Diseases of Banana, Abacá and Enset

Diseases of Banana, Abacá and Enset

Edited by

D.R. Jones

Consultant in International Agriculture
Droitwich Spa
Worcestershire
UK

CABI *Publishing*

CABI *Publishing* is a division of CAB *International*

CABI Publishing
CAB International
Wallingford
Oxon OX10 8DE
UK

Tel: +44 (0)1491 832111
Fax: +44 (0)1491 833508
Email: cabi@cabi.org

CABI Publishing
10 E 40th Street
Suite 3203
New York, NY 10016
USA

Tel: +1 212 481 7018
Fax: +1 212 686 7993
Email: cabi-nao@cabi.org

A catalogue record for this book is available from the British Library, London, UK.

Library of Congress Cataloging-in-Publication Data
Diseases of banana, abacá, and enset / edited by D.R. Jones.
p. cm.
Includes bibliographical references and indexes.
ISBN 0-85199-355-9 (alk. paper)
1. Bananas--Diseases and pests. 2. Abaca (Plant)--Diseases and pests. 3. Ensete--Diseases and pests. I. Jones, D.R. (David Robert), 1946– .
SB608.B16D57 1999
634'.7729--dc21 99-29320
CIP

ISBN 0 85199 355 9

Cover illustrations

Main picture: symptoms of black leaf streak disease on a leaf of 'Sucrier' (AA) in Western Samoa (photo: D.R. Jones, QDPI).
Front inserts (top to bottom): symptoms of Septoria leaf spot disease on 'Anamala' (AAA, syn. 'Gros Michel') in Sri Lanka (photo: D.R. Jones, INIBAP); symptoms of Fusarium wilt disease on 'Bluggoe' (ABB) in Honduras (photo: D.R. Jones, INIBAP); the almost total destruction of leaf tissue of 'Agbagba' (AAB, Plantain subgroup) in Nigeria by black leaf streak disease (photo: D.R. Jones, INIBAP); symptoms of leaf speckle disease on 'Pisang Awak' (ABB) in West Malaysia (photo: D.R. Jones, INIBAP).
Back cover insert: symptoms of bract mosiac disease on the bracts of *Musa ornata* in a germplasm collection in southern India (photo: D.R. Jones. INIBAP).
Spine insert: fruit of 'Kluai Khai' (AA, 'Sucrier') for sale at a roadside stall in north-western Thailand (photo: D.R. Jones, INIBAP).

Typeset in Melior by Columns Design Ltd, Reading
Origination, printing and binding in Singapore under the supervision of MRM Graphics Ltd, Winslow, Bucks.

Contents

Contributors

J. Carlier, Département amélioration des méthodes pour l'innovation scientifique (AMIS), Centre de coopération internationale en recherche agronomique pour le développement (CIRAD), BP 5035, 34032 Montpellier Cedex 1, France

R. Black, National Resources Institute (NRI), University of Greenwich, Central Avenue, Chatham Maritime, Kent ME4 4TB, UK

D. De Waele, Laboratory of Tropical Crop Improvement, Katholieke Universiteit Leuven, Kardinaal Mercierlaan 92, B-3001 Heverlee, Belgium

M. Diekmann, BEAF (Advisory Service for Development-Orientated Agricultural Research), Dottendorfer Straße 86, 53129 Bonn, Germany

S.J. Eden-Green, National Resources International (NRI), Central Avenue, Chatham Maritime, Kent ME4 4TB, UK

E. Fouré, Centre régional bananiers et plantains (CRBP), BP 832, Douala, Cameroon

F. Gauhl, Chiquita Brands, PO Box 217-1150, La Uruca, San José, Costa Rica

S.R. Gowen, Department of Agriculture, Earley Gate, PO Box 236, University of Reading, Reading, Berkshire RG6 2AT, UK

J.d'A. Hughes, International Institute of Tropical Agriculture, c/o L.W. Lambourn and Co., Carolyn House, 26 Dingwall Road, Croydon CR9 3EE, UK

S.C. Hwang, Taiwan Banana Research Institute, PO Box 18, Chiuju, Pingtung, Taiwan 90403

M.L. Iskra-Caruana, Mission défense des cultures (MIDEC), Centre de coopération internationale en recherche agronomique pour le développement (CIRAD), BP 5035, 34032 Montpellier Cedex 1, France

Y. Israeli, Jordan Valley Banana Experiment Station, Zemach 15132, Israel

D.R. Jones, 12 Charlotte Brontë Drive, Droitwich Spa, Worcestershire WR9 7HU, UK

E. Lahav, Institute of Horticulture, Agricultural Research Organization, The Volcani Center, Bet Dagan, 50250, Israel

P. Lepoivre, Unité de Phytopathologie, Passage des Déportés, 2, Faculté universitaire des Sciences agronomiques de Gembloux (FUSAGx), B-5030 Gembloux, Belgium

B.E.L. Lockhart, Department of Plant Pathology, College of Agricultural, Food and Environmental Sciences, University of Minnesota, 495 Borlaug Hall, 1991 Upper Buford Circle, St Paul, Minnesota 55108-6030, USA

E.O. Lomerio, Fiber Industry Development Authority (FIDA), Department of Agriculture, Bicol University Compound, Legazpi City, Philippines

L.V. Magnaye, Bureau of Plant Industry, Department of Agriculture, Bago-Oshiro, 8000 Davao City, Philippines

X. Mourichon, Département amélioration des méthodes pour l'innovation scientifique (AMIS), Centre de coopération internationale en recherche agronomique pour le développement (CIRAD), BP 5035, 34032 Montpellier Cedex 1, France

I.F. Muirhead, 12 Mondra Street, Kenmore Hills, Queensland 4069, Australia

C. Pasberg-Gauhl, c/o F. Gauhl, Chiquita Brands, PO Box 217-1150, La Uruca, San José, Costa Rica

K.G. Pegg, Plant Pathology Building, Department of Primary Industries, 80 Meiers Road, Indooroopilly, Queensland 4068, Australia

R.C. Ploetz, Tropical Research and Education Center, Institute of Food and Agricultural Sciences, University of Florida, 18905 SW 280 Street, Homestead, Florida 33031, USA

A.J. Quimio, Department of Plant Pathology, University of the Philippines at Los Baños, College, Laguna 3720, Philippines

R.A. Romero, Chiquita Brands, PO Box 217-1150, La Uruca, San José, Costa Rica

F.E. Rosales, Regional Office for Latin America and the Caribbean, International Network for the Improvement of Banana and Plantain (INIBAP), Apatardo Postal 60-7170, CATIE, Turrialba, Costa Rica

P.R. Rowe, Fundación Hondureña de Investigación Agricola (FHIA), Apartado Postal 2067, San Pedro Sula, Honduras

L. Sági, Laboratory of Tropical Crop Improvement, Katholieke Universiteit Leuven, Kardinaal Mercierlaan 92, B-3001 Heverlee, Belgium

J.L. Sarah, Département amélioration des méthodes pour l'innovation scientifique (AMIS), Centre de coopération internationale en recherche agronomique pour le développement (CIRAD), BP 5035, 34032 Montpellier Cedex 1, France

R.H. Stover, Apartado Postal 33, San Pedro Sula, Honduras

C.Y. Tang, Taiwan Banana Research Institute, PO Box 18, Chiuju, Pingtung, Taiwan 90403

M. Tessera, Institute of Agricultural Research, PO Box 2003, Addis Ababa, Ethiopia

J.E. Thomas, Plant Pathology Building, Department of Primary Industries, 80 Meiers Road, Indooroopilly, Queensland 4068, Australia

R. Thwaites, National Resources Institute (NRI), University of Greenwich, Central Avenue, Chatham Maritime, Kent ME4 4TB, UK

The Editor: David Jones

Dr David Jones grew up in Dudley in the English Midlands. He graduated in botany from the University of Hull in 1968, obtained an MSc in plant pathology from the University of Exeter in 1970 and was awarded a PhD by the University of Keele in 1973 for his work on rust fungi. He was then granted postdoctorate research fellowships with Agriculture Canada in London, Ontario, and the University of Sydney in New South Wales, Australia, to work on the physiology of host–pathogen interactions.

From 1977 to 1987, Dr Jones held the position of Quarantine Plant Pathologist with the Queensland Department of Primary Industries in Brisbane, helping to prevent the spread of black leaf streak of banana from the Torres Strait area and formulating guidelines for the introduction of banana germplasm into Australia. Subsequently, as Senior Plant Pathologist, he worked on banana leaf diseases and postharvest problems. During this period, he visited Papua New Guinea as part of an international mission to collect valuable *Musa* germplasm and participated in a banana improvement project in the South Pacific.

In 1992, Dr Jones joined the International Network for the Improvement of Banana and Plantain (INIBAP) in Montpellier, France, as Crop Protection Research Coordinator. Later, as Scientific Research Coordinator, he played a leading role in developing INIBAP's plan of action. During his time with INIBAP, he also served as coeditor of the magazine *Infomusa,* helping to improve its circulation and status as a citable publication, and led a global project to test and distribute new disease-resistant banana hybrids. Since 1996, he has been an international consultant based in Worcestershire, England, where he lives with his wife and two daughters.

Preface

In 1972, Dr R.H. Stover, who was Head of the Plant Pathology Section of the United Fruit Company based at La Lima in Honduras, published a book called *Banana, Plantain and Abaca Diseases.* As well as being concise, it was comprehensive and easy to use as a reference book. In time, it became the most acquired publication on banana and abacá diseases. Also in 1972, Profesor C.W. Wardlaw published the last of his three major works on banana and abacá diseases, earlier editions being published in 1935 and 1961. These two books were a compilation of all knowledge on banana and abacá problems up to that year.

Since 1972, there has been no one publication devoted solely to banana and abacá diseases. Dr Stover's and Professor Wardlaw's books have long been out of print and stocks exhausted. However, right up until the present day there has been a moderate demand for these old publications as there has been nothing new produced as a replacement. In 1993, CAB *International* considered the time to be right for an update of Dr Stover's book and I was approached in November of that year to act as editor of a modern version. As well as banana and abacá, CAB *International* considered a new text should also cover enset, another important plant in the *Musaceae* cultivated in Ethiopia. In 1995/96, I began the task of soliciting papers from the acknowledged experts of the various diseases that afflict the crops targeted for inclusion in the book.

The book is now at last complete and I would like to thank all those involved in its production. These include all the contributing authors, many of whom are personal friends, the staff of CABI, who provided me with scientific information, and offered advice and prepared the text for publication. Special thanks must go to E. Lahav and Y. Israeli, who prepared five chapters, E.O. Lomeiro for providing illustrations of abacá diseases and M. Tessera for slides of diseases of enset. I also thank A. Denham-Smith and L. Swan of Fyffes for their help in providing funds that have enabled most of the illustrations in this publication to be printed in colour. Fyffes' generous contribution has substantially increased the diagnostic value of this book and will be of much benefit to plant pathologists and other workers new to banana.

Over the last two decades, there has been an increasing awareness in developing countries of the potential and importance of banana as a food source. In the past, banana has been more or less 'taken for granted' and regarded as a crop that required little attention or input and virtually 'grew itself'. With the global spread of black leaf streak and the susceptibility of plantain, a staple in many countries, that attitude has changed. Governments, international research agencies and funding bodies now realize that the banana is under threat and conscious efforts must be made if the crop is to survive and flourish in the future. The battle against disease is a major issue. Diseases have to be correctly diagnosed so that the right measures can be taken to protect the crop. The international spread of serious diseases has to be slowed or stopped if possible. In these respects, I hope that the book will help provide valuable information for the identification of problems and in the fight against factors limiting production.

The book itself generally follows the format laid down by R.H. Stover. Diseases are described in chapters that are based on pathogen groups and sites of pathogen attack. Most diseases described are those that attack banana. No attempt has been made to separate the diseases of banana, abacá and enset because some pathogens can affect more than one crop and there would have been overlap and repetition. Instead, the different host crops affected are mentioned under disease headings. The general index should be consulted to find lists of names of diseases and pathogens that affect the three crops.

Much has been updated and new diseases have been described, but in some cases there has been little new information published so the text is basically that of R.H. Stover's original treatment. Fungal diseases of the foliage are described first, beginning with black leaf streak, which is the most economically important banana disease and the main constraint to commercial production. Fusarium wilt and other fungal diseases of the root, corm and pseudostem have been described in the next chapter. Chapters on fungal diseases of fruit, bacterial diseases, virus diseases and nematode pathogens follow. After diseases, there are five chapters devoted to non-infectious disorders, mineral deficiencies, injury caused by climatic conditions, injury caused by chemicals and toxic concentrations of elements and genetic abnormalities. The chapter on quarantine and the safe movement of *Musa* germplasm details the history of the development of guidelines to prevent the movement of exotic pathogens between countries on banana and abacá. This is particularly relevant at a time when more and more planting material is being exported from tissue-culture laboratories and germplasm is being exchanged around the world with increasing regularity. The book ends with chapters on breeding banana for disease resistance and the possibilities for genetically engineering disease resistance. These sections highlight the progress that has been made, is being made and will undoubtedly be made in this area.

Throughout the book, an effort has been made to give as much information as possible on the reaction of banana cultivars to disease because genetic resistance holds the key for the future, long-term, sustainable control of disease problems. So that this information can be put into perspective *vis-à-vis* the different types of banana, the classification and origins of edible banana have been explained in some detail in the introductory chapter. A list of host species within the *Musaceae* and cultivar names has been included as a separate index so that the available information on each clone can be accessed easily.

In the late 1970s, I was introduced to banana cultivation and banana diseases while Quarantine Plant Pathologist with the Queensland Department of Primary Industries in Australia. My keen interest in the crop later extended to banana classification and breeding. I take the opportunity here to thank all those persons who have helped, encouraged, guided, offered hospitality and provided companionship on missions since my early days with banana. Of these, K.G. Pegg, M. Smith, J.E. Thomas, I.F. Muirhead, J. Daniells, R.A. Fullerton, S.L. Sharrock, P.R. Rowe, E. Ostmark, R.C. Ploetz, M.L. Iskra-Caruana, M. Diekmann, Z.C. De Beer, S.H. Jamaluddin, D. Wattanachaiyingcharoen, X. Mourichon, F. Gauhl, C. Pasberg-Gauhl, R. Swennen and B.E.L. Lockhart deserve special mention. The years that I spent with the International Network for the Improvement of Banana and Plantain (INIBAP) from 1992 until 1996 were particularly rewarding and I appreciated the interest and friendship of the late T. Wormer and all the staff, especially N. Mateo, R.V. Valmayor, R.C. Jaramillo, J.P. Horry and H. Tezenas du Montcel.

My career has taken me to places as

varied as Western Samoa, Vietnam, Sri Lanka, Zanzibar, Rwanda, Nigeria, Cuba and Colombia studying banana and its problems. More recently, I have spent a year and a half on the banana-exporting island of St Vincent in the Caribbean, where I became known locally as 'Banana Doctor'. This book is for all banana doctors around the world.

David R. Jones
Droitwich Spa, Worcestershire
March 1999

1

Introduction to Banana, Abacá and Enset

D.R. Jones

The Genera *Musa* and *Ensete*

Banana and abacá belong to the genus *Musa* and enset to the genus *Ensete*. *Musa* and *Ensete* are the only two genera of the family *Musaceae* of the order *Zingiberales*. The wild *Musaceae* are distributed from the Pacific to West Africa, but are mainly found in the South-East Asian–New Guinea region. Characteristics of the *Musaceae* that differentiate this family from others in the same order are that the leaves and bracts are spirally arranged, male and female or hermaphrodite flowers are separated within one inflorescence and the fruit is a many-seeded berry (Stover and Simmonds, 1987). Some species within the *Musaceae* are utilized as ornamentals, but banana, abacá and enset are the three most economically important crop plants.

Species within *Musa* and *Ensete* are large herbs with pseudostems of leaf sheaths. New leaves are formed from a meristem near ground level and push up through the pseudostem in a tight roll. Lamina are large, usually oblong, with a stout midrib and numerous parallel veins extending to the margin. Flowering is initiated when the apical meristem stops producing leaves and forms an inflorescence. A peduncle develops which carries the inflorescence upwards and eventually emerges from the centre of the leaf crown at the top of the pseudostem. Flowers are borne in clusters, each cluster in the axil of a large spathaceous bract. The perianth consists of one compound lobed tepal and one narrow inner tepal. The ovary is inferior with three locula, each loculus with numerous ovules in axile placentation. Seeds have a thick, hard testa, a straight embryo and copious endosperm. Pseudostems die after flowering (Purseglove, 1972; Stover and Simmonds, 1987).

Musa spp. have tightly clasping leaf sheaths and a slightly swollen pseudostem base. Basal flowers are generally female only and distal flowers male. Flowers and bracts are inserted independently on the peduncle and, except for functional female ovaries in basal hands, are commonly deciduous by abscission. Bracts are usually reddish, purple or violet due to anthocyanins. Suckers arise freely from an underground rhizome. Pollen grains are finely granular and seed is 7 mm or less in diameter (Purseglove, 1972).

In contrast, *Ensete* spp. have lax leaf sheaths and a pseudostem base which is usually markedly swollen. Basal flowers are usually hermaphrodite. Flowers and bracts are integral with each other and with the axis and are persistent. Suckers are not produced unless induced by humans. Pollen grains are warty and seed is 6 mm or more in diameter (Purseglove, 1972).

The genus *Musa*, which has about 40 species, is divided into five sections:

- *Eumusa* and *Rhodochlamys* – contain species which have 22 (n = 11) chromosomes.

- *Australimusa* and *Callimusa* – contain species which have 20 (n = 10) chromosomes.
- *Ingentimusa* – contains one species which has 14 (n = 7) chromosomes.

Eumusa is the largest and most wide-ranging section of the genus and contains *Musa acuminata* Colla and *Musa balbisiana* Colla, which are the principal progenitors of most edible banana cultivars. Another, much smaller, series of edible *Musa*, comprising the Fe'i banana cultivars, is derived from wild species in the *Australimusa* section, representatives of which are found mainly on the island of New Guinea. Abacá is the species *Musa textilis* Née, which is also in the *Australimusa* section. *Musa textilis* is found in nature only in the Philippines, but has been introduced to other countries for cultivation.

The genus *Ensete* comprises six to seven species, which are divided between Asia and Africa. In Africa, *Ensete ventricosum* (Welw.) Cheesm. (syn. *Musa ensete* Gmel.; *Ensete edule* (Gmel.) Horan.), which is extremely variable, is the most widely distributed and is found between Ethiopia and the Central African Republic, Sudan and South Africa. In Ethiopia, edible cultivars of enset have been selected and developed from *E. ventricosum*.

Banana

Facts and figures

Banana is one of the most important, but undervalued, food crops of the world. It is grown predominantly in gardens and smallholdings in some 120 countries, mainly in the tropics and subtropics, and provides sustenance to millions of people. Roughly one-third of total world production of bananas, which in 1998 was estimated to be approximately 88 million tonnes (Anon., 1999), comes from the Latin American–Caribbean region, one-third from Africa and one-third from the Asian–Pacific region. Most fruit is consumed locally. It has been estimated that the highest consumption rates are on the island of New Guinea and in the Great Lakes region of East Africa, where bananas occupy a large proportion of the diet and amount to 200–250 kg person^{-1} year^{-1}. This compares with consumption figures of approximately 15–16 kg person^{-1} year^{-1} in Europe and North America (Anon., 1992). Bananas for export account for only about 10% of total production and are mostly grown in the Latin American–Caribbean region, though some are produced in Africa, Asia and the Pacific.

Banana is an attractive perennial crop for farmers in developing countries. The fruit can be produced all year round, thus providing a steady cash income or supply of nutritious food. Bananas for domestic consumption are harvested from a multitude of cultivars, which are grown on a wide variety of soils and in many different situations. These cultivars can be grown on holdings that range from small plantations to garden plots and jungle clearings. The number of different clones has been estimated to be 200–500 by Stover and Simmonds (1987) and 400–500 by Perrier and Tezenas du Montcel (1990). However, the New Guinea–Solomon Islands region, which is the centre of greatest banana diversity, has not yet been thoroughly explored and the number may be appreciably higher.

Bananas can be divided into two main categories: dessert bananas and cooking bananas. Dessert bananas, which constitute 43% of world production (Anon., 1992), are eaten raw when ripe, as they are sugary and easily digestible. There are many different types of dessert bananas, but fruit from cultivars in the Cavendish subgroup are the most common. In some tropical countries, it is the custom to cook dessert bananas. However, cooking bananas, which account for the other 57% of world production (Anon., 1992), are usually starchy when ripe and need to be boiled, fried or roasted to make them palatable. They are important in the diets of many peoples throughout the tropics and are mostly consumed locally. Plantains, the most well known of the cooking types, form 23% of the world's total production of bananas (Anon., 1992).

As well as being eaten as a fresh or cooked fruit, bananas can be sun- or oven-dried to make 'figs' or sliced and fried into 'chips'. Bananas can also be turned into flour, brewed to make beer and, in the Philippines, even form the basis of a 'tomato' sauce. In some countries in South-East Asia, the male bud of the banana is eaten after the removal of the fibrous outer bracts. Banana is often grown in association with other crops, affording protection and shade. Leaves of banana are utilized as packaging for other foods and serve as plates, tablecloths and decorative items for religious ceremonies. Chopped pseudostems, peduncles and fruit peel can be fed to animals. The banana is a very versatile plant.

Botany

The banana is a large, herbaceous plant consisting of a branched, underground stem (a *rhizome* in the strict botanical sense, but also often called a *corm*) with roots and vegetative buds, and an erect *pseudostem* composed of tightly packed leaf bases (Fig. 1.1). The apical meristem, which is located in the centre of the pseudostem at about soil level, gives rise to a succession of leaf primordia. Each primordium grows upwards and differentiates into a leaf base, a petiole and a lamina. The petiole and tightly rolled lamina emerge at the top of the pseudostem in the centre of the cylinder formed by older leaf bases. The lamina begins to unfold when fully emerged, the left-hand side, when viewed from above and from the petiole end, unrolling before the right-hand side. Initially, the leaf forms a tunnel where rainwater or dew can collect and it is at this stage when infection by leaf pathogens can occur. Leaves emerge at different rates depending on the cultivar and environmental conditions. With cultivars in the Cavendish subgroup growing in Honduras, this varies from 3.5–3.8 leaves month^{-1} in summer to 2.5–3.0 leaves month^{-1} in winter. In winter in the cool subtropics, the leaf emergence rate can fall to 0.1–1.2 leaves month^{-1}.

A banana plant generally has ten to 15 functional leaves at flowering and five to ten at harvest, but numbers can be less due to environmental and disease constraints. Unless removed, old, senescing leaves hang down the pseudostem.

At a certain stage of plant development, usually after about 25–50 leaves have been produced, the apical growing point stops initiating leaves and develops an inflorescence. The flower stalk or *peduncle* elongates and forces the inflorescence out through the top of the plant. The bracts lift to expose double layers of female nodes with tightly packed fruit called *hands*. Individual fruit that arise from each flower are called *fingers*. Bracts later dry and fall off. Hermaphrodite flowers, which usually abscise, are found next to developing fruit. At the end of the peduncle are the male flowers, which are tightly enclosed in bracts. This terminal arrangement is called the *bell* or *male bud*.

The *bunch* is the collective name for the hands of fruit. Bunch size and weight depend on plant vigour and health. In the tropics bunches can be harvested between 85 and 110 days after flowering, but in the cool subtropics fruit development can take up to 210 days. After harvest, the pseudostem dies back and is therefore usually cut down. The plant propagates by forming a *sucker*, which is an outgrowth of a vegetative bud on the rhizome. A young sucker emerging from the ground is called a *peeper*. A number of suckers can arise from each rhizome. One sucker is usually selected by the farmer as a *follower* to grow on and regenerate the plant. Other suckers are either removed physically or their apical meristems destroyed chemically. If all suckers are allowed to grow, a *clump* consisting of plants of various sizes all arising from the same connected rhizome system or *mat* eventually develops.

Primary roots, which are about 5–8 mm in diameter, usually arise in groups of three or four from the fleshy rhizome. When healthy and vigorous, primary roots are white in colour, but later turn grey or brown before dying. A system of secondary and tertiary roots develops from the primary

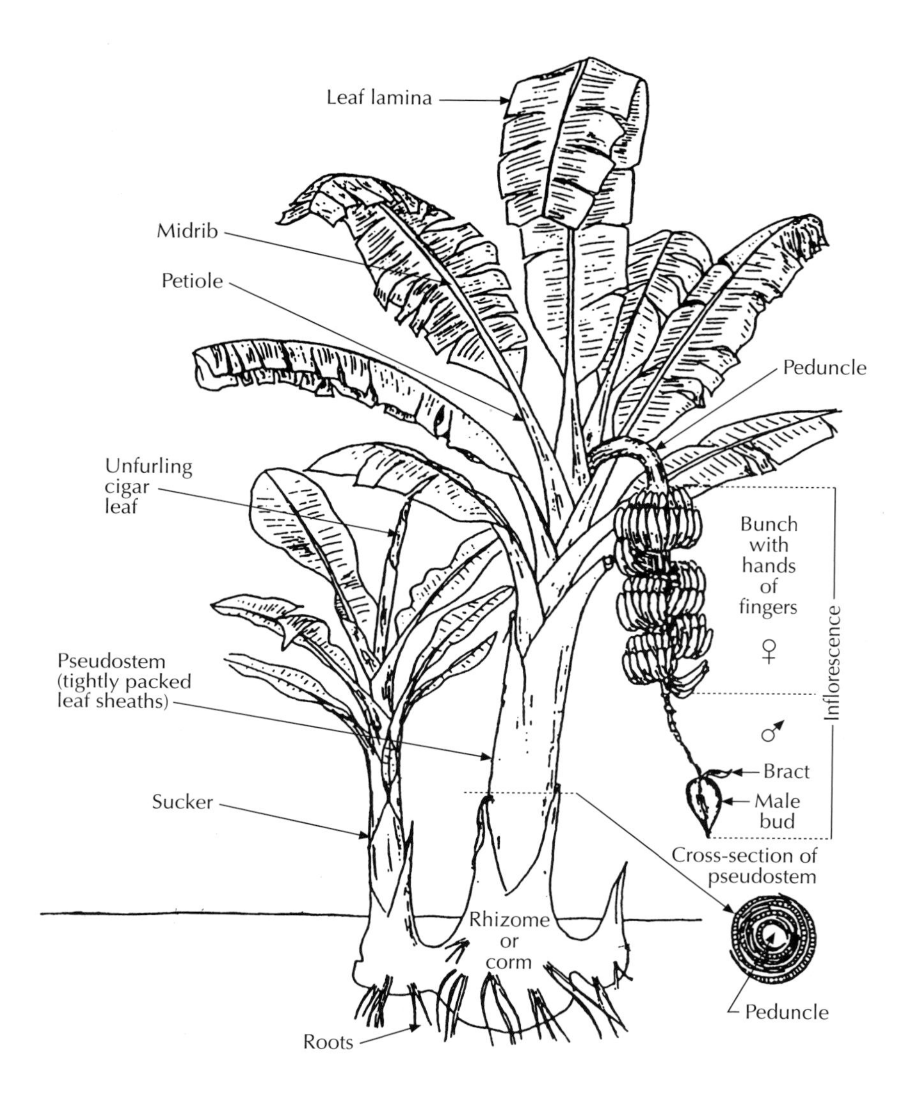

Fig. 1.1. Diagrammatic representation of a fruiting banana plant with suckers (from Champion, 1963).

roots. The root system is shallow, not penetrating much below 600 mm. Horizontal spread can be as far as 5 m, but more commonly 1–2 m. Banana roots are sensitive to adverse conditions, such as waterlogging, desiccation and an over-compact soil structure (Wardlaw, 1961; Robinson, 1996).

Cultivation and trade

Banana was initially grown in South-East Asia and Melanesia in small permanent gardens around settlements (Plate 1.1), in forest clearings by shifting cultivators (Plate 1.2) and in gardens with other crops

Plate 1.1. Banana plots close to houses in Kasaka village, East New Britain, Papua New Guinea (photo: D.R. Jones, QDPI).

Plate 1.2. Banana growing in a temporary, man-made clearing in the rain forest near Kiunga, Western Province, Papua New Guinea (photo: D.R. Jones, QDPI).

(Plate 1.3). Many cultivars with differing genetic backgrounds and different responses to pathogens and pests were most probably cultivated together. Disease problems could not have been severe enough under these agrosystems to have influenced the selection of disease-resistant cultivars over susceptible ones. Horticultural characteristics were probably the main considerations in the selection, propagation and spread of early landraces.

Traditional planting material is either the sucker (Plate 1.4) or *bit*, which is a piece of rhizome with a vegetative bud. More recently, plants derived from tissue culture have been utilized in commercial operations. Tissue-cultured plants have the advantage of a higher establishment rate producing a more uniform crop, which gives a higher yield in the first crop cycle, and they also have the potential of being completely free of pests and diseases when planted. However, this type of planting material needs extra care during establishment and plants can be affected by morphological changes caused by genetic mutations during *in vitro* multiplication. The first crop after planting is known as the *plant crop* and the second cycle as the *ratoon crop* or *first ratoon*. This is followed by the second ratoon, third ratoon, etc.

Banana plants need ample supplies of water if they are to yield well. This has been estimated at 25 mm per week for satisfactory growth in the tropics (Purseglove, 1972). Banana grown in drier areas has to be irrigated to maintain optimal growth. Plants also usually require inputs of nutrients, such as nitrogen and potassium, to prevent deficiency problems.

Banana cultivated as a monocrop on a commercial scale is only a relatively recent phenomenon. This was brought about 100 years ago, after the advent of steamships and cool storage techniques, by a great increase in demand for the fruit in Europe and North America. 'Gros Michel' was the first export banana cultivar to be planted over thousands of hectares, but fruit for export today is mainly produced by cultivars in the Cavendish subgroup of banana. In Latin America and the Philippines, export fruit is usually produced on vast plantations on relatively flat coastal plains with alluvial soils. However, export bananas in the Windward Islands are produced on smallholdings, which are often found on steeply sloping hillsides. In other parts of the world, large areas of country can also be covered with genetically similar banana plants, which are grown to produce fruit for domestic consumption (Plate 1.5).

Export bananas need to be as blemish-free as possible at markets to obtain premium prices. The bunch is therefore usually covered with a perforated, polyethylene *sleeve*, *cover* or *bag* (Plate 1.6) soon after emergence to protect the fruit from rubbing leaves, wind-blown debris and chemicals used in sprays to control leaf spot diseases. Bagging can also shorten the fruit maturation period by 3–4 days and increase the weight of the bunch by 1.8–2.3 kg. If impregnated with insecticide, the bag offers protection from thrips and other fruit-attacking insects that cause damage. At the time of bagging, the male bud is snapped off because in some circumstances this practice can also increase bunch weight. The bag is usually tied to the peduncle above the scar formed by the upper spathaceous bract. In windy areas, the bag is tied at the bottom to prevent the polyethylene rubbing on the fingers, which causes scarring. If the bottom of the bag is tied, the knot is positioned at one side to allow rainwater to drain freely from the bag. This is because a build-up of humidity within the bag can lead to fungal disease problems on the fruit (see Jones and Stover, Chapter 4, pp. 173–190).

Rough handling of bananas leads to bruising, scarring and other damage. Export plantations have developed methods of handling bananas to minimize this wastage by establishing cableways for moving bunches to packing sheds where fruit is dehanded, washed, treated with fungicide to reduce postharvest losses and placed into cardboard cartons for transport.

Banana fruit produces ethylene, which triggers ripening. Ripening involves the conversion of starch to sugar, a softening of the fruit texture and often a change in skin colour from green to yellow. These physiological changes can also stimulate the development of postharvest diseases. Locally consumed fruit can be left on the plant until ripe. However, bananas for export or consumption some distance away from growing areas are usually harvested green and immature. Timing of harvest is important in these circumstances. This can be determined by tagging bunches on emergence and then calculating approximate harvesting dates. Usually, coloured ribbons are tied around the peduncle, the colour of the ribbon signifying the particular week of emergence. When the bunch reaches harvesting age, maturity or grade of the fruit is checked by measuring the diameter of the middle finger on certain hands with a caliper. The bunch is harvested at the desired grade.

Export fruit has to be held and transported at cool temperatures to minimize ethylene release and ripening. Even so, premature ripening can occur if fruit is over-mature, damaged physically or harvested from plants with high levels of leaf spot. Mixed ripe fruit is often rejected at wholesale markets. Controlled atmosphere conditions on ships transporting bananas can reduce the incidence of ripe and turning fruit. At markets, bananas can be stored under cool conditions until needed and then ripened artificially with ethylene.

Disease and pest problems are always more serious when plants with an identical genetic make-up are cultivated on a large scale. With banana, this first happened when 'Gros Michel' was grown on plantations in the Latin American–Caribbean region. In the course of time, Fusarium wilt destroyed the viability of this first commercial cultivar. After 1960, the banana trade in the region was dominated by Cavendish cultivars and the problems associated with clones in this subgroup, such as Sigatoka and burrowing nematode, became limiting factors requiring much research.

Until fairly recently, banana scientists have worked predominantly for multinational companies or for governments supporting export growers. However, in the last few decades, banana has been recognized as an important local fruit crop by developing countries in the Americas, Africa and Asia and local improvement programmes have been initiated. Associated with these programmes has been a need to identify constraints to local banana production, including pests and diseases. In countries where Cavendish is not the most popular type of banana, new disease problems are emerging as local cultivars are beginning to be grown as a monocrop and on a scale much larger than before.

Origin of edible banana cultivars

Wild banana species are seedy, generally opportunistic and in nature grow in forest clearings and along watercourses. Early agriculturists are believed to have initially selected plants with high levels of parthenocarpy, which is the ability to form fruit without fertilization. This fruit would have had fewer seeds, thus increasing edibility. At the same time, selection would also have been for female sterility, which would also lower the number of seeds. Selected plants would have been propagated by suckers, thus conserving their genetic composition. The long-distance movement of vegetative planting material by humans enabled edible banana cultivars from Asia eventually to reach other regions of the world.

Evolution of the Eumusa *series of edible banana cultivars*

Simmonds (1962) postulated that the evolution of edibility in the *Eumusa* series of cultivars as outlined above most probably began with wild *M. acuminata* subspecies, which occur naturally in an area stretching from South Asia to Australasia, but are concentrated mainly in South-East Asia (Fig. 1.2). Another key event was believed to have been hybridization with *M. balbisiana*, a more drought-tolerant species. Hybridization is thought to have occurred as early landraces derived from *M. acuminata* began to be cultivated in peripheral, drier areas of South-East Asia where *M. balbisiana* is endemic (Fig. 1.3). This integration with *M. balbisiana* enabled banana to be grown in areas where rainfall is seasonal. Both *M. acuminata* and *M. balbisiana* are diploids. The genome of wild *M. acuminata* is represented by AAw and wild *M. balbisiana* by BBw. Diploid cultivars derived solely from *M. acuminata* are designated by AA and diploid hybrids with *M. balbisiana* by AB.

Triploidy, another important step forward in the evolution of banana, is thought to have arisen following the fertilization of viable, diploid egg cells, which can be

Plate 1.3. Banana growing together with taro, cassava, sweet potato and sugar cane in a garden plot on Badu Island in the Torres Strait region of Australia (photo: D.R. Jones, QDPI).

Plate 1.4. A sucker of 'Robusta' (AAA, Cavendish subgroup) selected for planting in St Vincent, Windward Islands (photo: D.R. Jones, SVBGA).

Plate 1.5. Large-scale cultivation of 'Kluai Namwa' (ABB, syn. 'Pisang Awak') near the Mekong River in Nong Khai Province, Thailand (photo: D.R. Jones, INIBAP).

Plate 1.6. A protective polyethylene bag covering a young bunch of 'Grand Nain' (AAA, Cavendish subgroup) in St Vincent, Windward Islands (photo: S. Vanloo, SVBGA).

formed when meiosis breaks down at the second division, with haploid pollen (Simmonds, 1962). Tripoid cultivars are found in the AAA, AAB and ABB genomic groups. Triploids are bigger, sturdier plants than diploids with increased fruit size (Fig. 1.4). Diploids can usually be distinguished from triploids because of their more slender peudostems and more upright leaves. A scheme outlining the possible evolution of edible banana cultivars in the AA, AB, AAA, AAB and ABB genomic groups is presented in Fig. 1.5.

Only a few, natural cultivars have been recognized as belonging to the AAAA, AAAB, AABB and ABBB genomic groups (Richardson *et al.*, 1965; Shepherd and Ferreira, 1984). These tetraploids, which have robust pseudostems and leaves that tend to droop, are believed to have arisen from the fertilization of triploid egg cells by haploid pollen. Recently, 'Kluai Teparot', a cultivar long described as a typical ABBB, has been found to be an ABB (Jenny *et al.*, 1997; Horry *et al.*, 1998). This calls in question the genomic status of other natural cultivars with typical tetraploid attributes. Tetraploid hybrids have been artificially bred for disease resistance and are becoming important in some countries (see Jones, Chapter 14, pp. 425–434; Rowe and Rosales, Chapter 14, pp. 435–449).

Classification of cultivars in the Eumusa *series of edible banana*

The genome of edible banana within the *Eumusa* series of cultivars is deduced from an analysis of morphological characters, such as pseudostem colour, shape of the petiolar canal and bract features (Simmonds and Shepherd, 1955). This taxonomic method has been proved over time to give a good approximation of the genetic composition and ploidy of cultivars, generally correlating well with the results of more modern molecular techniques.

Within each genomic group, morphological characters in addition to those used to define genomic group are used to identify clones. Many clones have mutated over time to form morphotypes that differ in certain characters, such as fruit and bunch morphology, pigmentation and height. Dwarfness is a common mutation and has occurred in many locations. Clones that are thought to have arisen originally from a base clone by mutation form subgroups. Major subgroups with large numbers of clones have formed in centres of secondary diversification. It is not always possible to identify the original base clone of major subgroups.

The same clone can have different names in different locations. This is especially true in Papua New Guinea, a country with over 700 languages, where the names of cultivars can vary between villages. In this publication, the synonym most frequently used by banana workers has been adopted to signify the clone. This 'best-known synonym' is in many cases the one

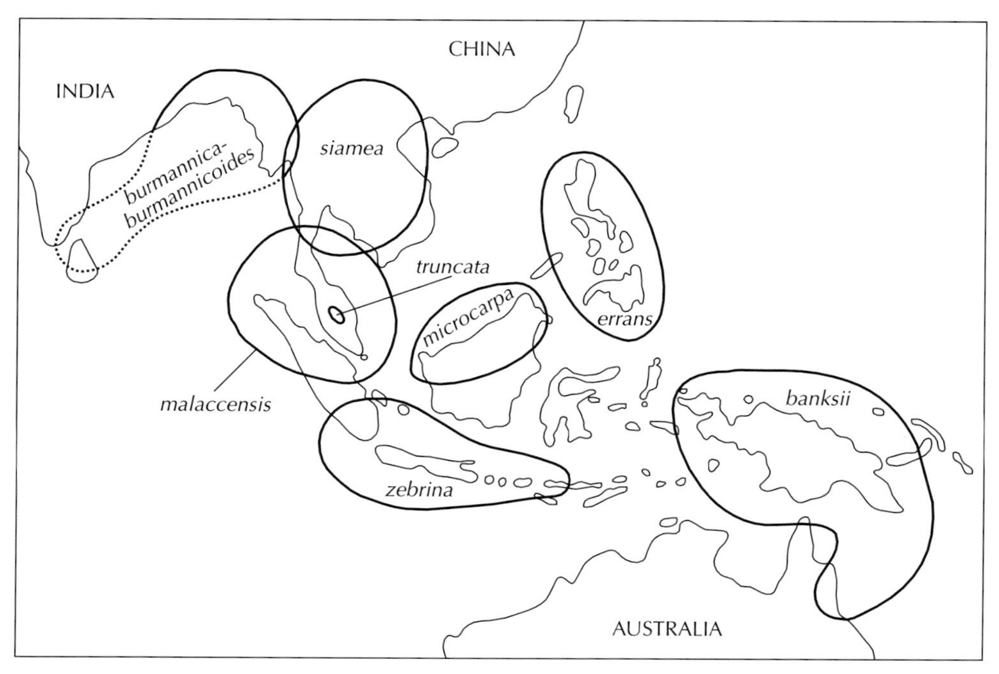

Fig. 1.2. Approximate natural distribution of *Musa acuminata* subspecies (adapted from Simmonds, 1962; Horry *et al.*, 1997).

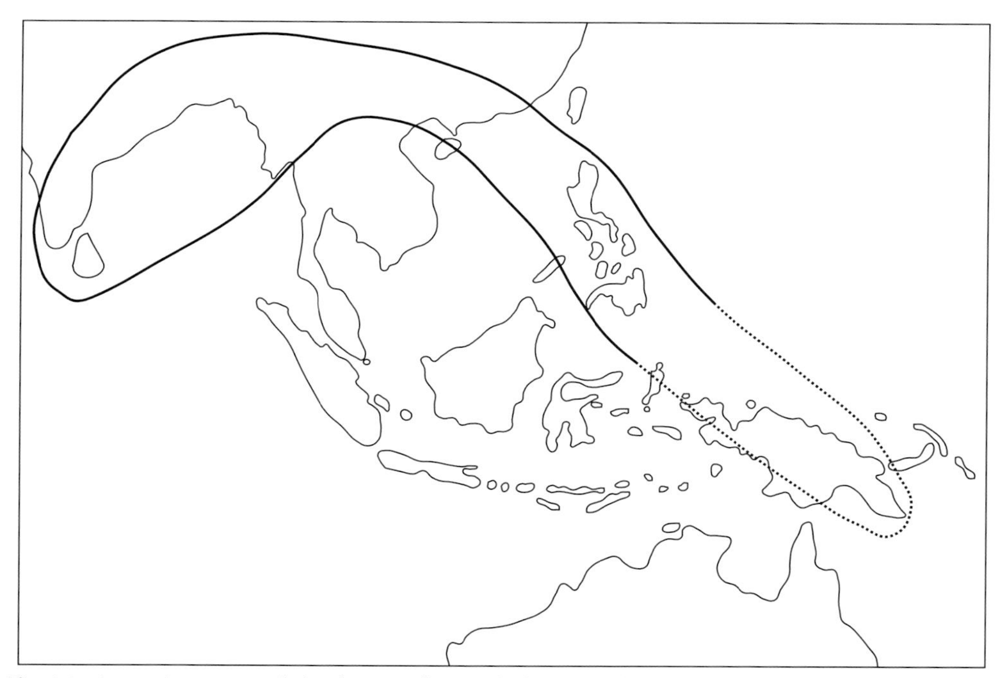

Fig. 1.3. Approximate natural distribution of *Musa balbisiana* (adapted from Simmonds, 1962). Although Simmonds (1962) considers *M. balbisiana* to be indigenous to New Guinea, Argent (1976) believes that the species may have been introduced, as in Thailand and Malaysia.

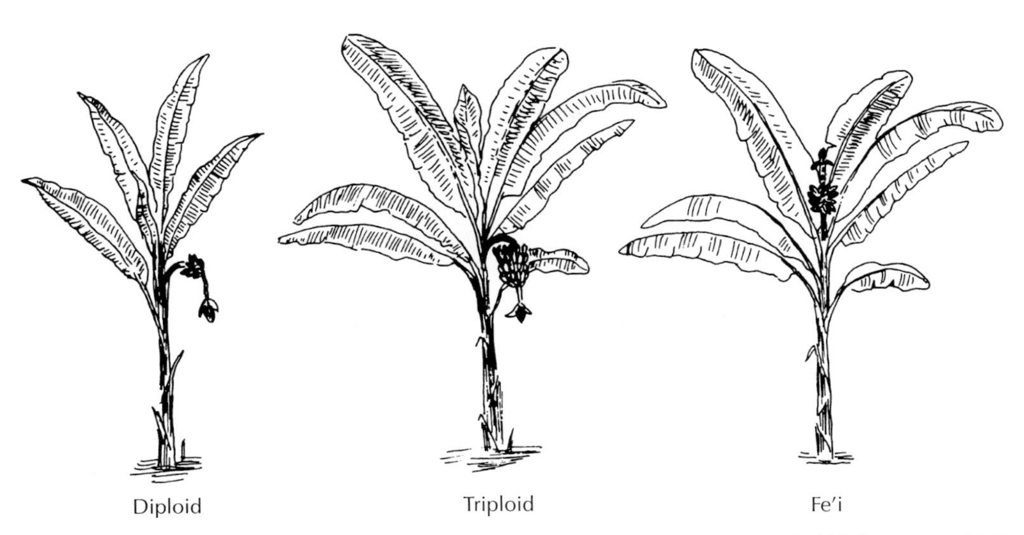

Fig. 1.4. General appearance of diploid and triploid cultivars in the *Eumusa* series of edible banana and Fe'i cultivars in the *Australimusa* series of edible banana (Bourke, 1976).

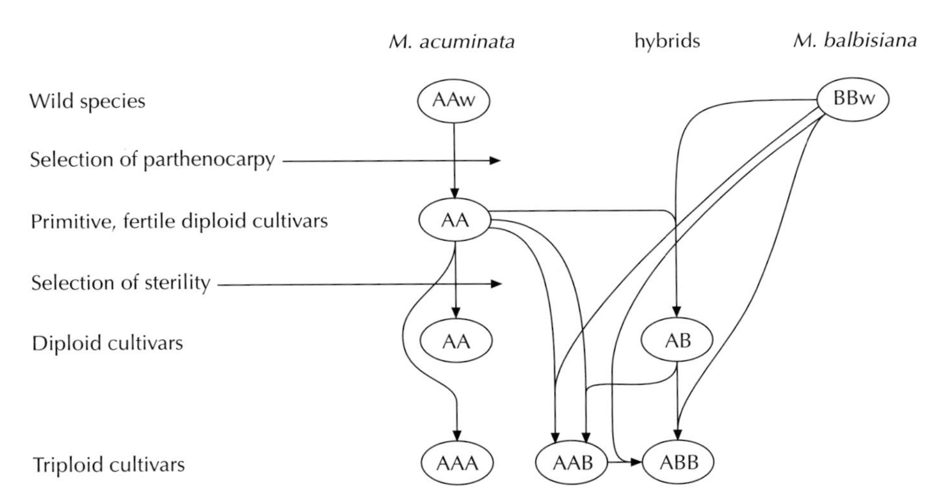

Fig. 1.5. Evolution of the main genomic groups of edible banana cultivars of the *Eumusa* series (after Simmonds, 1962; Carreel, 1995).

applied to the clone in the Caribbean where much of the early taxonomic work was undertaken. Thus 'Silk', a synonym of West Indian origin, gives its name to the clone that is known in India as 'Rasthali', in the Philippines as 'Latundan' and in Australia as 'Sugar'.

Cultivar names in some countries begin with the local word for banana. For example, names in Malaysia and Indonesia begin with 'Pisang', in Thailand with 'Kluai', in Vietnam with 'Chuoi', in Hawaii with 'Mai'a' and in Western Samoa with 'Fa'i'. Some workers believe this word to be redundant and accessions in germplasm collections have been registered without using the prefix. However, the local word for banana has been retained in this publication because it is considered an integral part of the cultivar name and also gives an indication as to the original source of the material.

Subgroups are named after the best-known synonym of the most important clone in the subgroup (Gros Michel, Pome and Bluggoe subgroups), the cultivars first described to identify the main characteristics of the subgroup (Iholena, Lujugira–Mutika and Maia Maoli–Popoulu subgroups) or a generic term (Cavendish and Plantain subgroups).

To help the reader relate the cultivars mentioned in this publication to the subgroup or clone, cultivar names are followed in parenthesis by the genomic group and then by either the subgroup or best-known synonym.

Subgroups and important clones in the **Eumusa** *series of edible banana*

AA and AB genomic groups

Some well-known AA and AB clones and synonyms are listed in Table 1.1.

The most important AA clone is 'Sucrier', which has exceptional fruit quality. It is quite common in South-East Asia and is present elsewhere in the world. 'Sucrier' is exported from the Philippines and Latin America. 'Pisang Lilin' is another popular AA clone, but is not grown widely outside Malaysia and Thailand. Fruit of both 'Sucrier' and 'Pisang Lilin' can often be seen on sale in Thailand (Plates 1.7 and 1.8). Other AA diploids are cultivated in South-East Asia, but are not as common. For example, 'inarnibal' is found in

Table 1.1. Important clones in the AA and AB genomic groups.

Name of clone*	Selected synonyms†	Location	Main use of fruit
AA genomic group			
'Sucrier'	'Sucrier'/'Figue Sucrée'	West Indies	Dessert
	'Pisang Mas'	Indonesia, Malaysia	
	'Kluai Khai'	Thailand	
	'Amas'	Philippines	
	'Lady's Finger'	Hawaii	
	'Orito'	Ecuador	
'Pisang Lilin'	'Pisang Lilin'	Malaysia	Dessert
	'Kluai Lep Mu Nang'	Thailand	
'Inarnibal'	'Inarnibal'	Philippines	Dessert
	'Pisang Lemak Manis'	Malaysia	
	'Pisang Lampung'	Indonesia	
'Senorita'	'Senorita'	Philippines	Dessert
'Pisang Jari Buaya'	'Pisang Jari Buaya'	Indonesia, Malaysia	Dessert
'Paka'	'Kipaka'	Zanzibar	Dessert
AB genomic group			
'Ney Poovan'	'Ney Poovan'/'Safet Velchi'	India	Dessert
	'Kisubi'	Uganda	Brewing
'Kunnan'	'Kunnan'	India	Dessert
'Sukari Ndizi'	'Sukali Ndizi'	Uganda	Dessert
	'Kamarangasenge'	Rwanda	

*The 'best-known synonym' gives its name to the clone.
†'Selected synonyms' indicate the most popular names for the clone in some of the different countries/locations where it is found.

Plate 1.7. Fruit of 'Kluai Khai' (AA, syn. 'Sucrier') for sale at roadside stall near Kamphaeng Phet, north-western Thailand (photo: D.R. Jones, INIBAP).

Plate 1.8. Fruit of 'Kluai Lep Mu Nang' (AA, syn. 'Pisang Lilin') for sale in Bangkok, Thailand (photo: D.R. Jones, INIBAP).

Malaysia, Indonesia and the Philippines and 'Senorita' is grown in the Philippines. There are also many AA clones in Papua New Guinea (Arnaud and Horry, 1997). In general, they are not very productive, but are maintained in village gardens as relics of a time when they were much more common and widespread (Stover and Simmonds, 1987).

'Pisang Lilin', 'Paka' from East Africa and 'Pisang Jari Buaya' (see Plate 14.6) from Malaysia have been used in breeding programmes because of their disease resistance attributes (see Jones, Chapter 14, pp. 425–434; Rowe and Rosales, Chapter 14, pp. 435–449).

The only AB clones of note are 'Ney Poovan', 'Kunnan' and 'Sakari Ndizi'. 'Ney Poovan' is not common, but is found in many countries. In Uganda, it is used for making beer. 'Kunnan' from South India has affinities to 'Ney Poovan' and may be a variant of this clone (Stover and Simmonds, 1987). 'Sakari Ndizi' is a popular dessert banana in East Africa. Fruit (Plate 1.9) is exported by air to Europe where it fetches high prices.

AAA genomic group

The AAA genomic group contains some of the most productive clones, which are widely grown and constitute about half of the world's total output of banana fruit (Anon., 1992). Major AAA subgroups are presented in Table 1.2.

'Gros Michel' is a tall clone and was the first banana of the export trades. Despite its susceptibility to Fusarium wilt, it is still grown for local consumption in many countries around the world because of its superior flavour. 'Gros Michel' produces a few seeds when pollinated. As a consequence, 'Gros Michel' and its shorter variants ('Cocos', 'Highgate' and 'Lowgate') have been utilized in breeding programmes (see Jones, Chapter 14, pp. 425–434); plant and bunch features of 'Lowgate', the shortest of the dwarf mutants of 'Gros Michel', are illustrated in Plate 14.9.

The Cavendish subgroup is responsible for 30% of the world's production of banana fruit. Stover and Simmonds (1987) recognize four major clone sets distinguished on height: 'Dwarf Cavendish' types are the shortest in stature, 'Grand Nain' types are medium dwarfs, 'Giant Cavendish' types are taller and 'Pisang Masak Hijau' types the tallest (Stover and Simmonds, 1987). This classification is somewhat arbitrary as in reality there is an almost continuous gradation in height from 'Extra-Dwarf Cavendish', which is shorter than 'Dwarf Cavendish' (see Israeli and Lahav, Chapter 12, pp. 396–397) upwards to 'Pisang Masak Hijau'. Each clone set contains different cultivars, some of which are listed in Table 1.2.

As well as height, clones in the Cavendish subgroup also differ in other morphological characters such as petiole length, bract persistence, bunch grade and pseudostem colour. Other characteristics, such as leaf emergence rate and length of crop cycle, have also been used to differentiate cultivars (Daniells, 1990). 'Dwarf Cavendish' (Plate 1.10) is better adapted to cooler climates than most other Cavendish clones and has been the basis of trade in subtropical areas. However, 'Williams' ('Giant Cavendish' type) is also suited to cooler environments and is gaining in popularity. The medium dwarf 'Grand Nain', which is resistant to toppling by wind, but more susceptible to drought than other Cavendish cultivars, has replaced 'Valery' ('Giant Cavendish' type) as the most popular cultivar in international trade in tropical areas. 'Robusta' ('Giant Cavendish' type), which is less susceptible to a water deficit than 'Grand Nain', is still grown extensively in the Windward Islands where irrigation is not widely available and seasonal droughts occur.

Cavendish cultivars may have originated in the South China–Vietnam area. In the 1820s, representatives of the subgroup were taken from this area to Europe as botanical specimens and may have been introduced to the Canary Islands and Guyane (Stover and Simmonds, 1987). As well as being the principal banana types grown for export fruit in the Latin America–Caribbean, African and South-East Asian regions, Cavendish cultivars also form the backbone of domestic banana industries in such places as Australia, China, Pakistan, India (Plate 1.11) and South Africa. They have higher yields than all other natural clones (Robinson, 1996).

East African highland cultivars form the Lujugira–Mutika subgroup. Approximately 18% of all banana fruit produced is from clones in this subgroup (Anon., 1992). 'Lujugira' and 'Mutika' are the cultivars that represent the two main subdivisions proposed for this subgroup by Shepherd (1957). This division is based on fruit morphology, with further differentiation related to other characteristics (Stover and Simmonds, 1987). A recent numerical taxonomic study of over 200 Ugandan accessions has distinguished five different clone sets (see Table 1.2) (Karamura, 1999).

Table 1.2. AAA genomic group – subgroups, clone sets, clones and important cultivars.

Subgroup	Name of clone*	Selected synonyms†	Location	Main use of fruit
Gros Michel	'Gros Michel'	'Gros Michel' 'Pisang Embun' 'Kluai Dok Mai' 'Anamala' 'Bluefields' 'Pisang Ambon Putih'	West Indies Malaysia Thailand Sri Lanka Hawaii Indonesia	Dessert
	'Cocos'	'Cocos'	Honduras	Dessert
	'Highgate'	'Highgate'	Jamaica	Dessert
	'Lowgate'	'Lowgate'	Honduras	Dessert
Subgroup	**Clone set**	**Some cultivars in clone set**	**Location**	**Main use of fruit**
Cavendish	'Extra-Dwarf Cavendish'	'Dwarf Parfitt' 'Dwarf Nathan'	Australia Israel	Dessert
	Dwarf Cavendish'	'Dwarf Cavendish' 'Basrai' 'Petite Naine'	Australia India, Pakistan French Antilles	Dessert
	'Grand Nain'	'Grand Nain' 'Grande Naine' 'Umalag'	Latin America French Antilles Philippines	Dessert
	'Giant Cavendish'	'Giant Cavendish' 'Robusta' 'Williams' 'Zhong Ba' 'Poyo' 'Pei-Chiao' 'Chuoi Tieu' 'Veimama' 'Nañicao' 'Valery'	General India, Windward Islands Australia, South Africa China West Africa Taiwan Vietnam Fiji Brazil Latin America	Dessert
	'Pisang Masak Hijau'	'Pisang Masak Hijau' 'Bungulan' 'Pisang Ambon Lumut' 'Lacatan'	Malaysia Philippines Indonesia Latin America	Dessert
Subgroup	**Clone set**	**Some cultivars in clone set‡**	**Location**	**Main use of fruit**
Lujugira–Mutika	Beer	'Bwara', 'Eridirra', 'Mende', 'Nametsi'	Uganda	Brewing
	'Musakala'	'Musakala', 'Enyoya', 'Mudwale', 'Mukazialanda'	Uganda	Cooking
	'Nakabululu'	'Nakabululu', 'Butobe', 'Kibuzi', 'Mukite'	Uganda	Cooking
	'Nakitembe'	'Nakitembe', 'Mbwazirume', 'Nasaala', 'Waikova'	Uganda	Cooking
	'Nfuuka'	'Nfuuka', 'Enyeru', 'Nakhaki', 'Kasenene'	Uganda	Cooking

*The 'best-known synonym' gives its name to the clone.
†'Selected synonyms' indicate the most popular names for the clone in some of the different countries/locations where it is found.
‡Cultivars listed represent different clones.

Plate 1.9. Fruit of 'Sakali Ndizi' (AB, syn. 'Sakari Ndizi') being transported to market near Jinja, Uganda (photo: D.R. Jones, INIBAP).

Plate 1.10. Dwarf Cavendish' (AAA) growing under adverse conditions in West Malaysia. Note the persistent bracts on the male axis and symptoms of freckle disease (see Jones, Chapter 2, pp. 120–125) on the leaf above the bunch (photo: D.R. Jones, INIBAP).

The fruit of the 'Lujugira–Mutika' subgroup, like fruit from other AAA clones, is sweet when ripe. However, East Africans either cook the fruit as 'matooke' or use it for brewing to make 'banana beer'. The pulp qualities of the fruit determine if a cultivar is a cooking or brewing type. Cooking cultivars can change spontaneously so that they become beer cultivars. This change occurs with more frequency above 1400 m. A check-list of East African highland types in Uganda lists 145 cooking cultivars and 88 beer cultivars together with synonyms (Karamura and Karamura, 1994). Fruit from these cultivars are very important in the diets of people in the eastern part of the Democratic Republic of Congo, Rwanda, Burundi, Uganda, Tanzania and western Kenya. Bunch features of this subgroup are illustrated in Plate 1.12.

The AAA group also contains a number of other clones, some of which are listed in Table 1.3. 'Red', together with its common sport 'Green Red', are widely distributed, tall, garden clones. The difference between the two cultivars is in their pigmentation, especially in the fruit peel. Immature fruit of 'Red' is red in colour while fruit of 'Green Red' is green. 'Lakatan', a banana with an excellent flavour, is popular in South-East Asia. 'Ibota Bota' is a clone highly resistant to disease. A semi-dwarf variant has been described (Daniells and Bryde, 1995).

Many other AAA clones, which are important locally, have been identified (Stover and Simmonds, 1987). 'Pisang Nangka' is found in Malaysia and Indonesia and in the Philippines as 'Nangka'. 'Pisang Susu' is grown in Indonesia. 'Paji' from Zanzibar is resistant

Plate 1.11. Bunches of 'Robusta' (AAA, Cavendish subgroup) for sale at the wholesale banana market in Tiruchirappalli, Tamil Nadu State, India (photo: D.R. Jones, INIBAP).

Plate 1.12. Bunches from cultivars in the Lujugira–Mutika subgroup (AAA) at a village in the Bushenyi district of Uganda awaiting transportation to market in Kampala (photo: D.R. Jones. INIBAP).

to black leaf streak disease. 'Pisang Ambon' and 'Orotava' accessions collected in Indonesia in the 1980s are different from clones in the Gros Michel and Cavendish subgroups (Carreel, 1995). Ten distinct AAA clones were collected in Papua New Guinea in 1988/89 (Arnaud and Horry, 1997).

AAB genomic group

Four subgroups, which are presented in Table 1.4, have been recognized within the AAB genomic group.

The Plantain subgroup is very important as plantain cultivars provide food for many millions of people in the West and Central

Table 1.3. AAA genomic group – other important clones.

Name of clone*	Selected synonyms†	Location	Main use of fruit
'Red' and 'Green Red'	'Red' – 'Green Red' 'Pisang Raja Udang' – 'Pisang Mundam' 'Morado' – 'Moradong Puti' 'Ratambala' – 'Galanamalu' 'Chenkadali' – 'Venkadali' 'Red Dacca' – 'Green Dacca'	West Indies Malaysia Philippines Sri Lanka India Australia	Dessert
'Lakatan'	'Lakatan' 'Pisang Berangan' 'Pisang Barangan'	Philippines Malaysia Indonesia	Dessert
'Ibota Bota'	'Ibota Bota' 'Kluai Khom Bao', 'Kluai Khai Thong Ruang' 'Pisang Saripipi' 'Yangambi Km 5'	Dem. Rep. of Congo Thailand Indonesia Widely distributed accession	Dessert

*The 'best-known synonym' gives its name to the clone.
†'Selected synonyms' indicate the most popular names for the clone in some of the different countries/locations where it is found.

Africa and Latin America–Caribbean regions. Cultivars are also found in East Africa and South and South-East Asia. A total of 23% of the world's production of banana fruit comes from plantain (Anon., 1992). The term plantain has been used in the past to describe all cooking banana types, but in this publication it refers only to clones belonging to the Plantain subgroup within the AAB genomic group of banana (Shepherd, 1990).

Cultivars in the plantain subgroup are placed in four main clone sets, which are distinguished on bunch and inflorescence characteristics. 'French' plantain types have many hands with comparatively small fingers and an inflorescence axis covered with persistent hermaphrodite and male flowers. The large male bud is also persistent. The bunch features of 'French' plantain are illustrated in Plate 14.11. At the other extreme, 'Horn' plantain types have few hands of very large fingers, no hermaphrodite flowers and no male axis (Plate 1.13). 'French Horn' and 'False Horn' plantain types are intermediate classification categories between 'French' and 'Horn' plantain. The male bud is absent at maturity in both of these types, but there are many hermaphrodite flowers on 'French Horn' cultivars and few on 'False Horn' cultivars (Tezenas du Montcel and Devos, 1978; Swennen, 1990). The bunch features of 'False Horn' plantain are shown in Plate 14.8). Tezenas du Montcel (1979) later proposed six subdivisions ('Giant French', 'Average French', 'Dwarf French', 'French Horn', 'False Horn' and 'True Horn') after an analysis of many different characteristics ranging from the colour and circumference of the pseudostem to orientation of the fingers and male bud characteristics. West and Central Africa is an important centre of secondary diversification of plantain. De Langhe (1961), who recognized 56 cultivars in the Democratic Republic of Congo, believes that they all arose from one clonal source. 'Agbagba', a 'False Horn' plantain, has been reported to revert occasionally to 'French' plantain (Tezenas du Montcel *et al.*, 1983). The Linnaean name *Musa paradisiaca* has been applied to 'French' plantain, but should not be used today (Stover and Simmonds, 1987).

Table 1.4. AAB genomic group – major subgroups, clone sets and important cultivars.

Subgroup	Clone set	Some cultivar names	Location	Main use of fruit
Plantain	'French'	'Obino l'Ewai'	Nigeria	Cooking
		'Njock Kon'	Cameroon	
		'Nendran'	India	
		'Dominico'	Colombia	
		'Bobby Tannap'	Cameroon	
	'French Horn'	'Mbang Okon'	Nigeria	Cooking
		Batard	Cameroon	
		3 Vert	Cameroon	
	'False Horn'	'Agbagba'	Nigeria	Cooking
		'Orishele'	Nigeria	
		'Dominico-Hartón'	Colombia	
		'Cuerno'	Central America	
		'Barraganete'	Ecuador	
	'Horn'	'Ihitisim'	Nigeria	Cooking
		'Pisang Tandok'	Malaysia	
		'Tanduk'	Philippines	
Subgroup	**Name of clone***	**Selected synonyms†**	**Location**	**Main use of fruit**
Pome	'Pome'	'Pome'	Canary Islands	Dessert
		'Prata'	Brazil	
		'Virupakshi', 'Vannan', 'Sirumalai'	India	
		'Kijakazi'	Zanzibar	
		'Brazilian'	Hawaii	
	'Prata Anã'	'Prata Anã'	Brazil	Dessert
		'Prata Santa Catarina'	Brazil	
	'Pacovan'	'Pacovan'	Brazil	Dessert
	'Pachanadan'	'Pachanadan'	India	Dessert
		'Lady Finger'	Australia	
Subgroup	**Subdivisions**	**Some cultivar names**	**Location**	**Main use of fruit**
Maia Maoli–Popoulu	'Maia Maoli'	'Mai'a Maoli Maoli', 'Mai'a Maoli Ele'ele'	Hawaii	Cooking
		'Mei'a Ma'ohi Hai', 'Mei'a Mao'i Maita'	French Polynesia	
		'Comino', 'Pompo', 'Maqueño'	Ecuador	
		'Pacific Plantain'	Australia	
	'Popoulu'	'Mai'a Popo'ulu Ka'io', 'Mai'a Popo'ulu Moa'	Hawaii	Cooking
		'Mei'a Po'u Hu'amene', 'Mei'a Po'upo'u'	French Polynesia	
		'Popoulou'	CRBP§	
Iholena		'Mai'a Iholena Ha'a', 'Mai'a Iholena Kapua', 'Mai'a Iholena Lele'	Hawaii	Cooking
		'Mei'a Ore'a 'Ute 'Ute'	French Polynesia	
		'Maritú'	Colombia	

*The 'best-known synonym' gives its name to the clone.
†'Selected synonyms' indicate the most popular names for the clone in some of the different countries/locations where it is found.
§Germplasm collection of the Centre régional bananiers et plantains, Cameroon.

Plate 1.13. 'Pisang Tandok' (AAB, 'Horn' plantain type) growing in West Malaysia (photo: D.R. Jones, INIBAP).

Cultivars in the Pome subgroup are important dessert banana types in India and Brazil, where their subacid flavour is much appreciated. The taste is also popular in Australia and Hawaii. However, cultivars are generally not very productive. 'Prata Anã' and 'Prata Santa Catarina' are dwarf variants, believed to be synonymous, from Brazil. A bunch of 'Prata Anã' can be seen in Plate 14.10. 'Pacovan' is a large fruited variant from Brazil. Both 'Prata Anã' and 'Pacovan' have been used in breeding programmes (see Jones, Chapter 14, pp. 425–434). 'Pachanadan' from India has blunt fruit. 'Lady Finger', which may be synonymous with 'Pachanadan', is grown commercially in Australia.

Traditional cooking banana cultivars found on islands in the Pacific Ocean have been placed in the Maia Maoli–Popoulu subgroup of the AAB genomic group. 'Maia Maoli' and 'Popoulu', which differ in fruit shape, are clones that represent the two major clone sets of the subgroup. Recent isozyme studies and other evidence suggest that the two types may have had independent origins (Lebot *et al.*, 1994), thus warranting their separation into two distinct subgroups. The base clones of the 'Maia Maoli' and 'Popoulu' types were probably carried by Polynesians on their migrations across the Pacific. Both may have originated in the New Guinea area (Lebot *et al.*, 1994). Fruit characteristic of 'Popoulu' types is illustrated in Plate 1.14.

Iholena is the name given to a third type of cooking banana found in the Pacific, but mainly in Hawaii. Cultivars, which differ in some morphological characteristics, are also thought to have arisen from a base clone carried by migrating Polynesians from the New Guinea area (Lebot *et al.*, 1994). The occurrence of a few representatives of the Iholena and Maia Maoli–Popoulu subgroups in the Andes region of South America has led to speculation that banana may have been introduced to the continent from the Pacific in pre-Columbian times (Langdon, 1993).

Some other important clones in the AAB genomic group are listed in Table 1.5.

'Silk', which has fruit with a sweet-acid taste, is a very popular dessert cultivar in South and South-East Asia, East Africa and Latin American–Caribbean regions (Plate 1.15). It is very popular in Brazil. 'Ayiranka Rasthali', which has a long tapering bunch and no male axis, is the only sport of 'Silk' that has been recorded. 'Silk' has been described under the Linnaean name *Musa sapientum*, but this and all other Latin names for banana should be disregarded (Stover and Simmonds, 1987).

Plate 1.14. Bunch of 'Popoulou' (AAB, Maia Maoli–Popoulu subgroup) in the CRBP germplasm collection at Nyombe, Cameroon (photo: D.R. Jones, INIBAP).

'Mysore' is grown on a large scale in South Asia (Plate 1.16) and like 'Silk' has sweet-acid fruit. However, outside India and Sri Lanka, it is usually only occasionally encountered. Exceptions are Trinidad, where it is used to shade cocoa, and Western Samoa. Only a few sports of 'Mysore' have been identified.

'Pisang Raja' is a clone found in Malaysia (Plate 1.17) and Indonesia, but is rare elsewhere. 'Pisang Kelat' is a dessert banana from Malaysia with poor quality fruit.

'Nendrapathi' bears sweet-acid fruit and persistent male flowers. 'Rajapuri' was originally thought to be its semi-dwarf mutant, but may be a distinct clone (Stover and Simmonds, 1987). 'Laknau' is a plantain-like cultivar that has been used for breeding (Plate 1.18).

One AAB clone is an oddity. 'Pisang Seribu' has numerous, small, half-developed fingers and has been described under the Latin name *Musa chiliocarpa* Backer (Plate 1.19).

ABB genomic group

ABB cultivars are generally hardy and disease-resistant. They produce starchy fruit, which is cooked.

Plate 1.15. Fruit of 'Rasthali' (AAB, syn. 'Silk') for sale in Tiruchirappalli, Tamil Nadu State, India (photo: D.R. Jones, INIBAP).

Table 1.5. AAB genomic group – other important clones.

Name of clone*	Selected synonyms†	Location	Main use of fruit
'Silk'	'Silk Fig' 'Rasthali' 'Kolikutt' 'Pisang Rastali' 'Pisang Raja Serah' 'Latundan' 'Maçã', 'Manzano' 'Pukusa' 'Sugar'	West Indies India Sri Lanka Malaysia Indonesia Philippines South America Zanzibar Australia	Dessert
'Mysore'	'Mysore' 'Poovan' 'Embul' 'Fa'i Misiluki' 'Pisang Keling' 'Kikonde Kenya'	West Indies, Australia India Sri Lanka Western Samoa Malaysia Zanzibar	Dessert
'Pisang Raja'	'Pisang Raja' 'Grindy'	Malaysia, Indonesia Windward Islands	Dessert and cooking
'Pisang Kelat'	'Pisang Kelat' 'King'	Malaysia Trinidad	Dessert
'Nendrapadathi'	'Nendrapadathi'	India	Dessert
'Laknau'	'Laknau' 'Pisang Raja Talong' 'Kune'	Philippines Malaysia Papua New Guinea	Cooking

*The 'best-known synonym' gives its name to the clone.
†'Selected synonyms' indicate the most popular names for the clone in some of the different countries/locations where it is found.

The popular clone 'Bluggoe' gives its name to a major ABB subgroup (Table 1.6). Other important clones in the subgroup are 'Silver Bluggoe', a common, waxy-fruited variant, and 'Nalla Bontha Bathees', a many-, small-fruited variant. 'Monthan', a cultivar with a pronounced cylindrical apex to its fruit instead of the tapering apex of 'Bluggoe', is recognized as a 'Bluggoe' variant by Shepherd (1990), but not all taxonomists are in agreement. 'Sambrani Monthan' and 'Pacha Monthan Bathees', are, respectively, the waxy-fruited and small-fruited sports of 'Monthan'. 'Bluggoe', which is regarded as the base clone, is found worldwide together with 'Silver Bluggoe'. Others may be restricted to India. Dwarfing to create the 'Dwarf Bluggoe' type is rare and may have only occurred in the western hemisphere. Fruit of 'Bluggoe' has a distinctive angular shape (see Plate 4.10).

Other important clones in the ABB genomic group are listed in Table 1.7.

There are a number of ABB cooking-banana cultivars in the Philippines, typified by 'Saba', which may form a distinct subgroup. It is possible that most of these cultivars arose from 'Saba' by mutation (Stover and Simmonds, 1987), but more taxonomic work needs to be undertaken to confirm this hypothesis (K. Shepherd, Portugal, 1998, personal communication). The cultivars include the waxy-fruited sport 'Abuhon', the larger-fruited 'Cardaba'

Table 1.6. ABB genomic group – clones in the Bluggoe subgroup.

Name of clone*	Selected synonyms†	Location	Main use of fruit
'Bluggoe'	'Bluggoe', 'Moko'	West Indies	Cooking
	'Pisang Abu Keling'	Malaysia	
	'Nalla Bontha'	India	
	'Fa'i Pata Samoa'	Western Samoa	
	'Kidhozi', 'Kivuvu'	East Africa	
	'Matavia'	Philippines	
	'Kluai Som'	Thailand	
	'Largo'	Hawaii	
	'Square Cooker', 'Mondolpin'	Australia	
'Silver Bluggoe'	'Silver Bluggoe', 'Silver Moko'	West Indies	Cooking
	'Kluai Hakmuk'	Thailand	
	'Thella Bontha'	India	
	'Katsila'	Philippines	
'Nalla Bontha Bathees'	'Nalla Bontha Bathees'	India	Cooking
'Monthan'	'Monthan'	India	Cooking
'Sambrani Monthan'	'Sambrani Monthan'	India	Cooking
'Pacha Monthan Bathees'	'Pacha Monthan Bathees'	India	Cooking
'Dwarf Bluggoe'	'Chamaluco Enano'	Puerto Rico	Cooking

*The 'best-known synonym' gives its name to the clone.
†'Selected synonyms' indicate the most popular names for the clone in some of the different countries/ locations where it is found.

and 'Gubao', the fused-fingered 'Inabaniko', the small-fruited 'Turangkog' and its waxy-fruited mutant 'Sabang Puti', and others. Fruit of 'Saba' is illustrated in Plate 1.20.

'Pisang Awak' is a widely disseminated, high-yielding, cooking cultivar, which is also used as a beer banana in East Africa. It is very common in Thailand (see Plate 1.5), Vietnam and elsewhere in the Indo-China region. Subclones in Thailand differ in fruit pulp colour. 'Kluai Namwa' (syn. 'Pisang Awak') has yellow pulp, 'Kluai Namwa Khao' has white pulp and 'Kluai Namwa Daeng' has pink pulp. 'Kluai Namwa Khom' is a dwarf form (Plate 1.21) (Silayoi and Chomchalow, 1987). One variant, which has been collected in West Malaysia, has sweeter fruit containing many more seed and an apparent increased susceptibility to freckle disease.

'Pelipita', which is known as 'Pelipia' in the Philippines, has been planted in Central America as a Moko-resistant cultivar (see Thwaites *et al.*, Chapter 5, p. 219). 'Ney Mannan', which is a clone from South Asia, is very popular in Sri Lanka, where it is known as 'Alukehel' or 'Ash Plantain'. 'Peyan' is another South Asian clone. 'Kalapua' is a common cultivar in Papua New Guinea and also has a dwarf form. 'Pitogo' is a cultivar from the Philippines with almost round fruit (Plate 1.22). 'Kluai Teparot' is now also included in the ABB group following the recent finding that it is not a tetraploid. The male axis is occasionally absent from this robust clone, which is

Plate 1.16. Bunches of 'Poovan' (AAB, syn. 'Mysore') for sale at the wholesale banana market in Tiruchirappalli, Tamil Nadu State, India (photo: D.R. Jones, INIBAP).

Plate 1.17. Bunch of 'Pisang Raja' (AAB) in the MARDI germplasm collection at Serdang, Selangor State, West Malaysia. Note the persistent male flowers (photo: D.R. Jones, INIBAP).

Plate 1.18. Bunch of 'Laknau' (AAB) in the CRBP germplasm collection at Nyombe, Cameroon. Note the lack of persistent bracts, one characteristic that separates this clone from 'French' plantain (photo: D.R. Jones, INIBAP).

Plate 1.19. Bunch of 'Kluai Roi Wi' (AAB, syn. 'Pisang Seribu'), which has numerous small fingers, at the Surat Thani Research Station in southern Thailand (photo: D.R. Jones, INIBAP).

Plate 1.20. Bunch of 'Kluai Hin' (ABB, syn. 'Saba') which has had the male axis removed, at the Surat Thani Research Station in southern Thailand (photo: D.R. Jones, INIBAP).

Plate 1.21. Bunch characteristics of 'Kluai Namwa Khom' (ABB), a dwarf variant of 'Pisang Awak' from Thailand, at the South Johnstone Research Station, North Queensland, Australia (photo: D.R. Jones, INIBAP).

Table 1.7. ABB genomic group – other important clones.

Name of clone*	Selected synonyms†	Location	Main use of fruit
'Saba'	'Saba' 'Pisang Kepok' 'Pisang Abu Nipah'	Philippines Indonesia Malaysia	Cooking
'Pisang Awak'	'Pisang Awak' 'Kluai Namwa' 'Katali' 'Chuoi Tay' 'Karpuravalli' 'Kayinja' 'Ducasse'	Malaysia Thailand Philippines Vietnam India East Africa Australia	Cooking
'Pelipita'	'Pelipita' 'Pelipia'	Central America Philippines	Cooking
'Ney Mannan'	'Ney Mannan' 'Alukehel', 'Ash Plantain' 'Blue Java' 'Ice Cream'	India Sri Lanka Fiji, Australia Hawaii	Cooking
'Peyan'	'Peyan'	India	Cooking
'Kalapua'	'Kalapua'	Papua New Guinea	Cooking
'Kluai Teparot'‡	'Kluai Teparot' 'Tiparot' 'Pisang Abu Siam'	Thailand Philippines Malaysia	Cooking

*The 'best-known synonym' gives its name to the clone.
†'Selected synonyms' indicate the most popular name for the clone in some of the different countries/locations where it is found.
‡'Kluai Teparot' was previously considered to be in the ABBB genomic group.

found in many countries in South-East Asia (Stover and Simmonds, 1987). Fruit of 'Kluai Teparot' is pictured in Plate 1.22.

Genomic groups with S and T components

Shepherd and Ferreira (1984) and Arnaud and Horry (1997) identified some cultivars in Papua New Guinea believed to contain genetic elements of *Musa schizocarpa*, another wild species in the *Eumusa* section. Genomic groups were designated AS, AAS, and ABBS where S indicates hybridization with *M. schizocarpa*. Other cultivars were thought to have genetic components from species in the *Australimusa* section. AAT, AAAT and ABBT genomic groups have been recognized where T indicates hybridization with an *Australimusa* species. Recent research using molecular taxonomic methods has indicated that *M. schizocarpa* and one or more species in the *Australimusa* section did play a role in the origin of some *Eumusa* cultivars present in Papua New Guinea (Carreel, 1995). However, cultivars with these unusual combinations of genomic constituents seem to be only occasionally found in cultivation.

BBB genomic group

There is no evidence to suggest that edibility occurred spontaneously in *M. balbisiana* as it did in *M. acuminata*. Although 'Saba'-type cultivars have been classified as having a BBB genome using the scoring method of Simmonds and Shepherd (1955) (Pascua and Espino, 1987), this has been challenged on morphological grounds (Shepherd, 1990) and on molecular evidence (Jarret and Litz, 1986; Carreel, 1995). However, 'Kluai Lep Chang Kut' from Thailand may be a BBB as it has no *M. acuminata* characteristics and resembles *M. balbisiana* morphologically. This clone may have arisen from a cross between an ABB cultivar, such as 'Kluai Teparot', and *M. balbisiana* (Shepherd, 1990; J.P. Horry, Montpellier, 1999, personal communication).

For more information on subgroups and clones, see Stover and Simmonds (1987). Consult Robinson (1996) for data on bunch weights and yields ha^{-1} of various banana cultivars.

Recent advances in knowledge on the origin of cultivars in the Eumusa *series*

Musa acuminata has a number of subspecies and each has its own area of distribution (see Fig. 1.2). The Malayan peninsula has been suggested by Simmonds (1962) as the location of the origin of edible banana because *M. acuminata* ssp. *malaccensis* was believed by him to be the primary source of edibility. However, this hypothesis has recently been challenged by Carreel (1995). With the aid of restriction fragment length polymorphism (RFLP) markers, she has compared DNA from chloroplasts (inherited through the female parent), mitochondria (inherited through the male parent) and the nucleus of many *Musa* species, subspecies and landraces. Her work has given an insight into the wild species and subspecies that have contributed to the genetic make-up of cultivars in the *Eumusa* series of edible banana. This in turn has given clues as to probable location of origin of cultivars.

The first step in the evolution of the *Eumusa* edible banana is generally believed to have been the development of parthenocarpic AA diploid cultivars (see Fig. 1.5) (Simmonds, 1962). Carreel (1995) analysed 128 AA diploid clones and found all contained DNA from *M. acuminata* ssp. *banksii* (Plate 1.23) and/or its close relative *M. acuminata* ssp. *errans*. She also analysed 115 clones in the other main genomic groups and found 111 contained genetic components from *M. acuminata* ssp. *banksii/errans*. These results indicate that the first edible banana landraces in the *Eumusa* series were most probably derived from *Musa acuminata* ssp. *banksii/errans*. *Musa acuminata* ssp. *banksii* occurs naturally in the New Guinea area and *M. acuminata* ssp. *errans* in the Philippines (see Fig. 1.2). Edibility may have evolved in one or both locations. As many more primitive AA diploid cultivars are found in Papua New Guinea than anywhere else in the world (Stover and Simmonds, 1987), rather than the last refuge of early diploids originating in Malaysia (Stover and Simmonds, 1987), this area may be the first centre of domestication of the edible banana.

Carreel (1995) has hypothesized that the genomic constituents of other *M. acuminata* subspecies were incorporated into edible banana as primitive, diploid clones spread westwards into South-East Asia. Early diploids derived from *M. acuminata* ssp. *banksii/errans* would have had starchy fruit. Carreel's work indicates that dessert cultivars probably arose by the integration of genetic material from *M. acuminata* ssp. *malaccensis* and ssp. *zebrina* in areas of present-day Indonesia and Malaysia, where these subspecies occur naturally. Hybridization with *M. balbisiana* may have occurred in the Philippines and/or when early cultivars spread to Indo-China, northern Burma and India. The wild *Musa* species and subspecies implicated in the ancestry of the cultivated *Eumusa* banana cultivars are listed in Table 1.8.

Other interesting information has emerged from Carreel's study. Cooking and beer-making cultivars in the Lujugira–Mutika subgroup (AAA), which are common in a secondary centre of diversity in the highlands of East Africa, contain genetic components of *M. acuminata* ssp. *banksii* and ssp. *zebrina*. One can speculate that the progenitor of the distinct East African highland banana types may have had its origin in western Indonesia where

Plate 1.22. Fruit of 'Kluai Teperot' (left), an ABB clone with angular fingers found in a number of South-East Asia countries and 'Pitogo' (right), an ABB clone with small rounded fingers grown in the Philippines (photo: D.R. Jones, INIBAP).

Plate 1.23. *Musa acuminata* ssp. *banksii*, pictured here growing wild in north-western Papua New Guinea, is the probable progenitor of the first edible banana cultivars in the *Eumusa* series (photo: D.R. Jones, QDPI).

M. acuminata ssp. *zebrina* is indigenous. Banana is considered by some historians to have reached the east coast of Africa at about the same time as other South-East Asian food crops and to have been transported by Indonesian voyagers (Simmonds, 1962). Carreel (1995) also confirmed that both the A genomes of AAB Plantain and Maia Maoli–Popoulu subgroups were derived from *M. acuminata* spp. *banksii*, as had been suggested by earlier workers (Horry and Jay, 1990; Lebot *et al.*, 1994).

From a plant-pathological perspective, it is interesting to note that *M. acuminata* ssp. *banksii/errans* is recognized as having considerably more disease problems than other *Musa* species and subspecies of *M. acuminata* (Vakili, 1965, 1968; Argent, 1976). If almost all edible banana cultivars in the *Eumusa* series have inherited some component of their genetic make-up from *M. acuminata* ssp. *banksii/errans*, it may explain their susceptibility to certain diseases to varying degrees. It is to be hoped that future work will throw much more light on the origin of today's cultivars. Perhaps with knowledge of their genetic background, it will be easier to explain the reaction of landraces and bred hybrids to disease.

Table 1.8. Wild *Musa* implicated in the ancestry of the *Eumusa* series of edible banana cultivars (Carreel, 1995).

Species	Subspecies	Geographical distribution*
Musa acuminata	*banksii*†	New Guinea, north-east Queensland (Australia), Western Samoa‡
	errans†	Philippines
Musa acuminata	*burmannica*§ (*burmannicoides*)	Burma
	siamea§	Thailand, Indo-China
Musa acuminata	*malaccensis*	Southern Thailand, West Malaysia, Sumatra (?)
Musa acuminata	*microcarpa*	North Borneo
Musa acuminata	*zebrina*	Java
Musa balbisiana		Indo-China, northern Burma, India, Sri Lanka, Philippines, New Guinea,‖ Malaysia,¶ Thailand¶
*Musa schizocarpa***		New Guinea
Australimusa species††		New Guinea

*Based on information from Simmonds (1962), Shepherd (1990) and Carreel (1995).
†The nuclear genomes of *M. acuminata* ssp. *banksii* and ssp. *errans* are similar, but the cytoplasmic genomes are different.
‡*Musa acuminata* ssp. *banksii* is believed to have been introduced to Western Samoa (Simmonds, 1962).
§The nuclear, chloroplastic and mitochondrial genomes of *M. acuminata* ssp. *burmannica, burmannicoides* and *siamea* accessions in international collections are similar (Carreel, 1995).
‖Simmonds (1962) believes *M. balbisiana* to be indigenous to Papua New Guinea, but Argent (1976) thinks the species was introduced.
¶*Musa balbisiana* was introduced to Malaysia and Thailand, where it was cultivated.
**This species has contributed to the genome of some cultivars found in Papua New Guinea.
††One or more species within the *Australimusa* section may be contributing to the genome of some cultivars in Papua New Guinea.

Australimusa *series of edible banana cultivars*

A second series of edible banana called Fe'i banana is found in eastern Indonesia, New Guinea and on islands in the Pacific. Cultivars in this group have upright fruit bunches (see Fig. 1.4), orange fruit when ripe and usually red sap like *Musa maclayi*, a wild *Australimusa* species found in Papua New Guinea and the Solomon Islands. Because of these shared characteristics, Simmonds (1962) believed that *M. maclayi* must have played a major role in the evolution of Fe'i cultivars, although he did not rule out an interspecific origin.

Cheesman (1947) noted similarities between the Fe'i banana and *Musa lolodensis*, another species in the *Australimusa* section from New Guinea. Close links have been demonstrated between three Fe'i cultivars and *M. lolodensis*, using RFLP analysis. This has led Jarret *et al.* (1992) to speculate that the series may be derived solely from this species. However, more recent work by Carreel (1995) has indicated that the Fe'i banana may have an interspecific origin as the nuclear genome of some of the cultivars she analysed were close to *M. maclayi*, some close to *M. lolodensis* and some close to *Musa peekelii*, which is yet another species in the *Australimusa* section found in Papua New Guinea.

It seems very likely that Fe'i cultivars originated in the New Guinea–Solomon Islands area and spread eastwards across the Pacific with migrating Polynesians (MacDaniels, 1947). The Fe'i cultivars seem highly resistant to diseases of the foliage, but little is known about their reaction to root diseases or their response to viruses and bacteria. This lack of knowledge is a reflection of the relatively unimportant role that these types now play as foodstuff in the countries where they are still cultivated.

Importance of banana classification to banana pathologists

Often in the past, research has been conducted on incorrectly identified germplasm, which has led to much confusion in the scientific literature. It is important that banana germplasm that is the subject of research be accurately identified by scientists and agriculturists. Only then can valid comparisons be made between the results of work undertaken at different times and at different locations.

One of the main challenges facing banana taxonomists today is to resolve synonymy among the many different names for banana clones and develop a system whereby cultivars can be easily identified. This would enable research workers to define their material accurately. A computerized *Musa* germplasm identification system based on morphological characters has been developed by the Centre de coopération internationale en recherche agronomique pour le developpement (CIRAD) in France and is being promoted by the International Network for the Improvement of Banana and Plantain (INIBAP) (Perrier and Tezenas du Montcel, 1990). Based on the available information, the software gives a probability rating of the identity of the clone. As more morphological information on a particular clone is introduced, the higher the probability of a correct identification. The programme is currently being tested on a range of clones, each derived from the same source, growing under diverse environmental conditions in a number of countries.

It is just as important for plant pathologists to know the correct identity of material being tested or screened for disease resistance as it is for them to know the correct identity of the pathogen. In this way, the reaction of a particular clone (with a particular genetic composition) to a disease can be put into perspective *vis-à-vis* the reaction of other clones. An attempt has been made in this volume to provide information on the reaction of the different types of banana to each disease. Inevitably, more information is available on the responses of cultivars to serious and widespread diseases that have been well studied than on host responses to minor diseases of limited distribution.

As mentioned previously, to help the reader place germplasm named in the text of this publication into perspective, the name of the genomic group followed by the best-known synonym or subgroup, has been placed in parentheses after the cultivar name at appropriate places throughout the text.

Abacá

Abacá is indigenous to the Philippines, where most is grown, and is often called 'Manila hemp', a name given to it by early European traders, who found it for sale in the Manila market. Abacá produces the strongest of the cordage fibres and is used to make the best grades of commercial cables and ropes. Because it has a high degree of resistance to sea water and a low degree of swelling when wet, it is particularly suited to marine cordage. Large amounts are also pulped and used to make high-quality paper and specialties such as tea bags, paper sacks and movable walls for Japanese houses (Purseglove, 1972).

Abacá grows best in the wet tropics where the annual precipitation is 2000–3200 mm, spread evenly throughout the year. The average annual temperature should be about 27°C, but not below 21°C, and the relative humidity around 80%. Abacá grows best in well-drained, deep, fertile soil, rich in humus and potash. Although these soils have relatively high fertility levels, continuous cropping of abacá leads eventually to yield declines because of the rapid rate of nutrient uptake by plants. In the Philippines, the crop is usually grown on land below 500 m (Berger, 1969; Purseglove, 1972).

Abacá closely resembles banana in appearance. However, abacá stalks are generally more slender and the leaves smaller, narrower and more tapered. The fruit is also smaller, being about 7.5 cm in length, and curves upwards as it ripens. Fingers are inedible, being hard green structures containing numerous black seed. Abacá has an underground rhizome with numerous small roots, which do not penetrate far into the soil. Erect pseudostems arise from the rhizome and reach a height of 5–8 m. As with banana, pseudostems consist of thickened, clasping leaf bases. The fibre from the outer sheaths is coarser, stronger and darker in colour than the innermost fibre, which is whiter and weaker.

Planting material may be either suckers or whole corms or pieces of corm with a vegetative bud. However, suckers are rarely used because of the difficulty of transport. True seed can also be planted, but plants take longer to mature. 'Seed' pieces, the equivalent of banana bits, are planted at a depth of 5–10 cm and 2.5–3.0 m apart. Young plants may be partially shaded for protection from excessive heat. The date of the first harvest depends on the cultivar, soil conditions and climate, but full-grown pseudostems can be harvested from 18 to 24 months after planting. After the first harvest, two or four pseudostems can be harvested from each mat every 4–6 months. Yields are initially small, but reach a maximum after 4–5 years, when 12.5 t of dry fibre ha^{-1} can be obtained. Yields decline after 7–8 years and the crop is replanted after 10–15 years.

Optimum cutting time is just before the *flag leaf* emerges. The flag leaf is the rudimentary and very small leaf that precedes the appearance of the inflorescence and its emergence can be anticipated because plant growth slows and leaf blades gradually shorten. Abacá is harvested by first trimming off the leaves and then cutting the pseudostem close to the ground. The percentage of fibre in a pseudostem is 1.5–3.0% of the weight. Fibre is extracted as soon after harvest as possible by hand-stripping, spindle stripping or using a decorticating machine. Extracted fibre is either sun-dried in the open or air-dried in sheds. It is then tied in bundles, graded and baled for export. Abacá fibre consists of collections of sclerenchyma cells about 2.5–12 mm long. Strands of fibre can reach a length of 1.5 to 3.0 m. They are three times as strong as cotton and twice as strong as sisal fibres. Low-quality fibre comes from plants harvested too young or too old. Abacá fibre deteriorates during prolonged storage. The main fungi responsible have been identified as *Aspergillus fumigatus* Fres. and *Chaetomium funicolum* Cke (Purseglove, 1972).

Most abacá is grown in the Philippines. Here, the bulk of the crop comes from the eastern Visayas and Bicol regions (E.O. Lomerio, Legazpi City, 1996, personal communication). In 1998, 65,000 million tonnes were produced on 95,000 ha (Anon., 1999). Ecuador was the next largest producer, with 16,380 million tonnes grown on 23,287 ha (Anon., 1999). Minor suppliers are Equatorial Guinea, with 1800 ha of abacá, Costa Rica, with 1100 ha, and Indonesia, with 620 ha. Total world production for 1998 came to approximately 90.5 million tonnes (Anon., 1999).

More than 100 cultivars of abacá have been identified in the Philippines, but only about 20 are of commercial significance (Anunciado *et al.*, 1977). These are listed in Table 1.9. Cultivars are distinguished by the colour and shape of their flowers and pseudostem, yield and quality of fibre and resistance to disease. Some, like 'Itom', are natural hybrids of *M. textilis* × *M. balbisiana* and are used to make 'Canton fibre' (Purseglove, 1972).

Enset

Enset (Fig. 1.6) is cultivated in south-western Ethiopia (Fig. 1.7), where it is important as a starchy foodstuff, animal feed and a source of fibre for making rugs, sacks, bags and ropes. In addition, fresh leaves are used as bread and food wrappers and as plates. Dried petioles and midribs are burnt as fuel and their pulp utilized as cleaning rags and brushes, baby nappies and cooking-pot stands. Parts of the plant also have medicinal properties (Brandt *et al.*, 1997).

Enset forms an important part of the diet

Table 1.9. Abacá cultivars recommended for growing in the Philippines (information from E.O. Lomerio, Legazpi City, 1996, personal communication).

Region	Cultivar
Visayas	'Inosa', 'Linawaan', 'Linino', 'Linlay'
Mindanao	'Tangoñgon', 'Maguindanao', 'Bongolanon'
Bicol	'Itom', 'Sogmad', 'Tinawagan Pula', 'Tinawagan Puti', 'Lausigon', 'Abaub', 'Casilihon', 'Lausmag 24', 'Itolaus 45', '*M. textilis* 52', '*M. textilis* 51', '*M. textilis* 50'
Southern Tagalog	'Tinawagan Pula', 'Tinawagan Put'i', 'Sinibuyas', 'Putian'

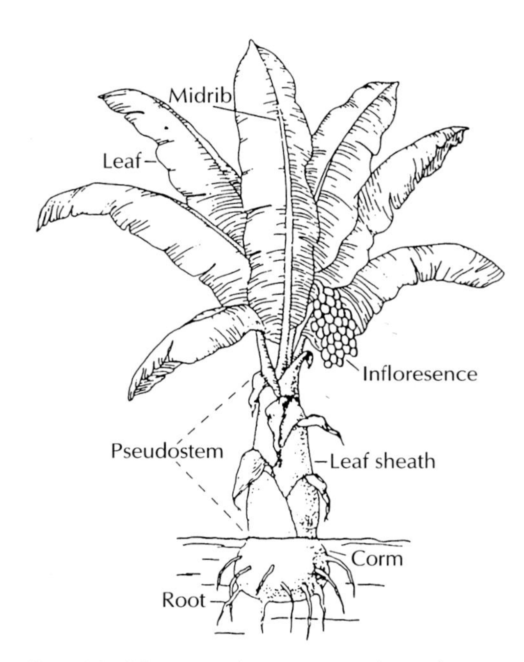

Fig. 1.6. Diagrammatic representation of enset (*Ensete ventricosum*) (from Brandt *et al.*, 1997).

of about 10 million people, which represents about 20% of the Ethiopian population. Enset farming is believed to be one of Africa's few surviving examples of indigenous, sustainable agriculture, which has evolved over hundreds of years. Enset is probably one of the oldest useful plants in Africa and has magic–religious significance in many areas. The Ari people preserve enset genetic resources by a highly ritualized, *ex situ* conservation of wild plants, which maintains the genetic diversity of landraces through a constant gene flow between wild and cultivated plants (Shigeta, 1990).

The system of cultivation enables farmers to grow enset in the same plot for generations without the aid of chemical fertilizer. Enset is resistant to climatic and environmental fluctuations and produces edible products that can be stored for months, if not years. It has been estimated that one mature enset plant will feed a family of four or five people for about 20 days (Alemu and Sandford, 1991).

Enset is planted at altitudes ranging from 1200 to 3100 m, but grows best at elevations of 2000 to 2750 m. The average temperature where it is grown varies from 10 to 21°C, with a relative humidity range of 63–80%. Like banana, enset does not tolerate freezing and frost damage is often observed above 2800 m. The annual rainfall of most growing areas is 1100–1500 mm, with the majority falling between March and September. Irrigation is used to supplement rainfall below 1500 m, where rainfall decreases and there is a greater evaporative demand. Enset grows in most soil types if they are sufficiently fertile and well drained. Cattle manure is used as the main organic fertilizer. Ideal soils for enset cultivation have a pH of 5.6–7.3 and contain 2–3% organic matter (Brandt *et al.*, 1997).

Enset development is similar to banana in that leaves are successively produced by an apical meristem until flowering. The plant is normally 5–7 m in height and the pseudostem has a circumference of

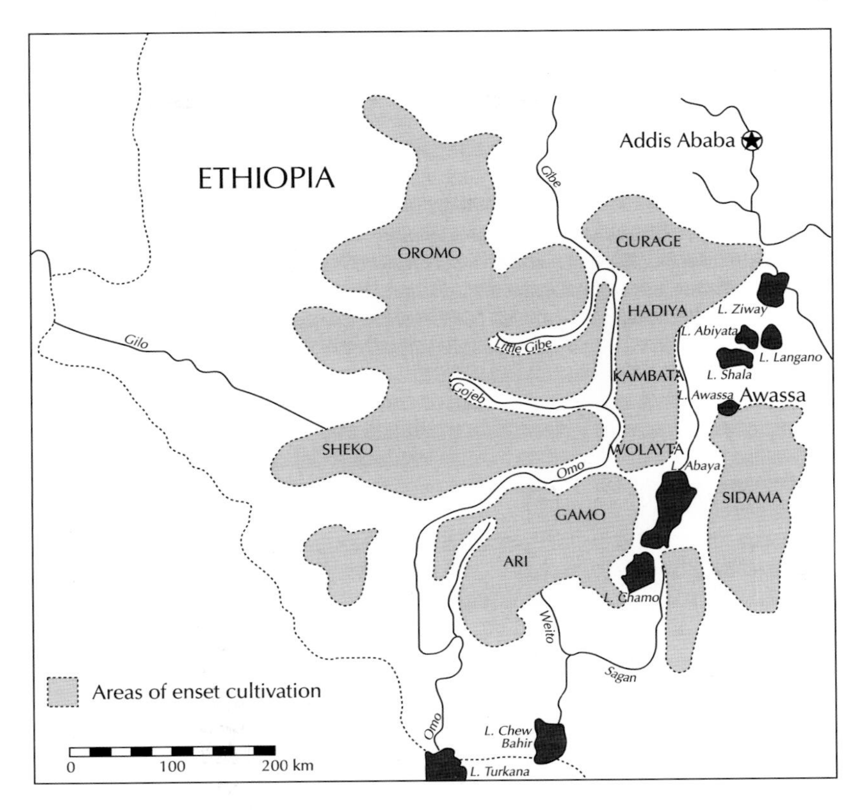

Fig. 1.7. Enset growing areas in south-west Ethiopia with names of enset-cultivating, ethnic groups (from Brandt *et al.*, 1997).

0.5–3.0 m at maturity. It has a shallow rooting system, which extends 2–3 m.

It takes about 6–7 years at the optimum altitude for growth for an enset plant to reach maturity and form an inflorescence. Several fruit, carrying 5–15 seeds, which vary in size and weight according to the enset clone, are produced in the inflorescence. The seedy, leathery fruit are inedible. After fruiting, the plant declines.

Although in nature enset only reproduces by seed, the Ethiopian people have developed a technique by which vegetative shoot formation is induced, thus enabling superior plants to be clonally propagated. The central bud at the top of the corm of a 4–6-year-old plant is hollowed out and the corm usually split into four equal parts, buried and covered with manure. About 40–200 buds are produced by this method and, after 4–6 weeks, the suckers emerge from the soil (Demeke, 1986). The suckers are later severed from the corm and planted in nurseries. Management varies considerably after this stage, depending on the ethnic community and household requirements. Plants may be transplanted up to four times at ever wider spacing. Some may be harvested young, while others are left for harvesting when mature. Generally, a leaf canopy is maintained that covers the soil for most of the year. Enset may grow alone or in mixed plantings with other crops, such as maize, cabbage, coffee and citrus (Brandt *et al.*, 1997).

Enset is harvested before the formation of an inflorescence, which uses up starch reserves in the plant (Purseglove, 1972). At harvest, the pseudostem and corm are decorticated with a bamboo scraper. Several different food products are made from the raw starch obtained. One is *kocho*, which is a dough made from the scraped leaf sheaths of the pseudostem and grated corm after fermentation, which takes 15–20 days. *Kocho* is stored in a pit lined with enset leaves for at least 1 month before consumption. However, it can be kept for many months or several years. Another is *bulla*, which is a porridge, pancake or dumpling made from dehydrated juice obtained from the scrapings of the leaf sheath, peduncle and grated corm. *Amicho* is the boiled corm, usually from a young plant (Brandt *et al.*, 1997). Yields of food products vary between 10 and 18 kg plant^{-1} (Demeke, 1986).

Acknowledgements

The advice given by Jean-Pierre Horry, Ken Shepherd, Rony Swennen and Jeff Daniells on some of the finer points of banana classification is much appreciated.

REFERENCES

Alemu, K. and Sandford. S. (1991) *Enset in North Omo Region.* FRP Technical Pamphlet No. 1, Farm Africa, Addis Ababa, Ethiopia, 49 pp.

Anon. (1992) Banana and plantain – food for thought. In: INIBAP (ed.) *Bananas, Plantains and INIBAP.* Annual Report 1992, INIBAP, Montpellier, France, pp. 7–11.

Anon. (1999) Statistics from FAO website http://apps.fao.org/default.htm

Anunciado, I.S., Balmes, L.O., Bawagan, P.Y., Benigno, D.A., Bondad, N.D., Cruz. O.J., Franco, P.T., Gavarra, M.R., Opeña, M.T. and Tabora, P.C. (1977) *The Philippines Recommends for Abaca, 1977.* Philippine Council for Agriculture and Resources Research, Los Baños, Philippines, 71 pp.

Argent, G.C.G. (1976) The wild bananas of Papua New Guinea. *Notes from the Royal Botanic Garden, Edinburgh* 35, 77–114.

Arnaud, E. and Horry, J.P. (eds) (1997) *Musalogue, a Catalogue of* Musa *Germplasm, Papua New Guinea Collecting Missions, 1988–1989.* INIBAP, Montpellier, France, 125 pp.

Berger, J. (ed.) (1969) Abacá or manila hemp (*Musa textilis*). In: *The World's Major Fibre Crops: Their Cultivation and Manuring.* Centre d'Etude de l'Azote, Zurich, pp. 233–235.

Bourke, R.M. (1976) Know your bananas. *Harvest (Papua New Guinea)* 3, 48–54.

Brandt, S.A., Spring, A., Hiebsch, C., McCabe, J.T., Tabogie, E., Diro, M., Wolde-Michael, G., Yntiso, G., Shigeta, M. and Tesfaye, S. (1997) *The 'Tree Against Hunger', Enset-based Agricultural Systems in Ethiopia.* American Association for the Advancement of Science, Washington, DC, USA, 56 pp.

Carreel, F. (1995) Etude de la diversité génétique des bananiers (genre *Musa*) à l'aide des marqueurs RFLP. PhD thesis, Institut National Agronomique, Paris-Grignan, France.

Champion, J. (1963) *Le Bananier.* Maisonneuvre and Larose, Paris, 214 pp.

Cheesman, E.E. (1947) Classification of the bananas. II The genus *Musa* L. *Kew Bulletin* 2, 106–117.

Daniells, J.W. (1990) The Cavendish subgroup: distinct and less distinct cultivars. In: Jarret, J.L. (ed.) *Identification of Genetic Diversity in the Genus* Musa, *Proceedings of a Workshop held at Los Baños, Philippines, 5–10 September 1988.* INIBAP, Montferrier-sur-Lez, France, pp. 29–35.

Daniells, J. and Bryde, N. (1995) Semi-dwarf mutant of Yangambi Km5. *Infomusa* 4(2), 16.

De Langhe, E. (1961) La taxonomie du bananier plantain en Afrique équatoriale. *Journal d'Agriculture Tropicale et de Botanique Appliquée (Brussels)* 8, 417–449.

Demeke, T. (1986) Is Ethiopia's *Ensete ventricosum* crop her greatest potential food? *Agriculture International* 38, 362–365.

Horry, J.P. and Jay, M. (1990) An evolutionary background for bananas as deduced from flavonoids diversification. In: Jarret, R.L. (ed.) *Identification of Genetic Diversity in the Genus* Musa, *Proceedings of an International Workshop held at Los Baños, Philippines, 5–10 September 1988.* INIBAP, Montferrier-sur-Lez, France, pp. 41–55.

Horry, J.P., Ortiz, R., Arnaud, E., Crouch, J.H., Ferris, R.S.B., Jones, D.R., Mateo, N., Picq, C. and Vulylsteke, D. (1997) Banana and plantain. In: Fuccillo, D., Sears, L. and Stapleton, P. (eds) *Biodiversity in Trust, Conservation and Use of Plant Genetic Resources in CGIAR Centres.* Cambridge University Press, Cambridge, UK, pp. 67–81.

Horry, J.P., Doležel, J., Doleželova, M. and Lysak, M.A. (1998) Do natural A × B tetraploid bananas exist? *Infomusa* 7(1), 5–6.

Jarret, R.K. and Litz, R.E. (1986) Determining the interspecific origins of clones within the 'Saba' cooking banana complex. *Horticultural Science* 21, 1433–1435.

Jarret, R.L., Gawel, N., Whittemore, A. and Sharrock, S. (1992) RFLP based phylogeny of *Musa* species, Papua New Guinea. *Theoretical and Applied Genetics* 84, 579–584.

Jenny, C.F., Carreel, F. and Bakry, F. (1997) Revision on banana taxonomy: 'Klue Tiparot' (*Musa* sp.) reclassified as a triploid. *Fruits* 52, 83–91.

Karamura, D.A. and Karamura, E.B. (1994) *A Provisional Checklist of Banana Cultivars in Uganda.* INIBAP, Montpellier, France, 28 pp.

Karamura, D.A. (1999) Numerical taxonomic studies of the East African Highland Bananas (*Musa* AAA – East Africa) in Uganda. Submitted PhD thesis, Department of Agricultural Botany, University of Reading, January 1998. INIBAP, Montpellier, France, 192 pp.

Langdon, R. (1993) The banana as a key to early American and Polynesian history. *Journal of Pacific History* 28, 15–35.

Lebot, V., Meilleur, B.A. and Manshardt, R.M. (1994) Genetic diversity in eastern Polynesian Eumusa bananas. *Pacific Science* 48, 16–31.

MacDaniels, L.H. (1947) A study of the 'Fe'i banana and its distribution with reference to Polynesian migrations. *Bernice P. Bishop Museum Bulletin* 190, 3–56.

Pascua, O.C. and Espino, R.R.C. (1987) Taxonomic classification of Philippine Bananas. In: Persley, G.J. and De Langhe, E.A. (eds) *Banana and Plantain Breeding Strategies. Proceedings of an International Workshop held at Cairns, Australia, 13–17 October 1986.* ACIAR Proceedings No. 21, Australian Centre for International Agricultural Research, Canberra, Australia, pp. 161–164.

Perrier, X. and Tezenas du Montcel, H. (1990) Musaid: a computerized determination system. In: Jarret, R.L. (ed.) *Identification of Genetic Diversity in the Genus* Musa, *Proceedings of an International Workshop held at Los Baños, Philippines, 5–10 September 1988.* INIBAP, Montferrier-sur-Lez, France, pp. 76–91.

Purseglove, J.W. (ed.) (1972) Musaceae. In: *Tropical Crops – Monocotyledons.* Longman, London, pp. 343–384.

Richardson, D.L., Hamilton, K.S. and Hutchison, D.J. (1965) Notes on bananas, I, Natural edible tetraploids. *Tropical Agriculture (Trinidad)* 42, 125–137.

Robinson, J.C. (1996) *Bananas and Plantains.* Crop Production Science in Horticulture 5, CAB International, Wallingford, UK, 238 pp.

Sheperd, K. (1957) Banana cultivars of East Africa. *Tropical Agriculture (Trinidad)* 34, 277–286.

Shepherd, K. (1990) Observations on *Musa* taxonomy. In: Jarret, R.L. (ed.) *Identification of Genetic Diversity in the Genus* Musa*: Proceedings of an International Workshop held at Los Baños, Philippines, 5–10 September 1988.* INIBAP, Montferrier-sur-Lez, France, pp. 158–165.

Shepherd, K. and Ferreira, F.R. (1984) The PNG Biological Foundation's Banana Collection at Laloki, Port Moresby, Papua New Guinea. *IBPGR Regional Committee for Southeast Asia Newsletter* 8(4), 28–34.

Shigeta, M. (1990) Folk *in-situ* conservation of ensete (*Ensete ventricosum* [Welw.] E.E. Cheesman): towards the interpretation of indigenous agricultural science of the Ari, southwestern Ethiopia. *Kyoto University, African Studies Monographs* 10(3), 93–107.

Silayoi, B. and Chomchalow, N. (1987) Cyotaxonomic and morphological studies of Thai banana cultivars. In: Persley, G.J. and De Langhe, E.A. (eds) *Banana and Plantain Breeding Strategies. Proceedings of an International Workshop held at Cairns, Australia, 13–17 October 1986.* ACIAR Proceedings No. 21, Australian Centre for International Agricultural Research, Canberra, Australia, pp. 157–160.

Simmonds, N.W. (1962) *The Evolution of the Bananas.* Longmans, Green, London, 170 pp.

Simmonds, N.W. and Shepherd, K. (1955) The taxonomy and origins of the cultivated bananas. *Journal of the Linnean Society of London* 55, 302–312.

Stover, R.H. and Simmonds, N.W. (1987) *Bananas*, 3rd edn. Longman Scientific and Technical, Harlow, UK, 468 pp.

Swennen, R. (1990) Limits of morphotaxonomy: names and synonyms of plantains in Africa and elsewhere. In: Jarret, R.L. (ed.) *Identification of Genetic Diversity in the Genus* Musa*: Proceedings of an International Workshop held at Los Baños, Philippines, 5–10 September 1988.* INIBAP, Montferrier-sur-Lez, France, pp. 172–210.

Tezenas du Montcel, H. (1979) Les plantains du Cameroon. Propositions pour leur classification et dénominations vernaculaires. *Fruits* 34, 83–97.
Tezenas du Montcel, H. and Devos, P. (1978) Proposal for establishing a plantain determination card. *Paradisiaca (Ibadan, Nigeria)* 3, 14–17.
Tezenas du Montcel, H., De Langhe, E. and Swennen, R. (1983) Essai de classification des bananiers plantain (AAB). *Fruits* 6, 461–474.
Vakili, N.G. (1965) Fusarium wilt resistance in seedlings and mature plants of *Musa* species. *Phytopathology* 55, 135–140.
Vakili, N.G. (1968) Response of *Musa acuminata* species and edible cultivars to infection by *Mycosphaerella musicola. Tropical Agriculture (Trinidad)* 45, 13–22.
Wardlaw, C.W. (1961) *Banana Diseases Including Plantains and Abaca.* Longmans, Green and Co. Ltd, London, 648 pp.

2

Fungal Diseases of the Foliage

SIGATOKA LEAF SPOTS

Black Leaf Streak

J. Carlier, E. Fouré, F. Gauhl, D.R. Jones, P. Lepoivre, X. Mourichon, C. Pasberg-Gauhl and R.A. Romero

Distribution and economic importance

C. Pasberg-Gauhl, F. Gauhl and D.R. Jones

Introduction

Black leaf streak, or black Sigatoka as it is known in the Latin American–Caribbean region, is regarded as the most economically important leaf disease of banana. It is a very similar leaf spot disease to Sigatoka (see Jones, this chapter, pp. 79–92), but it is more virulent and affects a wider range of banana genotypes. Black leaf streak develops much more rapidly than Sigatoka and causes more severe defoliation. In many coastal locations, it has replaced Sigatoka, which was present before the arrival of black leaf streak, as the dominant leaf spot within a few years of its introduction. Black leaf streak destroys banana leaves, which leads to a reduction in yield and premature ripening of the fruit. Since it was first recognized in the South Pacific, black leaf streak has spread to most of the major banana-growing regions of the world and is still spreading. Countries where the disease occurs are listed in Table 2.1.

Distribution in Asia and the Pacific

Black leaf streak was first noticed in February 1963 in the Sigatoka district of the island of Viti Levu in Fiji, where it was said to be spreading rapidly (Rhodes, 1964). The disease was also described as replacing Sigatoka wherever it became established (Leach, 1964a, b). Surveys conducted between 1964 and 1967 showed black leaf streak to be widespread in the Pacific and parts of the Pacific rim. According to Johnston (1965) and Graham (1969), the disease was present in the Federated States of Micronesia, New Caledonia, Papua New Guinea, Philippines, Western Samoa, Singapore, Solomon Islands, Tahiti, Taiwan, Tonga, Vanuatu and West Malaysia. Meredith and Lawrence (1969) identified black leaf streak in the Hawaiian Islands and Stover (1976) reported the presence of the disease in the additional Pacific locations of the Cook Islands and Niue.

The widespread distribution of black leaf streak in and around the Pacific suggested that it had been present in the region for some considerable time before its discovery in Fiji in 1963 (Meredith, 1970; Stover 1978; Long, 1979). Meredith and Lawrence (1969) thought that black leaf streak may have arrived in the Hawaiian Islands in 1958 and been mistaken for Sigatoka.

In 1970, black leaf streak was found throughout the Philippines on many different cultivars (Hapitan and Reyes, 1970). Evidence suggested that the disease had been present around Los Baños on the island of Luzon in the Philippines since 1964, but had most probably arrived even earlier.

Table 2.1. Countries where black leaf streak has been recorded and year of detection.

Region/country	Year of detection	Reference
West Africa		
Benin	1993	Jones and Mourichon (1993)
Cameroon		
South-east	1980	Tezenas du Montcel (1982)
South-west	1983	Wilson and Buddenhagen (1986)
Central African Republic	1996	X. Mourichon, Montpellier, 1996, personal communication
Congo	1985	Mourichon (1986)
Côte d'Ivoire	1985	Mourichon and Fullerton (1990)
Democratic Republic of Congo		
Highlands	1987	Sebasigari and Stover (1988)
Lowlands	1988	Mobambo and Naku (1993)
Gabon	1978	Frossard (1980)
Ghana	1986	Wilson (1987)
Nigeria	1986	Wilson and Buddenhagen (1986)
São Tomé	1980	Frossard (1980)
Togo	1988	Mourichon and Fullerton (1990)
East Africa		
Burundi	1987	Sebasigari and Stover (1988)
Comoros	1993	Jones and Mourichon (1993)
Kenya	1988	Kung'U *et al.* (1992)
Malawi	1990	Ploetz *et al.* (1992)
Mayotte	1993	X. Mourichon, Montpellier, 1996, personal communication
Rwanda	1986	Sebasigari (1990)
Tanzania	1987	Sebasigari and Stover (1988)
Pemba	1987	Dabek and Waller (1990)
Zanzibar	1987	Sebasigari and Stover (1988)
Uganda	1990	Tushemereirwe and Waller (1993)
Zambia	1973*	Raemaekers (1975)
Latin America and the Caribbean		
Belize	1975	Stover (1980b)
Bolivia	1996	Tejerina (1997)
Brazil	1998	Cordeiro *et al.* (1998)
Colombia	1981	Merchan (1990)
Costa Rica	1979	Stover (1980b)
Cuba	1990	Vidal (1992)
Dominican Republic	1996	X. Mourichon, Montpellier, 1996, personal communication
Ecuador	1986	Mourichon and Fullerton (1990)
El Salvador	1990	Mourichon and Fullerton (1990)
Florida (USA)	1998	Ploetz and Mourichon (1999)
Guatemala	1977	Stover (1980b)
Honduras	1972 (1969)†	Stover and Dickson (1976)
Jamaica	1995	A. Johanson, Chatham, 1996, personal communication
Mexico	1980	Stover and Simmonds (1987)
Nicaragua	1979	Stover (1980b)
Panama	1981	Stover (1987)
Venezuela	1991	Anon. (1994b)

Table 2.1. *Continued*

Region/country	Year of detection	Reference
Asia		
Bhutan	1985	Peregrine (1989)
China		
Hainan	1980	Stover and Simmonds (1987)
Guangdong	1990	Mourichon and Fullerton (1990)
Yunnan	1993	Jones and Mourichon (1993)
Indonesia		
Halmahera	1970	Stover (1978)
Java	1969	Reddy (1969)
Sumatra	1993	Jones and Mourichon (1993)
Kalimantan	1996	X. Mourichon, Montpellier, 1996, personal communication
Malaysia		
West Malaysia	1965	Graham (1969)
Johore	1993	Jones and Mourichon (1993)
Langkawi	1995	Anon. (1995)
East Malaysia (Sarawak)	1996	X. Mourichon, Montpellier, 1996, personal communication
Philippines		
Luzon	1964	Hapitan and Reyes (1970)
Mindanao	1965‡	Stover (1978)
Singapore	1964–1967	Graham (1969)
Taiwan	1927‡	Stover (1978)
Thailand	1969	Reddy (1969)
Vietnam	1993	Jones and Mourichon (1993)
Australasia/Oceania		
American Samoa	1975	Firman (1975)
Australia		
Torres Strait Islands	1981	Jones and Alcorn (1982)
Cape York Peninsula	1981	Jones and Alcorn (1982)
Cook Islands	1976	Firman (1975)
Fiji	1963	Rhodes (1964)
French Polynesia	1964–1967	Graham (1969)
Hawaii (USA)	1958‡	Stover (1978)
Micronesia	1964–1967	Graham (1969)
New Caledonia	1964–1967	Graham (1968)
Niue	1976	Stover (1976)
Norfolk Island	1980	Jones (1990b)
Papua New Guinea	1957	Stover (1976)
	(1951)§	Meredith (1970)
Solomon Islands	1957‡	Stover (1976)
	(1946)§	Meredith (1970)
Tonga	1965	Johnston (1965)
Vanuatu	1964–1967	Graham (1968)
Wallis and Futuna Islands	1996	X. Mourichon, Montpellier, 1996, personal communication
Western Samoa	1965	Johnston (1965)

* Authenticity of this record has been challenged.
† Symptoms of black leaf streak on colour transparencies.
‡ Date from herbarium material examined.
§ Date black leaf streak recognized in hindsight.

After examining old herbarium specimens of leaf spot, Stover (1976) believed that black leaf streak was present in Papua New Guinea in 1957 and in Taiwan as early as 1927. After analysing information on distribution and probable times of introduction to various countries, he believed that black leaf streak may have originated in the New Guinea–Solomon Islands area (Stover, 1978).

Black leaf streak was first described in Australia on islands in the Torres Strait and on the tip of Cape York Peninsula in 1981 during the first plant disease survey of the area (Jones and Alcorn, 1982; Jones, 1990a). However, it seems likely that black leaf streak may have been present in this remote part of Queensland, which is in close proximity to Papua New Guinea, for many years prior to its discovery. Since 1981, the disease has been detected at a few isolated locations further south on Cape York Peninsula, but has not appeared in commercial growing areas.

In 1980, black leaf streak was observed together with Sigatoka on Hainan Island in China (Stover and Simmonds, 1987). Mourichon and Fullerton (1990) isolated the fungal pathogen from leaf samples collected in the Guangdong region and later the disease was confirmed in Yunnan Province (Jones and Mourichon, 1993). These reports indicate that black leaf streak is now distributed throughout southern China. Black leaf streak was first identified in Vietnam in 1993 (Jones and Mourichon, 1993), but it may have been present much earlier. The disease has not yet been confirmed in Cambodia and Laos. The occurrence of black leaf streak was reported from Bhutan in 1985 (Peregrine, 1989), but it has not yet been recorded in Burma, India or Bangladesh.

In some areas of South-East Asia, the distribution of black leaf streak is uncertain. In Indonesia, the disease was first recognized in the late 1960s from symptoms seen on banana in the Bogor-Bandung area of Java (Reddy, 1969). However, the disease was not seen in the same area of Java in 1988 (Jones, 1990b). In 1996, although Sigatoka was still observed to be the dominant banana leaf spot disease in Java, the black leaf streak pathogen was also detected (X. Mourichon, Montpellier, 1999, personal communication). The causal agent of black leaf streak has also been identified from leaf spot specimens collected on Halmahera Island in the Maluku group (Stover, 1978), in West Sumatra (Anon., 1993) and in Kalimantan (X. Mourichon, Montpellier, 1996, personal communication). As in Java, Sigatoka was found to predominate in West Sumatra and Kalimantan (X. Mourichon, Montpellier, 1996, personal communication). Black leaf streak is almost certain to be present in Irian Jaya despite a lack of records because of its close proximity to Papua New Guinea where the disease is widespread.

Reports in the literature suggest that black leaf streak has been present in Singapore, West Malaysia and Thailand for some time (Reddy, 1969; Stover, 1978). However, black leaf streak was not seen near Kuala Lumpur and Melaka in west Malaysia or near Bangkok and Pakchong in Thailand in 1988 (Jones, 1990b). During a banana disease survey of West Malaysia in 1993, black leaf streak was identified in leaf spot specimens collected in Johore State (Anon., 1993b). The disease was later confirmed as present on Langkawi Island off the west coast of West Malaysia near the Thai border (Anon., 1995) and in Sarawak State in East Malaysia (X. Mourichon, Montpellier, 1996, personal communication). No specimens of black leaf streak were collected during a comprehensive banana disease survey of Thailand in 1994 (D.R. Jones, Montpellier, 1994, personal observation).

Where black leaf streak does occur in Asia, the disease has not become the dominant leaf spot, as has happened in the Pacific, Central America and West Africa. In northern Vietnam, black leaf streak is not seen on black leaf streak-susceptible cultivars in the Cavendish subgroup (AAA) in all areas (Anon., 1994a). In the Philippines, Sigatoka is still found many years after the introduction of black leaf streak. This phenomenon may be related to the varying degrees of resistance of the local banana cultivars and/or competition from other leaf spots. Septoria leaf spot (see Carlier *et al.*, this chapter, pp. 93–98) is probably the dominant leaf spot in Thailand, where it may have replaced Sigatoka. It is also common in West Malaysia, southern India and Sri Lanka. If this disease competes with black leaf streak, then this could explain the conflicting reports in the scientific literature on the distribution of black leaf streak in Asia (Jones, 1990b). Although black leaf streak may have been introduced to Asia many years ago, its relative visibility at any one time may depend on the locality, the environment and interactions with both the host cultivar and other leaf spots.

Distribution in the Latin American–Caribbean region

The history of the detection and spread of black leaf streak is well documented in the Americas. The first appearance of the disease outside Asia and the Pacific was in Honduras in 1972 (Stover and Dickson, 1976). It was found together with Sigatoka in the germplasm collection of the United Fruit Company at La Lima. This collection was used for Sigatoka studies and was always left unsprayed. Typical black leaf streak symptoms were subsequently recognized on colour transparencies taken in the collection in 1969. The disease may have been introduced earlier and remained undetected (Stover and Dickson, 1976).

After detection, a survey of local banana fields was undertaken. Symptoms of black leaf streak, which were always mixed with Sigatoka spots, were present in 100 ha of banana under cultivation close to La Lima (Stover and Dickson, 1976). The first severe outbreak of black leaf streak occurred in December 1973–February 1974 and covered an area of 1200 ha. By August 1974, the disease, which became known in Honduras as Sigatoka negra or black Sigatoka, was present in scattered locations for 51 km along the Ulúa River. Stover and Dickson (1976) and Buddenhagen (1987) postulated that, when black leaf streak first became established in the Ulúa valley, the pathogen did not have a sufficient level of virulence to replace Sigatoka. Only after an increased level of virulence was obtained was the pathogen able to become dominant and spread rapidly. As well as export banana cultivars in the Cavendish subgroup, cultivars in the Plantain subgroup (AAB) were attacked in coastal lowlands of Honduras. This characteristic distinguished black leaf streak from Sigatoka, which does not attack plantain at low altitudes.

Between 1973 and 1980, serious epidemics of the disease occurred throughout Latin America. Black leaf streak appeared in Belize in 1975, Guatemala in 1977 and El Salvador, Nicaragua and Costa Rica in 1979 (Stover, 1980a, b). In Costa Rica, the most probable source of inoculum for the black leaf streak outbreak in the San Carlos region was trash banana leaves from Honduras (Woods, 1980). At the time, plantains and green reject bananas from export plantations in Central America were regularly transported by road across international boundaries (Stover, 1980b). Trash banana and plantain leaves were used for padding and shading the shipments to reduce bruising and sunburn. Black leaf streak-affected leaves were thus moved from one country to another, spreading the disease.

By 1981, black leaf streak was present in all countries between southern Mexico and Panama (Stover, 1987). Also in 1981, the disease reached the Uraba export banana production area in the lowlands of Colombia, where all banana plantations were attacked (Merchan, 1990). From there, the disease spread along the Pacific and Atlantic coasts of Colombia to where plantain is grown in association with coffee or cocoa at altitudes of between 1000 and 1700 m above sea level. Black leaf streak was first observed in plantain districts in 1988 at 1500 m (Merchan, 1990) and later at 1600 m (Belalcazar, 1991). In 1986, black leaf streak was detected in northern Ecuador and within 4 years was found in the southern banana production areas (Mourichon and Fullerton, 1990). The disease reached western Venezuela in 1991

(Anon., 1994b) and by 1997 had spread to the eastern part of the country (Martinez *et al.*, 1998). Black leaf streak has also reached Bolivia (Tejerina, 1997) and Brazil (Cordeiro *et al.*, 1998).

In the Caribbean, black leaf streak has been found in Cuba (Vidal, 1992), Jamaica (A. Johansen, Chatham, 1995, personal communication) and the Dominican Republic (X. Mourichon, Montpellier, 1996, personal communication). It has not yet been detected in Haiti (T. Lescot, Santo Domingo, 1999, personal communication). Black leaf streak was recently detected in Florida (Ploetz and Mourichon, 1999). The disease will most probably spread to all Caribbean islands in due course, although progress may be slow due to prevailing westerly winds.

Distribution in Africa

The earliest report of black leaf streak in Africa was in Zambia in 1973 (Raemaekers, 1975). Symptoms resembled black leaf streak, but its identity could not be confirmed from specimens sent to the UK for positive identification (Dabek and Waller, 1990). This initial outbreak remains unconfirmed.

The first authenticated report was in Gabon in 1978, where it was believed that the disease might have been introduced on planting material from Asia (Frossard, 1980). Frossard (1980) also reported that symptoms of black leaf streak had been observed in São Tomé. In 1980, black leaf streak was observed in southern Cameroon, close to the borders of Equatorial Guinea and Gabon (Tezenas du Montcel, 1982). In 1985, the disease spread from Gabon to Congo (Mourichon, 1986; Wilson and Buddenhagen, 1986). It was also present in Côte d'Ivoire in the same year (Mourichon and Fullerton, 1990). In 1988, black leaf streak was reported from the lowlands of Zaire, now the Democratic Republic of Congo (Mobambo and Naku, 1993). The presence of black leaf streak was confirmed in Nigeria and Ghana in 1986 (Wilson and Buddenhagen, 1986; Wilson, 1987) and Togo in 1988 (Mourichon and Fullerton, 1990).

In East Africa, the first confirmed report of black leaf streak came from Pemba in 1987 (Dabek and Waller, 1990). This was most probably another independent introduction of the disease to the continent. From Pemba, the black leaf streak spread rapidly to the nearby island of Zanzibar and then to the Tanzanian mainland, reaching coastal Kenya in 1988 (Kung'U *et al.*, 1992). A survey the next year showed black leaf streak to be present in the Central and Eastern Provinces of Kenya (Kung'U *et al.*, 1992).

During a banana disease evaluation mission in the Great Lakes region of East Africa in 1987, Sebasigari and Stover (1988) observed black leaf streak in the highlands of eastern Democratic Republic of Congo, Rwanda, where symptoms had been seen in 1986 (Sebasigari, 1990), and Burundi. At that time, the disease was not observed in Uganda. However, in 1990, the disease was recorded in Uganda (Tushemereirwe and Waller, 1993) and in Malawi (Ploetz *et al.*, 1992). It is likely that black leaf streak spread to this region of East Africa from the Democratic Republic of Congo. Black leaf streak is now also present in the Comoros Islands (Jones and Mourichon, 1993).

Economic importance

Black leaf streak is a major constraint to banana production (Stover, 1983a, 1986a; Fouré, 1985; Fullerton, 1987; Stover and Simmonds, 1987). After the first occurrence of black leaf streak in an area, the disease builds up and often reaches epidemic level in a few years (Fullerton and Stover, 1990; Belalcazar, 1991). Chemical control costs and crop losses are well documented for industrially produced bananas (Stover, 1986a, 1990). Losses to smallholder's crops, which are consumed locally, are harder to estimate.

Black leaf streak does not kill plants immediately, but crop losses increase gradually with the age of plantings. The decrease in functional leaf area caused by the disease results in a reduction in the quality and quantity of fruit (Stover, 1983a; Stover and Simmonds, 1987; Pasberg-Gauhl, 1989; Mobambo *et al.*, 1993, 1996b). Fruit from infected plants ripens prematurely and does not fill properly. Bananas for export are sometimes harvested at a lower grade (younger age) in order to reduce the risks of premature ripening in transit to overseas markets (Stover and Simmonds, 1987).

Until the 1970s, the common leaf diseases of plantain were not considered economically important. This changed when black leaf streak spread to areas where the crop was extensively grown. All over the tropics, plantain is cultivated and fruit consumed by smallholders. In many areas, black leaf streak has caused a considerable decrease in the availability of fruit for local consumption and this has resulted in a substantial increase in their market price. Smallholders growing plantain in the Americas either go out of business, because they cannot cover the high costs of chemical control, or form cooperatives so that their limited resources can be pooled to fight the disease.

Black leaf streak is endangering the food security of resource-poor people. Africa alone contributes about 50% of the world plantain production and the demand for plantain is steadily increasing (Wilson, 1987). All known plantain cultivars (Fouré, 1985; Mobambo *et al.*, 1996a) are susceptible to black leaf streak and are severely defoliated by the disease. Plants in the ratoon crop are weaker than in the first cycle and thus more affected by wind damage. On poor sandy soils in West Africa, Mobambo *et al.* (1996b) estimated that yield losses due to black leaf streak are 33% and 76% during the first and second cropping cycle, respectively. However, in intensively cropped backyard or home garden systems, cultivation is not so seriously affected (Mobambo *et al.*, 1994b). Under marginal conditions, plantain production is often abandoned due to low yields.

Plantain is not the only smallholder banana to be affected in Africa. The disease also causes serious damage to East African Highland cultivars in the Lujugira–Mutika subgroup (AAA). In Uganda, Tushemereirwe (1996) reported yield losses of 37% due to the effects of a leaf spot complex consisting mainly of black leaf streak and Cladosporium speckle (see Jones, this chapter, pp. 107–110).

Black leaf streak has had a devastating effect on the production of export bananas in the South Pacific. Firman (1972) noted that only 49% of unsprayed Cavendish cultivars produced fruit that reached the export quality standard. Fiji ceased exporting bananas in 1974 and Western Samoa in 1984. Exports also dropped in Tonga and the Cook Islands, because producers had problems maintaining fruit quality standards for their markets in New Zealand. Black leaf streak control has become the single largest production cost (Fullerton, 1987).

In 1974, the production of dessert bananas and plantains in Central America was seriously affected by hurricane Fifi, which was also thought responsible for the wind-borne spread of black leaf streak to new areas. In many countries, production subsequently dropped substantially. Before 1974, Honduras exported 500,000 boxes of plantain each year, but afterwards exports dropped to below 1000 boxes (Stover, 1983a). In 1978, the export of plantain from Honduras to the USA was curtailed because of the shortage of fruit with the required quality (Bustamente, 1983). Plantain exports only resumed in 1985, when black leaf streak was controlled by the aerial application of fungicides (Stover, 1987).

After the identification of black leaf streak in Costa Rica in 1979, the government initiated a quarantine programme (Woods, 1980). This programme consisted of the destruction of host plants in the affected area and the establishment of roadside quarantine stations strategically located to stop movements of banana and plantain leaves, which were used for padding and shading fruit. About 3000 ha of plantain were destroyed in 1979 and early 1980. The Costa Rican government paid about $US3 million for the eradication programme, but the spread of the disease could not be stopped (Woods, 1980). By 1982, the Ministry of Agriculture in Costa Rica estimated that black leaf streak alone reduced plantain production by 40% (Romero, 1986). In general, yields of plantains in well-maintained fields on rich, fertile soils in Central America may have fallen by 20–50% (Stover, 1983a; Pasberg-Gauhl, 1989).

Black leaf streak was first detected in Panama in 1980. Bureau (1990) estimated that plantain production in Panama decreased by 69% between 1979 (100,910 t) and 1984 (31,134 t). During this period, the price of plantains rose by up to 50% in local markets. Jaramillo (1987) reported that between 1982 and 1985, the area planted with plantain decreased by 22%, from 7432 ha to 5800 ha. About 34% of growers were believed to have abandoned their holdings, leading to a decrease in production of 47%.

Colombia is one of the largest plantain producers in Latin America, with 400,000 ha under cultivation and an estimated yearly production of 2.5 million t. About 96% of plantains are consumed locally, the remainder being exported (Belalcazar, 1991). About 88% of the plantain is grown in association with coffee by smallholders. Only 12% of the crop is grown in monoculture in larger plantations. After the introduction of black leaf streak, this staple food became scarce and much higher prices were demanded in the market (Belalcazar, 1991). Due to the high cost of plantain, consumers changed to other, cheaper food crops. This in turn had a negative effect on plantain production. Black leaf streak has thus had a significant impact on Colombian agriculture and the eating habits of a nation.

Cost of control

Black leaf streak can be chemically controlled on plantations (see Romero, this chapter, pp. 72–79), but more applications are needed than are necessary to control Sigatoka (Stover, 1990). Up to 36 spray cycles per year may be required for plantations growing dessert bananas for export and up to 19 cycles for commercial plantings of plantain (Stover, 1980b, 1990; Fouré, 1983, 1988a, b; Belalcazar, 1991; Gauhl, 1994; Romero and Sutton, 1997a). The cost of black leaf streak control, therefore, is much higher than that of Sigatoka control. Stover and Simmonds (1987) reported that 27% of production costs in dessert banana plantations was spent controlling black leaf streak. From 1972 until 1985, the estimated cost of black leaf streak control in Central America, Colombia and Mexico was more than $US350 million (Stover and Simmonds, 1987). The cost of chemical control measures has been calculated to be $US400–1400 ha^{-1} year^{-1} (Anon., 1993b). During the 1980s, the cost of black leaf streak control in export banana crops in Costa Rica was estimated at approximately $US17.5 million year^{-1} (Stover and Simmonds, 1987). Between 1985 and 1994, the area under banana cultivation increased from approximately 21,000 ha to 52,737 ha (Serrano and Marín, 1998). As a consequence, the cost of black leaf streak control in 1995 was estimated to have increased to $US49 million year^{-1} (Romero and Sutton, 1997a).

Symptoms

D.R. Jones, C. Pasberg-Gauhl, F. Gauhl and E. Fouré

In Hawaii, Meredith and Lawrence (1969) observed the development of symptoms of black leaf streak on mature, preflowering plants of 'Dwarf Cavendish', a cultivar in the susceptible Cavendish subgroup (AAA). They described first symptoms, which were faint reddish-brown specks less than 0.25 mm in diameter, as being visible on the lower surface of the leaf at an 'initial speck stage'. These specks then elongated, becoming slightly wider, to form a narrow, reddish-brown streak with dimensions of 20 mm × 2 mm, the long axis of the streak being parallel to leaf veins. At this 'first streak stage', streaks were more clearly visible on the lower leaf surface than on the upper. The streaks could be densely aggregated in places and frequently overlapped to form larger, compound streaks. At the 'second streak stage', the colour of streaks, which were clearly visible on the upper leaf surface, changed to a very dark brown, almost black, colour. The entire leaf could blacken at this stage if streaks were numerous, but, if less densely congregated, streaks broadened and water-soaked borders, which were seen best after rain or dew, appeared. At this 'first spot stage', lesions became fusiform or elliptical in outline. At the 'second spot stage', the dark brown or black centres of spots became slightly depressed and leaf tissue immediately surrounding the more pronounced water-soaked border yellowed slightly. At the 'third or mature spot stage', the centre of the spot dried, becoming light grey or buff in colour. Each mature spot had a well-defined dark brown or black border and surrounding tissue was often bright yellow. Where spots coalesced, whole sections of leaves became necrotic. After the leaf withered, Meredith and Lawrence (1969) reported that spots remained visible because of their light-coloured centres and dark borders.

Fouré (1987), who worked with black leaf streak in West Africa, identified six main stages of symptom development. His 'stage 1', which precedes the initial speck stage of Meredith and Lawrence (1969), is when small whitish or yellow specks appear on the underside of the leaf (Plate 2.1). In the second phase of this stage, the specks turn rusty brown in colour (Plate 2.1). These specks are less than 1 mm long and are not visible in translucent light. The specks grow into characteristic narrow, reddish-brown or dark brown streaks, which Fouré calls 'stage 2'. Streaks, which are generally 2–5 mm long, are visible in translucent light and are easy to recognize from a distance of 1–2 m. On some clones, brown streaks are visible on the underside of the leaf (Plate 2.1) with corresponding yellow streaks on the upper surface (Plate 2.2). The colour of the upper surface streak changes through brown to black, but the brown colour is retained on the underside. The streaks become longer in 'stage 3' and can reach 20 or 30 mm in length (Plate 2.3). If conditions are favourable and inoculum potential high, streaks in stages 2 or 3 coalesce and cause leaf necrosis (Plate 2.4). Individual streaks that have not coalesced broaden to form a spot, which has a brown colour on the underside of the leaf (Plate 2.5) and is often black on the upper leaf surface. This is Fouré's 'stage 4'. An elliptical spot, which is black on both sides of the leaf and is usually surrounded by a yellow halo, defines 'stage 5' (Plate 2.5). The centre of the spot also begins to flatten out in 'stage 5'. During 'stage 6', the centre of the spot dries and fades to clear grey. The spot is usually surrounded by a well-defined black border, which in turn is surrounded by a water-soaked or yellow halo (Plates 2.6 and 2.7). The grey spot with dark border remained visible after the leaf had died and dried (Plate 2.8).

The pattern of first symptoms is determined by the stage at which the unfurling leaf is infected. As the unrolling leaf is a constantly expanding funnel, new susceptible tissue is being gradually exposed to the inoculum. If wind-dispersed spores are deposited on the lower surface of the apical third of the largely furled heart leaf, a distinct line of specks, which later develop into streaks and spots, appears along the left edge of the leaf, particularly towards the tip (as viewed looking down on the upper leaf surface from the base towards the apex). When conditions are extremely favourable for infection, these early speck symptoms may appear on leaf 2 (counting down the plant from the first fully opened leaf). However, they are usually first seen on leaves 3 and 4. Symptoms that develop due to later infections are generally observed at the distal end of the right lamina margin, followed by the middle section and successively towards the base of the leaf (Stover, 1980a; Gauhl, 1989).

On a growing plant, streaks are usually present on the third, fourth and fifth leaves and both streaks and spots on the fifth and older leaves. If the plant is stressed and slow-growing, advanced symptoms may be seen on the second or even the first leaf. On juvenile leaf tissue, such as on water-suckers and young plants derived from

Plate 2.1. Black leaf streak symptoms on the lower surface a leaf of a cultivar in the Cavendish subgroup (AAA) in Gabon. Yellow spots (stage 1, phase 1) can be seen upper left. Brown spots (stage 1, phase 2) and streaks (stage 2) can be seen elsewhere (photo: E. Fouré, CIRAD).

Plate 2.2. Black leaf streak symptoms on the upper leaf surface of 'Fougamou' (ABB, syn. 'Pisang Awak') in Gabon. Chlorotic streaks (stage 2) are visible (photo: E. Fouré, CIRAD).

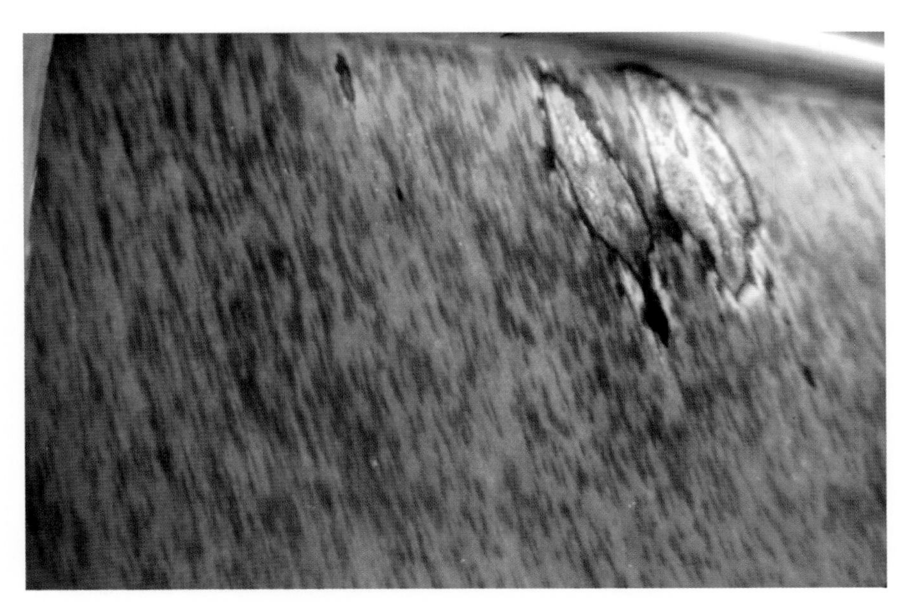

Plate 2.3. Black leaf streak symptoms on the lower surface of a leaf of an AAB cultivar in Gabon. Long brown streaks (stage 3) are visible (photo: E. Fouré, CIRAD).

Plate 2.4. Black leaf streak symptoms on the upper surface of 'Ovina' (AAB, Plantain subgroup) in Gabon. Streaks (stage 2) have coalesced to form necrotic areas at the lamina margin (photo: E. Fouré, CIRAD).

tissue culture, initial specks quickly develop into oval or circular brown spots without showing the streak stage. It usually takes 3–4 weeks after symptoms first appear for a leaf to die (Stover, 1972; Gauhl, 1994). However, when young streaks at 'stage 2' coalesce, large brown areas are formed, which quickly turn black and the whole leaf may die within 1 week. Sometimes, because of the intensity of infection, fewer than six living leaves may be present on growing plants in the vegetative stage. On more resistant cultivars, symptoms are only seen on the older, lower leaves, due to slower symptom development.

Black leaf streak looks more severe on plants with bunches because leaf production ceases and symptoms gradually appear on leaves higher and higher up the plant. If disease pressure is great, it is not uncommon for a very susceptible cultivar to have no viable leaves at harvest (Plate 2.9). The number of functional leaves at harvest has been used to gauge the resistance of cultivars to black leaf streak.

Symptoms have been noted as being different in respect of colour, size and shape of streaks and spots on different cultivars (Meredith and Lawrence, 1970a). Many accessions of wild *Musa acuminata* subspecies and some cultivars in the AA and AAA groups are highly resistant to *Mycosphaerella fijiensis* (see Table 2.4). These usually respond to infection by producing small, yellow or chlorotic spots (Plate 2.10), which do not develop further.

The causal agent

J. Carlier, X. Mourichon and D.R. Jones

Taxonomy

Mycosphaerella fijiensis Morelet, the fungus causing black leaf streak disease, was first described by Leach (1964a). It is a heterothallic ascomycetous species like *Mycosphaerella musicola*, the cause of Sigatoka disease (Stover, 1963; Mourichon and Zapater, 1990).

Conidiophores first develop in initial brown flecks or early streaks on the lower surface of the leaf and continue to be produced until the second spot stage of Meredith and Lawrence (1969). They emerge singly or in small groups from stomata within the boundary of the lesion. Most are produced on the lower surface, though a few arise on the upper. Conidiophores are pale to medium olive-brown, becoming slightly paler towards the tip. They are straight or bent, often with geniculations and sometimes with a basal swelling up to 8 μm in diameter, 0- to 5-septate, 16.5–62.5 μm × 4–7 μm, usually slightly narrower, but occasionally wider, at the tip. One or more scars are present near the tip of the conidiophore, either flat against the apex or on the side, or on a slightly sloping shoulder (Fig. 2.1).

Conidia are formed singly at the apex of the conidiophore, later becoming lateral as the conidiophore develops. Up to four mature conidia may be attached to a single conidiophore. Conidia are not quite colourless, being pale green or olivaceous. They are obclavate to cylindro-obclavate, 1- to 10-septate (commonly 5- to 7-septate), straight or curved, obtuse at the apex, truncate or rounded at the base, with a visible and slightly thickened hilum, 30–132 μm × 2.5–5 μm, the broadest point being at the base (Fig. 2.1).

Although the anamorphs of both *M. fijiensis* and *M. musicola* were first placed in the *Cercospora* genus, they are now separated as *Paracercospora fijiensis* and *Pseudocercospora musae* (Deighton, 1976, 1979; Pons, 1990). The two pathogens can be differentiated microscopically by the characteristics of their conidia and conidiophores (Meredith and Lawrence, 1969, 1970a; Mulder and Stover, 1976). Conidia of *M. fijiensis* are on average longer than those of *M. musicola* and are distinguishable because of a thickened basal hilum. Conidiophores of *M. fijiensis* possess conidial scars, which are absent from conidiophores of *M. musicola*.

The sexual stages of *M. fijiensis* and *M. musicola* are morphologically similar (Meredith and Lawrence, 1969, 1970a; Mulder and Stover, 1976). Spermogonia develop at the stage when streaks turn into

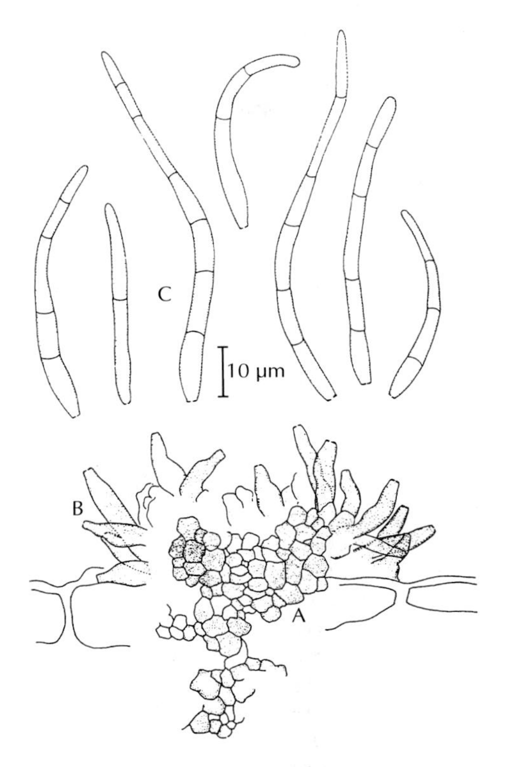

Fig. 2.1. Stroma (A), conidiogenous cells (B) and conidia (C) of *Mycosphaerella fijiensis* (from Pons, 1987).

spots and are more abundant on the lower surface of the leaf, being consistently associated with conidiophores. Spermogonia are hourglass-shaped, oval or almost globose and measure 55–88 μm × 35–50 μm. The ostiole is slighly prominent and protrudes through the stoma pore. Mature spermatia are hyaline, rod-shaped and measure 2.5–5.0 μm × 1.0–2.5 μm.

Perithecia vary in characteristics, but are mostly globose with a diameter of 47–85 μm. They are immersed in the leaf tissue with protruding ostioles and are found on both leaf surfaces, though more abundant on the upper. The wall of the perithecium is dark brown with three or more layers of polygonal-shaped cells. The numerous asci are bitunicate, obclavate and without paraphyses. The hyaline, biserate ascospores have dimensions of 12.5–16.5 μm × 2.5–3.8 μm and are two-celled, with the larger cell uppermost in the ascus. The ascospore is slightly constricted at the septum (Mulder and Holliday, 1974a). The ascostroma, asci and ascospores are illustrated in Fig. 2.2.

It is now generally accepted that the pathogens causing black leaf streak and Sigatoka are two different species. However, several authors in the past have questioned whether there are sufficient morphological differences between the anamorphs of the two fungi for their separation (Graham, 1969; Stover 1969; Meredith, 1970; Wardlaw, 1972). Recently, genetic divergence between *M. fijiensis* and *M. musicola* was detected using the restriction fragment length polymorphism (RFLP) technique (Carlier *et al.*, 1994) and sequence analysis of the internal transcribed spacer (ITS) region of nuclear ribosomal DNA (Johanson, 1993; J. Carlier, Montpellier, 1998, unpublished). This new evidence supports the classification of the two pathogens as separate species.

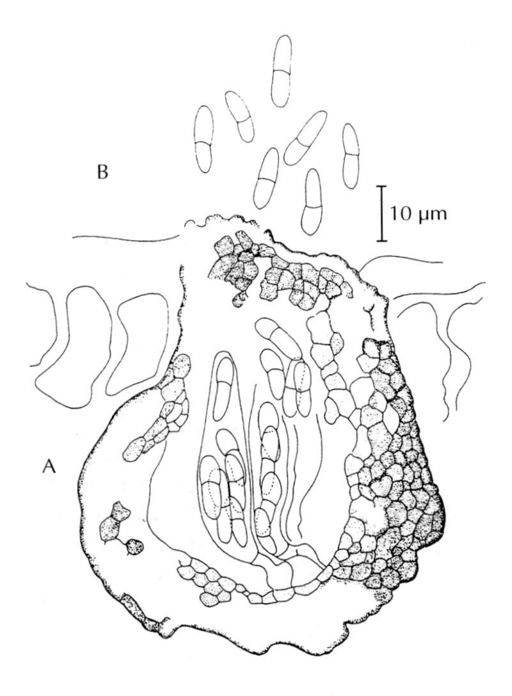

Fig. 2.2. Ascostroma with asci (A) and ascospores (B) of *Mycosphaerella fijiensis* (from Pons, 1987).

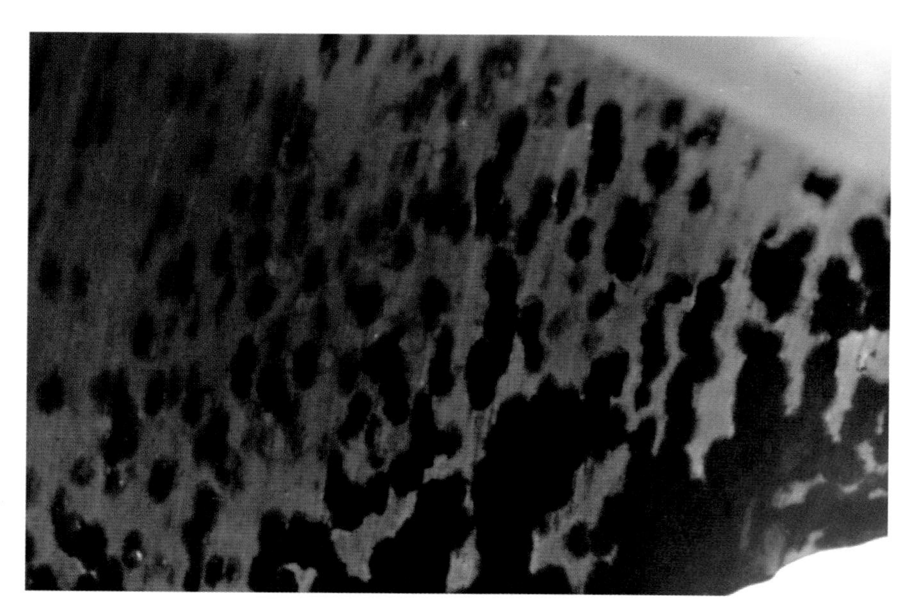

Plate 2.5. Black leaf streak symptoms on the lower surface of a leaf of a cultivar in the Cavendish subgroup (AAA) in Gabon. Brown spots (stage 4) (left) are developing into black spots (stage 5) (centre), which are coalescing (right) to form necrotic areas (photo: E. Fouré, CIRAD).

Plate 2.6. Black leaf streak symptoms on the upper surface of a leaf of 'Ovina' (AAB, Plantain subgroup) in Gabon. Mature spots with sunken, grey centres (stage 6) have coalesced to form a large necrotic area at the bottom of the picture. Individual spots are delimited by dark borders (photo: E. Fouré, CIRAD).

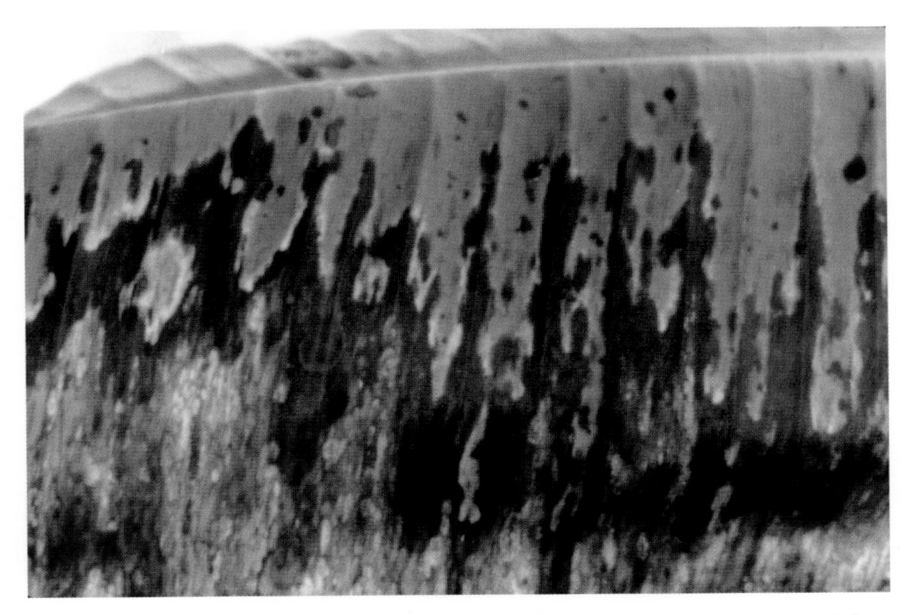

Plate 2.7. Black leaf streak symptoms on the upper surface of a leaf of a cultivar in the Cavendish subgroup (AAA) in Gabon. Numerous, mature spots (stage 6) with grey centres and dark borders surrounded by haloes of yellow leaf tissue, are present. The spots have coalesced at the lamina margin to form necrotic areas. Symptoms resemble those of Sigatoka (see Plate 2.13) (photo: E. Fouré, CIRAD).

Plate 2.8. Black leaf streak symptoms on the upper surface of a leaf of a cultivar in the Cavendish subgroup (AAA) in the Cook Islands. Mature spots (stage 6) are visible as light-coloured lesions in the brown, necrotic leaf tissue. The dark borders around the spots are indistinct where developing lesions have coalesced (photo: D.R. Jones, QDPI).

When black leaf streak was first found in Latin America, it was called black Sigatoka and the name *M. fijiensis* var. *difformis* assigned to the pathogen (Stover, 1974). This name was later validated on the basis of an examination of specimens from La Lima, Honduras (Mulder and Stover, 1976). The taxonomic characteristic used to separate *M. fijiensis* var. *difformis* from *M. fijiensis* was the presence of sporadic stroma, which gave rise to dense and loose fascicles of conidiophores. This feature had not been described for *M. fijiensis*. However, a later, comprehensive, taxonomic study of isolates of *M. fijiensis* and *M. fijiensis* var. *difformis* revealed that the two were synonymous (Pons, 1987, 1990). This synonomy was later confirmed by molecular studies (Johanson, 1993; Carlier *et al.*, 1994).

Cultural characteristics

Growth on culture media is slow. A single conidium seeded on to potato dextrose agar forms a colony about 1 cm in diameter after 38 days' incubation at 26°C (Meredith, 1970). The optimal temperature for growth ranges from 24°C to 28°C (Meredith, 1970; Stover, 1983b; Mouliom-Pefoura and Mourichon, 1990). Cultures are raised and stromatic and have a velvety surface coloured pale grey, pink–dark grey or grey-brown (Stover, 1976). As with *M. musicola*, mycelial characteristics in any one segment are not constant and may change as the colony grows. Colonies of *M. fijiensis* and *M. musicola* are very similar and cannot be used to differentiate the two species if conidia are not produced.

Ten- to 21-day-old colonies can produce conidia (Mourichon *et al.*, 1987; Jacome *et al.*, 1991). Sporulation has been reported to be optimal on cultures grown on V8-juice agar adjusted to pH 6 and incubated at 20°C under conditions of continuous light. *Mycosphaerella fijiensis* is heterothallic and perithecia, asci and ascospores can be produced in culture by crossing mating types (Mourichon and Zapater, 1990). The optimum temperature for the germination of conidia and ascospores in culture is around 26°C (Jacome *et al.*, 1991). Maximum germination occurs in water and decreases as the relative humidity (RH) lowers. No conidia have been observed germinating below 95% RH and no ascospores below 98% RH (Jacome *et al.*, 1991).

Diagnostic methods

Diseases caused by *M. fijiensis, M. musicola* (see Jones, this chapter, pp. 79–92). *Phaeoseptoria musae* (see Jones, this chapter, pp. 92–93) and the *Septoria* leaf spot pathogen (see Carlier *et al.*, this chapter, pp. 93–98) are difficult to distinguish on the basis of symptom expression. However, *M. fijiensis* and *M. musicola* can easily be separated from one another, *P. musae* and the pathogen that causes Septoria leaf spot on microscopic differences between asexual stages of the fungi on leaf samples and in culture (Meredith and Lawrence, 1969, 1970a; Mourichon and Fullerton, 1990; J. Carlier, X. Mourichon, M.F. Zapater, F. Lapeyre and D.R. Jones, 1999, unpublished results). Molecular methods have also been developed to distinguish *M. fijiensis, M. musicola* (Johanson and Jeger, 1993; Johanson *et al.*, 1994), *P. musae* and the causal fungus of Septoria leaf spot (J. Carlier, M.F. Zapater, F. Lapeyre, D.R. Jones and X. Mourichon, 1999, unpublished results).

Pathogenic variability

Isolates of *M. fijiensis* from different locations in Papua New Guinea and elsewhere have been found to vary in their pathogenicity (Fullerton and Olsen, 1995). This variability was detected when young plants of a standard set of *Musa* genotypes, which had been propagated in tissue culture, were inoculated with suspensions of conidia prepared from the isolates. The reactions of the genotypes to *M. fijiensis* isolates were differential, suggesting specific interactions. Some genotypes considered as highly resistant or hypersensitive in the field, such as the wild banana *M. acuminata* ssp. *burmannicoides* accession

'Calcutta 4' and 'Yangambi Km 5' (AAA, syn. 'Ibota Bota'), were attacked by several isolates.

In other inoculation experiments using young plants, significant differences in pathogenicity have been found between isolates of *M. fijiensis*. Jacome and Schuh (1993) reported that six isolates of *M. fijiensis* from Honduras induced different levels of disease severity, which was defined as the percentage of a leaf target area covered by black leaf streak lesions, on 'Grand Nain' (AAA, Cavendish subgroup). Romero and Sutton (1997b), who were working with isolates from different geographical regions, later showed that, as well as disease severity, the time between inoculation and the appearance of ten first streak symptoms could also differ. These results indicate that isolates of *M. fijiensis* vary in their aggressiveness.

Fullerton and Olsen (1995) reported the sudden breakdown of the resistance of the East African cultivar 'Paka' (AA) and the Jamaican bred tetraploid 'T8' (AAAA) on the island of Rarotonga in the Cook Islands after approximately 8 years exposure to black leaf streak. They suggested that the change from high resistance to complete susceptibility might have been because of the failure of a single gene conditioning resistance.

More recently in Cameroon, Mouliom-Pefoura (1999) has described the appearance of necrotic lesions with perithecia on leaves of 'Yangambi Km 5' (AAA, syn. 'Ibota Bota', a cultivar that has been reported as highly resistant to black leaf streak (see Fouré *et al.*, this chapter, p. 62). *Mycosphaerella fijiensis* has been isolated from ascospores discharged from perithecia formed in leaf lesions at 1250 m and its pathogenicity is being investigated (J. Carlier, 1999, Montpellier, personal communication). Lesions have been observed in the field to progress through all the developmental stages described by Fouré (see Jones *et al.*, this chapter, p. 45) (Mouliom-Pefoura, 1999, Cameroon, personal communication) and this may be as a result of the breakdown of a single gene that could control the hypersensitive-type response hitherto associated with the strong resistance of the cultivar. However, other necrotic areas on leaves seem to be derived from coalescing hypersensitive spots so the situation is still confused and needs further study. Even if the breakdown of a hypersensitivity gene is demonstrated, 'Yangambi Km 5' would still seem to have a high level of partial resistance (see Fouré *et al.*, this chapter, pp. 66–67) as the full development of lesions has been reported to take approximately 165 days (Mouliom-Pefoura, 1999, Cameroon, personal communication). These observations show that there is a need for a greater knowledge on genes for resistance so that resistance breeding can have a more scientific basis.

Genetic diversity

The genetic structure of the global population of *M. fijiensis* was studied by Carlier *et al.* (1996). As ascospores produced during sexual reproduction play a major role in the epidemiology of black leaf streak, the analysis of *M. fijiensis* required quantitative estimates of population genetic parameters based on allele frequencies at several loci. Geographical populations of *M. fijiensis* from the Philippines, Papua New Guinea, Africa, Latin America and islands in the Pacific were analysed, using 19 random single-locus probes (each probe defining an RFLP locus).

Results showed that populations of *M. fijiensis* can maintain a high level of genetic diversity (Table 2.2) compared with populations of other pathogens (McDonald and Martinez, 1990; McDonald *et al.*, 1994). Ecological conditions are favourable for black leaf streak development and banana cultivation almost year-round in most growing areas. Low genetic drift within large pathogen populations can maintain the high level of genetic diversity observed. Therefore, a high level of pathogenic variability might also be maintained in *M. fijiensis* populations, allowing an adaptation to newly introduced resistant host genotypes, such as has been observed with 'Paka' and 'T8' to Rarotonga in the Cook Islands (Fullerton and Olsen, 1995).

Plate 2.9. Black leaf streak has killed all leaves of 'Dwarf Cavendish' (AAA, Cavendish subgroup) in Gabon before harvest (photo: E. Fouré, CIRAD).

Recombination plays an important role in the population structure of *M. fijiensis*. For the 19 RFLP loci, 134 of the total 136 isolates studied corresponded to a single genotype. A very low level of gametic disequilibrium among RFLP loci was also detected in each population. These results are consistent with random mating. Since the genetic markers were statistically independent, pathogenic characteristics could not be related to RFLP genotypes. The important role that genetic recombination plays in *M. fijiensis* populations should also be considered in banana breeding programmes. Combining specific resistance genes within individual cultivars (pyramiding) may not be a durable deployment strategy. Mixing different cultivars with different resistance genes together in the same plantation or breeding for partial and non-specific resistance to black leaf streak might be more appropriate.

The results indicate that the Australasia–South-East Asia region may be the centre of *M. fijiensis* diversity. The level of genetic variation found in the Papua New Guinea and Philippine populations was much higher than in other geographical regions (Table 2.2). Most alleles detected in the Latin American, Pacific

Plate 2.10. Small, yellow spots on a leaf of *Musa acuminata* ssp. *siamea* in Cameroon after infection by *Mycosphaerella fijiensis*. These symptoms are typical of a highly resistant response to infection following an incompatible interaction (photo: E. Fouré, CIRAD).

Table 2.2. Allelic diversity in geographical populations of *Mycosphaerella fijiensis* at 19 nuclear RFLP loci (from Carlier *et al.*, 1996).

Location	Sample size	Mean no. of alleles per locus	% of polymorphic loci	Nei's measure of gene diversity
Total sample	136	10.1	100.0	0.59 (0.03)
Philippines	33	7.7	100.0	0.59 (0.05)
Tagum (Mindanao)	10	4.3	100.0	0.58 (0.05)
Papua New Guinea	26	4.8	100.0	0.57 (0.04)
Africa	33	1.9	68.4	0.25 (0.05)
Cameroon	16	1.7	57.9	0.22 (0.05)
Pacific Islands	26	2.1	73.7	0.30 (0.05)
New Caledonia	13	1.8	63.2	0.27 (0.05)
Latin America	18	2.3	78.9	0.40 (0.05)

Island and African populations were also found in the Papua New Guinea and Philippine populations. These results are consistent with the hypothesis that *M. fijiensis* originated in Australasia–South-East Asia and was recently introduced to other banana-producing regions (Mourichon and Fullerton, 1990).

From a study of the patterns of detection and distribution of *M. musicola* and *M. fijiensis* in South-East Asia and on islands in the Pacific, Stover (1978) suggested that *M. fijiensis* originally came from the New Guinea–Solomon Islands area. Because areas of host–pathogen co-evolution are thought to be favourable locations to find sources of resistance to disease (Harlan, 1976), this led to the collection of *Musa* germplasm, especially fertile diploids, which may have potential for use in breeding programmes, in Papua New Guinea (Sharrock, 1990). It now seems likely that

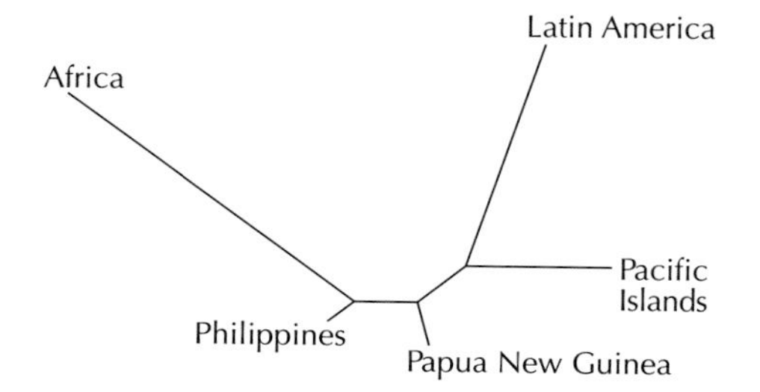

Fig. 2.3. Additive tree constructed from estimates of Wright's *Fst* among pairs of geographical populations of *Mycosphaerella fijiensis* (Carlier *et al.*, 1996).

germplasm in the Philippines and possibly other countries in South-East Asia may be equally valuable.

The founder-effect process (Hart and Clarke, 1989) accompanying introductions of *M. fijiensis* to Latin America, Africa and the Pacific Islands has led to a reduction of genetic diversity in these regions. However, the level of diversity detected remains relatively large and might be sufficient for rapid adaptations to newly introduced resistant host genotypes. Founder effects have also led to marked genetic differentiation between geographical populations (Fig. 2.3). The estimate of Wright's *Fst* parameter (Wright, 1951) for all loci and populations (Weir, 1990) was 0.32. The founder effects detected suggest occasional *M. fijiensis* migration events between the geographical regions studied. If appropriate quarantine measures can be enforced to prevent further movements of *M. fijiensis* between continents, the four different banana producing regions (Australasia–South-East Asia, Africa, Latin America–Caribbean, Pacific Islands) might be considered as separate epidemiological units requiring independent black leaf streak management strategies.

Müller *et al.* (1997), using oligonucleotide fingerprinting, have recently shown that genetic variation of *M. fijiensis* in Nigeria exists on a micro- and macroscale. Variability occurred within one lesion, between lesions on one plant and between plants and cultivars as well as between geographical locations. Research is now needed to estimate accurately the variability of *M. fijiensis* at plant, field, district, country and regional levels to determine if different banana-producing areas have only one or several epidemiological units. These studies should be conducted using molecular markers and undertaking inoculation tests to determine pathogenic differences.

Disease cycle and epidemiology

F. Gauhl, C. Pasberg-Gauhl and D.R. Jones

Infection

Spores usually germinate within 2–3 h of deposition on a moist leaf surface. The minimum, optimum and maximum temperatures for the development of ascospore germ tubes is 12°C, 27°C and 36°C, respectively, with no development taking place at 11°C and 38°C (Stover, 1983b; Jacome *et al.*, 1991; Porras and Peréz, 1997). However, ascospore germ-tube growth at 20°C is half the rate it is at 27°C. Germ tubes from both conidia and ascospores penetrate through stomata after 48–72 h above 20°C (Stover, 1980a; Fouré and Moreau, 1992). A film of water on the leaf surface is required for ascospore infection under controlled conditions. Water is not required for conidial infection provided the relative humidity is high (Jacome and Schuh, 1992). Although it seems that all leaves are equally susceptible to *M. fijiensis*, most infections occur on new leaves between emergence and unfurling (Stover and Simmonds, 1987; Gauhl, 1994).

Incubation period

Once infection is established, one or more vegetative hyphae of *M. fijiensis* emerge from stomata on the lower leaf surface, develop into conidiophores or grow across the leaf surface parallel to the veins for distances up to 3 mm to infect adjacent stomata (Stover, 1980a; Gauhl, 1989). These epiphyllic hyphae can anastomose to form a complex network. At various intervals along the hyphae, short side-branches develop, which terminate as appressoria over stomata (Meredith and Lawrence, 1969). Movement from one stoma to another is much more common with *M. fijiensis* than with *M. musicola* (see Jones, this chapter, pp. 79–92) and this eventually results in the development of black leaf streak lesions over entire leaves (Stover, 1980a). Disease severity increases the longer the period of leaf wetness (Jacome and Schuh, 1992) and this is most probably related to the epiphytic growth of the fungal mycelium.

The fastest incubation period, which is the time between infection and the appearance of the first speck symptoms on the leaf, is about 10–14 days under ideal conditions for disease development. The results of some field trials indicate that the incubation period varies with the cultivar (Meredith and Lawrence, 1970a; Firman, 1972; Fouré, 1991a; Jones and Tezenas du Montcel, 1994), but its length does not seem to correlate with overall resistance (Fouré, 1985). In contrast, no significant differences between incubation times in different cultivars have been found by other workers (Gauhl, 1994; Mobambo *et al.*, 1996a, 1997). The incubation period is faster on 3–5-month-old plants derived from tissue culture than on field-established plants (Fouré and Mouliom-Pefoura, 1988; Pasberg-Gauhl, 1993; Mobambo *et al.*, 1997).

Artificially inoculating young plants derived from tissue culture with conidia, Romero and Sutton (1997b) found the incubation period under laboratory conditions to be 4 days longer in 'FHIA-01®', an AAAB hybrid bred for resistance to black leaf streak, than in 'Grande Naine', a cultivar in the susceptible AAA Cavendish subgroup. In addition, they reported that the incubation period varied according to the isolate of *M. fijiensis* used in inoculations. Isolates from Cameroon were more aggressive than those from Central America and the South Pacific.

The incubation period is influenced by the season. During the rainy season in Nigeria, it has been recorded as 14 days on plantain (Mobambo *et al.*, 1996a), symptoms first being seen on leaf 2 on vegetatively growing plants. During the dry season, the incubation period was 24 days (Fig. 2.4). This difference probably relates to a delay in infection because of unfavourable conditions for ascospore discharge and germination, as well as temperature differences, which affect mycelial growth in leaf tissues.

Symptom evolution time

The number of days between the appearance of first symptoms and the appearance of mature spots with dry centres is the transition period or symptom evolution time. This varies according to the susceptibility of the cultivar and environmental conditions. Jones and Tezenas du Montcel (1994) reported transition periods ranging from 11 to 139 days.

Gauhl (1994) in Costa Rica noted that symptoms on 'Valery' (AAA, Cavendish subgroup) developed faster than those on 'Curraré' (AAB, Plantain subgroup) after the speck stage (Table 2.3) and this led to an earlier death of leaves. Gauhl (1994) also reported that first symptoms on 'Curraré' appeared, on average, about 8 days later on plants that had flowered compared with plants before flowering. Shade was also found to slow down symptom development, but only after speck formation. Shade has reduced the leaf area affected by black leaf streak by up to 50% (F. Gauhl and C. Pasberg-Gauhl, Costa Rica, 1999, personal observation).

Disease development time

A more useful criterion for measuring the rate of disease development is the disease development time, which is the period between infection and the formation of mature spots. Unlike the incubation and transition times, it does not rely on the detection of subtle first symptoms that can easily be overlooked.

The disease development time depends on the susceptibility of the cultivar, intensity of infection and environmental conditions, in much the same way as has been

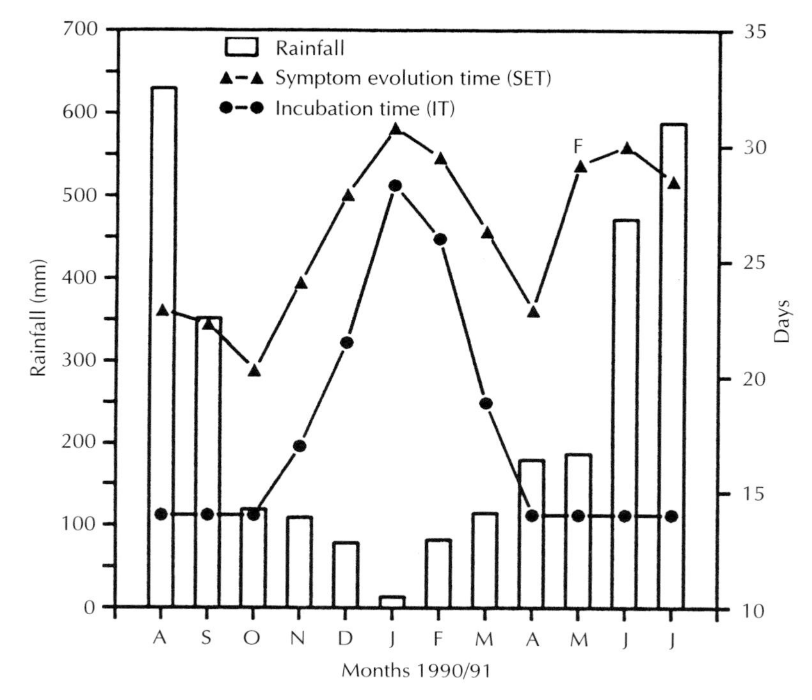

Fig. 2.4. The development of black leaf streak symptoms in relation to rainfall in Nigeria (from Mobambo *et al.*, 1996a). F = flowering.

Table 2.3. Comparison of the fastest black leaf streak symptom evolution times on 'Valery' (AAA, Cavendish subgroup) and 'Curraré' (AAB, Plantain subgroup) in Costa Rica between May and July 1986 when rainfall was fairly uniform (after Gauhl, 1994).

	'Valery'	'Curraré'
Time from leaf emergence to first speck symptom	24 days	24 days
Time from leaf emergence to first streak symptom	27 days	30 days
Time from leaf emergence to first spot symptom	30 days	38 days
Time from leaf emergence to first mature spot symptom	34 days	44 days

described for Sigatoka (Meredith, 1970). However, the rate of onset of mature spots is much faster with black leaf streak than with Sigatoka. After the introduction of black leaf streak to Honduras, spotting on unprotected plants in the susceptible Cavendish subgroup was reported to occur 8–10 days earlier after infection than with Sigatoka. As a consequence, spots appeared on leaves 3–4 and sometimes leaves 2–3 compared with leaves 4–5 with Sigatoka (Stover, 1980a). It is this increased speed of development, coupled with higher infection densities due to the earlier and more abundant formation of ascospores, that makes black leaf streak so much more difficult to control than Sigatoka.

The factors affecting the development of symptoms of black leaf streak and Sigatoka are presented in Fig. 2.5.

Conidiophores and conidia

Conidiophores with conidia are present in lesions beginning at the initial speck or first streak stages of Meredith and Lawrence (1969) and develop under conditions of high humidity. They are produced until the appearance of necrotic spots (Blanco, 1987; Fouré and Moreau, 1992; Gauhl, 1994). The production of conidiophores and conidia of *M. fijiensis* is, therefore, of short duration. Stover (1980a) estimated that about 1200 conidia are produced in the average 20 mm^2 streak on the lower leaf surface of a Cavendish cultivar. Conidiospore production is influenced by the weather. In Cameroon during 1987–1989, Fouré and Moreau (1992) counted a maximum of 44 conidiophores in each 1 mm^2 of stage 2 streak lesions in September (rainy season), whereas no conidiophores could be found in January (dry season). Blanco (1987) observed that the production of conidia was much higher on plants before flowering than on plants after flowering.

Conidia become dislodged by wind and water (Stover, 1980a), but, contrary to earlier work, which implicated water-splash in disease epidemiology (Meredith *et al.*, 1973), wind has been identified as the main agent that carries conidia to nearby plants (Rutter *et al.*, 1998). The scars on the conidia and conidiophores at the point of attachment are believed to facilitate the wind removal of conodiospores. However, conidia are not usually found in the air above the canopy of plantations (Gauhl, 1994). This and the fact that no distinct conidial infection patterns appear on leaves, as happens with Sigatoka (see Jones, this chapter, p. 86), suggest that conidia do not play a major role in disease spread.

Spermagonia and spermatia

Spermogonia with spermatia and perithecia are produced in substomatal cavities (Stover, 1980a). Although the cytological details of spermatization and ascospore

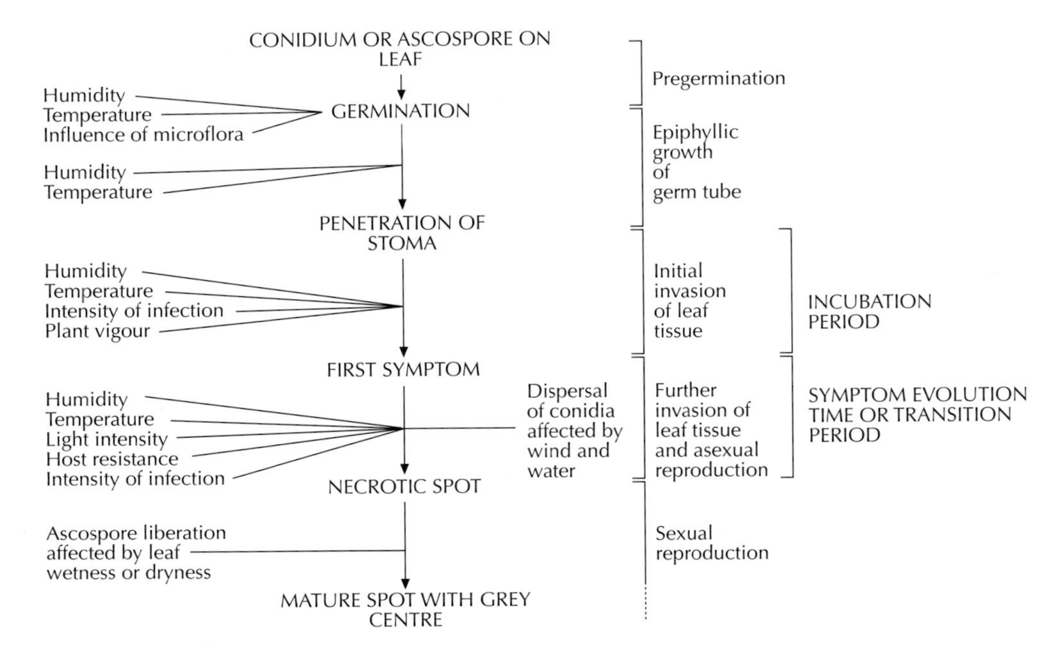

Fig. 2.5. Stages in the development of black leaf streak symptoms and factors that influence their duration (adapted from Meredith, 1970; Fouré and Moreau, 1992).

formation of *M. fijiensis* have not been described, the process is believed to be similar to that in *M. musicola* (see Jones, this chapter, pp. 79–92).

Perithecia and ascospores

Perithecia are abundant in lesions at the second and third or mature spot stage of Meredith and Lawrence (1969), which corresponds to stages 5 and 6 of Fouré (1987). Ascospores are forcibly ejected from the perithecia during periods of wet weather or, in the absence of rain, during the night, reaching a maximum before dawn as dew settles (Meredith *et al.*, 1973; Gauhl, 1994). Ascospores can be released from lesions up to 21 weeks after their formation and production is usually greater from lesions on the lower surface of the leaf (Gauhl, 1994). Only one crop of ascospores is produced in each perithecium (Stover, 1980a), but perithecia continue to mature in necrotic leaves. Therefore, dead leaves, which may hang on the plant or lie on the ground after pruning operations, are a source of inoculum. In Costa Rica, Gauhl (1994) monitored the discharge of ascospores from dead leaves lying on the ground. Large numbers were released during the first 3 weeks, with small numbers released during the next 6 weeks. The leaves disintegrated after 10 weeks. No differences were found in the numbers of ascospores discharged from the upper or lower leaf surfaces.

In long-term epidemiological studies on black leaf streak in Costa Rica (Blanco, 1987; Gauhl, 1994) and Nigeria (Gauhl and Pasberg-Gauhl, 1994), ascospores were shown to be the most common form of inoculum. These results confirm observations in Colombia (Mayorga, 1985), in Fiji (Firman, 1972) and in Hawaii (Meredith *et al.*, 1973).

Mature perithecia need to be impregnated with water before ascospores can be discharged (Stover, 1976; Fouré and Moreau, 1992). Therefore, spore release can follow a seasonal pattern, with only low numbers of ascospores being released during the dry season and high numbers during periods of higher rainfall (Gauhl, 1994; Gauhl and Pasberg-Gauhl, 1994). In the Atlantic lowlands of Costa Rica, Gauhl (1994) recorded the highest concentration of ascospores in December, which is the wettest month (Fig. 2.6). The highest ascospore concentration for one day was 6876 in 1 m^3 of air on 26 December 1985. Meredith *et al.* (1973) also found the highest daily ascospore concentrations during the rainy seasons in Hawaii and Fiji (Fig. 2.6). Ascospore counts in Nigeria also peaked during the rainy season, but were found to be much lower than in other areas of the world (Gauhl and Pasberg-Gauhl, 1994) (Fig. 2.6). The lower levels of inoculum recorded in Nigeria may have been because spores were trapped at a location remote from large banana plantations.

A film of water on the leaf surface is required for ascospore infection under controlled conditions. The minimum amount of rainfall necessary to induce ascospore release is as low as 0.1 mm h^{-1} (Gauhl, 1994). Rainfall can, therefore, always induce ascospore release if mature perithecia are present. Discharge can occur within 1 h or so after rain (Burt, 1993; Gauhl, 1994). Ascospores discharged during long periods of continuous rain can be washed out of the air and off foliage.

Another factor influencing ascospore discharge is the temperature. Less ascospore production was observed during periods with daily minimum temperatures below 20°C (Gauhl, 1994). Burt (1993) reported a decrease in spore release when temperature was under 19°C or over 33°C. In Cameroon, Fouré and Moreau (1992) found many perithecia in coalescing mature spots in the wet season, but very few perithecia in individual mature spots during the dry season. Spots probably did not coalesce in the dry season because low humidities and the absence of rain may have inhibited lesion spread by epiphyllic hyphae. This, coupled with restrictions on the movement of spermatia imposed by a lack of surface moisture, may have reduced the chances of cross-fertilization necessary for the production of perithecia and ascospores. In Costa Rica, Blanco (1987) counted 0.3–1.2 perithecia in each 1 mm^2

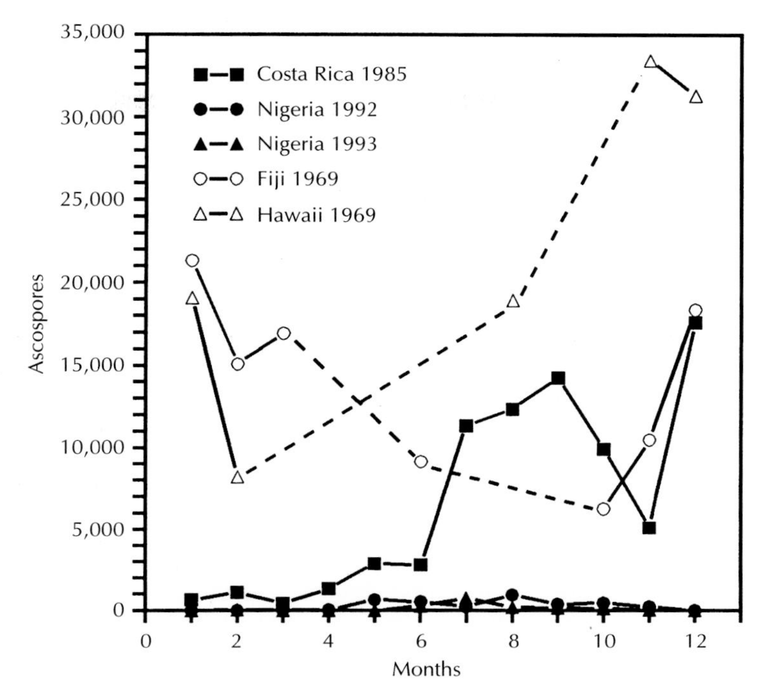

Fig. 2.6. Ascospore counts (total for month) in different regions of the world (after Meredith *et al.*, 1973; Gauhl, 1994; F. Gauhl and C. Pasberg-Gauhl, Costa Rica, 1999, unpublished results).

of mature lesion on the upper leaf surface of a Cavendish cultivar and 0.6–3.7 perithecia on the lower surface. He also demonstrated that more ascospores were discharged from the lower leaf surface than from the upper surface and that the proportion of ascospores discharged from two leaf surfaces in different seasons of the year varied considerably. Evaporation and relative humidity readings correlate best with ascospore discharge during daylight hours (Thal *et al.*, 1992).

Ascospores are commonly found in the air above the canopy of plantations following rainfall (Gauhl, 1994). Air currents carry the ascospores long distances to other plants within the plantation and to other plantations. Ascospores are killed after 6 h exposure to ultraviolet radiation and this indicates that the distance of dissemination of viable spores is determined not only by the speed of the wind, but also by cloud cover and the time of day that spore release occurs. Dispersal over a distance of a few hundred kilometres is considered unlikely (Parnell *et al.*, 1998).

Black leaf streak versus Sigatoka

Black leaf streak is well suited for the environmental conditions prevailing in the coastal tropics and has now virtually replaced Sigatoka as the dominant leaf spot disease in these areas. This process took 2–3 years in Honduras (Stover, 1980a) and less than 5 years in Costa Rica (Gauhl, 1994). Black leaf streak lesion expansion and ascospore production are greater in these areas than those for Sigatoka (Mouliom-Pefoura *et al.*, 1996) and this has probably given this disease a competitive advantage. However, the situation has been different at altitude. Sigatoka seems more adapted to cooler environments and lesions grow faster than those of black leaf streak in upland areas (Mouliom-Pefoura *et al.*, 1996). In Costa Rica in 1985, both Sigatoka and black leaf streak diseases were identified on the same banana leaf at 900 m, but only Sigatoka was found at 1200 m (Romero and Gauhl, 1988).

Black leaf streak is believed to be becoming more competitive at altitude as it gradually adapts to a cooler environment. In 1990, in the same area of Costa Rica as was surveyed in 1985, black leaf streak was found at altitudes up to 1500 m (F. Gauhl and C. Pasberg-Gauhl, Costa Rica, 1999, unpublished). In 1988, 7 years after its introduction to Colombia, black leaf streak was found for the first time in the plantain production areas in the Cauca valley at 1500 m (Merchan, 1990). In 1991, it had reached 1600 m (Belalcazar, 1991).

Host reaction

Evaluating germplasm for reaction to black leaf streak

E. Fouré, X. Mourichon and D.R. Jones

INTRODUCTION Numerous reviews have been published on the reaction of wild and cultivated banana to black leaf streak in the field. However, this information is sometimes contradictory. Variations in results could be due to the different methodologies used to evaluate levels of resistance and susceptibility (Laville, 1983; Tezenas du Montcel, 1990). Differences in virulence and aggressiveness of local populations of *M. fijiensis* (Fullerton and Olsen, 1995) or differences in environmental conditions at test sites (Vuylsteke *et al.*, 1993a) may also cause discrepancies.

Tests to determine the reaction of germplasm to black leaf streak have often been made within genetic resource collections containing wild species and banana cultivars belonging to different genomic groups. This work has been undertaken to identify genotypes that may have potential as breeding material and as replacement cultivars for traditionally cultivated clones in areas where they have succumbed to the disease.

The evaluation of resistance of germplasm is especially important for banana improvement programmes (Rowe and Richardson, 1975; Stover and Simmonds, 1987; Hahn *et al.*, 1989; Fouré, 1991b, 1992). *Musa* accessions obtained from gene banks or collecting expeditions need to be tested to determine their suitability as parental material (Bakry and Horry, 1994) and also the resistance of hybrids derived from breeding endeavours needs to be confirmed (Vuylsteke *et al.*, 1993a, 1995; Jenny *et al.*, 1994).

A review of the literature published over the last 30 years on the reaction of banana to *Mycosphaerella* spp., and *M. fijiensis* in particular, has made clear the need for standard testing procedures, which can characterize the different types of interaction between the host and pathogen.

METHODS FOR DETERMINING HOST REACTION Environmental conditions influence spore germination, subsequent growth of germ tubes and hence the number of successful infections. Studies have shown that, in black leaf streak– and Sigatoka–banana interactions, it is only after stomatal penetration that events occur determining resistance or susceptibility. Although high inoculation densities can result in a more rapid development of symptoms, for a given strain of pathogen, final disease expression is influenced by host factors and environmental conditions, which determine the number of lesions, the rate of development of necrotic lesions and the production of spores (Meredith, 1970).

Observations on the reaction of banana to leaf spots caused by *Mycosphaerella* spp. began in Trinidad. Cheesman and Wardlaw (1937) classified cultivars as either 'susceptible', 'mildly susceptible' or 'highly resistant or immune' to *M. musicola*. Symptoms were completely absent on cultivars in the highly resistant/immune category. Simmonds (1959) later proposed four classes of reaction of banana to *M. musicola*, ranging from 'highly susceptible' to 'highly resistant', and Brun (1962) used this same scale to assess the behaviour of several varieties towards *M. musicola* in Guadeloupe and Guinea.

Vakili (1968), also working with *M. musicola*, measured the time from the unfurling of a leaf to the development of advanced yellow streaks and the time from the appearance of advanced yellow streak to mature spot formation on different species and cultivars. He also noted the position of the youngest leaf with mature spots, counted the total number of mature spots on each plant on the first 14 leaves and sometimes estimated the number of sporodochia at different stages of lesion development. From this information, he proposed four levels of response to infection ('highly resistant', 'resistant', 'partially resistant', 'susceptible' and 'highly susceptible') and recommended four cultivars as Sigatoka resistance differentials. These differentials exemplified the four response categories identified. Vakili's studies also showed that, as susceptibility increases, the disease incubation period shortens and mature spot symptoms appear on younger leaves.

Meredith and Lawrence (1970a) in Hawaii were the first to undertake reaction evaluations of cultivars attacked by *M. fijiensis*. They collected eight different sets of data.

1. The rate of leaf emergence, expressed as the number of days taken for the heart leaf to unfurl completely.
2. The number of the youngest leaf with 'initial speck stage' symptoms. The youngest completely unfurled leaf was designated as leaf no. 1, the next youngest no. 2 and so on down the pseudostem.
3. The number of the youngest leaf bearing the 'first spot stage'.
4. The frequency of transition with which streaks changed to spots as measured by the percentage area on the oldest living leaf occupied by spots.
5. The intensity of streaking, as measured by the number of streaks in a 5 cm^2 area on the youngest leaf with the first spot stage. The area was selected where streaks were most numerous (usually near the edge of the left side of the leaf). Counts from five, randomly selected areas per leaf on each of five leaves per cultivar were averaged to get the final reading.
6. The average area of individual streaks on the second youngest leaf showing symptoms. This was calculated from length and width measurements of 300 streaks.
7. The number of functional leaves, which were defined as those with more than 50% of the lamina area free of spots, at the time of harvest.
8. The abundance of formation of conidia, as measured by the number of conidiophores mm^{-2} on the lower surface of 'second-stage streaks'.

The time taken for the appearance of first symptoms or 'incubation period' was determined from (1) and (2) by assuming that leaves became infected within 5 days of emergence from the top of the pseudostem. The time taken from the appearance of first stage streaks to the formation of the first spot stage or 'transition period' was determined from (1), (2) and (3). This information, together with (2), (3), (4), (5), (6) and (7) allowed the calculation of a score for each cultivar evaluated. The results obtained from (8) were thought by the authors to be of little value and were not taken into consideration. Various combinations of short and lengthy incubation and transition periods, frequent and infrequent transition from streak to spot and heavy and light streaking were observed on the different cultivars. Symptoms were found on all cultivars studied and none were rated as being highly resistant or immune. Although an almost continuous gradation of responses was recorded, cultivars were placed into three evaluation categories, which were 'very susceptible', 'moderately susceptible' and 'slightly susceptible' (Table 2.4). The evaluation indicated that there was much less natural resistance to *M. fijiensis* than there was to *M. musicola* (see Jones, this chapter, pp. 79–92).

Firman (1972) evaluated banana cultivars at the Koronivia Research Station in Fiji for reaction to *M. fijiensis*, using criteria similar to those of Meredith and Lawrence (1970a). In addition, the number of functional leaves at flowering was also noted. Difficulty was experienced in measuring the intensity of streaking and area of individual streaks, because streaks merged at

Table 2.4. Reaction of different banana cultivars to *Mycosphaerella fijiensis* (modification of Table 8 of Meredith and Lawrence, 1970a).

Genomic group	Very susceptible	Moderately susceptible	Slightly susceptible
AA		'Sucrier'	'Tuugia'*
AAA	'Gros Michel'†	'Green Red'	
	'Cocos'†		
	'Dwarf Cavendish'‡		
	'Robusta'‡		
	'Valery'‡		
	'Giant Cavendish'‡		
	'Lacatan'‡		
AAAA		I.C. 2§	
AAB	'Pome'	'Silk'	
	'Pisang Rajah'‖	'Horn' plantain	
	'Moongil'¶	'Maia Maoli'-types	
	'Platano Enano'**	'Popoulu'-types	
	'Walha'††	'Iholena'-types	
	'Eslesno'‡‡	'Father Leonore'‡‡	
ABB		'Monthan'§§	'Saba'
		'Ice Cream'‖‖	
		'Largo'§§	

* Cultivar from Vietnam.
† Cultivars in Gros Michel subgroup.
‡ Cultivars in Cavendish subgroup.
§ Tetraploid from the Trinidad breeding programme.
‖ Synonym of either 'Pisang Raja', 'Celat' (Stover and Simmonds, 1987) or 'Nendrapadathi' (see Table 2.5).
¶ 'Horn' plantain from southern India with one hand of very large fruit.
** Dwarf 'Horn' plantain from Puerto Rico (Stover and Simmonds, 1987).
†† Synonym of 'Rajapuri' (Stover and Simmonds, 1987).
‡‡ Believed to have been introduced from West Indies and may be ABB types (Stover and Simmonds, 1987).
§§ Cultivars in Bluggoe subgroup.
‖‖ Synonym of 'Ney Mannan'.

an early stage of development. Attempts to measure the production of conidia and ascospore discharge were unsuccessful, because results varied widely from leaf to leaf and not enough samples could be processed to make comparisons valid.

All cultivars tested in Fiji fell into the 'very susceptible' or 'moderately susceptible' categories. Results were similar to those of Meredith and Lawrence (1970a) in that commercially important triploid cultivars in the AAA group were rated as very susceptible and local cooking banana cultivars in the AAB group as moderately susceptible. Differences in readings between Hawaii and Fiji suggested that black leaf streak was more severe in Fiji (Firman, 1972).

Fouré (1982, 1985, 1987, 1989) and Fouré *et al.* (1984) evaluated the reaction of cultivars to black leaf streak in Gabon and Cameroon using the following four parameters.

1. The time taken from infection to the appearance of first symptoms (incubation period).
2. The time taken from the appearance of first symptoms to the appearance of the first mature necrotic spot (or necrosis due to the merging of earlier stages).

3. The intensity of asexual and sexual sporulation.
4. The number of functional leaves at harvest.

These studies also revealed a gradation of reactions to *M. fijiensis* from complete susceptibility to slight susceptibility or partial resistance. Cultivars with an AAB genome were found not to have the same level of resistance to *M. fijiensis* as they had to *M. musicola*. The work of Fouré and colleagues also confirmed that cultivars in the ABB genomic group were generally less susceptible than those in other groups. A few accessions evaluated were found to have a very high level of resistance. Spots never formed on these accessions and the development of symptoms was arrested at an early streak stage. The results and analysis of Fouré's work are presented in more detail later.

Jones (1993a) concluded that a close approximation of resistance to black leaf streak could be gauged from noting the position (number) of the youngest leaf with ten or more mature spots. Stover and Dickson (1970) first used this parameter for determining the prevalence and severity of Sigatoka in Cavendish banana plantations in Central America. They found that this leaf number, which was called the 'youngest leaf spotted' (YLS), correlated well with the amount of spotting present on a plant and disease intensity. The symptom marks the beginning of a stage where most damage to the leaf occurs and the health of leaves rapidly declines after this event.

In 1992, Mobambo *et al.* (1994a) evaluated young plants for reaction to black leaf streak in the field in Nigeria, beginning 2 months after planting. Infection was assumed to occur when the unfurling cigar leaf was at stage B, as described by Brun (1962) (Fig. 2.7). Five parameters were measured.

1. Incubation time (time from infection to appearance of first symptoms).
2. Symptom evolution time (time from appearance of first symptoms to appearance of spots with dry centres).
3. Disease development time (time from infection to appearance of spots with dry centres).
4. Number of the youngest leaf spotted.
5. Life of leaf (time from cigar leaf stage B to leaf death).

Results showed that the incubation time was similar for all clones tested, but other parameters were significantly different.

A guide to the evaluation of black leaf streak resistance, which gave details of trial layouts and readings taken to determine disease reaction at the International Institute of Tropical Agriculture (IITA) in Nigeria, was published by Gauhl *et al.* (1993). Parameters considered important

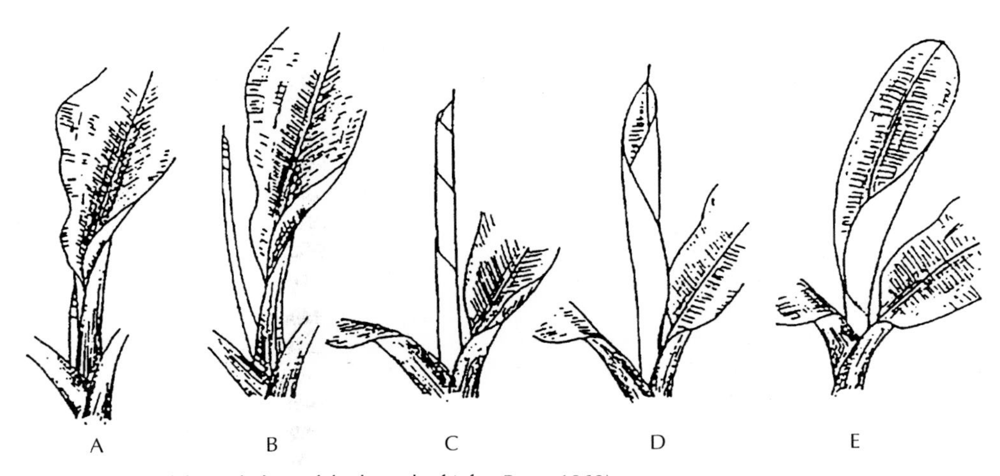

Fig. 2.7. Stages of the unfurling of the heart leaf (after Brun, 1962).

included the incubation period, the disease development time (time from infection to the formation of ten spots with a dry centre), percentage of leaf area necrotic and the lifespan of a leaf (time from emergence to leaf death).

Since 1990, the International Network for the Improvement of Banana and Plantain (INIBAP) has coordinated the worldwide testing of new hybrids from the major banana breeding programmes under its International *Musa* Testing Programme (IMTP). Hybrids were evaluated for black leaf streak resistance in plots with reference accessions of known disease resistance. As standard field plans and the same protocols for measuring disease reaction were employed at all test sites (Jones, 1994a), comparisons could be made on the response of the same germplasm to black leaf streak under different climatic conditions and inoculum pressures. It was hoped that this work, as well as determining the resistance of new material from genetic improvement centres and their agronomic suitability for further testing in different global environments, would also throw some light on the pathogenic variability of *M. fijiensis* at different locations. Readings employed in the latest series of trials have included the following.

1. Disease development time, which is the time from infection (assumed to occur at an early stage of leaf emergence) to the appearance of ten or more mature spots or large necrotic area with ten or more light-coloured dry centres.
2. Youngest leaf spotted, which is the position (number) of the first fully unfurled leaf with ten or more mature spots or large necrotic area with ten or more light-coloured dry centres counting down from the top of the plant.
3. Leaf emission rate expressed as the number of new leaves produced each week.
4. Disease severity expressed as the percentage of leaf area killed at 6 months after planting, bunch emergence and harvest.

INIBAP has recently published technical guidelines for evaluating germplasm for reaction to black leaf streak and Sigatoka (Orjeda, 1998).

Various field plans have been suggested to ensure that results of black leaf streak screening trials are statistically valid (Gauhl *et al.*, 1993; Jones, 1994a; Orjeda, 1998). Nokoe and Ortiz (1998) have recommended that about 13 (plus or minus three) plants of each clone should be used to assess reaction to black leaf streak and 15 (plus or minus two) plants to evaluate growth and yield characteristics.

Classification of Interactions From observations on the response of banana to black leaf streak, Fouré *et al.* (1990) and Fouré (1992, 1994) proposed three types of reaction to the disease.

Phenotype 1. The expression of very high resistance (HR), characterized by the apparent blockage of the development of symptoms at the early streak stage (see Plate 2.10). This reaction is believed to be close to hypersensitivity and the pathogen does not sporulate. Resistance of this type could be controlled by a single host resistance gene or a small number of host resistance genes giving the plant 'vertical' resistance, which may be easily overcome by mutations in the pathogen population. Most accessions of subspecies of *M. acuminata* and the cultivars 'Pisang Lilin' (AA), 'Tuugia' (AA) and 'Yangambi Km 5' (AAA, syn. 'Ibota Bota') react in this way. Studies have indicated that the defence response is activated soon after the pathogen penetrates the stoma.

Phenotype 2. The expression of partial resistance (PR), characterized by the normal, but slow, development of symptoms to necrosis. This reaction allows the pathogen to sporulate. Resistance of this type may be controlled by many genes giving the plant durable or 'horizontal' resistance. There is a large gradation of response with phenotype 2 from resistance to almost complete susceptibility. Plants may have a large number of functional leaves at harvest. Cultivars showing moderate partial resistance include 'Pisang Mas' (AA, syn. 'Sucrier'), while 'Pisang Ceylan' (AAB, syn. 'Mysore') and 'Fougamou' (ABB, syn. 'Pisang Awak') have strong partial resistance. These categories are the equivalent of the moderately susceptible and slightly susceptible divisions of Meredith and Lawrence (1970a).

Phenotype 3. The expression of pronounced susceptibility (S), characterized by the normal and rapid development of symptoms to necrosis (see Plate 2.8). The pathogen may sporulate profusely if environmental conditions are favourable. Few functional leaves, if any, remain on the plant at harvest (see Plate 2.9). All the important Cavendish cultivars (AAA), such as 'Dwarf Cavendish', 'Grand Nain', 'Williams' and 'Robusta', fall into this category. However, disease expression can be greatly influenced by environmental conditions and inoculum potential. This category is the equivalent of the very susceptible division of Meredith and Lawrence (1970a).

Fouré (1994) considers that these three phenotypes are expressions of two major types of interaction: a highly resistant or incompatible interaction (phenotype 1) and a compatible interaction, which has grades of expression ranging from strong partial resistance through to complete susceptibility (phenotypes 2 and 3).

REACTION OF *MUSA* IN FIELD TRIALS Symptoms of black leaf streak have only been identified on *Musa balbisiana*, subspecies of *M. acuminata* and cultivars in the *Eumusa* series of edible banana. The disease does not affect *Musa textilis* (abacá) or *Ensete ventricosum* (enset) (Gauhl, 1994).

Over 350 accessions, which belong mainly to the *Eumusa* series of cultivars, but also include wild diploids, have been tested for their reaction to black leaf streak in the field in Cameroon by Fouré. The response of these accessions is presented in Table 2.5.

The results reveal that *M. balbisiana* and cultivars belonging to genomic groups containing at least one B component (AB, AAB, ABB) all show a susceptible or partially resistant reaction (compatible interaction). The behaviour of cultivars within subgroups, with a few minor exceptions, is generally fairly uniform. The highly resistant or incompatible reaction is expressed in most wild *M. acuminata* subspecies and in some cultivars within the AA and AAA groups.

Although the results presented in Table 2.5 were obtained in Cameroon in one particular environment and may not necessarily reflect reactions observed elsewhere, they do generally confirm the findings made earlier by other authors in the Pacific (Meredith and Lawrence, 1970a; Firman, 1972; Pearson *et al.*, 1983; Gonsalves, 1987), in Latin America (Stover, 1983a; Pasberg-Gauhl, 1990) and in West Africa (Mobambo *et al.*, 1993, 1994a; Ortiz *et al.*, 1993, 1995; Vuylsteke *et al.*, 1993a, 1995).

Host–pathogen interactions

X. Mourichon, P. Lepoivre and J. Carlier

INTRODUCTION. The first serious studies on banana–black leaf streak interactions began when small banana plants propagated by tissue culture were inoculated with conidia and fragments of hyphae harvested from colonies of *M. fijiensis* grown in culture (Mourichon *et al.*, 1987). The results of this work showed that symptoms of compatible and incompatible interactions obtained under controlled environmental conditions were very similar to those observed in the field on mature plants. This led to further research on the microscopic events that take place in banana tissues after fungal penetration and culminate in the expression of resistance or susceptibility (Beveraggi, 1992; Beveraggi *et al.*, 1993, 1995).

GERMINATION AND STOMATAL PENETRATION. It was discovered that, after a 4–6-day phase that involves the germination of spores and subsequent epiphyllous growth of mycelium, appressoria develop on hyphae above stomatal openings (Beveraggi, 1992). Fine hyphal filaments then penetrate through the stomata to the cavity below, where they form sub-stomatal vesicles. Mycelia, which develop from these vesicles,

Table 2.5. Reaction of *Musa* species/subspecies, subgroups and clones to black leaf streak at Njombe, Cameroon (80 m altitude).

Genome	Species/subspecies	Accessions evaluated	Reaction*
AAw	*Musa acuminata* ssp. *banksii*	'Madang', 'Banksii AF'[†]	S and HR
	Musa acuminata ssp. *burmannica*	'Calcutta 4'	HR
	Musa acuminata ssp. *malaccensis*	'Pahang' and 'Kluai Pa (Musore) × '	HR
	Musa acuminata ssp. *microcarpa*	'Pisang Cici Alas'[‡]	HR
	Musa acuminata ssp. *siamea*	'Kluai Pa (Rayong)', 'Kluai Khae (Phrae)'[‡] and 'Kluai Pa (Songkla)'[‡]	HR
	Musa acuminata ssp. *truncata*		HR
	Musa acuminata ssp. *zebrina*		PR
BBw	*Musa balbisiana*	'Kluai Tani', 'Pisang Klutuk Wulung', 'Pisang Batu'	PR*

Genomic group	Subgroup/clone[§]	Cultivars evaluated	Reaction*
AA	'Sucrier'	'Kirun', 'Figue Sucrée'	PR
	'Pisang Lilin'	'Pisang Lilin'	HR
	Others	'Nzumoheli', 'Niyarma Yik', 'Sow Muk', 'Bie Yeng', 'Mambee Thu', 'Samba', 'Chicame', 'Pallen Berry', 'Pisang Bangkahulu', 'Akondro Mainty', 'Galeo', 'Guyod'	S
		'Pisang Berlin', 'Pisang Madu', 'Mak', 'Pisang Perecet', 'Pa Pathalong', 'Kluai Thong Det', 'Doumboumi', 'Pisang Tongat'	PR
		'Tuugia', 'Pisang Oli'	HR
AAA	Gros Michel subgroup	'Gros Michel', 'Cocos', 'Highgate'	S
	Cavendish subgroup	'Petite Naine', 'Grande Naine', 'Poyo', 'Robusta', 'Valery', 'Americani', 'Lacatan'	S
	Lujugira–Mutika subgroup	'Inokatoke', 'Intuntu', 'Igitsiri', 'Mbwazirume', 'Nakitengwa'	S
	'Red' and 'Green Red'	'Figue Rose', 'Figue Rose Verte', 'Figue Rose Naine'	S
	'Ibota Bota'	'Yangambi Km 5'[‖], 'Kluai Khom Bao', 'Kluai Khai Thong Ruang'	HR
	'Orotava'	'Orotava', 'Pisang Kayu'	S
	Others	'Kluai Hom', 'Pisang Sri', 'Kluai Hom Thong Mokho'	PR
AAAA		'Pisang Jambe', 'I.C.2', '1877'	PR
AB	'Ney Poovan'	'Safet Velchi', 'Lal Kelat'	PR
	Other	'Figue Pomme Ekona'	S
AAB	Plantain subgroup	'Amou', 'French Sombre', 'French Clair', 'Kelong Mekintu', 'Orishele', 'Msinyore'	S
	Pome subgroup	'Prata Anã', 'Foconah', 'Nadan', 'Guindy'	S
	Maia Maoli-Popoulu subgroup	'Iho U Maoli', 'Popoulou', 'Poingo'	S
	Iholena subgroup	'Mattui', 'Maritú'	PR
	'Silk'	'Figue Pomme', 'Muthia', 'Supari'	S and PR

Table 2.5. *Continued*

Genomic group	Subgroup/clone[§]	Cultivars evaluated	Reaction*
	'Mysore'	'Pisang Ceylan', 'Gorolo', 'Zabi'	PR
	'Pisang Raja'	'Pisang Rajah Bulu'	S
	'Pisang Kelat'	'Pisang Pulut', 'Pisang Kelat'	S and PR
	'Laknau'	'Laknao'	S
	'Nendrapadathi'	'Pisang Rajah' – from Rajapuri, India	S
ABB	Bluggoe subgroup	'Cacambou', 'Cachaco', 'Poteau', 'Monthan'	PR
	'Pisang Awak'	'Fougamou', 'Gia Hong', 'Foulah 4', 'Brazza 2'	PR* and PR
	'Pelipita'	'Pelipita'	PR
	'Peyan'	'Simili Radjah'	PR
	'Kluai Teparot'	'Kluai Tiparot'	PR*
	'Ney Mannan'	'Pisang Abu Perak', 'Ice Cream', 'Som'	PR
	'Saba'	'Saba'	PR
	'Inabaniko'	'Benedetta'	PR

* Classified using Cavendish cultivars as very susceptible reference clones. S, very susceptible; PR, partial resistance; PR*, very strong partial resistance (Fouré's compatible interactions); HR, high resistance (Fouré's incompatible interaction).
[†] AF, plant derived from the self-fertilization of a *Musa acuminata* ssp. *banksii* accession.
[‡] Hybrid with dominant characteristics of *Musa acuminata* subspecies.
[§] Name of best-known synonym signifies clone.
[||] This cultivar has recently succumbed to black leaf streak in Cameroon (Mouliom-Pefoura, 1999).

later spread through the intercellular spaces between the leaf parenchyma cells. This initial infection step does not differ markedly from that already described for *M. musicola* (Brun, 1963; Meredith, 1970; Stover, 1972; Wardlaw, 1972).

LESION DEVELOPMENT The first major event after penetration is necrosis of the stomatal cells. In compatible interactions, the necrosis slowly spreads to nearby cells while other necrotic areas develop some distance away from the initial microlesion. These clusters of dead lower hypodermal or spongy mesophyll cells are separated from the original microlesion by apparently healthy tissues. As the central lesions spread, the secondary necrotic areas increase in diameter and coalesce. Other necrotic areas regularly appear ahead of the margin of the main lesion.

After the initial penetration of the fungus, epiphyllous mycelia can infect other nearby stomata, leading to a succession of similar reactions. This phenomenon accelerates the colonization rate and leaf tissue at the centre of the expanding spot dries. When the inoculum is highly concentrated, lesions caused by penetrating hyphae originating from different conidia quickly coalesce and large areas of tissue rapidly become necrotic, as is often also noted in the field.

In adult banana leaves, secondary veins form a mechanical barrier to the lateral spread of *M. musicola* within the leaf (McGahan and Fulton, 1965). The pathogen can overcome these barriers by growing epiphytically on the leaf surface, but progress is not as rapid as in leaf tissue parallel to the veins and the lesion tends to become linear in shape. In young leaves, such as those found on small water suckers, the secondary veins are less developed and the lateral growth of the pathogen is not so limited. As a consequence, necrotic lesions become more elliptical. *Mycosphaerella fijiensis* invades banana leaf tissue in a similar manner to *M. musicola*. Young plants derived from tissue culture have limited vein development and black leaf streak lesions tend to be more spherical.

GROWTH OF *M. FIJIENSIS* IN BANANA TISSUE The course of infection has been followed in several susceptible and resistant banana cultivars inoculated with *M. fijiensis* (Beveraggi, 1992). In compatible interactions, hyphae were observed between living cells ahead of the necrotic zone, indicating that the pathogen is biotrophic during the initial phases of infection. The main difference between susceptible and partially resistant cultivars was the much faster growth rate of hyphae in the former. In incompatible interactions, as exemplified by the infection of the highly resistant cultivar 'Yangambi Km 5' (AAA, syn. 'Ibota Bota'), the pathogen stops growing 2–3 days after stomatal penetration and the death of the first plant cell.

COMPATIBLE INTERACTIONS The progress of infections in 'Fougamou' (ABB, 'Pisang Awak'), a cultivar with strong partial resistance, and 'Grande Naine' (AAA, Cavendish subgroup), a susceptible cultivar, were compared by Beveraggi *et al.* (1993, 1995). In 'Fougamou', a histological analysis of healthy tissues revealed the presence of many specialized parenchyma cells containing polyphenol-rich vacuoles in the palisade and spongy mesophyll layers. Far fewer of these specialized cells were found in uninfected 'Grande Naine'. An ultrastructural study of the interaction between *M. fijiensis* and 'Fougamou' revealed that, once the specialized cells became necrotic, the contents of their vacuoles were released into the intercellular spaces. This material had a high affinity for fungal cell walls and their appearance around hyphae seemed to be correlated with the slow spread of the mycelium in parenchyma tissues.

Soluble phenols present in the leaf tissues of several banana cultivars that have different levels of partial resistance were identified by Gire *et al.* (1994). They found flavonols, which were mainly quercetin derivatives, hydroxycinnamic derivatives, belonging to *p*-coumaric, ferulic and chlorogenic acid families, and flavanes, which have been identified as condensed tannins. In the cultivars studied, no significant differences in flavonol and hydroxycinnamic derivatives were detected. However, there was a close correlation between flavane content and the level of partial resistance of cultivars studied. These flavanes were found to be concentrated in the vacuoles of the specialized parenchyma cells descibed above and it is possible that they could be implicated in the expression of partial resistance.

INCOMPATIBLE INTERACTIONS Most of the information on incompatible reactions has been obtained from the studies of interactions involving 'Yangambi Km 5' (Beveraggi *et al.*, 1993, 1995). The first event that occurs following penetration is the very rapid death of about three to four cells around the stoma and the expulsion of electron-dense compounds through the walls of these cells into the intercellular space. The released compounds show a high affinity for the walls of the infection hypha and nearby host cells. The degeneration of hyphae in contact with these compounds suggests that they have a fungitoxic action. Deposits of these compounds on host cell walls could also explain the fluorescence observed at infection sites after analine-blue staining. The ability of 'Yangambi Km 5' cells nearest the pathogen to rapidly die seems to be a characteristic hypersensitive reaction. In this case, it appears that host resistance is determined by changes that occur in only a few cells either in contact with or in close proximity to the pathogen. A second event, which is characterized by further, more extensive, cell death in surrounding leaf tissue, follows the hypersensitive response.

Hypersensitive reactions typically take place in gene-for-gene relationships. There is no experimental evidence of such relationships in *Musa–M. fijiensis* interactions, as this requires genetic studies of the host and pathogen. However, some *M. fijiensis* isolates originating from the Cook Islands and Tonga are able to overcome the resistance of 'Yangambi Km 5' in laboratory tests (Fullerton and Olsen, 1993), thus suggesting that this type of host–pathogen relationship could occur. Such cultivar-specific hypersensitive reactions are usually explained by the presence of a specific elicitor or elicitors or by the coordinated action of non-specific elicitors, combined with the action of a specific suppressor (de Wit, 1992; Atkinson, 1993).

A wide range of fungal compounds have been proposed as elicitors of hypersensitive reactions in host–pathogen interactions: polysaccharides (Sharp et *al.*, 1984), glycoproteins (Schaffrath *et al.*, 1995), peptides (de Wit *et al.*, 1985) and hydrolytic enzymes (Boller, 1987). In *Musa–M. fijiensis* interactions, elicitation of necrosis was induced by high-molecular-weight compound(s), which appeared stable after high temperature treatment or extreme pH, and were not affected by protease treatments and sodium periodate oxidation. This suggests that polysaccharide compounds are not involved in eliciting activity (P. Lepoivre, Gembloux, 1997, unpublished).

The durability of resistance is crucial for efficient breeding for resistance to black leaf streak disease. The existence of *M. fijiensis* isolates able to attack 'Yangambi Km 5' indicates that this type of resistance is not sustainable and it may be more appropriate to utilize the strong partial resistance found in cultivars such as 'Fougamou'.

The Role of Toxins in Pathogenesis A number of phytotoxic compounds have been found in *M. fijiensis* culture filtrates (Upadhyay *et al.*, 1990; Stierle *et al.*, 1991; Strobel *et al.*, 1993). Juglone 2 was found to be the most biologically active, inducing necrotic lesions in all cultivars of banana tested. However, 2,4,8-trihydroxytetralone exhibited the greatest host selectivity, forming large necrotic lesions when applied to leaves of black leaf streak-susceptible cultivars, but showing little activity when applied to a resistant cultivar (Strobel *et al.*, 1993).

The role of toxins in pathogenesis is usually assessed by determining if a correlation exists between (i) toxin production and pathogenicity of the organism and (ii) the sensitivity of different banana genotypes to the toxin and their susceptibility to the disease (Yoder, 1980). Induction of necrosis by a leaf-puncture bioassay on detached banana leaves or the injection of toxic preparations into the leaf is simple to undertake, but has been found to be neither sensitive nor quantitative. An electrolyte leakage assay that was developed allowed a more sensitive assessment of the effect of *M. fijiensis* toxins, but was very time-consuming. The most sensitive and rapid assay was found to be the measurement of changes in the fluorescence of chlorophyll in suspensions of banana cells in contact with toxin (P. Lepoivre, Gembloux, 1997, unpublished results). This was considered a valid test, as the action of *M. fijiensis* toxins found in culture filtrates has been found to be light-dependent (Lepoivre and Acuna, 1989).

The method was used to assess the toxic effects of metabolites in culture filtrates on cultivars 'Grande Naine' and 'Pisang Glintong' (AAA), which are very susceptible to black leaf streak, and 'Fougamou', 'Pisang Berlin' (AA) and 'Pisang Madu' (AA), which vary in their partial resistance. The sensitivity of the different cultivars was compared with their reaction to black leaf streak in the field. The results showed that there was a significant correlation between sensitivity to the toxic metabolites and reaction to the disease. Cultivars that were more resistant were less sensitive to the toxins (P. Lepoivre, Gembloux, 1997, unpublished results). This correlation has also been found by other workers observing the effects of toxic filtrates on leaves (Molina and Krausz, 1988; Strobel *et al.*, 1993). However, this is not proof of the involvement of toxins in pathogenicity.

Daub (1986) has advised caution when using pathogen metabolites as selection agents to screen tissue cultures for resistance. Nevertheless, the use of fungal toxins has been proposed for screening banana *in vitro* for resistance to black leaf streak

(Escalant, 1990; Strobel *et al.*, 1993). There have been claims that resistant material has been produced by selecting callus tissues of banana that survive increasingly higher concentrations of *M. fijiensis* toxins (Okole and Schulz, 1993), but results of field tests, which would confirm resistance, have yet to be published.

The hypothesis of a role for *M. fijiensis* toxins as secondary, as distinct from primary, determinants of pathogenicity is in agreement with cytological studies, which do not suggest an early involvement of toxic compounds in the host–pathogen interaction. Indeed, cytological studies of compatible interactions has revealed that *M. fijiensis* behaves initially as a biotrophic parasite.

Control

R.A. Romero

Introduction

In commercial plantations of banana cultivars in the Cavendish subgroup, black leaf streak disease is currently controlled using a combination of cultural practices and chemical methods. Fungicides are used because non-chemical alternatives alone do not provide satisfactory control. Breeding programmes have provided important tetraploid hybrids of banana with resistance to black leaf streak (Jones, 1994a), but the fruit has not met the requirements of the export industries and has only been acceptable for local markets and as staple foods. At the moment, fruit from cultivars in the Cavendish subgroup are the export industry standards in terms of taste and green life, and are likely to remain so for some considerable time. Although advances in biotechnology in the last 5 years have led to the development of transformation systems for the introduction of genes that code for antifungal products into banana, a black leaf streak-resistant Cavendish cultivar has yet to be bred. Until a new black leaf streak-resistant commercial banana is obtained either by traditional breeding techniques or by genetic transformation of a Cavendish cultivar, current control methods will continue to be important.

The control of leaf spot diseases in Central America started with the use of Bordeaux mixture (suspension of copper sulphate, hydrated lime and water) against Sigatoka in the mid-1930s. Bordeaux was superseded by petroleum oil and dithiocarbamate fungicides, which started to be sprayed at almost the same time in the late 1950s. The conversion from ground spraying to aerial spraying was made possible by the change to oil. In the early 1960s, both oil and dithiocarbamates were used together in mixtures. The first systemic fungicide used in the control of Sigatoka disease was benomyl, which was in widespread use by the early 1970s. Benomyl was also found to be effective against black leaf streak after the appearance of the disease in Central America in 1972. Ten years later, tridemorph was used against black leaf streak and, in 1987, propiconazole, the first of the triazole fungicides, was introduced. The fungicides currently registered in Central America for the control of black leaf streak are the protectant-type fungicides, such as mancozeb and chlorothalonil, and systemic fungicides in the benzimidazole, triazole, morpholine and strobilurin groups.

Selection of the fungicide to be used to control black leaf streak must be based on an assessment of several factors. These include the cropping system, knowledge of the epidemiology of the disease, the sensitivity of the local pathogen population to fungicides, the management of the farm and a thorough knowledge of how all these factors interact with each other. Thus, the chemical control of black leaf streak needs to be viewed as part of an integrated disease management strategy, which involves good sanitation practices, adequate drainage systems, ideal plant densities and optimal nutritional conditions for plant growth.

Cultural control practices

The most common practices undertaken to help in the control of black leaf streak are related to the reduction of inoculum levels on farms. Leaves with extensive necrosis are removed from plants and small areas of necrotic leaf tissue excised as soon as they appear. The rationale behind this practice is to accelerate the decomposition of the necrotic sporulating tissue by rapid deposition on the ground. If the necrotic tissue is left on the plant, production of perithecia continues and ascospores can be released for several weeks.

Another important practice is the reduction of the relative humidity inside the crop. This is achieved through an efficient drainage system, which rapidly takes groundwater out of the plantation, and by maintaining an adequate balance between the distribution and number of plants to avoid overlapping of foliage. As well as increasing humidity levels, overlapping foliage also makes it difficult to cover all leaf surfaces with fungicide. In areas where irrigation is necessary, the use of under-tree or drip irrigation is preferred (Wielemaker, 1990). This is because overhead sprinkling systems maintain favourable humidity for spore germination in the upper leaves where most ascospores of *M. fijiensis* are deposited.

Chemical control

FUNGICIDE APPLICATION Aircraft are used to spray fungicides in most export banana plantations, but tractors or motorized backpack sprayers are used on small farms growing fruit for local markets. However, ground sprays are less efficient than aerial sprays. In most operations in Central America, the Ayres Trush Commander and the Ayres Turbo Jet, with capacity for 1135 l and 1514 l, respectively, are the most widely used aircraft (Stover and Simmonds, 1987). These aircraft are equipped with satellite-linked, geographical positioning systems, which direct the pilot to application targets and can be seen on an electronic screen inside the aircraft. This technology avoids unnecessary spray overlaps, missed paths and oversprays. The number of hectares that can be sprayed by an aircraft in 1 h depends on the configuration of the farms and the distance from the mixing and loading facility. However, it is usually in the range of 125–175 ha.

PROTECTANT FUNGICIDES Protectant fungicides remain on the surface of the leaf and do not penetrate underlying tissues. They have a broad spectrum of activity and exert a multi-site effect on the pathogen, mostly reacting with essential thiol groups (SH-groups) of enzymes, causing a non-selective toxicity (Gasztonyi and Lyr, 1995). The protectant fungicides approved for control of Sigatoka diseases are dithiocarbamates and chlorothalonil.

Dithiocarbamate fungicides available for use on banana are different formulations of mancozeb, which is a complex of zinceb and maneb (zinc and manganese salts of ethylene bisdithiocarbamate). Some formulations can be used in oil–water emulsion (Vondozeb SC®, Manzate 43SC®, Manzate 35SC®, Dithane MB®, Dithane OS®, Dithane SC®) or in straight oil (Dithane OS®, Dithane SC®, Vondozeb SC® and Manzate 35SC®). Others can be applied only in water (Vondozeb DG® and Manzate DF®). The rate of active ingredient ranges from 1000 to 1500 g ha^{-1}.

Chlorothalonil (Daconil®, Bravo®) has a low water solubility, which favours its fungitoxic activity and reduces the risk of phytotoxicity. This compound has to be sprayed in water alone. Care must be taken when leaves have a deposition of oil from previous treatments and when applying oil following a chlorothalonil treatment, because the interaction of oil and chlorothalonil causes damage to leaf tissues. The rate of active ingredient of chlorothalonil ranges from 1000 to 2000 g ha^{-1}.

SYSTEMIC FUNGICIDES Systemic fungicides have the ability to penetrate the leaf and exert a toxic effect on the pathogen after infections have taken place. Most systemic fungicides have a specific mode of action on the pathogen, which makes them non-phytotoxic to banana.

The benzimidazole group of systemic fungicides comprises several compounds, of which benomyl and thiophanate-methyl have been the most widely used in Central America. Benzimidazoles (Benlate®, Sigma® and Topsin®) act in preventing microtubule (spindle fibre) assembly, which is necessary for cell division during mitosis and meiosis. They bind to the β-tubulin, a subunit of the tubulin protein, which is a component of the microtubules (Davidse, 1988; Davidse and Ishii, 1995; Delp, 1995). Benomyl is used in oil or an oil–water emulsion at a rate of 1400–1500 g of active ingredient ha^{-1}.

Resistance to benzimidazole fungicides in populations of *M. fijiensis* in Honduras and the Philippines occurred 2–3 years after the fungicide was first used (Stover, 1980a). Resistance has also been reported in Western Samoa (Fullerton and Tracey, 1984) and in many other countries in Central America (Anon., 1993a). The build-up of resistance has been attributed to the intensive use of benzimidazoles for black leaf streak control when these fungicides first became available. Only a single mutation in the β-tubulin gene in *Venturia inaequalis* and other fungal pathogens results in resistance to benzimidazoles (Koenraadt *et al.*, 1992). It is likely that the same mutation occurs in *M. fijiensis*.

Tridemorph (Calixin®) is the only member of the morpholine group of systemic fungicides used for the control of black leaf streak. Although a systemic, it has been shown to have only limited penetrant properties in banana leaves (Cronshaw and Akers, 1990). Tridemorph is used at a rate of 450 g of active ingredient ha^{-1} and, like benomyl, can be applied in straight oil or in an oil–water emulsion. It inhibits the synthesis of ergosterol, which is an important component of the cell membrane in fungi, interfering mainly with sterol Δ^{8-7} isomerase (Köller, 1992), although an inhibition of sterol Δ^{14} reductase has also been identified (Kerkenaar, 1995).

There are several compounds in the triazole group of systemic fungicides registered for use on banana. All of them inhibit a cytochrome P-450 mono-oxygenase enzyme, which catalyses the C-14 demethylation reaction in the ergosterol biosynthesis pathway (Köller, 1992). Therefore, these fungicides are known as demethylation inhibitors (DMI).

In 1987, propiconazole (Tilt®) was the first DMI fungicide registered for use on banana in Central America. Due to its high efficacy against *M. fijiensis*, fewer applications of this fungicide were needed to achieve satisfactory standards of control. Other triazoles registered for use on banana are flusilazole (Punch®) (flusilazole is registered for use on bananas shipped to Europe, but not the USA), fenbuconazole (Indar®), bitertanol (Baycor®), tebuconazole (Folicur®, hexaconazole (Anvil®) and cyproconazole (Alto®). There are differences in the efficacy of these compounds (Guzmán and Romero, 1996a) and in their systemic movement inside the leaf. An indication of the amount of translocation is given by leaf infection patterns. Differences in infection patterns became evident in the field after the build-up to resistance to triazoles in the pathogen population. If a fungicide translocates mainly to the edges of the leaf, the areas along the midrib will have less fungicide concentration than necessary to arrest infections of fungicide-resistant isolates (sublethal dose), resulting in a characteristic pattern of infection along the midrib. Propiconazole translocates very rapidly from the midrib towards the edges of the leaf and this results in a pattern of infection along the midrib. Fenbuconazole application results in significantly less disease along the midrib, indicating less translocation from this point (M. Guzmán and R.A. Romero, Costa Rica, 1995, unpublished results). The use of bitertanol also results in less infection along the midrib of banana leaves as compared with propiconazole. All DMI fungicides are used at a rate of 100 g of active ingredient ha^{-1}, except bitertanol, which is applied at a rate of 150 g of active ingredient ha^{-1}. They can be applied in an oil–water emulsion or in straight oil. Resistance to propiconazole has been reported in many countries in Central America (Anon., 1993a; Romero and Sutton, 1997a). Other DMI fungicides have been shown to be less effective against propiconizole-resistant isolates, indicating cross-resistance (M. Guzmán and R.A. Romero, Costa Rica, 1996, unpublished results).

Systemic fungicides of the strobilurin group represent the newest chemicals registered for use against black leaf streak disease. Strobilurins are natural occurring compounds found in various genera of fungi belonging to the *Agaricaceae*, which live on decaying plant material in forests (Lange *et al.*, 1993). The natural products themselves have physical properties that make them unsuitable for use in agriculture, but analogues, such as the β-methoxyacrylates, have been developed with improved stability and physical properties. These compounds inhibit fungal respiration by binding strongly to a specific site on cytochrome b, preventing electron transfer between cytochromes b and c_1 (Clough *et al.*, 1994). In 1996, azoxystrobin (Bankit®) was the first strobilurin registered for use on banana in Central America, at a recommended rate of 100 g of active ingredient ha^{-1}. This fungicide has systemic activity on the leaf and it is applied in an oil–water emulsion or in straight oil (Guzmán and Romero, 1997).

OIL Petroleum oil has been used for the control of leaf spot diseases for many years (Stover, 1990). It has an effect on disease development, delaying the appearance of symptoms and retarding the expansion of lesions (Stover, 1990; Guzmán and Romero, 1996b). Oil does not affect the growth of *M. musicola in vitro* and it has been suggested that it acts by somehow increasing the resistance of banana (Meredith, 1970). Recent experimental work has shown that the therapeutic effect of oil is translocated to unsprayed leaves, indicating that it may act by stimulating the banana's natural defence mechanisms (Guzmán and Romero, 1996b). Oil application rates range from 5 to 15 l ha^{-1} and it can be applied straight or in an oil–water emulsion (Marín and Romero, 1992). From the results of work undertaken in Costa Rica, the rate of 10 l of oil ha^{-1} has been recommended for commercial plantations in Central America (Guzmán and Romero, 1996b).

Today, all systemic fungicides are applied in oil. Oil helps systemic fungicides penetrate leaves, thus optimizing their performance. When applied with a systemic fungicide, the application rate of oil and whether it should be applied straight or as an emulsion depend on several factors. In situations where the pathogen is highly sensitive to the fungicide and few applications are necessary for the control of disease, straight oil can be used. A rate of around 10 l of oil ha^{-1} is necessary to achieve the best results with systemic fungicides (D.H. Marín and R.A. Romero, Costa Rica, 1994, unpublished results; Guzmán and Romero, 1996b). However, in locations where a significant proportion of isolates in the pathogen population are resistant to the fungicide and more frequent applications of fungicide are required for adequate control, the amount of oil used in sprays should be reduced. This is because oil can accumulate on leaves and negatively affect yield by interfering with the exchange of gases and photosynthesis (Israeli *et al.*, 1993). In the final analysis, it is disease pressure that dictates the amount of oil used. The serious consequences of poor disease control in commercial plantations far outweigh yield losses that might be expected from a build-up of oil.

Oils should have an unsulphonated residue of 90% or more and an aromatic content of less than 12% (Stover and Simmonds, 1987; Israeli *et al.*, 1993). Those fractions distilling below 338°C at 760 mmHg do not control black leaf streak or Sigatoka. Fractions distilling above 365°C control the diseases, but are phytotoxic. Oils that combine both good disease control qualities and low phytotoxicity must have a 50% distillation range of 346–354°C (Calpouzos, 1968).

Chemical control strategies

FUNGICIDE APPLICATION BASED ON A DISEASE FORECASTING SYSTEM Forecasting systems depend on the ability of systemic fungicides applied in an oil suspension to arrest infections at early stages of development.

In the French Antilles, a disease forecasting system is used to time applications of systemic fungicides for the control of Sigatoka disease on Cavendish cultivars. It is based on regular measurements of temperature and evaporation, in conjunction with an assessment of early infection stages observed in the youngest leaves (Ganry and Laville, 1983). The method, with some modifications, has been used in Cameroon (Fouré, 1990a) and Central America (Marín and Romero, 1992) to time fungicide sprays for the control of black leaf streak. In general, this early warning system, which predicts optimum spraying times, is based on weekly disease evaluations of the second, third and fourth fully opened leaves of young banana plants in marked plots. Each plot has 25 plants and the number of plots varies according to the size of the plantation. The most advanced stage of symptom present on each of the three leaves from ten randomly selected plants in each plot is recorded. Symptom intensity is also recorded by assigning a plus (+) if more than 50 stages of the most advanced symptoms are present, or by a minus (−) if fewer than 50 stages are observed. Stages of symptoms follow the descriptions of Meredith and Lawrence (1969) and Fouré (1990b). The datum obtained for each leaf is adjusted by taking into account the age of the leaf. A correction factor for differences in the stage of unfurling of new leaves is also taken into consideration. Finally, the data are added and corrected by the rate of leaf emission to obtain a variable called 'the stage of evolution of the disease' (Ganry and Laville, 1983; Marín and Romero, 1992). If the disease is sufficiently advanced, the crop is sprayed with fungicide.

The early warning system allows fungicides to be used only when necessary, thus reducing the cost of control of the disease. The method has been very successful against Sigatoka in the French Antilles and black leaf streak in Cameroon, but has only had limited success against black leaf streak in Central America. The disappointing results in Central America have been attributed to the rapid build-up of resistance to benomyl and propiconazole fungicides in local populations of *M. fijiensis*. This resistance has significantly affected the efficacy of these fungicides to arrest early infections (Romero and Sutton, 1997a). It is not yet known why resistance to fungicides has developed in Central America and not Cameroon. However, the climate is more favourable for disease development in Central America and this allows the pathogen a greater number of sexual reproductive cycles. Also, farms are larger and there is a much greater concentration of farms in a given geographical area. The size of populations of the black leaf streak fungus being treated is also much bigger. All these factors are conducive to the build-up of resistance.

Forecasting systems for black leaf streak on Cavendish cultivars have also been proposed by Chuang and Jeger (1987a, b) and Wielemaker (1990). The former is based on disease severity, total precipitation and days with over 90% humidity and the latter solely on weekly observations of disease symptoms on young leaves. Bureau (1990) and Lescot *et al.* (1998) have developed forecasting systems for black leaf streak on plantain.

FUNGICIDE APPLICATION BASED ON CROP GROWTH AND WEATHER CONDITIONS Because of the build-up and persistence of resistance to the fungicide benomyl, which has been withdrawn from control programmes in Costa Rica, and to an increase in resistance to propiconazole (Romero and Sutton, 1997a, 1998), systemic fungicides alone are no longer able to provide satisfactory control of black leaf streak. A similar situation regarding resistance to benomyl and propiconazole has arisen in most other countries in Central America. The use of protectant fungicides is now necessary, both in alternation and in mixtures with systemic fungicides, to maintain adequate control of the disease.

To be able to give satisfactory control, protectant fungicides need to be sprayed at intervals that ensure that most unfurling and recently expanded leaves are covered. Therefore, during weather conditions favourable for infection, protectant fungicides are applied at intervals close to the rate of leaf emission. Emission rates vary according to the temperature, the nutritional condition of the crop and soil type. Although this strategy of application of protectants in alternation or in mixtures with systemic fungicides is referred to within the industry as a calendar-based schedule, in reality many factors are considered when making a decision on the time and type of fungicide treatment. These factors include a weekly analysis of disease data in plants of different ages on every farm, the rate of leaf emission, weather data on a weekly basis and the sensitivity of local populations of the pathogen to site-specific fungicides.

Fungicide Resistance Management As a consequence of the build-up of resistance to propiconazole, the fungicide has to be applied at shorter intervals than before and with protectants to achieve the levels of disease control required for commercial plantations. The efficacy of other DMI fungicides has also decreased significantly because of cross-resistance factors, but these too are still used to control black leaf streak in the same way as propiconazole. Because of the important economic consequences of resistance to systemic fungicides, action must be taken in those countries or areas where black leaf streak has recently spread or where shifts in sensitivities to fungicides have not yet occurred, to prevent or delay the build-up of resistance. The sensitivities of local pathogen populations to the different groups of systemic fungicides need to be established and regularly monitored so that any changes can be detected early. This has to be undertaken in conjunction with a fungicide application strategy. The strategies most commonly used to prevent or delay the development of resistance to fungicides in *M. fijiensis* are as follows.

1. Limit the use of fungicides, such as benomyl, that are at risk of developing resistance.
2. Always apply systemic fungicides in mixtures with a protectant fungicide, such as mancozeb, that has a non-specific mode of action.
3. Do not apply DMI, benzimidazole and strobilurin fungicides for 4 consecutive months during the season or year. They should be substituted for protectants and penetrants (tridemorph) during this period. This strategy is believed to decrease selection pressure in pathogen populations for resistance to these systemic fungicides. If only systemic fungicides are used to control the disease, it may be difficult to have a DMI-, benzimidazole- and strobilurin-free-period of the recommended duration. In this situation, the different systemics should be sprayed alternately. However, the systemic-free strategy has proved itself in Central America and should be attempted if possible.
4. Use a rate of oil close to 10 l ha^{-1} to enhance the activity of the systemic treatments and to avoid sublethal doses of the systemic fungicide inside the leaf.
5. Do not use reduced rates of systemic fungicides.
6. Always alternate fungicides, including protectants and tridemorph, that have different modes of action,
7. Maintain all cultural practices that will reduce the amount of inoculum.

The benefits of the use of DMI fungicides in mixtures with protectants have been demonstrated in Costa Rica. In experiments, the infection index of black leaf streak was found to be significantly lower after treatment with propiconazole and mancozeb in oil than after a treatment with propiconazole in oil alone. Further work has indicated that the effect of each fungicide is additive (M. Guzmán and R.A. Romero, Costa Rica, 1996, unpublished results).

The frequency of DMI-resistant strains in the pathogen population is likely to remain the same after a mixture with mancozeb is applied. The way in which a mixture helps in the prevention and management of resistance to DMI fungicides is thought to be by reducing the size of the population of the pathogen. This results in less sexual recombination during a season, thus retarding the flow of resistance genes within the population.

Although a DMI application-free period is advocated as an anti-resistance strategy, the value of this recommendation has yet to be demonstrated in Costa Rica (Jiménez *et al.*, 1996; Guzmán *et al.*, 1997). After DMI application-free periods of up to 6 months for 2 years, there has been no change in the sensitivity of the pathogen population. This could be because resistance to DMI fungicides does not affect the fitness of *M. fijiensis*. Certainly, the ability of resistant strains to produce conidia in culture is not affected (Jiménez *et al.*, 1997). However, a DMI application-free period is believed to be of great value in delaying the onset of resistance in those areas where DMI resistance is not present or where the pathogen population is composed of only a few resistant individuals.

Future developments in chemical control

It is expected that more fungicides of the strobilurin group will be registered for use against black leaf streak in the near future. Some may have greater efficacy than azoxystrobin, but they will all have a similar mode of action and share the same risk of becoming ineffective through a build-up of resistance in the pathogen population. A recent report has shown that *V. inaequalis* mutants obtained from the field are able to overcome the effect of a strobilurin fungicide by an alternative respiratory mechanism that circumvents the strobilurin inhibitor site (Köller, 1998). The multi-drug resistance of fungi that affect humans has been reported by other workers (de Waard *et al.*, 1996; Orth, 1998). It is not known if fungicide resistance is going to develop into a major problem for the banana industry in the future. However, more knowledge of the mechanisms of fungicide resistance in fungal pathogens is needed if progress is to be made in developing strategies to combat resistant strains.

New fungicides have been developed that have potential for use against black leaf streak. Acibenzolar-S-methyl, a compound in the benzothiadiazole (BTH) group that induces a systemically acquired resistance in several crops (Görlach *et al.*, 1996), is currently being tested (Madrigal *et al.*, 1998) and is likely to be registered. Pyrimethanil (anilinopyrimidine) is another example of a relatively new fungicide that may be effective. It has a novel mode of action interfering with the secretion of enzymes from the pathogen that are necessary for pathogenesis (Milling and Daniells, 1996). Molecular techniques are helping in the discovery of new fungicides and in acquiring a better understanding of target sites for fungicides. More fungicides with potential for use on banana will undoubtedly be developed in the future. Opportunities for improving the control of black leaf streak under an integrated pest management strategy will increase as more is learnt about host–pathogen interactions at the ecological and cellular level and if genetically modified plants with resistance to black leaf streak become available.

Introduction of resistant cultivars

Cultivars resistant or less susceptible to black leaf streak than local clones have been deployed in areas where the disease has been causing significant yield reductions to crops grown by subsistence farmers. In Central America, ABB cooking banana cultivars, such as 'Chato' (syn. 'Bluggoe'), 'Pelipita' and 'Saba', have been tested as substitutes for the local susceptible plantain (Jarret et al., 1985; Pasberg-Gauhl, 1989). In West Africa, IITA micropropagated AAA types from East Africa and ABB cooking-banana cultivars for distribution to local farmers whose own traditional plantain cultivars were severely attacked by black leaf streak (Vuylsteke *et al.*, 1993b). However, due to differences in palatability, these new introductions were not fully accepted. The introduction of certain black leaf streak-resistant plantain substitutes bred at IITA was also resisted for the same reason (Ferris *et al.*, 1996). The development of disease-resistant hybrids with fruit that suits local palates and cooking qualities that match traditionally grown cultivars remains a major challenge to banana breeders. However, in Cuba, the government has had more freedom to initiate change in the face of yield declines since the introduction of black leaf streak in 1990. Large areas are now planted with new black leaf streak-resistant hybrids developed by the Fundación Hondureña de Investigación Agrícola (FHIA) (see Rowe and Rosales, Chapter 14, pp. 435–449).

In some areas, black leaf streak-resistant clones have gradually been replacing more susceptible types through sheer necessity. Watson (1993) reported that the spectrum of cultivars grown on atolls in the western and central Pacific changed over the 20 years from 1973 to 1993. In the north, it was noted that the resistant clones 'Saba' and 'Mysore' (AAB) were much more widely grown in the Federated States of Micronesia and that 'Saba' had spread to the Marshall Islands. In the south, 'Bluggoe' and 'Rokua Mairana' ('Kalapua'-like clone (ABB), which probably originated in Papua New Guinea) had replaced other more susceptible cooking types on Tokelau and the northern Cook Islands. 'Mysore' was also more widespread.

Resistant cultivars, such as 'Mysore' and the bred hybrid 'T8' (AAAA), have also been used in Australia to reduce inoculum levels and thus buffer the advances made by black leaf streak in the Torres Strait and Cape York areas of Queensland (see Jones and Diekmann, Chapter 13, p. 411).

The development of black leaf streak-resistant hybrids is a major priority of banana breeding programmes (see Jones, Chapter 14, pp. 425–435; Rowe and Rosales, Chapter 14, pp. 435–449). This is because resource-poor people cannot afford chemicals for disease control, and fungicides that are now used by those that can afford them may not be effective in the future.

Acknowledgements

F. Gauhl and C. Pasberg-Gauhl are thanked for their contribution on the introduction of resistant cultivars as a control measure.

Sigatoka

D.R. Jones

Introduction

Until the discovery and spread of black leaf streak disease, Sigatoka or yellow Sigatoka, as it is now often called, was the most important foliar disease of banana. It was first recorded in Java (Zimmermann, 1902) and later in the Sigatoka valley on the island of Viti Levu, Fiji (Philpott and Knowles, 1913; Massee, 1914), the location giving its name to the disease. Herbarium specimens indicated that Sigatoka was present in Sri Lanka in 1919 and the Philippines in 1921. The disease was first found in Australia in 1924. In the 1930s, Sigatoka was reported throughout the Central American–Caribbean region. It was also recorded in Surinam, Guyana and Colombia in South America, Tanzania and Uganda in East Africa, China and West Malaysia. Reports from West Africa, India and Brazil began in the 1940s and the disease was recorded in most other banana-growing countries for the first time in the 1950s and 1960s (Stover, 1962).

Sigatoka's rapid global dissemination in the 1930s from original areas of distribution in the 1920s led to speculation that spores of the causal fungus were carried by air currents between continents (Stover, 1962). However, long-distance spread was more likely to have occurred by the uncontrolled movement of propagating material and/or diseased banana leaves used to wrap produce. It is highly likely that the disease may have been present at a low level in many countries for some time before it came to the attention of local plant pathologists. Today, Sigatoka is regarded as having a worldwide distribution, although it has not been recorded in the Canary Islands, Egypt or Israel (Meredith, 1970; Anon., 1981).

On a vegetatively growing banana plant, the most efficient leaves for photosynthesis are the second to fifth, counting down the profile (Robinson, 1996). Lower leaves are progressively less efficient as they are ageing and later senescing. Leaves are not static and, on a vigorous plant of a cultivar in the Cavendish subgroup (AAA) growing under optimal environmental conditions with one new leaf appearing every 7–8 days, the most efficient leaf area is renewed monthly. On a less vigorous plant growing under suboptimal environmental conditions, this renewal time can be extended to several months. Although a banana plant has the internal ability to partially compensate for lost photosynthetic assimilation, due to leaf area destruction, it is important that leaves 2–5 remain as free of excessive shade, severe leaf tearing and disease as possible; otherwise assimilation potential is greatly reduced (Robinson, 1996).

Sigatoka disease is very destructive if left uncontrolled. Leaf spots caused by the disease can coalesce, which leads to the premature death of large areas of leaf tissue, thus reducing the photosynthetic capabilities of the plant. However, Sigatoka appears to have little effect on vegetative growth in the tropics, as measured by the rate of leaf emergence, rate of increase in plant height and height of the plant at the time of shooting (Leach, 1946). This is because the effects of the disease are not sufficiently great on leaves 2–5 to cause a severe deficit of the assimilates used to power plant growth. Unfortunately, the effect of Sigatoka on fruit development is considerable. After shooting, leaf production ceases and the plant is unable to replace those leaves damaged by Sigatoka. Plants of 'Williams' (AAA, Cavendish subgroup) with fewer than five viable leaves at harvest – a viable leaf being defined as one with more than 30% healthy tissue – produce lighter bunches (Ramsey *et al.*, 1990). The greater the damage and the earlier it occurs after shooting, the greater the effects on yield. On 'Williams' plants with between zero and two viable leaves at harvest, yields are reduced on average by around 25–29% (Ramsey *et al.*, 1990). If Sigatoka is left uncontrolled, it is not uncommon for plants to have four or fewer leaves at shooting and none shortly afterwards (Meredith, 1970). Yield losses would be expected to be highest in these cases.

Sigatoka also disturbs the physiology of fruit, resulting in premature ripening (Meredith, 1970; Stover, 1972; Wardlaw, 1972). Premature ripening can occur in the field if plants are severely diseased or in transit to markets if moderately affected. Some field ripening of bunches occurs on 'Williams' plants with fewer than 11 viable leaves at harvest, which reduces marketable yields, and on all bunches on plants with fewer than four viable leaves at harvest, which results in total marketable yield loss (Ramsey *et al.*, 1990). When fruit ripens in transit, consignments are devalued because of uneven ripening and increased risk of postharvest problems. In Australia, badly affected shipments are often condemned because of the risk of infection by fruit fly, which is an interstate quarantine concern.

Sigatoka caused widespread disruption to the export trade when first introduced into the Latin American–Caribbean region. In Mexico, production fell from 525,000 t in 1937, the year after Sigatoka was first recorded, to 240,000 t in 1941. Exports from the State of Tabasco ceased entirely. In Honduras, production declined to less than one-third of the level prior to the introduction of the disease. Sigatoka caused crop losses of 25–50% in Guadeloupe in 1937. In Ecuador, out of 62 million bunches produced in 1954, only 19 million were fit for export because of uncontrolled Sigatoka on small farms. In subsequent years, the damage was reduced by the timely application of chemicals. However, 15–17 fungicide applications were required every year to control Sigatoka, which considerable increased the cost of production (Meredith, 1970). The cost of Sigatoka control in the North Queensland growing area of Australia has been estimated to be 14% of total production costs (Jones, 1991).

Because of its seriousness, developing a commercial cultivar with Sigatoka resistance became a priority in banana-breeding programmes (see Jones, Chapter 14, pp. 425–435). However, this endeavour was not successful. Although Sigatoka is still important in certain locations, such as

Australia, Brazil and the Windward Islands, black leaf streak is now attracting more of the banana breeder's attention because it has replaced Sigatoka in many banana-growing areas as the major leaf spot problem. Sigatoka is hard to find in most areas invaded by black leaf streak, but *M. musicola* is believed to survive at a low level in the leaf spot pathogen population. Indeed, it has been identified in coastal Nigeria on 'SH-3362' (AA), a hybrid that is susceptible to Sigatoka but has resistance to black leaf streak (C. Pasberg-Gauhl and F. Gauhl, 1994, Nigeria, personal communication). Sigatoka is also still present in the Philippines many years after the introduction of black leaf streak. Sigatoka is more adapted to cooler temperatures than black leaf streak and is dominant at altitudes over 1200–1500 m in tropical countries where both diseases occur.

Symptoms

Many descriptions of the development of symptoms have been published (Meredith, 1970). The earliest symptom is a light green, narrow speck about 1 mm in length on the upper surface of the leaf. The speck develops into a streak several millimetres long and 1 mm or less in width running parallel to the leaf veins. The streak elongates, expands laterally to become elliptical in shape and turns rusty red (Plate 2.11). A water-soaked halo forms around the lesion when the leaf is turgid. This is best seen in the early morning when dew is present on the leaf. This infiltrated tissue quickly turns brown and a young spot is formed (Plate 2.11). The dark brown centre of the spot later shrinks and appears sunken and the halo turns a darker brown. The sunken area becomes grey and the halo darker brown, forming a well-defined ring around the mature spot (Plate 2.12), which remains distinct even after the leaf tissue has died (Plate 2.13). On leaves of young plants, individual spots tend to be larger and more spherical. Mature spots are normally 4–12 mm in length. Symptom development from specks through streaks to spots has been divided into various stages by different authors (Leach, 1946; Klein, 1960; Brun, 1963). Stover (1972) and Stover and Simmonds (1987) have compared these stages.

Leaf tissue surrounding spots turns yellow (Plates 2.11–2.14). This is initially more pronounced on the leaf margin side of the spot (Plate 2.11). If the infection density is high, large areas of leaf tissue around spots yellow and eventually become necrotic, with dark brown borders (Plates 2.13). Where mass infection occurs, areas of necrotic leaf become whitish grey within a dark border and the outlines of individual lesions are not well defined.

Counting down from the first fully opened leaf, earliest symptoms are first seen on the third or fourth leaf of susceptible plants in an active state of growth, but sometimes appear on the second leaf if conditions are very favourable for infection. Mature spot symptoms are found on older leaves. The overall effect is that the severity of the disease increases in a descending order down the plant. On resistant plants, symptoms may appear only on the very oldest leaves or not at all. Plants with bunches appear more diseased because clean, new leaves are not being produced to displace old ones with symptoms.

Symptoms of Sigatoka disease are very similar to those of black leaf streak (*M. fijiensis*) (see Jones *et al.*, this chapter, pp. 44–48), Septoria leaf spot (*Mycosphaerella* sp.) see Carlier *et al.*, this chapter, pp. 93–98) and Phaeoseptoria leaf spot (*P. musae*) (see Jones, this chapter, pp. 92–93).

Causal agent

The fungus *Mycosphaerella musicola* Leach (anamorph *Pseudocercospora musae* [Zimm] Deighton) is the cause of Sigatoka leaf spot.

Conidiophores can be formed, if conditions are suitable, at the first brown spot stage. The sporodochia (mass of tightly aligned conidiophores on a dark stroma) develop in the substomatal air chamber and the conidiophores grow through the stoma pore in a tuft-like fashion. As more

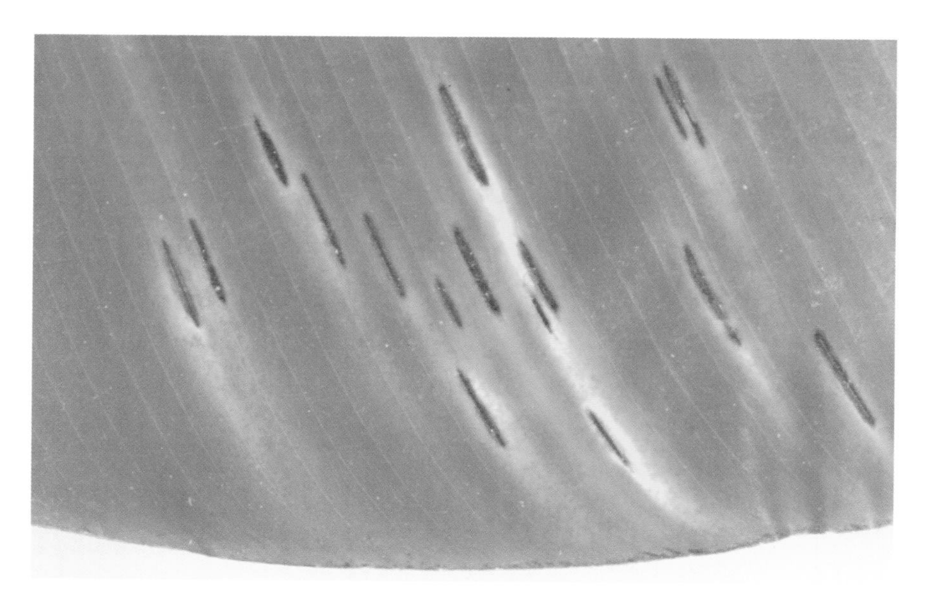

Plate 2.11. Sigatoka symptoms on the upper surface of a leaf of 'Williams' (AAA, Cavendish subgroup). Diffuse brown streaks (right) can be seen between developing spots. Note that the yellowing around the young spots is more pronounced between the spot and the leaf margin (photo: T. Cooke, QDPI).

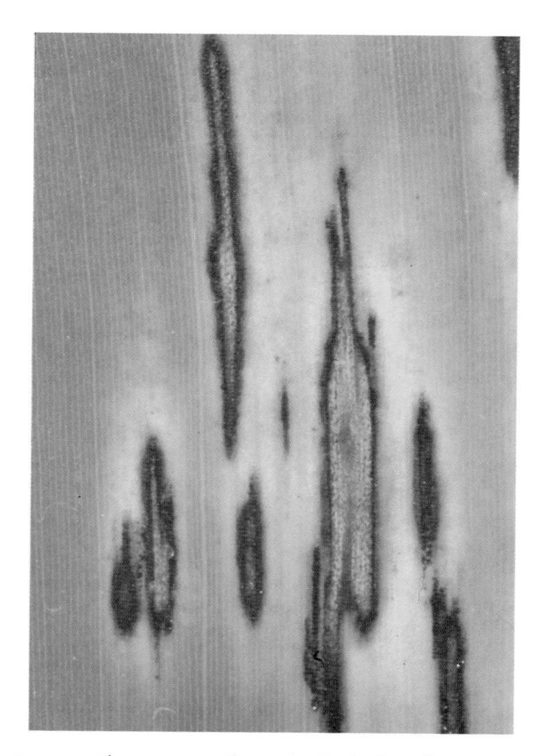

Plate 2.12. Sigatoka symptoms on the upper surface of a leaf of 'Williams' (AAA, Cavendish subgroup). Mature spots have sunken, grey centres with regular lines of sporodochia and dark brown borders. The leaf tissue surrounding spots is yellowing (photo: T. Cooke, QDPI).

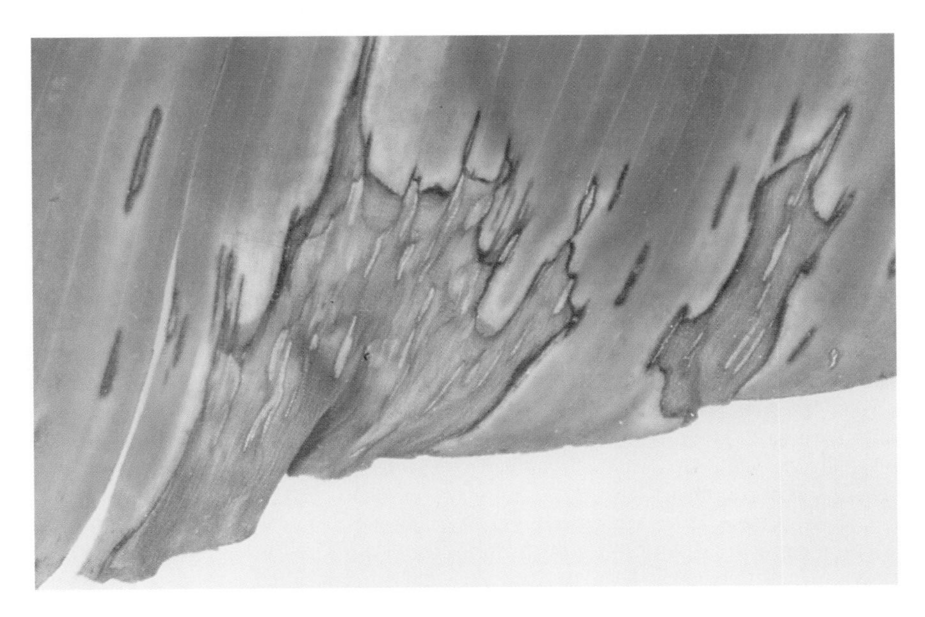

Plate 2.13. Advanced symptoms of Sigatoka on the upper surface of a leaf of 'Williams' (AAA, Cavendish subgroup). Individual spots with dark borders are discernible within the dead leaf tissue, which is surrounded by a dark brown border (photo: T. Cooke, QDPI).

Plate 2.14. Advanced symptoms of Sigatoka on the upper surface of a leaf of a cultivar in the Cavendish subgroup (AAA). Large areas of leaf tissue around spots are yellowing and becoming necrotic. Mature spots with grey centres can be seen within necrotic areas (photo: D.R. Jones, INIBAP).

and more conidiophores emerge, the sporodochia become erumpent, the guard cells becoming disrupted and the adjacent epidermis is pushed back. Sporodochia can easily be seen using a hand-lens (Plate 2.12) and are produced in spots on both sides of the leaf. Conidiophores are pale to very pale olivaceous brown, paler towards the apex, straight or slightly curved, only very rarely branched, without septa, not shouldered or geniculated, narrow towards the apex and without conidial scars. They measure 5–25 μm and are mainly bottle-shaped, with rounded or nearly truncate apices (Figs 2.8 and 2.9).

Conidia are borne terminally and singly on the conidiophore. They are pale to very pale olivaceous brown, smooth, straight or variously curved, occasionally undulate and almost perfectly cylindrical to obclavate–cylindrical (Fig. 2.9). The apex is obtuse or subobtuse and the basal cell is shortly attenuate and has no thickened basal hilum. Conidia are usually 2–5-septate or more and usually measure 10–80 μm × 2–6 μm (Meredith, 1970; Mulder and Holliday, 1974b).

Spermogonia, which also originate in the substomatal air chambers, develop before the mature spot stage. They are more abundant on the lower surface of the leaf. Microscopically, they appear as small, black, immersed flask-shaped structures, sometimes arising from the stomatic base of old conidial fructifications, and measure 46–77 μm × 37–63 μm. Spermatia are very minute, oblong hyaline cells, which form chains inside the spermogonium and ooze from the ostiole at its apex. They measure 2–5 μm × 0.8–1.4 μm (Meredith and Lawrence, 1970b).

Perithecia are often present on mature spots and are more numerous on the upper leaf surface. They are dark brown or black, erumpent, with a short protruding ostiole, and have a dark, well-defined wall. Diameters vary from 36.8 μm to 72.0 μm. Inside the perithecia, asci are septate, hyaline, obtuse–ellipsoidal, with the upper

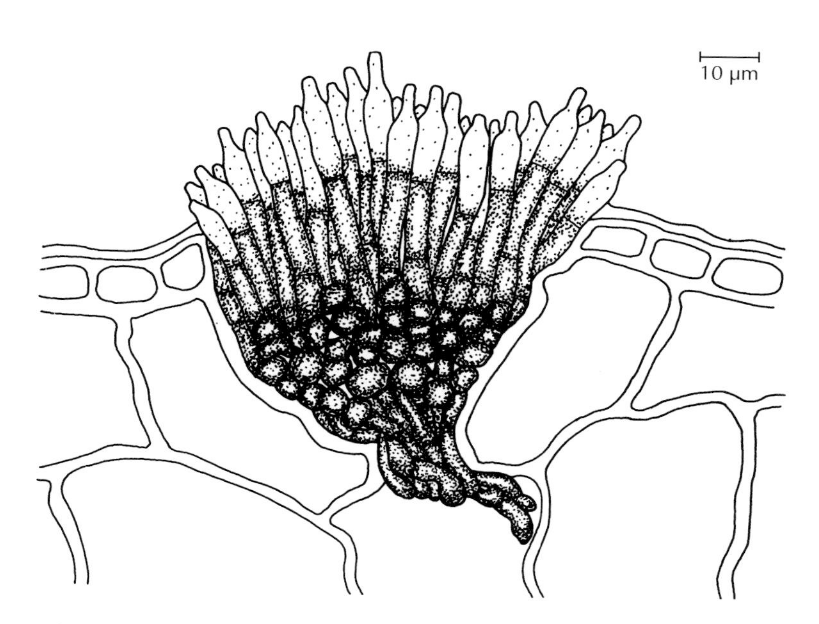

Fig. 2.8. Sporodochium of *Mycosphaerella musicola* in vertical section, showing bottle-shaped conidiophores borne terminally on stromatal hyphae (from Meredith, 1970).

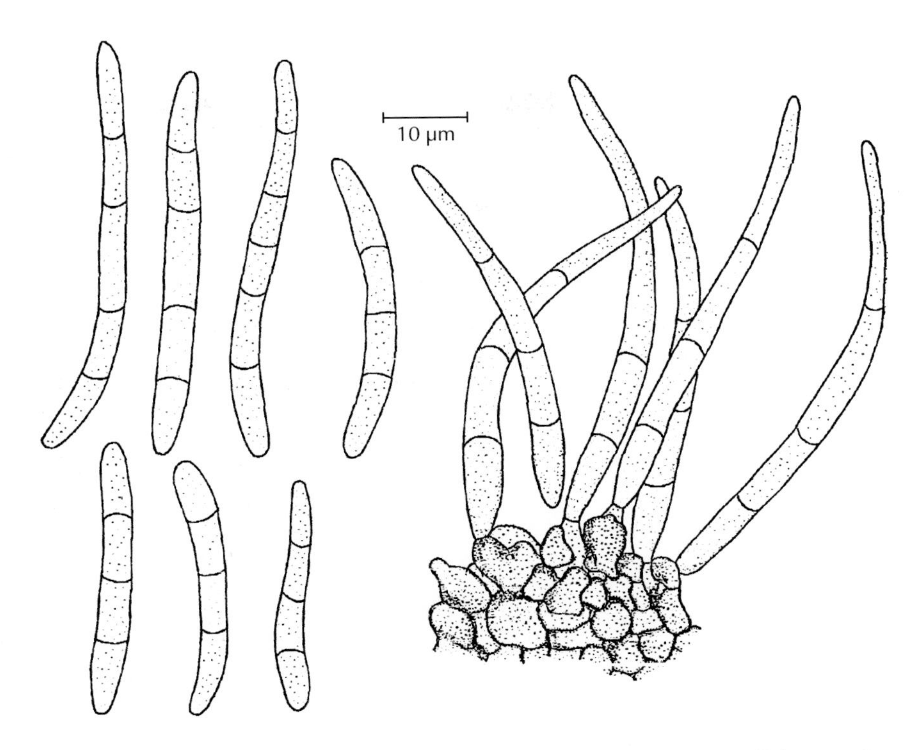

Fig. 2.9. Conidia and conidiophores of *Mycosphaerella musicola* (from Mulder and Holliday, 1974b).

cell slightly broader than the lower, and they measure 14.4–18.0 μm × 3–4 μm (Leach, 1941).

On most agar media, colonies first become visible 4–6 days after inoculation with ascospores or conidia. Growth is slow, but the centre of the colony soon becomes raised, almost hemispherical in side-view and unevenly folded. The surface of the colony sectors into distinct patches comprising aerial hyphae differing in texture and colour (Plate 2.15). Colours of colonies include white, light to dark grey, pink and dark green. Black stromatic masses with no aerial hyphae may occur. The submerged mycelium is very compact and dark green to almost black. The optimum temperature for growth is about 25°C. Colonies remain small and diameters of up to only 3 cm after 60 days' growth have been reported on potato dextrose agar (Meredith, 1970). Conidia may be produced on colonies and a profusely sporulating culture (Calpouzos, 1954) has been used to provide inoculum for infection experiments (Goos and Tschirch, 1962, 1963).

Overnight incubation of diseased leaves with early spot stage symptoms at 100% RH and at 25°C should lead to the production of abundant conidia and conidiophores for identification. Conidia closely resemble those produced by *M. fijiensis*, the cause of black leaf streak disease. However, they are, on average, shorter than those of *M. fijiensis* and lack a thickened basal hilum. The absence of scars on the conidiophores, which are usually bottle-shaped, also distinguishes *M. musicola* from *M. fijiensis*.

Mycosphaerella musicola and *M. fijiensis* can also be distinguished in leaf tissue and culture by a polymerase chain reaction technique (Johanson and Jeger, 1993; Johanson *et al.*, 1994).

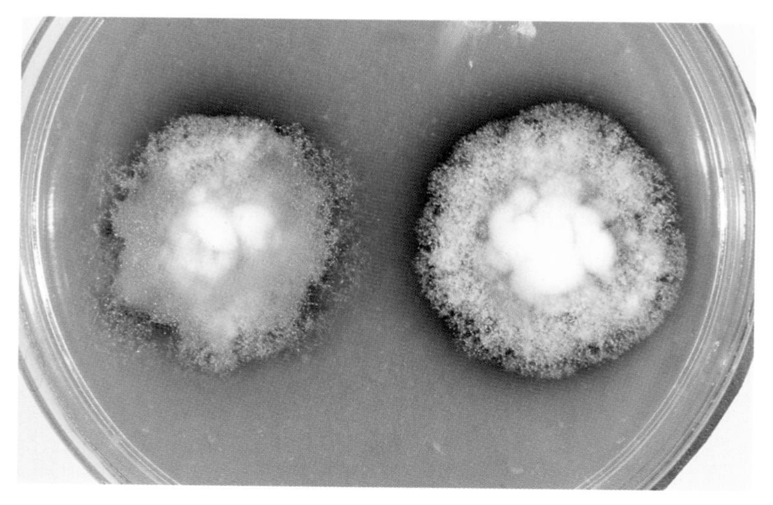

Plate 2.15. Colonies of *Mycosphaerella musicola* after 3 months' culture on potato dextrose agar at 25°C (photo: T. Cooke, QDPI).

Disease cycle and epidemiology

The disease develops as described previously under 'Symptoms', but the period between spore germination and the formation of mature spots depends on environmental conditions, the resistance or susceptibility of the cultivar and the intensity of infection (Meredith, 1970). Infection is believed to occur usually as a new leaf emerges from the pseudostem and unfurls (Stover and Fulton, 1966). If the cultivar is susceptible, the period from infection to mature spot formation can be short. If the cultivar has resistance, mature spots take longer to develop. As a consequence, mature spots are visible on younger leaves of cultivars that are susceptible. Some cultivars appear to be immune, but further studies are needed to determine if penetration occurs or not.

Conidiospores form in the presence of a film of dew or rainwater (Stahel, 1937b). Conidia are dislodged by rainwater to spread the disease to other leaves on the same plant or to leaves on nearby plants if wind-driven. Water drops laden with conidia can fall on unfurling heart leaves and run down the upright cylinder, depositing spores. As the leaf matures, spots develop in lines or other patterns which can be explained by infection of the leaf at different stages of unfurling (Stahel, 1937b; Stover and Fulton, 1966; Meredith, 1970). Recent studies have shown that conidia can also be airborne (Burt and Rutter, 1997).

Cultures of *M. musicola* derived from a single conidium or ascospore are sterile. The fungus is a heterothallic species and two compatible individuals must be mated before asci are formed. Spermatia, produced in the spermogonia, move in a film of moisture to protoperithecia of a compatible mating type. There is indirect evidence that one to three hyphae that emerge from protoperithecia and protrude through the stomata act as trichogynes. Spermatia of *M. musicola* are believed to be functional because spotted leaves that are not wetted by dew or rain, and thus do not have a film of water to facilitate mating, do not produce perithecia and ascospores (Stover, 1963).

As spots mature, fertilized protoperithecia develop into perithecia containing asci. Ascospores are forcibly ejected from perithecia during wet weather or after a heavy dew and can be carried in wind currents. The greatest number of ascospores is discharged after rain that follows a prolonged dry period. Discharge is also enhanced by alternative wetting and drying (Price, 1960). Ascospores are believed to be responsible for the long-distance spread of the disease and form characteristic apical-spotting symptoms on leaves after disease development. Stover (1980a) reported that leaves on the ground subjected to intermittent rainfall, dew and partial drying in the daytime continued to discharge ascospores in declining amounts for up to 4 weeks. Ascospores have survived for as long as 8 weeks in the shade on leaf tissue above the ground (Stover, 1971a).

Ascospores can germinate within 2–3 h of deposition provided that a film of water is present and temperatures are favourable (Brun, 1963). Conidia may take a little longer (Stover, 1965). The optimum temperature for germination is in the range 25–29°C for conidia (Meredith, 1970) and 25–26°C for ascospores (Brun, 1963; Stover, 1965). Germ tubes elongate only in moisture and collapse in hot, dry conditions. Growth can resume later in the presence of rainwater or dew. The optimum temperature for the growth of ascospore germ tubes is 25°C, which is 2°C less than for *M. fijiensis* (Porras and Pérez, 1997). Germ tubes of *M. musicola* ascospores also grow better at cooler temperatures than *M. fijiensis*. This physiological difference between the two pathogens may explain the dominance of *M. musicola* in upland plantations (Porras and Pérez, 1997). After a period of epiphyllic growth, which varies from 48–72 h for ascospore germ tubes (Brun, 1963) to 4–6 days for conidial germ tubes, an appressorium is produced above the pore of a stoma and the fungus penetrates by means of a fine infection hypha. Infection through the lower surface occurs more than through the upper surface. Later, hyphae grow out of stomata and can extend 2–3 mm over the leaf surface before forming new appressoria and re-entering the leaf.

Temperatures above 21°C favour the disease, with the formation of conidia being optimal at about 25–28°C (Pont, 1960a; Stover, 1965). Disease development is accelerated if infection densities are high and high light intensity also seems important, as shading can prevent symptom expression (Meredith, 1970). In general, conditions that favour plant growth also favour disease development.

The time interval between spore germination and the appearance of the first yellow streak symptoms (incubation period) varies on susceptible plants. It can be as short as 11 days under ideal environmental conditions for disease development (Klein, 1960) or as long as 105 days under unfavourable conditions (Simmonds, 1939). Similarly, the time taken for yellow streaks to change into brown spots (transition period) is variable and ranges from 2 to 103 days. Some streaks never develop into spots (Meredith, 1970).

In outdoor, artificial inoculation experiments in Surinam, Stahel (1937b) found that the first symptoms appeared after 15–17 days, rusty red streaks after 22–24 days, infiltration of tissue around streaks with water followed by necrosis after 26–27 days and the browning and death of the whole affected area after 30 days. In inoculation studies in misting cabinets in the glasshouse, Goos and Tschirch (1963) noted that conidia germinated after 24–48 h and stomata were penetrated after 4–6 days. The development of disease was favoured by nightly exposure to mist followed by 6–8 h exposure to low humidity during the day. Under this regime of alternating high and low humidity, leaf spots usually developed within 28–35 days. Jones (1995) inoculated young plants derived from tissue culture with mycelial fragments of *M. musicola* and observed that, at 25–26°C, faint chlorotic spot symptoms appeared after 12–16 days and mature lesions after 30 days. Drops of guttation water were frequently seen on water-soaked lesions at the streak to spot transition stage, as had been described by Stahel (1937b).

Several scales for measuring disease intensity (amount of damaged tissue on individual plants) have been proposed by different authors (Leach, 1946; Guyot and Cuillé, 1958; Brun, 1963; Kranz, 1965; Stover and Dickson, 1970; Stover, 1971b).

Methods have been based on allocating a disease rating for each leaf based on the percentage area of leaf damaged by Sigatoka. These leaf ratings can then be used to determine disease levels on the plant.

The intensity of disease on certain non-flowering plants selected at random in plantations has been used to estimate disease prevalence (amount of spotting in a population of plants). Results of such surveys are useful for monitoring changes in disease incidence in plantations over time. However, most systems proposed are time-consuming and cumbersome and have not proved suitable for commercial use.

As disease intensity increases, Sigatoka spots first appear on younger and younger leaves. Stover and Dickson (1970) showed that the age of the youngest leaf with ten or more spots correlates well with disease intensity. This leaf they called the 'youngest leaf spotted' or YLS. An estimation of the age of the YLS is obtained by counting down the plant from the first fully unfurled leaf until the YLS is reached. The higher the number, the older the leaf. Disease prevalence is obtained by averaging YLS data from a number of randomly selected medium-sized, non-flowering plants in a plantation. Changes in the average YLS between surveys give an indication of changes in disease prevalence. The average YLS in commercial plantations with regular spray programmes was found by Stover and Dickson (1970) to be 10–11, while the average YLS in unsprayed plots was 3–4.

Host reaction

There have been numerous published reports and reviews of the reaction of *Musa* species and banana cultivars to *M. musicola* in the field (e.g. Parham, 1935; Cheesman and Wardlaw, 1937; Brun, 1962; Simmonds, 1966; Vakili, 1968; Meredith, 1970; Laville, 1983; Tezenas du Montcel, 1990; Daniells *et al.*, 1996). In some instances, results have not been the same. Misidentified and mislabelled clones undoubtedly account for some differences. Another possibility is that populations of the pathogen in different locations may differ in their ability to cause disease. However, there is no proof yet that pathogenic variants of *M. musicola* exist, although the appearance of a more virulent form has been suggested to explain the more susceptible reaction of some cultivars in Brazil over recent years.

Trials to evaluate germplasm for reaction to Sigatoka have been undertaken in different locations with different environmental conditions. In addition, different methods have been used to measure disease severity and calculate host response. These range from ratings of reaction based on a limited number of observations to a detailed analysis of a variety of parameters (Vakili, 1968). Even under the same environmental conditions and using the same evaluation method, readings between plants of the same clone may differ considerably and different interpretations of reaction may be possible. Because of this variability, the layout of screening trials is extremely important, so that meaningful results can be obtained after statistical analysis. In nature, it is probable that there is an almost continuous gradation of response of *Musa* germplasm to Sigatoka from extremely susceptible to resistant. Therefore, even with statistically validated results, classification into artificial reaction categories is an arbitrary decision. In the following paragraphs, clones that are described as highly resistant to *M. musicola* are those that do not develop symptoms. Clones that have symptoms are divided into partially resistant and susceptible categories. These definitions may vary from those adopted by the authors of some papers cited.

Screening trials undertaken with wild species in the *Eumusa* section of *Musa* have shown that *M. schizocarpa, M. balbisiana* and *M. acuminata* ssp. *malaccensis, microcarpa, siamea* and *truncata* are on the whole highly resistant to Sigatoka (Cheesman and Wardlaw, 1937; Daniells *et al.*, 1966; Vakili, 1968). Of all the accessions of these wild types tested by Vakili (1968), only one of six accessions of

M. acuminata ssp. *siamea* was seen with symptoms.

The reaction of *M. acuminata* ssp. *banksii* and ssp. *errans* is more variable. Vakili (1968) recorded 14 accessions of *M. acuminata* ssp. *banksii* as susceptible, seven as partially resistant and five as highly resistant. He also found that one accession of *M. acuminata* ssp. *errans* was susceptible, one partially resistant and eight highly resistant. Daniells *et al.* (1996) screened 16 accessions of *M. acuminata* ssp. *banksii* for reaction to Sigatoka and found eight to be susceptible, five resistant and three highly resistant.

From the above, susceptibility to Sigatoka in wild banana populations involved in the evolution of the *Eumusa* series of banana (see Table 1.2) would seem to be the exception rather than the rule. However, Carreel (1995) has recently suggested *M. acuminata* ssp. *banksii* and/or *errans* played a major role in this evolution because genetic components of one or other or both of these subspecies are found in almost all banana cultivars of the *Eumusa* series. Therefore, the source of susceptibility to Sigatoka in edible banana may have been inherited from susceptible forms of *M. acuminata* ssp. *banksii/errans*.

All wild *Musa* in the *Australimusa* section tested in the field (*M. textilis*, *M. maclayi* ssp. *maclayi* var. *maclayi*, *M. peekelii* ssp. *peekelii* and *M. peekelii* ssp. *angustigemma*) have been classified as highly resistant (Cheesman and Wardlaw, 1937; Daniells *et al.*, 1966). It is not surprising, therefore, that all Fe'i cultivars in the *Australimusa* series of edible banana so far tested have also been highly resistant (Jones, 1995; Daniells *et al.*, 1996). *Musa ornata* in the *Rhodochlamys* section is also highly resistant (Cheesman and Wardlaw, 1937).

Cultivars in the AA genomic group of the *Eumusa* series of edible banana vary in their response to the disease. Vakili (1968) undertook the evaluation of 180 accessions in Honduras in 1963/64 and found that 45 were highly resistant, 25 partially resistant and 110 susceptible. His results also showed that resistance was present in a higher proportion of accessions originating from South-East Asia than from the New Guinea–Solomon Island area. Of 82 accessions collected in Papua New Guinea and rated for reaction to *M. musicola* in Australia by Daniells *et al.* (1996), only three remained free of Sigatoka symptoms, indicating high resistance. Twenty-nine fell into the partially resistant category and 50 were rated as susceptible.

The AA diploid cultivars 'Sucrier' and 'Inarnibal' are generally regarded as susceptible to Sigatoka disease and 'Pisang Lilin', 'Pisang Tongat' and 'Paka' as highly resistant (Cheesman and Wardlaw, 1937; Brun, 1962; Simmonds, 1966; Vakili, 1968; Laville, 1983). Vakili (1968) evaluated 16 accessions of 'Pisang Jari Buaya', a cultivar that has been used in breeding programmes as a source of nematode resistance, and found 14 to be susceptible and two partially resistant. An accession of 'Pisang Jari Buaya' from Papua New Guinea was also found to be partially resistant (Daniells *et al.*, 1996).

In the AAA genomic group, 'Gros Michel', 'Lakatan', 'Pisang Susu' and cultivars in the Cavendish subgroup are susceptible (Cheesman and Wardlaw, 1937; Simmonds, 1966; Vakili, 1968; Laville, 1983). 'Red' and 'Green Red' are regarded by some workers as susceptible (Laville, 1983), while others have noted degrees of partial resistance in some accessions (Simmonds, 1933; Simmonds, 1966; Vakili, 1968). Daniells *et al.* (1996) classified 'Mata Kun' (syn. 'Red') as susceptible, but it was less susceptible than a Cavendish cultivar, which was rated at the same trial. Cultivars in the Lujugira–Mutika subgroup have been rated as partially resistant (Simmonds, 1966) and 'Yangambi Km 5' (syn. 'Ibota Bota') as highly resistant (Laville, 1983).

'Ney Poovan' (AB) has been reported as highly resistant by Simmonds (1966) and 'Safet Velchi' (syn. 'Ney Poovan') was rated as highly resistant by Laville (1983). Reports of the susceptibility of some accessions of this clone (Brun, 1962; Laville 1983) may be erroneous.

Clones in the Plantain subgroup in the AAB genomic group are all highly resistant to *M. musicola* when tested at or near sea level (Simmonds, 1966; Laville, 1983). However, plantain has been found to be susceptible at elevations above 500 m in Puerto Rico and Colombia (Stover and Simmonds, 1987), and in upland areas of Cameroon (Mouliom-Pefoura and Mourichon, 1990). Most workers consider 'Silk' (AAB) to be partially resistant (Simmonds, 1933; Brun, 1962; Simmonds, 1966; Laville, 1983) and cultivars in the Pome subgroup (AAB) to be susceptible (Stover and Simmonds, 1987). Two cultivars in the Maia Maoli–Popoulu subgroup (AAB) have been rated as susceptible (Laville, 1983), but resistance may be present in some clones (Parham, 1935). The reaction of 'Mysore' (AAB) most probably varies between high resistance (Parham, 1935; Simmonds, 1966) and high partial resistance (Cheesman and Wardlaw, 1937; Brun, 1962; Laville, 1983), despite reports of susceptibility in some clones (Brun, 1962; Laville, 1983). 'Pisang Raja' (AAB) has partial resistance (Simmonds, 1966).

The situation is less confused for cultivars in the ABB genomic group. 'Bluggoe', 'Pisang Awak', 'Kluai Teparot' and others are known to be highly resistant (Simmonds, 1966; Stover and Simmonds, 1987). There are probably none that can be classified as susceptible.

The high resistance of cultivars in the ABB genomic group has been attributed to their high 'B' genome content. As mentioned previously, *M. balbisiana*, the source of the 'B' genome, is highly resistant to *M. musicola*. The general rule is that the greater the 'B' component in the genome of the cultivar, the greater the resistance of that cultivar to Sigatoka (Meredith, 1970). However, some *M. acuminata*-derived diploids and at least one triploid are also highly resistant. The AA clones 'Pisang Lilin' and 'Paka' have been used as pollen sources in breeding for resistance to Sigatoka (Stover and Simmonds, 1987). Resistance to the disease resides in both *M. acuminata* and *M. balbisiana*.

Evaluation techniques based on the inoculation of juvenile plants derived from tissue culture have also been shown to be useful (Jones, 1995), although differences with results in the field have been noted (Daniells *et al.*, 1996).

Black leaf streak attacks many cultivars resistant to Sigatoka. Although most cultivars that are resistant to black leaf streak are also resistant to Sigatoka, some hybrids bred at the FHIA in Honduras are more susceptible to Sigatoka than to black leaf streak. This indicates that the two pathogens may activate different resistance mechanisms in some instances.

Control

Chemical control

Meredith (1970) and Stover (1972, 1990) have reviewed the history of the chemical control of Sigatoka.

Knowles (1914) in Fiji first advocated the use of Bordeaux mixture to control Sigatoka. After the arrival of the disease in the Caribbean and Central America in the early 1930s, large pipeline systems were installed in commercial plantations to deliver Bordeaux, which was applied as a high-volume spray at intervals of 2–5 weeks (Wardlaw, 1941). The costs were high, but control was achieved and the industry saved from possible collapse. Bordeaux worked by controlling conidial production in the leaf spot and conidial infection, but was less effective in preventing infection of the unfurling cigar or heart leaf by ascospores (Leach, 1946). In Jamaica, the control strategy was to reduce the number of leaf spots before the rainy season, when ascospores would be released. Spraying with Bordeaux continued until 1957. During this period, zineb and copper oxychloride were found to give good control, but no exceptional substitutes for Bordeaux were found.

In the 1950s, the banana industry in the French Antilles was threatened because the hilly terrain made the application of high-volume sprays of Bordeaux impracticable. Experiments were conducted with zineb and copper oxychloride in petroleum oil, applied as a low-volume spray. This proved effective and even oil alone applied as a mist gave good control (Guyot and Cuillé, 1954). Oil phytotoxicity caused some initial problems, but improved oils were developed, which led to their widespread use. Oils and oil-based sprays were applied from the ground by various mist blowers or from the air by means of a helicopter or fixed-wing plane at the rate of between about 12 and 15 l ha^{-1}, according to disease severity and risks of phytotoxicity at different times of the year. In some years, it was necessary to spray every 10–12 days as new leaves appeared, whereas one application every 3–4 weeks maintained control in other years.

Klein (1960) discovered that the development of Sigatoka lesions was greatly retarded when oil was applied at the young yellow streak stage and that counts of these streaks on young leaves could be used to forecast when to spray. This method enabled oil to be applied strategically rather than regularly, with associated cost savings.

Soon after the adoption of oil spraying, the use of aqueous suspensions of dithiocarbamate fungicides, such as mancozeb (Dithane M-22®), was also shown to control Sigatoka. However, it soon became apparent that protectant fungicides were much more effective when applied as oil-in-water emulsions.

Benomyl (Benlate®), the first systemic fungicide of the benzimidazole group to be used on banana, was sent to Honduras for trial in 1967. By 1972, it was in widespread use. With the development of systemic fungicides, control was even more effective, as the chemicals could migrate in treated leaves and also from treated to untreated leaves. The penetrant fungicide tridemorph (Calixin®) was used soon after benomyl, but was found to be less effective.

A Sigatoka forecasting system based on climatic data (temperature and evaporation), as well as observations of disease symptoms on young leaves, was developed in the French Antilles (Ganry, 1986). The system, which relies on the therapeutic action of systemic fungicides, enabled the application of sprays only when necessary. *Mycosphaerella musicola* developed resistance to fungicides in the benzimidazole group in the French Antilles in the 1970s and this led to the temporary use of oil alone. Later, propiconizole (Tilt®), a systemic fungicide in the triazole group of DMI, was integrated into the forecasting system (Bureau and Ganry, 1987). Flusilazole (Punch®), another effective fungicide in the triazole group, and Imazalil (Fungaflor®), in the imidazole group of DMI of systemic fungicides, have also been used to control Sigatoka.

In northern Australia in the late 1980s, growers were advised to use mancozeb plus white oil every 14 days in the wet season and every 21–28 days during the dry season, but to use propiconizole in water-miscible oil during periods of high disease pressure. By using propiconizole only when the disease could not be held in check by protectant fungicides in oil, it was thought that the possibility of resistance developing to this valuable systemic fungicide would be lessened.

In the Windward Islands today, systemic/penetrant fungicides with different modes of action, such as benomyl, propiconizole or flusilazole and tridemorph, are alternated to reduce the risk of the development of resistance. The stages of disease development in young leaves of plants growing in different microclimates are monitored weekly to predict the most appropriate time to spray the crop from aircraft. Plants in locations where Sigatoka builds up more rapidly are also sprayed from the ground in between aerial applications. This management strategy developed from a forecasting system proposed by Cronshaw (1982).

Cultural methods

The removal and destruction of badly spotted leaves (trash) from banana plantations are recommended as a means of reducing inoculum levels. Heavily diseased leaves can also be buried within the plantation or piled on top of one another to prevent the effective discharge of ascospores.

Biological control

No effective biological control methods have yet been developed. However, epiphyllic mycelia of leaf-surface fungi have been reported to inhibit the germination of spores of *M. musicola* (Meredith, 1970).

SIGATOKA-LIKE LEAF SPOTS

Phaeoseptoria Leaf Spot

D.R. Jones

Introduction

Phaeoseptoria leaf spot was first described from Trivandrum, Kerala State, India, where it was the cause of a severe blight of banana leaves (Raghunath, 1963). The disease was not highlighted by Stover (1972), but Punithalingam (1983) reported that several samples of the disease had been received at the International Mycological Institute since the 1950s from different countries. Phaeoseptoria leaf spot has been found in Australasia–Oceania (Australia: Queensland), Asia (India, East Malaysia: Sabah), Africa (Cameroon, Ghana, Kenya, Tanzania: Zanzibar, Uganda) and the Latin American–Caribbean region (Colombia, Guyana, Honduras, Trinidad) (Punithalingam, 1983; Jones, 1991; B. Ritchie, CABI, 1999, personal communication).

Symptoms

Individual lesions mature into ellipsoidal, sometimes ovoid spots. Raghunath (1963) gives the dimensions as 15 mm × 7 mm, while Punithalingam (1983) has described them as 10–20 mm wide. Mature lesions are porcelain-white (Raghunath, 1963) or straw-yellow (Punithalingam, 1983) in the centre, with dark brown borders and yellow haloes. Often, infection density is high and spots coalesce as they enlarge to form irregular necrotic areas with white to straw-coloured centres. Young leaves are unaffected. Raghunath (1963), who published a photograph of mature lesions, thought that the symptoms of Phaeoseptoria leaf spot were reminiscent of Sigatoka (see Jones, this chapter, pp. 79–92).

Causal agent

The fungus *Phaeoseptoria musae* Punith. causes Phaeoseptoria leaf spot.

Pycnidia, which are immersed in leaf tissue and become erumpent, are found within the spots. They are yellowish brown to dark brown, subglobose and 85–145 μm in diameter. Conidiogenous cells lining the pycnidial cavity are hyaline, simple and subglobose to doliiform and they give rise to conidia. Conidia are hyaline to straw-yellow to pale brown, straight or slightly curved, cylindrical to slightly clavate, 2–4-septate and 22–30 μm × 2.5–3 μm in size. Micronidia are similarly coloured and shaped, but are 1–2-septate and have dimensions of 12 μm × 3 μm (Punithalingam, 1983). Although *Mycosphaerella* was identified as a possible perfect stage by Raghunath (1963), this was not reported by Punithalingam (1976, 1983). A *Leptosphaeria* sp. has been seen in association with one specimen of *P. musae* (G. Kinsey, CABI, 1999, personal communication).

On oat agar, colonies are floccose and greyish sepia, with abundant mycelium. Underneath, they appear colourless, with black blotches. Pycnidia are 80–140 μm in diameter and conidia are 25–33 μm × 2.5–3.5 μm and 1–4-septate (Fig. 2.10). Microconidia are uniseptate and measure 10–16 μm × 3–4 μm. Conidia produced in culture are generally broader than those within pycnidia found on the host. Conidiogenous cells lining the inner wall of the pycnidium are holoblastic, hyaline and subclavate to subobpyriform (Fig. 2.10) (Punithalingam, 1976).

Disease cycle and epidemiology

Little is known about the epidemiology of Phaeoseptoria leaf spot. It is assumed that conidia are dispersed by rain splash and ascospores are forcibly discharged during periods of wet weather.

Host reaction

In Kerala State in southern India, 'Nendran' (AAB, Plantain subgroup) has been described as highly susceptible and 'Poovan' (AAB, syn. 'Mysore') as being affected. The cultivars 'Annan' and 'Palayankodan' were also reported as susceptible. None of the cultivars growing in the area was believed to be immune (Raghunath, 1963).

Control

No information on control is available.

Septoria Leaf Spot

J. Carlier, X. Mourichon and D.R. Jones

Introduction

Between 1992 and 1995, as part of an INIBAP-initiated survey to determine the distribution of Sigatoka and black leaf streak in the South and South-East Asian regions, specimens of banana leaf spots

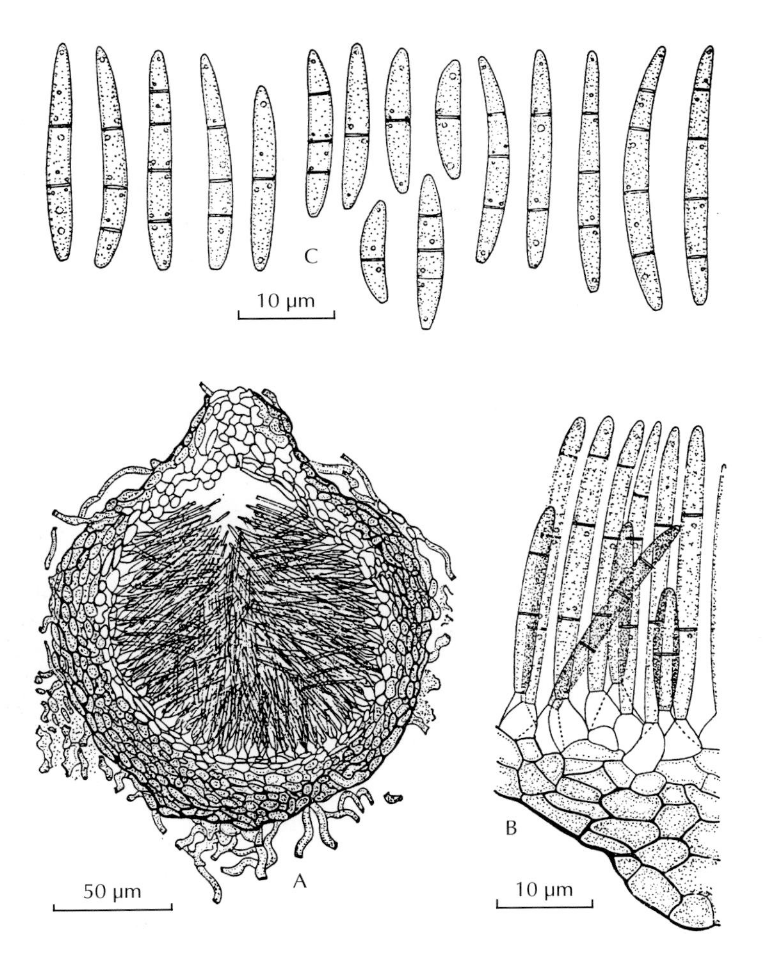

Fig. 2.10. Vertical section of pycnidium (A), part of the pycnidial wall with conidiogenous cells and conidia (B) and conidia (C) of *Phaeoseptoria musae* (from Punithalingam, 1976).

were collected in southern India, Sri Lanka, West Malaysia, Thailand and Vietnam. These were sent to the Centre de coopération internationale en recherche agronomique pour le développement (CIRAD) in Montpellier, France, for identification. *Mycosphaerella musicola*, the cause of Sigatoka disease, was not identified from any of the samples. Some specimens from Johore in West Malaysia and from Vietnam were found to be *M. fijiensis* and some could not be identified, but all others collected from southern India, Sri Lanka, West Malaysia, Thailand and southern Vietnam, were of an unknown banana pathogen. The fungus had *Septoria* as an imperfect stage and *Mycosphaerella* as a perfect stage (Anon., 1995). The disease was given the name Septoria leaf spot. A specimen of Septoria leaf spot was later collected in Mauritius in 1997 and an analysis of leaf spot isolates collected at Onne in Nigeria has revealed that the pathogen was present there in 1989.

Evidence suggests that Septoria leaf spot has most probably been mistaken for Sigatoka and perhaps black leaf streak on visual symptoms in the past. More specimens of Septoria leaf spot need to be collected and analysed to determine its true distribution. However, the disease appears to be the dominant leaf spot in Thailand, where it was collected from a number of different geographical areas. It also appears to be common in West Malaysia, Sri Lanka and southern India.

Symptoms

The symptoms of Septoria leaf spot are similar to those of Sigatoka (see Jones, this chapter, pp. 79–92), black leaf streak (see Jones *et al.*, this chapter, pp. 44–48) and Phaeoseptoria leaf spot (see Jones, this chapter, pp. 92–93). The first obvious symptom is a brown streak, which expands into a large spot and darkens (Plate 2.16). The spot later develops a dark brown border as the centre turns grey. When infection density is low, individual spots are ovoid or elliptical when mature (Plate 2.17). The size and shape of this type of spot distiguish it from most mature Sigatoka and black leaf streak spots, which are smaller and narrower. However, mature spots of Septoria leaf spot closely resemble those of Phaeoseptoria leaf spot (Raghunath, 1963). When the infection

Plate 2.16. Symptoms of Septoria leaf spot on 'Grand Nain' (AAA, Cavendish subgroup) near Phicit in northern Thailand. Grey centres are beginning to form in developing brown lesions, which are coalescing in places (photo: D.R. Jones, INIBAP).

Plate 2.17. Mature spots of Septoria leaf spot on a young leaf of 'Embul' (AAB, 'Mysore') in Sri Lanka. (photo: D.R. Jones, INIBAP).

density is high, lesions coalesce at an early stage of development and large areas of leaf tissue become necrotic (Plate 2.18). Grey spots and patches are visible in the necrotic areas. Leaf tissue around developing lesions turns yellow before dying (Plates 2.16 and 2.18).

Causal agent

When the disease was detected initially, it was thought that the pathogen may have been *Phaeoseptoria musae* (Anon., 1995), which has been reported in many locations including southern India (Raghunath, 1963) and East Malaysia (Punithalingam, 1983). However, further investigations, which are summarized below, proved otherwise (Carlier *et al.*, 1999, unpublished results).

Two distinct types of fruiting structures, which were more prevalent in lesions on the upper leaf surface, were observed to be closely associated in all the specimens collected. The first was identified as a flask-shaped pycnidium, which when mature measured 31–42 μm in width. Pycnidia were immersed, more or less erumpent,

Plate 2.18. Severe symptoms of Septoria leaf spot on 'Anamala' (AAA, syn. 'Gros Michel') in Sri Lanka. Developing lesions have coalesced to form large areas of necrotic tissue. Note lesions on the midrib (photo: D.R. Jones, INIBAP).

flask-shaped and ostiolated when young, but often acervular-like when mature. Conidia associated with the pycnidia were hyaline, filiform and measured 22–41.6 μm in length. Pycnidia and conidia helped identify the asexual stage as *Septoria*. The pycnidia were smaller and the conidia larger than those described for *P. musae* (Punithalingam, 1983). The other fruiting body was a perithecium. Perithecia were globose with a short protruding ostiole, 42–51 μm in diameter and dark brown in colour. Ascospores found in oblong asci were two-celled and measured 12.0–16.5 μm × 3.0–4.5 μm. The sexual stage was identified as *Mycosphaerella* and was indistinguishable from that of *Mycosphaerella musicola* and *M. fijiensis.*

Cultures of the pathogen were initiated from single ascospores ejected from the perithecia. All had black stroma-like structures and appeared similar to cultures of *M. fijiensis.* Pycnidia were observed 1 week after isolation. Conidia were three- to five-celled, hyaline, straight or slightly flexuous, with an average size of 40.9 μm × 2.1 μm (range 24.5–53.9 μm × 1.8–2.7 μm). These results confirmed the link between the *Mycosphaerella* and *Septoria* states of the fungus.

A number of micropropagated plants of 'Grande Naine' (AAA, Cavendish subgroup) were artificially inoculated with conidia produced by single ascospore isolates of the fungus. Symptoms on inoculated plants were similar to those observed in the field.

Phylogenetic analysis based on sequences of the ITS of ribosomal DNA from the *Mycosphaerella* sp. causing Septoria leaf spot, *M. musicola, M. fijiensis, M. musae and P. musae* has confirmed that all are different species.

Disease cycle and epidemiology

No research has yet been undertaken on this disease in the laboratory or field to determine optimum germination and growth temperatures and main means of dispersal and infection.

Host reaction

Little is known about the reaction of banana clones to Septoria leaf spot, but it has been isolated from 'Chuoi Ngu' (AA, syn. 'Sucrier'), 'Kluai Hom Thong' (AAA), 'Pisang Kapas' (AA), 'Anamala' (AAA, syn. 'Gros Michel'), cultivars in the Cavendish subgroup (AAA) and unidentified AAB clones from Nigeria, which are most probably cultivars in the Plantain subgroup. Symptoms have been recognized in southern Thailand on older leaves of 'Kluai Lep Mu Nang' (AA, 'Pisang Lilin'), a clone that is highly resistant to Sigatoka and black leaf streak diseases, and in Sri Lanka on young leaves of 'Mysore' (Plate 2.19), which also has resistance to these two diseases.

Control

There is some evidence that Septoria leaf spot may be controlled by the same chemicals that control Sigatoka leaf spot and black leaf streak.

Acknowledgement

The information presented in this disease description will form part of an article on Septoria leaf spot and its causal agent that has been prepared for publication in a scientific journal by J. Carlier, M.F. Zapater, F. Lepyre, D.R. Jones and X. Mourichon. The editor thanks these authors for allowing the results of some of their unpublished research to be presented in this book.

OTHER LEAF SPOTS

Black Cross Leaf Spot

D.R. Jones

Introduction

Black cross leaf spot is found in Australasia–Oceania (American Samoa, Australia, Fiji, New Caledonia, Niue, Papua New Guinea, Tonga, Vanuatu, Western Samoa) and Asia (Indonesia, Philippines) (Booth and Shaw, 1961; Meredith, 1969; Dingley *et al.*, 1981; Hyde, 1992). The disease is not regarded as serious, although in Western Samoa it can be a problem on 'Fa'i Misiluki' (AAB, syn. 'Mysore') when it is grown in shady areas under trees or coconuts. Under such conditions, black cross leaf spot symptoms can cover entire leaves (Gerlach, 1988).

Symptoms

Mature symptoms of the disease are found on older leaves and take the form of large, jet-black, four-pointed, diamond-shaped stars or crosses. The longer axis of the star runs parallel to leaf veins for up to 6 cm and the shorter arm or arms are at right angles and can extend for about 3 cm. The symptoms are clearly visible on the underside of the leaf (Plates 2.19 and 2.20), but, on the upper surface, symptoms are less pronounced, showing only a series of black dots (Booth and Shaw, 1961). In Western Samoa, a yellow, diamond-shaped spot, interspersed with dark brown lines, has been observed on upper leaf surfaces (Gerlach, 1988). Black cross leaf spot can serve as points of entry for *Cordana musae* (Plates 2.19 and 2.20), which can cause considerable damage on older leaves.

Causal agent

Black cross leaf spot is caused by the fungus *Phyllachora musicola* Booth & Shaw, the black crosses or stars being the distinctive mature stroma of the fungus. On the upper leaf surface, the fungus produces oval to globose perithecia 200–240 μm in diameter, which are immersed in the stroma with the ostiole visible only as a minute pore. Up to 40 perithecia can be found in each stroma. Asci develop from the base of the perithecia among abundant paraphyses. Asci are cylindrical or clavate in shape with the widest part near the centre of the ascus, measure 115–190 μm × 16–20 μm and contain eight ascospores. The apex is truncate, with a non-amyloid apical ring. Ascospores, which are arranged biserrately in the ascus and surrounded by a deliquescing mucilaginous sheath, measure 35–52 μm × 6.5–10 μm, are hyaline, smooth, usually aseptate and obovate to clavate in shape (Booth and Shaw, 1961; Hyde, 1992). The fungus has not been grown in culture.

An asexual state was not observed by Booth and Shaw (1961) or Hyde (1992), but Gerlach (1988) has implicated the genus *Scolecobasidium*, which he found covering the stroma on the lower side of leaves as a velvety layer. However, *Scolecobasidium humicola* Barron & Busch has been recorded as growing over black cross fructifications in Papua New Guinea (Shaw, 1984).

Disease cycle and epidemiology

White masses of spore exudates occur along the arms of the stroma in humid conditions (Booth and Shaw, 1961). During wet weather, it is presumed that ascospores are carried in water drops and rain splash to spread the disease along and between leaves. It is also likely that ascospores can become airborne when forcibly ejected from perithecia to disseminate the disease more widely.

Host reaction

Meredith (1969) has claimed that Knowles (1916) working in Fiji was the first to recognize black cross leaf spot because his description of the symptoms closely matches the disease attributed to *P.*

Plate 2.19. Symptoms of black cross leaf spot on the underside of a banana leaf on Murray Island in the Torres Strait region of Australia. Lesions caused by *Cordana musae* can be seen associated with black cross symptoms on the extreme right of the picture (photo: D.R. Jones, QDPI).

musicola. Knowles (1916) stated that the black cross leaf spot was not found on 'Dwarf Cavendish' (AAA, Cavendish subgroup), 'Gros Michel' (AAA) and other cultivars grown by Europeans, but did occur on 'Mysore' (AAB), 'Blue Java' (ABB, syn. 'Ney Mannan') and native cooking-banana cultivars. Meredith (1969) reported black cross leaf spot on members of the Maia Maoli–Popoulu subgroup (AAB), but not on 'Robusta' (AAA, Cavendish subgroup) in Fiji. Later, Firman (1972) took the opportunity to rate accessions growing at the Koronivia Research Station in Fiji for reaction to black cross leaf spot. His findings were similar to the previous reports, which indicated that black cross leaf spot was a disease of cultivars in the AAB and ABB genomic groups. Counting down non-flowering plants, symptoms were seen first on the third, fourth or fifth leaves, with the highest number of crosses being recorded on the fifth or sixth leaf.

Black cross leaf spot was observed on a wide range of genotypes (AA, AAA, AAB, AAS, ABB and AAAB) in the Papua New Guinea Biological Foundation Banana Collection at Laloki near Port Moresby in 1988. Although the disease was present on 75% of accessions, which included those identified as belonging to the Cavendish subgroup and clones 'Sucrier' (AA), 'Red' (AAA), 'Gros Michel' and 'Pisang Awak' (ABB), it was not significant on most cultivars. Very severe symptoms were only seen on seven local cultivars, which had been classified as belonging to the AAA, AAB, AAS and AAAB genotypes (D.R. Jones, Papua New Guinea, 1988, personal observation).

Other records include 'Lady Finger' (AAB, Pome subgroup) and 'Pacific Plantain' (AAB, Maia Maoli-Popoulu subgroup) in the Torres Strait region of Australia (Jones and Daniells, 1991) and 'Saba' (ABB) in the Philippines (Meredith, 1969).

Symptoms of black cross leaf spot have been seen on *M. balbisiana* in Papua New Guinea and on *M. acuminata* ssp. *banksii* in Papua New Guinea (Plate 2.20) and Queensland, Australia (D.R. Jones, Papua New Guinea, 1988 and 1989, personal observations; Jones and Daniells, 1991). The presence of the disease on many local cultivars in Papua New Guinea and Fiji may be because their genomes are mostly derived from one or both of these two species

Plate 2.20. Black cross leaf spot symptoms on the underside of a leaf of *Musa acuminata* ssp. *banksii* in the Oro Province of Papua New Guinea (photo: D.R. Jones, QDPI).

(Carreel, 1995). No other wild *Musa* species in Papua New Guinea have been seen with symptoms and introduced *M. textilis* also seems to be unaffected (Booth and Shaw, 1961; D.R. Jones, Papua New Guinea, 1988 and 1989, personal observation).

Control

Commercial cultivars of the AAA genotypes, although not immune, would seem to be fairly resistant to the disease. No control measures are practised on other cultivars, although, in Western Samoa, planting 'Fa'i Misiluki' in open, sunny places rather than in shade is recommended (Gerlach, 1998).

Cordana Leaf Spot

D.R. Jones

Introduction

Cordana leaf spot is usually of minor importance, but on occasions it can cause defoliation. Stover (1972) reported that it could be a serious problem in Central America on cultivars in the Plantain subgroup (AAB) during and following periods of wet weather. Epidemics have also occurred on 'Williams' (AAA, Cavendish subgroup) in the southernmost area of commercial banana production in New South Wales, Australia (Allen and Dettman, 1990). Most damage occurs when the pathogen gains entry to leaf tissue weakened because of age, adverse environmental conditions, nutritional deficiencies or wounds or through lesions caused by other disease organisms. Cordana leaf spot has a worldwide distribution.

Symptoms

Large, pale brown, oval to fusiform-shaped, necrotic lesions often with light grey centres and concentric ring patterns characterize Cordana leaf spot. A dark brown border surrounded by a bright yellow halo separates a lesion from normal leaf tissue (Plates 2.21 and 2.22). Often, lesions coalesce to form large areas of dead leaf tissue (Plate 2.22).

Invasion frequently occurs at leaf margins, where the lamina is more vulnerable to tearing and where the effects of senescence and nutritional deficiency problems are first felt. Symptoms are also often seen around lesions caused by other pathogens, such as *M. musicola* (Plate 2.21), *M. fijiensis*, the *Mycosphaerella* sp. responsible for Septoria leaf spot (Plate 2.22) and *P. musicola* (see Plates 2.19 and 2.20).

Causal agents

The cause of this distinctive leaf spot disease was first described as *Scolecotrichum musae* by Zimmerman (1902) from a specimen collected in Java, Indonesia. This fungal pathogen is now known as *Cordana musae* (Zimm.) Höhnel. A new species, *Cordana johnstonii* Ellis, which causes a disease very similar in appearance to the one caused by *C musae*, was described by Ellis (1971a) on banana from Irian Jaya in Indonesia, the Cameron Highlands of West Malaysia and Tonga. Two different species of *Cordana* are, therefore, responsible for Cordana leaf spot symptoms.

When herbarium specimens of Cordana leaf spot collected in New South Wales in Australia since the 1930s were examined, all were found to be of *C. johnstonii*. This species was also present on Lord Howe Island off the coast of New South Wales and on Norfolk Island between Australia and New Zealand. However, further studies revealed that *C. musae* was the cause of most Cordana leaf spot in Queensland, which lies to the north of New South Wales. *Cordana musae* was also identified as present at Darwin in the Northern Territory. In an area of Queensland close to the border with New South Wales, there was a zone where the distribution of both species overlapped (Priest, 1990). These findings indicate that *C. johnstonii* is more adapted to cooler environments than *C. musae*.

Priest (1990) has reported that leaf spots caused by *C. johnstonii* are generally smaller, more regular in outline and more tapered than the larger elliptical to oval spots caused by *C. musae*. It was estimated that individual leaf lesions of *C. johnstonii* measure up to 3 cm × 1 cm while those of *C. musae* are up to 7 cm × 2 cm.

Both fungi have erect, straight to flexuous, septate, denticulate conidiophores, which arise singly in small groups. However, they differ in colour and size. The pale brown conidiophores of *C. musae* are often nodose, up to 150 μm long and 4–6 μm in diameter, while the brown conidiophores of *C. johnstonii* are often twisted in the basal part, up to 300 μm long and 5–9 μm in diameter. The conidiophores of both fungi are paler towards the apex (Priest, 1990).

Cordana musae and *C. johnstonii* have conidia that are two-celled, smooth, pale brown to almost hyaline, with a visible thickened hylum. However, the conidia of *C. musae* are obclavate to pyriform and measure 14–18 μm × 8–10 μm on average, while those of *C. johnstonii* are broadly ellipsoidal to subglobose and measure 19–26 μm × 14–16 μm on average. Morphologically, the species can most easily be differentiated on conidial size and shape, *C. johnstonii* having longer and much wider conidia than *C. musae* (Fig. 2.11) (Priest, 1990).

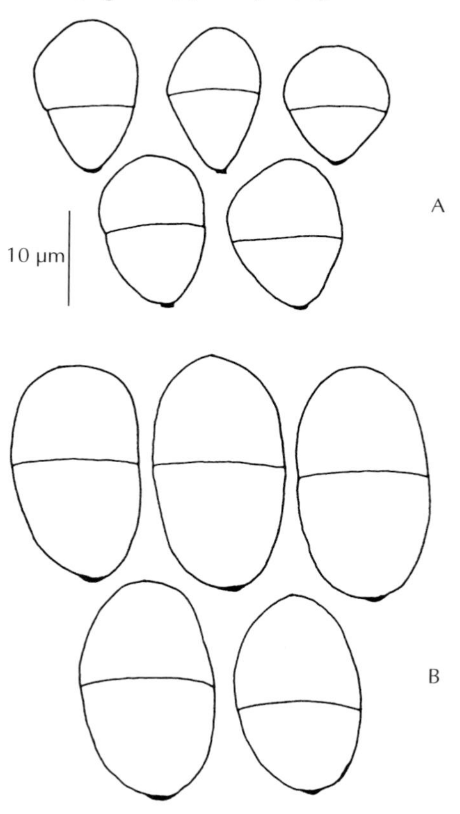

Fig. 2.11. Conidia of *Cordana musae* above and *C. johnstonii* below, showing shape and size differences (from Priest, 1990).

The colonies of both *C. musae* and *C. johnstonii* are hypophyllous, effuse to hairy and dark brown.

Disease cycle and epidemiology

Conidia of *C. musae* are formed at night during rainy periods or when dew is present and violently discharged from the conidiophores just after dawn when there is a sharp decline in humidity (Meredith, 1962a, b). They germinate in a film of moisture on the leaf surface and within 8 h have formed appressoria. Each appressorium produces an infection peg, which penetrates an epidermal cell. One or more fungal cells are then produced, which almost completely fill the host cell. Epidermal cells that are invaded colour and die as the infection peg swells. The fungus penetrates both living and dead tissue in the same manner. The rate of progress of the disease in living leaves seems to depend on the health of the tissue. *Cordana* is a weak pathogen of banana and is primarily a wound invader (Stahel, 1934).

The severe outbreaks of Cordana leaf spot that occur near Macksville in New South Wales are in localities with daily temperature ranges of 16–28°C in summer and 10–20°C in winter. This is relatively cold for banana cultivation and plants are extremely stressed during the winter months. In this area, *C. johnstonii* sporulates profusely on the adaxial surface of leaf spots in otherwise green leaves and, to a lesser extent, on spots in leaf litter, during cool, misty weather. Conidia germinate, produce germ tubes and appressoria and penetrate the leaf epidermis, as reported for *C. musae* by Stahel (1934). The time course for germination and infection was 6–48 h, depending on the temperature, and symptoms appeared after 4–10 days (Allen and Dettman, 1990).

Laboratory experiments with leaf lesions detached from living leaves and dead leaf litter showed that new conidia are produced 1 day after wetting and, under moist conditions, this is continued at a constant rate for another 5 days at least. The optimum temperature for production of conidia was 22°C. Using detached banana leaves, it was found that most appressoria were produced at 22–25°C, but some were formed at temperatures as low as 10–13°C (Allen and Dettman, 1990).

Host reaction

Cordana leaf spot occurs on a wide range of clones. In addition to cultivars in the Cavendish and Plantain subgroups, typical symptoms have also been seen on 'Sucrier' (AA), 'Inarnibal' (AA), 'Gros Michel' (AAA), 'Red' and 'Green Red' (AAA), 'Mysore' (AAB), 'Pome' (AAB), 'Silk' (AAB), 'Pisang Rajah' (AAB), 'Bluggoe' (ABB), 'Pisang Awak' (ABB), 'Kluai Teparot' (ABB) and 'Saba' (ABB) in South-East Asia (Jones and Daniells, 1988; Jones, 1993, 1994b).

A disease survey of the Papua New Guinea Biological Foundation's Banana Collection at Laloki near Port Moresby in 1988 revealed that 81% of the 238 accessions inspected were affected by Cordana leaf spot. However, only four accessions were very severely diseased. During a mission to collect *Musa* germplasm in Papua New Guinea, 86% of cultivars collected were observed to have symptoms of the disease. Cordana leaf spot was also seen on a Fe'i banana cultivar (Jones, 1988).

Cordana leaf spot has been reported on *M. acuminata* ssp. *banksii*, *M. acuminata* ssp. *banksii* × *M. schizocarpa*, *M. balbisiana*, *Musa boman*, *M. maclayi*, *M. schizocarpa* and *Ensete glaucum* in Papua New Guinea (Jones, 1988; Sharrock and Jones, 1989).

Control

Cordana leaf spot is usually of no economic importance and does not warrant control. However, when it has become a problem, fungicides used to control Sigatoka and black leaf streak have also been found to control Cordana leaf spot.

Plate 2.21. Symptoms of Cordana leaf spot on a leaf of 'Williams' (AAA, Cavendish subgroup) in Queensland, Australia. *Cordana musae* has invaded Sigatoka spots, causing large lesions, some with concentric ring patterns (photo: QDPI).

Deightoniella Leaf Spot

D.R. Jones, E.O. Lomerio, M. Tessera and A.J. Quimio

Introduction

The fungus that causes Deightoniella leaf spot is a saprophytic colonizer of dead *Musa* leaves and flowers. It is also a weak parasite of the older foliage of banana and has been reported on young leaves of *Musa* seedlings (Stover, 1972). The leaf spot is more prevalent if plants are growing under poor conditions and humidity is high. Senescing or injured leaves are more prone to the disease.

Deightoniella leaf spot is more of a problem on abacá than on banana. When abacá was grown in Central America, as much as 14% of the total leaf area of 'Bungolanon', the most widely planted cultivar, was destroyed (Stover, 1972). However, this was not thought to affect plant growth (Lopez and Loegering, 1953).

The disease has also recently been recorded on both young and old enset plants and is common in the Sidamo and North Omo regions of Ethiopia (Quimio and Tessera, 1996).

Plate 2.22. Cordana leaf spot symptoms on 'Novaria', a radiation-induced mutant of 'Grand Nain' (AAA, Cavendish subgroup), which is being grown commercially in Perak State, West Malaysia. *Cordana musae* has invaded the leaf through Septoria leaf spot lesions (photo: D.R. Jones, INIBAP).

Symptoms

On banana, lesions are more prevalent along the edges of the leaf blade and on the older, lower leaves. They first appear as small, black necrotic spots between 1 and 2 mm in diameter. These increase in size, becoming oval in shape with a black border. Mature spots can be 25 mm or more in diameter and may unite to form bands of necrotic tissue along leaf edges. Deightoniella leaf spot can easily be confused with Cordana leaf spot. However, there is a dark, smoky colouring over a tan background in Deightoniella spots (Stover, 1972). The fungus can cause a speckle-like spotting of the petiole and a pin-spotting disease of preharvest fruit (see Jones, Chapter 4, pp. 181–182).

Leaf spot of abacá first manifests itself as small pinpoints on the leaf lamina, which are yellowish at first, then brownish and finally black. Lesions increase in size, becoming longer in length than in width. Finally, blotches of various sizes form, which sometimes run the whole length of the leaf. Well-defined black bands, 1–2 mm in width, and bright yellow haloes surround lesions. Fully developed lesions become brown and dry, but the original small black spots remain distinct. As with banana, a speckle-like spotting can also be seen on petioles. The causal agent also attacks the pseudostem of abacá (see Lomerio, Chapter 3, pp. 161–162).

Symptoms on the leaves of enset look like large tar-spot lesions (Plate 2.23). Areas of dead tissue are evident in places where many spots coalesce.

Causal agent

Deightoniella torulosa (Syd.) Ellis causes Deightoniella leaf spot. Conidiophores of this fungus arise singly or in small groups. They are brown and swollen at the apex, with up to six successive proliferations. Conidia are produced singly as expanded ends of the conidiophores and their successive proliferations. They are 35–70 μm × 13–25 μm, straight or slightly curved, obpyriform to obclavate, subhyaline to olive in colour with three to 13 pseudosepta (Fig. 2.12). *Deightoniella torulosa* has been described by Ellis (1957a), Meredith (1961a, b) and Subramanian (1968).

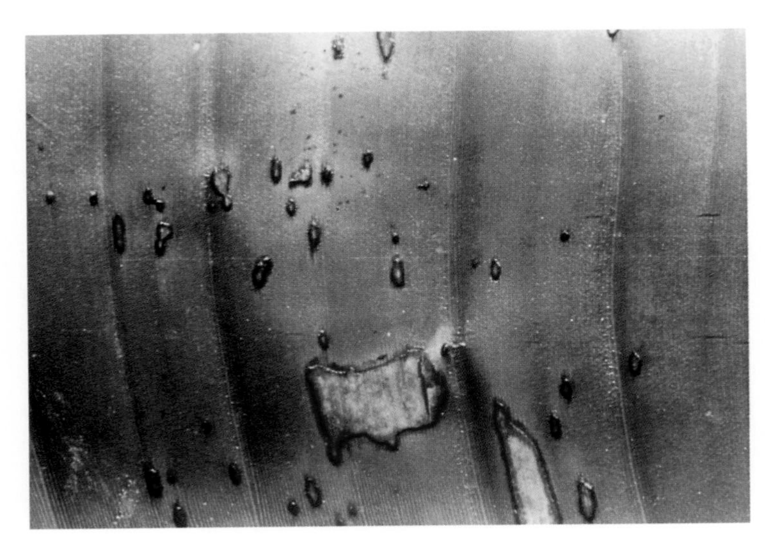

Plate 2.23. Tar-spot-like lesions on a leaf of enset caused by *Deightoniella torulosa* (photo: M. Tessera and A.J. Quimio, IAR).

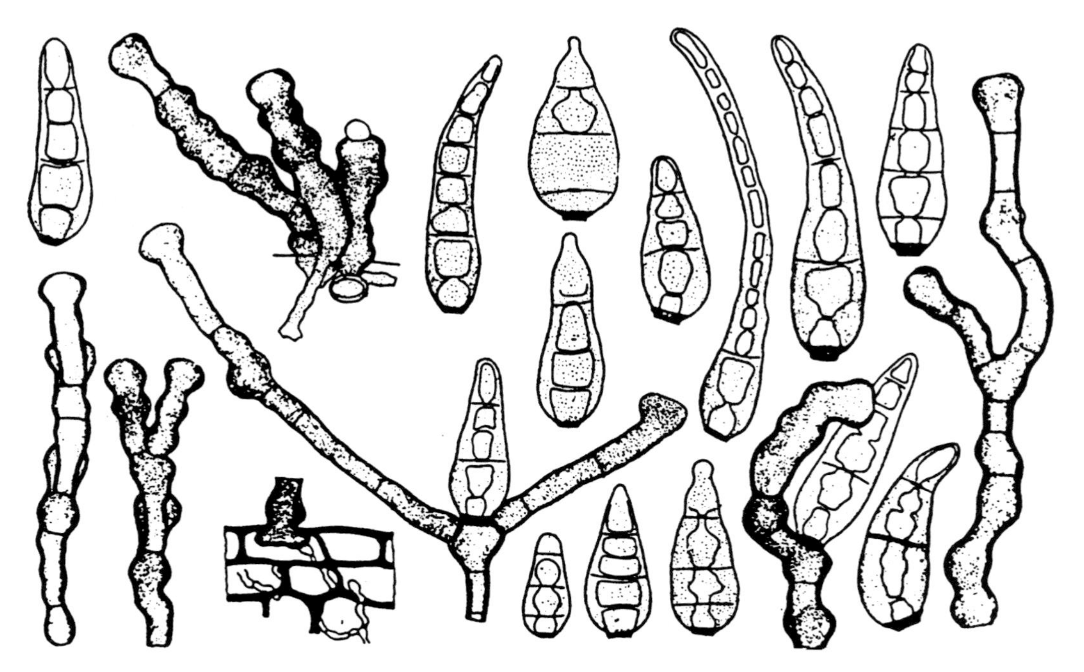

Fig. 2.12. Conidia and conidiophores of *Deightoniella torulosa* (from Ellis, 1957a). Conidia are 35–700 μm × 13–25 μm.

Disease cycle and epidemiology

Deightoniella torulosa is present in dead banana leaves and inoculum is produced during periods of rain or dew (Meredith, 1961a). Spores are violently discharged, when the humidity declines, and become airborne (Meredith, 1961b). It has been reported that spores of *D. torulosa* are present in the air throughout the year in Jamaica (Meredith, 1961c). However, they are not transmitted long distances and viability is lost within 4 days at humidities less than 95% (Meredith, 1961a, b). Spores land on plants and germinate in water. Studies with abacá show that germ tubes produce appressoria and leaves are directly infected through the epidermis. As the wall of the epidermis is pierced, the epidermal cell is immediately killed and the contents transformed into a reddish granular substance. The infection tube swells up and fills the infected cell.

Host reaction

Little is known on the reaction of different banana cultivars to the disease. The problem is mainly reported from cultivars of the Cavendish subgroup (AAA) growing under plantation conditions. The abacá cultivar 'Bongalonan' is more susceptible than 'Libuton', 'Maguindanao', 'Tangoñgon' and 'Putian'. Most enset clones have been reported as susceptible.

Control

No control measures are required on banana. However, the removal of dead and diseased leaves will reduce the amount of inoculum in the field.

It is recommended that only resistant abacá cultivars should be planted in areas where the disease causes problems. Other control measures include reducing planting densities, harvesting plants before rotting becomes severe and cutting down and burning seriously affected plants (Ela and San Juan, 1954).

Drechslera Leaf Spot

D.R. Jones, M. Tessera and A.J. Quimio

Drechslera leaf spot, also known as eyespot, is a minor disease of banana suckers less than 2 m tall and occurs during periods of wet weather and heavy dew. In Central America and Jamaica, it only affects banana where Bermuda grass (*Cynodon dactylon*) is present beneath the banana canopy (Stover, 1972). Meredith (1963a) first reported the disease in Jamaica. Drechslera leaf spot is common on enset seedlings and young transplants in the Sidamo and North Omo regions of Ethiopia (Tessera and Quimio, 1994). Older enset plants are not seriously affected.

On banana, the first symptoms are minute, slightly sunken, reddish spots with a pale green or faintly yellow border. The spots become oval or lens-shaped in the direction of the leaf veins and the centre turns dark brown. The centre of the spot later dries to a bleached white or grey colour, with a narrow, well-defined, dark brown margin and pale green or yellowish-green halo (Plate 2.24). Spots, which may be as large as 16 mm × 8 mm on leaves, also occur on the midrib and petiole (Stover, 1972). On enset, spots usually coalesce to form large blighted and dead areas (Plate 2.25). Severe blighting may be seen on the unfurling leaf and the first and second expanded leaves of succulent, rapidly growing plants.

The causal agent of Drechslera leaf spot on banana has been identified as *Drechslera gigantea*. It is a pathogen that causes a zonate eyespot symptom on grasses (Drechsler, 1928, 1929). Sporulation occurs on eyespot lesions of Bermuda grass when the grass is wet. Conidia are forcibly expelled from the conidiophore as the humidity decreases and can be picked up by a Hirst spore trap between 8.00 and 14.00 h (especially after rain). Moisture on the leaf surface favours germination and infection. When conidia were placed on the heart leaves of banana suckers, reddish brown spots appeared after 24 h at 26.7°C. After 5 days, these spots measured 4–5 mm, but no fructifications formed (Meredith, 1963a). The fungus causing leaf spot on enset has been shown to be a *Drechslera* species. It has been isolated and found to attack leaves of both enset and the banana cultivar 'Dwarf Cavendish' (AAA, Cavendish subgroup). Symptoms have also been observed on cultivars in the Lujugira-Mutika subgroup in Uganda. Most cultivars of enset are susceptible. Early removal of diseased leaves has been recommended to minimize spread.

Malayan Leaf Spot

D.R. Jones

Introduction

Malayan leaf spot, known also as diamond leaf spot, was first observed in Fiji by Knowles (1916) and was later described from leaves collected at altitude in the Cameron Highlands of West Malaysia (Ellis, 1957b). The disease has also been reported in Tonga and Western Samoa (Anon., 1990b) and is present in the highlands of Papua New Guinea, where some local cultivars are very susceptible (P. Kokoa, Papua New Guinea, 1988, personal communication; D.R. Jones, Papua New Guinea, 1989, personal observation). Malayan leaf spot has been reported as severe in Fiji during the cool season, particularly in the upper Waidina valley, where it can replace black leaf streak as the most significant problem (Firman, 1971).

Symptoms

In Fiji, symptoms on the upper surface of the leaf appear as diamond-shaped, greyish-white spots with dimensions of 2–4 mm × 3–5.5 mm, the longer axis being parallel to the leaf veins. These spots, which sometimes have brown centres, are surrounded by a black border about 0.5 mm wide. On the undersurface, the lesion can be covered with a dense, velvety, brown mass (Knowles, 1916).

From specimens collected in the highlands of West Malaysia, ellipsoid and

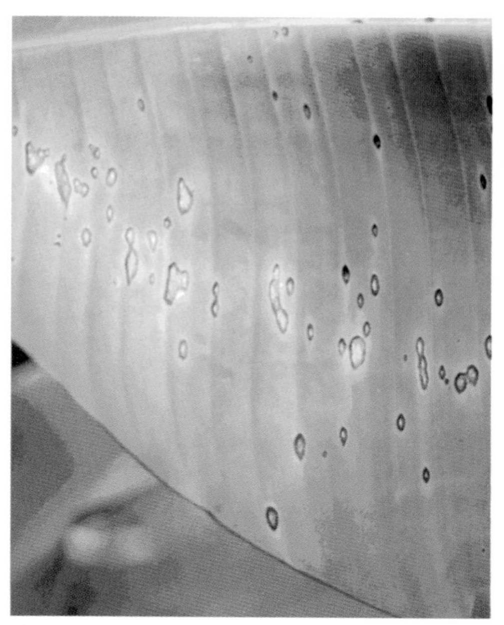

Plate 2.24. Symptoms of Drechslera leaf spot on an East African highland cultivar (AAA, Lujugira-Mutika subgroup) near Kampala, Uganda (photo: D.R. Jones, INIBAP).

Plate 2.25. Symptom of Dreschlera leaf spot on the first expanded leaf of enset (photo: M. Tessera and A.J. Quimio, IAR).

Plate 2.26. Symptoms of Malayan leaf spot on the leaf lamina and midrib of 'Mala' (AA) in the Southern Highland Province of Papua New Guinea (photo: D.R. Jones, QDPI).

round spots were described as well defined, white, grey or pale brown. The dimensions of the ellipsoid spots were recorded as 2–4 mm × 3–12 mm and those of the round spots as 2–5 mm in diameter. The spots were often very pale on the upper surface and darker on the lower, with dark purple borders (Ellis, 1957b).

In the highlands of Papua New Guinea, round and ellipsoid-shaped lesions of various sizes, with well-defined, dark brown borders and grey centres, are seen on the leaf lamina and midrib (Plate 2.26). Very few could be described as having a true diamond shape. When infection density is high, surrounding tissue yellows slightly and large necrotic areas develop. The dark brown borders and grey centres of lesions can still be distinguished in dead leaf tissue (Plate 2.26). First symptoms were seen on the second and third fully expanded leaf on 'Mala', a highly susceptible local AA cultivar.

Causal agent

The fungus *Haplobasidion musae* Ellis causes Malayan leaf spot (Ellis, 1957b; Ellis and Holliday, 1976). Microscopical examination of the velvety mass on the lesion on the undersurface of the leaf shows many conidiophores (Fig. 2.13). These arise singly or in groups of two to six at the ends of hyphae or as lateral branches. They emerge through the epidermal wall and cuticle and can be straight or flexuous. Condiophores are pale brown in colour, 0–3-septate, 50–110 μm × 4–6 μm in size, with an apical cell swollen at the end into a subglobose apical proliferation 9–12 μm in diameter. Spherical sporogenous cells, which are 4–8 μm in diameter and are at first hyaline and smooth and later brown and verrucose, are formed on the surface of each successive apical vesicle. Conidia are borne either singly or in simple or branched chains of two to five. Conidia are spherical, brown, verrucose and 4–6 μm in diameter (Ellis, 1957b).

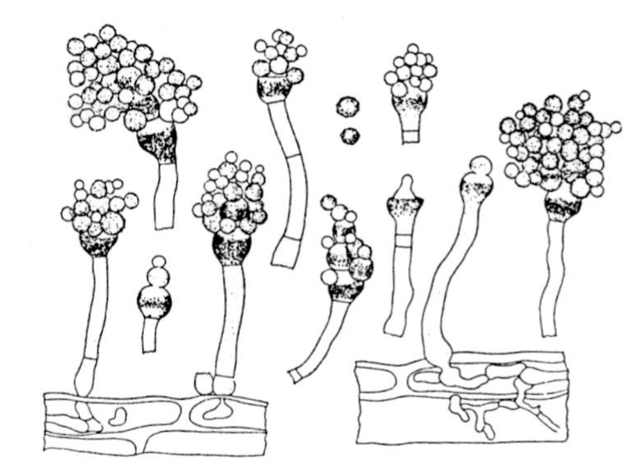

Fig. 2.13. Conidia and conidiophores of *Haplobasidion musae* (× 500) (from Ellis, 1957b).

Disease cycle and epidemiology

In Fiji, shade and cool temperatures (monthly means below 23.8°C) favour Malayan leaf spot. The disease is only serious in areas with low sunshine hours, long periods of high humidity and high rainfall (> 2500 mm year^{-1}). Leaves seem more susceptible when plants are near to flowering. Plants flowering in the cool season suffer most (Firman, 1971). In West Malaysia, Malayan leaf spot is found at between 1372 and 1525 m (Firman, 1971) and in Papua New Guinea at well over 1000 m. Firman (1971) noticed that, in Western Samoa, incidence increases with altitude. In Tonga, which has a cooler climate, Malayan leaf spot occurs on banana growing at sea level.

Not much is known about the disease cycle, but infection would seem to occur on young leaves soon after emergence (Ellis and Holliday, 1976).

Host reaction

No studies have been undertaken on cultivar reaction, but clones in the Cavendish subgroup (AAA) have been reported as susceptible. Some cultivars growing at altitude in Papua New Guinea seem to be extremely susceptible.

Control

Leaves sprayed with oil seem more prone to Malayan leaf spot in Fiji and it has been suggested that this is because oil inhibits photosynthesis and transpiration, similar to when plants are shaded. Maneb applied in water delayed the appearance of the disease and symptoms were never severe (Firman, 1971).

Pestalotiopsis Leaf Spot

D.R. Jones

This disease has been described as naturally occurring on 2–6-month-old seedlings of *M.*

acuminata ssp. *banksii*, *M. acuminata* ssp. *zebrina* and *M. balbisiana* in Honduras. *Pestalotiopsis palmarum* (Cke.) Steyaert (*Pestalotia palmarum* Cke.) was identified as the causal agent, which formed brownish-yellow, circular spots between leaf veins. Spots developed until they reached vascular tissue, which then appeared as linear extensions of the lesion. Dimensions of 6-week-old spots were 1–4 cm × 0.3–0.9 cm. The fungus penetrated the epidermis of the *Musa* seedlings directly by means of appressoria and infection pegs. Growth was intracellular. Inoculum was believed to have originated from sporulating spots on Manaca palm (*Orbignya cohune*) used as shade. Sporulation did not occur on *Musa* even after 8 months (Vakili, 1963).

Spotting attributed to *Pestalotiopsis leprogena* (Speg.) Kausar (*Pestalotia leprogena* Speg.) has occurred on wounded leaves of *M. balbisiana* and 'Bluggoe' (ABB) in Jamaica (Meredith, 1963b) and Central America (Stover, 1972). Grey or fawn spots develop around abrasions or tears, which are usually found at the leaf margin. Delicate concentric zonation and acervuli of the pathogen are present on the upper leaf surface. Acervuli, which are sparse on the lower surface, increase in size and number if the leaf section is incubated for 24 h. The centre of the spot is surrounded by a narrow, dark brown band and a conspicuous, bright orange or yellow halo. Crescentic areas of necrosis develop on one or both sides of the tear and decay may extend to the midrib. Conidia on *M. balbisiana* measure 17.5–24.5 μm × 6.0–7.5 μm and are four to five-celled, spindle-shaped and slightly constricted at the septa. The basal cell is hyaline, long and cone-shaped with a short hyaline pedicel. The apical cell is also hyaline and cone-shaped and bears three knobless and widely spread setae (Meredith, 1963b: Stover, 1972).

Vakili (1963) considered *P. palmarum* and *P. leprogena* to be synonymous.

Pyricularia Leaf Spot of Enset

M. Tessera and A.J. Quimio

This disease is severe on young enset suckers and can lead to premature leaf death. It is common in the Sidamo and North Omo enset-growing areas of Ethiopia (Tessera and Quimio, 1994). Circular, oblong and spindle-shaped lesions with dark borders are found on leaves, midribs (Plate 2.27), petioles and leaf sheaths (Plate 2.28). These can coalesce to cause large areas of necrosis. A *Pyricularia* sp. isolated from lesions was shown to be pathogenic on enset, but not on the 'Dwarf Cavendish' (AAA, Cavendish subgroup) banana cultivar (Quimio and Tessera, 1996). Most enset cultivars appear to be susceptible, but older plants are not affected. The removal of affected leaves is advocated to minimize the spread of the disease in the sucker production field.

SPECKLES, FRECKLE AND RUST

Cladosporium Speckle

D.R. Jones

Introduction

Cladosporium speckle is usually found on the older leaves of banana plants growing in humid environments. Although it has been regarded as a minor problem, it can be serious on certain cultivars in certain locations. Badly diseased leaves tend to dry out and fall prematurely (Frossard, 1963) and thus lower the photosynthetic ability of the plant. However, no serious attempts have been made to calculate yield losses. Cladosporium speckle has been reported in Australasia–Oceania (Papua New Guinea, Solomon Islands, Western Samoa), Asia (Bangladesh, Hong Kong, Indonesia, Malaysia, Nepal, Sri Lanka, Thailand, Vietnam), Africa (Burundi, Cameroon, Côte d'Ivoire, Democratic Republic of Congo, Egypt, Ethiopia, Ghana, Guinea, Rwanda, Sierra Leone, South

Africa, Sudan, Togo, Uganda, Zimbabwe) and the Latin American–Caribbean region (Cuba, Ecuador, Honduras, Jamaica) (Frossard, 1963; Anon, 1988; David 1988; Sebasigari and Stover, 1988; Jones, 1993, 1994b; D.R. Jones, Ecuador, 1999, personal observation).

The disease has been reported as significant on plant crops of 'Petite Naine', 'Grande Naine', 'Poyo' and 'Lacatan' (AAA, Cavendish subgroup) grown on commercial, export plantations in West Africa (Frossard, 1963). It is also a serious problem on 'Kluai Khai' (AA, syn. 'Sucrier') on smallholdings near Kamphaeng Phet in Thailand (Jones, 1994b) and on 'Pisang Berangan' (AAA, syn. 'Lakatan') in a plantation in Perak State, West Malaysia (Jones, 1993), where symptoms are present on young as well as old leaves.

A Cladosporium leaf spot disease has also been reported on enset in Ethiopia. It is said to be a minor problem of old plants in the Awasa Zuria, Sidamo and Welayita regions (Tessera and Quimio, 1994).

Symptoms

In West Africa, symptoms on banana first appear as pale brownish spots like pencil marks, which measure 0.3 mm × 1.5 mm. These are easily distinguished by transmitted light and can be seen 3–4 weeks after the youngest leaf unfurls. The spots elongate to streaks, which increase in size and eventually become lesions measuring 15 mm × 30 mm. These can coalesce (Plate 2.29) and yellow, later becoming a violet-black colour (Frossard, 1963).

Similar leaf symptoms have also been seen in Uganda, Malaysia and Thailand. Early symptoms are small spots, which elongate into grey and later brown streaks, which can be mistaken for the early stages of black leaf streak disease. If infection densities are high, expanding grey streaks coalesce, causing affected areas to turn orange and surrounding tissue yellow (Plate 2.30). Coalescing lesions later change colour to dark brown or violet-black before the leaf tissue is killed (Plate 2.31). Whole leaves can eventually turn necrotic. Lesions have also been observed on leaf midribs.

In Central America, symptoms have been described as consisting of a diffuse, greyish-brown blotching on the upper surface of the oldest leaves, which later become orange-yellow and then brown and necrotic (Stover, 1972).

On enset, the disease affects the middle and lower leaves (Plate 2.32). When spotting is severe, large areas of dead tissue can be seen along leaf margins.

Causal agent

The pathogen causing Cladosporium speckle of banana is the fungus *Cladosporium musae* Mason. It produces conspicuous erect, dark brown, conidiophores, which can be seen easily with a hand-lens. These are 6–8 μm wide and an average of 610 μm in length, with a thick-walled basal cell (Fig. 2.14). Conidiogenous cells, which are found in branches at the apex of the conidiophore, can be terminal or intercalary (Fig. 2.14). Conidia have dimensions of 6–22 μm × 3–5 μm. They are ovate cylindrical, ellipsoidal or fusiform in shape, tending to be restricted in the middle, with one or more protuberant scars at each end. Conidia are thin-walled, almost colourless and mainly aseptate or with one septum (Fig. 2.14).

Colonies growing in culture are first white and then olive. Thin-walled, colourless hyphae, 1–2 μm in diameter and slightly constricted at the septa, form the mycelium, which is superficial. Conidiophores may occur singly or in groups of six (Martyn, 1945; Stover, 1972; David, 1988).

The pathogen affecting enset has been identified as a *Cladosporium* species.

Disease cycle and epidemiology

Conidia of *C. musae*, which are carried by air currents, germinate in moisture on leaf tissue. The development of the disease is favoured by high humidity.

Host reaction

Cultivars in the Cavendish subgroup (AAA) and 'Gros Michel' (AAA) were

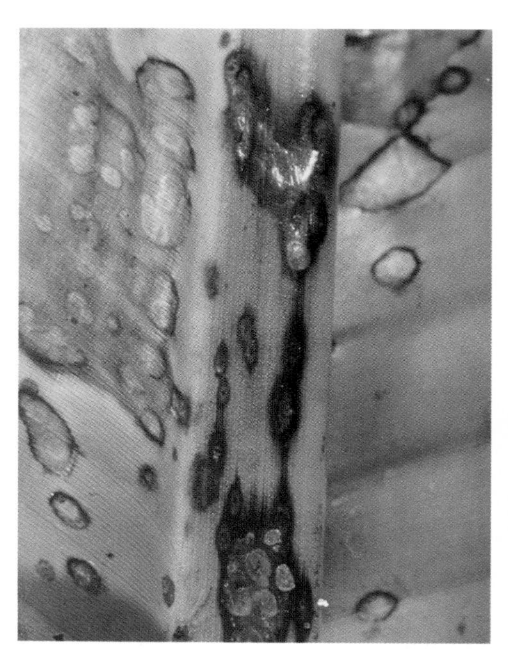

Plate 2.27. Symptoms of Pyricularia leaf spot on the leaf midrib and lamina of enset in Ethiopia (photo: M. Tessera and A.J. Quimio, IAR).

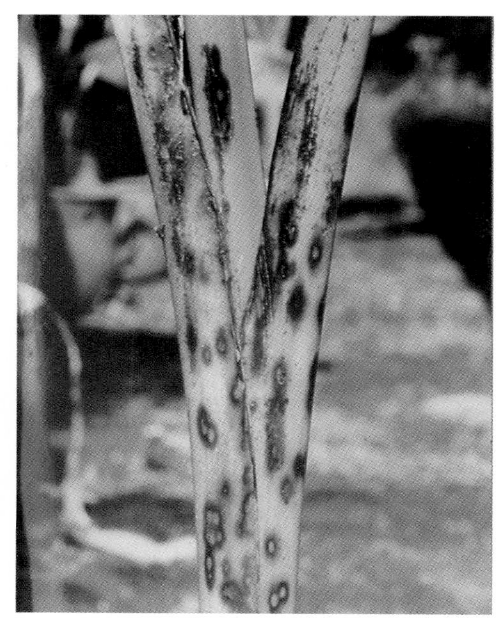

Plate 2.28. Symptoms of Pyricularia leaf spot on the petioles and leaf sheaths of enset in Ethiopia (photo: M. Tessera and A.J. Quimio, IAR).

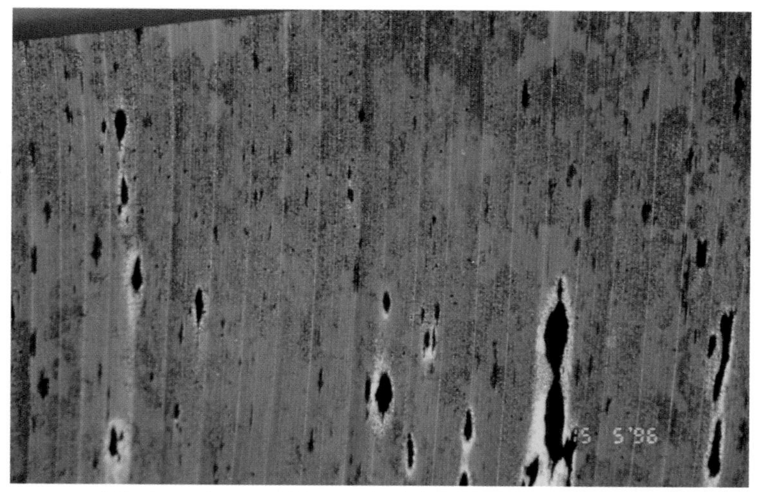

Plate 2.29 Symptoms of Cladosporium speckle on a leaf of 'Essong' (AAB, Plantain subgroup) in Cameroon. Dark brown lesions of black leaf streak in various stages of development, the more mature ones surrounded by yellow haloes, are also present (photo: C. Pasberg-Gauhl and F. Gauhl, IITA).

affected in Côte d'Ivoire, but cultivars in the AB, AAB and ABB genomic groups were not regarded as susceptible (Frossard, 1963). Nevertheless, symptoms have been seen on cultivars in the Plantain subgroup (AAB) in Cameroon (Plate 2.29). Frossard (1963) regarded 'Figue Sucrée' (AA, syn. 'Sucrier') as resistant in Côte d'Ivoire, but 'Kluai Khai' (AA, syn. 'Sucrier') is thought to be the most susceptible clone in

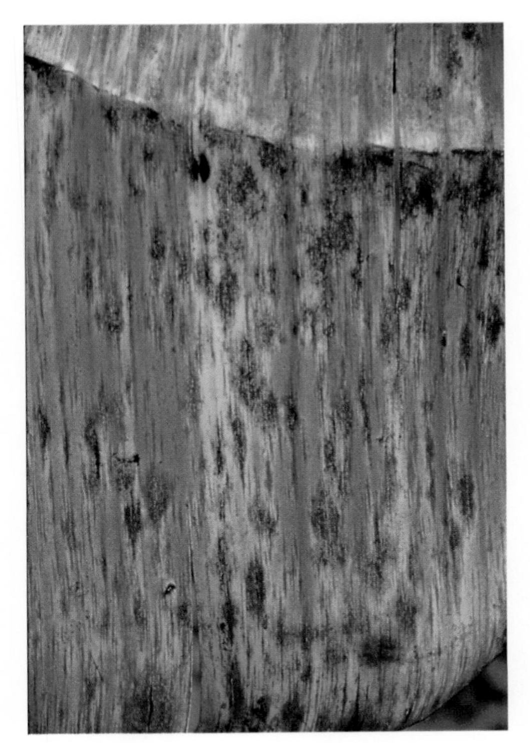

Plate 2.30. Symptoms of Cladosporium speckle on a leaf of an East African highland cultivar (AAA, Lujugira–Mutika subgroup) in the Bushenyi district of Uganda. Pencil-thin streaks coalesce and change from grey through orange to brown as the disease progresses and leaf tissue surrounding developing lesions turns yellow (photo: D.R. Jones, INIBAP).

Thailand. The disease on 'Kluai Khai' is particularly severe between August and November during the rainy season (Plate 2.31). Symptoms have also been seen on 'Orito' (AA, syn. 'Sucrier') in Ecuador.

In addition to 'Pisang Berangan' (AAA), 'Pisang Nangka' (AAA) has been seen with symptoms of Cladosporium speckle in Malaysia (Jones, 1993) and East African highland cultivars (AAA, Lujugira–Mutika subgroup) in Africa are also affected (Plate 2.30) (Sebasigari and Stover, 1988). In Côte d'Ivoire, no symptoms were seen on *M. balbisiana* or *M. textilis* (Frossard, 1963).

Most enset cultivars are susceptible to Cladosporium leaf spot.

Control

Fungicides used to control Sigatoka and black leaf streak are believed to control Cladosporium speckle on banana. In Thailand, the disease is controlled by applying benomyl every 2 weeks during the rainy season at a rate of 0.5 g l^{-1}. The early removal of diseased enset leaves is recommended in Ethiopia.

Acknowledgement

The information provided by M. Tessera and A.J. Quimio on *Cladosporium* affecting enset is gratefully acknowledged.

Plate 2.31. Symptoms of Cladosporium speckle on a leaf of 'Kluai Khai' (AA, syn. 'Sucrier') on a smallholding near Kamphaeng Phet, Thailand. High infection densities cause diseased leaf tissue to eventually turn necrotic (photo: D.R. Jones, INIBAP).

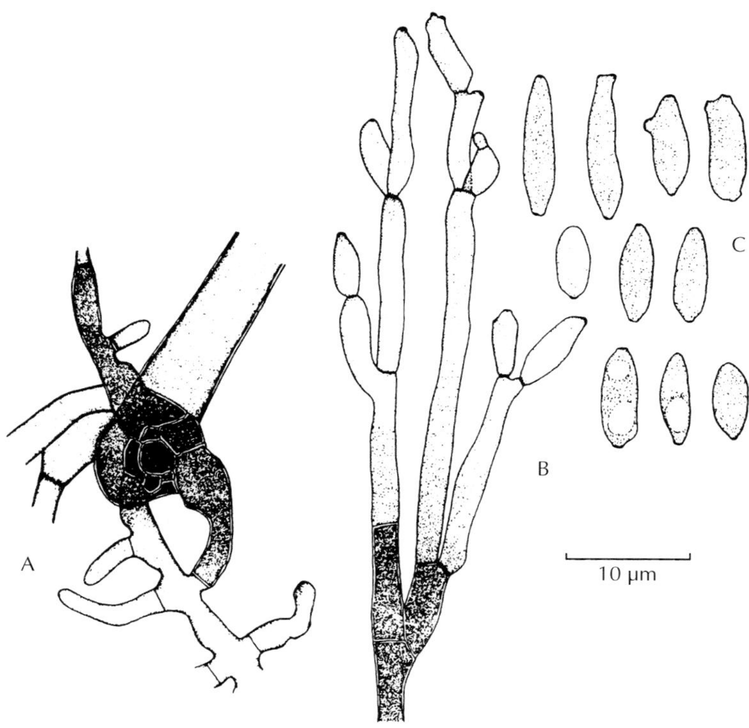

Fig. 2.14. Base of a conidiophore (a), a conidiophore with conidia (b) and conidia (c) of *Cladosporium musae* (from David, 1988).

Leaf Speckle

D.R. Jones

Introduction

Leaf speckle is a minor disease of banana in South-East Asia, but it has been described as a major problem on commercially grown cultivars in the Cavendish subgroup (AAA) in Taiwan since 1981 (Hwang and Chen, 1986; Hwang, 1991). Symptoms of leaf speckle have also been recognized in Australia, Malaysia, Thailand and Vietnam.

Symptoms

Lesions first appear as small brown to dark brown specks, densely distributed on the lower leaf surface. These develop into fine, dark brown streaks up to 4 mm long and 0.3 mm wide, running parallel to veins (Plate 2.33). On the upper surface of old leaves where there are concentrations of streaks, the disease appears as tan-coloured blotches (Plate 2.34). Affected areas yellow and eventually become necrotic.

Causal agent

The fungal pathogen has been identified as *Acrodontium simplex* (Mangenor, de Hoog) by the Central Bureau voor Shimmecultures, Baarn, the Netherlands (Hwang and Chen, 1986). However, *A. simplex*, as *Beauveria simplex* Mangenot, was first described from a decaying tree trunk in France (de Hoog, 1972) and seems an unlikely pathogen of banana.

The fungus produces erect, brownish-coloured, septate conidiophores, which are single or sparingly branched and approximately 41–90 µm in length. Conidia are ovoid, 2.9–3.8 µm × 1.9 µm and cover the entire terminal part of the conidiophore (Plate 2.35) (Hwang and Chen, 1986).

The optimum temperature for growth

in culture is 25°C. Dark grey colonies attain a diameter of only 7 mm after 1 week on potato dextrose agar. Conidia produced in culture have been used to inoculate plants and typical symptoms obtained (Hwang and Chen, 1986).

Symptoms identical to leaf speckle in Asia have been found in North Queensland, Australia (Plate 2.36). The fungal pathogen responsible has been identified as *Periconiella musae* Stahel ex Ellis (syn. *Ramichloridium musae* Stahel) (Ellis, 1967). Samples of leaf speckle sent to Australia from Vietnam have also been tentatively identified as caused by *P. musae* (J. Alcorn, Australia, 1995, personal communication).

If leaf speckle is caused by *R. musae*, then this pathogen has been implicated as the causal agent of three different types of symptom on banana (see Jones, this chapter, pp. 114–116). Further studies on the taxonomy of the fungus or fungi involved in the tropical leaf speckle seem warranted.

Disease cycle and epidemiology

In the south of Taiwan, disease development is greatly affected by climatic conditions. The incubation period is about 35 days in the hot, rainy summer and about 60 days in the cool, dry winter (Hwang, 1991). Leaf speckle builds up to epidemic levels on unsprayed Cavendish cultivars after shooting.

Host reaction

Cultivars of the Cavendish subgroup (AAA) are affected in Taiwan and Vietnam. *Musa balbisiana* and cultivars in the AAB genomic group are also reported susceptible in Taiwan (Hwang, 1991). Symptoms have also been seen on other cultivars in Australia and South-East Asia, including 'Chuoi Bom' (AAA, syn. 'Lakatan'), 'Kluai Namwa' (ABB, syn. 'Pisang Awak'), 'Kluai Namwa Khom' (ABB, dwarf 'Pisang Awak'), 'Pisang Abu Keling' (ABB, Bluggoe subgroup), 'Pisang Abu Siam' (ABB, syn. 'Kluai Teparot') and 'Pisang Abu Nipah' (ABB, syn. 'Saba') (D.R. Jones, 1993–1994, personal observations).

Control

Only in Taiwan has this disease been a significant problem. After the initiation of a forecasting system to control black leaf streak, banana plants were not sprayed with fungicide after bunch emergence. Under these conditions, leaf speckle built up to epiphytotic levels during the period of fruit maturation (Hwang and Chen, 1986). Fungicidal sprays containing dithiocarbonates in oil provide effective control (Hwang, 1991).

Mycosphaerella Speckle

D.R. Jones

Introduction

Mycosphaerella speckle is a minor disease of banana leaves. It is found worldwide (Pont, 1960b; Stover, 1969), but has been reported as a problem only in subtropical areas of Australia, where it is a common cause of leaf death. The disease has also been recorded on abacá (Anunciado *et al.*, 1977).

Symptoms

The first symptoms, which are rarely seen above the fifth fully opened leaf, are water-soaked patches, which exude moisture droplets. These become visible during rain or early in the morning when dew is present. In the absence of moisture, affected areas are first noticeable on the lower leaf surface as light brown or tan-coloured, irregular blotches. These may show on the upper surface as smoky patches. Blotches on the lower surface darken in colour and eventually become dark purple to black, irregularly shaped, speckled areas, which are visible on both leaf surfaces. At this stage of disease development, leaf tissue in and around the speckled areas is yellow (Plates 2.37 and 2.38). Later, necrotic tissue within the speckled areas coalesces and bleaches with age to become grey on the lower surface and straw-coloured on the upper as it dries out.

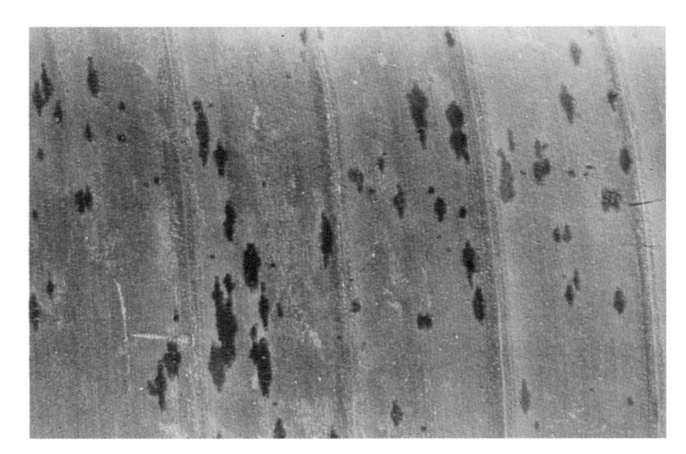

Plate 2.32 Symptoms of Cladosporium leaf spot on an enset leaf in Ethiopia (photo: M. Tessera and A.J. Quimio, IAR).

Plate 2.33 Dark brown streak symptoms of leaf speckle on 'Chuoi Bom' (AAA, syn. 'Lakatan') in Vinh Phu Province, Vietnam. Lesions are also noticeable on the midrib (photo: D.R. Jones, INIBAP).

The extensive death of leaf tissue is not often seen above the eighth leaf of an actively growing plant. However, when leaf production ceases with bunch emergence, extensive defoliation can occur before the fruit is ready for harvest (Pont, 1960b).

Causal agent

Speckle is caused by *Mycosphaerella musae* (Spreg.) Syd. The perithecia, which measure 45–99 μm × 34–81 μm, are highly scattered, globose, immersed in leaf tissue and black in colour. Asci, which have dimensions of 24–44 μm × 8–12 μm are eight-spored and obclavate in shape. Ascospores measure 9–16 μm × 2–3 μm and are hyaline, obtuse to cylindrical and two-celled, with one cell broader than the other. No conidial stage has been observed *in vivo* on banana leaves, but conidia of the *Cercospora* type have been produced by ascospore cultures on plain agar after 4–5 days (Stover, 1994). Conidia have dimensions of 55–200 μm × 2.6–3.2 μm (average 127 μm × 2.9 μm) and are usually verrucose, with a basal scar. Conidiophores

Plate 2.34. Tan-coloured blotch symptoms of leaf speckle on 'Pisang Awak' (ABB) in Terengganu State, West Malaysia (photo: D.R. Jones, INIBAP).

Plate 2.35. Conidiophores and conidia of the leaf speckle pathogen in Taiwan (photo: TBRI).

are 24–46 μm long (average 35 μm), with one septum. Although about one culture in 20 produces a few spermogonia with spermatia on plain agar, perithecia and ascospores have not been found (Stover, 1994). However, perithecia of *M. musae* may have been produced on V8 juice agar (Van der Berg-Lorida, 1989, reported by Stover, 1994).

On potato dextrose agar, colonies consist of a thin layer of compact, light grey aerial mycelium on a hard, black, sclerotium-like hump. The outline of colonies becomes more irregular with age and droplets of dark brown or black fluid may exude from parts of its elevated surface. The surface mycelium later turns pink. No fruiting bodies have been detected on potato dextrose agar. The optimum temperature for growth of mycelium is 26°C, but the fungus grows well between 20° and 30°C (Pont, 1960b).

Disease cycle and epidemiology

Perithecia develop in necrotic leaf tissue as it dries out. They are numerous on the lower leaf surface, but sparse on the upper surface. Ascospores discharge from perithecia when leaves become wet. Germ tubes grow out of the ends of each cell of the ascospores, often simultaneously. The germ tubes give rise to an extensive, branched, epiphytic mycelium. After about 5–6 weeks, lobed appressoria are formed on or near stomatal pores on the undersides of leaves and guard cells turn light brown. After penetration of the stomata, coarse intercellular hyphae up to 3 μm thick, with finger-like lateral branches, grow through the leaf tissue, killing cells of the spongy parenchyma. The success in control when protectant fungicides are sprayed on the lower surfaces of leaves 4–6, indicates that infection does not take place on young leaves.

High humidity favours disease development. Symptoms can appear within 45 days if inoculated plants are held in a saturated atmosphere during the night. Without saturation, the incubation period is 80–102 days. Speckle is always more prevalent in sheltered locations, hollows or wet spots in a plantation.

Disease development is more rapid in senescing or injured leaves. Advanced stages of the disease are usually seen only on older leaves. Temperatures below 20°C retard the development of disease, but cooler weather also slows plant growth. As a consequence, speckle is more obvious in winter in subtropical areas, because leaf production does not keep pace with symptom expression, as it does in summer (Pont, 1960b).

Stover (1994) has shown that, in tropical America, *M. musae* is an endophyte on early streak lesions of both Sigatoka and black leaf streak.

Host reaction

In Australia, cultivars 'Mons Mari' and 'Williams' (AAA, Cavendish subgroup), 'Lady Finger' (AAB, Pome subgroup), 'Sugar' (AAB, syn. 'Silk'), 'Ducasse' (ABB, syn. 'Pisang Awak') and 'T8' ('61–882–1', an AAAA tetraploid from the Jamaican breeding programme) are susceptible. The disease is also found on the wild banana *M. acuminata* ssp. *banksii* (Pont, 1960b).

Control

In the past, copper oxychloride and the dithiocarbomate fungicide zineb gave good control when applied to the underside of leaves (Stover, 1972). Today, fungicides that are applied to control Sigatoka disease in Australia (mancozeb, propiconazole) give adequate control of Mycosphaerella speckle. Removing leaves killed by Mycosphaerella speckle can reduce inoculum levels.

Tropical Speckle

D.R. Jones

Introduction

Tropical speckle is a collective name given here to two types of symptom which have been observed on banana leaves in hot, humid tropical environments around the world (Stahel, 1937b; Martyn, 1945; Pont, 1960b; Stover, 1972). Some authors contend that the disease is caused by two different fungal pathogens and this is supported by taxonomic studies undertaken by Ellis (1967, 1976). However, de Hoog (1977) is of the opinion, after examining specimens of both types from many locations, that only one fungus is involved.

The disease is not regarded as serious and no effect on yield or growth has been documented. Tropical speckle has been described in many countries in the Asian–Pacific, African and Latin American–Caribbean regions and probably has a worldwide distribution.

Symptoms

In northern Australia, tropical speckle has been reported on young foliage, even as high on the plant as the third fully expanded leaf, where symptoms of Sigatoka or Mycosphaerella speckle are not yet evident. However, no extensive necrosis or breakdown of leaf tissue, which could lead to premature defoliation, has been associated with the disease.

Pont (1960b) has described two types of leaf symptom. The first appears as roughly circular, chlorotic blotches up to 4 cm in diameter on the upper surface of the leaf and as tan-coloured blotches underneath. Thickly distributed, dark brown or black pinpoint-sized specks are clearly visible on the upper leaf surface. Densely packed, bristle-like conidiophores of the causal fungus can be seen on the lower surface, even with the naked eye, if the leaf is wrapped around a finger and held against the light. Blotches can merge to form extensive tan-coloured areas on leaves. Plate 2.39 illustrates this type of symptom.

The second type of symptom is obvious as irregular circular, dark grey to black patches on the lower leaf surface (Plate 2.40). With a hand-lens, these appear as densely aggregated, minute black specks. Individual speckled blotches are smaller than those characteristic of the first symptom, but often merge into extensive discoloured areas. These blotches are less distinct on the upper surface. The conidiophores are clearly visible on the lower surface as a dense, almost velvety coating. Symptoms have also been seen on midribs (Plate 2.40) and peduncles (Plate 2.41). Sometimes, in commercial plantations of 'Umalag' (AAA, Cavendish subgroup) in the Philippines, the ends of fingers are affected, which can render fruit unacceptable for export.

Both types of symptom can occur together on the same leaf in northern Australia (Pont, 1960b).

Causal agents

Two species of fungus, *Veronaea musae* Stahel ex Ellis (syn. *Chloridium musae* Stahel) (Ellis, 1976) and *Periconiella musae* Stahel ex Ellis (syn. *Ramichloridium musae* Stahel) (Ellis, 1967, 1971b) have been distinguished mainly on conidiophore morphology. *Veronaea musae* (as *C. musae*), with unbranched conidiophores (Fig. 2.15), was associated with the first type of symptom described and *P. musae* (as *R. musae*) with branched conidiophores (Fig. 2.16), was associated with the second type of symptom (Stahel, 1937a; Pont, 1960b; Stover, 1972). However, de Hoog (1977) claimed that many specimens of *V. musae* have some branched conidiophores and regarded the two fungi as conspecific. If this analysis is correct, then the name *Ramichloridium musae* (Stahel ex Ellis) de Hoog (syns. *Chloridium musae* Stahel, *Veronaea musae* Stahel ex Ellis, *Periconiella musae* Stahel ex Ellis) should be applied to both.

Pont (1960b) reported that conidiophores of *V. musae* (as C. *musae*) were shorter than those of *P. musae* (as *R. musae*) and conidia of the former were oval and the latter elliptical. Conidia of *V. musae* (as *C. musae*) were also observed to have minute papillae at the point of attachment to the conidiophore (Stahel, 1937a; Pont, 1960b).

More taxonomic work seems necessary to resolve the issue of whether one or two fungi cause tropical speckle symptoms. *Ramichloridium musae* (as *P. musae*) has also been implicated in leaf speckle disease (see Jones, this chapter, pp. 110–111). The description of *R. musae* below is that of de Hoog (1977).

The conidia (5.5 μm × 8.5–2 μm × 2.6 μm) are hyaline to subhyaline, thin-walled, ellipsoidal and have inconspicuous basal scars. Conidiogenous cells arise terminally from conidiophores. They are pale brown in colour, cylindrical and variable in length. Conidiophores arise vertically from epiphyllic mycelium on the lower leaf surface and can be up to 500 μm high. They are composed of four to ten intercalary cells and in the apical region are often branched just below the septa. Stalks are 1.8–2.5 μm wide, thick-walled and a golden brown in colour.

On oatmeal agar, colonies grow to a diameter of 21–26 mm in 14 days. At this stage, the culture consists of compact, flat, submerged mycelium, with some central aerial hyphae. The colour of the colony is pale greyish green when viewed from above and pale orange underneath. Conidiophores are usually unbranched, up to 250 μm in height and contain two to six intercalary cells. On potato dextrose agar, colonies are dark grey with flat or slightly elevated centres. The optimum temperature for growth is 26–28°C (de Hoog, 1977).

Pont (1960b) reported that *P. musae* (as *R. musae*) grew slower than *V. musae* (as *C. musae*) on potato dextrose agar, but had similar colony characteristics. The optimum growth temperature for the former was recorded as 27–28°C and for the latter 26°C.

Disease cycle and epidemiology

Conidia germinate on lower leaf surfaces after 24 h and produce fine hyphae.

Plate 2.36. Symptoms of leaf speckle on 'Kluai Namwa Khom' (ABB, a dwarf 'Pisang Awak') in north Queensland, Australia (photo: D.R. Jones, INIBAP).

Plate 2.37. Symptoms of Mycosphaerella speckle on the upper leaf surface of a cultivar in the Cavendish subgroup (AAA) in Vinh Phu Province, Vietnam (photo: D.R. Jones, INIBAP).

Eventually a loose network of epiphytic hyphae forms on the leaf surface. The hyphae, which are 1–2 μm thick, have side-branches that develop over about every sixth stoma. These side-branches, called stomatopodia, are club-shaped and a darker brown than the other hyphae. An infection tube from the stomatopodia enters the leaf through the stoma. Infection hyphae grow across the air space beneath the stoma to the palisade tissue and side-branches enter cells, which then die. However, the fungus spreads no further and necrosis is confined to the tissue adjacent to invaded stoma. This gives the lesion its speckled appearance. Conidiophores arise from epiphytic mycelium. The base of the conidiophore is attached to the cuticle.

The causal fungi are weak parasites and in Australia symptoms are only seen in unsprayed plantations located in rainforest clearings in high-rainfall areas. In Central America, the disease is only noticeable on lower leaves during the high-rainfall months. In Papua New Guinea, it is present on banana plants grown in shady, moist locations.

Plate 2.38. Close-up of symptoms of Mycosphaerella speckle on the upper leaf surface of a cultivar in the Cavendish subgroup (AAA) in southern Queensland, Australia (photo: QDPI).

Plate 2.39. Tan blotch symptoms of tropical speckle on the underside of a leaf of 'Embul' (AAB, syn. 'Mysore') in Sri Lanka (photo: D.R. Jones, INIBAP).

Host reaction

Not much is known about cultivar response. Tropical speckle has been recorded on cultivars in the AA, AAA, AAB, ABB genomes in South-East Asia, including members of the Cavendish subgroup (D.R. Jones, 1993, personal observation). The disease has also been seen on *M. schizocarpa* and *M. acuminata* ssp. *banksii* in Papua New Guinea (D.R. Jones, Papua New Guinea, 1989, personal observation) and on *M. acuminata* ssp. *banksii* in Queensland, Australia (Pont, 1960b).

Control

The disease is of little economic significance, except when it occasionally affects fruit, and control is not usually warranted. Fungicide sprays used to control leaf spot diseases of banana reduce the incidence of tropical speckle.

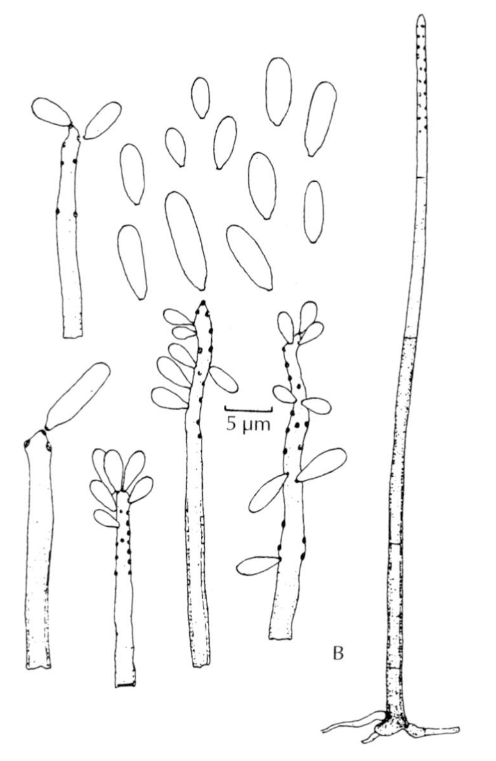

Fig. 2.15. Conidia and conidiophores of *Veronaea musae* (× 650 except where indicated by the scale) (from Ellis, 1976).

Freckle

D.R. Jones

Introduction

Freckle is a leaf- and fruit-spotting disease of banana common in South and East Asia (Bangladesh, Bhutan, Brunei, China, Hong Kong, India, Indonesia, Malaysia, Nepal, Pakistan, Philippines, Sri Lanka, Taiwan, Thailand, Vietnam) and Australasia–Oceania (American Samoa, Australia, Cook Islands, Fiji, Hawaii, New Caledonia, New Zealand, Nuie, Papua New Guinea, Solomon Islands, Tonga, Western Samoa) (Dingley *et al.*, 1981; Zhou and Xie, 1990; Anon, 1990a, 1994a; Jones, 1994b). Records from Africa (Congo, Zambia) and the Caribbean (Dominican Republic, Jamaica and St Lucia) need confirmation because of confusion in the taxonomy of the fungus causing the disease and because typical symptoms have not been seen in these regions.

Freckle is regarded as more serious than black leaf streak on 'Pei-Chiao' (AAA, Cavendish subgroup) in Taiwan (Tsai *et al.*, 1993) and it is emerging as a significant disease on plantations of 'Grand Nain' (AAA, Cavendish subgroup) in the Philippines, where fruit can be severely blemished thus affecting its marketability. Together with Cladosporium speckle, freckle also causes a serious disease problem on leaves of 'Pisang Berangan' (AAA, 'Lakatan') grown on a plantation scale in West Malaysia. On banana grown in smallholdings, freckle is not usually considered a major problem, as blemished fruit is acceptable at local markets. However, heavy infections do lead to the premature death of older leaves of some cultivars. Freckle has also been recorded on leaves and fruit of abacá in the Philippines, although it is not regarded as a serious disease (Anunciado *et al.*, 1977).

Symptoms

Two types of leaf spotting have been described, which occur mainly on the upper surfaces of older banana leaves. One consists of very small, dark brown to black spots less than 1 mm in diameter. These give the leaf a sooty appearance and rough texture. Sometimes, spots may cluster in lines and appear as streaks running diagonally or horizontally across the leaf (Plate 2.42) or from the midrib to the leaf edge along veins (Plate 2.43). Large, individual, dark brown to black spots up to 4 mm in diameter characterize the other type of spotting (Plate 2.44). These spots may have a grey or fawn centre and can aggregate to form large blackened areas or streaks. Because the pycnidia are prominent under a raised epidermis, diseased leaves and fruit feel rough to the touch. Severely affected leaves yellow, wither and die prematurely in both cases (Meredith, 1968). Spots also form on petioles, midribs (Plate 2.44) and transition leaves. The longevity of freckle-affected leaves has been esti-

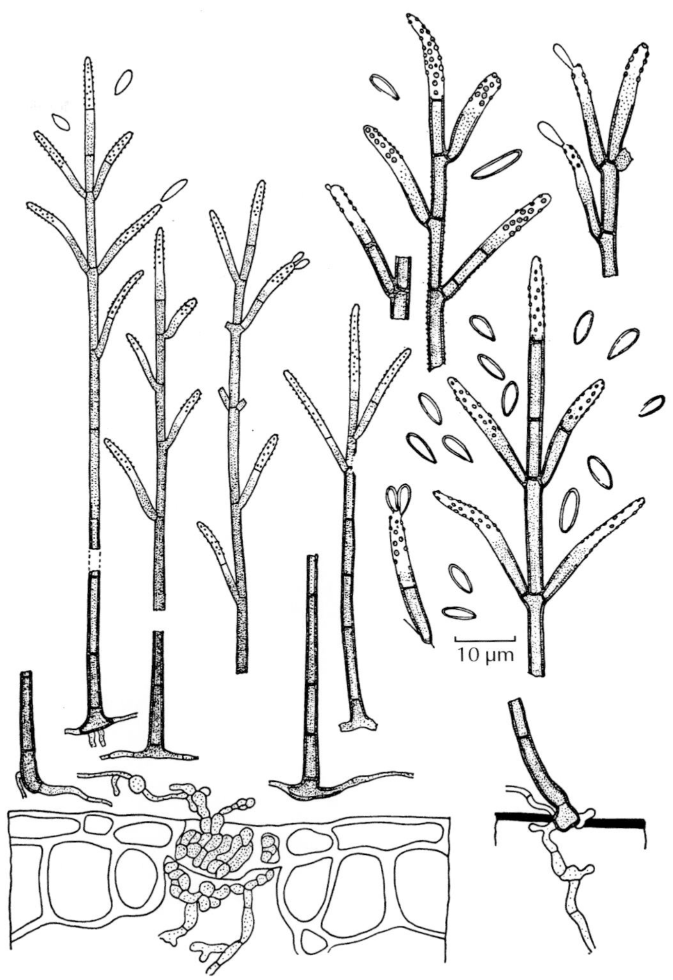

Fig. 2.16. Conidia and conidiophores of *Periconiella musae* (× 650 except where indicated by the scale) (from Ellis, 1967).

mated to be half that of healthy leaves in Taiwan (Chuang, 1984). For a description of symptoms on fruit, see Jones and Stover, Chapter 4, p. 185).

Causal agent

The fungus *Guignardia musae* Racib (anamorph *Phyllosticta musarum* [Cooke] van der Aa, synonyms *Phyllostictina musarum* [Cooke, Petr.] and *Macrophoma musae* [Cooke] Berl. & Volg) causes freckle. Usually, the pycnidial stage is present on the host. Herbarium specimens from the Philippines, India and Indonesia have been described by van der Aa (1973). Pycnidia were found to be 60–170 μm in diameter (but mostly 135 μm), globose and brown to black and could occur singly or in groups. Meredith (1968) reported how small spots in the first type of spotting described in 'Symptoms' usually contain fewer than five pycnidia, whereas the larger spots in the second type of spotting could contain up to 70 pycnidia. Conidia are one-celled, obovoidal, ellipsoidal or short cylindrical, with a truncate base, broadly rounded apically and conspicuously indented. Sometimes a distinctive apical appendage, up to 8–16 μm long, is present. The

Plate 2.40. Grey to black patch symptoms of tropical speckle on the underside of a leaf lamina and midrib of 'Umalag' (AAA, Cavendish subgroup) in a commercial plantation on Mindanao Island in the Philippines (photo: D.R. Jones, INIBAP).

Plate 2.41. Symptoms of tropical speckle on the peduncle of 'Umalag' (AAA, Cavendish subgroup) in a commercial plantation on Mindanao Island in the Philippines (photo: D.R. Jones, INIBAP).

dimensions of conidia are usually 15–18 μm × 9–10 μm, but can be 10–20 μm × 7–13 μm. Conidia are surrounded by a 1–3-μm-thick gelatinous envelope and can germinate within 12 h to form lobed appressoria 15 μm in diameter (Stover, 1972). Conidiogenous cells are cylindrical or conical, with dimensions of 4–11 μm × 2.5–5 μm.

Perithecia are globose or somewhat depressed, 70–220 μm in diameter and distinctly papillate. Their walls are 1–2 μm thick, composed of dark brown cells. Asci have eight ascospores, are clavate or cylindrical, usually with a short stalk, and measure 35–85 μm × 20–25 μm. Ascospores are single-celled, ovoid or oblong ovoid and 17–22 μm × 8–10 μm in size. Spermatia have similar dimensions to pycnidia with a 20-μm-wide pore. Spermatia cells are aseptate, cylindrical or dumb-bell-shaped, 6–10 μm long and 0.5–2 μm wide.

A description of *G. musae* in culture (Punithalingam, 1974; Punithalingam and Holliday, 1975) is believed by Chuang (1981) to be a saprophyte associated with diseased tissue.

In culture on V8 juice agar, the fungus produces minute, blackish-coloured colonies, which grow slowly, attaining a diameter of only 5 mm in 2 months (Chuang, 1981).

Plate 2.42. Symptoms of freckle on a leaf of 'Pisang Rastali' (AAB, syn. 'Silk') in Kelantan State, West Malaysia. Aggregates of spots in horizontal lines indicate that water drops laden with spores may have run down the leaf as it was unfurling in an upright position (photo: D.R. Jones, INIBAP).

Plate 2.43. Symptoms of freckle on a leaf of a cultivar in the Cavendish subgroup (AAA) in Vinh Phu Province, Vietnam. Aggregates of spots are often associated with leaf veins, as illustrated in this picture (photo: D.R. Jones, INIBAP).

Disease cycle and epidemiology

Ascospores are discharged from perithecia and conidia are exuded from pycnidia *en masse* as white, gelatinous tendrils during wet weather or heavy dews. The importance of ascospores in the epidemiology of the disease is unclear, but conidia play a significant role.

Conidia adhere to each other because of the mucilaginous sheath that surrounds each conidium. However, water readily separates them and dissemination occurs when drops run across the leaf or fruit. On immature fruit of 'Dwarf Cavendish' (AAA, Cavendish subgroup), germination can begin as early as 2–3 h at 24°C and, after 12 h, a lateral swelling proliferates into an irregular, hyaline appressorium. After 18–30 h, the appressorium is distinct, being light or dark grey with a thickened wall and a septum where it joins the conidium. Most appressoria are formed in depressions between epidermal cells. Penetration is thought to occur after 24–72 h at 24°C, when single epidermal cells become reddish brown. After 96 h, more than 60% of appressoria are associated with discoloured host cells. This response is more rapid and extensive in areas where appressorium density is high. Fine penetration hyphae enter the cell from the underside of the appressoria. These swell to a diameter of 3–5 mm on entering the cell lumen. The surrounding tissue is subsequently invaded by intercellular and intracellular hyphae. Lesions are superficial and tissue below five cell layers is rarely invaded. Pycnidia can develop as early as 3 weeks after inoculation (Meredith, 1968).

In Taiwan, the incubation period varies between about 60 days in the dry, cool season to 20 days in the warm, wet season. The susceptibility of leaves increases as they age (Chuang, 1984).

Pycnidia develop in lesions of all sizes, some as small as five dead cells. Streaks of pycnidia are found where water drops loaded with conidia have run across the leaf surface. Fruit becomes diseased when water runs off infected leaves on to bunches. Often, infection occurs near the pycnidium, where the conidia are formed, which leads to an increase in intensity of the disease. On leaves, continuous day-to-day infection results in an overlap of lesions, which results in extensive areas of dead tissue (Meredith, 1968).

Host reaction

In Hawaii, clones in the Cavendish subgroup (AAA) and Plantain subgroup (AAB) are susceptible, as are 'Pome' (AAB), 'Silk' (AAB), 'Sucrier' (AA), 'Gros Michel' (AAA), 'Red' and 'Green Red' (AAA), 'Bluggoe' (ABB), 'Monthan' (ABB) and 'Blue Java' (ABB, syn. 'Ney Mannan'). Cultivars in the AAB Maia Maoli–Popoulu subgroup are resistant (Meredith, 1968). The situation in Taiwan is similar (Hwang *et al.*, 1984; Tsai *et al.*, 1993), but the situation is the opposite in Australasia and the South Pacific, where cultivars in the Cavendish subgroup are resistant and 'Bluggoe' is susceptible.

In the South Asia and South-East Asia region, freckle has been found affecting both clones in the Cavendish subgroup (AAA) and 'Bluggoe' (ABB). 'Pome' (AAB), 'Silk' (AAB), 'Lakatan' (AAA), 'French' plantain (AAB), 'Horn' plantain (AAB), 'Mysore' (AAB), 'Saba' (ABB), 'Pisang Awak' (ABB), 'Pelipita' (ABB) and 'Kluai Teparot' (ABB) are also susceptible. Less seriously attacked are 'Sucrier' (AA), 'Gros Michel' (AAA), 'Red' and 'Green Red' (AAA), 'Pisang Nangka' (AAA) and 'Pisang Raja' (AAB). Freckle symptoms have not yet been recorded on the AA cultivars 'Pisang Lilin', 'Inarnibal' and 'Pisang Jari Buaya' (Jones and Daniells, 1988; Jones, 1993, 1994b).

More than one type of banana pathogen recorded as *G. musae* may exist. One, which attacks clones in the Cavendish subgroup, but not 'Bluggoe', may have spread from South and South-East Asia to Taiwan and the North Pacific, while the other, which attacks 'Bluggoe', but not Cavendish clones, may have spread to Australasia and the South Pacific. Research needs to be undertaken to clarify whether the two types, which have Cavendish and Bluggoe as differential hosts, can be distinguished taxonomically on morphological or molecular grounds. This would reveal if more than one species is involved or if there are two different races of the same pathogen. Of particular interest would be whether the large-spot symptom (Plate 2.44) is connected to taxonomic differences. The disease found in the African and Caribbean regions also needs careful study to determine if the pathogens involved fit the description of van der Aa (1973).

In Papua New Guinea, freckle is present on *M. acuminata* ssp. *banksii* and *M. schizocarpa*. Symptoms are especially severe on hybrids between the two wild species. No symptoms have been seen on *M. balbisiana* in Papua New Guinea, Taiwan, Hawaii and South-East Asia.

Control

Freckle symptoms are present on fruit in local markets in South-East Asia, Papua New Guinea and the South Pacific, but are tolerated by most consumers. Control becomes important when fruit is exported to more discerning buyers. In Taiwan, control measures include the removal of lower, diseased leaves to reduce inoculum levels and bagging bunches to help prevent fruit infection. Mancozeb has been shown to be more effective than benomyl in controlling freckle, but mineral oil was proved to be ineffective (Chuang, 1983). Propiconazole (Tilt®) is now the preferred fungicide. In the Philippines, propiconazole and flusilazole (Punch®), which control black leaf streak, also control freckle.

Rust

D.R. Jones

Introduction

Rust is usually a minor disease of banana. It has been described in Australasia–Oceania (Australia, Fiji, Papua New Guinea, Wallis Islands, Western Samoa), Asia (Malaysia, Philippines) and Africa (Congo, Nigeria) (Wardlaw, 1961; Mulder and Holliday, 1971; Dingley *et al.*, 1981; Shaw, 1984). Epiphytotics have occurred on commercial plantations of cultivars in the Cavendish subgroup (AAA) in the Philippines and Malaysia. In Fiji, Cavendish cultivars are quite susceptible to rust when oil-sprayed (Firman, 1972).

Symptoms

Uredosori appear as small brown linear lesions, mainly on the lower surfaces of older leaves. These elongate, broaden and can coalesce, causing leaf yellowing (Plate 2.45) and necrosis if the density of infection is great. Powdery, light brown masses of uredospores can often be seen covering developing lesions. Uredosori can act as points of entry for secondary infections caused by *Cordana musae*.

Causal agents

Uromyces musae P. Henn and *Uredo musae* Cummins are the causal agents of banana rust.

Uromyces musae has powdery brown to black, elongate, erumpent sori. The uredosorus is generally hypophyllous, round and pulvinate. Uredospores are globose to subglobose or ellipsoidal, light brown, finely echinulate and 20–28 μm × 17–24 μm in size with 2.5-μm-thick walls. The telia are hypophyllous, scattered or aggregated, oblong or ellipsoidal, 0.5–1 mm in length or up to 3 mm when joined in series. Teliospores are subglobose, ovoid or oblong, brown with dimensions of 23–35 μm × 17–25 μm on persistent pedicels about 60 μm long (Mulder and Holliday, 1971). Authenticated specimens of this rust fungus are only known from Africa (Firman, 1976).

Uredo musae has uredia that are more crowded. Uredospores have a thinner cell wall (1.5 μm) than those of *Uromyces musae* and more pronounced echinulations (Mulder and Holliday, 1971). All collections of rust on banana on Pacific Islands have been of this fungus (Firman, 1976).

Disease cycle and epidemiology

Little has been reported on the life cycle of *U. musae*. The existence of a host other than banana has not been determined. Wind-blown uredospores carry the disease from banana to banana. These germinate in moisture and invade leaf tissue.

Host reaction

Many cultivars seem to be susceptible to rust. In Fiji, where rust is common, symptoms have been seen on representatives of the AA, AAA, AAAA, AAB and ABB genomes, including 'Dwarf Cavendish', 'Veimama' and 'Poyo' (AAA, Cavendish

Plate 2.44. Symptoms of freckle on a leaf of 'Mondolpin' (ABB, syn. 'Bluggoe') on Badu Island in the Torres Strait region of Australia. Note the larger size of spots, which are also present on the leaf midrib, compared with Plates 2.42 and 2.43 (photo: D.R. Jones, QDPI).

Plate 2.45. Symptoms of rust disease on a leaf of 'Williams' (AAA, Cavendish subgroup) in North Queensland, Australia (photo: D.R. Jones, QDPI).

subgroup), 'Gros Michel' (AAA) and 'Blue Java' (ABB, syn. 'Ney Mannan'). About half the cultivars examined in a screening trial at Koronivia Research Station were diseased, but in most cases symptoms were slight and very few leaves were affected. Rust appeared on two unaffected cultivars in the Cavendish subgroup after spraying with oil (Firman, 1972). Rust has been seen on *M. acuminata* ssp. *banksii* in Western Samoa (Dingley *et al.*, 1981).

Control

No control is generally warranted. However, severe outbreaks that have occurred in some commercial Cavendish plantations appear to be related to the continual use of either oil or benomyl alone to control leaf spot diseases (Gerlach, 1988).

REFERENCES

Allen, R.N. and Dettman, E.B. (1990) Production of conidia and appressoria by *Cordana johnstonii* on banana. *Mycological Research* 94, 815–818.

Anon. (1981) Mycosphaerella musicola *Mulder*. Distribution Maps of Plant Diseases, No. 7, edn 7, Commonwealth Mycological Institute, Kew, Surrey, UK.

Anon. (1988) Cladosporium musae *Mason*. Distribution Maps of Plant Diseases, No. 594, Commonwealth Mycological Institute, Kew, Surrey, UK.

Anon. (1990a) Guignardia musae *Racib*. Distribution Maps of Plant Diseases, No. 263, CAB International Mycological Institute, Kew, Surrey, UK.

Anon. (1990b) Haplobasidian musae *M.B. Ellis*. Distribution Maps of Plant Diseases, No. 474, CAB International Mycological Institute, Kew, Surrey, UK.

Anon. (1993a) *Fungicide Resistance Action Committee (FRAC) Banana Working Group. Conclusions and Recommendations of the Third Meeting, Orlando, Florida, USA, February 1993.* Ciba-Geigy, Switzerland, 14 pp.

Anon. (1993b) *Musa*news. *Infomusa* 2(2), 21–23.

Anon. (1994a) *Musa*news. *Infomusa* 3(2), 25–29.

Anon. (1994b) *The Global Banana and Plantain Network, INIBAP Annual Report 1994.* INIBAP, Montpellier, France, 29 pp.

Anon. (1995) *Musa*news. *Infomusa* 4(2), 26–30.

Anunciado, I.S., Balmes, L.O., Bawagan, P.Y., Benigno, D.A., Bondad, N.D., Cruz, O.J., Franco, P.T., Gavarra, M.R., Opeña, M.T. and Tabora, P.C. (1977) *The Philippines Recommends for Abaca, 1977.* Philippine Council for Agriculture and Resources Research, Los Baños, Philippines, 71 pp.

Atkinson, M.M. (1993) Molecular mechanisms of pathogen recognition by plants. *Advances in Plant Pathology* 10, 35–59.

Bakry, F. and Horry, J.P. (1994) *Musa* breeding at CIRAD–FLHOR. In: Jones, D.R. (ed.) *The Improvement and Testing of Musa: A Global Partnership, Proceedings of the First Global Conference of the International Musa Testing Program held at FHIA, Honduras, 27–30 April 1994.* INIBAP, Montpellier, pp. 169–175.

Belalcazar, S.L. (1991) *El cultivo del plátano en el trópico.* Manual de Asistencia Técnica No. 50, Instituto Colombiano Agropecuario (ICA), Cali, Colombia, 376 pp.

Beveraggi, A. (1992) Etude des interactions hôte–parasite chez des bananiers sensibles et résistants inoculés par *Cercospora fijiensis* responsable de la maladie des raies noires. Thèse de 3ème cycle, Université de Montpellier II, USTL.

Beveraggi, A., Mourichon, X. and Salle, G. (1993) Study of host–parasite interactions in susceptible and resistant bananas inoculated with *Cercospora fijiensis*, pathogen of black leaf streak disease. In: Ganry, J. (ed.) *Breeding Banana and Plantain for Resistance to Diseases and Pests, Proceedings of the International Symposium on Genetic Improvement of Bananas to Diseases and Pests organized by CIRAD–FLHOR, Montpellier, France, 7–9 September 1992.* CIRAD, Montpellier, France, pp. 171–192.

Beveraggi, A., Mourichon, X. and Salle, G. (1995) Etude des interactions hôte-parasite chez des bananiers sensibles et résistants inoculés par *Cercospora fijiensis* (*Mycosphaerella fijiensis*) responsable de la maladie des raies noires. *Canadian Journal of Botany* 73, 1328–1337.

Blanco, M.V. (1987) Relación de los factores climaticos con los niveles de esporulación e infección de *Mycosphaerella fijiensis* var. *difformis* en el cultivo de banano en la zona de Rio Frio. Tesis de ingeniero agronomo, Universidad de Costa Rica, 60 pp.

Boller, T. (1987) Hydrolytic enzymes in plant disease resistance. In: Kosuge, T. and Nester, E.W. (eds) *Plant–Microbe Interactions*, Vol. 2, Macmillan, New York, pp. 385–413.

Booth, C. and Shaw, D.E. (1961) Black cross, caused by a new species of *Phyllachora* on banana. *Papua New Guinea Agricultural Journal* 13, 157–159.

Brun, J. (1962) Etudes préliminaires sur l'utilisation des variétés de bananiers résistants dans la lutte contre la cercosporiose. *Fruits* 17, 113–119.

Brun, J. (1963) La cercosporiose du bananier en Guinée. Etude de la phase ascoporée du *Mycosphaerella musicola* Leach. Thèse de Docteur de Science, Orsay, Paris, France, 190 pp.

Buddenhagen, I.W. (1987) Disease susceptibility and genetics in relation to breeding of bananas and plantain. In: Persley, G.J. and De Langhe, E.A. (eds) *Banana and Plantain Breeding Strategies. Proceedings of an International Workshop held at Cairns, Australia, 13–17 October 1986.* ACIAR Proceedings 21, Australian Centre for International Agricultural Research (ACIAR), Canberra, Australia, pp. 95–109.

Bureau, E. (1990) Adaption of a forecasting system to control black Sigatoka (*Mycosphaerella fijiensis* Morelet) in plantain plantations of Panama. *Fruits* 45, 329–338.

Bureau, E. and Ganry, J. (1987) A climatic forecasting system to control banana Sigatoka (*Mycosphaerella musicola*) using sterol-biosynthesis inhibiting fungicides. *Fruits* 42, 199–205.

Burt, P.J.A. (1993) Windborne dispersal of the pathogens causing Sigatoka leaf spots in banana and plantain. In: *Abstracts of 6th International Plant Pathology Congress, Montreal, Canada.* International Plant Pathology Society, Montreal, p. 104.

Burt, P.J.A. and Rutter, J. (1997) The role of wind in the short-distance transport of fungal pathogens causing Sigatoka diseases in banana and plantain. In: Agashe, S.N. (ed.) *Aerobiology.* Oxford and IBH, New Delhi, India, pp. 307–320.

Bustamente, M. (1983) Impact du Cercospora noir sur la production du plantain au Honduras. *Fruits* 38, 330–332.

Calpouzos, L. (1954) Controlled sporulation of *Cercospora musae* Zimm. in pure culture. *Nature* 173, 1084–1085.

Calpouzos, L. (1968) Oils. In: Totgason, D.C. (ed.) *Fungicides,* Vol. 2. Academic Press, New York, USA, pp. 367–393.

Carlier, J., Mourichon, X., Gonzales de Leon, D., Zapater, M.F. and Lebrun, M.H. (1994) DNA restriction fragment length polymorphisms in *Mycosphaerella* species causing banana leaf spot diseases. *Phytopathology* 84, 751–756.

Carlier, J., Lebrun, M.H., Zapater, M.F., Dubois, C. and Mourichon, X. (1996) Genetic structure of the global population of banana black leaf streak fungus *Mycosphaerella fijiensis. Molecular Ecology* 5, 499–510.

Carreel, F. (1995) Etude de la diversité génétique des bananiers (genre *Musa*) à l'aide des marqueurs RFLP. PhD thesis, Institut National Agronomique, Paris-Grignan, France.

Cheesman, E.E. and Wardlaw, C.W. (1937) Specific and varietal susceptibility of bananas to Cercospora leaf spot. *Tropical Agriculture (Trinidad)* 12, 335–336.

Chuang, T.Y. (1981) Isolation of *Phyllosticta musarum*, causal organism of banana freckle. *Transactions of the British Mycological Society* 77, 670–671.

Chuang, T.Y. (1983) Chemical control of banana freckle caused by *Phyllosticta musarum. Plant Protection Bulletin (Taiwan)* 25, 15–22.

Chuang, T.Y. (1984) Ecological studies of banana freckle caused by *Phyllosticta musarum. Plant Protection Bulletin (Taiwan)* 26, 335–345.

Chuang, T.Y. and Jeger, M.J. (1987a) Relationship between incidence and severity of banana leaf spot in Taiwan. *Phytopathology* 77, 1537–1541.

Chuang, T.Y. and Jeger, M.J. (1987b) Predicting the rate of development of black Sigatoka (*Mycosphaerella fijiensis* var. *difformis*) disease in southern Taiwan. *Phytopathology* 77, 1542–1547.

Clough, J.M., Evans, D.A., De-Fraine, P.J., Fraser, T.E.M., Godfrey, C.R.A. and Youle, D. (1994) Role of natural products in pesticide discovery: the beta-methoxyacrylate fungicides. *ACS Symposium Series (USA)* 551, 37–53.

Cordeiro, Z.J.M., de Matos, A.P. and de Oliveira e Silva, S. (1998) Black Sigatoka confirmed in Brazil. *Infomusa* 7(1), 31.

Cronshaw, D.K. (1982) Management of banana leaf spot (Sigatoka) diseases in the Windward Islands. *Tropical Pest Management* 28, 136–146.

Cronshaw, D.K. and Akers, A. (1990) Mode of action of tridemorph and sensitivity of *Mycosphaerella fijiensis.* In: Fullerton, R.A. and Stover, R.H. (eds) *Sigatoka Leaf Spot Diseases of Bananas, Proceedings of an International Workshop held at San José, Costa Rica, March 28–April 1, 1989.* INIBAP, Montpellier, France, pp. 84–89.

Dabek, A.J. and Waller, J.M. (1990) Black leaf streak and viral streak: new banana diseases in East Africa. *Tropical Pest Management* 36(2), 157–158.

Daniells, J.W., Peterson, R.A., Reid, D.J. and Bryde, N.J. (1996) Screening of 165 Papua New Guinea banana accessions for resistance to yellow Sigatoka (*Mycosphaerella musicola* Leach) in North Queensland. *Infomusa* 5(1), 31–34.

Daub, M.E. (1986) Tissue culture and the selection of resistance to pathogens. *Annual Review of Phytopathology* 24, 159–186.

David, J.C. (1988) *Cladosporium musae.* CMI Descriptions of Pathogenic Fungi and Bacteria, No. 958. *Mycopathologia* 103, 119–120.

Davidse, C.L. and Ishii, H. (1995) Biochemical and molecular aspects of the mechanisms of action of benzimidazole fungicides, *N*-phenylcarbamates and *N*-phenylformamidoximes and the mechanisms of resistance to these compounds in fungi. In: Lyr, H. (ed.) *Modern Selective Fungicides: Properties, Applications, Mechanisms of Action.* Gustav Fischer Verlag, Jena, Germany, pp. 305–322.

Davidse, L.C. (1988) Benzimidazole fungicides: mechanism of action and resistance. In: Delp, C.J. (ed.) *Fungicide Resistance in North America.* American Phytopathological Society, St Paul, Minnesota, pp. 25–27.

de Hoog, G.S. (1972) *Acrodontium simplex.* In: *The Genera* Beauveria, Isaria, Tritirachium *and* Acrodontium. *Gen. Nov. Studies in Mycology* 1, 33–34.

de Hoog, G.S. (1977) *Rhinocladiella* and allied genera. *Studies in Mycology* 15, 1–140.

Deighton, F.C. (1976) Studies on *Cercospora* and allied genera. VI. *Pseudocercospora* Speg., *Pantospora* Cif. and *Cercosetporia petri. Mycological Papers* 140, 1–168.

Deighton, F.C. (1979) Studies on *Cercospora* and allied genera. VII. New species and redisposition. *Mycological Papers* 144, 1–56.

Delp, C. (1995) Benzimidazole and related fungicides. In: Lyr, H. (ed.) *Modern Selective Fungicides: Properties, Applications, Mechanisms of Action.* Gustav Fischer Verlag, Jena, Germany, pp. 291–303.

de Waard, M.A., Van Nistelrooy, J.G.M., Langeveld, C.R., Van Kan, J.A.L. and Del Sorbo, G. (1996) Multidrug resistance in filamentous fungi. In: Lyr, H., Russel, P.E. and Sisler, H.D. (eds) *Modern Fungicides and Antifungal Compounds.* Intercept Ltd, Andover, UK, pp. 293–300.

de Wit, P.G.J.M. (1992) Molecular characterization of gene-for-gene systems in plant–fungus interactions and the application of avirulence genes in control of plant pathogens. *Annual Review of Phytopathology* 30, 391–418.

de Wit, P.G.J.M., Hofman, E.E., Velthius, G.C.M. and Kuc, J. (1985) Isolation and characterization of an elicitor of necrosis isolated from intercellular fluids of compatible interactions of *Cladosporium fulvum* (syn. *Fulvia fulva*) and tomato. *Plant Physiology* 77, 642–647.

Dingley, J.R., Fullerton, R.A. and McKenzie, E.H.C. (1981) *Survey of Agricultural Pests and Diseases, Technical Report Volume 2, Records of Fungi, Bacteria, Algae and Angiosperms Pathogenic on Plants in Cook Islands, Fiji, Kiribati, Niue, Tonga, Tuvalu and Western Samoa.* South Pacific Bureau of Economic Cooperation, United Nations Development Programme, Food and Agriculture Organization of the United Nations, Rome, Italy, 485 pp.

Drechsler, C. (1928) Zonate eyespot of grasses caused by *Helminthosporium giganteum. Journal of Agricultural Research* 37, 473–492.

Drechsler, C. (1929) Occurence of the zonate eyespot fungus *Helminthosporium giganteum* on some additional grasses. *Journal of Agricultural Research* 39, 129–136.

Ela, V.M. and San Juan, M.O. (1954) Leaf spot and stem rot of abaca. *Philippine Journal of Agriculture* 38, 251–271.

Ellis, M.B. (1957a) Some species of *Deightoniella. Mycological Papers* 66, 12 pp.

Ellis, M.B. (1957b) *Haplobasidion, Lacellinopsis* and *Lacellina. Mycological Papers* 67(3), 13 pp.

Ellis, M.B. (1967) Dematiaceous Hyphomycetes. VIII. *Periconiella, Trichodochium,* etc. *Mycological Papers* 111, 1–46.

Ellis, M.B. (1971a) Dematiaceous Hyphomycetes. X. *Mycological Papers* 125, 1–30.

Ellis, M.B. (1971b) *Periconiella musae.* In: *Dematiaceous Hyphomycetes.* Commonwealth Mycological Institute, Kew, Surrey, UK, pp. 296–297.

Ellis, M.B. (1976) *Veronea musae.* In: *More Dematiaceous Hyphomycetes.* Commonwealth Mycological Institute, Kew, Surrey, UK, pp. 209–210.

Ellis, M.B. and Holliday, P. (1976) Haplobasidion musae. Descriptions of Pathogenic Fungi and Bacteria, No. 496, Commonwealth Mycological Institite, Kew, Surrey, UK.

Escalant, J.V. (1990) Using tissue culture techniques for production of new varieties of plantain and early screening against black Sigatoka. In: Fullerton, R.A. and Stover, R.H. (eds) *Sigatoka Leaf Spot Diseases of Bananas: Proceedings of an International Workshop held at San José, Costa Rica, March 28–April 1, 1989.* INIBAP, Montpellier, France, pp. 338–348.

Ferris, R.S.B., Adenji, T., Chukwu, U., Akalumhe, Y.O., Vuylsteke, D. and Ortiz, R. (1996) Postharvest quality of plantains and cooking bananas. In Ortiz, R. and Akoroda, M.O. (eds) *Plantain and Banana; Production and Research in West and Central Africa, Proceedings of a Regional Workshop held at High Rainfall Station, Onne, Nigeria, September 24–28, 1995.* IITA, Ibadan, Nigeria, pp. 15–22.

Firman, I.D. (1971) Banana leaf spot caused by *Haplobasidion musae. PANS* 17, 315–317.

Firman, I.D. (1972) Susceptibility of banana cultivars to fungus leaf diseases in Fiji. *Tropical Agriculture (Trinidad)* 49, 189–196.

Firman, I.D. (1975) *Plant Diseases in the Area of the South Pacific Commission. 1. Banana Diseases.* South Pacific Commission, Noumea, New Caledonia, 9 pp.

Firman, I.D. (1976) Banana rust in Fiji and other Pacific Islands. *Fiji Agricultural Journal* 38, 85–86.

Fouré, E. (1982) Les cercosporioses du bananier et leurs traitements. Comportement des variétés. Etude de la sensibilité variétale des bananiers et des plantains à *Mycosphaerella fijiensis* MORELET au Gabon (I) – I Incubation et évolution de la maladie – II Etude de quelques paramètres. *Fruits* 37, 749–771.

Fouré, E. (1983) Les Cercosporioses du bananier et leurs traitements. Sélection de molécules fongicides nouvelles. Activités comparées de différentes molécules fongicides sur *Mycosphaerella fijiensis* MORELET, agent de la 'maladie des raies noires' des bananiers et plantains du Gabon. *Fruits* 38, 21–34.

Fouré, E. (1985) Les Cercosporioses du bananier et leurs traitements. Comportement des variétés. Étude de la sensibilité variétale des bananiers et plantains à *Mycosphaerella fijiensis* MORELET au Gabon (maladie des raies noires) (suite III). *Fruits* 40, 393–399.

Fouré, E. (1987) Varietal reactions of bananas and plantains to black leaf streak disease. In: Persley, G.J. and De Langhe, E.A. (eds) *Banana and Plantain Breeding Strategies, Proceedings of an International Workshop held in Cairns, Australia, 13–17 October 1986.* ACIAR Proceedings No. 21, ACIAR, Canberra, Australia, pp. 110–113.

Fouré, E. (1988a) Les Cercosporioses du bananier et leurs traitements. Efficacité comparée de différentes molécules fongicides sur *Mycosphaerella fijiensis* MORELET, agent de la maladie des raies noires des bananiers et plantains au Cameroun (I). *Fruits* 43, 15–19.

Fouré, E. (1988b) Les Cercosporioses du bananier et leurs traitements. Efficacités comparées du pyrazophos et du triadimenol sur *Mycosphaerella fijiensis* MORELET (agent de la Cercosporiose noire des bananiers et des plantains au Cameroun) lors de traitements sur grandes surfaces. *Fruits* 43, 143–147.

Fouré, E. (1989) Contribution to genetic control of banana and plantain against Sigatoka leaf spot disease in Cameroon, Studies on varietal susceptibility and early inoculation trials on plantlets produced by *in vitro* culture. In: Fullerton, R.A. and Stover, R.H. (eds) *Sigatoka Leaf Spot Diseases of Bananas, Proceedings of an International Workshop held at San José, Costa Rica, March 28–April 1, 1989.* INIBAP, Montpellier, France, pp. 290–305.

Fouré, E. (1990a) La lutte integrée contre le cercosporiose noire des bananiers au Cameroun. L'avertissement biologique et son évolution de 1985 à 1988. In: Fullerton, R.A. and Stover, R.H. (eds) *Sigatoka Leaf Spot Diseases of Bananas, Proceedings of an International Workshop held in San José, Costa Rica, March 28–April 1, 1989.* INIBAP, Montpellier, France, pp. 124–134.

Fouré, E. (1990b) Contribution to genetic control of banana and plantain Sigatoka leaf spot in Cameroon: studies on varietal susceptibility and early inoculation trials on plantlets produced by *in vitro* culture. In: Fullerton, R.A. and Stover, R.H. (eds) *Sigatoka Leaf Spot Diseases of Bananas,*

Proceedings of an International Workshop held in San José, Costa Rica, March 28–April 1, 1989. INIBAP, Montpellier, France, pp. 290–305.

Fouré, E. (1991a) Contribution à la lutte génétique contre les cercosporioses des bananiers et des plantains au Cameroun. Etudes de sensibilité variétale. In: Añez, B., Nara, C., Sosa, L. and Jaramillo, R. (eds) *ACORBAT, memoriãs IX Réunion*, Septiembre 24–29 1989, Maracaibo, Venezuela, pp. 245–257.

Fouré, E. (1991b) Les cercosporioses des bananiers et des plantains au Cameroun, *M. fijiensis* et *M. musicola*: amélioration des stratégies de lutte intégrée par des études épidémiologiques et la lutte génétique. In: Gold, C.S. and Gemmill, B. (eds) *Biological and Integrated Control of Highland Banana and Plantain Pest and Diseases, Proceedings, Research Coordination Meeting.* IITA, Cotonou, Benin, 12–14 November 1991, pp. 290–304.

Fouré, E. (1992) Characterization of the reactions of banana cultivars to *Mycosphaerella fijiensis* MORELET in Cameroon and genetics of resistance. In: Ganry, J. (ed.) *Breeding Banana and Plantain for Resistance to Diseases and Pests, Proceedings of the International Symposium on Genetic Improvement of Bananas for Resistance to Diseases and Pests organized by CIRAD–FLHOR, Montpellier, France, 7–9 September 1992.* CIRAD–FLHOR, Montpellier, pp. 159–170.

Fouré, E. (1994) Leaf spot diseases of banana and plantain caused by *M. musicola* and *M. fijiensis.* In: Jones, D.R. (ed.) *The Improvement and Testing of Musa: A Global Partnership, Proceedings of the First Global Conference of the International Musa Testing Program held at FHIA, Honduras, 27–30 April 1994.* INIBAP, Montpellier, France, pp. 37–46.

Fouré, E. and Moreau, A. (1992) Contribution à l'étude épidémiologique de la cercosporiose noire dans la zone de Moungo au Cameroon de 1987 à 1989. *Fruits* 47, 3–16.

Fouré, E. and Mouliom-Pefoura, A. (1988) La cercosporiose noire des bananiers et des plantains au Cameroun (*Mycosphaerella fijiensis*). Contribution à l'étude des premières phases de l'infection parasitaire. Mise au point de tests précoces d'inoculation sur plants issus de vitro-culture. *Fruits* 43, 339–348.

Fouré, E., Grisoni, M. and Zurfluh, R. (1984) Les cercosporioses du bananier et leurs traitements. Comportement des variétés. Etude de la sensibilité variétale des bananiers et plantains B *Mycosphaerella fijiensis* MORELET et de quelques caractéristiques biologiques de la maladie des raies noires au Gabon (II). *Fruits* 39, 365–377.

Fouré, E., Mouliom-Pefoura, A. and Mourichon, X. (1990) Etude de la sensibilité variétale des bananiers et des plantains à *M. fijiensis* MORELET au Cameroun. Caractérisation de la résistance au champ de bananiers appartenant à divers groupes génétiques. *Fruits* 45, 339–345.

Frossard, P. (1963) Une cladosporiose du bananier en Côte d'Ivoire. *Fruits* 18, 443–453.

Frossard, P. (1980) Apparition d'une nouvelle et grave maladie foliaire des bananiers et plantains au Gabon: la maladie des raies noires: *Mycosphaerella fijiensis* MORELET. *Fruits* 35, 519–527.

Fullerton, R.A. (1987) Banana production in selected Pacific islands. In: Persley, G.J. and De Langhe, E.A. (eds) *Banana and Plantain Breeding Strategies, Proceedings of an International Workshop held at Cairns, Australia, 13–17 October 1986.* ACIAR Proceedings 21, Australian Centre for International Agricultural Research, Canberra, Australia, pp. 57–62.

Fullerton, R.A. and Olsen, T.L. (1993) Pathogenic diversity in *Mycosphaerella fijiensis* Morelet. In: Ganry, J. (ed.) *Breeding Banana and Plantain for Resistance to Diseases and Pests, Proceedings of the International Symposium on Genetic Improvement of Bananas for Resistance to Diseases and Pests Organized by CIRAD–FLHOR, Montpellier, France, 7–9 September 1992.* CIRAD, Montpellier, France, pp. 201–211.

Fullerton, R.A. and Olsen, T.L. (1995) Pathogenic variability in *Mycosphaerella fijiensis* MORELET, cause of black Sigatoka in banana and plantain. *New Zealand Journal of Crop and Horticultural Science* 23, 39–48.

Fullerton, R.A. and Stover, R.H. (eds) (1990) *Sigatoka Leaf Spot Diseases of Bananas, Proceedings of an International Workshop held at San José, Costa Rica, March 28–April 1, 1989.* INIBAP, Montpellier, France, 374 pp.

Fullerton, R.A. and Tracey, G.M. (1984) Tolerance of *Mycosphaerella fijiensis* to benomyl and carbendazim in the Pacific Islands. *Tropical Agriculture (Trinidad)* 61, 133–136.

Ganry, J. (1986) Basic principles of a forecasting system for regulated control of Sigatoka disease in banana when using a systemic oil fungicide. In: *Improving Citrus and Banana Production in the Caribbean through Phytosanitation, Seminar Proccedings, 2–5 Dec. 1986, St Lucia, West Indies.* CTA–CARDI, Trinidad, pp. 34–56.

Ganry, J. and Laville, E. (1983). Les cercosporiosis du bananier et leurs traitements. Evolution des méthodes de traitement. 1. Traitements fongicides. 2. Avertissement. *Fruits* 38, 3–20.

Gasztonyi, M., and Lyr, H. (1995) Miscellaneous fungicides. In: Lyr, H. (ed.) *Modern Selective Fungicides: Properties, Applications, Mechanisms of Action.* Gustav Fischer Verlag, Jena, Germany, pp. 389–414.

Gauhl, F. (1989) *Untersuchungen zur Epidemiologie und Ökologie der Schwarzen Sigatoka-Krankheit (*Mycosphaerella fijiensis *MORELET) an Kochbananen (*Musa *sp.) in Costa Rica.* Göttinger Beiträge zur Land- und Forstwirtschaft in den Tropen und Subtropen, Heft 42, Verlag Erich Geltze, Göttingen, Germany, 128 pp.

Gauhl, F. (1994) *Epidemiology and Ecology of Black Sigatoka (*Mycosphaerella fijiensis *MORELET) on Plantain and Banana (*Musa *spp.) in Costa Rica, Central America.* INIBAP, Montpellier, France, 120 pp.

Gauhl, F. and Pasberg-Gauhl, C. (1994) Epidemiology of black sigatoka disease on plantain in Nigeria. *Phytopathology* 84, 1080.

Gauhl, F., Pasberg-Gauhl, C., Vuylsteke, D. and Ortiz, R. (1993) *Multilocational Evaluation of Black Sigatoka Resistance in Banana and Plantain.* IITA Research Guide 47, International Institute of Tropical Agriculture, Ibadan, Nigeria, 60 pp.

Gerlach, W.W.P. (1988) *Plant Diseases of Western Samoa.* Samoan-German Crop Protection Project, Apia, Western Samoa, 215 pp.

Gire, A., Beveraggi, A., Macheix, J.J. and Mourichon, X. (1994) *Evidence of a constitutive polyphenolic component in resistance of banana to* Mycosphaerella fijiensis. Société Francaise de Phytopathologie, Montpellier, France.

Gonsalves, R.A. (1987) Reactions of breeding lines of banana from Jamaica to black leaf streak. In: Jaramillo, R. and Mateo, N. (eds) *Memoria de la reunion regional de INIBAP para America Latina y el Caribe, San José, Costa-Rica, 1986*, INIBAP, LACNET, pp. 224–226.

Goos, R.D. and Tschirch, M. (1962) Establishment of Sigatoka disease of bananas in the greenhouse. *Nature* 194, 887–888.

Goos, R.D. and Tschirch, M. (1963) Greenhouse studies on the *Cercospora* leaf spot of banana. *Transactions of the British Mycological Society* 46, 321–330.

Görlach, J., Volrath, S., Knauf-Beiter, G., Beckhove, U., Kogel, K.H., Oostendorp, M., Staub, T., Ward, E., Kessman, H. and Ryals, J. (1996) Benzothiadiazole, a novel class of inducers of systemic acquired resistance, activates gene expression and disease resistance in wheat. *Plant Cell* 8, 629–643.

Graham, K.M. (1969) A simple way to distinguish black leaf streak from Sigatoka disease on bananas. *Bulletin of the Department of Agriculture, Fiji* 49, 4.

Guyot, H. and Cuillé, J. (1954) Les formules fongicides huileuses pour le traitement des bananeraies. *Fruits* 9, 289–292.

Guyot, H. and Cuillé, J. (1958) Essai de prevision des attaques de *Cercospora* en Guadeloupe. *Fruits* 13, 85–94.

Guzmán, M. and Romero, R.A. (1996a) Evaluación de la eficacia biológica de cinco fungicida inhibores de la biosíntesis de ergosterol en el combate de la Sigatoka negra del banano. In: *Informe Annual 1995.* Dirección de Investigaciones Agrícolas, CORBANA, Costa Rica, pp. 40–42.

Guzmán, M. and Romero, R.A. (1996b) Eficacia de cuatro dosis de aceite agrícola en el control de la Sigatoka negra (*Mycosphaerella fijiensis* Morelet) en banano (*Musa* AAA). *CORBANA* 21(46), 129–139.

Guzmán, M. and Romero, R.A. (1997) Comparación de los fungicidas azoxistrobina, propiconazole y difenoconazole en el control de la Sigatoka negra (*Mycosphaerella fijiensis* Morelet) en banano (*Musa* AAA). *CORBANA* 22(47), 49–59.

Guzmán, M., Jiménez, A., Vargas, R. and Romero, R.A. (1997) Evaluación de tres períodos libres de fungicidas triazoles sobre la sensibilidad de la Sigatoka negra (*Mycosphaerella fijiensis* Morelet) en banano. In: *Informe Annual 1997.* Dirección de Investigaciones Agrícolas, CORBANA, Costa Rica, pp. 94–95.

Hahn, S., Vuylsteke, D. and Swennen, R. (1989) First reactions to ABB cooking bananas distributed in south eastern Nigeria. In: Fullerton, R.A. and Stover, R.H. (eds) *Sigatoka Leaf Spot Diseases of Bananas, Proceedings of an International Workshop held in San José, Costa Rica, March 28–April 1, 1989.* INIBAP, Montpellier, France, pp. 306–315.

Hapitan, J.C. and Reyes, T.T. (1970) Black leaf streak disease of bananas in the Philippines. *Philippine Agriculturist* 54, 47–54.

Harlan, J.R. (1976) Diseases as a factor in plant evolution. *Annual Review of Phytopathology* 14, 31–51.

Hart, D.L. and Clarke, A.G. (1989) *Principles of Population Genetics.* Sinauer Associates, Sunderland, Massachusetts, 682pp.

Hwang, S.C. (1991) Status of banana diseases in Taiwan. In: Valmayor, R.V., Umali, B.E. and Bejosano, C.P. (eds) *Banana Diseases in Asia and the Pacific: Proceedings of a Technical Meeting on Diseases Affecting Banana and Plantain in Asia and the Pacific, Brisbane, Australia, 15–18 April 1991.* INIBAP, Montpellier, France, pp. 73–83.

Hwang, S.C. and Chen, C.L. (1986) A new leaf speckle disease of banana caused by *Acrodontium simplex* in Taiwan. *Plant Protection Bulletin (Taiwan)* 28, 413–415.

Hwang, S.C., Chen, C.L. and Wu, F.L. (1984) An investigation on susceptibility of banana clones to Fusarium wilt, freckle and marginal scorch disease in Taiwan. *Plant Protection Bulletin (Taiwan)* 26, 155–156.

Hyde, K.D. (1992) *Phyllochora musicola.* IMI Descriptions of Fungi and Bacteria No. 1127. *Mycopathologia* 119, 57–58.

Israeli, Y., Shabi, E. and Slabaugh, W.R. (1993) Effect of banana spray oil on banana yield in the absence of Sigatoka (*Mycosphaerella* sp.). *Scientia Horticulturae* 56, 107–117.

Jacome, L.H. and Schuh, W. (1992) Effects of leaf wetness duration and temperature on development of black Sigatoka disease on banana infected by *Mycosphaerella fijiensis* var. *difformis. Phytopathology* 82, 515–520.

Jacome, L.H. and Schuh, W. (1993) Effect of temperature on growth and conidia production *in vitro*, and comparison of infection and aggressiveness *in vivo* among isolates of *Mycosphaerella fijiensis* var. *difformis. Tropical Agriculture (Trinidad)* 70, 51–59.

Jacome, L.H., Schuh, W. and Stevenson, R.E. (1991) Effect of temperature and relative humidity on germination and germ tube development of *Mycosphaerella fijiensis* var. *difformis. Phytopathology* 81, 1480–1485.

Jaramillo, R. (1987) Banana and plantain production in Latin America and the Carribean. In: Persley, G.J. and De Langhe, E.A. (eds) *Banana and Plantain Breeding Strategies, Proceedings of an International Workshop held at Cairns, Australia, 13–17 October 1986.* ACIAR Proceedings 21, Australian Centre for International Agricultural Research, Canberra, Australia, pp. 29–35.

Jarret, R.L., Rodriguez, W. and Fernandez, R. (1985) Evaluation, tissue culture propagation, and dissemination of 'Saba' and 'Pelipita' plantains in Costa Rica. *Scientia Horticulturae* 25, 137–147.

Jenny, C., Auboiron, E. and Beveraggi, A. (1994) Breeding plantain-type hybrids at CRBP. In: Jones, D.R. (ed.) *The Improvement and Testing of Musa: A Global Partnership, Proceedings of the First Global Conference of the International Musa Testing Program held at FHIA, Honduras, 27–30 April 1994.* INIBAP, Montpellier, France, pp. 176–187.

Jiménez, A., Guzmán, M. and Romero, R.A. (1996) Dynamics of *Mycosphaerella fijiensis* fungicide-resistant strains in banana plantations in Costa Rica. *Phytopathology* 86(11), S32 (Suppl.).

Jiménez, A., Guzmán, M. and Romero, R.A. (1997) Parasitic fitness and ability to produce conidia of propiconazole-less sensitive isolates of *Mycosphaerella fijiensis*, causal agent of black Sigatoka disease of banana. In: *Proceedings of the XXXVII Annual Meeting of the Caribbean Division of the American Phytopathological Society, San José, Costa Rica*, Abstract 100.

Johanson, A. (1993) Molecular methods for the identification and detection of the *Mycosphaerella* species that cause Sigatoka leaf spots of banana and plantain. PhD thesis, University of Reading, United Kingdom.

Johanson, A. and Jeger, M.J. (1993) Use of PCR for detection of *Mycosphaerella fijiensis* and *M. musicola*, the causal agents of Sigatoka leaf spots in banana and plantain. *Mycological Research* 97, 670–674.

Johanson, A., Crowhurst, R.N., Rikkerink, E.H.A., Fullerton, R.A. and Templeton, M.D. (1994) The use of species-specific DNA probes for the identification of *Mycosphaerella fijiensis* and *M. musicola*, the causal agents of Sigatoka disease of banana. *Plant Pathology* 44, 701–707.

Johnston, A. (1965) *Spread of Banana Black Leaf Streak.* FAO Plant Protection Committee for the South East Asia and Pacific Region. Information Letter No. 41, Bangkok, Thailand, 2 pp.

Jones, D.R. (1988) *Report of an IBPGR Banana Germplasm Collecting Mission to Papua New Guinea, February/March 1988.* Queensland Department of Primary Industries, Brisbane, Australia, 14 pp.

Jones, D.R. (1990a) Black Sigatoka – a threat to Australia. In: Fullerton, R.A. and Stover, R.H. (eds) *Sigatoka Leaf Spot Diseases of Bananas. Proceedings of an International Workshop held at San José, Costa Rica, 28 March–1 April 1989.* INIBAP, Montpellier, France, pp. 29–37.

Jones, D.R. (1990b) Black Sigatoka in the Southeast Asian–Pacific region. *Musarama* 3(1), 2–5.

Jones, D.R. (1991) Status of banana diseases in Australia. In: Valmayor, R.V., Umali, B.E. and Besjosano, C.P. (eds) *Banana Diseases in Asia and the Pacific: Proceedings of a Technical Meeting on Diseases Affecting Banana and Plantain in Asia and the Pacific, Brisbane, Australia, 15–18 April 1991.* INIBAP, Montpellier, France, pp. 21–37.

Jones, D.R. (1993a) Evaluating banana and plantain for reaction to black leaf streak disease in the South Pacific. *Tropical Agriculture (Trinidad)* 70, 39–44.

Jones, D.R. (1993b) *Banana Disease Survey of West Malaysia, 16–26 August 1993.* INIBAP, Montpellier, France, 11 pp.

Jones, D.R. (1994a) *The Improvement and Testing of* Musa*: A Global Partnership, Proceedings of the First Global Conference of the International* Musa *Testing Program held at FHIA, Honduras, 27–30 April 1994.* INIBAP, Montpellier, France, 303 pp.

Jones, D.R. (1994b) *Banana Disease Survey of Thailand, 26 August–10 September 1994.* Report to INIBAP, Montpellier, France, 9 pp.

Jones, D.R. (1995) Rapid assessment of *Musa* for reaction to Sigatoka disease. *Fruits* 50, 11–22.

Jones, D.R. and Alcorn, J.L. (1982) Freckle and black Sigatoka diseases of banana in Far North Queensland. *Australasian Plant Pathology* 11, 7–9.

Jones, D.R. and Daniells, J.W. (1988) *Report of an ACIAR Banana Germplasm Collecting Mission to Southeast Asia, June 1988.* Queensland Department of Primary Industries, Brisbane, Australia, 33 pp.

Jones, D.R. and Daniells, J.W. (1991) Black cross disease in Queensland. *Infomusa* 1(1), 9.

Jones, D.R. and Mourichon, X. (1993) *Black Leaf Streak/Black Sigatoka Disease. Musa* Disease Fact Sheet No. 2, INIBAP, Montpellier, France, 2 pp.

Jones, D.R. and Tezenas du Montcel, H. (eds) (1994) *International* Musa *Testing Program Phase 1.* INIBAP, Montpellier, France, 495 pp.

Kerkenaar, A. (1995) Mechanism of action of cyclic amine fungicides: morpholines and piperidines. In: Lyr, H. (ed.) *Modern Selective Fungicides: Properties, Applications, Mechanisms of Action.* Gustav Fischer Verlag, Jena, Germany, pp. 185–204.

Klein, H.H. (1960) Control of *Cercospora* leaf spot of bananas with applications of oil sprays based on the disease cycle. *Phytopathology* 50, 488–490.

Knowles, C.H. (1914) *Disease in Bananas.* Pamphlet of the Department of Agriculture, Fiji 8, Suva.

Knowles, C.H. (1916) *Visit to Upper Rewa to Investigate Leaf Diseases of the Banana.* Pamphlet of the Department of Agriculture, Fiji 24, Suva.

Koenraadt, H., Somerville, S.C. and Jones, A.L. (1992) Characterization of mutations in the β-tubuline gene of benomyl-resistant field strains of *Venturia inaequalis* and other plant pathogenic fungi. *Phytopathology* 82, 1348–1354.

Köller, W. (1992) Antifungal agents with target sites in sterol functions and biosynthesis. In: Köller, W. (ed.) *Target Sites of Fungicide Action.* CRC Press, Boca Raton, Florida, pp. 119–206.

Köller, W. (1998) Mode of action of fungicides in real plant pathogens. In: *Proceedings of the 7th International Conference in Plant Pathology, Edinburgh, Scotland, 9–16 August 1998.* International Society for Plant Pathology, Abstract 5.5.1S.

Kranz, J. (1965) Feldrersuche zur bekampfung der Sigatoka-krankheit der banane (*Mycosphaerella musicola* Leach) in Guinea. *Phytopathologische Zeitung* 52, 335–348.

Kung'U, J.N., Seif, A.A. and Waller, J.M. (1992) Black leaf streak and other foliage diseases of bananas in Kenya. *Tropical Pest Management* 38, 359–361.

Lange, L., Breinholt, F., Nielsen, R. and Nielse, R. (1993) Microbial fungicides – the natural choice. *Pesticide Science* 39, 155–160.

Laville, E. (1983) Les cercosporioses du bananier et leurs traitements. Comportement des variétés. Généralités. *Fruits* 38, 147–151.

Leach, R. (1941) Banana leaf spot *Mycosphaerella musicola*, the perfect stage of *Cercospora musae* Zimm. *Tropical Agriculture (Trinidad)* 18, 91–95.

Leach, R. (1946) *Banana Leaf Spot (*Mycosphaerella musicola*) on the Gros Michel Variety in Jamaica. Investigations into the Aetiology of the Disease and the Principles of Control by Spraying.* Bulletin, Government Printer, Kingston, Jamaica.

Leach, R. (1964a) *Report on Investigations into the Cause and Control of the New Banana Disease in Fiji, Black Leaf Streak.* Council Papers Fiji 38, Suva.

Leach, R. (1964b) A new form of banana leaf spot in Fiji, black leaf streak. *World Crops* 16, 60–64.

Lepoivre, P. and Acuna, C.P. (1989) Production of toxins by *Mycosphaerella fijiensis* and induction of antimicrobial compounds in banana: their relevance in breeding for resistance to black Sigatoka. In: Fullerton, R.A. and Stover, R.H. (eds) *Sigatoka Leaf Spot Diseases of Bananas: Proceedings of an International Workshop held at San José, Costa Rica, March 28–April 1, 1989.* INIBAP, Montpellier, France, pp. 201–207.

Lescot, T., Simonot, H., Fages, O. and Escalant, J.V. (1998) Developing a biometeorological forecasting system to control leaf spot disease of plantains in Costa Rica. *Fruits* 53, 3–16.

Long, P.G. (1979) Banana black leaf streak disease (*Mycosphaerella fijiensis*) in Western Samoa. *Transactions of the British Mycological Society* 72, 299–310.

Lopez, R. and Loegering, W.Q. (1953) Resistencia de variedades de abaca (*Musa textilis* Nee) a la mancha de la hoja y perdidas ocasionadas por la enfermedad. *Turrialba* 3, 159–162.

McDonald, B.A. and Martinez, J.P. (1990) DNA restriction fragment length polymorphisms among *Mycosphaerella graminicola* (anamorph *Septoria tritici*) isolates collected from a single wheat field. *Phytopathology* 80, 1368–1373.

McDonald, B.A., Miles, J., Nelson, L.R. and Pettway, R.E. (1994) Genetic variability in nuclear DNA in field populations of *Stagonospora nodorum. Phytopathology* 84, 250–255.

McGahan, M.W. and Fulton, R.H. (1965) Leaf spot of bananas caused by *Mycosphaerella musicola*: a comparative anatomical study of juvenile and adult leaves in relation to lesion morphology. *Phytopathology* 55, 1179–1182.

Madrigal, A., Ruess, A. and Staehle-Csech, U. (1998) CGA 245704, a new plant activator to improve natural resistance of banana against black Sigatoka (*Mycosphaerella fijiensis*). In: Arizago, L.H. (ed.) *Proceedings of the 8th ACORBAT meeting, Guayaquil, Ecuador, 23–27 November 1998.* CONABAN, Ecuador, pp. 266–274.

Marín, D.H. and Romero, R.A. (1992) *El combate de la Sigatoka negra.* Boletín No. 4, Departamento de Investigaciones, CORBANA, San José, Costa Rica, 22 pp.

Martinez, G., Pargas, R., Manzanilla, E. and Mu, D. (1998) Report on black Sigatoka status in Venezuela in 1997. *Infomusa* 7(1), 31–32.

Martyn, E.B. (1945) A note on banana leaf speckle in Jamaica and some associated fungi. *Mycological Papers* 13, 1–5.

Massee, G. (1914) Fungi exotici: XVIII. *Kew Bulletin* 1914, 159.

Mayorga, M.H. (1985) *Informe anual de actividades.* Programa Fitopatológia–ICA, Regional 4, Tulenapa, 33 pp.

Merchan, V.M. (1990) Update of research on *Mycosphaerella* spp. in Colombia. In: Fullerton, R.A. and Stover, R.H. (eds) *Sigatoka Leaf Spot Diseases of Bananas, Proceedings of an International Workshop held at San José, Costa Rica, 28 March–1 April 1989.* International Network for the Improvement of Banana and Plantain (INIBAP), Montpellier, France, pp. 50–55.

Meredith, D.S. (1961a) Fruit spot ('speckle') of Jamaican bananas caused by *Deightoniella torulosa* (Syd.) Ellis. I–III. *Transactions of the British Mycological Society* 44, 95–104, 265–284, 391–405.

Meredith, D.S. (1961b) Spore discharge in *Deightoniella torulosa* (Syd.) Ellis. *Annals of Botany* 25, 271–278.

Meredith, D.S. (1961c) Fruit spot ('speckle') of Jamaican bananas caused by *Deightoniella torulosa* (Syd.) Ellis. IV. Further observations on spore dispersal. *Annals of Applied Biology* 49, 488–496.

Meredith, D.S. (1962a) Spore discharge in *Cordana musae* (Zimm.) and *Zygosporium oscheoides* Mont. *Annals of Botany* 26, 233–241.

Meredith, D.S. (1962b) Dispersal of conidia of *Cordana musae* (Zimm.) Höhnel in Jamaican banana plantations. *Annals of Applied Biology* 50, 263–267.

Meredith, D.S. (1963a) 'Eyespot', a foliar disease of bananas caused by *Drechslera giganteum. Annals of Applied Biology* 51, 29–40.

Meredith, D.S. (1963b) *Pestalotia leprogena* on leaves of *Musa* spp. in Jamaica. *Transactions of the British Mycological Society* 46, 537–540.

Meredith, D.S. (1968) Freckle disease of banana in Hawaii caused by *Phyllostictina musarum* (Cke) Petr. *Annals of Applied Biology* 62, 328–340.

Meredith, D.S. (1969) A note on black cross disease of banana caused by *Phyllachora musicola. Transactions of the British Mycological Society* 53, 324–325.

Meredith, D.S. (1970) *Banana Leaf Spot Disease (Sigatoka) caused by* Mycosphaerella musicola *Leach.* Phytopathological Papers, No. 11, Commonwealth Mycological Institute, Kew, Surrey, UK, 147 pp.

Meredith, D.S. and Lawrence, J.S. (1969) Black leaf streak disease of bananas (*Mycosphaerella fijiensis*):

Symptoms of disease in Hawaii, and notes on the conidial state of the causal fungus. *Transactions of the British Mycological Society* 52, 459–476.

Meredith, D.S. and Lawrence, J.S. (1970a) Black leaf streak disease of bananas (*Mycosphaerella fijiensis*); susceptibility of cultivars. *Tropical Agriculture* 47, 275–287.

Meredith, D.S. and Lawrence, J.S. (1970b) Morphology of the conidial state of *Mycosphaerella musicola* Leach. *Transactions of the British Mycological Society* 54, 265–281.

Meredith, D.S., Lawrence, J.S. and Firman, I.D. (1973) Ascospore release and dispersal in black leaf streak disease of bananas (*Mycosphaerella fijiensis*). *Transactions of the British Mycological Society* 60, 547–554.

Milling, R.J. and Daniells, A. (1996) Effect of pyrimethanil on the infection process and secretion of fungal cell wall deagrading enzymes. In: Lyr, H., Russel P.E. and Sisler, H.D. (eds) *Modern Fungicides and Antifungal Compounds*. Intercept, Andover, UK, pp. 53–59.

Mobambo, K.N. and Naku, M. (1993) Situation de la cercosporiose noire des bananiers et plantains (*Musa* spp.) sous différents systèmes de culture à Yangambi, Haut-Zaïre. *Tropicultura* 11, 7–10.

Mobambo, K.N., Gauhl, F., Vuylsteke, D., Ortiz, R., Pasberg-Gauhl, C. and Swennen, R. (1993) Yield loss in plantain from black Sigatoka leaf spot and field performance of resistant hybrids. *Field Crops Research* 35, 35–42.

Mobambo, K.N., Pasberg-Gauhl, C., Gauhl, F. and Zuofa, K. (1994a) Early screening for black leaf streak/black Sigatoka disease resistance under natural inoculation conditions. *Infomusa* 3(2), 14–15.

Mobambo, K.N., Zuofa, K., Gauhl, F., Adeniji, M.O. and Pasberg-Gauhl, C. (1994b) Effect of soil fertility on host response to black leaf streak of plantain (*Musa* spp., AAB group) under traditional farming systems in southeastern Nigeria. *International Journal of Pest Management* 40, 75–80.

Mobambo, K.N., Gauhl, F., Pasberg-Gauhl, C. and Zuofa, K. (1996a) Season and plant age affect evaluation of plantain for response to black Sigatoka disease. *Crop Protection* 15, 609–614.

Mobambo, K.N., Gauhl, F., Swennen, R. and Pasberg-Gauhl, C. (1996b) Assessment of the cropping cycle effects on black leaf streak severity and yield decline of plantain and plantain hybrids. *International Journal of Pest Management* 42, 1–8.

Mobambo, K.N., Pasberg-Gauhl, C., Gauhl, F. and Zoufa, K. (1997) Host response to black sigatoka in *Musa* germplasm of different ages under natural conditions. *Crop Protection* 16, 359–363.

Molina, G. and Krausz, J.P. (1988) A phytotoxic activity in extracts of broth cultures of *Mycosphaerella fijiensis* var. *difformis* and its use to evaluate host resistance to black Sigatoka. *Plant Disease* 73, 142–143.

Mouliom-Pefoura, A. (1999) First observation of the breakdown of high resistance in Yangambi Km 5 (*Musa* sp.) to the black leaf streak disease in Cameroon. *Plant Disease* 83, 78.

Mouliom-Pefoura, A. and Mourichon, X. (1990) Développement de *Mycosphaerella musicola* (maladie de Sigatoka) et *M. fijiensis* (maladie des raies noires) sur bananiers et plantains. Etude du cas particulier des productions d'altitude. *Fruits* 45, 17–24.

Mouliom-Pefoura, A., Lassoudiere, A., Foko, J. and Fontem, D.A. (1996) Comparison of development of *Mycosphaerella fijiensis* and *Mycosphaerella musicola* on banana and plantain in the various ecological zones in Cameroon. *Plant Disease* 80, 950–954.

Mourichon, X. (1986) Mise en évidence de *Mycosphaerella fijiensis* Morelet, agent de la maladie des raies noires (black leaf streak) des bananiers plantains au Congo. *Fruits* 41, 371–374.

Mourichon, X. and Fullerton, R.A. (1990) Geographical distribution of the two species *Mycosphaerella musicola* Leach (*Cercospora musae*) and *M. fijiensis* Morelet (*C. fijiensis*), respectively agents of Sigatoka disease and black leaf streak disease in bananas and plantains. *Fruits* 45, 213–218.

Mourichon, X. and Zapater, M.F. (1990) Obtention *in vitro* du stade *Mycosphaerella fijiensis* forme parfaite de *Cercospora fijiensis*. *Fruits* 45, 553–557.

Mourichon, X., Peter, D. and Zapater, M.F. (1987) Inoculation expérimentale de *Mycosphaerella fijiensis* Morelet sur de jeunes plantules de bananiers issues de culture *in vitro*. *Fruits* 42, 195–198.

Mulder, J.L. and Holliday, P. (1971) Uromyces musae. Descriptions of Pathogenic Fungi and Bacteria No. 295, Commonwealth Mycological Institute, Kew, Surrey, UK.

Mulder, J.L. and Holliday, P. (1974a) Mycosphaerella fijiensis. Descriptions of Pathogenic Fungi and Bacteria No. 413, Commonwealth Mycological Institute, Kew, Surrey, UK.

Mulder, J.L. and Holliday, P. (1974b) Mycosphaerella musicola. Descriptions of Pathogenic Fungi and Bacteria No. 414, Commonwealth Mycological Institute, Kew, Surrey, UK.

Mulder, J.L. and Stover, R.H. (1976) *Mycosphaerella* species causing banana leaf spot. *Transactions of the British Mycological Society* 67, 77–82.

Müller, R., Pasberg-Gauhl, C., Gauhl, F., Ramser, J. and Kahl, G. (1997) Oligonucleotide fingerprinting detects genetic variability at different levels in Nigerian *Mycosphaerella fijiensis. Journal of Phytopathology* 145, 25–30.

Nokoe, S. and Ortiz, R. (1998) Optimum plot size for banana trials. *Hortscience* 33, 130–132.

Okole, B.N. and Schulz, F.A. (1993) Selection of banana and plantain (*Musa* spp.) resistant to toxins produced by *Mycosphaerella* species using in-vitro culture techniques. In: Ganry, J. (ed.) *Breeding Banana and Plantain for Resistance to Diseases and Pests, Proceedings of the International Symposium on Genetic Improvement of Bananas to Diseases and Pests Organized by CIRAD–FLHOR, Montpellier, France, 7–9 September 1992.* CIRAD, Montpellier, France, p. 378.

Orjeda, G. (1998) *Evaluation of Musa Germplasm for Resistance to Sigatoka Diseases and Fusarium Wilt.* INIBAP Technical Guidelines 3, INIBAP, Montpellier, France, 63 pp.

Orth, A.B. (1998) Molecular genetics of fungicide resistance. In: *Proceedings of the 7th International Conference in Plant Pathology, Edinburgh, Scotland, 9–16 August 1998.* International Society for Plant Pathology, Abstract 5.5.4S.

Ortiz, R., Vuylsteke, D., Okoro, J., Ferris, S., Hemeng, O.B., Yeboah, D.F., Anojulu, C.C., Adelaja, B.A., Arene, O.B., Agbor, A.N., Nwogu, A.N., Kayode, G.O., Ipinmoye, I.K., Akele, S. and Lawrence, A. (1993) Host response to black Sigatoka across West and Central Africa. *Musafrica* 3, 8–10.

Ortiz, R., Vuylsteke, D., Okoro, J., Ferris, S., Dumpe, B., Apanisile, R., Fouré, E., Jenny, C., Hemeng, O.B., Yeboah, D.K., Anojulu, C.C., Adelaja, B.A., Arene, O.B., Ikiediugwu, F.E.O., Agbor, A.N., Nwogu, A.N., Kayode, G.O., Ipinmoye, I.K., Akele, S. and Lawrence, A. (1995) Genotypic responses of *Musa* germplasm to black Sigatoka disease in West and Central Africa. *Musafrica* 6, 16–18.

Parham, B.E.V. (1935) Annual report of banana disease investigations, 1934. In: *Report of the Department of Agriculture of Fiji for 1934.* Department of Agriculture, Fiji, Suva.

Parnell, M., Burt, P.J.A. and Wilson, K. (1998) The influence of exposure to ultraviolet radiation in simulated sunlight on ascospores causing black Sigatoka disease of banana and plantain. *International Journal of Biometeorology* 42, 22–27.

Pasberg-Gauhl, C. (1989) *Untersuchungen zur Symptomenttwicklung und Bekämpfung der Schwarzen Sigatoka-Krankheit (*Mycosphaerella fijiensis *MORELET) an bananen (*Musa *sp.)* in vitro *und im Freiland.* Göttinger Beiträge zur Land- und Forstwirtschaft in den Tropen und Subtroen, Heft 40, 142 pp.

Pasberg-Gauhl, C. (1990) Development of black Sigatoka disease on different banana and plantain clones propagated by rhizomes and shoot-tip culture in Costa Rica. In: *Report of the First Research Coordination Meeting on Mutation Breeding of Bananas and Plantains.* FAO/AIEA, Vienna, Austria, pp. 41–55.

Pasberg-Gauhl, C. (1993) Symptom development of black Sigatoka leaf spot on young or adult banana and plantain plants after natural inoculation. In: Gold, C. and Gemmill, B. (eds) *Biological and Integrated Control of Highland Banana and Plantain Pests and Diseases. Proceedings of a Research Coordination Meeting, Cotonou, Benin, 12–14 November 1991.* International Institute of Tropical Agriculture, Ibadan, Nigeria, pp. 263–275.

Pearson, M.N., Bull, P.B. and Shepherd, K. (1983) Possible sources of resistance to black Sigatoka in the Papua New Guinea Biological Foundation Banana Collection. *Tropical Pest Management* 29, 303–308.

Peregrine, W.T.H. (1989) Black leaf streak of banana *Mycosphaerella fijiensis* Morelet. *FAO Plant Protection Bulletin* 37(3), 130.

Philpott, J. and Knowles, C.H. (1913) *Report on a Visit to Sigatoka.* Pamphlet of the Department of Agriculture, Fiji 3, Suva.

Ploetz, R.C. and Mourichon, X. (1999) First report of black Sigatoka in Florida. *Plant Disease* 83, 300.

Ploetz, R.C., Channer, A.G., Chizala, C.T., Banda, D.L.N., Makina, D.W. and Braunworth, W.S., Jr (1992) A current appraisal of banana and plantain diseases in Malawi. *Tropical Pest Management* 38, 36–42

Pons, N. (1987) Notes on *Mycosphaerella fijiensis* var. *difformis. Transactions of the British Mycological Society* 89, 120–124.

Pons, N. (1990) Taxonomy of *Cercospora* and related genera. In: Fullerton, R.A. and Stover, R.H. (eds) *Sigatoka Leaf Spot Diseases of Bananas, Proceedings of an International Workshop held in San José, Costa Rica, March 28–April 1, 1989.* INIBAP, Montpellier, France, pp. 360–370.

Pont, W. (1960a) Epidemiology and control of banana leaf spot (*Mycosphaerella musicola* Leach) in North Queensland. *Queensland Journal of Agricultural Science* 17, 273–309.
Pont, W. (1960b) Three leaf speckle diseases of the banana in Queensland. *Queensland Journal of Agricultural Science* 17, 271–309.
Porras, A. and Peréz, L. (1997) The role of temperature in the growth of the germ tubes of ascospores of *Mycosphaerella* spp. responsible for leaf spot diseases of banana. *Infomusa* 6(2), 27–32.
Price, D. (1960) Climate and control of banana leaf spot. *SPAN* 3, 122–124.
Priest, M.J. (1990) Distribution of *Cordana* ssp. on *Musa* in Australia. *Mycological Research* 94, 861–863.
Punithalingam, E. (1974) Studies on Sphaeropsidales in culture. II. *Mycological Papers* 136, 1–63.
Punithalingam, E. (1976) A new species of *Phaeoseptoria* on *Musa*. *Kew Bulletin* 31, 469–470.
Punithalingam, E. (1983) Phaeoseptoria musae. Description of Pathogenic Fungi and Bacteria No. 772, Commonwealth Mycological Institute, Kew, Surrey, UK.
Punithalingam, E. and Holliday, P. (1975) Guignardia musae. Descriptions of Pathogenic Fungi and Bacteria No. 467, Commonwealth Mycological Institute, Kew, Surrey, UK.
Quimio, A.J. and Tessera, M. (1996) Diseases of enset. In: Abate, T., Hibisch, C., Brandt, S.A. and Gebremariam, S. (eds) *Proceedings of the International Workshop on Enset.* Institute of Agricultural Research, Addis Ababa, Ethiopia, pp. 188–203.
Raemaekers, R. (1975) Black leaf streak like disease in Zambia. *PANS* 21, 396–400.
Raghunath, T. (1963) A new leaf spot of banana from India. *Plant Disease Reporter* 47, 1084–1085.
Ramsey, M.D., Daniells, J.W. and Anderson, V.J. (1990) Effects of Sigatoka leaf spot (*Mycosphaerella musicola* Leach) on fruit yields, field ripening and greenlife of bananas in North Queensland. *Scientia Horticulturae* 41, 305–313.
Reddy, D.B. (1969) Black leaf streak (*M. fijiensis*). *Quarterly Newsletter of the Plant Protection Committee for South East Asia and Pacific Region* 12(1–3). FAO, Bangkok, Thailand.
Rhodes, P.L. (1964) A new banana disease in Fiji. *Commonwealth Phytopathological News* 10, 38–41.
Robinson, J.C. (1996) *Bananas and Plantains.* Crop Production Science in Horticulture 5, CAB International, Wallingford, UK, 238 pp.
Romero C., R. (1986) Impacto de Sigatoka Negra y Roya del Cafeto en actividad platanera nacional. *Revista de la Asociación Bananera Nacional (ASBANA), San José, Costa Rica* 9(26), 10–11.
Romero, R. and Gauhl, F. (1988) Determinación de la severidad de la Sigatoka negra (*Mycosphaerella fijiensis* var. *difformis*) en bananos a diferentes altitudes sobre el nivel del mar. *Revista de la Asociación Banaanera Nacional (ASBANA), San José, Costa Rica* 12(29), 7–10.
Romero, R.A., and Sutton, T.B. (1997a) Sensitivity of *Mycosphaerella fijiensis*, causal agent of black Sigatoka of banana, to propiconazole. *Phytopathology* 87, 96–100.
Romero, R.A., and Sutton, T.B. (1997b) Reaction of four *Musa* genotypes at three temperatures to isolates of *Mycosphaerella fijiensis* from different regions. *Plant Disease* 81, 1139–1142.
Romero, R.A., and Sutton, T.B. (1998) Characterization of benomyl resistance in *Mycosphaerella fijiensis*, cause of black Sigatoka of banana, in Costa Rica. *Plant Disease* 82, 931–934.
Rowe, P.R. and Richardson, D.L. (1975) *Breeding Bananas for Disease Resistance, Fruit Quality and Yield, Bulletin No. 2 – December, 1975.* Tropical Agriculture Research Services (SIATSA), La Lima, Honduras, 41 pp.
Rutter, J., Burt, P.J.A. and Ramirez, F. (1998) Movement of *Mycosphaerella fijiensis* spores and sigatoka disease development on plantain close to an inoculum source. *Aerobiologia* 14, 201–208.
Schaffrath, U., Schelinpflug, H. and Reisner, H.J. (1995) An elicitor from *Pyricularia oryzae* induces resistance response in rice: isolation, characterization and physiological properties. *Physiological and Molecular Plant Patholo*gy 46, 293–307.
Sebasigari, K. (1990) Effects of black Sigatoka (*Mycosphaerella fijiensis* Morelet) on bananas and plantains in the Imbo Plain in Rwanda and Burundi. In: Fullerton, R.A. and Stover, R.H. (eds) *Sigatoka Leaf Spot Diseases of Bananas. Proceedings of an International Workshop held at San José, Costa Rica, 28 March–1 April 1989.* International Network for the Improvement of Banana and Plantain (INIBAP), Montpellier, France, pp. 61–65.
Sebasigari, K. and Stover, R.H. (1988) *Banana Diseases and Pests in East Africa. Report of a Survey in November 1987.* INIBAP, Montpellier, France, 15 pp.
Serrano, E. and Marín, D.H. (1998) Disminución de la productividad bananera en Costa Rica. *CORBANA* 23, 85–96.

Sharp, J.K., McNeil, M. and Albersheim, P. (1984) The primary structures of one elicitor-active and seven elicitor-inactive hexa(-D-glucopyranosyl)-D-glucitols isolated from mycelial walls of *Phytophthora megasperma* f. sp. *glycinea*. *Journal of Biological Chemistry* 259, 11321–11336.
Sharrock, S. (1990) Collecting *Musa* in Papua New Guinea. In: Jarret, R.L. (ed.) *Identification of Genetic Diversity in the Genus* Musa, *Proceedings of an International Workshop held in Los Banös, Philippines, 5–10 September 1988*. INIBAP, Montpellier, France, pp. 140–157.
Sharrock, S.L. and Jones, D.R. (1989) *Report on a Third IBPGRI/QDPI Banana Germplasm Collecting Mission to Papua New Guinea, 15 February to 12 March 1989*. Queensland Department of Primary Industries, Brisbane, Australia, 7 pp.
Shaw, D.E. (1984) *Microorganisms in Papua New Guinea*. Research Bulletin No. 33, Department of Primary Industries, Port Moresby, Papua New Guinea, 344 pp.
Simmonds, J.H. (1933) Banana leaf spot. Progress report. *Queensland Agricultural Journal* 39, 21–40.
Simmonds, J.H. (1939) Influence of seasonal conditions on the development of *Cercospora* leaf spot of the banana with special reference to the control programme. *Queensland Agricultural Journal* 52, 633–647.
Simmonds, N.W. (1959) *Bananas*. Longman, Green, London, 466 pp.
Simmonds, N.W. (1966) *Bananas,* 2nd edn. Longman, Green, London, UK, 512 pp.
Stahel, G. (1934) The banana leaf disease in Surinam. *Tropical Agriculture (Trinidad)* 11, 138–142.
Stahel, G. (1937a) The banana leaf speckle in Surinam caused by *Chloridium musae* Nov. Spec. and another related banana disease. *Tropical Agriculture (Trinidad)* 14, 42–45.
Stahel, G. (1937b) Notes on Cercospora leaf spot of bananas (*Cercospora musae*). *Tropical Agriculture (Trinidad)* 14, 257–264.
Stierle, A.A., Upadhyay, R., Hershenhorn, J., Strobel, G.A. and Molina, G. (1991) The phytotoxins of *Mycosphaerella fijiensis*, the causative agent of black Sigatoka disease of bananas and plantains. *Experientia* 47, 853–859.
Stover, R.H. (1962) Intercontinental spread of banana leaf spot (*Mycosphaerella musicola* Leach) *Tropical Agriculture (Trinidad)* 29, 327–338.
Stover, R.H. (1963) Sexuality and heterothallism in *Mycosphaerella musicola*. *Canadian Journal of Botany* 41, 1531–1532.
Stover, R.H. (1965) Leaf spot of bananas caused by *Mycosphaerella musicola*: effect of temperature on germination, hyphal growth and conidia production. *Tropical Agriculture (Trinidad)* 42, 351–360.
Stover, R.H. (1969) The *Mycosphaerella* species associated with banana leaf spots. *Tropical Agricultural (Trinidad)* 46, 325–332.
Stover, R.H. (1971a) Ascospore survival in *Mycosphaerella musicola*. *Phytopathology* 61, 139–141.
Stover, R.H. (1971b) A proposed international scale for estimating intensity of banana leaf spot. *Tropical Agriculture (Trinidad)* 48, 185–196.
Stover, R.H. (1972) *Banana Plantain and Abaca Diseases*. Commonwealth Mycological Institute, Kew, Surrey, UK, 316 pp.
Stover, R. H. (1974) Pathogenic and morphologic variation in *Mycosphaerella fijiensis* (*M. musicola*). *Proceedings of the American Phytopathological Society* 1, 123.
Stover, R.H. (1976) Distribution and cultural characteristics of the pathogens causing banana leaf spot. *Tropical Agriculture (Trinidad)* 53, 111–114.
Stover, R.H. (1978) Distribution and probable origin of *Mycosphaerella fijiensis* in Southeast Asia. *Tropical Agriculture (Trinidad)* 55, 65–68.
Stover, R.H. (1980a) Sigatoka leaf spots of bananas and plantains. *Plant Disease* 64, 750–755.
Stover, R.H. (1980b) Sigatoka leaf spots of banana and plantain. In: Krigsvold, D.T. and Woods, T.L. (eds) *Proceedings of the Sigatoka Workshop, 18–19 February 1980, La Lima, Honduras*. United Fruit Company, La Lima, Honduras, pp. 1–18.
Stover, R.H. (1983a) Effet du Cercospora noir sur les plantains en Amérique centrale. *Fruits* 38, 326–329.
Stover, R.H. (1983b) The effect of temperature on ascospore germinative tube growth of *Mycosphaerella musicola* and *Mycospaerella fijiensis* var. *difformis*. *Fruits* 38, 625–628.
Stover, R.H. (1986a) Disease management strategies and the survival of the banana industry. *Annual Review of Phytopathology* 24, 83–91.
Stover, R.H. (1986b) Measuring response of *Musa* cultivars to the Sigatoka pathogens and proposed screening procedures. In: Persley, G.J. and De Langhe, E.A. (eds) *Banana and Plantain Breeding*

Strategies, Proceedings of an International Workshop held at Cairns, Australia, 13–17 October 1986. ACIAR Proceedings No. 21, ACIAR, Canberra, Australia, pp. 114–118.

Stover, R.H. (1987) Produccion de platano en presencia de la Sigatoka Negra. *Union de Paises Exportadores de Banano (UPEB), Informe Mensual* 82, 50–56.

Stover, R.H. (1990) Sigatoka leaf spots: thirty years of changing control strategies: 1959–1989. In: Fullerton, R.A. and Stover, R.H. (eds) *Sigatoka Leaf Spot Diseases of Bananas, Proceedings of an International Workshop held in San José, Costa Rica, March 28–April 1, 1989.* INIBAP, Montpellier, France, pp. 66–74.

Stover, R.H. (1994) *Mycosphaerella musae* and *Cercospora* 'non-virulentum' from Sigatoka leaf spots are identical. *Fruits* 49, 187–190.

Stover, R.H. and Dickson, J.D. (1970) Leaf spot of bananas caused by *Mycosphaerella musicola*: methods of measuring spotting prevalence and severity. *Tropical Agriculture (Trinidad)* 47, 289–302.

Stover, R.H. and Dickson, J.D. (1976) Banana leaf spot caused by *Mycosphaerella musicola and M. fijiensis* var. *difformis*: a comparison of the first Central American epidemics. *FAO Plant Protection Bulletin* 24, 36–42.

Stover, R.H. and Fulton, R.H. (1966) Leaf spot of bananas caused by *Mycosphaerella musicola*: the relation of infection sites to leaf development and spore type. *Tropical Agriculture (Trinidad)* 43, 118–129.

Stover, R.H. and Simmonds, N.W. (1987) *Bananas*, 3rd edn. Longman Scientific and Technical, Harlow, Essex, UK, 468 pp.

Strobel, G.A., Stierle, A.A., Upadhyay, R., Hershenhorn, J. and Molina, G. (1993) The phytotoxins of *Mycosphaerella fijiensis*, the causative agent of black Sigatoka disease, and their potential use in screening for disease resistance. In: *Proceedings of the Workshop on Biotechnology Applications for Banana and Plantain Improvement held in San José, Costa Rica, 27–31 January 1992.* INIBAP, Montpellier, France, pp. 93–103.

Subramanian, C.V. (1968) *Deightoniella torulosa. CMI Descriptions of Pathogenic Fungi and Bacteria* 165, 2 pp.

Tejerina, J.C. (1997) First report of black Sigatoka in Bolivia. *Plant Disease* 81, 1332.

Tessera, M. and Quimio, A.J. (1994) Research on ensat pathology. In: Herath, E. and Desalegn, L. (eds) *Proceedings of the 2nd National Horticultural Workshop of Ethiopia.* Institute of Agricultural Research, Addis Ababa, pp. 217–225.

Tezenas du Montcel, H. (1982) *Propositions d'études pour la lutte contre la cercosporiose noire sur les plantains d'Hevecam au Cameroun. Rapport d'activités 1980–82.* IRA, Centre de recherches d'Ekona, Ekona, Cameroon.

Tezenas du Montcel, H. (1990) The susceptibility of various cultivated bananas to Sigatoka diseases. In: Fullerton, R.A. and Stover, R.H. (eds) *Sigatoka Leaf Spot Diseases of Bananas: Proceedings of an International Workshop held in San José, Costa Rica, March 28–April 1, 1989.* INIBAP, Montpellier, France, pp. 272–289.

Thal, W.M., Sauter, H.P., Spurr, H.W. Jr and Arroyo, T. (1992) Factors influencing airborne ascospore counts of *Mycosphaerella fijiensis*, cause of black sigatoka on banana. *Phytopathology* 82, 1163.

Tsai, Y.P., Chen, H.P. and Liu, S.H. (1993) Freckle disease of banana in Taiwan. In: *Proceedings: International Symposium on Recent Developments in Banana Cultivation Technology, Chiuju, Pingtung, Taiwan, 14–18 December 1992.* INIBAP–ASPNET, Los Baños, Laguna, Philippines, pp. 298–307.

Tushemereirwe, W.K. (1996) Factors influencing the expression of leaf spot diseases of highland bananas in Uganda. PhD thesis, University of Reading, UK, 197 pp.

Tushemereirwe, W.K. and Waller, J.M. (1993) Black leaf streak (*Mycosphaerella fijiensis*) in Uganda. *Plant Pathology* 42, 471–472.

Upadhyay, R., Strobel, G.A. and Coval, C. (1990) Some toxins of *Mycosphaerella fijiensis.* In: Fullerton, R.A. and Stover, R.H. (eds) *Sigatoka Leaf Spot Diseases of Bananas: Proceedings of an International Workshop held at San José, Costa Rica, March 28–April 1, 1989.* INIBAP, Montpellier, France, pp. 231–236.

Vakili, N.G. (1963) A leaf spotting disease of *Musa* seedlings incited by *Pestalotia palmarum* Cke. *Plant Disease Reporter* 47, 644–646.

Vakili, N.G. (1968) Responses of *Musa acuminata* species and edible cultivars to infection by *Mycosphaerella musicola. Tropical Agriculture (Trinidad)* 45, 13–32.

van der Aa, H. (1973) Studies in Phyllosticta I. *Studies in Mycology* 5, 1–110.

Vidal, A. (1992) Outbreaks and new records. Cuba. Black Sigatoka. *FAO Plant Protection Bulletin* 40, 1–2.

Vuylsteke, D., Ortiz, R. and Fouré, E. (1993a) Genotype by environment interaction and black Sigatoka resistance in the humid forest zone of West and Central Africa. *Musafrica* 2, 6–7.

Vuylsteke, D., Ortiz, R., Pasberg-Gauhl, C., Gauhl, F., Gold, C., Ferris, S. and Speijer, P. (1993b). Plantain and banana research at the International Institute of Tropical Agriculture. *HortScience* 28(9), cover story, 3 pp.

Vuylsteke, D., Ortiz, R., Ferris, S. and Swennen, R. (1995) Pita-9: a black Sigatoka resistant hybrid from the false horn plantain gene pool. *Hortscience* 30, 395–397.

Wardlaw, C.W. (1941) The banana in Central America. II. The control of *Cercospora* leaf disease. *Nature* 147, 344–349.

Wardlaw, C.W. (1961) *Banana Diseases Including Plantains and Abaca.* Longman, Green, London, 648 pp.

Wardlaw, C.W. (1972) *Banana Diseases Including Plantains and Abaca*, 2nd edn. Longman, London, UK, 878 pp.

Watson, B. (1993) Major banana cultivars in Pacific atoll countries. *Infomusa* 2(2), 19–20.

Weir, B.S. (1990) *Genetic Data Analysis.* Sinauer Associates, Sunderland, UK, 377 pp.

Wielemaker, F. (1990) Practical notes on black Sigatoka control. In: Fullerton, R.A. and Stover, R.H. (eds) *Sigatoka Leaf Spot Diseases of Bananas, Proceedings of an International Workshop held at San José, Costa Rica, March 28–April 1, 1989.* INIBAP, Montpellier, France, pp. 107–114.

Wilson, G.F. (1987) Status of bananas and plantains in West Africa. In: Persley, G.J. and De Langhe, E.A. (eds) *Banana and Plantain Breeding Strategies, Proceedings of an International Workshop held at Cairns, Australia, 13–17 October 1986.* ACIAR Proceedings 21, Australian Centre for International Agricultural Research (ACIAR), Canberra, Australia, pp. 29–35.

Wilson, G.F. and Buddenhagen, I. (1986) The black Sigatoka threat to plantain and banana in West Africa. *IITA Research Briefs* 7, 3.

Woods, T.L. (1980) The black Sigatoka situation in Costa Rica. In: Krigsvold, D.T. and Woods, T.L. (eds) *Proceedings of the Sigatoka Workshop, 18–19 February 1980, La Lima, Honduras.* United Fruit Company, La Lima, Honduras, pp. 19–20.

Wright, S. (1951) The genetical structure of populations. *Annals of Eugenics* 15, 323–354.

Yoder, O.C. (1980) Toxins in pathogenesis. *Annual Review of Phytopathology* 18, 103–129.

Zhou, Z. and Xie, L. (1990) Status of banana diseases in China. *Fruits* 47, 715–721.

Zimmerman, A. (1902) Uber einige tropischer Kulturpflanzen beobachtete Pilze. *Zentralblatt für Bakteriologie, Parasitenkunde, Infektionskrankheiten und Hygiene* 8, 219 (abstract).

3

Fungal Diseases of the Root, Corm and Pseudostem

Fusarium Wilt

R.C. Ploetz and K.G. Pegg

Introduction

Fusarium wilt is one of the most destructive and notorious diseases of banana. It is also known as Panama disease, in recognition of the extensive damage it caused in export plantations in this Central American country (Pegg and Langdon, 1987). By 1960, Fusarium wilt had destroyed an estimated 40,000 ha of 'Gros Michel' (AAA), causing the export industry to convert to cultivars in the Cavendish subgroup (AAA). Based on these losses, Simmonds (1966) believed Fusarium wilt to be one of the most catastrophic of all plant diseases.

The pathogen that causes Fusarium wilt systemically colonizes the xylem of susceptible banana cultivars and causes a lethal vascular wilt. Abacá is also affected, but damage on this crop is usually restricted to a reduction in fibre quality (Waite, 1954). Fusarium wilt was the most serious disease to affect abacá in the western hemisphere between 1900 and 1960.

History and dispersal

Fusarium wilt is thought to have originated in South-East Asia, but was first recognized elsewhere. Bancroft's (1876) initial description from Australia was quickly followed by reports from tropical America (Costa Rica and Panama in 1890) (Stover, 1962). A dramatic increase in the number of new records occurred in the early 1900s (Fig. 3.1), most of which described damage in export plantations (Ploetz, 1992). Currently, the disease is found in virtually all areas where banana is grown. Exceptional locations where it is not known include islands in the South Pacific, parts of Melanesia, countries around the Mediterranean Sea and Somalia (Stover and Simmonds, 1987; Pegg *et al.*, 1993).

Humans have been the key factor in the global distribution of Fusarium wilt. Since infected rhizomes are usually free of visual symptoms, it was not uncommon for the pathogen to be introduced to new areas in conventional planting material.

Stover (1962) speculated that the dissemination of susceptible clones into new areas closely paralleled the movement of the disease. He indicated that the highly susceptible 'Silk' (AAB) was introduced into the West Indies before 1750, and that this cultivar was widely planted as shade for cacao in the region. Since the dissemination of 'Silk' in the West preceded that of 'Gros Michel', he believed that its presence set the stage for widespread epidemics in the early export plantations.

The expansion of the export trade was linked to the further dispersal of the pathogen (Stover, 1962). Before the 1960s, the trade was wholly dependent upon susceptible 'Gros Michel'. In the western hemisphere, 'Gros Michel' was among the first cultivars that was reported to be affected in 22 of 25 locations, a situation

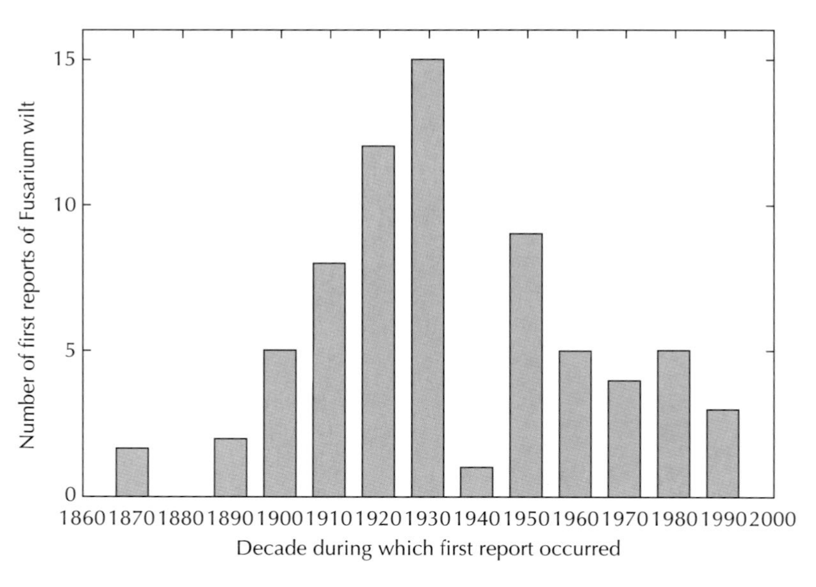

Fig. 3.1. Number of first reports of Fusarium wilt on banana during a given decade in the 19th and 20th centuries. Data are from Stover (1962), Anon. (1977), Ploetz *et al.* (1990), Pegg *et al.* (1994), Shivas and Philemon (1996) and Shivas *et al.* (1996).

that was repeated in West Africa (Ploetz, 1992). Although an increasing demand for bananas in the importing countries caused more land to be planted to 'Gros Michel', increased planting was also due to the effects of Fusarium wilt. Since the disease reduced production in many plantations and caused the abandonment of others, the trade needed to continually replant in new, disease-free areas. This planting and abandonment cycle hastened the spread of the disease (Stover, 1962).

Elsewhere, the pathogen appears to have been spread on susceptible cultivars used for smallholder production. For example, the first reports in East Africa occurred on the ABB cultivars 'Bluggoe' and 'Pisang Awak', which were introduced to the region around the Second World War (Wallace, 1952; Jameson, 1953; Anon., 1954). Fusarium wilt is now widespread in East Africa, but is still found primarily on the recently introduced susceptible cultivars (Ploetz, 1993). Recent reports of the disease on the East African highland cultivars of the Lujugira–Mutika subgroup (AAA) appear to be associated with the spread of the introduced clones (Ploetz *et al.*, 1995).

Recently, Fusarium wilt was reported in New Guinea, an island that had been historically free of the disease. Emigrants from Java are thought to have carried infected rhizomes to a newly infested site at Manokweri in Irian Jaya, with subsequent movement to a neighbouring site at Bewani in the West Sepik Province of Papua New Guinea (Shivas and Philemon, 1996; Shivas *et al.*, 1996). The Indonesian government has encouraged transmigration to ease population pressures on Java. As this activity continues, it is probable that additional outbreaks of Fusarium wilt will be detected in Irian Jaya. Since the Irian Jaya and Papua New Guinea sites are 800 km apart and because native populations move freely across the intervening border, the disease may already be widely disseminated on the island. The appearance of Fusarium wilt in Melanesia emphasizes the need to observe quarantine laws in the few areas that remain free of this disease.

Table 3.1. Vegetative compatibility among strains of *Fusarium oxysporum* f. sp. *cubense*.

VCG*	Genomic group/cultivar(s)†	Origin(s)
0120–01215	*Musa* sp.; AA: 'SH-3142', 'SH-3362' AAA: 'Gros Michel', 'Highgate', 'Pisang Ambon Putih', 'Pisang Ambon', 'Dwarf Cavendish', 'Williams', 'Mons Mari', Grand Nain', 'Lacatan'; AAB: 'Prata', 'Lady Finger', 'Pacovan', 'Huamoa', 'Silk'	Australia, Brazil, Costa Rica, France (Guadeloupe, Guyane), Honduras, Indonesia (Java), Jamaica, Malaysia (Sarawak), Nigeria, Portugal (Madeira), South Africa, Spain (Canary Islands), Taiwan, USA (Florida)
0121	AAA: 'Gros Michel', Cavendish clone	Indonesia (Sumatra), Taiwan
0122	AAA: Cavendish clone ABB: 'Saba'	Philippines
0123	*Musa textilis*; AAA: 'Gros Michel', 'Grand Nain'; AAB: 'Silk', 'Latundan', 'Pisang Keling'; ABB: 'Pisang Awak', 'Kluai Namwa'	Malaysia (West and Sarawak), Philippines, Taiwan, Thailand
0124–0125–0128–01220	AA: 'Niyarma Yik'; AAA: 'Gros Michel', 'Highgate', 'Mons Mari', 'Williams', 'Grand Nain', 'Mbwazirume'; AAAA: 'Jamaica 1242'; AB: 'Ney Poovan', 'Sukali Ndizi', 'Kamaramasenge'; AAB: 'Lady Finger', 'Maçã', 'Manzano', 'Maqueño'; ABB: 'Pisang Awak', 'Ducasse', 'Kayinga', 'Zambia', 'Kluai Namwa', 'Bluggoe', 'Harare', 'Kholobowa', 'Dwarf Bluggoe', 'Mbufu', 'Burro Criolla', 'Pelipita', 'Ice Cream'	Australia, Brazil, Burundi, China, Cuba, Democratic Republic of Congo, Haiti, Honduras, India, Jamaica, Malawi, Malaysia (West), Mexico, Nicaragua, Rwanda, Tanzania, Thailand, Uganda, USA (Florida)
0126	AA: 'Pisang Berlin'; AAA: 'Highgate'; AAB: 'Maqueño'; AAB: 'Pisang Manurung'	Honduras, Indonesia (Irian Jaya, Sulawesi), Papua New Guinea, Philippines
0129	AAA: 'Mons Mari'; AAB: 'Lady Finger'	Australia
01210	AAA: 'Gros Michel'; AAB: 'Manzano'	Cayman Islands, Cuba, USA (Florida)
01211	AA: 'SH-3142'	Australia
01212	AB: 'Ney Poovan'; AAB: 'Silk'; ABB: 'Pisang Awak', 'Kisubi', 'Bluggoe'	Tanzania
01213–01216	AA: 'Pisang Lilin', 'Pisang Mas'; AAA: 'Pisang Ambon', 'Valery', 'Williams', 'Grand Nain', 'Novaria','Pisang Nangka', 'Red', 'Pisang Raja Udang', 'Pisang Susu', 'Pisang Barangan'; AAB: 'Pisang Raja Serah', 'Pisang Rastali', 'Pisang Raja', 'Relong'; ABB: 'Pisang Awak', 'Pisang Awak Legor', 'Saba', 'Pisang Kepok', 'Pisang Caputu', 'Pisang Kosta'; Unknown: 'Pisang Batan'	Australia (Northern Territory), Indonesia (Halmahera, Irian Jaya, Java, Sulawesi, Sumatra), Malaysia (West), Taiwan
01214	ABB: 'Harare', 'Mbufu'	Malawi
01217	AAB: 'Pisang Rastali'	Malaysia
01218	AAB: 'Pisang Rastali', 'Pisang Raja Serah'; ABB: 'Pisang Awak', 'Kluai Namwa', 'Pisang Kepok', 'Pisang Siam'	Indonesia (Java, Sumatra), Malaysia (West), Thailand
01219	AAA: 'Pisang Ambon', 'Pisang Ambon Putih'; Unknown: 'Pisang Raja Garing'	Indonesia (Java, Sumatra)
01221	ABB: 'Kluai Namwa'	Thailand

* VCGs are according to Hernandez *et al.* (1993), Pegg *et al.* (1994), Ploetz (1994), Ploetz *et al.* (1997), N.Y. Moore and K.G. Pegg, Brisbane, 1998, unpublished and R.C. Ploetz, Florida, 1998, unpublished, and were determined with nitrate-auxotrophic mutants, as described previously (Correll *et al.*, 1987). VCG complexes (e.g. 0120–01215) are those in which isolates in different VCGs are capable of infrequently forming heterokaryons (see Ploetz, 1990a, and Ploetz *et al.*, 1997, for details).

†Refer to Tables 1.1–1.7 for more details of clone or subgroup of affected cultivars. Information in addition to Tables 1.1–1.7: 'SH-3142' and 'SH-3362' (breeding diploids from the Honduran Programme); 'Mons Mari' (Cavendish clone in Australia); 'Novaria' (radiation-induced mutant of 'Grand Nain' reputed to have a shorter cycling time); 'Highgate' (dwarf mutant of 'Gros Michel'); 'Pisang Nangka' and 'Pisang Susu' (local cultivars from Malaysia–Indonesia region); 'Jamaica 1242' (tetraploid bred in early 1970s with 'Highgate' as the female parent); 'Huamoa' (Maia Maoli–Popoulu subgroup); 'Zambia', 'Pisang Awak Legor' ('Pisang Awak' types); 'Harare', 'Kholobowa', 'Pisang Kosta', 'Mbufu', 'Burro Criolla' (Bluggoe subgroup); 'Pisang Capatu', 'Pisang Kepok' ('Saba'); 'Pisang Siam' (syn. 'Kluai Teparot').

Plate 3.1. Yellowing of the oldest leaves of 'Bluggoe' (ABB) affected by Fusarium wilt in Honduras. The plant visible in the right background has more advanced symptoms (photo: D.R. Jones, INIBAP).

Plate 3.2. Splitting of the lower portion of the pseudostem can be an early indication of Fusarium wilt (photo: D.R. Jones, INIBAP).

Plate 3.3. Collapse of lower leaves of 'Kluai Namwa' (ABB, syn. 'Pisang Awak') affected by Fusarium wilt in Thailand (photo: D.R. Jones, INIBAP).

Plate 3.4. 'Rasthali' (AAB, syn. 'Silk') dying of Fusarium wilt near Bangalore in India. Note that the young plant in the foreground is also affected. 'Silk' is extremely susceptible to Fusarium wilt (photo: D.R. Jones, INIBAP).

Impact

Fusarium wilt ceased to be a concern for the export trade once the plantations converted to cultivars in the Cavendish subgroup (Stover and Simmonds, 1987). Although these clones continue to perform well in the western hemisphere, they are affected in other areas.

The first reports of the disease on Cavendish clones were from the subtropics in the eastern hemisphere. Damage to 'Dwarf Cavendish' was reported in the Canary Islands in the 1920s (Ashby, 1926) and thereafter in Australia, South Africa and Taiwan (Su *et al.*, 1986). In the latter locations, very serious efforts have been made to identify new, disease-resistant clones (Jones, 1994).

Recently, Cavendish plantations in the tropics have also been affected (Pegg *et al.*, 1994; Ploetz, 1994; Bentley *et al.*, 1998). Although minor problems had been observed in the tropics for many years, the recent outbreaks on Cavendish clones have caused extensive damage in Halmahera, Sumatra and Java in Indonesia and in Johore State in West Malaysia. Unlike the subtropical cases, the outbreaks do not appear to involve predisposing factors (see below). Instead, a unique population of the pathogen is implicated. Although this population has not been found outside South-East Asia and its immediate environs, these developments have ominous implications about the durability of the western export trade, which is based on the Cavendish subgroup.

Currently, the most important effect of Fusarium wilt is on local and subsistence production (Ploetz, 1994). The affected cultivars have diverse genotypes and uses and are widespread geographically (Table 3.1).

On abacá, the disease is found wherever the crop is grown commercially. In the Philippines, serious damage occurs only on a few plantings at high altitude and is said to be more severe in the dry season. Fusarium wilt on abacá became more important when major production shifted to Central America during the Second World War (Waite, 1954). Outbreaks were recorded in Costa Rica, Guatemala, Honduras and Panama (Waite, 1954).

Symptoms

The characteristic internal symptom of Fusarium wilt is a reddish to dark brown discoloration of the host's vascular system. The first internal symptoms occur in feeder roots, which are the initial sites of infection. These symptoms progress to the rhizome and are most pronounced where the stele joins the cortex (Stover, 1962). Eventually, the pseudostem is colonized. In the latter organ, symptoms are often evident as faint brown streaks and/or flecks when outer portions of older leaf sheaths are examined (Stover, 1972; Ploetz, 1992).

The first external symptoms of Fusarium wilt in banana are a yellowing of the oldest leaves (Plate 3.1) or a longitudinal splitting of the lower portion of the outer leaf sheaths on the pseudostem (Plate 3.2). This is followed by a wilt and collapse of leaves at the petiole base (Plate 3.3) (Stover, 1972). In some cases, these leaves remain green. As the disease progresses, younger and younger leaves collapse, until the entire canopy consists of dead or dying leaves (Plate 3.4) (Ploetz, 1994). At this stage, a pronounced, red-brown discoloration of the vascular tissues is usually evident if the pseudostem is cut (Plate 3.5). After plants die, they usually remain standing for 1–2 months before they decay and topple.

Two key features help distinguish Fusarium wilt from Moko disease (see Thwaites *et al.*, Chapter 5, pp. 213–221). External symptoms of Fusarium wilt do not usually develop on plants and suckers that are less than about 4 months old, whereas plants that are affected by Moko disease will wilt and become chlorotic at a very early stage of development. In addition, internal portions of fruit are discoloured brown by Moko disease, but not by Fusarium wilt.

In the Philippines, the first noticeable symptom on abacá is the inward curling of the leaf blade at or near the tip of the lower leaves, which then gradually droop, wilt and turn from pale yellow to yellowish brown. Newly emerged leaves are reduced in size and also wilt. A reddish-violet discoloration is noticeable in the vascular system when infected rhizomes are cut open. In Central America, external symptoms on abacá were said to be less conspicuous than on banana. Leaves dried and were broken by the wind. The fibre from decorticated pseudostems of diseased abacá was discoloured and graded low in quality (Stover, 1972).

Causal agent

The soil-borne fungus *Fusarium oxysporum* Schlect. f. sp. *cubense* (E.F. Smith) Snyder and Hansen is the cause of Fusarium wilt.

Host range

Fusarium oxysporum f. sp. *cubense* is one of around 100 formae speciales (special forms) of *F. oxysporum* that cause vascular wilts of flowering plants (Gerlach and Nirenberg, 1982). Hosts of the various formae speciales are usually restricted to a limited and related set of taxa. As currently defined, *F. oxysporum* f. sp. *cubense* affects the following species in the order *Zingiberales*: in the family *Musaceae*, *Musa acuminata* Colla, *Musa balbisiana* Colla, *Musa schizocarpa* Simmonds and *Musa textilis* Née; and in the family *Heliconeaceae*, *Heliconia caribaea* Lam, *Heliconia chartacea* Lane ex Barreiros, *Heliconia crassa* Griggs, *Heliconia collinsiana* Griggs, *Heliconia latispatha* Benth., *Heliconia mariae* Hook, *Heliconia rostrata* Ruiz and Pavon, and *Heliconia vellerigera* Poeppig (Stover, 1962; Waite, 1963). Additional hosts include hybrids between *M. acuminata* and *M. balbisiana* and between *M. acuminata* and *M. schizocarpa.*

Physiological races

Traditionally, four races of *F. oxysporum* f. sp. *cubense* are recognized (Stover and Buddenhagen, 1986; Stover and Simmonds, 1987; Stover, 1990) (Table 3.2). Race 1 was responsible for the epidemics on 'Gros Michel' and also affects 'Maqueño' (AAB, Maia Maoli–Popoulu subgroup), 'Silk' (AAB), 'Pome' (AAB) (Plate 3.6), 'Pisang Awak' (ABB) (see Plate 3.3) and 'I.C.2' (AAAA). Race 2 affects ABB cooking bananas, such as 'Bluggoe' (ABB) (see Plate 3.2), and some bred AAAA tetraploids.

Race 3 was reported to affect *Heliconia* spp. and was weakly pathogenic on 'Gros Michel' and seedlings of *M. balbisiana* (Waite, 1963). However, Fusarium wilt on *Heliconia* has not been reported since Waite's (1963) work more than 30 years ago and the disease has not been found in recent surveys in Central America (R.C. Ploetz, Florida, 1998, personal observation; R.H. Stover, Honduras, 1998, personal communication). For example, in 1991 no symptomatic plants of *Heliconia caribaea*,

Table 3.2. Susceptiblity of various species and clones to different races of *Fusarium oxysporum* f. sp. *cubense.*

Race	Susceptible species and clones
1	*Musa textilis* (abacá), 'Gros Michel' (AAA), 'Maqueño' (AAB, Maia Maoli-Popoulu subgroup), 'Silk' (AAB), Pome subgroup (AAB), 'Pisang Awak' (ABB), 'I.C.2' (bred AAAA)
2	'Bluggoe' (ABB), bred AAAAs
3	*Heliconia caribea, H. chartacea, H. crassa, H. collinsiana, H. latispatha, H. mariae, H. rostrata, H. vellerigera, M. balbisiana* (seedlings),* 'Gros Michel'* (AAA)
4	Cultivars in the AAA Cavendish subgroup (except 'Dwarf Parfitt') in the subtropics, plus cultivars susceptible to race 1 and race 2. Race 4 in the tropics (T4) also attacks 'Sucrier' (AA) and 'Lakatan' (AAA)

*Waite (1963) indicated that seedlings of *M. balbisiana* and 'Gros Michel' were slightly susceptible to race 3.

Plate 3.5. Pseudostem of banana affected by Fusarium wilt cut to reveal a dark brown discoloration of the vascular tissue (photo: QDPI).

Heliconia latispatha and *H. vellerigera* were found in the Quepos area of Costa Rica (R.C. Ploetz, Costa Rica, 1991, personal observation), an area in which Waite reported finding the disease (Waite, 1961, 1963). It is not clear why this once widely spread disease has not been found on heliconia in recent years.

Finally, race 4 affects race 1- and race 2-susceptible clones in addition to the Cavendish cultivars (Su *et al.*, 1986; Stover and Simmonds, 1987). Before recent outbreaks in South-East Asia, Cavendish clones had only been affected in the subtropical production areas in the Canary Islands, South Africa (Plate 3.7), Taiwan and Australia. In these areas, cold winter temperatures are presumed to predispose Cavendish to damage that would not normally develop. Work in Australia indicates that the photosynthetic abilities of 'Williams', a Cavendish clone, are compromised by cold temperatures (Moore *et al.*, 1993b). In contrast, a unique population of the pathogen, vegetative compatibility group (VCG) 01213–01216, is responsible for the affected Cavendish monocultures in tropical South-East Asia (Pegg *et al.*, 1994;

Plate 3.6. Plantation of 'Lady Finger' (AAB, Pome subgroup) in northern New South Wales, Australia, devastated by Fusarium wilt (photo: D.R. Jones, INIBAP).

Plate 3.7. A cultivar in the Cavendish subgroup (AAA) with symptoms caused by race 4 of *Fusarium oxysporum* f. sp. *cubense* (VCG 0120) near Nelspruit in South Africa (photo: D.R. Jones, INIBAP).

Ploetz, 1994; Bentley *et al.*, 1998; R.C. Ploetz, Florida, 1998, personal observation). Also affected in these areas are 'Pisang Mas' (AA, syn. 'Sucrier'), 'Pisang Berangan' (AAA, syn. 'Lakatan') and other cultivars that are resistant to Fusarium wilt in other locations.

Work is needed to clarify the race structure in *F. oxysporum* f. sp. *cubense*. In addition to tropical race 4, there are other situations which suggest that more than the four current races exist (Ploetz, 1994). For example, strains of the pathogen from East Africa and Florida (USA) damage both 'Gros Michel' and 'Bluggoe', but not Cavendish clones (R.C. Ploetz, Florida, 1998, personal observation; R.H. Stover, Honduras, 1998, personal communication).

Determining the extent of pathogenic variation in this fungus may ultimately rely on experiments with diverse isolates and banana clones under controlled conditions (Ploetz, 1989, 1990b). The differentiation of races of *F. oxysporum* f. sp. *cubense* with restriction digests of mitochondrial DNA was recently reported by Thomas *et al.* (1994), but this technique has been shown to distinguish different VCGs, rather than races, of the fungus (D'Alessio, 1998).

Morphology

Fusarium oxysporum is in Wollenweber's section *Elegans* (Wollenweber, 1913). The species contains pathogenic and saprophytic strains that cannot be distinguished morphologically. In culture, colonies are fast-growing (4–7 mm day^{-1} radial growth on potato dextrose agar (PDA) at 24°C), with sparse to abundant aerial mycelium, and white, pink, salmon or purple pigmentation (Gerlach and Nirenberg, 1982; Nelson *et al.*, 1983). When formed, sporodochia are tan to orange and sclerotia are blue and submerged. The characteristic, sporodochial, wild-type isolate of the species mutates easily in culture to either mycelial types, which produce cottony, aerial mycelium, or to slimy, pionnatal types, which produce depressed aerial mycelium and macroconidia in pionnates; the latter attribute lends colonies a wet, yellow to orange appearance. Some strains of *F. oxysporum* produce strong odours in culture (Domsch *et al.*, 1980) and these have been used to classify isolates of *F. oxysporum* f. sp. *cubense* (Brandes, 1919; Moore *et al.*, 1991).

Microscopic features of the species include the production of micro- and macroconidia on branched and unbranched monophialides (Nelson *et al.*, 1983). Microconidia are one- or two-celled and oval to kidney-shaped and are borne in false heads. Macroconidia are four- to eight-celled, sickle-shaped, thin-walled and delicate, with foot-shaped basal and attenuated apical cells. Dimensions of the micro- and macroconidia typically are in the range of 5–16 μm × 2.4–3.5 μm and 27–55 μm × 3.3–5.5 μm, respectively (Gerlach and Nirenberg, 1982). Terminal and intercalary chlamydospores are usually globose and are formed singly (7–11 μm) or in pairs in hyphae or conidia. Although chlamydospore production is a diagnostic character for the species, they are not produced by isolates of *F. oxysporum* f. sp. *cubense* in VCG 01214 (Ploetz and Pegg, 1997).

Origins, genetic diversity and ancestry

Fusarium oxysporum f. sp. *cubense* is thought to have co-evolved with edible banana in South and South-East Asia and to have been subsequently dispersed to other regions by humans (Stover, 1962). Vakili (1965) believed that the resistance to Fusarium wilt that he observed in *M. acuminata* sspp. *burmannica*, *malaccensis*, *microcarpa* and *siamea* indicated that these subspecies had been exposed to selection pressures exerted by *F. oxysporum* f. sp. *cubense*. He noted that their geographical distributions in Burma (Myanmar), Thailand, West Malaysia and Vietnam coincided with that of the disease. In contrast, the susceptible *M. acuminata* ssp. *banksii* and *M. schizocarpa* are native to New Guinea, an island that, until recently, had been free of the disease (Stover, 1962; Simmonds, 1966). Thus, these subspecies were thought to have evolved in the absence of the pathogen.

The resistance of two taxa from the Philippines provides additional, circumstantial support for Vakili's (1965) hypothesis. *Musa textilis* (abacá) is susceptible to Fusarium wilt, but resistance is known among various accessions of the species (Waite, 1954). Likewise, *M. acuminata* ssp. *errans* possesses partial resistance to this disease; Vakili (1965) indicated that it was resistant to race 2, but susceptible to race 1. Thus, both *M. textilis* and *M. acuminata* ssp. *errans* may have coexisted with *F. oxysporum* f. sp. *cubense* long enough to allow selection for resistance to Fusarium wilt to occur. The partial resistance of these taxa may be the result of the relatively narrow genetic diversity of the pathogen that has been found in the Philippines. For example, only three VCGs, 0122, 0123 and 0126, are known in the Philippines, whereas at least 11 VCGs or VCG complexes have been described in the areas in which *M. acuminata* sspp. *burmannica*, *malaccensis*, *microcarpa* and *siamea* evolved.

Since a teleomorph (sexual stage) for *F. oxysporum* has not been found, the pathogen appears to rely solely on asexual reproduction. Genetic variation that occurs in the species is assumed to arise by neutral mutations that are maintained by mitosis (see Kistler and Miao (1992) for a discussion on the generation and maintenance of genetic variation in asexual filamentous fungi). Although a parasexual cycle in *F. oxysporum* f. sp. *cubense* has been described in the laboratory (e.g. Buxton, 1962; Kuhn, *et al.*, 1995), it has not been shown to occur in the field.

During the last decade, isolates of *F. oxysporum* f. sp. *cubense* have been genetically characterized in several different ways. These results are summarized in Table 3.3.

Somatic compatibility has been used to identify genetically isolated populations of the pathogen, the VCGs (Correll and Leslie, 1987; Ploetz and Correll, 1988; Brake *et al.*, 1990; Hernandez *et al.*, 1993; Moore *et al.*, 1993a; Rutherford *et al.*, 1998). Usually, genetic complementation (heterokaryon formation) between nitrate-non-utilizing auxotrophic (*nit*) mutants is used to identify compatible isolates (Puhalla, 1985; Correll *et al.*, 1987; Leslie, 1993). To date, 16 VCGs or VCG complexes have been reported in *F. oxysporum* f. sp. *cubense* (Ploetz, 1990a; Ploetz *et al.*, 1997b) (see Table 3.1). This is an unusually large number of VCGs for a forma speciales of *F. oxysporum*, and may be the result of the diverse banana host and the presumed old age of this pathosystem.

The distribution of and relationships among the various VCGs provide a wealth of information on the dissemination and evolution of this pathogen. Two of the VCG complexes, 0120–01215 and 0124–0125–0128–01220, are found both in and outside Asia (see Table 3.1). Their worldwide distribution is probably the result of frequent dispersal of *F. oxysporum* f. sp. *cubense* in

Table 3.3. Groupings of *Fusarium oxysporum* f. sp. *cubense* based on various genetic and phenetic data.

VCG*	VCG complex†	Race‡	DFG§	EK‖	Aldehyde formation¶	Clonal lineage**	Sequence data††
0120	0120–01215	1, 4	I	II	Odoratum	2, 3	C2
0121	None	4	III	II	Odoratum	3	C2
0122	None	4?	II	II	Odoratum	6	C2
0123	0123a–0123b	1	V	II	Inodoratum	7, 10	C3
0124	0124–0125–0128–01220	1, 2	IV	I	Inodoratum	1	C1
0125	0124–0125–0128–01220	1, 2	IV	I	Inodoratum	1	C1
0126	None	1	II	nt	Odoratum	2	C4
0128	0124–0125–0128–01220	1, 2	IV	I	Inodoratum	1	C1
0129	None	4	I	II	Odoratum	2	nt
01210	None	1	II	I	Inodoratum	4	C4
01211	None	4	I	nt	Odoratum	9	nt
01212	None	?	IV	nt	Inodoratum	8	C1
01213	01213–01216	T4	III	nt	Odoratum	2	C5
01214	None	2	VII	I	Inodoratum	5	C2
01215	0120–01215	1, 4	I	nt	Odoratum	2, 3	C2
01216	01213–01216	T4	III	nt	Odoratum	nt	nt
01217	None	?	V	nt	Inodoratum	nt	nt
01218	None	?	VI	nt	Odoratum	nt	nt
01219	none	?	II	nt	Inodoratum	nt	nt
01220	0124–0125–0128–01220	4?	IV	nt	Inodoratum	nt	nt
01221	None	?	nt	nt	nt	nt	nt

nt, Not tested.
* VCGs, vegetative compatibility groups as defined in the text.
† VCG complex, broad groups of VCGs in which cross-compatibility or bridging occurs (0123a–b refers to two distinct subgroups of VCG 0123 that were evident in a recent study with isolates from Thailand (Ploetz *et al.*, 1997)).
‡ Race, physiological race as defined in Table 3.2.
§ DFG, DNA fingerprint group, determined by RAPD banding patterns, as reported by Bentley *et al.* (1998).
‖ EK, groups based on electrophoretic karyotypes (Boehm *et al.*, 1994).
¶ Aldehyde formation, groups determined by the presence (odoratum) or absence (inodoratum) of volatile aldehydes produced over the head space of rice cultures of isolates (Pegg *et al.*, 1994).
** Clonal lineages, those determined by RFLP analysis, as described by Koenig *et al.* (1997).
†† Sequence data, clonal lineages determined by nucleotide sequences of nuclear (transition elongation factor 1α) and mitochondrial (small subunit ribosomal RNA) genes (O'Donnell *et al.*, 1998).

banana rhizomes during the movement of this food crop to new production areas. Only five VCGs have not been found in Asia.

Boehm *et al.* (1994) determined the chromosome number and genome size for 118 isolates in 12 VCGs/VCG complexes. Mean variation in both attributes was consistently associated with VCG ($P < 0.0001$). Two major groups of isolates were distinguished by the karyotype data – those with large genomes (40–59 Mbp) and high chromosome numbers (11–14) and those with smaller genomes (32–45 Mbp) and fewer chromosomes (9–12); this distinction was also highly significant ($P < 0.0001$). Notably, of the globally disseminated complexes, 0124–0125–0128–01220 was found in the former and 0120–01215 in the latter major groups.

Bentley and her colleagues (Bentley *et*

al., 1995, 1998; Bentley and Bassam, 1996) used the randomly amplified polymorphic DNA (RAPD) technique and a modified DNA amplification fingerprinting (DAF) system to study some of the same isolates that were examined by Boehm *et al.* (1994). They noted similar or identical banding patterns for isolates within a VCG that were independent of their geographical and host origins. Phenetic (UPGMA cluster) comparisons of isolates among the different VCGs generally recognized the same major groups that were outlined by Boehm *et al.* (1994). Group 1 contained isolates in VCGs 0120–01215, 0121, 0122, 0126, 0129, 01210, 01211, 01213–01216 and 01219, whereas group 2 contained isolates in VCGs 0123, 0124–0125–0128–012120, 01212, 01214, 01217 and 01218. When banding patterns were compared, these groups were about 60% similar. These matched, respectively, the odoratum and inodoratum groups of Moore *et al.* (1991) and Pegg *et al.* (1994), which were determined by whether or not isolates formed volatile aldehydes in culture. Bentley *et al.* (1998) further subdivided their major groups into subgroups that contained isolates which were 71 to 100% similar. Group 1 contained subgroups 1-A (VCGs 0120–01215, 0129 and 01211), 1-B (VCGs 0126, 01210 and 01219), and 1-C (VCGs 0121, 0122 and 01213–01216), whereas group 2 contained subgroups 2-A (VCGs 0124–0125–0128–01220 and 01212) and 2-B (VCGs 0123, 01214, 01217 and 01218).

Boehm *et al.* (1994) and Pegg *et al.* (1993) observed that, in general, isolates in the two major groups differed by host genotype: those represented by the 0124–0125–0128–01220 complex were usually recovered from *M. acuminata* × *M. balbisiana* hybrids, whereas those associated with the 0120–01215 complex were recovered from *M. acuminata* hybrids. These relationships caused Pegg *et al.* (1993) to propose that the two groups may have evolved in the respective centres of diversification for the *M. acuminata* × *M. balbisiana* hybrids and *M. acuminata* hybrids, i.e. the monsoonal areas of Thailand, Burma and India versus the rain forests of Malaysia and Indonesia (Simmonds, 1962).

Results from the diversity studies are consistent with, but do not verify, the hypothesis that *F. oxysporum* f. sp. *cubense* is not monophyletic. To construct an evolutionary history for this pathogen, unambiguous data are needed (Bruns *et al.*, 1991; Clark and Lanigan, 1993). A recent study was the first to meet the necessary criteria for this objective. It provides the most complete phylogeny for *F. oxysporum* f. sp. *cubense* to date, as well as compelling evidence that the pathogen evolved in Asia (Ploetz and Pegg, 1997).

Koenig *et al.* (1997) studied 165 isolates in 13 VCGs/VCG complexes with restriction fragment length polymorphisms (RFLPs) of genomic DNA and 19 anonymous, single-copy probes. Parsimony analyses of the RFLP data generated trees, which usually had a very significant bifurcation. One of the major groups that was defined was genetically variable and contained isolates in the 0120–01215 complex, whereas the other group was relatively homogeneous and contained isolates in the 0124–0125–0128–01220 complex.

The midpoint-rooted tree based on bootstrap analysis delineated several clonal lineages of the pathogen. The two largest, lineages I and II, contained 65 and 43 isolates, respectively. Lineages I and II corresponded to the VCG 0124–0125–0128–01220 and VCG 0120–01215 complexes noted above and, together, represented 65% of the isolates which were tested. Lineages I and II were genetically more similar to *F. oxysporum* f. sp. *niveum*, a pathogen of watermelon, than to each other, and as closely related to each other as they were to *F. oxysporum* f. sp. *lycopersici*, a pathogen of tomato. DNA sequence data from these and other members of section *Elegans* verified the occurrence of a large dichotomy between the two populations that cause Fusarium wilt of banana. Thus, there is strong molecular endorsement for the independent evolutionary origins of the two major populations of this pathogen (O'Donnell *et al.*, 1998).

For most of the VCGs studied by Koenig

et al. (1997), it was possible to infer descent from lineage I or lineage II, even for those that have been found exclusively outside South-East Asia. For example, VCG 0129 from Australia was a member of lineage II, signifying its descent from the 0120–01215 complex. For three other VCGs/lineages outside South-East Asia, similarity coefficients indicated common ancestry with VCGs from the region. Isolates in VCG 01210/lineage IV, which is found in Cuba, the Cayman Islands and the United States, shared 89–92% of their alleles with isolates in VCG 0120. Thus, lineage IV may represent diversification of lineage I in the New World. In the Old World, similar relationships could be proposed for two of the lineages outside South-East Asia: isolates in VCG 01212/lineage VIII (Tanzania) were 92–95% similar to isolates in lineage I, whereas isolates in VCG 01211/lineage IX (Australia) were 87–93% similar to isolates in lineage II.

Isolates from Lineage V were members of VCG 01214. This VCG has been found only in the Misuku Hills in Malawi, a small area of about 300 km^2 on the country's northern border with Tanzania (Ploetz *et al.*, 1992). Isolates in lineage V were genetically distinct from isolates in all other lineages, due to numerous lineage-specific alleles. Coefficients of similarity between it and the other lineages averaged 75%, or about the same values noted for two outgroups that were used in the study, *F. oxysporum* f. sp. *niveum* and *F. oxysporum* f. sp. *lycopersici*. As indicated above, isolates in this lineage do not form chlamydospores, which is atypical for *F. oxysporum*. Furthermore, sequence data also illustrate the uniqueness of this VCG (Table 3.3) (O'Donnell *et al.*, 1998). These data provide rare support for the hypothesis that pathogenicity to banana in *F. oxysporum* has also developed outside banana's centre of origin.

Isolates in VCG 0123 were unusual in that they were heterogeneous and did not fall into a single clade. Two lineages, VII and X, were identified in the VCG. Isolates in lineage VII shared 43% of their polymorphic alleles with isolates in lineage II and 36% with isolates in lineage I. Lineage VII also had several lineage-specific alleles. These data suggest that genetic exchange may have occurred in VCG 0123 between lineages I and II. Alternatively, VCG 0123 may represent the parental line from which lineages I and II arose.

In summary, the genetic and phenetic data usually recognize the same major groups within *F. oxysporum* f. sp. *cubense* (Table 3.3). In all of these studies, there has been a clear demarcation between isolates in the VCG 0120–01215 and VCG 0124–0125–0128–01220 complexes. There has been incomplete agreement, however, over the identity of minor groups within the taxon. Variation detected in VCGs 0120–01215 and 0123 by the RFLP analyses was not discerned with RAPDs, nor were differences that were detected between VCG 0120–01215 and VCG 0126 with RAPDs observed with RFLPs (Table 3.3). These differences may be due to the less than identical set of isolates that were used in these studies or may be an indication that the various techniques that were used provided varying degrees of discrimination. In any event, the most profound conclusion from this work is that significant genetic variation exists within this pathogen. Work to identify resistance in banana to Fusarium wilt would clearly benefit by taking into account this variation.

Disease cycle and epidemiology

The response of banana to infection by *F. oxysporum* f. sp. *cubense* has been studied by Beckman and his colleagues (Beckman and Keller, 1977; Vander Molen *et al.*, 1977, 1987; Beckman and Talboys, 1981; Beckman, 1987, 1990). In a series of experiments, the pathogen's colonization of the host and associated interactions was followed. Root systems of banana plants were grown in large mist chambers, which encouraged the production of long, easily manipulated roots. These were severed, dipped in conidial suspensions of the pathogen and subsequently observed over time for infection and disease development.

During this work, race 1 of the pathogen was observed to form abundant microconidia in xylem vessels of the race1-susceptible 'Gros Michel'. These propagules moved acropetally in vessels via the plant's transpirational flux and were trapped at the first scalariform vessel end with which they came in contact. As the fungus continued to grow in the infected vessel, it colonized the vessel end and, within 2–3 days, produced microconidia on its adaxial side, thus allowing the pathogen to move through another vessel.

Although this process continued unabated in 'Gros Michel', it was stopped in a Cavendish cultivar shortly after it was inoculated with race 1. In the latter case, gels formed in infected vessels after 24–48 h, followed by the growth of vascular parenchyma into vessels after 48–96 h. These pathogen-induced activities in the host effectively trapped spores of the pathogen and denied further colonization of the host's vascular system. Ultimately, the host released phenolic compounds, which infused and lignified the occluding structures. Thus, in the resistant host, there was a clear and rapid orchestration of mechanical host defences to ensure that systemic colonization of the xylem did not occur.

The most common means by which *F. oxysporum* f. sp. *cubense* is disseminated is in infected rhizomes. The pathogen can also be spread in soil and running water and on farm implements and machinery. Work in the early export plantations indicated that susceptible clones could not be successfully replanted in an infested site for up to 30 years, due to the long-term survival of *F. oxysporum* f. sp. *cubense* in soil and as a parasite of non-host weed species (Waite and Dunlap, 1953; Stover, 1962).

Root tips are the natural, initial sites of infection (Beckman, 1990). Wounded rhizome surfaces are apparently minor infection courts. Chlamydospores germinate in response to exudates that originate from behind the zone of elongation on root tips. In most cases, root-tip infections are stopped shortly after the pathogen reaches the host xylem, due to the formation of gels and tyloses and vascular collapse. However, some of these infections are not recognized in susceptible cultivars, and the colonization of the xylem and associated parenchymal tissues continues.

Eventually, chlamydospores are formed in dead and dying host tissues. Chlamydospores are the most significant survival structure of the pathogen and allow it to endure periods of desiccation and the absence of a host.

Host reaction

Current knowledge on the susceptibility of banana cultivars to VCGs is summarized in Tables 3.1 and 3.2.

In trials undertaken as part of the International *Musa* Testing Programme organized by the International Network for the Improvement of Banana and Plantain, guidelines for rating germplasm for reaction were issued to participating agencies. Parameters proposed to compare the susceptibility of cultivars, somaclonal variants and bred hybrids to Fusarium wilt included the rate of onset of pseudostem splitting, leaf yellowing, vascular discoloration in leaf bases, malformation of new leaves (i.e. the appearance of pale margins to the lamina, narrowing of the lamina, burning plus ripping of the lamina and degree of erectness), shortened internodes, wilting and petiole buckling. It was recommended that the first observations should be made 3 months after planting and then at monthly intervals until harvest. Three categories of reaction ('no symptoms', 'mild symptoms' and 'severe symptoms') were considered sufficient for most of the parameters, except for malformation of leaves, which was either present or absent. Another important measurement, which proved the most useful, was an assessment of the internal discoloration of the vascular tissue of the rhizome as viewed after the rhizome was cut transversely about one-quarter of the way up from the bottom. This was to be undertaken immediately after harvest and illustrations indicating varying degrees of vascular discoloration in the corm were published to aid evaluation (Jones, 1994). Results from screening trials

Table 3.4. Rhizome vascular discoloration ratings at harvest of a wild *Musa* subspecies, banana cultivars, selected somaclonal mutants and bred hybrids growing in plots containing *Fusarium oxysporum* f. sp. *cubense* race 1 (VCG 0124) and race 4 (VCG 0120, 0129) in subtropical Australia (K.G. Pegg, Brisbane, 1999, unpublished).

Banana tested*	Average discoloration rating in race 1 plot†	Average discoloration rating in race 4 plot†
Musa acuminata ssp. *burmannicoides* ('Calcutta 4' accession)	1.00	1.00
'Pisang Jari Buaya' (AA)	1.20	1.00
'Pisang Lilin' (AA)	1.00	1.30
'Pisang Mas' (AA)	1.10	5.20
'Cultivar Rose' (AA)	1.00	1.00
'Gros Michel' (AAA)	5.80	5.62
'Williams' (AAA, Cavendish subgroup)	1.00	2.44
'Yangambi Km 5' (AAA, syn. 'Ibota Bota')	1.00	5.50
'Pisang Nangka' (AAA)	1.50	1.01
'GCTCV 119' (AAA, Cavendish subgroup)	1.10	1.06
'GCTCV 215' (AAA, Cavendish subgroup)	1.00	1.60
'FHIA-02' (AAAA)	3.40	–
'FHIA-17' (AAAA)	1.00	1.86
'FHIA-23' (AAAA)	1.70	4.20
'Ney Poovan' (AB)	2.80	–
'Lady Finger' (AAB, Pome subgroup)	3.80	–
'Pisang Ceylan' (AAB, syn. 'Mysore')	3.00	1.75
'Bluggoe' (ABB, Bluggoe subgroup)	2.30	5.45
'FHIA-01®' (AAAB, syn. 'Goldfinger')	1.00	1.00
'PA 03–22' (AAAB)	1.00	2.29
'PV 03–44' (AAAB)	1.10	2.24
'PC 12–05' (AAAB)	1.00	–
'FHIA-03' (AABB)	–	5.74

* GCTCV (Giant Cavendish tissue culture variant) prefix indicates clones are somaclonal variants of 'Pei-Chiao' from the Taiwan Banana Research Institute (see Hwang and Tang, Chapter 14, pp. 449–458); FHIA prefix indicates clones bred by the Fundación Hondureña de Investigación Agrícola (see Rowe and Rosales, Chapter 14, pp. 435–449); PA, PV and PC prefixes indicate clones are bred by the Empresa Brasiliera de Pesquisa Agropecuaria–Centro Nacional de Pesquisa de Mandioca e Fruticultura (EMBRAPA-CNPMF) and are derived from 'Prata Anã', 'Pacovan' and common 'Prata', respectively.

† Plants were completely randomized in plots; rating is an average of readings of 20 plants clone^{-1} and is based on the following scores: 1, no vascular discoloration; 2, isolated points of discoloration in vascular tissue; 3, discoloration of up to one-third of vascular tissue; 4, discoloration of between one-third and two-thirds of vascular tissue; 5, discoloration of more than two-thirds of vascular tisssue, but not total discoloration; 6, total discoloration of the vascular tissue.

in subtropical Australia using rhizome discoloration readings for rating reaction to Fusarium wilt are shown in Table 3.4.

Technical guidelines for the evaluation of *Musa* germplasm to Fusarium wilt, which include illustrations of categories of rhizome discoloration for rating purposes (Jones, 1994), have recently been published by INIBAP (Orjeda, 1998).

Of the abacá cultivars most commonly grown in Central America, 'Bongolanon' was the most susceptible to Fusarium wilt, with 'Maguindanao' seldom affected. 'Tangoñgon' was either unaffected or only slightly affected (Waite, 1954).

Control

In general, effective chemical control measures do not exist. Herbert and Marx (1990)

Plate 3.8. A Fusarium wilt quarantine plot in a commercial plantation in southern Mindanao, Philippines. Sawdust and rice hulls are being burnt where a diseased plant in the Cavendish subgroup (AAA) has been removed in an attempt to sterilize the soil. The chopped remains of the plant lie on banana leaves in the background awaiting removal in polyethylene bags (photo: D.R. Jones, INIBAP).

reported a ninefold reduction in disease incidence on 'Williams' 26 months after soil was fumigated with methyl bromide. However, after 3 more years, the fumigated areas were thoroughly reinvaded by the pathogen. In India, Lakshmanan *et al.* (1987) reported that rhizome injection with 2% carbendazim (Bavistin 50 WP®) protected 'Rasthali' (AAB, syn. 'Silk') for one crop cycle. This treatment had no effect in South Africa (Herbert and Marx, 1990). In Australia, pseudostem injections of 20% potassium phosphonate gave some control on 'Williams', but results were quite erratic (K.G. Pegg, Brisbane, 1998, personal observation). This may be due to the fact that the sensitivity of *F. oxysporum* f. sp. *cubense* to phosphonate is reduced at phosphate concentrations that occur naturally in the banana plant (Davis *et al.*, 1994). Heat sterilization of the soil has been attempted in one commercial plantation in southern Mindanao in the Philippines to control the spread of race 4 (Plate 3.8).

Disease-suppressive soils are found in several different locations (Stover, 1962, 1990; Alvarez *et al.*, 1981; Chuang, 1988). In general, these soils are recognized by the length of time that high levels of production can be maintained in the presence of the pathogen. Stolzy and coworkers (reviewed by Toussoun, 1975) associated disease suppression with chemical and physical factors and found the closest association between suppression and soils in which a clay fraction of the montmorillonoid type was found.

Susceptible clones can be grown if pathogen-free propagation material is used in uninfested soil. Micropropagated plantlets are the most reliable source of clean material. Since they are also free of bacterial, nematode and other fungal pathogens, plantlets should be used whenever possible. It should be noted, however, that plants grown from tissue-cultured plantlets have been shown to be more susceptible to Fusarium wilt than those that are grown from conventional planting material (Smith *et al.*, 1998). Moreover, the expense of plantlets may make their use in subsistence agriculture impractical. In the latter situations, plantlets may be better utilized to establish disease-free nurseries, which could produce clean rhizome pieces or suckers.

Clearly, the use of resistant genotypes is the best way to combat this disease. Resistant cultivars exist for several different kinds of banana (Buddenhagen, 1990), and these should be used when they are available. In other situations, new hybrids could be used to replace susceptible clones. Readers who are interested in recent progress in breeding disease-resistant banana hybrids should refer to Jones (1994).

Acknowledgement

The authors thank Ms Editha O. Lomerio for information on Fusarium wilt of abacá in the Philippines.

Armillaria Corm Rot

D.R. Jones and R.H. Stover

Armillaria corm rot of banana is uncommon and has only been recorded in Australia (Blake, 1963), Kenya (Anon, 1964) and Florida (Rhoads, 1942). It is usually associated with new plantations that have been established on recently cleared land. In Australia, the disease was worst where hardwoods were cleared and was rare on bush land (Magee and Eastwood, 1935). The disease has also been called dry rot or stump rot (Stover, 1972).

Armillaria mellea and *Armillaria tabescens* in the order *Agaricales* are the causes of the disease (Blake, 1963; Anon., 1964). These pathogens are woodland inhabitants and colonize tree stumps or large roots left behind after clearing. Rhizomorphs, which are compacted masses of hyphal strands covered by a cortex, often form networks under the bark of trees. The fungi penetrate the banana's roots and corm at or below ground level. When the corm is fully invaded, leaves turn yellow and die from the base upwards. If only part of the corm is invaded, only the leaves on the affected side might die. The plant is easily pushed over or breaks off at ground level in the advanced stage of the disease. Rhizomorphs and white fungal strands can be seen permeating the dry, brown corm. Sometimes, fruiting bodies of the fungi are found at the base of the plant (Stover, 1972).

To avoid the disease, all tree stumps and large roots should be burnt or cleared from the plantation site before planting (Blake, 1963). If corm rot is a problem, diseased corms should be dug up, cut into pieces and burnt. Replanting should take place some distance away from the position of the affected plant.

Cylindrocladium Root Rot

D.R. Jones

In Martinique and Guadeloupe, the main pathogens associated with necrosis on the roots of 'Poyo' and 'Grande Naine' (AAA, Cavendish subgroup) are *Radopholus similis* (see J.L. Sarah, Chapter 7, pp. 295–303) and fungi in the genus *Cylindrocladium*. The intensity of necrosis caused by each component of this disease complex varies in respect to climate and soil characteristics. Both *R. similis* and the *Cylindrocladium* spp. cause a typical necrotic response, but symptoms are enhanced when these two pathogens interact (Loridat and Ganry, 1991). This nematode–fungus interaction is believed responsible for a decline in the vigour of root systems, which has led to a drop in production (Loridat, 1989).

Risède (1994) has partially characterized the main species of *Cylindrocladium* present in Martinique. The fungus produces tan to brownish colonies with irregular margins on 1% malt extract agar. Numerous chlamydospores arranged in chains and brown microsclerotia are produced. Conidia are straight, cylindrical, hyaline and uniseptate and measure 69–86 μm × 5.6–6.2 μm (Plate 3.9). The fungus is believed to be *Cylindrocladium pteridis* Wolf, but this needs confirmation by molecular and mating studies. Investigations also showed that the fungus could be trapped in soil with tissue-cultured banana plants. Microsclerotia have been detected in naturally infected roots in the field and also in artificially inoculated banana roots. Microsclerotia have been shown to be capable of infecting roots and may be the main survival structure of the fungus in soil.

Cylindrocladium spp. have been isolated from the roots and corms of many different cultivars of banana (Loridat and Ganry, 1991) in Martinique and *Cylindrocladium* is also present in Costa Rica (Semer *et al.*, 1987). More recently, the fungus has been found affecting banana in Côte d'Ivoire (Kobenan, 1991) and Cameroon (Castaing *et al.*, 1996).

In Cameroon, *Cylindrocladium spathyphilli* is the most common species (J.M. Risède, Guadeloupe, 1999, personal communication).

Investigations are being conducted to correctly identify other species of *Cylindrocladium* involved, assess their pathogenic potential and devise biological control methods (J.M. Risède, Guadeloupe, 1999, personal communication).

Rosellinia Root and Corm Rot

D.R. Jones

Rosellinia bunodes (Berkeley and Bromme) Saccardo was found by Smith (1929) attacking the root system of banana in Jamaica. However, the existence of another species of *Rosellinia* in the disease complex was suspected. *Rosellinia bunodes*, which causes a black root rot disease of many plant species, but mainly of tropical and subtropical woody hosts, is widespread in tropical America and has been found in Central Africa, South Asia and South-East Asia (Sivanesan and Holliday, 1972).

A *Rosellinia* sp. has also been implicated in a 'stinking root' disease of banana in coastal areas of southern Brazil in 1937 by R.D. Gonçalves (Wardlaw, 1961). As in Jamaica, other pathogens were associated with the problem. *Fusarium* spp. were also isolated and unusual sclerotial formations were seen in affected plantations.

More recently, *Rosellinia pepo* Pat. has been identified as causing 'star-like wound' disease of 'Dominico-Hartón' (AAB, Plantain subgroup) and other clones in Colombia (Belalcazar, 1991). Symptoms on banana include an internal reddish necrosis of the corm, necrosis of leaf margins, lower yields and the eventual toppling of plants following the decay of the root system. White, star or fan-shaped patterns of mycelium, which can be seen underneath the outer cortex of the corm and roots, helps distinguish this disease from Fusarium wilt.

Rosellinia pepo is a soil inhabitant and is found in organic litter and woody debris. It attacks many plant species including avocado, breadfruit, cocoa and coffee, but is thought to have a narrower host range than *R. bunodes* and a more restricted distribution (Booth and Holliday, 1972). In Colombia, it is common on plantain grown as an intercrop with coffee or cocoa and its incidence is greatest where plant residues are highest, such as on land recently cleared of trees. Cultivars in the AAA genomic group are not as susceptible as plantain (Belalcazar, 1991).

Damping-off of *Musa* Seedlings

D.R. Jones and R.H. Stover

A rot that attacks the lower leaves, pseudostem and roots where they join the corm has been reported to affect young, succulent seedlings of *Musa* species. The plants were frequently killed and fell over, giving the appearance of damping-off (Stover, 1972). A necrosis of the tip of the lower leaf, often followed by the death of the leaf, was the first symptom of the disease. The rot then progressed to the pseudostem and caused a dark brown discoloration of the leaf sheaths. The rot extended to the inner sheaths and the seedling collapsed and died. Sometimes the discoloration of the leaf sheaths began at or below ground level and moved down to where the leaf sheaths were attached to the developing corm and up to the crown. On occasion, the roots were affected. The disease could also affect the tips and base of the coleoptile of seedlings only a few centimetres long. These became brown as they emerged from the soil. When conditions were not

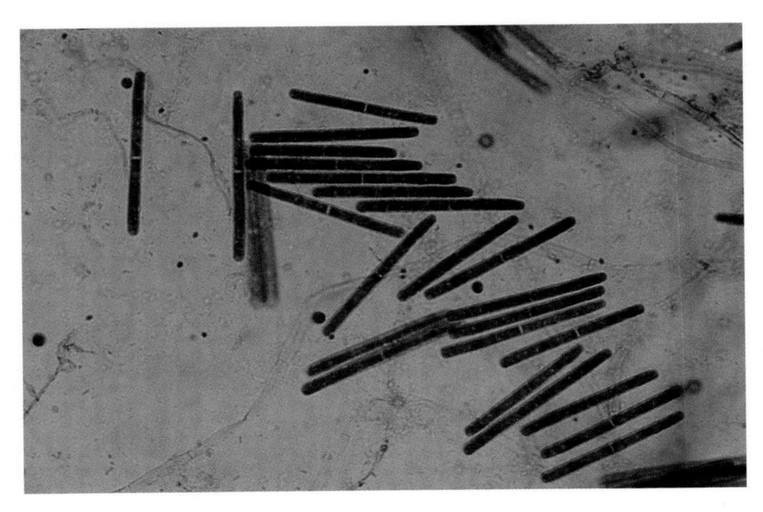

Plate 3.9. Conidia of a *Cylindrocladium* sp. isolated from banana roots in the French Antilles (photo: J.M. Risède, CIRAD).

favourable for disease development, only the older leaves died and one or two leaf sheaths became discoloured. The seedling continued to grow, but was retarded. In some instances, the disease remained below ground and the plant was stunted as a result of root and rhizome injury associated with the necrosis of adjacent leaf sheaths. Infection was greatest during periods of cool, wet weather. As seedlings became older and hardier, only the outer leaf sheaths were affected (Stover, 1963).

This damping-off disease was caused by *Deightoniella torulosa* (Syd.) Ellis (Stover, 1963), which also affects banana, abacá and enset leaves (see Jones *et al.*, Chapter 2, pp. 102–104), abacá pseudostems (see Lomerio, this chapter, pp. 161–162) and banana fruit (see Jones and Stover, Chapter 4, pp. 178–179, 181–182). Seedlings of the wild species *M. acuminata* ssp. *banksi*, *M. balbisiana*, *M. schizocarpa* and *M. textilis* were reported as affected, as were seedlings arising from crosses between *M. acuminata* and edible banana cultivars (Stover, 1972). The problem was controlled by thinning the stands of seedlings, reducing humidity levels and maintaining conditions that allowed the foliage of seedlings to dry rapidly after watering. Banana trash and banana plants with Deightoniella leaf spot were kept well away from seedlings.

The disease was of little economic importance and occurred only in the nurseries of banana breeding programmes where hybrid seedlings were grown in large quantities.

Deightoniella Pseudostem Rot of Abacá

E.O. Lomerio

Introduction

The disease was first described on the island of Luzon in the Phillipines in 1927 and was reported on the island of Mindanao in 1931 (Agati *et al.*, 1934). It has been serious in plantations of the cultivar 'Bongolanon' and has been reported as causing severe damage to seedlings, but is generally regarded as of minor economic importance. The quality of abacá fibre is lowered because the disease causes an increase in the percentage of dark strands. Severely diseased pseudostems become useless for stripping. In the Philippines, the disease is also known as trunk rot and red-sheath rot.

Symptoms

The disease is more severe on the pseudostem of abacá than on the leaves. Lesions about the size of a pinhead form first on the outer leaf sheaths. These have dark brown centres, surrounded by a lighter area, which is delimited by a brown marginal line. As lesions increase in size, they coalesce and become depressed in the middle. Enlargement is more rapid along the axis of the pseudostem so that, in advanced stages, a dark brown to black oval–oblong blemish is produced. Green-brown, superficial mycelium of the fungus can be seen in the centre. The disease also progresses inwards, affecting one leaf sheath after another until the stem is girdled and sufficiently weakened for it to bend over under the weight of the foliage. Affected leaf sheaths become desiccated and curl backwards and downwards (Ela and San Juan, 1954; Wardlaw, 1961; Stover, 1972).

Causal agent

Pseudostem rot of abacá is caused by *Deigtoniella torulosa* (Syd.) Ellis. This fungus has been described under Deightoniella leaf spot of banana, abacá and enset (see Jones *et al.*, Chapter 2, pp. 102–104). It also causes a damping-off of *Musa* seedlings (see Jones and Stover, this chapter, pp. 160–161) and affects banana fruit (see Jones and Stover, Chapter 4, pp. 178–179, 181–182)

Disease cycle and epidemiology

In the Philippines, the disease is more severe in the rainy season. *Deightoniella torulosa* is a common saprophyte on dead and dying abacá leaves. Spores of the fungus are also found in association with leaf spots and diseased pseudostems. The fungus sporulates at night under conditions of 100% relative humidity and conidia are violently discharged the next morning as the humidity falls and temperatures rise (Meredith, 1961). Germination takes place in water or at 95% relative humidity and above. Wounding the leaf sheath favours infection. Wounds on the pseudostem occur naturally when leaves buckle with age, when sheaths are loosened as older leaves fall and when collapsed leaves are blown by the wind. If wounds are inoculated, the first symptoms appear 15 days after inoculation (Arce, 1954, reported by Stover, 1972).

Host reaction

All cultivars become infected, but some show field resistance. 'Bongolanon' is the most susceptible cultivar (Stover, 1972).

Control

Serious damage is usually associated with poor cultural practices involving spacing, pruning and weed control. The disease spreads rapidly in high-density plantings and proper spacing reduces incidence. Seriously affected plants should be cut down and burned. Losses are minimized if mature plants are harvested before the rot becomes severe (Ela and San Juan, 1954; Stover, 1972).

Fungal Root Rot

D.R. Jones and R.H. Stover

Many genera of fungi have been isolated from rots in the roots of banana. However, there has usually been no proof that the fungi found have been pathogenic. From 1964 to 1966, studies were made of the fungi present in lesions on banana roots and of their pathogenic potential (Laville, 1964; Blake, 1966; Mallessard, 1966; Stover, 1966).

Root rot consists of the rotting and dieback of the tips of the main root and the blackening, necrosis and death of the smaller, lateral, feeding roots. It is common even in the absence of nematode injury (Stover, 1972).

Fungi consistently isolated from banana root rots have included *Fusarium solani, F. oxysporum* and *Rhizoctonia* sp. However, there has been no proof that *F. solani* and *F. oxysporum* (other than f. sp. *cubense*) are important pathogens by themselves. It is likely that they extend cortical necrosis in nematode lesions, which may reach the stele. In Australia, root lesions caused by the burrowing nematode (see Sarah, Chapter 7, pp. 295–303) were found to be more extensive when *F. oxysporum* was present (Blake, 1966). In Central America, where *F. oxysporum* is rare in root lesions, *F. solani* is believed to enlarge lesions caused by nematodes. *Rhizoctonia* is capable of invading young, intact roots and causing necrosis, but damage is greater if the roots are wounded (Mallesard, 1966; Stover, 1966). Some isolates of *Rhizoctonia* have been shown to be highly pathogenic on *Musa* seedlings (Stover, 1966), but not all isolates are pathogenic on banana roots. Stover (1972) has reported that all *Rhizoctonia* isolates from banana roots in Central America were binucleate and therefore not the same species as *R. solani,* which is common on other crops. *Phythium* and *Phytophthora* are isolated infrequently from banana root lesions and are thought not to play an important role in banana root rot (Stover, 1972). However, a *Pythium* sp. has been implicated in a damping-off disease of abacá seedlings (Roldan, 1933). A species of *Zythia* has been found to be pathogenic on banana roots in Côte d'Ivoire (Kobenan *et al.*, 1997).

Root rots are common when conditions are unsuitable for the optimum growth of the banana plant, such as when soils are waterlogged, compacted or nematode-infested or have a 40% or more clay content (Stover, 1972). In a healthy root system, root rots only affect a very small proportion of the roots. Large, extensive and deep root systems develop in well-drained, deep, friable loams. Root rot is less likely to occur if nematodes are controlled and if soil structure and moisture content permit plants to grow under optimum conditions.

Marasmiellus Pseudostem and Root Rot

D.R. Jones and E.O. Lomerio

Introduction

This disease affects banana and abacá and has been reported from Australasia–Oceania (Australia, Fiji, Hawaii), Asia (Philippines, Sri Lanka, East Malaysia), Africa (Mauritius, Uganda, Sierra Leone, Côte d'Ivoire, Ghana) and the Latin America–Caribbean region (Jamaica, Trinidad, Windward Islands, Brazil). The causal agent is a saprophyte of decaying vegetation, including leaf and pseudostem trash. It attacks banana and abacá that have been weakened by drought or poor cultural conditions and is rare in well-managed plantations. The disease is reported to be more common on sandy or gravelly areas, but is also reported in inadequately drained soils (Wardlaw, 1961; Stover, 1972).

Symptoms

The outer leaf sheaths and leaves of both banana and abacá become dry. Growth is stunted and new leaves are either slow to emerge or fail to emerge. Invaded leaf sheaths turn brown or grey and either stick to underlying layers or rot and become easy to remove. Layers, patches and strands (rhizomorphs) of white fungal mycelium are often visible on and between the dead leaf sheaths. The mycelium is conspicuous on water-soaked, brown lesions, which form on underlying leaf sheath layers (Plates 3.10 and 3.11). The rot spreads to the inner leaf sheaths, diminishing in size as it advances. The disease penetrates to the centre of the pseudostem in severe cases. The dry, diseased, outer leaf sheaths at the base of the pseudostem become moist and can be torn off easily. There is usually a characteristic odour of mushrooms. Fungal fruiting bodies (basidiocarps) develop on the pseudostem in wet weather (Plate 3.11) (Ramos, 1941; Wardlaw, 1961; Stover, 1972).

Plate 3.10. Symptoms of Marasmiellus pseudostem rot on the pseudostem of a cultivar in the Cavendish subgroup (AAA) in Queensland, Australia. The dead, outer leaf sheaths have been stripped away to reveal white mycelium present on brown lesions (photo: QDPI).

Plate 3.11. Fruiting bodies of *Marasmiellus inoderma* (arrow) visible on the pseudostem of a diseased abacá plant in the Philippines (photo: E.O. Lomerio, FIDA).

Causal agent

The disease is caused by *Marasmiellus inoderma* (Berkeley) Singer (formerly *Marasmius semiustus* Berkeley and Curtis), a member of the order *Agaricales*. Its mushroom-like fructifications are at first brownish or salmon-yellow in colour on the upper surface, later turning pale. The cap, which measures 5–15 mm in diameter, frequently turns up, exposing widely separated gills. A white glabrous stalk, which is about 7–9 mm in length and flattened oval in cross-section, arises from a sclerotic body and is attached to the cap off-centre. The spores, which are hyaline, oval and papillate at the point of attachment to the basidium, measure 7–8.5 μm × 5–6 μm. The stalk and cap shrivel readily, but recover quickly when placed in water (Wardlaw, 1961).

Disease cycle and epidemiology

The plant is infected at soil level. As well as the pseudostem, the fungus may also invade and kill the roots. Young roots first become covered with white mycelium and develop a blackened tip. Later roots shrivel and the cortex turns brown and soft. Eventually the vascular cylinder darkens and decays. The tissues of the corm are not invaded. Plants that have had their root system partially destroyed can be easily pushed over. In abacá plantations, the fungus has been reported to be present in old pseudostem stumps (Wardlaw, 1961).

The fruiting bodies of *M. inoderma* develop abundantly at soil level, but are also produced on the surface of the pseudostem (Plate 3.11). The basidiospores are formed in the gills of the fruiting body and disseminated by the wind (Wardlaw, 1961).

Plate 3.12. Yellowing and decaying leaves of abacá affected by pseudostem heart rot in the Philippines (photo: E.O. Lomerio, FIDA).

Plate 3.13. Symptoms of Sclerotium root and corm rot of enset in Ethiopia. Leaves are dying on the stunted plant in the foreground (photo: M. Tessera and A.J. Quimio, IAR).

Plate 3.14. Plot of enset at Areka Experimental Station, Ethiopia, devastated by Sclerotium root and corm rot (photo: M. Tessera and A.J. Quimio, IAR).

Host reaction

The fungus is probably capable of attacking a wide range of banana and abacá cultivars.

Control

Cultural practices need improving if the disease is present. Affected plants, particularly stumps, should be removed and trash eliminated from the surroundings. Weeds also need to be kept under control. Plants should not be overcrowded, as high humidity favours infection.

Pseudostem Heart Rot

D.R. Jones and E.O. Lomerio

Introduction

Pseudostem heart rot of banana is a minor disease and is usually associated with plant injury. However, it was considered to be serious on young plantations of 'Gros Michel' (AAA) following severe storms and floods when this cultivar was grown extensively in Central America and the West Indies. After 'Gros Michel' was replaced in the export trades by cultivars in the Cavendish subgroup (AAA), the disease was rarely seen (Stover, 1972).

Heart rot also affects abacá, occurring mainly in weakened plants growing in dense stands in damp locations under conditions of high humidity. The disease has been observed to spread from centres of infection (Wardlaw, 1961).

Other names for pseudostem heart rot are heart-leaf disease, Fusarium stalk rot and stalk heart rot.

Symptoms

The external symptoms of pseudostem heart rot of banana are the dying and rotting of the inner leaves of the crown. These leaves can become nearly fully unrolled or remain furled as they emerge. Diseased leaves are bright yellow or brown and may collapse at the petiole base. In severe cases, all leaves collapse and the pseudostem is killed. Usually, no more than one pseudostem in a mat may be affected, but several mats may be affected in one area (Wardlaw, 1961).

The initial internal symptom is a blackening and decay of the furled heart leaf within the pseudostem. The rot may be firm and odourless at this stage. The decay begins in the upper part of the pseudostem and may extend down to the base, while surrounding leaf sheaths remain unaffected. The central tissue breaks down to a foul-smelling liquid, as a result of a secondary bacterial infection, and healthy sheaths in contact with the diseased core often become streaked brown or purple. Diseased leaves that develop in the pseudostem become twisted and abortive. This obstructs the upward movement of following leaves, which are damaged and decay. Sometimes, median rather than central sheaths are decayed, resulting in a ring of rot in the pseudostem. In these cases, the plant can recover and the decay is sloughed off during growth. Rots may develop in scattered pockets of infection throughout the pseudostem. However, these usually remain free of secondary rots. The corm is not invaded (Wardlaw, 1961; Stover, 1972).

If plants are young and actively growing, central sheaths can recover from heart rot. In these cases, the growing point has either not been damaged or growth through the decayed area has been rapid. However, the lamina of emerged leaves may be ragged, with missing portions (Waite, 1956). Plants damaged by windstorms, floods and pruning operations frequently recover (Stover, 1972).

Symptoms of the disease in abacá are similar. Plants can have their central cylinder completely rotted. Leaves become yellowed, with dry margins and stiff petioles (Plate 3.12).

Causal agent

Pseudostem heart rot is initiated by *Fusarium moniliforme* Sheldon. Bacteria follow and cause secondary infections. There is no evidence that bacteria alone can cause the disease. Other *Fusarium* spp.

and *Cephalosporium* have later been found in association with *F. moniliforme* attacking banana.

Disease cycle and epidemiology

Fusarium moniliforme sporulates profusely on diseased sheaths that have not been affected by secondary rots. Microconidia are mainly produced in false heads by most isolates, but sometimes chains are formed. The fungus has no chlamydospores. Banana plants could not be infected by pouring spore suspensions into crowns between leaf sheaths or by the application of spores or soil. Infections have only occurred when the fungus has been applied to wounded furled leaves inside the pseudostem (Waite, 1956). This indicates that banana tissues have to be injured for infection to take place (Stover, 1972).

Ocfemia and Mendiola (1932) and Ramos (1933) reported that water-soaked, light to dark brown lesions could be induced in abacá when *F. moniliforme* var. *subglutinans* was applied to unrolled, heart leaves kept under moist conditions. The effects were most severe when the tissues were wounded. Abacá affected by bunchy top, nematode attacks on roots and weevil borer infestations in the corm are more prone to pseudostem heart rot than healthy plants. Ramos (1933) showed that abacá could be predisposed to infection by pruning the root system. In plants affected by bunchy top, the decay is said to begin on the thin, chlorotic margins of the youngest leaves. This extends to the midrib and then to the petiole. During wet weather, the decay then progresses down to the heart.

Host reaction

As mentioned above, 'Gros Michel' seems more prone to pseudostem heart rot than Cavendish cultivars. The disease has also been recorded on 'Latundan' (AAB, syn. 'Silk') in the Philippines (Lee, 1921, and Serrano, 1925, reported by Wardlaw, 1961). Pseudostem heart rot is reported as rare on cultivars in the AAB and ABB genomic groups in Central America (Stover, 1972).

Control

The disease is rare in banana plantations that are well maintained. Badly diseased plants should be eliminated (Stover, 1972). In abacá plantations, other disease problems need to be kept under control and plants should be well aerated.

Ceratocystis Corm Rot

D.R. Jones

Ceratocystis corm rot, which has been reported on banana in Australia, is caused by *Chalara paradoxa* (de Seynes) Hohn. (*Ceratocystis paradoxa* [Dade] Moreau). Extensive brown or dark water-soaked areas develop in corms after planting and cause a breakdown of the tissue. A dirty-white mould is generally visible in cavities in the affected areas. The pathogen is a soil inhabitant, especially in old pineapple land, and enters the corm through wounds. It is regarded as a minor problem and is controlled by planting healthy corms (Persley, 1993).

Sclerotium Root and Corm Rot of Enset

M. Tessera and A.J. Quimio

In 1992, the disease was observed for the first time at Areka Experimental Station in Welayita, Ethiopia, where it was killing young seedlings and new transplants. Roots and leaf sheaths at soil level became rotten and the growth of plants was stunted (Plate 3.13). Severely attacked plants died and enset plots were devastated (Plate 3.14). The causal agent was identified as a species of *Sclerotium*, which gained entry to the enset seedlings and transplants through damaged roots and corms. The pathogen was found to survive in disintegrating root and corm tissue present in the soil. Sclerotial bodies and mycelia isolated from diseased specimens were demonstrated to be pathogenic and initiated root rots and wilt when inoculated into the crown of young, healthy enset seedlings of different cultivars. The disease was controlled effectively by dipping corms in a fungicide suspension (Quimio and Tessera, 1996).

Cephalosporium Infloresence Spot of Enset

M. Tessera and A.J. Quimio

This minor disease, which causes extensive necrosis of flower bracts and necrotic spots on leaf sheaths of mature plants, was recorded for the first time at Areka Experimental Station, Welayita, Ethiopia, in 1991 (Tessera and Quimio, 1994). The causal agent is believed to be a *Cephalosporium* species. This fungus was isolated from diseased tissue and then injected into the stems of enset seedlings where it was found to initiate a necrotic reaction (Quimio and Tessera, 1966).

REFERENCES

Agati, J.A., Calinisan, M.M. and Aldaba, V.C. (1934) Further studies on the stem rot of abaca in the Philippines. *Philippine Journal of Agriculture* 5, 191–222.

Alvarez, C.E., Garcia, V., Robles, J. and Diaz, A. (1981) Influence des caractéristiques du sol sur l'incidence de la maladie de Panama. *Fruits* 36, 71–81.

Anon. (1954) *Annual Report*, Vol. 1. Department of Agriculture, Nairobi, Kenya.

Anon. (1964) *Annual Report*. Department of Agriculture, Nairobi, Kenya.

Anon. (1977) *Fusarium oxysporum* f. sp. *cubense*. Distribution Maps of Plant Diseases, Map No. 31, edn 4, Commonwealth Mycological Institute, Kew, Surrey, UK.

Ashby, S.F. (1926) Panama disease in the Canaries and West Africa. *Tropical Agriculture (Trinidad)* 3, 8.

Bancroft, J. (1876) Report of the board appointed to enquire into the cause of disease affecting livestock and plants. Queensland, 1876. *Votes and Proceedings* 1877(3), 1011–1038.

Beckman, C.H. (1987) *The Nature of Wilt Disease of Plants*. APS Press, St Paul, Minnesota, USA, 175 pp.

Beckman, C.H. (1990) Host responses to the pathogen. In: Ploetz, R.C. (ed.) *Fusarium Wilt of Banana*. APS Press, St Paul, Minnesota, USA, pp. 93–105.

Beckman, C.H. and Keller, J.L. (1977) Vessels do end! *Phytopathology* 52, 954–956.

Beckman, C.H. and Talboys, P.W. (1981) Anatomy of resistance. In: Mace, M.E., Bell, A.A. and Beckman, C.H. (eds) *Fungal Wilt Diseases of Plants*. Academic Press, New York, USA, pp. 487–521.

Belalcazar, S.L. (1991) *El Cultivo del Plátano en el Trópico*. Manual de Assistencia Técnica No. 50. Instituto Colombiano Agropecuario, Armenia, Colombia, pp. 288–289.

Bentley, S. and Bassam, B.J. (1996) A robust DNA amplification fingerprinting system applied to analysis of genetic variation within *Fusarium oxysporum* f. sp. *cubense*. *Journal of Phytopathology* 144, 207–213.

Bentley, S., Pegg, K.G. and Dale, J.L. (1995) Genetic variation among a world-wide collection of isolates of *Fusarium oxysporum* f. sp. *cubense* analyzed by RAPD–PCR fingerprinting. *Mycological Research* 99, 1378–1384.

Bentley, S., Pegg, K.G., Moore, N.Y., Davis, R.D. and Buddenhagen, I.W. (1998) Genetic variation among vegetative compatibility groups of *Fusarium oxysporum* f. sp. *cubense* analysed by DNA fingerprinting. *Phytopathology* 88, 1283–1293.

Blake, C.D. (1963) Root and corm diseases of banana. *Agricultural Gazette of New South Wales* 24, 526–528.

Blake, C.D. (1966) The histological changes in banana roots caused by *Radopholus similis* and *Helicotylenchus multicinctus*. *Nematologia* 12, 129–137.

Boehm, E.W.A., Ploetz, R.C. and Kistler, H.C. (1994) Statistical analysis of electrophoretic karyotype variation among vegetative compatibility groups of *Fusarium oxysporum* f. sp. *cubense*. *Molecular Plant–Microbe Interactions* 7, 196–207.

Booth, C. and Holliday, P. (1972) *Rosellinia pepo*. CMI Descriptions of Pathogenic Fungi and Bacteria No. 354. Commonwealth Mycological Institute, Kew, Surrey, UK, 2 pp.

Brake, V.M., Pegg, K.G., Irwin, J.A.G. and Langdon, P.W. (1990) Vegetative compatibility groups within Australian populations of *Fusarium oxysporum* f. sp. *cubense*, the cause of Fusarium wilt of bananas. *Australian Journal of Agricultural Research* 41, 863–870.

Brandes, E.W. (1919) Banana wilt. *Phytopathology* 9, 339–389.

Bruns, T.D., White, T.J. and Taylor, J.W. (1991) Fungal molecular systematics. *Annual Review of Ecology and Systematics* 22, 525–564.

Buddenhagen, I.W. (1990) Banana breeding and fusarium wilt. In: Ploetz, R.C. (ed.) *Fusarium Wilt of Banana.* APS Press, St Paul, Minnesota, USA, pp. 107–113.

Buxton, E.W. (1962) Parasexual recombination in the banana-wilt *Fusarium. Transactions of the British Mycological Society* 45, 274–279.

Castaing, V., Beveraggi, A., Foure, E. and Fogain, E. (1996) Detection of a *Cylindrocladium* sp. in Cameroon. *Infomusa* 5(1), 4–7.

Chuang, T.Y. (1988) Studies on the soils suppressive to banana fusarium wilt II. Nature of suppression to race 4 of *Fusarium oxysporum* f. sp. *cubense* in Taiwan soils. *Plant Protection Bulletin (Taiwan)* 30, 125–134.

Clark, A.G. and Lanigan, C.M.S. (1993) Prospects for estimating nucleotide divergence with RAPDs. *Molecular Biology and Evolution* 10, 1096–1111.

Correll, J.C. and Leslie, J.F. (1987) Genetic diversity in the Panama disease pathogen, *Fusarium oxysporum* f. sp. *cubense*, determined by vegetative compatibility. *Mycological Society of America Newsletter* 38, 22.

Correll, J.C., Klittich, C.J.R. and Leslie, J.F. (1987) Nitrate-nonutilizing mutants of *Fusarium oxysporum* and their use in vegetative compatibility tests. *Phytopathology* 77, 1640–1646.

D'Alessio, N. (1998) Mitochondrial inheritance during a parasexual cycle in *Fusarium oxysporum* f. sp. *cubense.* PhD dissertation, Florida International University, Miami, USA.

Davis, A.J., Say, M., Snow, A.J. and Grant, B.R. (1994) Sensitivity of *Fusarium oxysporum* f. sp. *cubense* to phosphonate. *Plant Pathology* 43, 200–205.

Domsch, K.H., Gams, W. and Anderson, T.H. (1980) *Compendium of Soil Fungi*, Vol. 1. Academic Press, New York, USA, 859 pp.

Ela, V.M. and San Juan, M.O. (1954) Leaf spot and stem rot of abaca. *Philippine Journal of Agriculture* 38, 251–271.

Gerlach, W. and Nirenberg, H. (1982) *The Genus* Fusarium – *A Pictorial Atlas.* Paul Parey, Berlin, Germany, 406 pp.

Herbert, J.A. and Marx, D. (1990) Short-term control of Panama disease of bananas in South Africa. *Phytophylactica* 22, 339–340.

Hernandez, J.M., Freitas, G., Ploetz, R.C. and Kendrick, C. (1993) Fusarial wilt of banana in the Canary Islands with some data regarding the Madeira Archipelago. In: Valmayor, R.V., Hwang, S.C., Ploetz, R.C., Lee, S.W. and Roa, V.N. (eds) *Proceedings: International Symposium on Recent Developments in Banana Cultivation Technology, Taiwan Banana Research Institute, Chiuju, Pingtung, Taiwan, 14–18 December 1992.* INIBAP/ASPNET, Los Baños, Laguna, Philippines, pp. 247–254.

Jameson, J.D. (1953) Outbreaks and new records. Uganda. *FAO Plant Protection Bulletin* 1, 62.

Jones, D.R. (ed.) (1994) *The Improvement and Testing of* Musa*: A Global Partnership, Proceedings of the First Global Conference of the International* Musa *Testing Program held at FHIA, Honduras, 27–30 April 1994.* INIBAP, Montpellier, France, 303 pp.

Kistler, H.C. and Miao, V. (1992) New modes of genetic change in filamentous fungi. *Annual Review of Phytopathology* 30, 131–152.

Kobenan, K. (1991) Parasites et ravageurs du système souterrain des bananiers en Côte d'Ivoire. *Fruits* 46, 633–641.

Kobenan, K., Ake, S., Kone, D. and Kehe, M. (1997) *Zythia* sp. (Ascomycotina), causal agent of root necrosis in bananas (*Musa* sp., AAA) cv. Grande Naine. *Cahiers Agricultures* 6, 91–96.

Koenig, R., Ploetz, R.C. and Kistler, H.C. (1997) *Fusarium oxysporum* f. sp. *cubense* consists of a small number of divergent and globally distributed lineages. *Phytopathology* 87, 915–923.

Kuhn, D.N., Cortes, B., Pinto, T. and Weaver, J. (1995) Parasexuality and heterokaryosis in *Fusarium oxysporum* f. sp. *cubense. Phytopathology* 85, 1119.

Lakshmanan, P., Selvaraj, P. and Mohan, S. (1987) Efficacy of different methods for the control of Panama disease. *Tropical Pest Management* 33, 373–374.

Laville, E. (1964) Etude de la mycoflore des racines du bananier 'Poyo'. *Fruits* 20, 435–449, 521–528.

Leslie, J.F. (1993) Vegetative compatibility in fungi. *Annual Review of Phytopathology* 31, 127–151.

Loridat, P. (1989) Etude de la microflore fongique et des nématodes associés aux nécroses de l'appareil souterrain du bananier en Martinique. Mise en evidence du pouvoir pathogene du genre *Cylindrocladium. Fruits* 44, 587–598.

Loridat, P. and Ganry, J. (1991) Mise en évidence d'une interaction nématode–champignon (*Radopholus similis*/*Cylindrocladium*), comme composante du parasitisme tellurique du bananier en culture industrielle aux Antilles. In: Anez, B., Nava, C., Sosa, L. and Jaramillo, R. (eds) *Proceedings of the Ninth Meeting of ACORBAT, Merida, Venezuela, September 24–29, 1989.* ACORBAT, Maracaibo, Venezuela, pp. 283–303.

Magee, C.J. and Eastwood, H.W. (1935) Corm rot of bananas. *Agricultural Gazette of New South Wales* 46, 631–632.

Mallessard, R. (1966) Etude de la mycoflore des racines du bananier 'Poyo'. *Fruits* 21, 543–552.

Meredith, D.S. (1961) Fruit spot ('speckle') of Jamaican bananas caused by *Deightoniella torulosa* (Syd.) Ellis. IV. Further observations on spore dispersal. *Annals of Applied Biology* 49, 488–496.

Moore N.Y., Hargreaves, P.A., Pegg, K.G. and Irwin, J.A.G. (1991) Characterisation of strains of *Fusarium oxysporum* f. sp. *cubense* by production of volatiles. *Australian Journal of Botany* 39, 161–166.

Moore, N., Pegg, K.G., Allen, R.N. and Irwin, J.A.G. (1993a) Vegetative compatibility and distribution of *Fusarium oxysporum* f. sp. *cubense* in Australia. *Australian Journal of Experimental Agriculture* 33, 797–802.

Moore, N., Pegg, K.G., Langdon, P.W., Smith, M.K. and Whiley, A.W. (1993b) Current research on Fusarium wilt of banana in Australia. In: Valmayor, R.V., Hwang, S.C., Ploetz, R.C., Lee, S.W. and Roa, V.N. (eds) *Proceedings: International Symposium on Recent Developments in Banana Cultivation Technology, Taiwan Banana Research Institute, Chiuju, Pingtung, Taiwan, 14–18 December 1992.* INIBAP/ASPNET, Los Baños, Laguna, Philippines, pp. 270–284.

Nelson, P.E., Toussoun, T.A., and Marasas, W.O. (1983) Fusarium *Species. An Illustrated Guide for Identification.* Pennsylvania State University Press, University Park, USA, 193 pp.

Ocefemia, G.O. and Mendiola, V.B. (1932) The *Fusarium* associated with some field cases of heart rot of abaca. *Philippine Agriculturist* 21, 296–308.

O'Donnell, K.O., Kistler, H.C., Cigelnik, E. and Ploetz, R.C. (1998) Multiple evolutionary origins of the fungus causing Panama disease of banana: concordant evidence from nuclear and mitochondrial gene genealogies. *Proceedings of the National Academy of Science (USA)* 95, 2044–2049.

Orjeda, G. (1998) *Evaluation of Musa Germplasm for Resistance to Sigatoka Diseases and Fusarium Wilt.* INIBAP Technical Guidelines 3, INIBAP, Montpellier, France, 63 pp.

Pegg, K.G. and Langdon, P.W. (1987) Fusarium wilt (Panama disease): a review. In: Persley, G. and DeLanghe, E.A. (eds) *Banana and Plantain Breeding Strategies, Proceedings of an International Workshop held at Cairns, Australia, 13–17 October 1986.* ACIAR Proceedings No. 21, Australian Centre for International Agricultural Research, Canberra, Australia, pp. 119–123.

Pegg, K.G., Moore, N.Y. and Sorensen, S. (1993) Fusarium wilt in the Asian Pacific region. In: Valmayor, R.V., Hwang, S.C., Ploetz, R.C., Lee, S.W. and Roa, V.N. (eds) *Proceedings: International Symposium on Recent Developments in Banana Cultivation Technology, Taiwan Banana Research Institute, Chiuju, Pingtung, Taiwan, 14–18 December 1992.* INIBAP/ASPNET, Los Baños, Laguna, Philippines, pp. 255–269.

Pegg, K.G., Moore, N.Y. and Sorensen, S. (1994) Variability in populations of *Fusarium oxysporum* f. sp. *cubense* from the Asia/Pacific region. In: Jones, D.R. (ed.) *The Improvement and Testing of* Musa*: A Global Partnership, Proceedings of the First Global Conference of the International* Musa *Testing Program held at FHIA, Honduras, 27–30 April 1994.* INIBAP, Montpellier, France, pp. 70–82.

Persley, D. (ed.) (1993) *Diseases of Fruit Crops.* Queensland Department of Primary Industries, Brisbane, Australia, 178 pp.

Ploetz, R.C. (1989) Factors influencing the development of fusarial wilt of banana (Panama disease). *Phytopathology* 79, 1181.

Ploetz, R.C. (1990a) Population biology of *Fusarium oxysporum* f. sp. *cubense.* In: Ploetz, R.C. (ed.) *Fusarium Wilt of Banana.* APS Press, St Paul, Minnesota, USA, pp. 63–76.

Ploetz, R.C. (1990b) Roundtable discussions. In: Ploetz, R.C. (ed.) *Fusarium Wilt of Banana.* APS Press, St Paul, Minnesota, USA, pp. 135–138.

Ploetz, R.C. (1992) Fusarium wilt of banana (Panama disease). In: Mukhopadhyay, A.N., Chaube, H.S., Kumar, J. and Singh, U.S. (eds) *Plant Diseases of International Importance,* Vol. III. Prentice Hall, Englewood Cliffs, New Jersey, USA, pp. 270–282.

Ploetz, R.C. (1993) Fusarium wilt (Panama disease) in Africa: Current status and outlook for smallholder Agriculture. In: Gold, C.S. and Gemmill, B. (eds) *Biological and Integrated Control of Highland Banana and Plantain Pests and Diseases.* IITA, Ibadan, Nigeria, pp. 312–323.

Ploetz, R.C. (1994) Panama disease: return of the first banana menace. *International Journal of Pest Management* 40, 326–336.

Ploetz, R.C. and Correll, J.C. (1988) Vegetative compatibility among races of *Fusarium oxysporum* f. sp. *cubense. Plant Disease* 72, 325–328.

Ploetz, R.C. and Pegg, K.G. (1997) Fusarium wilt of banana and Wallace's line: was the disease originally restricted to the Indo-Malayan region? *Australasian Plant Pathology* 26, 239–249.

Ploetz, R.C., Herbert, J., Sebasigari, K., Hernandez, J.H., Pegg, K.G., Ventura, J.A. and Mayato, L.S. (1990) Importance of fusarium wilt in different banana-growing regions. In: Ploetz, R.C. (ed.) *Fusarium Wilt of Banana.* APS Press, St Paul, Minnesota, USA, pp. 9–26.

Ploetz, R.C., Braunworth, W.S., Jr, Hasty, S., Gantotti, B., Chizala, C.T., Banda, D.L.N., Makina, D.W. and Channer, A.G. (1992) Fusarium wilt of banana (Panama disease) in Malawi. *Fruits* 47, 503–508.

Ploetz, R.C., Jones, D.R., Sebasigari, K., and Tushemerirewe, W. (1995) Fusarium wilt (Panama disease) on East African highland cultivars of banana. *Fruits* 49, 253–260.

Ploetz, R.C., Vázquez, A., Nagel, J., Benscher, D., Sianglew, P., Srikul, S., Kooariyakul, S., Wattanachaiyingcharoen, W., Lertrat, P. and Wattanachaiyingcharoen, D. (1997) Current status of Panama disease in Thailand. *Fruits* 51, 387–395.

Puhalla, J.E. (1985) Classification of strains of *Fusarium oxysporum* on the basis of vegetative compatibility. *Canadian Journal of Botany* 63, 179–183.

Quimio, A.J. and Tessera, M. (1996) Diseases of enset. In: Abate, T., Hiebsch, C., Brandt, S.A. and Gebremarium, S. (eds) *Proceedings of the International Workshop on Enset.* Institute of Agricultural Research, Addis Ababa, Ethiopia, pp. 188–203.

Ramos, M.M. (1933) Mechanical injuries to roots and corms of abaca in relation to heart rot disease. *Philippine Agriculturist* 22, 322–337.

Ramos, M.M. (1941) Dry sheath rot of abaca caused by *Marasmius* and suggestions for its control. *Philippine Journal of Agriculture* 12, 31–39.

Rhoads, A.S. (1942) Notes on *Clitocybe* root rot of bananas and other plants in Florida. *Phytopathology* 32, 487–496.

Risède, J.M. (1994) Eléments de caractérisation de *Cylindrocladium* sp., agent de nécroses racinaires du bananier en Martinique. *Fruits* 49, 167–178.

Roldan, E.F. (1933) Four new diseases of Philippine economic plants caused by species in the family *Pythiaceae. Philippine Agriculturist* 21, 541–546.

Rutherford, M.A., Kangire, A., Kung'u, J.N. and Mabagala, R.B. (1998) Fusarium wilt in East Africa. In: *7th International Congress of Plant Pathology, Edinburgh, 9–16 August 1998,* Abstract 3.7.63.

Semer, C.R., Mitchell, D.J., Mitchell, M.E., Martin, F.R. and Alfenas, A.C. (1987) Isolation, identification and chemical control of *Cylindrocladium musae* sp. nov. associated with toppling disease of banana. *Phytopathology* 77, 1729.

Shivas, R.G. and Philemon, E. (1996) First record of *Fusarium oxysporum* f. sp. *cubense* on banana in Papua New Guinea. *Australasian Plant Pathology* 25, 260.

Shivas, R.G., Suyuko, S., Raga, N. and Hyde, K.D. (1996) Some disease-associated microorganisms on plants in Irian Jaya, Indonesia. *Australasian Plant Pathology* 25, 36–49.

Simmonds, N.W. (1962) *The Evolution of the Bananas.* Longmans, Green, London.

Simmonds, N.W. (1966) *Bananas*, 2nd edn. Longmans, Green, London, UK, 512 pp.

Sivanesan, A. and Holliday, P. (1972) *Rosellinia bunoides.* CMI Descriptions of Pathogenic Fungi and Bacteria No. 351. Commonwealth Mycological Institute, Kew, Surrey, UK, 2 pp.

Smith, F.E.V. (1929) Report of the Government Microbiologist. Annual Report of the Department of Agriculture, Jamaica for 1928. Department of Agriculture, Kingston, Jamaica, pp. 17–20.

Smith, M.K., Whiley, A.W., Searle, C., Langdon, P.W., Schaffer, B. and Pegg, K.G. (1998) Micropropagated bananas are more susceptible to Fusarium wilt than plants grown from conventional material. *Australian Journal of Agricultural Research* 49, 1133–1139.

Stover, R.H. (1962) *Fusarial Wilt (Panama Disease) of Bananas and other* Musa *species.* Commonwealth Mycological Institute, Kew, Surrey, UK, 117 pp.

Stover, R.H. (1963) Leaf spot and damping-off of *Musa* seedlings. *Tropical Agriculture (Trinidad)* 40, 9–14.

Stover, R.H. (1966) Fungi associated with nematode and non-nematode lesions on banana roots. *Canadian Journal of Botany* 44, 1703–1709.

Stover, R.H. (1972) *Banana, Plantain and Abaca Diseases.* Commonwealth Mycological Insititute, Kew, Surrey, UK, 316 pp.

Stover, R.H. (1990) Fusarium wilt of banana: some history and current status of the disease. In: Ploetz, R.C. (ed.) *Fusarium Wilt of Banana.* APS Press, St Paul, Minnesota, USA, pp. 1–7.

Stover, R.H. and Buddenhagen, I.W. (1986) Banana breeding: polyploidy, disease resistance and productivity. *Fruits* 41, 175–191.

Stover, R.H. and Simmonds, N.W. (1987) *Bananas*, 3rd edn. Longmans Scientific and Technical, Harlow, Essex, UK, 468 pp.

Su, H.J., Hwang, S.C. and Ko, W.H. (1986) Fusarial wilt of Cavendish bananas in Taiwan. *Plant Disease* 70, 814–818.

Tessera, M. and Quimio, A.J. (1994) Research on enset pathology. In: Herath, E. and Desalegn, L. (eds) *Proceedings of the 2nd National Horticultural Workshop of Ethiopia.* Institute of Agricultural Research, Addis Ababa, pp. 217–225.

Thomas, V., Rutherford, M.A. and Bridge, P.D. (1994) Molecular differentiation of two races of *Fusarium oxysporum* special form *cubense. Letters in Applied Microbiology* 18, 193–196.

Toussoun, T.A. (1975) Fusarium-suppressive soils. In: Bruehl, G.W. (ed.) *Biology and Control of Soil-Borne Plant Pathogens.* APS, St Paul, Minnesota, USA, pp. 145–151.

Vakili, N.G. (1965) Fusarium wilt resistance in seedlings and mature plants of *Musa* species. *Phytopathology* 55, 135–140.

Vander Molen, G.E., Beckman, C.H. and Rodehorst, E. (1977) Vascular gelation: a general response phenomenon following infection. *Physiological Plant Pathology* 11, 95–110.

Vander Molen, G.E., Beckman, C.H. and Rodehorst, E. (1987) The ultrastructure of tylose formation in resistant banana following inoculation with *Fusarium oxysporum* f. *cubense. Physiological and Molecular Plant Pathology* 31, 185–200.

Waite, B.H. (1954) Vascular disease of abaca or Manila hemp in Central America. *Plant Disease Reporter* 38, 575–578.

Waite, B.H. (1956) Fusarium stalk rot of bananas in Central America. *Plant Disease Reporter* 40, 309–311.

Waite, B.H. (1961) Variability and pathogenesis in *Fusarium oxysporum* f. *cubense.* PhD thesis, University of California, Berkeley, USA, 147 pp.

Waite, B.H. (1963) Wilt of *Heliconia* spp. caused by *Fusarium oxysporum* f. sp. *cubense* Race 3. *Tropical Agriculture (Trinidad)* 40, 299–305.

Waite, B.H. and Dunlap, V.C. (1953) Preliminary host range studies with *Fusarium oxysporum* f. sp. *cubense. Plant Disease Reporter* 37, 79–80.

Wallace, G.B. (1952) Wilt or Panama disease of banana. *East African Agricultural Journal* 17, 166–175.

Wardlaw, C.W. (1961) *Banana Diseases Including Plantains and Abaca.* Longmans, Green, London, 648 pp.

Wollenweber, H.W. (1913) Studies on the *Fusarium* problem. *Phytopathology* 3, 24–50.

4

Fungal Diseases of Banana Fruit

PREHARVEST DISEASES

D.R. Jones and R.H. Stover

Introduction

Preharvest diseases of banana fruit caused by fungi are important because they produce unsightly blemishes, which are unacceptable to consumers, on the fruit peel. Although eating quality is seldom affected, because the pulp is only occasionally invaded, the fruit is rejected because its value is significantly lowered. From time to time in the past, severe outbreaks of various preharvest diseases have occurred, which have resulted in substantial losses. These diseases were controlled by implementing good sanitation practices in plantations, covering fruit with a polyethylene bag (see Plate 1.6) and spraying fruit with fungicide when necessary. Although many of the problems reported here, such as brown spot and pitting disease, have declined in significance, others, such as freckle and speckle, have increased. In the past, fungicides were applied to fruit on commercial plantations to control diseases when necessary. Today, because of health concerns, options for chemical control are limited.

Bacterial diseases, such as Moko, blood disease, finger-tip rot and bugtok, also affect preharvest fruit. In contrast to fungal problems, these invariably affect the pulp and make the fruit inedible. Bacterial diseases of fruit are described in the next chapter (Thwaites *et al.*, Chapter 5, pp. 213–239). Other problems with preharvest fruit are covered under virus diseases (Chapter 6, pp. 241–293), non-infectious disorders (Lahav *et al.*, Chapter 8, pp. 325–338) and genetic abnormalities (Israeli and Lahav, Chapter 12, pp. 395–408).

Anthracnose Fruit Rot

Anthracnose is a common and widespread disease of postharvest fruit (see Muirhead and Jones, this chapter, pp. 199–203), but, in the Asia–Pacific region, it can also affect immature fruit on the plant. It has been reported that up to 16% of unharvested bunches have been diseased in the Philippines (Wardlaw, 1961). Infections begin as small, black circular specks on the flowers and peel at the distal ends of the hands. Salmon-pink-coloured spore masses of the fungal pathogen appear as the lesions increase in size and coalesce to form large sunken, blackened areas. In severe cases, fingers can be covered in lesions. Affected fruit ripens prematurely, rots and eventually shrivels (Plate 4.1). The disease may spread to crowns and attack other fingers via pedicels (Plate 4.1). The fruit of abacá has also been reported to be attacked (Ocfemia, 1924; Agati, 1925).

Anthracnose fruit rot is caused by *Colletotrichum musae* (Berk. and Curtis) Arx (syn. *Gloeosporium musarum* Cooke and Massee). This fungus is described in the section on postharvest fruit (see Muirhead and Jones, this chapter, pp. 199–203). High humidity, high temperatures and the presence of bruises promote infection, which probably occurs mainly at the remains of the perianth (Wardlaw, 1961). The conidia of *C. musae* are abundant in banana plantations because the fungus is a common inhabitant of decaying banana leaves and old fruit. They are spread by wind and may also be carried by insects that frequent banana flowers. Germination occurs readily in water, but conidia quickly die in the absence of moisture. The disease is favoured, therefore, by conditions of high humidity and rainfall. In inoculation experiments, the disease was initiated when the fungus was applied to slight wounds in the peel. Very immature fruit, with flowers still attached, were more readily attacked than older fruit (Agati, 1922). Field inoculations carried out in the West Indies have been unsuccessful (Wardlaw, 1931).

Thirty different cultivars showed susceptibility in the Philippines. Cavendish cultivars were the most severely attacked, with cooking-banana types being less susceptible (Wardlaw, 1961). However, immature bunches of 'Blue Java' (ABB, syn. 'Ney Mannan') have been completely destroyed in Fiji (Parham, 1938). In Sri Lanka, 'Alukehel' or 'Ash Plantain' (ABB, syn. 'Ney Mannan') is the main cultivar affected (Plate 4.1) (Park, 1933).

Brown Spot

Introduction

Brown spot was a common blemish in some areas of the Americas on fruit developing during warm, rainy weather. Incidence varied greatly, but the most severe outbreaks occurred in Mexico, Guatemala and Honduras. In the latter two countries, entire bunches were rejected and up to 20% of hands lost due to the disease. It also caused losses in Costa Rica, Panama, Colombia, Ecuador, Surinam and the Caribbean Islands. Although it was first described only in 1965, it may have been present earlier, because the pathogen is a common saprophyte on hanging banana-leaf trash and on leaves of dead and dying weeds (Stover, 1972). Brown spot is not an important disease today.

Symptoms

Symptoms of brown spot occur on fingers, crowns and the peduncle. Spots are pale to dark brown with an irregular margin, surrounded by a halo of water-soaked tissue (Plates 4.2 and 4.8). The size of the spots, which are centred over stomata, varies, but they average 5–6 mm in diameter. Brown spot can be distinguished from Deightoniella speckle because brown spot lesions are much larger than those caused by *Deightoniella torulosa* (see Jones and Stover, this chapter, pp. 181–182). They differ from those of pitting disease because they are not sunken, as are lesions caused by *Pyricularia grisea* (see Jones and Stover, this chapter, pp. 186–189). Also, they do not increase in size or number during ripening, as do those of pitting disease. Brown spot occurs on fruit that is 50 days old or older. The pathogen does not sporulate in the fruit spots.

Causal agent

Brown spot is caused by *Cercospora hayi* Calpouzos. This fungus is readily isolated from lesions, but does not produce spores on most media. However, *C. hayi* does sporulate on propylene oxide-sterilized banana leaf tissue after 7 days. Conidia are hyaline, measure 90–150 μm × 3–4 μm, contain five to 15 cells and have truncate bases and acute tips (Kaiser and Lukezic, 1965).

Disease cycle and epidemiology

Cercospora hayi is present in dead banana and weed leaf tissue. Mycelium of the fungus survives in dried leaf tissue for 15 weeks and spores remain viable for at least 5 weeks when exposed to fluctuating temperatures and humidity levels (Stover, 1972). Conidia form within 16 h at 23–26°C in a saturated atmosphere and are released in air currents with velocities as low as 2.4 km h^{-1}. Spore release peaks in the afternoon between 14.00 and 16.00 h at a time that usually coincides with the highest daily wind speeds. Conidia germinate within 3 h on green banana peel. Infection mainly occurs on fruit less than 21 days old, but spots do not appear until the fruit is approaching maturity, at about 90 days (Kaiser and Lukezic, 1966).

In Honduras, conidia were present throughout the year, but numbers were highest in July, September and October, when most rain fell. Brown spot incidence was highest during the rainy season, from July to January.

Host reaction

Fruit of accessions of *Musa acuminata* ssp. *malaccensis* and ssp. *microcarpa* were seen with brown spot, but not fruit of *Musa balbisiana*. Fruit of all banana cultivars was susceptible, but the clones in the Cavendish subgroup (AAA) seemed to be more susceptible than 'Gros Michel'. Dessert banana cultivars in the AB and AAB genomic groups were also affected, but cultivars in the Plantain subgroup (AAB) were not seriously spotted (Kaiser and Lukezic, 1966).

Control

When brown spot was serious, it was controlled in the same way as pitting disease (see Jones and Stover, this chapter, pp. 186–189).

Cigar-end Rot

Introduction

Cigar-end rot is important in West and Central Africa. A loss of 10,000 bunches was reported in Cameroon in 1958 (Stover, 1972). The disease has also been reported from Australia, the Middle East, Egypt, the Canary Islands, the West Indies and South America (Wardlaw, 1961; Stover, 1972; Slabaugh, 1994b).

Symptoms

A black necrosis spreads from the perianth into the tip of immature fingers. The pulp undergoes a dry rot. The corrugated necrotic tissue becomes covered with the mycelia of fungi and resembles the greyish ash of a cigar-end (Plate 4.3). Diseased tissue is delimited from healthy tissue by a black band and a line of chlorosis. The rot spreads slowly and seldom affects more than the first 2 cm of the finger tip (Plate 4.4). However, if the fingers are attacked soon after emergence, the entire finger sometimes becomes rotten. In Australia, cigar-end rot is often associated with 'choke throat' (see Israeli and Lahav, Chapter 10, pp. 354–356) (Simmonds, 1966).

Causal agents

Cigar-end rot is caused by either *Verticillium theobromae* (Turconi) Mason and Hughes or *Trachysphaera fructigena* Tabor and Bunting. However, *V. theobromae* is more widespread than *T. fructigena*, which is not known in the western hemisphere. In Central and West Africa, both fungi are responsible for the disease, but *T. fructigena* is the main agent. In Queensland, only *V. theobromae* has been isolated from cigar-end rots (Simmonds, 1966).

Conidiophores of *V. theobromae*, which measure 150–400 μm × 4–6 μm, are solitary or in small groups on the surface of the peel. Conidia are borne at the ends of tapering phialides and are aggregated into rounded, mucilaginous, translucent heads. Conidia are 4–9 μm × 2–3.5 μm and phialides 15–30 μm (Meredith, 1961b).

Conidiophores of *T. fructigena* are simple or bear a terminal vesicle, to which a whorl of pedicellate conidia is attached (Fig. 4.1). The conidia are spherical, strongly echinulate, with an average diameter of 35 μm. They are borne on pedicels, which vary in length up to 30 μm. Oogonia characteristic of the *Peronosporales* are produced in diseased tissue. They are small, averaging 40 μm × 24 μm, thick-walled and characterized by the presence of irregular, sac-like outgrowths (Fig. 4.1). The

Plate 4.1. Symptoms of anthracnose fruit rot on fingers of 'Alukehel' or 'Ash Plantain' (ABB, syn. 'Ney Mannan') in Sri Lanka. Note the disease spreading from the crown into a finger via the pedicel (photo: D.R. Jones, INIBAP).

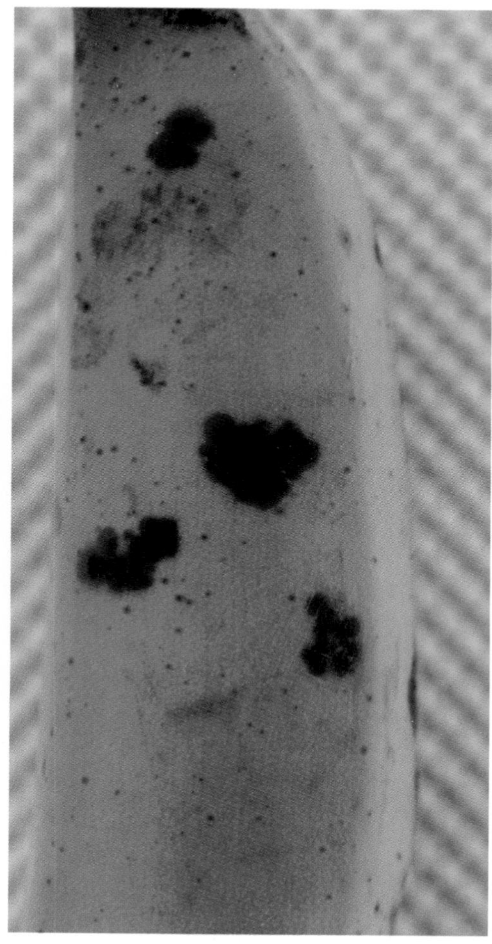

Plate 4.2. Symptoms of brown spot in Honduras. Note the irregular outline of the spots and water-soaked haloes (photo: J. Rivera, FHIA).

antheridia are amphigynous, completely surrounding the stalk of the oogonium. Pure cultures can easily be established from conidia. *Trachysphaera fructigena* was first described on cacao and coffee in West Africa (Tabor and Bunting, 1923).

Disease cycle and epidemiology

In Cameroon, cigar-end rot was found to be at its worst during the rainy season, virtually disappearing in the dry season (Brun, 1954). Later, a direct correlation was shown between rainfall and incidence of the disease on Cavendish cultivars (Beugnon *et al.*, 1970). The problem was more serious at higher elevations and around the borders of plantations next to native vegetation (Brun, 1954). In Egypt, cigar-end rot was rare in the winter months, reaching a peak in May and June (El-Helaly *et al.*, 1955). In general, poorly maintained plantations had the highest levels of disease.

Verticillium theobromae is a common colonizer of banana-leaf trash and flowers. Conidia are disseminated in air currents and infect dying flower parts. The optimum growth of *V. theobromae* in culture is 25°C.

Spores of *T. fructigena* are not believed to remain viable for long under field conditions (Maramba and Clerk, 1974). The

Plate 4.3. Symptoms of cigar-end rot in Australia caused by *Verticillium theobromae* (photo: QDPI).

Plate 4.4. Symptoms of a cigar-end rot on fruit of 'Yangambi Km 5' (AAA, syn.' Ibota Bota') in Nigeria. *Trachysphaera frucigena* was one of several fungi isolated from affected fruit. (Photo: C. Pasberg-Gauhl and F. Gauhl, IITA.)

fungus grows best in culture at 24°C, but cigar-end rot caused by *T. fructigena* is accentuated when moderate temperatures of around 20°C are followed by higher temperatures of above 27°C (Tezenas du Montcel and Laville, 1977). In inoculation experiments, advanced symptoms appeared on immature, wounded fruit after 5–7 days' incubation at 18–24°C (Meredith, 1960).

Unlike cigar-end rot caused by *V. theobromae*, the disease caused by *T. fructigena* can lead to premature ripening of fruit. This pathogen also continues to attack and rot fruit after harvest. It can invade freshly cut crown surfaces and through wounds in the peel caused by improper handling. Inoculum for postharvest infections originates from fruit harvested with the disease and may be present in dehanding and delatexing tanks. In the UK, bananas from Jamaica that were stored in ripening rooms formerly occupied by diseased fruit from Cameroon became infected (Meredith, 1960).

Host reaction

Cigar-end rot affects fruit of 'Gros Michel' (AAA) and cultivars in the Cavendish subgroup (AAA).

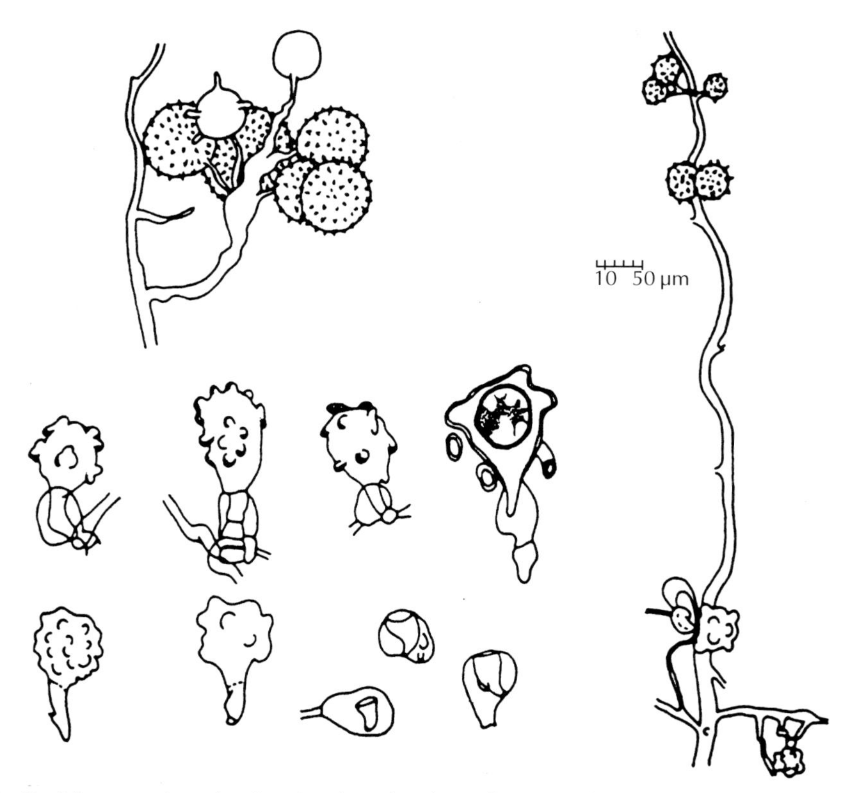

Fig. 4.1. Conidia, oogonia and antheridia of *Trachysphaera fructigena* (from Tabor and Bunting, 1923).

Control

Field sanitation, such as the removal of dead, hanging leaves from plants, will reduce inoculum levels of *V. theobromae*. The early removal of dead flowers, which may eliminate the source of infection before it reaches the finger, followed by the use of polyethylene bags on bunches, is also advocated as a control measure (Beugnon *et al.*, 1970). Any trash accumulating in the bags should be removed after a few weeks. Packing stations and ripening rooms need to be kept clean to minimize the chances of postharvest infection by *T. fructigena*. Fingers affected by cigar-end rot should be cut from hands in the field to avoid contamination of water in dehanding and delatexing tanks. It may be necessary to apply fungicide when cigar-end rot is prevalent. Brun (1970) found benzimidazole fungicides to be ineffective in controlling cigar-end rot. Tezenas du Montcel (1981) tested different fungicides and found that metalaxyl (Ridomil®) gave the best control.

Tip-end Rot

A number of tip-end rots have been reported on unripe banana fruit. They are not economically significant, but have caused some problems on occasions.

A black-tip disease was first reported from Bermuda by Ogilvie (1927, 1928) and was subsequently found in the Philippines (Reyes, 1934), Trinidad, Guinea, Somalia, Papua New Guinea (Wardlaw, 1961) and Jamaica (Stover, 1972). The disease is confined to the peel in the early stages. The skin turns black and has a dry crumpling consistency. An internal dry rot of the pulp sometimes occurs later, probably as a result of secondary invasion. The lesion, which frequently has an irregular outline, forms more on one side of the finger than on the other. Under conditions of high humidity, the surface of the affected tip is often covered with fungal mycelia. The disease does not have the ashy grey appearance associated with cigar-end rot and rarely affects more than the apical third of the finger. It is often associated with an injury, such as sunburn (see Israeli and Lahav, Chapter 10, pp. 364–365). Fingers in all stages of development can be affected. *Deightoniella torulosa* (Syd.) Ellis, which is the agent responsible for speckle disease (see Jones and Stover, this chapter, pp. 181–182), is usually isolated from banana fruit with black-tip disease (Meredith, 1961a). However, *Verticillium theobromae* and *Fusarium moniliforme* Sheldon have also have been isolated from black-tip lesions (Meredith, 1965). Both are pathogenic on banana. *Verticillium theobromae* is the cause of cigar-end rot (see Jones and Stover, this chapter, pp. 175–178) and *F. moniliforme* causes a minor, dark brown rot of the pulp of unripe fruit in Israel, known as black heart disease, which can also induce premature ripening (Chorin and Rotem, 1961). All the fungi implicated are common saprophytes of dead flower parts of banana. Although attempts to reproduce symptoms of black tip in the field have failed, symptoms were induced in fingers of 'Lacatan' (AAA, Cavendish subgroup) by placing *D. torulosa* in incisions in the corky tissue at the tip of the fingers and incubating the fruit in a saturated atmosphere for 7 days at 28–30°C. Spraying suspensions of conidia of *D. torulosa* into young fruit and then sealing the bunch in a polyethylene bag also occasionally induced black tip (Meredith, 1961d).

Dothiorella tip-end rot has been reported from Israel (Reichert and Hellinger, 1938; Chorin and Rotem, 1961), Guinea and Côte d'Ivoire (Roger and Mallamaire, 1937, in Reichert and Hellinger, 1938) and Australia (Wardlaw, 1961). The disease progresses more rapidly in the peel than in the pulp. Blackening is preceded by a narrow, brown, watery margin, which forms a sharp line of demarcation between healthy and diseased tissues. In Australia, the whole fruit may be affected, with the distal end becoming tapered as the diseased tissues shrink, dry and become mummified (Wardlaw, 1961). Dothiorella tip-end rot is usually attributed to *Dothiorella gregaria* Sacc. This fungus may have *Botryosphaeria ribis* Grossenbacher and Duggar, which has been implicated in a postharvest soft rot of bananas in India (Kachhwaha *et al.*, 1992), as the teleomorph stage (Reichert and Hellinger, 1938). Pycnidia of *D. gregaria* are usually produced in the superficial layers of the peel and burst through the epidermis during development. They may occur singly or together in groups, but with little stromatic material. The black, thick-walled, globose pycnidia have a papillate ostiole and measure 141–321 μm × 141–242 μm. The fusoid or ovoid conidia are thin-walled and measure 13–21 μm × 4–7 μm. The conidia are forcibly ejected *en masse* and each pycnidium becomes surmounted by a little mass of white powder, which is a characteristic of the disease. Insects and wind may spread the conidia. In Israel, *D. gregaria* also causes a black rot of orange fruit, and an isolate of the fungus from oranges caused decay on banana fruit in inoculation experiments (Reichert and Hellinger, 1938). Dothiorella tip-end rot increases in the second and third year after planting and then decreases (Rotem and Chorin, 1961). Fruit and plants in plots irrigated by sprinklers are more prone to this disease than on plants in surface-irrigated plots. Conditions that favour the maintenance of high humidity levels, such as are prevalent in Israel in the winter, favour the problem. Disease levels were reduced in Israel by spraying with zineb and cuprous oxide. Control was also achieved by removing flowers immediately after formation. However, the practice of breaking off flowers 20–30 days after formation gave no control (Chorin and Rotem, 1961).

In Jamaica, a similar disease to Dothiorella tip-end rot is reported on

'Dwarf Cavendish' (AAA, Cavendish subgroup) at an elevation of around 800 m, especially in damp localities. The large abaxial tepal or sometimes the smaller abaxial tepal are believed to be infected first and then finger tips are invaded. The withered flower parts remain attached to diseased fruit and are a characteristic feature. Often the disease does not progress more than a few millimetres, but decay can sometimes spread 6–8 cm along the finger after 20–30 days, usually advancing more quickly on one side than on the other. In the early stages, the decayed pulp is reddish brown and firm. Later it turns black and is hollowed out by insects. Occasionally, the whole finger is affected and is reduced to a dry, sunken, mummified state. *Hendersonula toruloidea* Nattrass (Punithalingam and Waterston, 1970) has been provisionally adopted as the causal agent (Meredith, 1963a). Meredith (1963b) has suggested that this fungus may be the same pathogen as was described by Reichert and Hellinger (1938) as *D. gregaria*, the main difference between the two being the presence of a few septate conidia in pycnidia of *H. toruloidea*. These pycnidia are formed singly or together in groups on the blackened peel and withered flowers. The pycnidium is immersed, globose and black and measures 110–340 μm × 100–260 μm, with a papillate ostiole. Most conidia are non-septate, with hyaline granular contents, and measure 14–24 μm × 4–8 μm. They are produced at the ends of short, thin-walled cells, which form the innermost layer of the pycnidium wall. They are tightly packed and exude from the ostiole in a mucilaginous sheath. Conidia germinate in 1–2 h at 27°C in a film of water, becoming 1–3-septate in 12 h. Finger tips have been artificially inoculated by inserting mycelium from cultures in wounds at the junction of the perianth and finger tip in detached fingers. Symptoms appeared after 6 days on fruit of both 'Dwarf Cavendish' and 'Lacatan' (AAA, Cavendish subgroup). Decay was slow at 10°C, but fingers could be completely decayed within 12 days at 27°C (Meredith, 1963a). 'Lacatan' fruit is not affected in locations where fruit of 'Dwarf Cavendish' is diseased. This may be because 'Dwarf Cavendish' has a more persistent perianth than taller members of the Cavendish subgroup (Simmonds, 1959). A corky abscission layer, which may preclude further fungal invasion, forms early in 'Lacatan' fruit at the junction of the finger tip and perianth (Meredith, 1963a).

Plate 4.5. Severe symptoms of a tip-end rot on fruit of 'Fougamou' (ABB, syn. 'Pisang Awak') in Nigeria. The small, necrotic fruit on the bunch on the right were most probably infected at an early stage of development. (Photo: C. Pasberg-Gauhl and F. Gauhl, IITA.)

Sclerotinia fruit rot is an uncommon, low-temperature disease and has been reported from Israel (Reichert and Hellinger, 1930), Bermuda (Waterston, 1947) and Costa Rica (Laguna and Salazar, 1984). A rot, caused by *Sclerotinia sclerotiorum* (Lib.) de Bary, begins at the flower end and spreads down the finger. The normally white pulp tissue becomes reddish-black and then black and is finally broken

Plate 4.6. Symptoms of Deightoniella speckle caused by *Deightoniella torulosa* on a cluster of fingers from 'Robusta' (AAA, Cavendish subgroup) in St Vincent, Windward Islands (photo: D.R. Jones, SVBGA).

down and hollowed out (Wardlaw, 1961). Unlike tip rots, the entire finger becomes rotten and shrivelled, and abundant sclerotia are found both internally and on the surface of the peel. In Israel, the disease only occurred in the winter. The problem was controlled by the removal of affected fruit and the application of copper sprays.

Botrytis tip-end rot, caused by *Botrytis cinerea* Pers., has been found in Israel and was reported as fairly common during the wet, winter months. It discolours the peel, which is covered by a fairly dense, brown mycelium, and rapidly rots the pulp (Reichert and Hellinger, 1932).

Fusarium and *Lasiodiplodia* species have also been isolated from tip-end rots (Wardlaw, 1961; C. Pasberg-Gauhl and F. Gauhl, Nigeria, 1995, personal communication). *Glomerella cingulata* (Stonem.) Spauld. and Schrenk, which has been found in association with brown blotches on green fruit in Queensland, Australia (Simmonds, 1966), has also been isolated from tip-end rots in Nigeria. This fungus, together with *T. fructigena*, was isolated from the tip-end rots illustrated in Plate 4.5 (C. Pasberg-Gauhl and F. Gauhl, Nigeria, 1995, personal communication).

Deightoniella Speckle

Introduction

Speckle, pin-spotting or 'swamp spot', as it is known in Jamaica, is a common blemish of banana peel. Most fruit has a low incidence of speckle, but it is not regarded as a serious problem as symptoms usually become much less obvious after ripening, when they are masked by the yellow colour of the peel. However, speckle can cause export fruit to be downgraded when severe. High levels of the disease can occur following prolonged higher-than-normal rainfall and when a high humidity builds up inside the polyethylene bunch covers because of impeded drainage or lack of sufficient perforations.

Symptoms

Symptoms consist of minute, reddish brown to black spots with small green haloes (Plate 4.6). The spots are usually less than 2 mm in diameter, but some can be up to 4 mm if the disease is severe. The distribution of speckle within a bunch may be uneven. Stover (1972) reports that it is more severe on the outer face of inner-whorl fingers and towards the tips of fingers in the outer whorls. Symptoms can increase as the fruit ages, but resistance to infection also increases with age. Meredith (1961a) found that 10–30-day-old fingers were more easily infected than 70–100-day-old fingers. Speckling does not increase and spots do not enlarge after harvest. Speckle can be distinguished from spots caused by flower thrips because the latter are raised and bumpy to the touch.

Causal agent

Deightoniella speckle is caused by *Deightoniella torulosa* (Syd.) Ellis, which also causes a leaf spot of banana, abacá and enset (see Jones *et al.*, Chapter 2, pp. 102–104) and a pseudostem rot of abacá (see Lomerio, Chapter 3, pp. 161–162). *Deightoniella torulosa* can also incite black-tip disease on preharvest banana fruit (see Jones and Stover, this chapter, pp. 178–179). The fungus is a weak pathogen of old or injured banana leaves and is commonly found on dead banana leaves. It has been described by Ellis (1957), Meredith (1961a, d) and Subramanian (1968) (see also Jones *et al.*, Chapter 2, pp. 102–104).

Disease cycle and epidemiology

Inoculum is generated during periods of rain or dew on dead banana leaves (Meredith, 1961a). Conidia are violently discharged as the humidity falls, and they become airborne (Meredith, 1961d). They are not disseminated long distances and lose their viability within 4 days at a humidity of less than 95% (Meredith, 1961a, c). Conidia germinate in moisture and form appressoria after about 6 h (Meredith, 1961a). Penetration can occur within 12–20 h, as indicated by a reddish-brown discoloration in the invaded epidermal cell (Meredith, 1961a). Adjacent cells then become discoloured as the fungus spreads intercellularly. Usually, no more than 15 cells are affected in width and depth. Speckles become visible to the naked eye on green banana peel within 72 h (Meredith, 1961a). The fungus does not sporulate in lesions on fruit.

Host reaction

Stover (1972) has reported that all cultivars of banana are affected by speckle.

Control

The disease is more prevalent in badly maintained and poorly drained plantations that have abundant leaf trash. High-density planting, inadequate weed control and standing water on the ground all help to maintain a moist atmosphere, which encourages speckle. In well-maintained and properly drained plantations, where fruit is covered with a perforated polyethylene bag, speckle is rarely a problem. However, the bag should have sufficient perforations to prevent the build-up of a high humidity level around the bunch. Also, if the bag is knotted at the bottom to stop the polyethylene rubbing on fruit in the wind, the knot should not impede drainage.

As leaves decay more rapidly on the ground than hanging on the plant, inoculum levels in the air can be reduced by removing dead leaves (Meredith, 1961a, c). Spraying the young fruit with dithiocarbamate fungicides, as was practised for the control of pitting disease, also reduces the incidence of speckle.

Central American speckle

Recently, research into speckle has been resumed because the problem has increased, particularly in the wet season, in some areas in Central America (Plate 4.7). In Costa Rica, up to 70% of fruit from one farm has been rejected for export. It was discovered that, although symptoms were very similar to Deightoniella speckle, *D. torulosa* was rarely isolated from lesions. Young, unprotected fruit before bagging is vulnerable to damage caused by agrochemicals, such as fungicides and leaf fertilizers, which may cause speckle-like symptoms. Physiological reactions of the fruit leading to small necrotic flecks may also be induced by insects feeding or laying eggs and other fungi penetrating the peel. In total, 16 different sporulating fungi have been isolated from 'speckle' lesions during recent investigations. Ten of these fungi were selected for artificial inoculation experiments. These were a *Cephalosporium* sp., *Colletotrichum musae*, a *Cylindrocarpon* sp., *D. torulosa*, *Fusarium dimerum*, *Fusarium fusarioides*, *Fusarium semitectum*, *Fusarium tricinctum*, *Penicillium roseum* and a *Trichoderma* sp. Typical speckle symptoms could be reproduced with *D. torulosa*, the *Cephalosporium* sp., *C. musae*, *F. fusarioides*, *F. semitectum* and *F. tricinctum*. Very faint speckling, similar to some symptoms seen in the field, formed after inoculation with the *Cylindrocarpon* sp., *P. roseum* and *Trichoderma* sp. Speckle is less common on farms where banana plants are treated with fungicide than on farms where banana plants are untreated. The disease still continues to cause problems and research is continuing (C. Pasberg-Gauhl, Costa Rica, 1999, personal communication).

Diamond Spot

Introduction

Diamond spot was common in Honduras and Guatemala following the expansion of the cultivation of clones in the Cavendish subgroup (AAA) in the 1960s. It was also a problem in Mexico and frequently encountered on unsprayed bunches on the Pacific coasts of Nicaragua, Costa Rica and Panama and in the Philippines. It was not common on the Atlantic coasts of Costa Rica and Panama or in Ecuador, Colombia, Surinam and the Caribbean (Stover, 1972).

The disease caused serious losses of fruit in Honduras and Guatemala during the rainy season from July to January. In these countries, up to 10% of hands were discarded. However, diamond spot is no longer important.

Symptoms

Spotting begins to appear as fruit approaches maturity. The first symptom on green peel is a slightly raised, inconspicuous, yellow spot between 3 and 5 mm in diameter. The cells colonized by the pathogen do not expand and a longitudinal crack, which is surrounded by a yellow halo, develops as the surrounding tissue grows. The crack increases in length beyond the halo and widens at the centre. The tissue exposed by the crack and the yellow halo becomes necrotic, collapses and turns black. The spot then appears as a black, sunken, diamond-shaped lesion with dimensions 1.0–3.5 cm × 0.5–1.5 cm (Plates 4.8 and 4.9). Small spots rarely extend below the peel, but large spots occasionally expose the pulp. The size and number of spots can increase after harvest and inconspicuous spots can become unsightly blemishes as fruit is transported and ripened (Berg, 1968). During the early and late stages of spot development, the symptoms are distinctive. However, the middle stage can be confused with pitting disease (see Jones and Stover, this chapter, pp. 186–189).

Causal agents

Diamond spot is caused by *Fusarium solani*, *Fusarium pallidoroseum* and possibly other fungi invading lesions initiated by *Cercospora hayi* (Stover, 1972). Originally, it was thought that *F. pallidoroseum*, possibly in conjunction with other fungi common on the banana peel, was the cause of the disease (Berg, 1968).

The strain of *C. hayi* that causes diamond spot may be different from the one that causes brown spot (see this chapter, pp. 174–175). *Cercospora hayi* was consistently isolated from diamond-spot lesions together with *Fusarium* spp. When near-mature bananas were inoculated with *C. hayi* and incubated in a saturated atmosphere, yellow spots up to several mm in diameter appear 2–3 weeks later. Certain isolates of *F. solani* and *F. pallidoroseum*, if inoculated at the same time as *C. hayi*,

Plate 4.7. Symptoms of the recent 'speckle' problem on fruit of 'Grand Nain' (AAA, Cavendish subgroup) in Costa Rica. Spots closely resemble those of Deightoniella speckle, but are caused by a complex of different fungi (photo: C. Pasberg-Gauhl and F. Gauhl, CB).

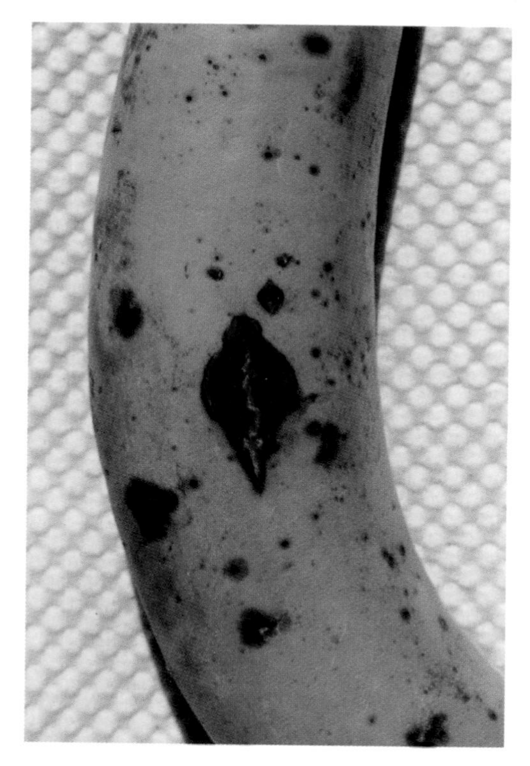

Plate 4.8. Close-up of a diamond spot lesion on fruit in Honduras (photo: J. Rivera, FHIA).

could invade and extend these initial lesions into typical diamond spots. The exact sequence of invasion and the stage of fruit maturity when this can take place in the field have not been determined.

Fusarium spp. alone cannot penetrate uninjured banana peel, although they can invade artificial wounds and cause various types of lesions. The failure of fruit, in many instances, to produce typical diamond spot symptoms after inoculation with spores of *C. hayi* and *Fusarium* spp. is an indication that highly specific conditions may be necessary for disease development. Sporulation does not occur in diamond-spot lesions.

Disease cycle and epidemiology

Cercospora hayi and *Fusarium* spp. are common inhabitants of dead and decaying banana trash and produce abundant spores under moist conditions. *Fusarium pallidoroseum* also sporulates on the dead pistils of banana flowers. Air currents carry the spores, which are common components of the air flora of banana plantations, to fruit (Kaiser and Lukezic, 1966; Lukezic

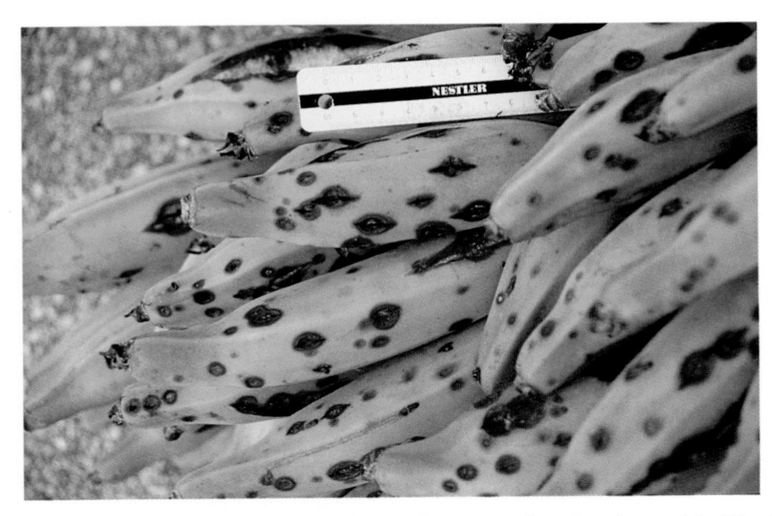

Plate 4.9. Symptoms of diamond spot on fruit of 'Agbagba' (AAB, Plantain subgroup) in Nigeria (photo: C. Pasberg-Gauhl and F. Gauhl, IITA).

and Kaiser, 1966; Berg, 1968). Diamond spot was only prevalent following prolonged periods of rain.

Host reaction

'Gros Michel' (AAA) was less seriously affected than cultivars in the Cavendish subgroup (AAA) (Stover, 1972).

Control

Measures used to control pitting disease (see Jones and Stover, this chapter, pp. 186–189) also controlled diamond spot.

Fruit Freckle

Freckle disease of banana, which is caused by the fungus *Guignardia musae* Racib., affects banana and abacá fruit as well as leaves. As early as 2–4 weeks after bunch emergence, a few widely scattered spots or, in some cases, dense aggregates, in the form of streaks or circles, can appear on banana fingers. On green fruit, individual spots first appear as minute, reddish-brown flecks, surrounded by a halo of dark green, water-soaked tissue. Large areas of the peduncle and fruit surface sometimes become black, due to dense aggregations of spots (Plate 4.10). Bracts can also be affected. The severity of the disease increases as the fruit matures. Freckle is particularly severe where fruit is in contact with diseased leaves. During ripening, the centre of individual spots darkens and each spot is surrounded by a halo of green tissue up to 3 mm in diameter. Although this discoloration detracts from the appearance of the fruit, eating qualities are not affected (Meredith, 1968).

Fruit spots are caused by concentrations of pycnidia. Consumers who buy fruit from local markets in South and South-East Asia tolerate these symptoms. However, for discerning buyers in Japan, the blemishes caused by freckle are unacceptable. The disease is, therefore, a problem for export industries based in Taiwan and the Philippines.

Information on the causal agent, leaf symptoms, epidemiology, host reaction and control has been presented in an earlier report (see Jones, Chapter 2, pp. 120–125).

Peduncle Rot

In commercial plantations and in many smallholders' plots, the male flower bud is broken off after the last female hand has emerged because this may result in a slight increase in bunch weight and eliminates a habitat for thrips. This practice also reduces the incidence of Moko disease, caused by the insect transmission of *Ralstonia solanacearum*, race 2, biovar 1, to male flower bract scars in areas where this pathogen is endemic. However, under certain conditions, fungi can enter the wound and rot the peduncle (Plate 4.11). Usually, the rot only extends a few centimetres and is checked in the vicinity of the *false hands*, small transition hands between completely female and male flowers. However, during long periods of high humidity, the rot occasionally extends further and prevents the development of lower hands. Sometimes, rots originate in the old male flower bud if it is not removed (Stover, 1972).

In northern Australia, fungi isolated from the interface between necrotic and living tissue of affected peduncles include *Fusarium oxysporum*, *F. pallidoroseum*, *F. solani* and *Verticillium theobromae* (D.R. Jones, Brisbane, 1990, personal observation). In Central America, *Lasiodiplodia theobromae* is frequently isolated from rotten peduncles and sometimes a soft rot caused by *Erwinia carotovora* is present (Stover, 1972). The main agents responsible for peduncle rot in the Philippines are *Colletotrichum musae*, *L. theobromae* and *Fusarium* spp. (Quimio, 1986).

Control measures are not usually warranted. However, in commercial banana plantations, it is common practice to leave one finger of a small apical hand attached to the peduncle when removing the false and small apical hands (Plate 4.11). This is believed to prevent any peduncle rot from reaching the remaining hands by keeping the vascular tissue in the peduncle active, thus enabling the peduncle tissue to resist infection.

Pitting Disease

Introduction

Wastage caused by pitting disease was first described by Tomkins (1931) in England on fruit of 'Dwarf Cavendish' (AAA, Cavendish subgroup) imported from Brazil. Pitting disease was subsequently investigated in Central America (Johnston, 1932), where it was a serious problem on 'Gros Michel' (AAA) and became known there as 'Johnston spot', and in Brazil (Wardlaw and McGuire, 1932). In the 1930s, it was also recorded in Australia, as 'black pit'. Here the incidence was reported as usually slight, but occasionally severe. Pitting disease disappeared in Central America from 1936 onwards, when plantations were routinely treated with high-volume sprays of Bordeaux mixture to control Sigatoka leaf spot. However, it reappeared around 1958–1960 with the conversion to low-volume, aerial sprays of oil-based fungicides. It was thought at the time by Stover (1972) to be the most important of the fruit spotting diseases in Central America.

Pitting disease was a serious problem in Mexico, Taiwan and the Philippines in the 1970s. It was also important seasonally in Ecuador and Colombia, but rare in Surinam and the Caribbean. In Central America, the disease was more severe on the Pacific coast than on the Atlantic coast. It has not been reported as a problem in most of Africa, although hands exported from Cameroon in late 1970 had a 14% incidence. It was also thought to be common in Madagascar (Stover, 1972).

The disease caused an unsightly blemish, which was unacceptable to consumers of export fruit. In the 1960s, up to 50% of fruit arriving at some packing stations in Central America were diseased, causing significant losses due to the rejection. Also, because the disease could develop while fruit was in transit and during ripening, consignments with less than 2% incidence at the packing station were known to have a 60% incidence at markets. However, these high incidences only occurred following long periods of heavy rainfall and on plantations where control measures were inadequate (Stover, 1972). Since 1980, the disease has declined considerably in significance.

Symptoms

Pitting disease gets its name from the round, sunken pits that appear on fruit as it approaches maturity or after harvest (Plate 4.12). The pits average 4–6 mm in diameter and are surrounded by a reddish-brown zone, which in turn is surrounded by a greenish, narrow, water-soaked halo. Although the pit centres sometimes split, the damage does not extend to the pulp. Under favourable conditions during transit and ripening, the number of pits on fruit may increase fivefold or more. The pathogen does not sporulate in lesions on fruit (Stover, 1972).

Fruit on the side of the bunch facing away from the pseudostem is more severely affected than fruit facing towards the pseudostem. The larger proximal hands have more pits than the smaller distal hands. Small pits can also occur on pedicels and crowns, which can increase finger drop. After periods of abundant rainfall, larger, shallower pits appear on spathaceous bracts (transition leaves) and on young watersuckers. The pathogen can sporulate in these tissues as they decay (Stover, 1972; Slabaugh, 1994c).

Causal agent

Pitting disease is caused by *Pyricularia grisea* (Cooke) Sacc. *Magnaporthe grisea* (Herbert) Barr., the sexual stage, is rarely seen. Conidia of this fungus occur singly and are attached to the conidiophore, which is mainly unbranched, at their broader end (Fig. 4.2). They are ovate to pyriform, with a small basal apiculus, hyaline or nearly so, measuring 17.0–19.0 μm × 6.5–8.5 μm and having two septa (Meredith, 1963b). *Pyricularia grisea* closely resembles *Pyricularia oryzae*, which affects rice. However, both pathogens are host-specific (Halmos, 1969). *Pyricularia grisea* sporulates poorly on most agar media, but sporulates well on autoclaved leaves of *Commelinia erecta* within 3–4 days.

Disease cycle and epidemiology

Pyricularia grisea is a common inhabitant of hanging, banana-leaf trash, including the transition leaf and bracts, and is found on grasses (Halmos, 1970). However, the form of the pathogen that is found on senescing leaves of *C. erecta* has been reported as non-pathogenic on banana (Stover, 1972). Conidia are carried to banana fruit by air currents. Although these spores are found in the air throughout the year in Central America, they are much more common during periods of high rainfall. In the field, conidia germinate and form appressoria on green fruit within 4–8 h in a saturated

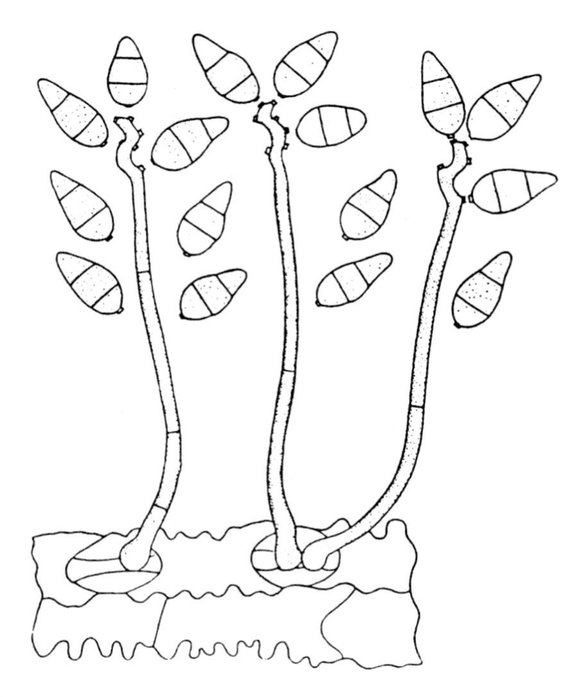

Fig. 4.2. Conidia and conidiophores of *Pyricularia grisea* (from Ellis, 1971).

Plate 4.10. The peduncle and fruit of 'Mondolpin' (AAB, syn. 'Bluggoe') affected by freckle on Badu Island in the Torres Strait region of Queensland, Australia (photo: D.R. Jones, QDPI).

Plate 4.11. Rot affecting the broken end of the peduncle of 'Grand Nain' (AAA Cavendish subgroup) in a commercial plantation in Ecuador. Note the single finger that has been left attached after the removal of the apical hands. The blue ribbon is impregnated with chlorpyrifos (Dursban®) to control insects under the polyethylene bag covering the bunch (photo: D.R. Jones, INIBAP).

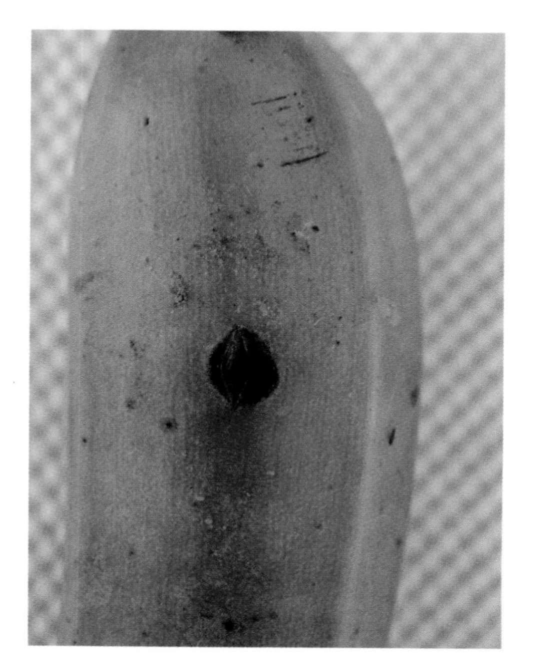

atmosphere. The optimum temperature for infection is 24–26°C. Although pits were produced on young fruit 2–3 weeks after inoculation in the laboratory, symptoms are rarely seen in the field until 70 days after bunch emergence. This is because the fungus remains dormant until the fruit is nearing maturuity and then spreads through the tissue intercellularly (Meredith, 1963b). In Central America, the disease was only severe from 6 weeks after the beginning of the rainy season until the beginning of the dry season (Stover, 1972).

Plate 4.12. Symptom of pitting disease in Honduras (photo: J. Rivera, FHIA).

Plate 4.13. Symptoms of sooty mould in Central America. The fungus is growing on the honeydew of the banana aphid (photo: R.D. Caid, CB).

More recently, the incidence of pitting disease has been associated with the presence of rotting wood, which can support *P. grisea*. The disease has been found to be more of a problem in areas where forest has been cut to establish new plantations and close to forested mountain ranges (M. Jiminez, Costa Rica, personal communication with C. Pasberg-Gauhl).

Host reaction

Fruit of Cavendish types (AAA) was more severely pitted than the fruit of 'Gros Michel' (AAA) or cultivars in the Plantain subgroup (AAB) (Stover, 1972). 'Silk' (AAB) was almost always affected (Wardlaw, 1961).

Control

Collapsed and dying banana leaves, the transition leaf and bracts need to be removed from plants at regular intervals during the rainy season to reduce inoculum levels. Dithiocarbamates were more efficacious for control than benomyl and other fungicides in Nicaragua (Guyon, 1970). Two spray applications were found to be better than one and mist spraying was not effective. Dusting the polyethylene bag with a fungicide has also been shown to be effective (Slabaugh, 1994c). Today, disease incidence is slight and chemical control measures are seldom necessary (Stover and Simmonds, 1987).

Sooty Mould and Sooty Blotch

Sooty mould and sooty blotch are the names given to the superficial peel blemish caused by various saprophytic fungi growing on honeydew from banana aphids and mealybugs. It is usually only a problem on bunches covered with polyethylene bags that are not insecticide-impregnated.

Plate 4.14. Symptoms of sooty blotch in Australia (photo: QDPI).

Sooty mould is seasonal in appearance, being most common during cool, rainy weather. The condition gets its name from the black, soot-like mycelium of the causal fungi on the fingers, pedicels, crowns and peduncles (Plate 4.13). Discoloration is greatest near the pedicels and between the fingers on the inner whorl of fingers. When rubbed, some of the fungal mycelium is removed, but a black or dark brown stain remains. Fungi associated with sooty mould in Latin America have been identified as *Chaetothyrina musarum* (Speg.) Theiss, *'Fumago vagans'* Pers., *Cladosporium atriellum* Cooke, *Microxyphium* sp. and *Penicillium citrinum* Thom. (Stover, 1972, 1975). However, *Cladosporium cladosporiodes* (Fres.) de Vries may be the main agent (Stover, 1975). Sooty mould is controlled by applying insecticide sprays to the bunch or using polyethylene bags impregnated with insecticide to protect fruit.

In Australia, sooty blotch, caused by the fungus *C. musarum*, is a light brown, superficial blemish (Plate 4.14). It is considered a minor disease, most likely to occur in cool, wet weather. Control is not warranted (Jones *et al.*, 1993).

POSTHARVEST DISEASES

I.F. Muirhead and D.R. Jones

Introduction

After harvest, the banana fruit is subject to a range of fungal diseases with potential to cause significant damage. These diseases and their importance in international trade were described in the earlier texts of Wardlaw (1961), Meredith (1971) and Stover (1972) and more recently by Slabaugh and Grove (1982), Stover and Simmonds (1987), Jeger *et al.* (1995) and Thompson and Burden (1995). These authors trace the history of postharvest disease, beginning at the time when whole bunches ripened and rotted during long sea voyages from Central America to Europe. They reveal how losses were reduced by improved management of ripening through better temperature control and reduced transit times, and how crown rot emerged as a problem when bananas were exported in boxes rather than on the bunch in the early 1960s. They also describe the use and effectiveness of fungicides to control postharvest diseases.

With some exceptions, fungicides are generally effective in controlling postharvest diseases when other parts of the fruit-handling system are well managed. However, in recent times, there has been a worldwide trend away from the routine use of pesticides, particularly after harvest (Eckert, 1990). This trend has been due to consumers becoming aware of the damage that some agricultural chemicals cause to the environment and to concerns that chemical residues on produce could lead to health problems. However, at the same time, the consumer is demanding higher-quality fruit with fewer blemishes caused by pests and diseases.

Today, except for a few organically grown bananas, most banana fruit that is exported is treated with fungicide to reduce the incidence of postharvest disease. Achieving postharvest disease control in bananas in international trade without using fungicides will require an appreciation and detailed knowledge of all the factors affecting both ripening and fungal development.

One of the keys to minimizing losses from postharvest disease is to make use of the natural resistance of green banana fruit by preventing the initiation of ripening. Another is to control the postharvest environment both before and after ripening begins. The close association between the activity of pathogenic fungi and the physiological state of the banana fruit is highlighted in this chapter.

The ripening process in bananas is initiated by ethylene gas, which is produced naturally by fruit tissues. Most postharvest disease problems of banana occur during ripening. The international trade in Cavendish bananas makes effective use of temperature control to prevent fruit ripening during transit. Many ships transporting bananas also have controlled atmospheric conditions to further reduce the incidence of ripening. However, banana is a staple food crop in many tropical and subtropical countries where refrigerated road and rail trucks and sophisticated storage facilities are not available. Fruit is also more likely to be damaged in these countries as a result of rough handling during harvest and transit to urban markets. Postharvest diseases are a major cause of wastage under these conditions. Although it has been shown that it is possible to hold fruit for long periods without refrigeration in simple controlled atmospheres generated within polyethylene bags (Scott *et al.*, 1971), diseases are one of the factors limiting the success of this technology.

Most scientific investigative work on postharvest diseases of bananas has been undertaken on fruit from Cavendish cultivars, the mainstay of the international export trade. However, the principles underlying the control of postharvest diseases apply equally to bananas from other cultivars that may be marketed in the country of production. Non-infectious problems associated with postharvest fruit are described later (Lahav *et al.*, Chapter 8, pp. 325–338).

Crown Rot

Introduction

Crown rot is a disease of the pad of tissue severed when the hand of fruit is cut from the bunch. Crown rot rose to prominence in international trade after Fusarium wilt in Central and South America led to the replacement of 'Gros Michel' (AAA) with the wilt-resistant cultivar 'Valery' (AAA, Cavendish subgroup) in the late 1950s and early 1960s (Slabaugh and Grove, 1982). Unfortunately, fruit of 'Valery' was more susceptible to scarring and bruising than fruit of 'Gros Michel', which had been exported as whole bunches. The trade therefore moved to dehanding and marketing hands or clusters of fruit in boxes, a procedure that also allowed greater efficiencies in packing and handling. However, the process of cutting the hand from the bunch, dividing it into clusters and pulling off individual rejected fingers exposed wounded crown tissue to infection by fungi and the development of crown rot.

Symptoms

Crown rot is a firm, dark brown or black rot, which spreads through the crown and may penetrate into the pedicels of individual fingers (Plate 4.15). A layer of fluffy white, grey or pink fungal mycelium may cover the cut surface of the crown (Plate 4.16). The mycelium and the rot itself spoil the fresh, clean appearance of the ripening fruit. Individual fingers may fall from the weakened crown if the rot penetrates deeply. In severe cases, the decay reaches the pulp itself and the entire fruit is lost. Disease development stimulates ripening (Slabaugh and Grove, 1982), which is another cause of wastage in international trade.

Causal agents

Unlike most other banana diseases, a single microorganism does not cause crown rot. The fungi associated with the disease in boxed bananas exported from the Windward Islands to the United Kingdom have been studied thoroughly. Johanson and Blazquez (1992) showed that the main pathogens were species of *Fusarium*, principally *F. pallidoroseum* (Cooke) Sacc, and *Colletotrichum musae* (Berk. and Curt.) Arx. A series of other known crown-rot-causing fungi were also isolated, including species of *Nigrospora*, *Lasiodiplodia theobromae* (Pat.) Griffon and Maubl. and *Verticillium theobromae* (Turc.) Mason and Hughes. This was consistent with the earlier studies in the Windward Islands by Wallbridge and Pinegar (1975), Griffee and

Plate 4.15. Fruit of a Cavendish cultivar affected by crown rot. The decay has spread from the crown to the pedicels and fingers. Note the anthracnose spots that develop during ripening on the fingers (photo: D.R Jones, QDPI).

Plate 4.16. A white fungal mycelium is visible on the surface of a crown affected by crown rot (photo: D.R. Jones, QDPI).

Burden (1976) and Wallbridge (1981). However, Finlay and Brown (1993) carefully examined the relative importance of the different pathogens in the Windward Islands and considered *C. musae* to be by far the most aggressive species.

Fusarium pallidoroseum and a *Verticillium* sp. were the most common pathogens isolated from the crowns of fruit cut in North Queensland, Australia (Jones, 1991). Of other fungi isolated infrequently, only an *Acremonium* sp. caused severe crown rot in inoculation experiments. *Colletotrichum musae* was not isolated, but was commonly seen sporulating on the surface of necrotic pedicels of fruit in the advanced stages of crown rot initiated by other pathogens.

Reports from other banana-growing regions reveal similar patterns. *Colletotrichum musae* and *F. pallidoroseum* or *F. moniliforme* Sheld., acting alone or together, are the dominant pathogens in Mauritius (Lutchmeah and Santchurn, 1991) and Malaysia (Sepiah and Nik Mohd, 1987). *Fusarium* spp. feature prominently in China (Wang Bisheng *et al.*, 1991) and Central and South America (Stover, 1972). However, other pathogens, which include a *Cephalosporium* sp. and *Chalara paradoxa* (de Seynes) Hohn. (*Ceratocystis paradoxa* [Dade] Moreau), are also frequently cited as additional causes of crown rot.

Thus, a number of well-defined pathogenic fungi, which sometimes act alone, but more usually in combination, cause crown rot. In any given situation, it is important to know which pathogenic fungi are dominant, as the effectiveness of control treatments may depend upon a correct diagnosis.

Several authors (see Johanson and Blazquez, 1992) have mentioned the possibility that enzymes produced by saprophytic bacteria may contribute to tissue breakdown, but this has not been confirmed by detailed research. What is known is that *C. musae* on its own is able to produce polygalacturonase isozymes capable of effectively macerating green banana tissue (Stanley and Brown, 1994).

Disease cycle and epidemiology

The fungi causing crown rot are ubiquitous components of the microflora of banana plantations (Stover, 1972; Stover and Simmonds, 1987). They live saprophytically in dead banana leaves, flower bracts, discarded fruit and bunch stems. In the field, airborne spores of fungi, including species of *Fusarium* and *Verticillium*, settle on banana bunches while water-borne spores, such as the conidia of *C. musae*, are splashed on to fruit by rain or irrigation water. Spores of some of the crown-rot pathogens may remain viable for months in the field under extremes of relative humidity and temperature (Stover, 1972). In Guadeloupe, the peak of inoculum of both *C. musae* and *Fusarium* species in close proximity to plants occurred between 25 and 40 days after bunch emergence (de Lapeyre de Bellaire and Mourichon, 1997). The flowers and the last bracts were the main sources of inoculum. *Colletotrichum musae* establishes latent infections, which survive in green banana fruits and probably crown tissue for several months (Muirhead and Deverall, 1981). Thus, at harvest, fruit often carries a heavy load of crown-rot inoculum.

Cutting the hand from the bunch results in a number of opportunities for crown-rot fungi. Firstly, it creates a substantial wound, which is directly accessible to airborne or water-borne fungal spores. Spores present in water used to wash bananas can be drawn into this wounded tissue to a depth of 5–7 mm (Greene and Goos, 1963). Secondly, the dehanding knife itself can drag inoculum from the surface of the crown across the wounded tissue (Finlay *et al.*, 1992). Thirdly, wounding may activate previously established latent infections of *C. musae* (Jones, 1991).

Spores that reach the surface of the cut crown germinate in response to the favourable environmental and nutritional stimuli. The progress of disease from that point depends on several factors, principally the amount of inoculum, the physiological state of the fruit, the length of time before ripening commences and the environmental conditions before and after ripening (see Finlay and Brown, 1993: Chillet and de Lapeyre de Bellaire, 1996b). In the Windward Islands, crown rot is a problem all year round but is more severe during the more humid, wetter and hotter summer months. This is thought to be because conditions favour increased activity of the crown-rot fungi in leaf trash and discarded fruit.

Host reaction

Stover (1972) has reported that all commercial dessert bananas, which are produced on cultivars of the AAA genome, are susceptible and that the problem is rare on fruit from cultivars in the Plantain subgroup (AAB). In the Cavendish subgroup (AAA), 'Robusta' and 'Valery' were found to be more susceptible to natural crown rot infections than 'Lacatan' (Shillingford and Sinclair, 1977). More recently, it has been suggested that fruit of 'FHIA-01'® (AAAB, syn. 'Goldfinger') from the Honduran breeding programme is less susceptible to crown rot (Jones, 1994).

Control

Preventing infection

The amount of inoculum in the environment is thought to affect levels of crown rot. It is now standard practice on many plantations to remove dead banana leaves and bracts from plants and keep areas where fruit is packed free of banana trash. Water used in washing and delatexing tanks is also treated with a sanitizer, such as chlorine. This latter procedure recommended for crown-rot control is sound, but in practice may have a limited effect. Slabaugh and Grove (1982) noted that maintaining the desirable concentration of chlorine in wash water did not reduce crown rot, suggesting that sources of inoculum other than contaminated water were more important. The importance of field inoculum is highlighted in studies in the Windward Islands, where fruit was cut and packed in the field in an attempt to reduce the physical damage caused during the transport of bunches to boxing plants (Thompson and Burden, 1995). This fruit was not washed and, even though a cellulose pad containing the fungicide thiabendazole was used on the crown to stem the flow of sap, crown rot still caused losses (Johanson and Blazquez, 1992).

Finlay *et al.* (1992) investigated the importance of different sources of crown-rot inoculum in the Windward Islands and the effects of alternative handling systems on crown-rot development. In experiments, they reduced the disease by more than 50% by dehanding (without washing) in the relatively clean packing shed instead of the field. Although the benefit was not significant in commercial shipping trials (in which disease levels were naturally low), the work provides a good argument for keeping the packing area free of leaf trash, floral remnants and discarded fruit.

Reducing inoculum levels in the field and the packing area should be seen as only the first steps in controlling crown rot. Other methods are generally needed as well.

Dehanding technique

The way in which the hands are removed from the bunch is another factor that affects development of crown rot in green tissue. The standard texts recommend a clean cut with a sharp tool and neat trimming. Finlay *et al.* (1992) showed that rough trimming and breaking crowns from the bunch stalk instead of cutting increased crown rot. The action of breaking crowns left fragments of tissue, which readily dehydrated and senesced, providing conditions favouring the establishment of crown-rot fungi. The crowns of the broken hands also yellowed more quickly, suggesting premature ripening of the tissue, which also favours fungal development. Smoothly cut, large crowns are known in the banana trade to be more resistant to crown rot than irregularly cut, shallow crowns. It is always prudent to trim hands leaving as much crown tissue as possible. As well as being more resistant to rot, any rot that may initiate at the cut surface has less chance of spreading to the pedicels and fingers before sale and consumption.

Managing the ripening process

Bananas harvested green pass through three physiological stages (John and Marchal, 1995).

- The preclimacteric or preripening phase during which the fruit's metabolic activity is relatively low.
- Ripening, which begins with an intense burst of respiration, called the climacteric.
- Senescence, during which metabolism slows once more.

Fruit grown for international and some domestic markets must be prevented from ripening while in transit. This is achieved by transporting the fruit under controlled temperature conditions (12–13°C). Sometimes the relative humidity and composition of the storage atmosphere is also controlled. The transit period may be 1 week or more and fruit must be harvested with sufficient 'green life' to ensure arrival before ripening commences. Applying ethylene in special rooms under controlled conditions then synchronizes ripening. If ripening begins *en route*, the fruit arrives at its destination 'mixed ripe', with hands at different stages of yellowing. Mixed-ripe fruit cannot be marketed successfully, often develops crown rot and is wasted. In commercial practice, premature ripening is prevented primarily by harvesting at the appropriate stage of maturity, which is judged by the age of the bunch, the size or 'grade' of the fruit and the internal colour of the pulp. Less mature fruit with a longer green life is used for more distant markets.

There is a very close relationship between the stage of ripening and the resistance of the banana to postharvest diseases, including crown rot. During the preclimacteric phase, the fruit retains much of the resistance of the unharvested fruit. This resistance is lost once ripening commences and the ripe fruit very rapidly becomes fully susceptible. The factors responsible for mediating resistance to *C. musae* are discussed under anthracnose (see Muirhead and Jones, this chapter, pp. 199–203) and are likely to be similar for other crown-rotting fungi.

Incidence of crown rot also depends on how the fruit is managed after harvest. When the period between harvest and artificial ripening is short and temperatures in transit are controlled, fruit reaches the consumer quickly and crown rot is rarely a problem. This is commonly the case in Australia, where the transit time from farm to market is as short as 1–3 days. In Australia, crown rot is associated with longer transit times, attempts to hold fruit before ripening to control market supply or suboptimal environmental conditions. In these cases, the resistance of the green fruit is insufficient to prevent disease development and other factors need to be considered.

Effects of modified atmosphere

The main reason for modifying the atmosphere around green bananas during the preclimacteric period is to extend the green life and hence prevent the premature initiation of ripening during transit and storage. Ripening is delayed by reducing the oxygen level to 3–7%, increasing carbon dioxide to 10–13% and absorbing ethylene (Wade *et al.*, 1993). The modified-atmosphere technique, in combination with refrigeration, is now being used in international trade and is steadily gaining popularity. Usually, the atmosphere in the holds of the ships containing the banana cartons is modified using sophisticated equipment. However, the benefits of modified-atmosphere storage can be achieved in other ways. Fruit can be packed in polyethylene bags and a vacuum applied to each bag before it is sealed and the carton closed. The atmosphere within the bag is modified by the respiratory activity of the fruit. If the fruit is then transported under refrigeration, the system combines the beneficial effects of reduced temperature and modified atmosphere. As well as extending green life, it is claimed that a modified atmosphere also reduces the incidence of crown rot (Slabaugh and Grove, 1982; Stover and Simmonds, 1987). Chillet and de Lapeyre de Bellaire (1996a) found that the beneficial effect of a modified atmosphere on crown rot caused by *C. musae* was more noticeable over a long period (28 days at 13.5°C), as compared with a short, storage period. The modified atmosphere, as opposed to reduced water loss, controlled the crown rot.

Once ripening is initiated, the process is irreversible, shelf-life is limited and storage is impractical. The possibility of holding green bananas in a preclimacteric state without refrigeration for 40 days or even longer (Satyan *et al.*, 1992) is potentially important in tropical and subtropical countries where refrigeration is not available. It has been known since at least 1966 that green life can be extended to such periods without refrigeration by enclosing preclimacteric bananas in sealed polyethylene bags with an ethylene absorbent (Scott and Roberts, 1966). However, under these conditions, postharvest disease is a major constraint and Scott *et al.* (1971) considered the use of fungicides to be essential.

More recent work (Wade *et al.*, 1993) on long-term modified atmosphere storage reveals a complex pattern of different fungal species causing different amounts of damage, depending on the atmospheric composition, the part of the bunch or hand affected and the fungicidal treatment. While some pathogens were unaffected or suppressed by modified atmosphere, others, including *Alternaria alternata* (Fr.) Keissler and *F. pallidoroseum*, were stimulated. However, the main pathogens affecting crown tissue were still *C. musae* and *Fusarium* species.

Effects of temperature

Fungi causing crown rot come from tropical environments and are favoured by tropical temperatures. For example, the optimum temperature for growth of *C. musae* is 28°C. Storage, transport and ripening temperatures are determined by the requirements of the fruit. As bananas from Cavendish cultivars are damaged at 12°C or below, the fruit is stored and shipped at 13–14°C. During ripening, the fruit is held at the optimum of 16–18°C, with some adjustments of temperature to control the ripening speed. These temperatures are well below the optimum for fungal growth. After harvest, fruit should be cooled as soon as possible and certainly within 48 h if transit times are going to be longer than 10 days (Stover, 1972). Cooling soon after harvest maintains green life as well and, to some extent, limits the development of crown rot. The longer that fruit is held at ambient temperatures before refrigeration, the more chance there is of crown rot developing.

Effects of a high humidity level

Stover and Simmonds (1987) note that any pack that maintains high humidity and keeps crowns turgid and fresh helps to reduce crown rot, but there appear to have been no specific studies undertaken on the effects of different humidity levels on crown-rot development during storage. One would expect the growth of superficial mycelia on the surface of decaying crowns to be encouraged at a high humidity. However, the overriding effect of high humidity levels must be that they prevent the desiccation of the crown, which favours tissue death and fungal colonization.

Fungicidal treatments

When the world export trade moved from shipping bananas in boxes instead of on the bunch, the cut crown tissue became a major site of fungal infection. However, it was also easy to treat the exposed crowns with a fungicide to protect them during transit and ripening. The major breakthrough in chemical control occurred when the first of the highly effective systemics benzimidazole fungicides (thiabendazole and then benomyl) became available in the late 1960s and early 1970s. These systemic fungicides did not penetrate far into the fruit, but the effect was sufficient to control crown rot well. Over the last 30 years, many fungicides from other chemical groups have been tested and some, including imazalil and prochloraz, have been approved for use in different countries. In Central America, imazalil is often used for the distant European market and, although more expensive, is regarded as more effective than the benzimidazoles (Slabaugh and Grove, 1982). Imazalil is also now used on most Windward Island fruit, with good results. Prochloraz was recommended for use in Australia in the months of glut when bananas were held longer at markets and crown rot tended to be more of a problem (Jones, 1991).

Before a fungicide can be registered for use on bananas, its efficacy must be proved and authorities satisfied that it will not endanger public health. The regulations of the country where the fruit is to be marketed must also be taken into account. The necessary trials and tests are usually conducted or sponsored by the chemical company promoting its use.

Fungicides are applied in many different ways (Thompson and Burden, 1995). The simplest, now in use in the Windward Islands, is to immerse fruit in a solution of fungicide in a polyethylene tub at the farmer's packing shed. As well as fungicide, the dip solution also contains alum to counteract latex staining. The cut surfaces are allowed to drain latex for a maximum of 10 min before immersion. The time between cutting and dipping is regarded as critical, as the risk of crown rot is believed to increase the longer the cut surface is exposed without protection. In commercial packing houses in Central America, treatment is integrated into a mechanized packing operation. Although dip tanks are common, fungicide solutions can also be sprayed or cascaded on to hands or clusters placed crown up on slowly moving trays. More recently, fungicides have been applied to crowns using a brush.

There are some practical problems associated with the use of fungicides. For health reasons, personnel in contact with fungicides, fungicide solutions or fungicide-treated fruit should wear appropriate protective clothing (Plate 4.17). However, such practices are not always adopted in hot, tropical environments, where protective clothing may be uncomfortable. If the mixture of fungicide is used continually or recycled, its strength declines with time and the passage of fruit. Dirt, sap and spores of fungi tolerant to the fungicide may also build up and reduce efficacy. To maintain the fungicide at an effective concentration, it becomes necessary either to use a topping-up procedure or to change the mixture regularly. The correct disposal of the used fungicide mixture is also now seen as an important environmental issue. Some producers in Costa Rica absorb the fungicide on to clay before disposal. In the Windward Islands, the use of charcoal-

Plate 4.17. Banana worker in protective clothing dipping fruit in a fungicide–alum–water mixture in St Vincent, Windward Islands (photo: D.R. Jones, SVBGA).

lined disposal pits is advocated. Applying the fungicide in pads, which are placed directly on to the crown of hands, overcomes some of these problems (Johanson and Blanquez, 1992). Unfortunately, pads are unsightly and do not appeal to consumers. In addition, pads do not work as well on clusters, the form of fruit now commonly exported. The more irregular crown of the cluster means that the pad with the fungicide does not come into contact with all of the cut tissue.

If a fungicide used to control crown rot is also used regularly to control leaf spot diseases on the same farms, there is a danger that the population of crown-rot fungi in the field may become less sensitive to that fungicide and related fungicides (Johanson and Blazquez, 1992). In Guadeloupe, de Lapeyre de Bellaire and Dubois (1997) reported that an average of 23% of isolates of *C. musae* were tolerant of thiabendazole. Their presence was correlated with the exclusive use of benomyl as an aerial spray to contol Sigatoka (see Jones, Chapter 2, pp. 79–92) in the field.

Future control methods

Innovations that improve control over fruit maturity, reduce physical injury and shorten transit times will continue to be introduced by the banana industries of the world. Improvements in these areas will also reduce the impact of crown rot. However, we are also likely to see an increase in public opposition to the use of postharvest pesticides of all kinds. Du Pont has withdrawn benomyl (Benlate®) as a postharvest treatment for fruit and vegetables. With chemical use on the decline, there is renewed interest in alternative means of control. Recent work has shown that hot water may have the potential to replace chemical fungicides. In tests in Hawaii, a 45°C hot water treatment for 20 min given 15–20 min after dehanding reduced the incidence of crown rot caused by *C. paradoxa* from 100% to less than 15% (Reyes *et al.*, 1998). Hot-water dips may also be effective against *Fusarium* and *Colletotrichum* (López-Cabrera and Marrero-Domínguez, 1998)

Biological control using microorganisms, such as yeasts or bacteria, is an attractive prospect and progress has been made in this area (Chuang and Yang, 1993; Kraus, 1996; de Costa *et al.*, 1997; Postmaster *et al.*, 1997). While there are significant difficulties in developing and commercializing biological control agents, there are some factors that would favour this technology for crown-rot control. Firstly, the freshly cut crown provides nutrients and moisture suited to microbial growth. Secondly, the cut crown could be treated within minutes of exposure to most crown-rot inocula. Thirdly, most packing systems would allow a controlled dose of microbial agent to be delivered easily and efficiently. Fourthly, the temperature and humidity of the environment while bananas are in transit can be regulated. Fifthly, control is required only for a limited period, perhaps 1–3 weeks. Finally, the concerns of consumers who influence supermarket chains buying bananas and the size of the international trade might attract commercial investment.

Antagonistic microorganisms would need to be adapted to temperatures from 13 to 18°C, the range used during commercial transit and ripening. However, a much wider temperature range would be needed for storage and handling without temperature control.

Another alternative to using conventional fungicides is to apply edible film-forming polymers or other acceptable food additives, such as organic acids and salts, e.g. potassium sorbate and sodium benzoate. These materials are inhibitory to *C. musae*, one of the major crown-rot pathogens, both *in vitro* and *in vivo* (Al Zaemey *et al.*, 1993). Although not spectacularly effective in its own right, this technique might be useful in combination with low levels of fungicide or as part of an integrated control package. Another interesting possibility is treatment with papaya latex, which has given control of crown rot under experimental conditions (Indrakeerthi and Adikaram, 1996).

Control of postharvest diseases is one of the benefits cited for the postharvest irradiation of fruit. While the doses required for fruit-fly control (75–300 Gy) might be tolerated by the banana fruit (Akamine and Moy, 1983; Burditt, 1994), those required for disease control are 1000 Gy or higher (Moy, 1983) and are likely to be damaging (Thomas, 1986). The opportunities for controlling crown rot with irradiation, therefore, appear limited.

Cultivars in the Cavendish subgroup (AAA) have been selected for improvement through genetic-engineering techniques (see Sági, Chapter 15, pp. 465–466), because they have many desirable horticultural attributes and they cannot be improved easily by conventional breeding techniques. Resistance to black leaf streak, Fusarium wilt race 4 and bunchy top are likely to be the main targets initially, but control of crown rot may be possible in due course. In this respect, it is interesting that Gomes *et al.* (1996) have demonstrated that defence proteins from cowpea, an unrelated host crop, inhibited *C. musae* in culture.

Anthracnose

Introduction

Anthracnose is the name given to the disease that appears as sunken, brown-black lesions found on the peel of bananas during transport, storage and ripening. Anthracnose arises when dormant infections of the causal fungus in the green peel reactivate as fruit ripens. This results in the gradual necrosis of the yellowing peel from initial brown spots. Anthracnose symptoms also appear when green fruit is wounded. Lesions develop to form scars, which spread and become more serious on ripening. Scars detract from the appearance of fruit and lower quality.

Symptoms

Brown spots form on the peel as the fruit ripens to cause a cosmetic blemish, which can deter potential consumers (Plates 4.15 and 4.18). The spots increase in size and often coalesce to form extensive areas of sunken, brown-black tissue (Plate 4.18). Orange or salmon-coloured spore masses develop under the right conditions (Plate 4.18). Usually, the pulp remains unaffected until the fruit becomes overripe. When green fruit is wounded, black scars are formed. The shape of the scar depends on the type of injury. Injury caused when the ends of fingers impinge on the curved shoulder of fingers on lower hands results in diamond-shaped lesions (Stover, 1972). Lesions are sunken and can have a yellow halo. Sometimes the outer layer of pulp may be invaded. When the fruit is ripened or placed at high temperatures, the lesion increases in size and the fungus invades the pulp.

Causal agent

Colletotrichum musae (Berk. and Curtis) Arx (syn. *Gloeosporium musarum* Cooke and Massee) is the cause of anthracnose. *Colletotrichum musae* is also implicated in crown rot (see Muirhead and Jones, this chapter, pp. 191–199), main stalk rot (see Muirhead and Jones, this chapter, p. 203), stem-end rot (see Muirhead and Jones, this chapter, p. 203), anthracnose rot of preharvest fruit (see Jones and Stover, this chapter, pp. 173–174) and is a specialized pathogen of bananas.

This fungus produces conidia on conidiophores, which arise in acervuli. Acervuli are primarily found on fruit, but can also occur on peduncles, petioles and occasionally leaves. They are usually rounded, elongated, erumpent, up to 400 μm in diameter and are composed of epidermal and subepidermal pale brown pseudoparenchyma, which becomes subhyaline towards the conidiophore region. Setae are absent. Conidiophores, which are formed from the upper pseudoparenchyma, are cylindrical, hyaline, septate, branched, subhyaline towards the base and tapered towards the apex. They measure 30 μm × 3–5 μm, with a single phialidic aperture. Conidia are hyaline, aseptate, oval to elliptical or cylindrical and often have a flattened base. They measure 11–17 μm × 3–6 μm, have an obtuse apex, are variably guttulate and are produced in yellow to salmon-coloured masses (Sutton and Waterston, 1970). An acervulus, conidiophores and conidia are illustrated in Fig 4.3.

Colonies on potato dextrose agar have sparse to abundant white, grey or olive-coloured, floccose, aerial mycelium. Acervuli, which are dark brown to black in colour, are abundant, irregularly scattered or sometimes confluent and are produced throughout the culture, particularly in areas devoid of aerial mycelium and around the point of inoculation. Setae are rarely formed. Masses of conidia are salmon-orange and can also be formed from separate phialides on the vegetative mycelium. On potato carrot agar, aerial mycelium is sparse and scattered acervuli are not abundant. As the cultures age, the colonies become grey underneath, due to appressorium formation. Appressoria are readily formed from vegetative hyphae, where they are navicular to ovate, becoming

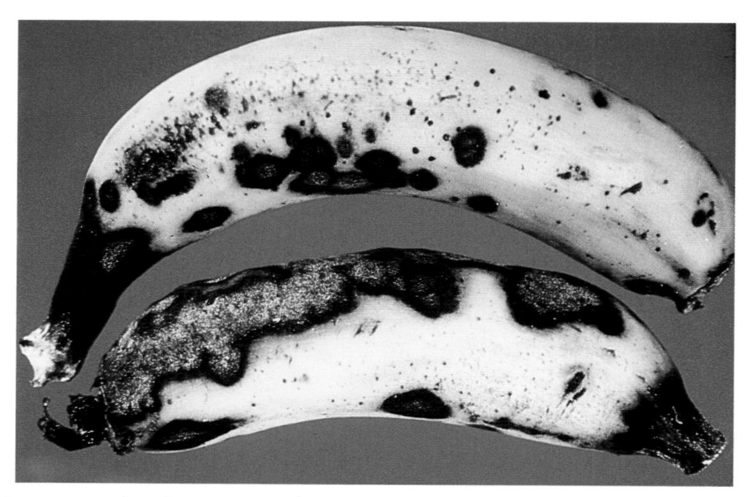

Plate 4.18. Symptoms of anthracnose on fingers of a Cavendish cultivar. The small, brown spots that develop during ripening can be seen at the centre of the top finger. Larger, black lesions caused following damage to green fruit are evident elsewhere. The white mycelium and salmon-pink-coloured masses of conidia of *Colletotrichum musae* can be seen on the very large lesion on the lower finger. (Photo: QDPI.)

irregularly lobed and dark brown, and they measure 6–12 μm × 5–10 μm (Fig. 4.3). Appressoria formed from germinating conidia are pale brown, subglobose to irregular and up to 8 μm in diameter (Fig. 4.3) (Sutton and Waterston, 1970).

The fungus (as *G. musarum*), isolated from fruit of 'Gros Michel' (AAA) in Trinidad, has been reported to produce *Glomerella cingulata* (Stonem.) Spauld and Schrenk in culture (Wardlaw, 1934). However, later work indicated that the anamorph of *G. cingulata* may have been *C. gloeosporioides* Penz., which can also be isolated from rotted bananas (Greene, 1967). Von Arx (1957) noted that *C. musae* could be distinguished from *C. gloeosporioides* by slightly longer and broader conidia and its faster growth in culture at 24°C (Sutton and Waterston, 1970).

Disease cycle and epidemiology

Spores of the fungus are produced on senescing banana tissue, including leaves, bracts, discarded fruit and fruit stems. They are dislodged primarily by water. They reach the fruit in the field in rain splash or irrigation water and in the packing shed in recirculated washing water, which accumulates spores from plant debris.

The optimum growth and sporulation temperature for *C. musae* is 27–30°C and there is little growth below 15°C. Conidia adhere to the intact fruit surface (Sela Buurlage *et al.*, 1991), germinate and form appressoria within 24–48 h. The fungus remains latent until the fruit starts to ripen. Infection also occurs directly through wounds, which stimulate conidial germination and mycelial growth but inhibit appressorial formation (Yang and Chuang, 1996).

Anthracnose lesions develop on the main body of the fruit. In commercial trade, the main location is on the angular corners of fruit, which are often scraped, scratched or bruised during packing and general handling. The injuries caused by the tips of the outer whorl of fruit on the inner whorl are also common starting-points. These injuries are either invaded by hyphae growing from spores that lodge in the wound or by hyphae growing from latent or quiescent infection that existed before the injury.

The banana–*C. musae* system has been

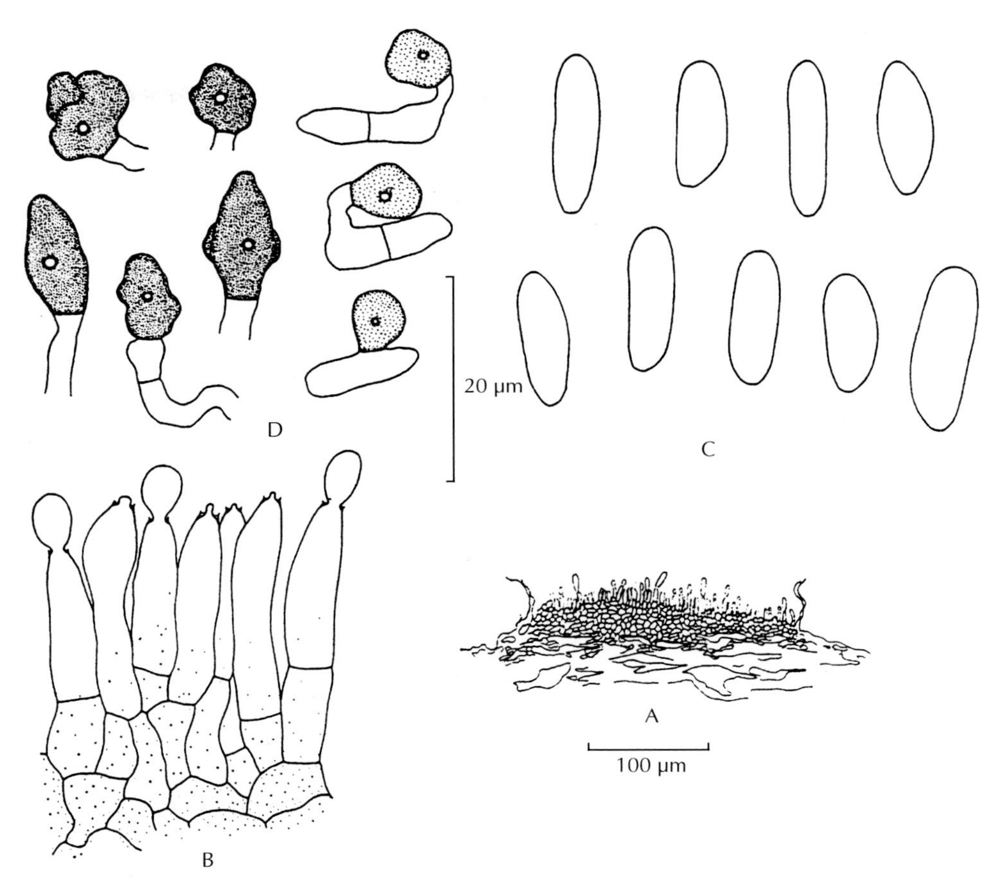

Fig. 4.3. Acervulus (A), conidiophores (B), conidia (C) and appressoria formed from hyphae (left) and germinating conidia (right) (D) of *Colletotrichum musae* (from Sutton and Waterston, 1970).

extensively studied in attempts to understand how latent infection by *Colletotrichum* species in tropical fruits is controlled. Early work by Simmonds (1941) showed that *C. musae* could remain inactive in green bananas for several months, resuming activity during ripening. He found that some appressoria produced infection threads and then subcuticular hyphae before fungal growth ceased. Simmonds (1941) believed that these hyphae were the latent structures. Using light microscopy, Muirhead and Deverall (1981) noted that these subcuticular hyphae induced hypersensitive reactions in green fruit, visible to the eye as red-brown flecks. The hyphae above cells that had died hypersensitively appeared to play no further role in initiating disease. In contrast, new subcuticular hyphae emerging from appressoria during ripening proceeded to invade the fruit, causing disease. They thus considered the appressoria to be the latent structures.

Electron-microscopy studies with *C. gloeosporioides* in avocado fruit show that the fungus remains latent in the cuticle as an infection peg of variable length (Coates *et al.*, 1993). Similar studies may reveal the same to be true for the banana–*C. musae* system. These histopathological studies are important because they provide a guide to the kinds of physiological mechanisms that control latency.

To investigate the physiological processes controlling latency, researchers have followed several hypotheses as initially proposed by Simmonds (1963). The first is that a preformed antifungal factor existing in green fruit declines in concentration or effectiveness during ripening, thus allowing the fungus to resume growth. Several such compounds, including 3,4-dihydroxybenzaldehyde (Mulvena *et al.*, 1969) and 3,4-dimethoxybenzaldehyde (Abdel-Sattar and Nawwar, 1986), have been identified. Dopamine, the main phenolic substrate responsible for the browning of wounded, green banana peel, has also been implicated (Jiang, 1997). However, it remains to be shown that preformed compounds of this kind are present at antifungal concentrations at the sites of latent infection for the duration of the latent period (Muirhead and Deverall, 1984).

The second type of investigation involves antifungal compounds or phytoalexins formed in response to infection. Brown and Swinburne (1981) showed that five inhibitory compounds were produced in necrotic tissue formed when green bananas were challenged by *C. musae*. One of the major compounds has been identified by Harai *et al.* (1994) as 2-(4′-hydroxyphenyl)-napthalic anhydride.

The host–pathogen interaction is confounded by other factors, such as the presence of iron and chelating agents at the fruit surface. Iron inhibits the germination of conidia of *C. musae*. On the other hand, chelating agents, such as 2,3-dihydroxybenzoic acid (formed from anthranilic acid present in green bananas) and siderophores from epiphytic bacteria, stimulate germination and appressorial formation (McCracken and Swinburne, 1979; Harper *et al.*, 1980). Brown and Swinburne (1981) cultured *C. musae* on media that were either iron-replete or iron-deficient and showed that, when conidia from these colonies were used to inoculate green fruit, there were variations in necrosis, the formation of antifungal chemical compounds and the speed of disease development during ripening. These interactions are complex and the role of antifungal chemicals in latent infection has not been fully resolved.

If antifungal compounds are not involved, stimulation of the fungus by ethylene (Flaishman and Kolattukudy, 1994) or nutrients (Yang and Chuang, 1996) during ripening or changes in the susceptibility of cell walls to enzyme attack (Shillingford and Sinclair, 1980) may provide the answer. More than one factor may operate simultaneously.

When anthracnose develops in wounds in green fruit, the extent of disease development depends, as it does for crown rot, on whether the fruit is held for a long period before ripening and under what conditions. While the fruit remains green, the lesion is limited but the longer the pre-ripening period, the larger the lesion. Disease development accelerates during ripening.

If the fruit is not injured after harvest, latent infections resume activity during ripening and similar lesions form. These lesions develop later in the ripening process and are less damaging and of little concern commercially.

Host reaction

Fruit from all dessert cultivars is affected, but fruit from cultivars in the Plantain subgroup (AAB) is said to be resistant. Of the dessert cultivars, fruit of 'Gros Michel' (AAA) is less susceptible to green-fruit wound anthracnose than fruit from cultivars in the Cavendish subgroup (AAA). Fruit of a high caliper grade, which is more susceptible to wounding than fruit of a lower caliper grade, is also more likely to develop wound anthracnose (Stover, 1972).

Shillingford and Sinclair (1977) compared the susceptibility of fruit of three Cavendish cultivars to postharvest infections. 'Robusta' and 'Valery' developed more wound anthracnose than did 'Lacatan'. When inoculated with *C. musae*, the Jamaican-bred tetraploid 'Calypso' ('65–3405–1') was resistant to wound anthracnose, 'Lacatan' was intermediate in reaction and both 'Valery and 'Giant Cavendish' were equally susceptible.

Control

The control of anthracnose depends on the same principles discussed for crown rot; minimizing damage to fruit, reducing inoculum levels by using fresh or clean water to wash fruit, using fungicides to protect fruit and maintaining fruit in a hard, green state until ripening. In the international banana trade, injured bananas are rejected during packing operations and subsequent quality checks. Severe cases of wound anthracnose are, therefore, rarely seen. However, scarring is still the main quality defect of green fruit and is a cause for consumer rejection. Anthracnose may cause significant wastage in domestic markets in developing countries, where damage levels are higher and storage temperatures are not controlled.

Fungal Scald

This disease, which superficially resembles anthracnose, can occur in fruit in controlled atmosphere packs in transit for more than 14 days. Fungal scald develops where the tips of fruit touch polyethylene liners in the presence of moisture. Reddish-brown, sunken spots form on green fruit near the finger tips and on the bottom clusters of the carton. The spots increase in size during ripening.

The problem is initiated by *Colletotrichum musae*, the cause of anthracnose, but is followed by *Fusarium pallidoroseum*, which causes a secondary infection. Fungicides used to protect bananas against crown rot also control fungal scald.

Stem-end Rot

If the pedicel or stem of green banana fruit is injured, a rot similar in appearance to crown rot often develops (Plates 4.19 and 4.20). Stem-end rot is the name given to the decay that advances from the cut end when individual fingers are cut from the bunch and marketed as single fruit, rather than in clusters or hands. The practice was once common in Australia and stem-end rot, or 'black-end', was a significant problem. When bananas were shipped as whole bunches in the international trade, the pedicel was often creased and bruised when the fingers flexed on the bunch. The ensuing decay, described by Stover (1972), was called Santa Marta stem-end rot and was important because individual fingers often fell from the bunch during transit and handling. *Colletotrichum musae* (Plate 4.19) was generally the cause of decay, but other crown-rot fungi, such as *Chalara paradoxa* (Plate 4.20), were also involved. These diseases are not important now, because handling procedures have changed. The principles and procedures described above for crown rot apply equally to stem-end rot.

Main-stalk Rot

Main-stalk rot refers to the decay of the peduncle that occurred when bananas were exported on the bunch. Spreading from the proximal end, the rot had a characteristic sweetish odour and often engulfed the lower hands, causing finger drop (Stover, 1972). The main cause was the fungus *Chalara paradoxa*, but the fungi *Lasiodiplodia theobromae* and *Colletotrichum musae* were also implicated. Although mainly of historical interest, the disease has recently emerged as a threat to a new method for storing green bananas on the bunch in sealed polyethylene tubes without refrigeration (Satyan *et al.*, 1992). Under modified atmospheres, *Fusarium moniliforme* var. *subglutinans* Wolenw. and Reinking was a major cause of this new stalk rot (Wade *et al.*, 1993).

Botryodiplodia Finger Rot

Stover (1972) reported that Botryodiplodia finger rot was one of the most common rots of bananas that were in transit in boxes for more that 14 days. However, it rarely occurs in fruit transported for 10 days or less. The disease has been reported on fruit

Plate 4.19. Symptoms of stem-end rot caused by *Colletotrichum musae.* Salmon-pink-coloured fruiting bodies of the fungus can be seen near the tip of the diseased Cavendish finger. (Photo: QDPI.)

Plate 4.20. Symptoms of stem-end rot caused by *Chalara paradoxa.* The white mycelium of this fungus can be seen at the pedicel end of the Cavendish finger. (Photo: QDPI.)

from Central and South America, the Caribbean, India, Taiwan and the Philippines (Slabaugh, 1994).

Botryodiplodia finger rot is a soft rot that advances from the tip of the fruit below the flower remnants. The pulp is converted into a black mass and the entire fruit can decay. The skin is black and wrinkled, ripens prematurely and becomes encrusted with pycnidia. Grey to black, woolly mycelium on the surface of the fruit under conditions of high humidity is diagnostic (Stover, 1972). Fully mature fruit is more susceptible to infection and affected clusters ripen earlier. The disease develops faster during ripening and can spread to adjacent fingers (Slabaugh, 1994a).

The disease is caused by *Lasiodiplodia theobromae* (previously *Botryodiplodia theobromae*), which is a common inhabi-

Plate 4.21. Internal symptoms of squirter caused by *Nigrospora sphaerica* (photo: QDPI).

Plate 4.22. Symptoms of ring rot caused by an unknown phycomycete in Australia (photo: QDPI).

tant of decaying banana vegetation in the banana plantation. The fungus is easily cultured. Pycnidia are black, flask-shaped with short necks and 250–300 μm in diameter and contain short conidiophores accompanied by paraphyses. Conidia are exuded as single, hyaline cells, which mature to two-celled, brown, unconstricted, longitudinally striate spores measuring 15 μm × 25 μm. Wind and water disseminate the conidia. The fruit is infected at the flower end of fingers and through wounds.

Control measures used for crown rot are recommended. The wash water should be clean, the fruit should have sufficient green life to reach its destination without ripening, temperature and humidity should be controlled and fungicides may be used (Stover, 1972).

Squirter

Squirter, like stem-end rot, is a disease of single fruit. It is caused by *Nigrospora sphaerica* (Sacc.) Mason, which has been placed in synonymy with *Nigrospora musae* under *Nigrospora maydis* by Jechová (1963). The pathogen enters through the cut pedicel and advances into the fruit as it ripens. The discoloured pulp (Plate 4.21) becomes a liquid mass and may squirt from the end of a fruit if squeezed. External symptoms become visible as a bluish tan discoloration of the peel as the fruit ripens (Stover, 1972). Conidia of *N. sphaerica* in culture are jet black when mature and measure 18 μm × 15 μm. They form on dead vegetation, including many grasses and banana trash. Airborne conidia are common in banana plantations (Meredith, 1961e) and in wash water in packing stations (Stover, 1972). Like stem-end rot, squirter is not a disease of hands or clusters and is now of little economic importance. Dipping fruit in a solution of fungicide easily controls the problem.

Other Diseases

Ring rot has been reported in single bananas from Australia (Jones and Wade, 1995). A dark, water-soaked, 20–40-mm-diameter ring develops on the peel as the fruit yellows during ripening (Plate 4.22). Internally, a dark, watery rot extends along the centre of the fruit pulp. The symptoms may be confused with squirter (see above). The disease is caused by a phycomycete fungus, which appears to infect from zoospores in the soil or wash tank. Ring rot is more often found after prolonged wet, windy weather. It can be avoided by rejecting fruit from bunches which have been in contact with the ground and by using clean wash water.

Trachysphaera finger rot, described by Stover (1972), is a rare disease of 'Gros Michel' fruit from the Cameroons. 'Lacatan' fruit from Jamaica ripened in rooms in England after fruit from Cameroon have also been affected. The cause, *Trachysphaera fructigena* Tabor and Bunting, is an inhabitant of the floral end of African bananas and may invade through wounds in the green peel, causing a dry black rot, which becomes fibrous (see Jones and Stover, this chapter, pp. 175–176).

REFERENCES

Abdel-Sattar, M.A. and Nawwar, M.A. (1986) 3-4 Dimethoxybenzaldehyde, a fungistatic substance in peels of green banana fruits in relation to resistance at different degrees of maturity. *HortScience* 21, 812.

Agati, J.A. (1922) Banana stem and fruit rot. *Philippine Agriculturist* 10, 411–422.

Agati, J.A. (1925) The anthracnose of abaca and Manila hemp. *Philippine Agriculturist* 13, 337–344.

Akamine, E.K. and Moy, J.H. (1983) Delay in postharvest ripening and senescence of fruits. In: Josephson, E.S. and Peterson, M.S. (eds) *Preservation of Food by Ionising Radiation.* CRC Press, Boca Raton, Florida, pp. 129–158.

Al Zaemey, A.B., Magan, N. and Thompson, A.K. (1993) Studies on the effect of fruit-coating polymers and organic acids on growth of *Colletotrichum musae in vitro* and on post-harvest control of anthracnose of bananas. *Mycological Research* 97, 1463–1468.

Berg, L.A. (1968) Diamond spot of bananas caused by *Fusarium roseum* 'Gibbosum'. *Phytopathology* 58, 388–389.

Beugnon, M., Brun, J. and Melin, P. (1970) La lutte contre *Trachysphaera fructigena*, parasite de bananes au Cameroon. *Fruits* 25, 187–197.

Brown, A.E. and Swinburne, T.R. (1981) Influence of iron and iron chelators on formation of progressive lesions by *Colletotrichum musae* on banana fruits. *Transactions of the British Mycological Society* 77, 119–124.

Brun, J. (1954) La pourriture des bananes au Cameroun Français. *Fruits* 9, 311–313.

Brun, J. (1970) Un nouvel aspect des pourritures de bananes 'Poyo' en cours de transport. *Fruits* 25, 781–784.

Burditt, A.R. (1994) Irradiation. In: Sharp, J.L. and Hallman, G.J. (eds) *Quarantine Treatments for Pests of Food Plants.* Westfield Press, San Francisco, pp. 101–117.

Chillet, M. and de Lapeyre de Bellaire, L. (1996a) Conditionnement en polybag pour le contrôle de l'anthracnose de blessures des bananes. *Fruits* 51, 162–172.

Chillet, M. and de Lapeyre de Bellaire, L. (1996b) Elaboration de la qualité des bananes au champ. Détermination de critères de mesure. *Fruits* 51, 317–326.

Chorin, M. and Rotem, J. (1961) Experiments on the control of tip rot in banana fruits. *Israel Journal of Agricultural Research (Ktavim)* 11, 185–188.

Chuang, T.Y. and Yang, H.R. (1993) Biological control of banana anthracnose. *Plant Pathology Bulletin (Taiwan)* 2, 71–77.

Coates, L.M., Muirhead, I.F., Irwin, J.A.G. and Gowanlock, D.H. (1993) Initial infection processes by *Colletotrichum gloeosporioides* in avocado fruit. *Mycological Research* 97, 1363–1370.

de Costa, D.M., Amaradasa, B.S. and Wegiriya, R.N. (1997) Antagonists of *Colletotrichum musae* associated with banana fruit skin. *Journal of the National Science of Sri Lanka* 25, 95–104.

de Lapeyre de Bellaire, L. and Dubois, C. (1997) Distribution of thiabendazole-resistant *Colletotrichum musae* isolates from Guadeloupe banana plantations. *Plant Disease* 81, 1378–1382.

de Lapeyre de Bellaire, L. and Mourichon, X. (1997) The pattern of fungal contamination of the banana bunch during its development and potential influence on incidence of crown rot and anthracnose diseases. *Plant Pathology* 46, 481–489.

Eckert, J.W. (1990) Recent developments in the chemical control of postharvest diseases. *Acta Horticulturae* 269, 477–494.

El-Helaly, A.F., Ibrahim, I.A. and El Arosi, H.M. (1955) Studies of some factors affecting the prevalence and distribution of cigar-end disease of bananas in Egypt. *Alexandria Journal of Agricultural Research* 2, 9–24.

Ellis, M.B. (1957) Some species of *Deightoniella. Mycological Papers* 66, 12 pp.

Ellis, M.B. (1971) *Dematiaceous Hyphomycetes.* Commonwealth Mycological Institute, Kew, Surrey, UK, pp. 206–207.

Finlay, A.R. and Brown, A.E. (1993) The relative importance of *Colletotrichum musae* as a crown rot pathogen on Windward Island bananas. *Plant Pathology* 42, 67–74.

Finlay, A.R., Lubin, C. and Brown, A.E. (1992) The banana stalk as a source of inoculum of fungal pathogens which cause crown rot. *Tropical Science* 32, 343–352.

Flaishman, M.A. and Kolattukudy, P.E. (1994) Timing of fungal invasion using host's ripening hormone as a signal. *Proceedings of the National Academy of Sciences of the United States of America* 91, 6579–6583.

Gomes, V.M., Olivera, A.E.A. and Filho, J.X. (1996) A chitinase and beta-1,3 glucanase isolated from seeds of cowpea (*Vigna unguiculata* L Walp.) inhibit the growth of fungi and insect pests of the seed. *Journal of Science of Food and Agriculture* 72, 86–90.

Greene, G.L. (1967) Effect of 2-deoxy-D-glucose on the respiration of tropical anthracnose fungi. *Physiologia Plantarum* 20, 580–586.

Greene, G.L. and Goos, R.D. (1963) Fungi associated with crown rot of boxed bananas. *Phytopathology* 53, 271–275.

Griffee, P.J. and Burden, O.J. (1976) Fungi associated with crown rot of boxed bananas in the Windward Islands. *Phytopathologische Zeitschrift* 85, 206–216.

Guyon, M. (1970) Essais de lutte chimique contre la 'Johnston fruit spot' au Nicaragua. *Fruits* 25, 685–691.

Halmos, S. (1969) Rice blast in Honduras and host specificity of *Pyricularia oryzae* and *Pyricularia grisea. Plant Disease Reporter* 53, 878–882.

Halmos, S. (1970) Inoculum sources of *Pyricularia grisea*, the cause of pitting disease of bananas. *Phytopathology* 60, 183–184.

Harai, N., Ishida, H. and Koshimuzu, K. (1994) A phenalenone-type phytoalexin from *Musa acuminata. Phytochemistry* 37, 383–385.

Harper, D.B., Swinburne, T.R., Moore, S.K., Brown, A.E. and Graham, H. (1980) A role for iron in the germination of conidia of *Colletotrichum musae*, a pathogen responsible for anthracnose lesions on bananas. *Journal of General Bacteriology* 121, 169–174.

Indrakeerthi, S.R.P. and Adikiram, N.K.B. (1996) Papaya latex, a potential postharvest fungicide. In: *Proceedings of the Australian Postharvest Horticulture Conference 'Science and Technology for the Fresh Food Revolution', Melbourne, Australia, 18–22 September 1995.* Institute for Horticultural Development, pp. 423–427.

Jechová, V. (1963) (New species of the genus *Nigrospora* causing rots of southern fruits). *Ceská Mykologie* 17, 12–20. (Czech.)

Jeger, M.J., Eden-Green, S., Thresh, J.M., Johanson, A., Waller, J.M. and Brown, A.E. (1995) Banana diseases. In: Gowen, S. (ed.) *Bananas and Plantains.* Chapman and Hall, London, pp. 317–402.

Jiang, Y.M. (1997) Occurrence of anthracnose inhibited by dopamine and induced by sugars in banana fruit after harvest. *Acta Phytopathologica Sinica* 27, 157–160.

Johanson, A. and Blazquez, B. (1992) Fungi associated with banana crown rot on field-packed fruit from the Windward Islands. *Crop Protection* 11, 79–83.

John, P. and Marchal, J. (1995) Ripening and biochemistry of the fruit. In: Gowen, S. (ed.) *Bananas and Plantains.* Chapman and Hall, London, pp. 434–467.

Johnston, J.R. (1932) The fruit spot of banana. *Research Department Bulletin* 43. United Fruit Company, Boston.

Jones, D.R. (1991) Chemical control of crown rot in Queensland bananas. *Australian Journal of Experimental Agriculture* 31, 693–698.

Jones, D.R. (1994) International *Musa* Testing Program Phase 1. In: Jones, D.R. (ed.) *The Improvement and Testing of* Musa *: A Global Partnership, Proceedings of the First Global Conference of the International* Musa *Testing Program held at FHIA, Honduras, 27–30 April 1994.* INIBAP, Montpellier, France, pp. 12–20.

Jones, D.R. and Wade, N.L. (1995) Bananas. In: Coates, L., Coke, T., Persley, D., Beattie, B., Wade, N. and Ridgeway, R. (eds) *Postharvest Diseases of Horticultural Produce. Tropical Fruits.* Department of Primary Industries, Brisbane, Queensland, Australia, p. 22.

Jones, D.R., Pegg, K.G. and Thomas, J.E. (1993) Banana. In: Persley, D. (ed.) *Diseases of Fruit Crops.* Queensland Department of Primary Industries, Brisbane, Australia, pp. 25–35.

Kachhwaha, M., Chile, A., Khare, V., Mehta, A. and Mehta, P. (1992) A new fruit rot disease of banana. *Indian Journal of Mycology and Plant Pathology* 22, 211.

Kaiser, W.J and Lukezic, F.L. (1965) Brown spot disease of banana fruit caused by *Cercospora hayi. Phytopathology* 55, 977–980.

Kaiser, W.J. and Lukezic, F.L. (1966) Influences of certain environmental conditions on spore dispersal and survival of *Cercospora hayi* from banana. *Phytopathology* 56, 1290–1293.

Kraus, U. (1996) Establishment of a bioassay for testing control measures against crown rot of banana. *Crop Protection* 15, 269–274.

Laguna, I.G. and Salazar, L.G. (1984) Pudricion del fruto del banano causada por *Sclerotinia sclerotiorum* (Lib.) de Bary en Costa Rica. *Turrialba* 34, 105–106.

López-Cabrera, J.J. and Marrero-Domínguez, A. (1998) Use of hot water dips to control the incidence of banana crown rot. *Acta Horticulturae* 490, 563–569.

Lukezic, F.L. and Kaiser, W.J. (1966) Aerobiology of *Fusarium roseum* 'Gibbosum' associated with crown rot of boxed bananas. *Phytopathology* 56, 545–548.

Lutchmeah, R.S. and Santchurn, D. (1991) Chemical control of common postharvest diseases of banana cv. 'Naine' in Mauritius. *Revue Agricole et Sucriere de L'Ile Maurice* 70, 8–11.

McCracken, A.R. and Swinburne, T.R. (1979) Siderophores produced by saprophytic bacteria as stimulants of germination of conidia of *Colletotrichum musae* causing anthracnose of bananas. *Physiological Plant Pathology* 15, 331–340.

Maramba, P. and Clerk, G.C. (1974) Survival of conidia of *Trachysphaera fructigena. Transactions of the British Mycological Society* 63, 391–393.

Meredith, D.S. (1960) Some observations on *Trachyshaera fructigena* Tabor and Bunting with particular reference to Jamaican bananas. *Transactions of the British Mycological Society* 43, 100–104.

Meredith, D.S. (1961a) Fruit spot ('speckle') of Jamaican bananas caused by *Deightoniella torulosa* (Syd.) Ellis. I–III. *Transactions of the British Mycological Society* 44, 95–104, 265–284, 391–405.

Meredith, D.S. (1961b) *Deightoniella torulosa* (Syd.) Ellis and *Verticillium theobromae* (Turc.) Mason and Hughes associated with a tip rot of banana fruits. *Transactions of the British Mycological Society* 44, 487–492.

Meredith, D.S. (1961c) Fruit spot ('speckle') of Jamaican bananas caused by *Deightoniella torulosa* (Syd.) Ellis. IV. Further observations on spore dispersal. *Annals of Applied Biology* 49, 488–496.

Meredith, D.S. (1961d) Spore discharge in *Deightoniella torulosa* (Syd.) Ellis. *Annals of Botany* 25, 271–278.

Meredith, D.S. (1961e) Atmospheric content of *Nigrospora* spores in Jamaican banana plantations. *Journal of General Microbiology* 26, 343–349.
Meredith, D.S. (1963a) Tip rot of banana fruits in Jamaica. I. *Hendersonula toruloidea* on Dwarf Cavendish bananas. *Transactions of the British Mycological Society* 46, 473–481.
Meredith, D.S. (1963b) *Pyricularia grisea* (Cooke) Sacc. Causing pitting disease of banana in central America. I. Preliminary studies on pathogenicity. *Annals of Applied Biology* 52, 453–463.
Meredith, D.S. (1965) Tip rot of banana fruits in Jamaica. II. *Verticillium theobromae* and *Fusarium* spp. *Transactions of the British Mycological Society* 48, 327–336.
Meredith, D.S. (1968) Freckle disease of banana in Hawaii caused by *Phyllostictina musarum* (Cke) Petr. *Annals of Applied Biology* 62, 328–340.
Meredith, D.S. (1971) Transport and storage diseases of bananas, biology and control. *Tropical Agriculture (Trinidad)* 48, 35–50.
Moy, J.H. (1983) Radurisation and radicidation: fruits and vegetables. In: Josephson, E.S. and Peterson, M.S. (eds.) *Preservation of Food by Ionising Radiation.* CRC Press, Boca Raton, Florida, pp. 83–108.
Muirhead, I.F. and Deverall, B.J. (1981) Role of appressoria in latent infections of banana fruits by *Collelototrichum musae. Physiological Plant Pathology* 19, 77–84.
Muirhead, I.F. and Deverall, B.J. (1984) Evaluation of 3,4-dihydroxybenzaldehyde, dopamine and its oxidation products as inhibitors of *Colletotrichum musae* (Berk. & Curt.) Arx in green banana fruits. *Australian Journal of Botany* 32, 575–582.
Mulvena, D., Webb, E.C. and Zerner, B. (1969) 3-4 Dihydroxybenzaldehyde, a fungistatic substance from green Cavendish bananas. *Phytochemistry* 8, 393–395.
Ocfemia, G.O. (1924) Notes on some ecomomic plant diseases new in the Philippine Islands. *Philippine Agriculturist* 13, 163–166.
Ogilvie, L. (1927) The black tip disease of banana. *Bulletin of the Department of Agriculture of Bermuda* 6(9), 4–5.
Ogilvie, L. (1928) 'Black tip', a finger disease of Chinese banana in Bermuda. *Phytopathology* 18, 531–538.
Parham, B.E.V. (1938) Central Agricultural Division, Annual Report for 1937. In: *Annual Bulletin of the Department of Agriculture of Fiji*, pp. 37–49.
Park, M. (1933) Anthracnose of plantains. *Tropical Agriculturist* 81, 330–333.
Postmaster, A., Kuo, J., Sivasithamparam, K. and Turner, D.W. (1997) Interaction between *Colletotrichum musae* and antagonistic microorganisms on the surface of banana leaf discs. *Scientia Horticulturae* 71, 113–125.
Punithalingam, E. and Waterston, J.M. (1970) Hendersonula toruloidea. Descriptions of Pathogenic Fungi and Bacteria No. 274, Commonwealth Mycological Institute, Kew, Surrey, UK, 2 pp.
Quimio, A.J. (1986) Postharvest diseases of bananas and their control in the Philippines. In: Umali, B.E. and Lantican, C.M. (eds) Banana and Plantain Research and Development, *Proceedings of the International Seminar-Workshop on Banana and Plantain Research and Development 25–27 February 1985, Davas City, Philippines.* ACIAR and PCARRD, Los Baños, Laguna, Philippines, pp. 130–135.
Reichert, I. and Hellinger, E. (1930) A Sclerotinia disease new to banana fruits, and its relation to 'Citrus'. *Hadar* 3, 14.
Reichert, I. and Hellinger, E. (1932) On 'Botrytis' tip-end of banana fruits in Palestine. *Hadar* 5(7), 162–163.
Reichert, I. and Hellinger, E. (1938) *Dothiorella* rot of bananas and oranges in Palestine. *Palestine Journal of Botany (Rehovot Series)* 2, 79–88.
Reyes, G.M. (1934) Banana black-tip disease in the Philippines. *Philippine Journal of Agriculture* 2, 117–119.
Reyes, M.E.Q., Nishijima, W. and Paull, R.E. (1998) Control of crown rot in 'Santa Catarina Prata' and 'Williams' banana with hot water treatments. *Postharvest Biology and Technology* 14, 71–75.
Rotem, J. and Chorin, M. (1961) Grove aeration and mode of irrigation as factors in the development of *Dothiorella* rot of banana fruits. *Israel Journal of Agricultural Research (Ktavim)* 11, 189–192.
Satyan, S., Scott, K.J. and Graham, D. (1992) Storage of banana bunches in sealed polyethylene tubes. *Journal of Horticultural Science* 67, 283–287.
Scott, K.J. and Roberts, E.A. (1966) Polyethylene bags to delay ripening of bananas during transport and storage. *Australian Journal of Experimental Agriculture and Animal Husbandry* 6, 197–199.

Scott, K.J., Blake, J.R., Strachan, G., Tugwell, B.L. and McGlasson, W.B. (1971) Transport of bananas at ambient temperatures using polyethylene bags. *Tropical Agriculture (Trinidad)* 48, 245–254.
Sela Buurlage, M.B., Epstein, L. and Rodriguez, R.J. (1991) Adhesion of ungerminated *Colletotrichum musae* conidia. *Physiological and Molecular Plant Pathology* 39, 345–352.
Sepiah, M. and Nik Mohd, N.A.M. (1987) Effects of benomyl and prochloraz on postharvest diseases of banana. *ASEAN Food Journal* 3, 101–104.
Shillingford, C.A. and Sinclair, J.B. (1977) Susceptibility of five banana cultivars to anthracnose and crown rotting fungi. *Plant Disease Reporter* 61, 797–801.
Shillingford, C.A. and Sinclair, J.B. (1980) Macerating enzyme production by *Colletotrichum musae* and *Fusarium semitectum*, incitants of banana fruit decay. *Phytopatholgische Zeitschrift* 97, 127–135.
Simmonds, J.H. (1941) Latent infection in tropical fruits discussed on relation to the part played by species of *Gloeosporium* and *Colletotrichum*. *Proceedings of the Royal Society of Queensland* 52, 92–120.
Simmonds, J.H. (1963) Studies in the latent phase of *Colletotrichum* species causing ripe rots of tropical fruits. *Queensland Journal of Agricultural Science* 20, 373–425.
Simmonds, J.H. (1966) *Host Index of Plant Diseases in Queensland*. Queensland Department of Primary Industries, Brisbane, Australia, 111 pp.
Simmonds, N.W. (1959) *Bananas*. Longman, Green and Co. Ltd, London, 466 pp.
Slabaugh, W.R. (1994a) Botryodiplodia finger rot. In: Ploetz, R.C., Zentmayer, G.A., Nishijima, W.T., Rohrbach, K.G. and Ohr, H.D. (eds) *Compendium of Tropical Fruit Diseases*. American Phytopathological Society Press, St Paul, Minnesota, pp. 5–6.
Slabaugh, W.R. (1994b) Cigar-end rot. In: Ploetz, R.C., Zentmayer, G.A., Nishijima, W.T., Rohrbach, K.G. and Ohr, H.D. (eds) *Compendium of Tropical Fruit Diseases*. American Phytopathological Society Press, St Paul, Minnesota, pp. 6–7.
Slabaugh, W.R. (1994c) Pitting disease (Johnston spot). In: Ploetz, R.C., Zentmayer, G.A., Nishijima, W.T., Rohrbach, K.G. and Ohr, H.D. (eds) *Compendium of Tropical Fruit Diseases*. American Phytopathological Society Press, St Paul, Minnesota, pp. 11–12.
Slabaugh, W.R. and Grove, M.D. (1982) Postharvest diseases of bananas and their control. *Plant Disease* 66, 746–750.
Stanley, M.S. and Brown, A.E. (1994) Separation and degradative action of polygalacturonase isozymes produced by *Colletotrichum musae*. *Journal of Phytopathology* 2, 217–224.
Stover, R.H. (1972) *Banana, Plantain and Abaca Diseases*. Commonwealth Mycological Institute, Kew, Surrey, UK, 316 pp.
Stover, R.H. (1975) Sooty mould of bananas. *Transactions of the British Mycological Society* 65, 328–330.
Stover, R.H. and Simmonds, N.W. (1987) *Bananas*, 3rd edn. Longman Scientific and Technical, Harlow, Essex, UK, 468 pp.
Subramanian, C.V. (1968) Deightoniella torulosa. Descriptions of Pathogenic Fungi and Bacteria No. 165. Commonwealth Mycological Institute, Kew, Surrey, UK, 2 pp.
Sutton, B.C. and Waterston, J.M. (1970) Colletotrichum musae. Descriptions of Pathogenic Fungi and Bacteria No. 222, Commonwealth Mycological Institute, Kew, Surrey, UK, 2 pp.
Tabor, R.J. and Bunting, R.H. (1923) On a disease of cocoa and coffee fruits caused by a fungus hitherto undescribed. *Annals of Botany* 37, 153–157.
Tezenas du Montcel, H. (1981) Perspectives nouvelles dans la lutte chimique contre *Trachysphaera fructigena* du bananier au Cameroon. *Fruits* 36, 3–8.
Tezenas du Montcel, H. and Laville, E. (1977) Influences des conditions climatiques sur développement du *Trachysphaera fructigena* sur bananier dans le sudouest du Cameroon. *Fruits* 32, 77–85.
Thomas, P (1986) Radiation preservation of foods of plant origin. III. Tropical fruits: bananas, mangoes. *CRC Critical Reviews of Food Science* 23, 147.
Thompson, A.K. and Burden, O.J. (1995) Harvesting and fruit care. In: Gowen, S. (ed.) *Bananas and Plantains*. Chapman and Hall, London, pp. 403–433.
von Arx, J.S. (1957) Die arten der gaitung *Colletotrichum*. *Phytopathologische Zeitschrift* 29, 413–468.
Wade, N.L., Kavanagh, E.E. and Sepiah, M. (1993) Effects of modified atmosphere storage on banana postharvest diseases and control of bunch main-stalk rot. *Postharvest Biology and Technology* 3, 143–154.

Wallbridge, A. (1981) Fungi associated with crown rot disease from the Windward Islands during a two-year survey. *Transactions of the British Mycological Society* 77, 567–577.

Wallbridge, A. and Pinegar, J.A. (1975) Fungi associated with crown rot disease of bananas from St Lucia and the Windward Islands. *Transactions of the British Mycological Society* 64, 247–254.

Wang Bisheng, Liu Chaozhen and Qi Peikun (1991) Studies on the Fusarium crown rot of bananas. *Acta Phytophylactica Sinica* 18, 133–137.

Wardlaw, C.W. (1931) Banana diseases. 3. Notes on the parasitism of *Gloeosporium musarum* Cke. and Mass. *Tropical Agricuture* 8, 12.

Wardlaw, C.W. (1934) Banana diseases. 6. The nature and occurrence of pitting disease and fruit spots. *Tropical Agriculture (Trinidad)* 11, 8–13.

Wardlaw, C.W. (1961) *Banana Diseases Including Plantains and Abaca.* Longmans, Green, London, UK, 648 pp.

Wardlaw, C.W. and McGuire, L.P. (1932) Pitting disease of bananas. Its nature and control. *Tropical Agriculture (Trinidad)* 9, 193–195.

Waterston, J.M. (1947) The fungi of Bermuda. *Bulletin of the Department of Agriculture of Bermuda* 23, 1–305.

Yang, H.R. and Chuang, T.Y. (1996) The nutrient effect on the appressorial behaviour and latent infection of *Colletotrichum musae. Plant Protection Bulletin (Taiwan)* 3, 247–259.

5

Diseases Caused by Bacteria

R. Thwaites, S.J. Eden-Green and R. Black

VASCULAR WILT DISEASES

Moko

Introduction

Moko disease takes its name from the cooking banana 'Moko' (ABB, syn. 'Bluggoe'), which suffered significant losses as a result of an outbreak of the disease in Trinidad in the 1890s (Rorer, 1911). The disease has probably been present in South America for over a century (Buddenhagen, 1961), and symptoms characteristic of Moko were observed some 70 years before the first description of the causal agent by Rorer (1911). Rorer named the causal agent *Bacillus musae* and noted its similarity to *Bacillus solanacearum* (now *Ralstonia solanacearum*), although, unlike strains of the latter, it apparently did not cause symptoms on tomato. Subsequent investigations showed that banana isolates were pathogenic to solanaceous plants, and the bacterium was reclassified as a member of the species then known as *B. solanacearum* (Ashby, 1926). Several decades elapsed before researchers recognized that banana is infected by a distinct subgroup of *R. solanacearum*, designated race 2, which had evolved from pathogens of wild *Heliconia* species (Buddenhagen, 1960; Sequeira and Averre, 1961). Independent appearances of such new pathogens, with differing pathological and epidemiological characteristics on cultivated banana, have been noted in regions extending from Costa Rica to Peru (French and Sequeira, 1970).

The outbreak in Trinidad in the 1890s, described by Rorer (1911), apparently eliminated the most susceptible host, 'Moko' cooking banana, so subsequent appearances of the disease were less serious. The commercial significance of Moko disease became apparent with the development of large-scale dessert banana plantations in Central America after the Second World War. An improved understanding of the disease and the development of effective control measures resulted partly from research into Moko disease in the late 1950s and 1960s and also from experience in controlling outbreaks of bacterial wilt in other crops, notably tomato, tobacco and potato (Kelman, 1953). Serious losses are now rare in commercial plantations because of the implementation of control measures. However, the disease still strikes susceptible plants grown on smallholdings and has continued to spread in Latin America and the southern Caribbean. Records now exist for Mexico, Guatemala, Belize, Honduras, El Salvador, Nicaragua, Costa Rica, Panama, Colombia, Venezuela, Guyana, Surinam, Trinidad, Grenada, Brazil, Peru and Ecuador (Stover, 1972; Phelps, 1987). In Nicaragua, over 20% of 'Bluggoe' plantings in some areas were found to be infected with Moko (Lehmann-Danziger, 1987) and, in Grenada, annual losses stabilized at 1.6% of mats (Hunt,

1987). An outbreak in Trinidad in the early 1960s virtually wiped out the island's export banana trade (Phelps, 1987). Moko disease has been recorded as a continuing problem in Mexico and Belize (Black and Delbeke, 1991). Significant losses have also occurred in South America. Reductions in yield due to Moko disease of up to 74% have been reported in Guyana (Phelps, 1987), while smallholder crops of 'Bluggoe' have been badly affected in the Magdalena valley of Colombia (Buddenhagen and Elsasser, 1962) and the Amazon valley in Peru (French and Sequeira, 1970). Reports of the disease in the Dominican Republic and Haiti (Smith *et al.*, 1997) are unconfirmed, but records from Jamaica and Guadeloupe are erroneous (Anon., 1998).

Moko also occurs on cultivars in the Cavendish subgroup (AAA) in commercial plantations in the Davao region of Mindanao in the Philippines. It is believed to have been introduced on propagating material imported from Honduras in 1968 (Rillo, 1979; Buddenhagen, 1994). Although Moko has been reported from India (Gnanamanickam *et al.*, 1979; Sivamani and Gnanamanickam, 1987), its presence is doubtful (Buddenhagen, 1994). There are also records of Moko from several African countries (Smith *et al.*, 1997), but many of these should be viewed circumspectly as they date from several decades ago and are not substantiated by authoritative descriptions of the symptoms or authenticated by identification of the Moko pathogen in culture. The record from Malawi, for example, is very doubtful (Anon., 1998).

Symptoms

All parts of the plant may be invaded by *R. solanacearum*, but the route of infection and the nature of the causal strain determine the type of symptoms (Table 5.1). Infection of dessert banana cultivars typically occurs via the roots or rhizome. In these cases, the first signs of the disease are yellowing and wilting of the oldest leaves, which eventually become necrotic and collapse. The symptoms spread to the younger leaves, which develop pale green or whitish panels before becoming necrotic. The disease may cause the suckers to wilt (Plate 5.1), possibly without exhibiting the more localized foliar symptoms of yellowing and necrosis. Fruit development is arrested and fingers may ripen prematurely or split. Internally, fruit become discoloured and eventually rot (Plate 5.2). Vascular tissues become progressively discoloured (Plate 5.3) and, when cut, exude bacterial ooze, from which the causal organism may readily be isolated. In the later stages of infection, the vascular discoloration may be brown or black. In fruit-bearing plants, the internal symptoms are most evident in the fruit stem and at the bases of younger leaves. Moko disease in cooking-banana cultivars, such as 'Bluggoe', commonly results from infection with strains of the small fluidal round (SFR) colony form via insect vectors. Stingless bees (*Trigona* spp.) and other generalist feeders are attracted to bacterial ooze, which exudes from diseased peduncle cushions for 15–25 days after infection. Insect transmission of the B (banana) strain may also occur, but to a much lesser extent as this strain causes far less bacterial oozing than do SFR isolates (Buddenhagen and Elsasser, 1962; Stover, 1972). Symptoms of insect-transmitted Moko disease are first seen in the flower buds and peduncles, which become blackened and shrivelled (Plate 5.4). The bacterium spreads to the fruit, which may ripen prematurely (Plate 5.4), and initiates a rot. The infection continues towards the pseudostem, causing blackening of the vascular tissue. Eventually the bacterium becomes systemic and the whole mat is diseased. In 'Bluggoe', however, systemic invasion of the rhizome is typically incomplete. Suckers that become infected will rapidly die, but those that do not will survive to produce flower buds, which can be infected by insects. In this way, infected 'Bluggoe' mats may persist indefinitely (Black and Delbeke, 1991). Such mats act as sources of inoculum for further spread of the pathogen.

Other strains of the pathogen cause less

Table 5.1. Characteristics of strains of *Ralstonia solanacearum* associated with Moko disease.

Strain*	Distribution	Colony characteristics†	Host range and ecology‡
B[1, 5] (banana rapid wilt)	Central and South America	Large, elliptical/irregular, fluidal, lace-like slime, slight formazan[3,4]	AAA/AAB/ABB: high *Heliconia*: moderate Insects: low/moderate Soil: moderate
D[1, 5] (distortion)	Costa Rica, Surinam, Guyana	As B strain	AAA/AAB/ABB: low/mod. *Heliconia*: low Insects: no report Soil: low
H[3] (Heliconia)	Costa Rica	As B strain,[3] smooth slime and intense formazan[4]	AAA/AAB: low ABB: moderate *Heliconia*: low Insects: probably high Soil: probably low
R[5] (red colony form)	Costa Rica	As B strain,[3, 4] intense formazan[5]	AAA/AAB: low ABB: no report *Heliconia*: no report Insects: no report Soil: no report
SFR,[2] SFR-C[3, 4] (small fluidal round)	C. America, S. Caribbean, Venezuela, Colombia	Small, round, little (SFR) or moderate (SFR-C) slime, smooth or initially lace-like, slight formazan	AAA/ABB: high AAB: moderate/high *Heliconia*: no report Insects: high Soil: low
A[3, 4] (Amazon)	Amazon basin (Brazil, Colombia, Peru)	Medium size, near round, fluidal, faintly lace-like slime, slight helical formazan	AAA/AAB/ABB: high *Heliconia*: no report Insects: high Soil: no report
AFV[6] (afluidal virulent)	Honduras	Small, round, no slime, intense formazan	AAA: low AAB/ABB: probably low Insects: no report Soil: no report
T[1, 5] (tomato = race 1, biovar 1)	Americas	As H strain, but brown pigment on L-tyrosine medium[4]	Pathogenic on some *Musa* species Not pathogenic on banana cultivars and *Heliconia* Insects: no report Soil: high

* Descriptions according to [1]Buddenhagen (1960); [2]Buddenhagen and Elsasser (1962); [3]French (1986); [4]French and Sequeira (1970); [5]Sequeira and Averre (1961); [6]Woods (1984). Additional data from Lehmann-Danziger (1987) and Stover (1972).

† Comparative sizes and colony morphologies after 2 days' incubation on Kelman's (1954) medium.

‡ Virulence to AAA, AAB and ABB genomic groups and *Heliconia* spp.; frequency of transmission by insects; persistence in soil.

Plate 5.1. Collapsed and withered heart leaf of a sucker of 'Robusta' (AAA, Cavendish subgroup) caused by Moko in Grenada. The necrotic margin on the next leaf is also reputed to be a symptom of the disease (photo: D.R. Jones, INIBAP).

severe symptoms (Sequeira and Averre, 1961; French and Sequeira, 1970). *Heliconia* (H) strains do not affect dessert bananas and cause localized symptoms on 'Bluggoe', affecting only the flower buds. Conversely, isolates of the red colony form (R strains) do not wilt 'Bluggoe' and are weakly pathogenic on dessert banana. Afluidal colony forms (AFV) cause mild wilting (Woods, 1984), while infection of young plants by the distortion (D) strain results in a slow wilt and stunting.

Causal agent

Ralstonia solanacearum (Yabuuchi *et al.*, 1995; syn. *Pseudomonas solanacearum*, *Burkholderia solanacearum*) is an aerobic, Gram-negative, non-fluorescent rod belonging to the rRNA homology group II (Palleroni, 1984). The species is large and highly diverse, but Hayward (1964) has described bacteriological characteristics common to all strains. The bacterium does not grow in the presence of 2% sodium chloride (NaCl), accumulates poly-β-hydroxybutyric acid, is arginine dihydrolase-negative and oxidase-positive. A casamino acid-based indicator medium containing 0.1 g l^{-1} 2,3,5-triphenyltetrazolium chloride (Kelman, 1954) is useful for distin-

Plate 5.2. Internal rot of fruit of 'Grand Nain' (AAA, Cavendish subgroup) caused by Moko on a commercial plantation in Honduras (photo: D.R. Jones, INIBAP).

Plate 5.3. Brown discoloration of the vascular tissues caused by Moko in a pseudostem of 'Grand Nain' (AAA, Cavendish subgroup) on a commercial plantation in Honduras (photo: D.R. Jones, INIBAP).

Plate 5.4. Premature ripening of fruit of 'Robusta' (AAA, Cavendish subgroup) affected by Moko in Grenada. The bunch also has a blackened and withered peduncle, indicating insect transmission (photo: D.R. Jones, INIBAP).

guishing between various colony types, as well as allowing for tentative discrimination between *R. solanacearum* and contaminating organisms in primary isolations from diseased material. Most strains are readily isolated from infected plants, and various selective media, based on Kelman's medium, are effective in reducing contamination (Granada and Sequeira, 1983; Engelbrecht, 1994).

Spontaneous avirulent mutants, which often arise during repeated subculturing, produce non-fluidal, butyrous colonies on Kelman's medium. Spontaneous avirulence has been termed phenotype conversion (Brumbley and Denny, 1990), because of the wide range of phenotypic alterations involved, including loss of extracellular polysaccharide production and changes in the activity of various exported enzymes. Motility by means of single polar flagella may not be evident in freshly isolated virulent cultures. Motility is also associated with spontaneous avirulence (Kelman and Hruschka, 1973).

Ralstonia solanacearum is a highly heterogeneous species and has been subdivided into five biovars on the basis of carbohydrate catabolism (Hayward, 1964; He *et al.*, 1983) and five races designated by host range (Buddenhagen *et al.*, 1962;

He *et al.*, 1983). Strains pathogenic to banana cultivars are designated race 2 and belong to biovar 1. Considerable variation exists within race 2 and a number of distinct colony forms, with different host ranges and ecological characteristics, have been observed (Table 5.1). In addition to the phenotypes more normally associated with disease in the field, a number of aberrant forms have been reported, either on first isolation from diseased plants (Woods, 1984) or as a result of subculture or storage (French and Sequeira, 1970; Wardlaw, 1972). Race 2 strains may be distinguished from generalist pathogens of solanaceous and non-solanaceous hosts (race 1) and potato pathogens (race 3) by their ability to cause a hypersensitive response when infiltrated into tobacco leaves (Lozano and Sequeira, 1970).

Race 1 strains are common in banana-growing areas, even where race 2 is not present (Stover, 1972). While race 1 strains have not been reported to cause bacterial wilt of banana cultivars, they have been implicated in a wilt disease of the wild diploids *Musa acuminata* ssp. *banksii*, *M. acuminata* ssp. *zebrina* and *Musa schizocarpa* in Honduras (Buddenhagen, 1962). Race 1 strains may infect wild diploid plants via the soil, by mechanical inoculation (Buddenhagen, 1962) and by insect transmission between inflorescences (Vakili and Baldwin, 1966). Symptoms resemble those of Moko disease. The 'V-10' accession of *M. acuminata* ssp. *banksii* was found to be resistant and used in studies on the inheritance of resistance (Vakili, 1965).

Disease cycle and epidemiology

Strains of *R. solanacearum* that cause Moko disease have probably evolved from pathogens of close wild relatives of banana. The appearance of the disease in some newly planted areas has been closely linked with the presence of infected *Heliconia* species. Sequeira and Averre (1961) found that isolates from *Heliconia* became more pathogenic on banana after repeated inoculation and reisolation from that host. Despite the ubiquitous presence of other races of the bacterium, only strains isolated from *Musa* and its close relative *Heliconia* appear to be capable of infecting banana (Buddenhagen, 1960, 1961; Sequeira and Averre, 1961). These facts suggest that the widespread presence of cultivated banana has selected for virulence in the endemic population of *Heliconia*-attacking strains of *R. solanacearum*.

Weed hosts have also been implicated in new outbreaks of Moko disease. Infection of endemic weeds may have a role in the Moko disease cycle, although the persistence of the bacterium in such hosts has been questioned (Buddenhagen, 1986). Race 2 strains have, however, been isolated from commonly occurring weed species. Berg (1971) isolated SFR strains from several weeds, unrelated to banana, which occurred frequently on Honduras plantations, and B strains can survive in the rhizosphere if certain natural flora are present (Wardlaw, 1972). Elimination of potential weed hosts is therefore considered an important factor in Moko disease control. *Ralstonia solanacearum* survives less well in soils where no natural hosts are present. In studies on plantations abandoned due to Moko infestation, fallowing for 18 months reduced the infective potential of the soil to negligible levels (Sequeira, 1962). The length of fallow period required for effective disease control was reduced still further if the soil was well drained (Lozano *et al.*, 1969) or shallow-tilled in dry weather (Sequeira, 1958). Wardlaw (1972) reported that the bacterium cannot survive for 6 months in soil maintained at or below 65% water-holding capacity.

Race 1 strains of *R. solanacearum* from infested soil may infect roots following wounding by soil-inhabiting insects and nematodes (Kelman, 1953), although the importance of this pathway in the infection of either banana or solanaceous hosts is unclear. Lehmann-Danziger (1987) observed that banana growing in close proximity to plants with Moko disease often remain free of the disease, despite the presence of nematodes. Both race 1 and race 2 can spread from root to root without mechanical injury (Kelman and Sequeira, 1965), but most plant-to-plant spread results from either insect or human activities. The bacterium may be spread in contaminated irrigation water and on pruning knives, the shoes of workers and animal hooves. Insect-mediated dissemination of over 90 km has been recorded (Wardlaw, 1972) and epidemics which occur via this route, such as those in Honduras (Buddenhagen and Elsasser, 1962) and Peru (French and Sequeira, 1970), spread very rapidly. This is a result not only of the rapid plant-to-plant spread afforded by insect dissemination, but also of the rapidity with which symptoms ensue following infection of the inflorescence. Bacterial ooze starts to appear on peduncles and bract bases approximately 15 days after infection of flowers by SFR strains (Buddenhagen and Elsasser, 1962), providing a focus for new infections. B strains produce less bacterial ooze than do SFR strains and so are less frequently transmitted by insects. Both SFR and B strains are capable of inflorescence infection (Stover, 1972), but banana cultivars with dehiscent bracts, such as 'Bluggoe', are particularly susceptible. Following infection, the bacterium spreads to the vascular tissue and the disease usually becomes systemic, eventually reaching the rhizome. Bacteria may remain viable in the rhizome for several months and any suckers or other propagating material removed for planting will also carry the disease.

Host reaction

Very few studies on resistance to Moko disease have been published. Stover (1972) reported that all clones of commercial importance are susceptible to B and SFR strains. Cultivars with dehiscent bracts, notably 'Bluggoe', are particularly susceptible to infection by insect-transmitted strains. The cooking-banana cultivar 'Pelipita' (ABB), which possesses indehiscent male bracts and has strong field resistance, has been recommended for replanting in Moko-infested areas in Central America as a replacement for 'Bluggoe' (Stover and Richardson, 1968). 'Horn' plantain (AAB) is said to have some degree of inherent field resistance (Stover, 1972). Stover (1993) reported that East African highland cultivars of the Lujugira–Mutika subgroup (AAA) are susceptible to the SFR strain.

In artificial inoculation tests in Honduras, bacteria were injected into the pseudostems of 345 accessions of banana (Stover, 1972). 'Pelipita' was highly resistant to the SFR strain, as were accessions of *Musa balbisiana*. The 'Manang' accession of *M. acuminata* ssp. *banksii* showed moderate resistance. In all, about 10% of the accessions showed some degree of resistance to the SFR strain. Clones of 'Sucrier' (AA), 'Inarnibal' (AA), 'Laknau' (AAB), 'Saba' (ABB), 'Pitogo' (ABB) and others were rated as susceptible. In tests with the B strain, only *M. balbisiana* showed resistance. Some *M. acuminata* accessions have also been shown to exhibit a degree of resistance to race 1 strains and race 2 distortion strains of *R. solanacearum* (Vakili, 1965; Wardlaw, 1972).

More recently, the bred tetraploid 'FHIA-03' (ABBB) has been observed to remain healthy after planting in Moko-infested areas of Grenada (see Rowe and Rosales, Chapter 14, pp. 435–449). However, this apparent 'resistance' should be used with caution. Bacteria injected into 'resistant' plants in the Honduran breeding programme were found to survive for up to 6 months without causing symptoms. As such symptomless plants could, in theory, serve as a hidden source of inoculum, this tolerance to infection was considered to be an undesirable trait and is no longer exploited (P. Rowe, Honduras, 1997, personal communication).

Control

The systemic nature of infection by the Moko disease bacterium means that the pathogen cannot be destroyed without also killing the host plant. Where the pathogen is present, the only means of effective control is the eradication of infected plants, coupled with the adoption of improved cultural practices to limit pathogen dispersal.

However, since the distribution of Moko remains fairly limited, the primary means of Moko control is exclusion from areas where the disease does not occur. As Moko strains also affect *Heliconia*, which is traded internationally as an ornamental, quarantine authorities must be alert to the possibility of spread on this alternative host (see Jones and Diekmann, Chapter 13, pp. 409–423). The importation of vegetative *Heliconia* into banana-growing countries should be treated with extreme caution.

Eradication

Successful management of Moko in areas where the disease occurs relies on early detection, followed by the destruction of any infected mats. Experienced Moko inspectors inspect banana plants in commercial plantations in advance of pruning operations. Regular inspections, every 2–4 weeks, depending on local disease incidence, are required for early identification of symptoms. All parts of diseased plants are destroyed *in situ* by injections with a systemic herbicide, such as glyphosate. Healthy mats surrounding the diseased plants are also destroyed by herbicide injection and weeds in this 'buffer zone' are treated with herbicide in order to reduce the chances of the bacterium surviving. Buffer zones of 5 m and 10 m have been recommended for SFR and B strain infections, respectively (Stover, 1972), and this area may either be left fallow or planted with wilt-suppressive crops, such as maize. Recommendations vary for the length of the fallow period required to eliminate the bacterium, but all are based on investigations and observations conducted under field conditions. Fallow periods of 6 and 12 months for SFR and B strains, respectively, have been suggested (Stover, 1972), although a banana-free (but not weed-free) period of up to 2 years has been recommended in Grenada, where Moko is caused by SFR strains (Hunt, 1987). Where weeds are effectively controlled, the fallow period may be considerably reduced. Lehmann-Danziger (1987) reported that a weed-free period of only 5 months was necessary for Moko control in Nicaragua. Shallow ploughing or disc-harrowing serves to desiccate the soil and has been shown to reduce the necessary fallow period (Sequeira, 1958). The same study also investigated treatment of infected soil with bactericidal chemicals, but these did not reduce disease incidence in subsequent crops. Soil fumigation with methyl bromide is, however, effective in dramatically reducing the fallow period (Stover and Simmonds, 1987), although this method is impractical for small-scale farmers and environmental concerns have arisen over the role of the fumigant in atmospheric ozone depletion.

In Belize, systematic surveys of 'Bluggoe' and smallholder dessert banana cultivars, coupled with glyphosate treatment of all banana plants within a 5-m radius, effectively eradicated Moko. The only outbreaks recorded now are possible incursions of the disease over the border from Guatemala. In Belize, eradication was achieved with the cooperation of the public and without the need for compensation for healthy plants (mostly 'Bluggoe') destroyed. In the eradication programme in Grenada, compensation had to be given for healthy mats of 'Bluggoe' destroyed, as the fruit is highly valued as a food (P. Hunt, UK, 1998, personal communication).

Cultural practices

It has been estimated that, without sufficient precautions against contamination, almost 97% of the dissemination of Moko type B strains may be due to cultural practices alone (Wardlaw, 1972). Disinfection of digging and cutting tools is therefore considered to be of great importance in disease control, and sterilization systems suitable for routine field use have been designed (Buddenhagen and Sequeira, 1958; Sequeira, 1958; Stover, 1972; Wardlaw, 1972). Pruning knives may be disinfected by contact with a 10% solution of formaldehyde for 10 s or a 5% solution for 30 s. Plantation workers undertaking pruning operations usually carry two knives. One knife is sterilized in a scabbard lined with felt soaked in a 5% or 10% formaldehyde solution while the other is in use. Tools need to be sterilized between each plant and a dye, such as crystal violet, is often added to the disinfectant in order to monitor the operation. Since the bacterium may also be spread on contaminated footwear, the same sterilizing solution is used to treat the shoes of workers when they leave infested areas.

Removal of the male bud from 'Bluggoe' as soon as the last female fingers had emerged was said to prevent the spread of the SFR strain from plant to plant via male bract scars in Honduras (Stover and Richardson, 1968). The practice of breaking the peduncle with the bud by hand has been recommended for all cultivars with dehiscent bracts (Stover, 1972).

Plant quarantine

Ralstonia solanacearum race 2 is currently listed as an important threat to banana cultivation by the Asian and Pacific Plant Protection Commission, the Caribbean Plant Protection Commission, the Pacific Plant Protection Organization and China. It should be excluded from any Moko-free area where *Musa* is grown for local consumption or as a cash crop. Importation of *Musa* planting material into an area considered at risk should only be undertaken after a thorough assessment of the risks. Information on the safe international movement of *Musa* germplasm is provided by Jones and Diekmann, Chapter 13, pp. 409–423.

Blood Disease

Introduction

The first report of 'Penyakit Darah', or blood disease, was from the Saleiren Islands off the south coast of Sulawesi in Indonesia in 1906. The outbreak was so serious that it forced the abandonment of newly established dessert banana plantations (Rijks, 1916). Subsequent investigations by Gäumann (1921b, 1923) showed that the disease was widespread in southern Sulawesi, where it was frequently found both in local banana cultivars and in wild *Musa* species. The disease was apparently not present in Java or on other islands, and quarantine restrictions were placed on the export of banana plants and fruit from Sulawesi to prevent further spread of the disease. Gäumann (1923) traced the cause of the disease to a Gram-negative bacterium, which he named *Pseudomonas celebensis*. Blood disease was later regarded as an anomalous appearance of Moko disease. In the late 1980s, the presence of a vascular bacterial disease of banana was noted on the Indonesian Island of Java (Eden-Green and Sastraatmadja, 1990). Investigations revealed that the causal bacterium was not the same as the Moko pathogen. Blood disease is now spreading rapidly among both dessert and cooking-banana cultivars in Java and constitutes a substantial limiting factor in banana cultivation. More recently, the disease has been reported from Sumatra, Kalimantan, the Moluccan Islands and Irian Jaya (Baharuddin *et al.*, 1994). To date, there have been no records of blood disease outside Indonesia.

Symptoms

Blood disease affects both dessert and cooking-banana cultivars. The latter are more widely grown in Sulawesi, perhaps due to the higher susceptibility of dessert varieties, and the symptoms of blood disease are described from this ABB group. In mature leaves, the disease causes a conspicuous yellowing, which is followed by wilting, necrosis and collapse near the junction with the pseudostem (Plate 5.5). Younger leaves turn bright yellow, before becoming necrotic and dry, and the emergence of the youngest leaf is arrested. The male flower bud and youngest fruit bunches may be blackened and shrivelled (Plate 5.6), although often they appear outwardly unaffected. However, internally the fruit exhibit a reddish-brown discoloration and are rotten or dry (Plate 5.7). The name blood disease derives from the reddish discoloration visible in the vascular tissues of infected plants. This discoloration extends from affected fruit bunches down into the corm via the peduncle and pseudostem (Plate 5.8). Cut vascular tissue exudes droplets of bacterial ooze, which may vary in colour from cream to reddish-brown to black. The vascular discoloration may extend through the mat and into the suckers, although this is not always the case and uninfected suckers are sometimes produced.

Causal agent

The symptoms of blood disease are very similar to those of insect-transmitted Moko disease caused by *Ralstonia solanacearum*. However, bacteriological characteristics of the blood disease bacterium differ from those of *R. solanacearum* in several respects and the precise taxonomic status of the organism currently remains undetermined. Characteristics of recent blood-disease isolates agree reasonably well with Gäumann's description of *P. celebensis*, although none of his original cultures have survived. *Pseudomonas celebensis* was subsequently reclassified as a member of the genus *Xanthomonas* (Dowson, 1943), but the name was never validated by inclusion in the approved bacteriological nomenclature. A pathovar of *Xanthomonas campestris*, isolated from leaf spots on banana, has since been assigned the name *celebensis* (Rangaswamy and Rangajaran, 1965; Young *et al.*, 1978), although this bacterium is entirely unrelated to the blood-disease organism.

Eden-Green *et al.* (1988) have determined characteristics of the blood-disease bacterium. The bacteria are straight, Gram-negative rods, 0.5 μm × 1.0–1.5 μm in size, which grow rather slowly on TZC medium (Kelman, 1954), producing non-fluidal colonies with smooth margins and dark red centres. Colonies are non-fluorescent on King's B medium and do not produce brown pigment on tyrosinase medium, although brown pigmentation does accumulate after prolonged incubation on these media. Bacteria grow at 37°C, but not at 41°C or in the presence of 2% NaCl. No motility, flagella or intracellular sudanophilic inclusions have been observed. Isolates are oxidase-positive and arginine dihydrolase-negative, and do not reduce nitrate or hydrolyse gelatin, starch or lecithin. On first isolation, all strains produce acid from galactose and glycerol, but none utilize glucose, fructose or sucrose, and hence isolates are not compatible with the biovar system used to delineate biochemical groups within *R. solanacearum*. However, the ability to utilize glucose, sucrose and other sugars was acquired by some strains during culture. In experimental inoculations, all isolates were found to be pathogenic to young banana plants in the AAA and AAB genomic groups, causing wilting 7–14 days after introduction into the corm. No symptoms were observed following inoculation of tomato, eggplant, sweet pepper, groundnut or tobacco and a hypersensitive response was elicited after infiltration of the bacteria into tobacco leaves.

Many of the above bacteriological features distinguish the blood-disease bacterium from race 2 of *R. solanacearum*, although the similarity between the two taxa has been established not only on the basis of symptomatology, but also serologically (Eden-Green *et* al., 1988; Robinson, 1993) and by a number of molecular studies. Seal *et al.* (1992) found that blood-disease isolates and some race 2 strains of *R. solanacearum* gave very similar fingerprints when subjected to polymerase chain reaction (PCR) amplification with primers targeted to tRNA genes. Furthermore, analysis of 16S rRNA gene sequences has grouped blood-disease isolates with some race 1 strains of *R. solanacearum* and isolates of *Pseudomonas syzygii*, a bacterium related to *R. solanacearum* that causes a vascular disease of clove trees in Indonesia (Roberts *et al.*, 1990; Seal *et al.*, 1993; Taghavi *et al.*, 1996). Both the tRNA and the 16S rRNA regions of the bacterial genome are considered to be relatively stable over time and analyses of these regions provide good indicators to the evolutionary relationships between bacterial taxa.

Randomly primed PCR amplification has recently been used to produce a collection of fingerprints more representative of the whole genome, as well as PCR amplification with primers targeted to specific intergenic DNA repeat sequences (Thwaites *et al.*, 1999). The results are in close agreement with those of previous studies, in that both PCR fingerprinting and analysis of tRNA and 16S rRNA regions linked blood-disease isolates with the same subset of *R. solanacearum* strains, although the repetitive primers grouped all blood-disease strains in a distinct group. Despite the wealth of information that has accumulated on the relationship between the causal agent of blood disease and strains of *R. solanacearum*, the blood-disease bacterium has not been officially classified. A formal description, therefore, is required to resolve this uncertainty.

Disease cycle and epidemiology

Our understanding of the disease cycle of blood disease stems largely from the investigations carried out by Gäumann (1921b, 1923) and, apart from the recognition of obvious similarities to Moko disease, more modern studies have added relatively little. The causal bacterium can infect healthy plants via the soil, and can survive in excess of 1 year in the rhizosphere if contaminated plant residues are present. Like Moko, blood disease is readily spread on pruning tools and may be carried in suckers removed to seed new plants. The disease may have been exported from Sulawesi to Java in infected fruit intended for local consumption since diseased fruit often display no external symptoms.

Transmission of blood disease between plants is rapid and dissemination at the rate of at least 25 km $year^{-1}$ has recently been observed. Gäumann noted this capacity for rapid spread. He found that bacteria applied to the stigma of female flowers could infect the developing fruit via the style. Some naturally infected flowers were observed and, some 40 years before insect transmission of Moko disease was reported by Buddenhagen and Elsasser (1962), Gäumann postulated that the disease may be transmitted between flowers by the action of wind or insects. More recent observations suggest that infection via the male flower bud may play a significant role in the blood-disease cycle.

Host reaction

Gäumann (1921b) examined over 100 banana cultivars, but found none to be resistant to blood disease. However, *Musa* spp. in Sulawesi and blood disease have probably undergone a period of co-evolution, so natural sources of resistance or tolerance to the disease may well exist.

Control

The unavailability of either resistant cultivars or effective chemical treatment means that, like Moko, vigilance and strict cultural practices are needed to control blood disease. Procedures for managing outbreaks of blood disease in the field are similar to those developed for Moko, although attempts to reduce the incidence of insect transmission by removal of the male flower bud will be ineffective if the bacterium enters via female flowers. Blood disease still has a fairly limited distribution, so restrictions imposed on the movement of plant material will continue to curb the spread of the causal bacterium. The use of disease-free planting material is perhaps the most effective strategy for blood-disease control.

Javanese Vascular Disease

According to Gäumann (1921a), this disease was first observed in 1915 and was subsequently found on almost all cultivated bananas in Java, Indonesia. Affected plants exhibited widespread vascular discoloration, although the disease was not considered to be particularly harmful to its host and 90% of infected plants displayed no external symptoms whatsoever. Where external symptoms were visible, they resembled those of Fusarium wilt and, as

Plate 5.5. Withering and collapse of leaves caused by blood disease on 'Pisang Kepok' (ABB, syn. 'Saba') in North Sulawesi, Indonesia (photo: S. Eden-Green, NRI).

many vascular diseases had been ascribed to fungal pathogens, Gäumann examined six species of *Fusarium*, which were isolated from diseased banana plants. None of these fungi was pathogenic to banana, but a bacterium was identified which caused vascular discoloration when introduced into the rhizome, pseudostem or leaf veins. The bacterium was not specific to banana and was also pathogenic on *Ravenala*, *Strelitzia* and possibly *Heliconia*. Gäumann named the bacterium *Pseudomonas musae*, but the name is no longer valid and none of the original isolates survive. The bacterium was described as a Gram-positive, non-spore-forming rod, 0.8–1.2 μm long with one to three polar flagella.

The original 1921 study asserted that external symptoms of Javanese vascular disease were caused by secondary infection by *Fusarium* spp. following primary vascular invasion by *P. musae*. Gäumann suggested that this was true also of Fusarium wilt and, in so doing, criticized the work of several of his contemporaries. Wardlaw's (1935) account of the disease details the debate that ensued over the role of bacteria in such infections. No subsequent observations of Javanese vascular disease have been made and the original reports have been ascribed to Fusarium wilt.

Plate 5.6. Shrivelling and blackening of the male flower bud and lower hands on a bunch of 'Pisang Kepok' (ABB, syn. 'Saba') affected by blood disease in West Java, Indonesia (photo: S. Eden-Green, NRI).

Plate 5.7. Internal rot of fruit of 'Pisang Kepok' (ABB, syn. 'Saba') affected by blood disease in West Java, Indonesia (photo: S. Eden-Green, NRI).

Plate 5.8. Brown discoloration of vascular tissues in a pseudostem of 'Pisang Kepok' (ABB, syn. 'Saba') affected by blood disease in West Java, Indonesia (photo: S. Eden-Green, NRI).

Bacterial Wilt of Abacá

A disease of abacá, referred to variously as 'abacá wilt', 'vascular disease' and 'banana wilt-like disease', was reported by Palo and Calinisan (1939) to be the most serious disease of abacá in the Davao region of Mindanao in the Philippines, where over 1300 ha were found to be infected. The symptoms of the disease were very similar to those of Moko, except for the appearance of distinctive rusty-brown streaks along the leaf veins of infected plants. These streaks could extend from the midrib to the leaf margin. Several abacá cultivars were affected by the disease, principally 'Maguindanao', 'Tangoñgon', 'Balindag' and 'Bongolanon'. Wilting was not observed in 'Tangoñgon'. The disease had previously been considered to be fungal in origin, but Palo and Calinisan (1939) were consistently able to isolate bacteria with characteristics similar to *Ralstonia solanacearum* from diseased tissue. These isolates induced wilt in abacá plants when introduced into the centre of the pseudostem. More recent investigations have found the pathogenicity of isolates from abacá to be inconsistent when mechanically inoculated (Zehr, 1970; Rillo, 1981), although these studies were carried out on the 'Tangoñgon' cultivar, which was reportedly unaffected by bacterial wilt in the field. The financial significance of the disease is now slight, since abacá is no longer an economically important crop on Mindanao.

Bacterial Wilt of Enset

The symptoms of bacterial wilt of enset, which is widely distributed in all growing districts in Ethiopia, were first recorded nearly 60 years ago. Initially, the spear leaf or one of the neighbouring inner leaves loses turgor and wilts. Necrotic panels develop and a slimy secretion may be apparent. Leaves of younger plants are distorted and wilted and, at an advanced stage of infection, older leaves wilt and collapse (Plate 5.9). Internally, vascular bundles become discoloured, although this symptom is not as conspicuous as the internal discoloration observed in banana infected with *Ralstonia solanacearum*. This pale grey, yellow or pinkish discoloration extends from the rhizome to the corm and into the lower leaf bases. A bacterial slime oozes from cut vascular tissue. Pockets of yellow or cream-coloured bacterial matter may develop within the pseudostem (Plate 5.10). Nematodes, which can be found within the roots of both diseased and healthy plants, may further damage the rhizome. Total yield loss is expected once the disease takes hold.

Yirgou and Bradbury (1968) reported that the disease was caused by the bacterium *Xanthomonas musacearum* (now *Xanthomonas campestris* pv. *musacearum*). This bacterium is a motile, Gram-negative rod possessing a single polar flagellum, and has the following characteristics: oxidase- and tyrosinase-negative, does not reduce nitrate or hydrolyse starch or gelatin and does not accumulate poly-β-hydroxybutyric acid. It is non-fluorescent on King's B medium and produces typical yellow, convex, mucoid colonies on nutrient agar and other media. Experimental inoculation of solanaceous and other plants does not induce symptoms of the disease. However, isolates have been found to be pathogenic to banana and natural infections have been diagnosed in Ethiopia (M. Tessera, Ethiopia, 1999, personal communication).

There have been few studies aimed at developing control strategies for enset wilt, and the techniques employed in enset culture clearly aggravate the problem. Enset does not produce suckers naturally, so lateral bud development is induced by removal of the central growing point from a mature corm. Daughter rhizomes are excised to seed new plants and, since a single parent corm can yield 150–300 such explants, the potential for dissemination of the disease via infected material is enormous. The crop is usually grown in small gardens close to the cultivator's dwelling, and new seed pieces are planted in the area of land furthest from the home. As the plants develop they are dug up and moved progressively closer to the home, a process that may be repeated several times before the plants are harvested. Dissemination of the pathogen throughout the garden probably occurs during transplantation and pruning. Animals, such as aardvark and porcupine, which feed on the corm, have also been implicated in disease spread in gardens. Spread between gardens is through the movement of diseased planting material and other human activity.

There are too few suitable crops available to provide an alternative where enset cannot be grown, so severely affected gardens are sometimes abandoned for up to 5 years. However, the 'Genticha' cultivar found in the Sidamo area and the 'Mazae' cultivar from the North Omo area have recently been shown to have resistance to the disease, as they were found to recover after showing symptoms of wilting in the field and after artificial inoculation. It has been recommended that these cultivars be planted in gardens where the disease is prevalent.

DISEASES OF FRUIT

Bacterial Finger-tip Rot

Finger-tip rot affects fruit of dessert banana cultivars in the AAA genomic group. It is sometimes known as mokillo, on account of the similarity between the symptoms of finger-tip rot and the fruit symptoms of Moko disease. Initially, parts of the pulp within the finger appear slightly gelatinous and yellow. Often, a brown discoloration is evident at the tip of the flower end. Later, sections of the pulp can become rusty red in colour and degenerate to a gummy, sap-like constituency (Plate 5.11). However, symptoms never spread further and vascular discoloration within the floral stem, of the type associated with Moko disease, does not occur. Losses from the disease are not severe and only a few fingers from the youngest hands are affected. Diseased fruit can be distinguished from unaffected fingers because they taper towards the tip and remain straighter. As a consequence, they stick out of the line of the other fingers. Internal fruit discoloration has been attributed to invasion by bacteria, but more severe symptoms may result from subsequent infection by *Fusarium moniliforme*. Stover (1972) reported that mechanical inoculation of a *Pseudomonas* species into the pulp reproduced the symptoms of finger-tip rot within 14 days. More recently, Luna *et al.* (1988) isolated a non-fluorescent pseudomonad from bananas exhibiting similar symptoms in Mexico. Isolates produced discoloration on inoculation into the fruit, but not the pseudostem, and did not elicit a hypersensitive reaction in tobacco. The bacterium was oxidase-positive and arginine dihydrolase-negative, did not reduce nitrate and accumulated poly-β-hydroxybutyric acid.

The disease has a widespread distribution, with symptoms observed in Central America, Trinidad, Israel, Taiwan and Australia (Stover, 1972). In Australia, the disease is known as 'gumming' and is most common on November-dump fruit (see Israeli and Lahav, Chapter 10, p. 356), which emerges in November and is harvested the following February–March. About 2–3% of fingers in a bunch can be affected.

Stover (1972) also described a bacterial soft rot of green fruit in Honduras and Nicaragua. This description was largely based on an account of a severe mokillo infection, published in an earlier edition of Wardlaw's (1972) work, and may in fact refer to finger-tip rot.

The precise details of the disease cycle are not known, but insect transmission of the pathogen may be involved and the disease seems to be less common in dry weather. Distorted fingers are removed from hands during packing operations and discarded.

Bugtok

Introduction

A disease affecting fruit of 'Saba' (ABB) and 'Cardaba' (ABB) in the Philippines was first reported in the mid-1960s by Roperos (1965). This disease, which may have been present for some considerable time prior to its discovery, is known as bugtok in the southern Philippines and as tapurok in central Visayas (Zehr and Davide, 1969). It has also been described as bacterial finger rot of 'Saba' by Stover (1972). Early reports drew particular attention to the symptoms on fruit, which becomes discoloured and eventually rotten. In this respect, the disease resembles bacterial finger-tip rot, although bugtok is associated with more extensive vascular symptoms. Although the causal agent appears to be identical to that of Moko disease, bugtok is treated here separately, as the disease is distinct in terms of symptoms and host range. Bugtok is not recognized in commercially grown dessert banana cultivars, but in many areas of the southern Philippines the disease has

Plate 5.9. Enset in Ethiopia showing advanced symptoms of bacterial wilt (photo: M. Tessera and A.J. Quimio, IAR).

become a major limiting factor in the cultivation of cooking banana.

Symptoms

Symptoms of bugtok disease have been described by Soguilon *et al.* (1994, 1995) and Jeger *et al.* (1995). Infection seems to occur via the male bud, and early symptoms resemble those caused by insect-transmitted strains of Moko disease. Drops of bacterial ooze are sometimes seen at the bases of bracts, where localized blackening also occurs. Peduncles become dry and distorted, and the whole male axis may eventually dry up. Bacterial ooze from a cut peduncle and the male inflorescence of an infected plant are illustrated in Plate 5.12.

If the male flower is removed, external symptoms may be entirely absent. Internally, the fruit contains pockets of dry or gelatinous tissue, which sometimes extends to the entire pulp and may be either greyish black or yellowish red in colour (Plate 5.13). The distribution of discoloured fruit within a bunch is random. In severe cases, all fruit can be discoloured. Reddish-brown vascular discoloration extends upwards from the male axis, but is rarely visible beyond the peduncle. However, in severely diseased plants, vascular strands in the pseudostem

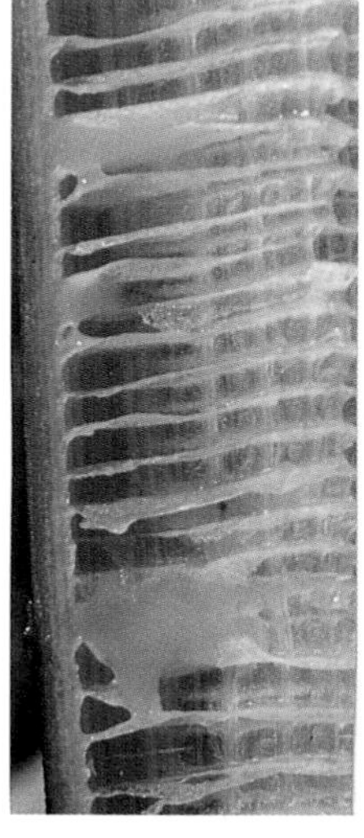

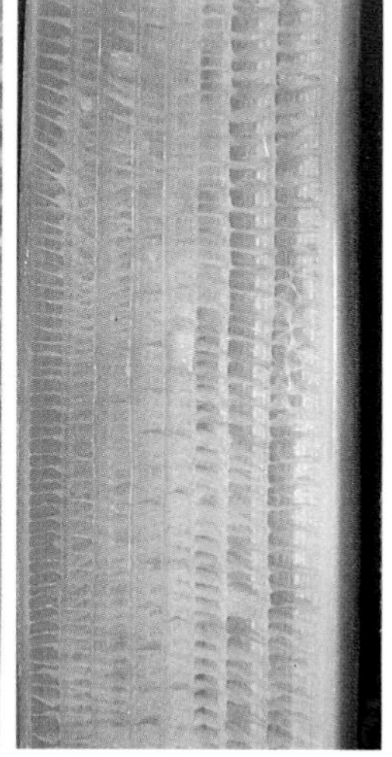

Plate 5.10. Yellow-orange bacterial matter in the air pockets of leaf sheaths of enset affected by bacterial wilt (photo: M. Tessera and A.J. Quimio, IAR).

Plate 5.11. Symptoms of gumming in fruit of 'Mons Mari' (AAA, Cavendish subgroup) in Queensland, Australia. Initial symptoms are seen in the top finger. A brown discoloration is evident at the tip of the flower end and the pulp is slightly gelatinous and yellow. More advanced sympoms are evident in the bottom finger. The pulp has become rusty red in colour and has degenerated to a gummy, sap-like constituency (photo: QDPI).

Plate 5.12. Bacterial exudation associated with bugtok disease. Cream-coloured bacterial exudate on the cut surface of a peduncle after 2 weeks' incubation (left) and an infected male inflorescence (arrow) (right). Note also the symptoms of bract mosaic on the male bracts (right) (photo: M.P. Natural, UPLB).

and corm may also be discoloured (Soguilon *et al.*, 1994). Symptoms do not progress into suckers and affected plants do not wilt.

External diagnosis of bugtok is possible in cultivars with an undetached male axis. Old bracts do not dehisce in diseased plants, but dry and remain attached to the male axis (Plate 5.14). Healthy plants have a long, clean male axis bearing the terminal male bud.

Causal agent

Zehr and Davide (1969) isolated a bacterium from diseased plants, the cultural characteristics of which resembled those of *Ralstonia* (then *Pseudomonas*) *solanacearum*. Inoculation of 'Saba' cooking-banana plants with the bacterium reproduced the characteristic fruit symptoms of bugtok, but *M. balbisiana* and tomato were apparently not affected. More recently, Soguilon (1990) was able to demonstrate pathogenicity of bugtok isolates on both tomato and triploid banana, and pathogenicity on eggplant has since been confirmed. The bacterium is, therefore, a strain of *R. solanacearum*.

Recent observations of bugtok and Moko diseases in the Philippines have led to the suggestion that the causal strains of the two diseases should be considered identical (Eden-Green and Seal, 1993). The differences in symptoms observed between Moko and bugtok diseases stem from the fact that the former occurs on dessert banana in plantations, while the latter is confined to cooking varieties, usually grown in smallholdings. Occurrences of Moko and bugtok are, therefore, separated geographically and by host genotype. DNA-fingerprinting data confirm this close relationship (Thwaites *et al.*, 1999). Genetic fingerprinting by PCR amplification, using random primers, consistently indicated a high level of genetic relatedness between bugtok and South-East Asian strains of Moko. Moreover, amplification with conserved repetitive primers could not distinguish at all between bugtok and South-East Asian strains of Moko. Similar investigations recently carried out in the Philippines on Moko and bugtok isolates support these observations (Raymundo, 1998). The close relationship between Moko and bugtok strains raises an interesting question, since reports of bugtok predate the first introduction of Moko into the Philippines, reported by Rillo (1979). There are two possible explanations; either bugtok strains derive from earlier undocumented introductions of Moko into the Philippines, or some Moko strains from Central America share a common South-East Asian ancestor with bugtok.

Disease cycle and epidemiology

The similarities between the symptoms of bugtok and those of insect-transmitted Moko suggest that the two diseases may share common dispersal and infection mechanisms. Evidence indicates that bugtok is transmitted by insects, probably thrips, which infect plants on contact with floral bract scars. The infection spreads along the upper part of the floral stem and into the fruit, but the bacteria rarely proceed further and the disease does not become systemic. When plants from bugtok areas were grown in isolation in an area where bugtok had not been reported, all fruit harvested after three cycles were healthy (Soguilon *et al.*, 1994). Dissemination of the causal agent in planting material is, therefore, not considered of great importance in the disease cycle. Bugtok, in common with insect-transmitted Moko disease, causes bacterial ooze to appear on peduncles and bract bases. This exudate probably attracts insects and serves as a reservoir for further dissemination.

Host reaction

No true resistance to bugtok has yet been identified in ABB clones, even in 'Pelipia' (syn. 'Pelipitia'), which is resistant to Moko in Central America (Soguilon *et al.*, 1995). However, cultivars that are affected may be considered tolerant of the disease, as the bacterium does not adversely affect either the survival or propagation potential of the host.

Other ABB clones reported with bugtok symptoms in the germplasm collection at the Davao National Crops Research and Development Centre in Mindanao include the cultivars 'Maduranga', 'Katsila' (syn. 'Silver Bluggoe'), 'Mundo', 'Turangkog', 'Pa-a Dalaga', 'Bigihan', 'Inabaniko' and 'Gubao' (Soguilon *et al.*, 1994). These observations indicate that susceptibility in these triploids may be linked to the possession of the 'BB' genome, although Zehr and Davide (1969) reported that *M. balbisiana* was not susceptible in its own right.

Control

Attempts to limit the spread of bugtok have centred mainly on protection of the inflorescence from infection. Effective control can be achieved by enclosing the inflorescence in an insecticide treated polyethylene bag, polypropylene bag, nylon net or cheesecloth bag on emergence (Soguilon *et al.*, 1994). Debudding is not effective, and injection of male inflorescences with insecticide to reduce thrips infestation seems to be less effective than bagging.

ROTS OF THE CORM AND PSEUDOSTEM

Bacterial Head Rot

Introduction

Bacterial head rot, sometimes referred to as rhizome rot, was first noted in Honduras in 1948 (Stover, 1959) and has since been reported elsewhere in Central America (Wardlaw, 1972) and in Israel (Volcani and Zutra, 1967), Jamaica (Shillingford, 1974) and Papua New Guinea (Tomlinson *et al.*, 1987). Head rot characteristically affects newly established plantings of diploid (AA) and some dessert (AAA) banana cultivars and also causes germination failure in recently planted corm pieces. In mature plants, it may cause the condition known as 'tip-over' or 'snap-off', where the pseudostem weakens and breaks at the top of the corm. Although the disease may be prevalent for many years in some localized areas, losses are generally greatest in younger plants. Stover (1959) reported incidences of up to 50% in newly planted 'Gros Michel' (AAA), although losses of between 10% and 15% were more normally observed. Cultivars in the AAA Cavendish subgroup were generally less affected (Stover, 1972), but up to 30% of corms planted in wet, cool weather were infected and losses higher that 90% have been reported in Guatemala. In Papua New Guinea, head rot was most prevalent on experimental plantings of AA diploid cultivars and Cavendish cultivars. In Jamaica, the disease caused 50% losses of the Cavendish cultivar 'Valery'.

A bacterial corm rot has been reported affecting enset in Ethiopia, killing both young and mature plants. Severity of the disease has been said to be high in the lowlands and mid-altitude locations up to 1800 m (Tessera and Quimio, 1994).

Symptoms

The external appearance of affected banana plants depends on the age of the plant and the severity of infection. Newly planted rhizomes fail to sprout and young plants may exhibit stunted growth and yellowing. External symptoms in mature plants are often absent, but, where decay is extensive (Plate 5.15), the rhizome is weakened. In such cases, the wind or the weight of fruit may easily fell the plant. Suckers developing from diseased rhizomes may appear healthy and bear normal fruit, although the bunches are sometimes abnormally small and deformed. Internally, the cortex contains regions of brown or yellow water-soaked material, with well-defined dark margins. In severe infections, these pockets of decay may be observed throughout the rhizome, and tissue may rot sufficiently to form cavities, bordered by spongy tissue.

Plate 5.13. Discoloration of the pulp of fruit from a bugtok affected plant (photo: M.P. Natural, UPLB).

Plate 5.14. Bugtok-affected plant with old bracts remaining attached to the male axis (left). Old bracts dehisce from healthy plants leaving a long, clean male axis (right) (photo: M.P. Natural, UPLB).

Plate 5.15. Bacterial head rot symptoms at the base of a pseudostem of a cultivar of the Cavendish subgroup (AAA) in a plantation in central Cuba. This particular problem was reported to be caused by *Erwinia chrysanthemi* and to be absent in plantations established using tissue-cultured plants (photo: D.R. Jones, INIBAP).

Plate 5.16. Wilting of the older leaves of enset caused by bacterial corm rot in Ethiopia (photo: M. Tessera and A.J. Quimio, IAR).

This symptom can easily be confused with the tunnels made by weevil borers (*Cosmopolites sordidus*). If the decay is widespread, the emerging daughter suckers may by affected. If these suckers are used as planting material, the disease can be spread to new areas.

The bacterial corm rot of enset is reported to disintegrate the tissues in the corm, forming a hollow area, which predisposes the plant to toppling by the wind. Wilting of the old leaves is said to be a distinguishing feature of the disease (Plate 5.16).

Plate 5.17. Leaf sheath of enset affected by bacterial leaf sheath rot in Ethiopia. The watery rot has been invaded by secondary infections (photo: M. Tessera and A.J. Quimio, IAR).

Causal agent

Erwinia carotovora, now known as *E. carotovora* ssp. *carotovora*, was isolated from affected banana plants by Stover (1959). The bacterium could reproduce the symptoms of rhizome rot when an agar plug containing the inoculum was placed inside the rhizome. Subsequent isolates from diseased material in Jamaica (Shillingford, 1974) and Central America (Dickey, 1979) were identified as *Erwinia chrysanthemi*. Both species of *Erwinia* were present in material from Papua New Guinea examined by Tomlinson *et al.* (1987), but only *E. chrysanthemi* caused rotting on test plants. However, the inoculation method employed in this study was not as invasive as that used by Stover (1959). The isolates are not specific to banana and, under experimental conditions, can cause rotting on potato, tomato and onion, although the natural host ranges of the organisms are not known.

The causal agent of the bacterial corm rot of enset has not been identified.

Disease cycle and epidemiology

The causal bacteria probably invade the rhizome directly after planting via wounds or decaying leaf sheathes around the central bud. It is likely that *E. chrysanthemi* persists in the soil, although the bacterium could not be isolated from soil where diseased plant material was not present (Tomlinson *et al.*, 1987). The disease is more common in wet weather, particularly if the ground is poorly drained and plants become waterlogged.

In enset, infection is believed to occur through damaged roots, the disease then spreading into the corm.

Host reaction

Information on susceptible and resistant banana cultivars is not extensive, but clones containing the B genome are rarely affected by head rot. Of cultivars in the AAA genomic group, 'Lacatan' (Cavendish subgroup) is less susceptible than 'Valery' (Cavendish subgroup) and 'Gros Michel'.

The 'Halla' enset cultivar has been described as highly susceptible to bacterial corm rot.

Control

Head rot is rarely a severe problem and control measures are only deemed necessary if the risk of infection is high. If susceptible cultivars must be planted in high-rainfall areas, then seed pieces should be carefully selected and inspected. Large rhizomes (more than 500 g in weight) are less affected, and suckers derived from axillary buds, induced by removal of the central growing point, are usually free of the disease (Loos, 1962). However, removal of the heart bud of 'Valery' serves to increase germination failure, particularly if rhizomes are planted during rainy periods (Stover, 1972).

Bacterial Soft Rot of Rhizome and Pseudostem

This disease is of relatively minor importance and therefore has not been investigated in great detail. An extensive cone-shaped rot in the centre of the pseudostem, which destroys the growing point and surrounding tissues, characterizes the symptoms. Older leaves collapse and younger leaves turn yellow and wilt. If the rot has progressed up the pseudostem, the emerging spear leaf may cease to be attached to the corm and will become flaccid and necrotic. Stover (1972) observed that the disease usually affects only large, non-fruiting plants and does not invade daughter suckers. The condition is thought to be caused by soft-rotting *Erwinia* species, although no empirical proof of this or any studies relating the agent to head-rot pathogens have been reported. The disease is of relatively low incidence and control measures are not practised. Common occurrences have, however, been reported in Costa Rica, Panama and Columbia (Stover, 1972). A similar condition has been observed on *Ensete ventricosum* (enset) in Ethiopia.

Pseudostem Wet Rot

Stover (1972) records that this disease has been responsible for severe losses of cultivars in the Plantain subgroup (AAB) in Colombia, Venezuela and Ecuador. More recently, the problem has been recorded in the Marigot and Petit Goave areas of Haiti (T. Lescot, Santo Domingo, 1999, personal communication). Dessert banana clones can also be infected experimentally, but are more resistant than plantain. The disease is most serious in young, preflowering plants. When rotting is severe, the pseudostem may be weakened to such an extent that the plant collapses, although leaves and fruit appear unaffected. A soft-rotting *Erwinia* species enters the leaf sheaths via pruning wounds and propagates throughout the pseudostem, but does not enter the rhizome. The causal bacterium differs slightly from isolates of the head-rot pathogen and was named *Erwinia chysanthemi* pv. *paradisiaca* by Dickey and Victoria (1980). However, the status of *E. chrysanthemi* pathovars has been brought into question (Dickey, 1981) and insufficient evidence is available to state whether the various banana soft-rotting diseases share a common causal agent. Stover (1972) reports that control measures consist of improving the aeration of plantations, avoiding green or yellow-green leaves when pruning and the frequent sterilization of pruning tools.

A similar disease, called bacterial leaf sheath rot, has been reported on enset plants over 3 years old. It is said to be common in the Guraghe, Welayita and Sidamo regions of Ethiopia and occurs during the dry time of the year. Yield loss has not been studied, but nothing can be consumed from diseased leaf sheaths. A watery rot starts at the base of the petiole of the outer leaf sheath and then spreads downward (Plate 5.17). The affected leaf becomes yellowish and eventually wilts and dies. The rot spreads inwards until only two or three leaves are left on the diseased plant. The causal agent is an unidentified bacterium. Diseased plants are usually harvested early before the rot penetrates deeply into the leaf sheath layers.

ACKNOWLEDGEMENTS

The authors would like to thank S. Seal for helpful and critical comments during the preparation of the manuscript. The information provided on the bacterial problems of enset by M. Tessera and A.J. Quimio is also gratefully acknowledged.

REFERENCES

Anon. (1998) *Moko disease (*Ralstonia solanacearum *race 2) is not present in Jamaica and its presence in Malawi is doubtful.* EPPO Reporting Service 98/025.

Ashby, S.F. (1926) A wilt disease of bananas. *Tropical Agriculture (Trinidad)* 3, 127–129.

Berg, L.A. (1971) Weed hosts of the SFR strain of *Pseudomonas solanacearum*, causal organism of bacterial wilt of bananas. *Phytopathology* 61, 1314–1315.

Black, R. and Delbeke, A. (1991) Moko disease (*Pseudomonas solanacearum*) of *Musa* in Belize. *Tropical Science* 31, 347–353.

Brumbley, S.M. and Denny, T.P. (1990) Cloning of wild-type *Pseudomonas solanacearum phcA*, a gene that when mutated alters expression of multiple traits that contribute to virulence. *Journal of Bacteriology* 172, 5677–5685.

Buddenhagen, I. (1960) Strains of *Pseudomonas solanacearum* in indigenous hosts in banana plantations of Costa Rica, and their relationship to bacterial wilt of banana. *Phytopathology* 50, 660–664.

Buddenhagen, I.W. (1961) Bacterial wilt of bananas: history and known distribution. *Tropical Agriculture (Trinidad)* 38, 107–121.

Buddenhagen, I.W. (1962) Bacterial wilts of certain seed-bearing *Musa* spp. caused by the tomato strain of *Pseudomonas solanacearum*. *Phytopathology* 52, 286 (abstract).

Buddenhagen, I.W. (1986) Bacterial wilt revisited. In: Persley, G.J. (ed.) *Bacterial Wilt Disease in Asia and the South Pacific, Proceedings of an International Workshop held at PCARRD, Los Baños, Philippines, 8–10 October 1985.* ACIAR Proceedings No. 13, Australian Centre for International Agricultural Research, Canberra, Australia, pp. 126–143.

Buddenhagen, I.W. (1994) Banana diseases caused by bacteria. In: Ploetz, R.C., Zentmyer, G.A., Nishijima, W.T., Rohrbach, K.G. and Ohr, H.D. (eds) *Compendium of Tropical Fruit Diseases*, APS Press, St Paul, Minnesota, pp. 15–17.

Buddenhagen, I.W. and Elsasser, T.A. (1962) An insect-spread bacterial wilt epiphytotic of Bluggoe banana. *Nature* 194, 164–165.

Buddenhagen, I.W. and Sequeira, L. (1958) Disinfectants and tool disinfection for prevention of spread of bacterial wilt of bananas. *Plant Disease Reporter* 42, 1399–1404.

Buddenhagen, I.W., Sequeira, L. and Kelman, A. (1962) Designation of races in *Pseudomonas solanacearum*. *Phytopathology* 52, 726.

Dickey, R.S. (1979) *Erwinia chrysanthemi*: a comparative study of phenotypic properties of strains from several hosts and other *Erwinia* species. *Phytopathology* 69, 324–329.

Dickey, R.S. (1981) *Erwinia chrysanthemi*: reaction of eight plant species to strains from several hosts and to strains of other *Erwinia* species. *Phytopathology* 71, 23–29.

Dickey, R.S. and Victoria, J.I. (1980) Taxonomy and emended description of strains of *Erwinia* isolated from *Musa paradisiaca* Linnaeus. *International Journal of Systematic Bacteriology* 30, 129–134.

Dowson, W.J. (1943) On the generic names *Pseudomonas*, *Xanthomonas* and *Bacterium* for certain bacterial plant pathogens. *Transactions of the British Mycological Society* 26, 4–14.

Eden-Green, S.J. and Sastraatmadja, H. (1990) Blood disease of banana present in Java. *FAO Plant Protection Bulletin* 38, 49–50.

Eden-Green, S.J. and Seal, S.E. (1993) Bacterial diseases of banana and plantain in Southeast Asia. In: Ganry, J. (ed.) *Breeding Banana and Plantain for Resistance to Diseases and Pests, Proceedings of the International Symposium on Genetic Improvement of Bananas for Resistance to Diseases and Pests, organized by CIRAD–FLHOR, Montpellier, France, 7–9 September 1992.* CIRAD, Montpellier, France, pp. 115–121.

Eden-Green, S.J., Supriadi and Sri Yuni Hartati (1988) Characteristics of *Pseudomonas celebensis*, the cause of blood disease of bananas in Indonesia. In: *Fifth International Congress on Plant Pathology, Kyoto, August 1988*, p. 389 (abstract).

Engelbrecht, M.C. (1994) Modification of a semi-selective medium for the isolation and quantification of *Pseudomonas solanacearum*. *ACIAR Bacterial Wilt Newsletter* 10, 3.

French, E.R. (1986) Interaction between strains of *Pseudomonas solanacearum*, its hosts and the environment. In: Persley, G.J. (ed.) *Bacterial Wilt Diseases in Asia and the South Pacific, Proceedings of an International Workshop held at PCARRD, Los Baños, Philippines, 8–10 October 1985.* ACIAR Proceedings No. 13, Australian Centre for International Agricultural Research, Canberra, Australia, pp. 99–104.

French, E.R. and Sequeira, L. (1970) Strains of *Pseudomonas solanacearum* from Central and South America: a comparative study. *Phytopathology* 60, 506–512.

Gäumann, E. (1921a) *Over een bacterieele vaatbundelziekte der bananen in Nederlandisch-Indië* (On a vascular bacterial disease of the banana in the Dutch East Indies). Mededeelingen van het Instituut voor Plantenziekten No. 48.

Gäumann, E. (1921b) *Onderzoekingen over de bloedziekte der bananen op Celebes* I (Investigations on the blood-disease of bananas in Celebes I). Mededeelingen van het Instituut voor Plantenziekten No. 50.

Gäumann, E. (1923) *Onderzoekingen over de bloedziekte der bananen op Celebes II* (Investigations on the blood-disease of bananas in Celebes II). Mededeelingen van het Instituut voor Plantenziekten No. 59.

Gnanamanickam, S.S., Lokeswari, T.S. and Nandini, K.R. (1979) Bacterial wilt of banana in southern India. *Plant Disease Reporter* 63, 525–528.

Granada, G.A. and Sequeira, L. (1983) A new selective medium for *Pseudomonas solanacearum*. *Plant Disease* 67, 1084–1087.

Hayward, A.C. (1964) Characteristics of *Pseudomonas solanacearum*. *Journal of Applied Bacteriology* 27, 265–277.

He, L.Y., Sequeira, L. and Kelman, A. (1983) Characteristics of strains of *Pseudomonas solanacearum* from China. *Plant Disease* 67, 1357–1361.

Hunt, P. (1987) Current strategies for Moko control in Grenada: technical and logistical constraints.

In: *Seminar Proceedings, Improving Citrus and Banana Production in the Caribbean Through Phytosanitation, St Lucia, W.I., 2–5 September 1986.* CTA/CARDI, Wageningen, pp. 121–129.

Jeger, M.J., Eden-Green, S., Thresh, J.M., Johanson, A., Waller, J.M. and Brown, A.E. (1995) Banana diseases. In: Gowen, S. (ed.) *Bananas and Plantains.* Chapman and Hall, London, pp. 317–381.

Kelman, A. (1953) *The bacterial wilt caused by* Pseudomonas solanacearum. Technical Bulletin No. 99, North Carolina Agricultural Experiment Station, Raleigh, 194 pp.

Kelman, A. (1954) The relationship of pathogenicity in *Pseudomonas solanacearum* to colony appearance on a tetrazolium medium. *Phytopathology* 44, 693–695.

Kelman, A. and Hruschka, J. (1973) The role of motility and aerotaxis in the selective increase of avirulent bacteria in still broth cultures of *Pseudomonas solanacearum. Journal of General Microbiology* 76, 177–188.

Kelman, A. and Sequeira, L. (1965) Root-to-root spread of *Pseudomonas solanacearum. Phytopathology* 55, 304–309.

Lehmann-Danziger, H. (1987) The distribution of Moko disease in Central and South America and its control on plantains and bananas. In: *Seminar Proceedings, Improving Citrus and Banana Production in the Caribbean Through Phytosanitation. St Lucia, W.I., 2–5 September 1986.* CTA/CARDI, Wageningen, pp. 130–152 and pp. 153–155.

Loos, C.A. (1962) Factors affecting incidence of rhizome-rot caused by *Erwinia carotovora* of Gros Michel and Lacatan bananas. *Phytopathology* 52, 110–114.

Lozano, J.C. and Sequeira, L. (1970) Differentiation of races of *Pseudomonas solanacearum* by a leaf infiltration technique. *Phytopathology* 60, 833–838.

Lozano, J.C., Thurston, H.D. and Galvez, G.E. (1968) Control del 'moko' del plátano y banano causado por la bacteria *Pseudomonas solanacearum. Agricultura Tropicale* 25, 315–324.

Luna, L., Fucikovsky, L., Cardenas, M., Orozco, M. and Lopez, L. (1988) Pudricion seca del fruto del platano en Tecoman, Colima. In: *Memorias XV Congreso Nacional de Fitopatologia, Agosto, 1988, Xalapa, Veracruz.* p. 57.

Palleroni, N.J. (1984) Pseudomonadaceae. In: Krieg, N.R. and Holt, J.G. (eds) *Bergey's Manual of Systematic Bacteriology* 1. Williams and Wilkins, Baltimore, pp. 141–219.

Palo, M.A. and Calinisan, M.R. (1939) The bacterial wilt of the abacá (manila hemp) plant in Davao: I. Nature of the disease and pathogenicity tests. *Philippine Journal of Agriculture* 10, 373–395.

Phelps, R.H. (1987) The status of Moko disease in the Caribbean. In: *Seminar Proceedings, Improving Citrus and Banana Production in the Caribbean Through Phytosanitation, St Lucia, W.I., 2–5 September 1986.* CTA/CARDI, Wageningen, pp. 100–107.

Rangaswami, G. and Rangarajan, M. (1965) A bacterial leaf spot disease of banana. *Phytopathology* 55, 1035–1036.

Raymundo, A.K. (1998) Analysis of genetic variation of a population of banana infecting strains of *Ralstonia solanacearum.* In: Prior, P., Allen, C. and Elphinstone, J. (eds) *Bacterial Wilt Disease: Molecular and Ecological Aspects.* INRA, Paris/Springer, Heidelberg, pp. 56–60.

Rijks, A.B. (1916) *Rapport over een onderzoek naar de pisangsterfte op de saleiereilanden.* Mededeelingen van het Laboratorium voor Plantenziekten No. 21, 19 pp.

Rillo, A.R. (1979) Bacterial wilt of banana in the Philippines. *FAO Plant Protection Bulletin* 27, 105–108.

Rillo, A.R. (1981) Differences of *Pseudomonas solanacearum* EFS isolates in abaca and banana. *Philippine Agriculturist* 64, 329–334.

Roberts, S.J., Eden-Green, S.J., Jones, P. and Ambler, D.J. (1990) *Pseudomonas syzygii,* sp. nov., the cause of Sumatra disease of cloves. *Systematic and Applied Microbiology* 13, 34–43.

Robinson, A. (1993) Serological detection of *Pseudomonas solanacearum* by ELISA. In: Hartman, G.L. and Hayward, A.C. (eds) *Bacterial Wilt, Proceedings of an International Conference held at Kaohsiung, Taiwan, 28–31 October 1992.* ACIAR Proceedings No. 45, Canberra, pp. 54–61.

Roperos, N.I. (1965) Note on the occurrence of a new disease of cooking banana in the Philippines. *Coffee and Cacao Journal* 8, 135–136.

Rorer, J.B. (1911) A bacterial disease of bananas and plantains. *Phytopathology* 1, 45–49.

Seal, S.E., Jackson, L.A. and Daniels, M.J. (1992) Use of tRNA consensus primers to indicate subgroups of *Pseudomonas solanacearum* by polymerase chain reaction amplification. *Applied and Environmental Microbiology* 58, 3759–3761.

Seal, S.E., Jackson, L.A., Young, J.P.W. and Daniels, M.J. (1993) Differentiation of *Pseudomonas solanacearum, Pseudomonas syzygii, Pseudomonas pickettii* and the blood disease bacterium by

partial 16S rRNA sequencing: construction of oligonucleotide primers for sensitive detection by polymerase chain reaction. *Journal of General Microbiology* 139, 1587–1594.
Sequeira, L. (1958) Bacterial wilt of bananas: dissemination of the pathogen and control of the disease. *Phytopathology* 48, 64–69.
Sequeira, L. (1962) Control of bacterial wilt of bananas by crop rotation and fallowing. *Tropical Agriculture (Trinidad)* 39, 211–217.
Sequeira, L. and Averre, C.W. (1961) Distribution and pathogenicity of strains of *Pseudomonas solanacearum* from virgin soils in Costa Rica. *Plant Disease Reporter* 45, 435–440.
Shillingford, C.A. (1974) Bacterial rhizome rot of banana in Jamaica. *Plant Disease Reporter* 58, 214–218.
Sivamani, E. and Gnanamanickam, S.S. (1987) Occurrence of Moko wilt in 'Poovan' banana in Pudukottai district of Tamil Nadu. *Indian Phytopathology* 40, 233–236.
Smith, I.M., McNamara, D.G., Scott, P.R. and Holderness, M. (eds) (1997) *Quarantine Pests for Europe*, 2nd edn. EPPO/CABI, Wallingford, 1440 pp.
Soguilon, C.E. (1990) Survey, etiology and control of 'Bugtok' disease of cooking bananas. MSc thesis, University of the Philippines at Los Baños, 65 pp.
Soguilon, C.E., Magnaye, L.V. and Natural, M.P. (1994) Bugtok disease of cooking banana in the Philippines. *Infomusa* 3(2), 21–22.
Soguilon, C.E., Magnaye, L.V. and Natural, M.P. (1995) *Bugtok Disease of Banana. Musa* Disease Fact Sheet 6. INIBAP, Montpellier, France.
Stover, R.H. (1959) Bacterial rhizome rot of bananas. *Phytopathology* 49, 290–292.
Stover, R.H. (1972) *Banana, Plantain and Abaca Diseases*. Commonwealth Mycological Institute, Kew, 316 pp.
Stover, R.H. (1993) The insect-transmitted SFR strain of *Pseudomonas solanacearum* destroys East African AAA cultivars in Honduras. *Infomusa* 2(1), 7.
Stover, R.H and Richardson, D.L. (1968) Pelipita, an ABB Bluggoe-type plantain resistant to bacterial and fusarial wilts. *Plant Disease Reporter* 52, 901–903.
Stover, R.H. and Simmonds, N.W. (1987) *Bananas*, 3rd edn. Longmans, Harlow, 468 pp.
Taghavi, M., Hayward, C., Sly, L.I. and Fegan, M. (1996) Analysis of the phylogenetic relationships of strains of *Burkholderia solanacearum*, *Pseudomonas syzygii*, and the blood disease bacterium of banana based on 16S rRNA gene sequences. *International Journal of Systematic Bacteriology* 46, 10–15.
Tessera, M. and Quimio, A.J. (1994) Research on ensat pathology. In: Herath, E. and Desalegn, L. (eds) *The Proceedings of the 2nd National Horticultural Workshop of Ethiopia*. Institute of Agricultural Research, Addis Ababa, pp. 217–225.
Thwaites, R., Mansfield, J., Eden-Green, S. and Seal, S. (1999) RAPD and rep PCR-based fingerprinting of vascular bacterial pathogens of *Musa* spp. *Plant Pathology* 48, 121–128.
Tomlinson, D.L., King, G.A. and Ovia, A. (1987) Bacterial corm and rhizome rot of banana (*Musa* spp.) in Papua New Guinea caused by *Erwinia chrysanthemi*. *Tropical Pest Management* 33, 196–199.
Vakili, N.G. (1965) Inheritance of resistance in *Musa acuminata* to bacterial wilt caused by tomato race of *Pseudomonas solanacearum*. *Phytopathology* 55, 1206–1209.
Vakili, N.G. and Baldwin, C.H. (1966) Insect dissemination of the tomato race of *Pseudomonas solanacearum*, the cause of bacterial wilt of certain *Musa* species. *Phytopathology* 66, 355–356.
Volcani, Z. and Zutra, D. (1967) Bacterial soft rot: *Erwinia carotovora* in Israel. *Review of Applied Mycology* 46, 288.
Wardlaw, C.W. (1935) *Diseases of the Banana and of the Manila Hemp Plant*. Macmillan, London, 615 pp.
Wardlaw, C.W. (1972) *Banana Diseases*, 2nd edn. Longman, London, 878 pp.
Woods, A.C. (1984) Moko disease: atypical symptoms induced by afluidal variants of *Pseudomonas solanacearum* in banana plants. *Phytopathology* 74, 972–976.
Yabuuchi, E., Kosako, Y., Yano, I., Hotta, H. and Nishiuchi, Y. (1995) Transfer of two *Burkholderia* and an *Alcaligenes* species to *Ralstonia* Gen. Nov.: proposal of *Ralstonia pickettii* (Ralston, Palleroni and Doudoroff 1973) Comb. Nov., *Ralstonia solanacearum* (Smith 1896) Comb. Nov. and *Ralstonia eutropha* (Davis 1969) Comb. Nov. *Microbiology and Immunology* 39, 897–904.
Yirgou, D. and Bradbury, J.F. (1968) Bacterial wilt of enset (*Ensete ventricosum*) incited by *Xanthomonas musacearum* sp. n. *Phytopathology* 58, 111–112.

Young, J.M., Dye, D.W., Bradbury, J.F., Panagopoulos, C.G. and Robbs, C.F. (1978) A proposed nomenclature and classification for plant pathogenic bacteria. *New Zealand Journal of Agricultural Research* 21, 153–177.

Zehr, E.I. (1970) Isolation of *Pseudomonas solanacearum* from abaca and banana in the Philippines. *Plant Disease Reporter* 54, 516–520.

Zehr, E.I. and Davide, R.G. (1969) An investigation of the cause of the 'Tapurok' disease of cooking banana in Negros Oriental. *Philippine Phytopathology* 5, 1–5.

6

Diseases Caused by Viruses

Bunchy Top

J.E. Thomas and M.L. Iskra-Caruana

Introduction

Of the virus diseases affecting banana worldwide, bunchy top is by far the most serious and can have a devastating effect on crops (Dale, 1987). The disease also affects abacá and is one of the major constraints to production of this crop in the Philippines (Anunciado *et al.*, 1977). The centre of origin of the genus *Musa* is the South and South-East Asian–Australasian region (Simmonds, 1962) and it is probable that bunchy top also originated somewhere in this broad geographical area.

History and losses

The first recorded outbreaks of bunchy top were on banana in Fiji in 1889, though reports and photographs taken around that time leave little doubt that the disease was present there as early as 1879 (Darnell-Smith, 1924; Magee, 1953). Interest in the disease may have only arisen due to its effects on the new export industry, which commenced production in 1877 and was based on cultivars in the Cavendish subgroup (AAA). Although the outbreaks were first observed in commercial plantations of Cavendish cultivars, the pattern of spread of the disease and the severe symptoms in local cooking-banana cultivars led Magee (1953) to consider it unlikely that bunchy top originated in Fiji. Cavendish cultivars were not traditionally grown in Fiji and must have been imported at some stage, thus providing the opportunity to introduce the disease. The industry reached a production peak of 788,000 bunches in 1892, but by 1895 had declined to 147,000 bunches, mainly because of bunchy top. Temporary recoveries in the industry occurred around 1896, possibly due to the planting of 'Gros Michel' (AAA), and from 1912 to 1916, to a large extent due to the opening up of new plantations. Surprisingly, a survey in 1937 showed that, although bunchy top incidence was high, with 5–30% of plants in all plantations and gardens affected, no plantings had an incidence of 100% (Magee, 1953).

Other early records of the disease include Egypt in 1901 (Fahmy, 1924, in Magee, 1927), the origin of infection unknown, and Australia and Sri Lanka in 1913, both outbreaks probably resulting from the importation of infected planting material from Fiji (Magee, 1953).

Bunchy top was first reported in Australia in the Tweed District of New South Wales (NSW) (Magee, 1927). Rapid expansion of the banana industry took place over the next decade and, with it, the inadvertent spread of the disease. By 1927, bunchy top was present along a 300 km stretch of the east coast of Australia from Grafton in central NSW to Yandina in southern Queensland. An isolated outbreak at Innisfail, a further 1300 km to the north, was detected in 1926 but subsequently eradicated. Production in the expanding Australian banana industry peaked in 1922, but 3 years later had collapsed. In the Tweed and Brunswick districts of NSW, every plantation was affected with bunchy top by 1925, with a disease incidence of 5–90%. In addition, the total area of production had been decreased by 90% (Magee, 1927). A similar situation existed in southern Queensland, where, for example, production in the Currumbin District was reduced by over 95%. C.J.P. Magee noted at the time: 'It would be difficult for anyone who has not visited these devastated areas to visualize the completeness of the destruction wrought in such a short time by a plant disease' (Magee, 1927).

In more recent times, a severe outbreak of banana bunchy top occurred in Pakistan (Kahlid and Soomro, 1993). In the early 1990s, land under production fell by 55% in 1 year as a direct result of the disease. In some plantations, the incidence of bunchy top was observed to be 100% (Anon., 1992). The disease has also appeared in Hawaii, causing much damage (Ferreira *et al.*, 1989).

Bunchy top of abacá was first noticed in the Philippines in 1915 and was causing serious economic damage by 1923 (Wardlaw, 1961). Many production areas were replanted with other crops.

Early investigations

Initial attempts to attribute a cause to bunchy top implicated a wide variety of agents, including fungi (Knowles and Jepson, 1912, in Magee, 1927), nematodes (Nowell, 1925 and Fahmy, 1924, cited in Magee, 1927) and bacteria (Darnell-Smith, 1924). Soil and nutritional factors were also considered (Darnell-Smith and Tryon, 1923; Darnell-Smith, 1924), though none of these investigations was conclusive.

Although some studies had suggested an association between the banana aphid and the disease (Darnell-Smith, 1924), it was not until the ground-breaking work of Magee (1927) that the viral nature of the disease was established. In the space of 3 years, he established that the causal agent was a virus transmitted by the banana aphid and in infected planting material. He also proposed management strategies that still form the basis of Australia's control programmes today.

Geographical distribution

Bunchy top of banana occurs in many, though not all, countries in the South and South-East Asia/Pacific region and in various African countries. Significantly, the banana-exporting countries of the Latin American–Caribbean region are free from the disease, though the aphid vector is present.

The countries for which authenticated records of banana bunchy top exist are shown in Table 6.1, together with records of the detection of banana bunchy top virus (BBTV). In addition to the Philippines (Ocfemia, 1926), bunchy top of abacá has also been observed in Sri Lanka, and this is likely to be a true record, as bunchy top of banana is reliably recorded in this country (Magee, 1953). However, reports from East Malaysia (Sabah), West Malaysia and Papua New Guinea (Magee, 1953; Wardlaw, 1961) need to be authenticated, as they were associated with atypical symptoms, they were not confirmed by aphid transmission tests and bunchy top of banana was not present.

Symptoms

The typical symptoms of bunchy top of banana are very distinctive, readily distinguished from those caused by other viruses of banana, and have been described in detail by Magee (1927). Plants can become infected at any stage of growth and there are some initial differences between the symptoms produced in aphid-infected plants and those grown from infected planting material.

In aphid-inoculated plants, symptoms usually appear in the second leaf to emerge after inoculation and consist of a few dark

Table 6.1. Countries where banana bunchy top disease has been recorded.

Country	First reference	BBTV detected*
Pacific		
Australia	Magee, 1927	+
Fiji	Magee, 1927	+
Guam	Beaver, 1982	
Hawaii (USA)	Dietzgen and Thomas, 1991	+
Kiribati (formerly Gilbert Is.)	Shanmuganathan, 1980	
Samoa (American)	Magee, 1967	
Samoa (Western)	Magee, 1967	+
Tonga	Magee, 1967	+
Tuvalu (formerly Ellice Is.)	Campbell, 1926	
Wallis Island	Simmonds, 1933	
Asia		
China	Diekmann and Putter, 1996	+
India	Magee, 1953	+
Indonesia	Sulyo *et al.*, 1978	+
Japan: Bonin Islands	Gadd, 1926	
Japan: Okinawa	Kawano and Su, 1993	+
Malaysia (Sarawak)	H.J. Su, personal communication	+
Pakistan	Kahlid *et al.*, 1993	+
Philippines	Castillo and Martinez, 1961	+
Sri Lanka	Bryce, 1921	+
Taiwan	Sun, 1961	+
Vietnam	Vakili, 1969	+
Africa		
Burundi	Sebasigari and Stover, 1988	+
Central African Republic	Diekmann and Putter, 1996	
Congo	Wardlaw, 1961	+
Democratic Republic of Congo (formerly Zaire)	Manser, 1982	
Egypt	Magee, 1927	+
Gabon	Manser, 1982	+
Malawi	Kenyon *et al.*, 1997	
Rwanda	Sebasigari and Stover, 1988	+

*Banana bunchy top virus detected using enzyme-linked immunosorbent assay (ELISA) or polymerase chain reaction (PCR) tests.

green streaks or dots on the minor veins on the lower portion of the lamina. The streaks form 'hooks' as they enter the midrib and are best seen from the underside of the leaf in transmitted light (Plate 6.1). The 'dot–dash' symptoms can sometimes also be seen on the petiole (Plate 6.2). The following leaf may display whitish streaks along the secondary veins when it is still rolled. These streaks become dark green as the leaf unfurls. Successive leaves become smaller, both in length and in width of the lamina, and often have chlorotic, upturned margins. The leaves become harsh and brittle and stand more erectly than normal, giving the plant a rosetted and 'bunchy top' appearance (Plate 6.3).

Suckers from an infected stool can show severe symptoms in the first leaf to emerge. The leaves are rosetted and small, with very chlorotic margins, which tend to turn

Plate 6.1. Dark green streak ('dot and dash') symptom of bunchy top in the minor leaf veins of a cultivar in the Cavendish subgroup (AAA). Note 'hooking' as streaks enter the midrib (photo: J.E. Thomas, QDPI).

Plate 6.2. Dark green streaks in the petiole of a cultivar in the Cavendish subgroup (AAA) affected by bunchy top (photo: J.E. Thomas, QDPI).

necrotic (Plate 6.4). Dark green streaks are usually evident in the leaves.

Infected plants rarely produce a bunch after infection and do not fruit in subsequent years. Plants infected late in the growing cycle may fruit once, but the bunch stalk and the fruit will be small and distorted. On plants infected very late, the only symptoms present may be a few dark green streaks on the tips of the flower bracts (Plate 6.5) (Thomas *et al.*, 1994a).

Mild strains of BBTV, which produce only limited vein clearing and dark green flecks, and symptomless strains have been reported in Cavendish plants from Taiwan (Su *et al.*, 1993). Mild disease symptoms are expressed in some banana cultivars and *Musa* species. The dark green leaf and peti-

Plate 6.3. Progressive shortening and bunching of the leaves of a cultivar in the Cavendish subgroup (AAA) affected by bunchy top (photo: J.E. Thomas, QDPI).

ole streaks, so diagnostic and characteristic of infection of cultivars in the Cavendish subgroup, can be rare or absent (Magee, 1953). Some plants of 'Veimama' (AAA, Cavendish subgroup), after initial severe symptoms, have been observed to recover and to display few if any symptoms. These reports and findings are reported in more detail under 'Causal agent' (Magee, 1948).

The symptoms of abacá bunchy top include a reduction in leaf size and lamina area, rosetting of leaves, upcurling and yellowing of leaf margins and stunting of the pseudostem. Chlorotic areas on the leaves sometimes collapse, becoming necrotic, and 'heart rot' often occurs in the pseudostem. No fruit is produced and infected plants usually die within 1 or 2 years (Magee, 1953; Wardlaw, 1961). Dark green 'dots and dashes' on the minor veins, midribs and petioles occur in only a few abacá cultivars. Magee (1953) concluded that in abacá 'the green streak symptom occurs so rarely as to lose nearly all its value as an aid in diagnosis'. Symptoms described for abacá bunchy top in the field are very close to those that develop on abacá inoculated with the bunchy top virus from banana. When enset is experimentally infected with BBTV, symptoms are similar to those described for abacá (Magee, 1953).

Plate 6.4. Symptoms of bunchy top in a sucker of 'Basrai' (AAA, Cavendish subgroup) in Pakistan. Note the yellowing margins to the leaves (photo: D.R. Jones, INIBAP).

Causal agent

Evidence for a viral agent

BBTV is presumed to be the causal agent of bunchy top of banana, though unequivocal evidence, by reproduction of the disease through inoculation of purified virions or cloned genomic components, is lacking. The virions are intimately associated with the disease (Harding *et al.*, 1991; Thomas and Dietzgen, 1991) and have been detected in all symptomatic plants tested (Dietzgen and Thomas, 1991; Thomas, 1991; Thomas and Dietzgen, 1991; Karan *et al.*, 1994). Dale *et al.* (1986) isolated disease-specific double-stranded RNA (dsRNA), suggestive of luteovirus infection, from Cavendish cultivars and Iskra-Caruana isolated similar dsRNAs from five of six bunchy-top samples (M.L. Iskra-Caruana, Montpellier, 1997, unpublished results). However, neither these nor any subsequent studies have identified or established a clear role for any virus other than the single-stranded DNA (ssDNA) BBTV in banana bunchy-top disease.

Particle and genome properties

The virions of BBTV are icosahedra, *c.* 18–20 nm in diameter (Plate 6.6), have a coat protein of *c.* 20,000 M_r, a sedimentation coefficient of *c.* 46*S* and a buoyant density of 1.29–1.30 g cm^{-3} in caesium sulphate (Wu and Su, 1990c; Dietzgen and Thomas, 1991; Harding *et al.*, 1991; Thomas and Dietzgen, 1991). Purified preparations have an $A_{260/280}$ of 1.33 (Thomas and Dietzgen, 1991). The virus possesses a multi-component genome, consisting of at least six circular ssDNA components, each *c.* 1000–1100 nucleotides long (Wu *et al.*, 1994; Yeh *et al.*, 1994; Burns *et al.*, 1995; Xie and Hu, 1995; Karan *et al.*, 1997). Component 1 encodes two proteins and components 2–6 each encode one protein (Burns *et al.*, 1995; Dale, 1996; Beetham *et al.*, 1997). Two areas of the non-coding regions are highly conserved between the six components (Burns *et al.*, 1995). The first is a stem-loop common region of up to 69 nucleotides. It contains a nonanucleotide loop sequence conserved among ssDNA plant viruses and which may be involved in rolling circle replication and initiation of viral-strand DNA synthesis. The second, 5′ to the stem-loop common region, is a major common region, varying in size between components from 65 to 92 nucleotides and which may have a promoter function. The initiation factor for endogenous DNA primers is also located within the major common region (Hafner *et al.*, 1997a). Component 1 encodes a putative replication initiation protein and contains a second functional open reading frame (ORF) internal to this, while component 3 codes for the coat protein (Harding *et al.*, 1993; Dale, 1996; Hafner *et al.*, 1997b; Wanitchakorn *et al.*, 1997). The function of the other ORFs is not known.

Strains of BBTV

Most isolates of BBTV are associated with typical severe disease symptoms. However, as mentioned earlier, mild and symptomless isolates have been reported from Taiwan (Su *et al.*, 1993) and may occur more widely, though apparently not in Australia (J.E. Thomas and A.D.W. Geering, Brisbane, 1998, unpublished results). BBTV has been confirmed in specimens of mild and symptomless infections from Taiwan by both enzyme-linked immunosorbent assay (ELISA) and polymerase chain reaction (PCR) (H.J. Su, J.L. Dale and J.E. Thomas, Brisbane, 1996, unpublished) and the isolates can be transmitted by *Pentalonia nigronervosa* (H.J. Su, Taipei, 1996, personal communication). Genomic differences, which correlate with these biological variants, have not yet been determined.

Two broad groups of isolates have been identified, based on nucleotide sequence differences between some, possibly all, of the six recognized genome components (Karan *et al.*, 1994; J.L. Dale, Brisbane, 1996, personal communication). The 'South Pacific' group comprises isolates from Australia, Fiji, Western Samoa, Tonga, Burundi, Egypt and India, while the

'Asian' group comprises isolates from the Philippines, Taiwan and Vietnam. These differences are present throughout the genomes of components 1 and 6, but are most striking in the untranslated major common region. No biological differences have been associated with these sequence differences.

As reported in 'Symptoms', Magee (1948) noted that certain plants of 'Veimama', a cultivar originally from Fiji and growing then in NSW, showed a 'partial recovery' from bunchy-top symptoms and produced bunches. After an initial flush of typical severe symptoms in three or four leaves, subsequent leaves showed few, if any, dark green flecks. Suckers derived from these partially recovered plants also displayed a flush of typical symptoms followed by partial recovery. The origin of the infection, whether from Australia or Fiji, was uncertain. This partial recovery was noted for some infected plants of 'Veimama' only, and in Fiji was noted for one sucker only on a single infected stool from among hundreds of infected stools of 'Veimama' observed. Magee was not able to transmit the virus from partially recovered plants and was only able to superinfect them, with difficulty, with high inoculum pressure. This may be an example of a mild strain of BBTV, possibly a non-aphid-transmitted one, propagated vegetatively, reaching only a low titre and conferring a degree of cross-protection. Alternatively, 'Veimama' may not be uniform and individual plants with a degree of resistance may exist. The complete explanation for this phenomenon is unclear.

The inability to transmit bunchy top from abacá to banana (Ocfemia and Buhay, 1934) was originally considered evidence that two distinct strains of the virus existed. However, as noted by Magee (1953), technical deficiencies in these experiments mean that the results must be viewed with caution. As first voiced by Magee (1953), the many similarities between the bunchy-top diseases of banana and abacá, including transmission properties with the vector *P. nigronervosa*, suggests that both are caused by the same virus. ELISA has recently detected BBTV in a sample of abacá bunchy top from the Philippines (J.E. Thomas and N. Bajet, Brisbane and Los Baños, 1993, unpublished).

Detection of BBTV

Prior to 1990, the only assays available for banana bunchy top were visual assessment of symptoms and aphid transmission to a sensitive banana cultivar. Subsequently, both serological and nucleic acid-based assays have become available.

Polyclonal and monoclonal antibodies are now routinely used in ELISA to detect BBTV in field and tissue-culture plants and can detect the virus in single viruliferous aphids (Wu and Su, 1990b; Dietzgen and Thomas, 1991; Thomas and Dietzgen, 1991; Thomas *et al.*, 1995). Triple antibody sandwich-ELISA was found to be the optimum assay format for routine virus indexing (Geering and Thomas, 1996). All isolates tested from Africa, Australia, Asia and the Pacific region are serologically related (Thomas, 1991).

A variety of nucleic acid-based assays have been applied to the detection of BBTV in plant tissue and viruliferous aphids, including DNA and RNA probes, labelled either non-radioactively or with ^{32}P (Hafner *et al.*, 1995; Xie and Hu, 1995). PCR has proved to be about 1000 times more sensitive than ELISA or dot blots with DNA probes (Xie and Hu, 1995). Substances in banana sap inhibitory to PCR can be circumvented by simple extraction procedures (Thomson and Dietzgen, 1995) or by immunocapture PCR (IC-PCR) (Anceau, 1996; M. Sharman, J.E. Thomas and R.G. Dietzgen, Brisbane, 1997, unpublished).

Disease cycle and epidemiology

BBTV is transmitted by an aphid vector (*P. nigronervosa*) and in vegetative planting material, but not by mechanical inoculation (Magee, 1927).

Plate 6.5. Dark green streaks on the tips of the male flower bracts from a banana plant infected with banana bunchy top virus (photo: J.E. Thomas, QDPI).

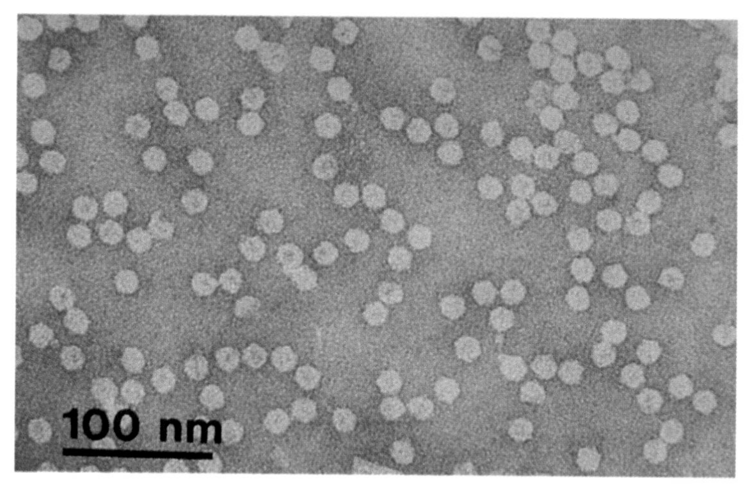

Plate 6.6. Virions of banana bunchy top virus contrasted with ammonium molybdate. Bar represents 100 nm (photo: J.E. Thomas, QDPI).

Distribution and movement within the plant

Magee (1927) showed that banana bunchy top was systemic. Following aphid inoculation, symptoms generally do not appear until a further two or more leaves have been produced (Magee, 1927). This period can vary between 19 days, in summer, and 125 days, in winter (Allen, 1978a). The virus can only be recovered by aphids from the first symptom leaf or those formed subsequently (Magee, 1940b). Suckers produced on an infected stool generally develop symptoms before reaching maturity (Magee, 1927).

Magee (1939) also concluded that the virus was restricted to the phloem tissue. Microscopic examination revealed hypertrophy and hyperplasia of the phloem tissue and a reduction in the development of the fibrous sclerenchyma sheaths surrounding the vascular bundles. The cells surrounding the phloem contained abnormally large numbers of chloroplasts giving rise to the macroscopic dark green streak symptom.

Subsequent investigation, using RNA probes and PCR (Hafner *et al.*, 1995), has demonstrated that BBTV replicates for a short period at the site of aphid inoculation, and then moves down the pseudostem to the basal meristem and finally to the corm, roots and newly formed leaves. Trace levels of virus were eventually detected by PCR in leaves formed prior to inoculation, but replication was not demonstrated. This latter observation is consistent with the inability to transmit the virus by aphids from such leaves (Magee, 1940b) and with the sequential development of single, new leaves from the basal meristem.

BBTV has been detected by ELISA and/or PCR in most parts of the plant, including leaf lamina and midrib, pseudostem, corm, meristematic tissues, roots, fruit stalk and fruit rind (Thomas, 1991; Wu and Su, 1992; Hafner *et al.*, 1995; A.D.W. Geering and J.E. Thomas, Brisbane, 1996, unpublished).

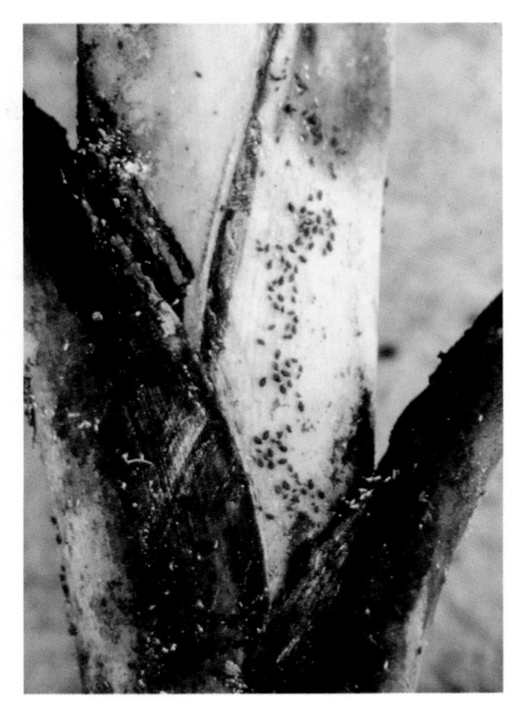

Plate 6.7. Black banana aphids (*Pentalonia nigronervosa*) on the basal portion of abacá leaf sheaths (photo: E.O. Lomerio, FIDA).

Aphid transmission

In Australia, the black banana aphid (*P. nigronervosa*) had been under suspicion as having a role in the aetiology of banana bunchy top (Darnell-Smith, 1924) and in 1925 was conclusively demonstrated to be the vector (Magee, 1927). Banana aphids have a worldwide distribution, with a host range that includes *Musa textilis* (Plate 6.7) and other species in the *Musaceae*. Species in several closely related plant families, including the *Araceae* (*Alocasia* sp., *Calladium* spp., *Dieffenbachia* spp., *Xanthosma* sp.), *Cannaceae* (*Canna* spp.), *Heliconiaceae* (*Heliconia* spp.), *Strelitzeaceae* (*Strelitzia* spp.) and *Zingiberaceae* (*Alpinia* spp., *Costus* sp., *Hedychium* spp.), are also colonized (Wardlaw, 1961; R.N. Allen, Brisbane, 1996, personal communication). However, a degree of host preference is displayed and some difficulty can be experienced transferring them between host species.

On banana plants in NSW, aphids are found at the base of the pseudostem at soil level and for several centimetres below the soil surface, beneath the outer leaf sheaths and on newly emerging suckers. Aphid numbers decrease during periods of drought (Wardlaw, 1961).

Transmission of BBTV, the probable causal agent, is of the circulative, non-propagative type. The transmission parameters reported from Hawaii (Hu *et al.*, 1996) and Australia (Magee, 1927), respectively, are: minimum acquisition access period 4 h/17 h; minimum inoculation access period 15 min/30 min–2 h; retention of infectivity after removal from virus source 13 days/20 days. No evidence was found for transmission of BBTV to the parthenogenetic offspring (Magee, 1940b; Hu *et al.*, 1996) or for multiplication of BBTV in the aphid vector (Hafner *et al.*, 1995).

Transmission efficiency for individual aphids has been reported as ranging from 46 to 67% (Magee, 1927; Wu and Su, 1990a; Hu *et al.*, 1996) and the virus is more efficiently acquired by nymphs than by adults (Magee, 1940b).

Colonies of *P. nigronervosa* from Australia (where bunchy top occurs) and from Réunion (where bunchy top does not occur) both transmitted each of six isolates of BBTV with similar efficiency (M.L. Iskra-Caruana, Montpellier, 1994, unpublished).

Vegetative propagation

Bunchy top is efficiently transmitted through conventional planting material, including corms, bits and suckers. All suckers from an infected stool will eventually become infected. Magee (1927) demonstrated 100% transmission of bunchy top through new 'eyes' (meristematic growing points), even in a plant that had only been expressing symptoms for 2–3 weeks.

Bunchy top is also transmitted in micropropagated banana plants (Drew *et al.*, 1989; Ramos and Zamora, 1990; Wu and Su, 1991), though not always at rates of 100%. From time to time, apparently virus-free meristems producing apparently virus-free plants can arise from an infected clone (Thomas *et al.*, 1995).

In Pakistan in the early 1990s, much of the available planting material of 'Basrai' (AAA, Cavendish subgroup) was infected with BBTV. Plantations established from infected suckers and corms were completely unproductive (Plate 6.8).

Epidemiology of banana bunchy top

The epidemiology of banana bunchy top in Australia is simplified by the presence of a single susceptible host and a single vector species (*P. nigronervosa*; Magee, 1927). Spread over long distances is by infected planting material and it is by this means that new plantings in isolated areas usually become infected. Dissemination over short distances from these infection foci is by the banana aphid.

In studies of actual outbreaks of bunchy top in commercial banana plantations, Allen (1978b, 1987) showed that the average distance of secondary spread of the disease by aphids was only 15.5–17.2 m. Nearly two-thirds of new infections were within 20 m of the nearest source of infection and 99% were within 86 m. Allen and Barnier (1977) showed that, if a new plantation was located adjacent to a diseased plantation, the chance of spread of bunchy top into the new plantation within the first 12 months was 88%. This chance was reduced to 27% if the plantations were separated by 50–1000 m and to less than 5% if they were 1000 m apart. On average, the interval between infection of a plant and movement of aphids from this plant to initiate new infections elsewhere (the disease latent period) was equivalent to the time taken for 3.7 new leaves to emerge. The rate of leaf emergence varied seasonally, with a maximum in summer (Allen, 1987). Based on the above studies, a computer program that simulates epidemics of bunchy top has been developed (R.N. Allen, Brisbane, 1994, personal communication). This program, which is commercially available, allows epidemiological factors to be varied and their effect on the progress of the epidemic and disease control to be monitored.

In the Philippines, Opina and Milloren (1996) also demonstrated that most new infections were adjacent to or in close proximity to primary sources of infection.

Host reaction

General host range

In the *Musaceae*, BBTV is known to infect a range of *Musa* species, cultivars in the *Eumusa* and *Australimusa* series of edible banana and *Ensete ventricosum*. Susceptible *Musa* include *M. balbisiana* (Magee, 1948; Espino *et al.*, 1993), *M. acuminata* ssp. *banksii* and *M. textilis* (Magee, 1927), *M. velutina* (Thomas and Dietzgen, 1991), *M. coccinea*, *M. jackeyi*, *M. ornata* and *M. acuminata* ssp. *zebrina* (A.D.W. Geering and J.E. Thomas, Brisbane, 1998, unpublished results).

There is some evidence for hosts outside the *Musaceae*, though reports have been conflicting. Su *et al.* (1993) obtained positive ELISA reactions from BBTV-inoculated *Canna indica* and *Hedychium coronarium*, and recovery of the virus to banana, though not reported here, was demonstrated (H.J. Su, Taipei, 1996, personal communication). Ram and Summanwar (1984) reported *Colocasia esculenta* as a host of BBTV. However, Hu *et al.* (1996) were unable to demonstrate *C. esculenta* or *Alpinia purpurata* as experimental (E) or natural (N) hosts of BBTV in Hawaii. Geering and Thomas (1997) also found no evidence for the following species as hosts of BBTV in Australia: *Strelitzia* sp. (N), *C. indica* (E, N), *Canna* × *generalis* (N), *Canna* × *orchiodes* (N), *H. coronarium* (E), *Heliconia psittacorum* (E), *Alpinia caerulea* (E, N), *Alpinia arundinelliana* (E), *Alpinia zerumbet* (E), *Alocasia brisbanensis* (E, N), *C. esculenta* (E, N). Magee (1927) was unable to infect *Strelitzia* sp., *Ravenala* sp., *Canna* spp. (including *C. edulis*), *Solanum tuberosum* and *Zea mays*.

Cultivar susceptibility

To date, there are no confirmed reports of immunity to BBTV in any *Musa* species or cultivar. However, differences in susceptibility between cultivars subject to either experimental or field infection have frequently been noted (Magee, 1948; Jose, 1981; Muharam, 1984; Espino *et al.*, 1993).

Espino *et al.* (1993) evaluated a total of 57 banana cultivars for their reaction to bunchy top, both by experimental inoculation and field observations. All cultivars in the AA and AAA genomic groups were highly susceptible. However, low levels of infection (as assessed by symptom expression) or total absence of symptoms following aphid inoculation was noted in some cultivars containing the B genome. These included 'Radja' (AAB, syn. 'Pisang Raja' – 12.5% of inoculated plants with symptoms), 'Bungaoisan' (AAB – 0%), 'Pelipia' (ABB, syn. 'Pelipita' – 10%), 'Pundol' (ABB – 0%), 'Katali' (ABB, syn. 'Pisang Awak' – 0%), 'Abuhon' (ABB – 0%) and 'Turangkog' (ABB – 0%).

These cultivars were not back-indexed by aphid transmission to a susceptible banana cultivar or biochemically induced (e.g. by ELISA), so the presence of symptomless infection cannot be ruled out. Also, greater numbers of aphids than the 15 used here might have resulted in infection. Cultivars 'Abuhon' and 'Bungaoisan' are susceptible to BBTV by experimental aphid inoculation (A.D.W. Geering and J.E. Thomas, Brisbane, 1998, unpublished). Nevertheless, it appears that real differences exist in cultivar reaction to bunchy top and the time taken before symptoms are expressed.

Cultivars within the Cavendish subgroup form the basis of the international banana export trade and are generally highly susceptible to bunchy top. However, it appears that not all cultivars with an AAA genome are similarly susceptible. 'Gros Michel' exhibits resistance to the disease under both experimental inoculation and field conditions and Magee (1948) considered that the introduction of this cultivar to Fiji in the early 1900s contributed to partial rehabilitation of the bunchy-top-devastated industry. Compared with 'Williams' (AAA, Cavendish subgroup), the concentration of virions of BBTV in infected plants of 'Gros Michel' and the proportion of plants infected by aphid inoculation are lower. Symptoms are also slower to develop and are less severe (A.D.W. Geering and J.E. Thomas, Brisbane, 1997, unpublished results). These factors may contribute to a reduced rate of aphid transmission and field spread in plantations of 'Gros Michel'.

Control

On the basis of his research in the early 1920s, Magee (1927) proposed a range of measures for the control of banana bunchy top. These recommendations involved two major components.

- Exclusion of the disease from unaffected and lightly affected areas.
- Eradication of infected plants from both lightly and heavily-affected areas.

Plate 6.8. Bunchy-top-affected plantation of 'Basrai' (AAA, Cavendish subgroup) in the Sindh Province of Pakistan (photo: D.R. Jones, INIBAP).

Magee considered that the measures would fail if left entirely to the goodwill of the growers (Magee, 1936), so legislation enforcing these control measures was gradually introduced by the state governments of NSW (Plant Diseases Act and Regulations commencing 1927) and Queensland (Diseases in Plants Act and Banana Industry Protection Act commencing 1929). The legislation in the two states was similar, the major difference being the onus of detection on the growers in Queensland and on the government-funded inspectors in NSW. These measures, which are listed below, still form the basis of control today.

- Registration of all banana plantations.
- Establishment of quarantine zones.
- Restrictions on the movement and use of planting material.
- Regular inspections of all banana plantations for bunchy top.
- Prompt destruction of all infected plants.
- Ongoing education and extension programmes for growers.

When adopted, these measures allowed the complete rehabilitation of the Australian banana industry. A vital aspect of eradication was the complete removal of infected stools to prevent the re-emergence of infected suckers. When identified, the infected plant was first sprayed with paraffin to kill any aphids. The plant was then chopped down, the corm and all associated suckers dug out and all plant material chopped into small pieces to prevent regrowth. Nowadays, herbicide injections have replaced physical removal and destruction.

A steady rehabilitation and expansion of the Australian banana industry followed in the late 1920s and early 1930s. However, a period of complacency, due to the success of the control programme, combined with a period of overproduction and resultant neglect of unprofitable plantations from about 1933 to 1935, led to a resurgence of bunchy top (Magee, 1936). An intensification of the control programme and increased vigilance by the disease inspectors brought the situation under control again, where it has remained ever since. Allen (1987) evaluated control measures in NSW by computer simulation and showed that bunchy top could be controlled because of the long interval between first appearance of disease symptoms (*c.* two new leaves) and first transmission to other plants (a further 1.7 new leaves). He recommended that inspection intervals be varied according to the growth rate of the plants, with successive inspections made before the growth of four additional leaves.

In a few cases, total eradication of BBTV from districts has been achieved (e.g. Yarrahappini and Richmond Valley areas of NSW and Innisfail, Queensland), but, even with a large input of resources, the disease is usually just reduced to very low, manageable levels. Few other countries have experienced such successful control of bunchy top. In most cases, this has been due to an inability to enforce an organized control programme across whole districts.

In addition to plantations, feral banana plants and plants in home gardens must be taken into consideration if bunchy top is to be successfully controlled. Smith *et al.* (1998) studied a bunchy-top epidemic in a commercial banana plantation in Mindanao, Philippines. They found that, despite regular inspection and roguing of infected plants, a clear gradient of infection was observed from the edge of the plantation towards the centre. This could only be explained by a significant, continuing source of infection, probably from banana plots on smallholdings and feral banana plants, in close proximity to the plantation.

Attempts to control bunchy top by controlling the aphid vector with insecticide spray programmes have usually met with limited success.

Bract Mosaic

J.E. Thomas, M.L. Iskra-Caruana, L.V. Magnaye and D.R. Jones

Introduction

Bract mosaic is one of the more recently described virus diseases of banana. Symptoms of the disease were first noted on several banana cultivars on the Philippine island of Mindanao in 1979 and thought to be different from all other recognized viruses of banana (Magnaye and Espino, 1990). Many accessions in the South-East Asian Regional Germplasm Collection near Davao were also seen with symptoms. Later, the disease was found to be widespread throughout the Philippines, particularly in plantings of the ABB cooking-banana cultivars 'Saba' and 'Cardaba'. In Mindanao, cultivars in the Cavendish subgroup (AAA) on commercial plantations were also affected. The disease was given the name bract mosaic at a meeting of banana virologists that took place in Los Baños in 1988 and included in a list of viruses of quarantine importance (Frison and Putter, 1989). Filamentous virus particles were later associated with the disease.

In 1992, banana cultivars with bract mosaic symptoms were noticed in the germplasm collections at the Indian Insitute of Horticultural Research near Bangalore and at the Tamil Nadu Agricultural University in Coimbatore in India. The disease was also found in field plots near Coimbatore. Filamentous virus particles were found in specimens from affected plants (Anon., 1993). In 1995 in Kerala State, bract mosaic symptoms were seen on 'Nendran' (AAB, Plantain subgroup) affected by a disease of hitherto unknown aetiology known locally as 'kokkan' (Shanmugavelu *et al.*, 1992). Other banana cultivars and *M. ornata* were seen with symptoms at the Kerala Agricultural University's Banana Research Station at Kannara near Trichur. Bract mosaic was confirmed from specimens sent to Brisbane, Australia (Anon., 1995).

'Embul' (AAB, syn. 'Mysore'), seen with bract mosaic symptoms at the Horticultural Research and Development Institute's field station at Walgolla, Sri Lanka, was also found to be affected by the disease following diagnosis. Other 'Embul' plants at the field station had double infections of bract mosaic and bunchy top or banana streak (Anon., 1995). Symptoms of bract mosaic were also seen in a germplasm collection and in back gardens at Gannoruwa near Kandy (D.R. Jones, INIBAP, 1995, personal observation).

In summary, bract mosaic disease has been recorded from the Philippines, India and Sri Lanka (Diekmann and Putter, 1996). Additional records from Western Samoa and Vietnam have been from banana plants containing the virus, but displaying symptoms more typical of banana mosaic (see Lockhart and Jones, this chapter, p. 258) than of bract mosaic (Rodoni *et al.*, 1996).

Very limited data are available on the economic impact of bract mosaic. In one study conducted in Mindanao, yield losses of up to 40% were noted in 'Cardaba' and 'Lakatan' (AAA) (Kenyon *et al.*, 1996; Thomas and Magnaye, 1996). In commercial plantations in this region, a correlation between high incidence of bract mosaic and high fruit rejection rates due to misshapen fingers has been noted (A. Pedrosa, Davao, 1993, personal communication). Streaks on fruit can also result in rejection. The failure of fruit to fill on infected plants has been noted in India (Diekmann and Putter, 1996).

Symptoms

Typical symptoms of bract mosaic are distinctive. Mosaic patterns on bracts are diagnostic and distinct from symptoms caused by all other known viruses of banana (see Plates 5.12, 6.9 and 6.15). Mosaic patterns, stripes and spindle-shaped streaks may also be visible on pseudostem bases when the outer leaf sheaths are removed and can extend up the petiole bases. Infection is often associated with an increase in pseudostem pigment. Sometimes the symptoms are chlorotic on a red background (Plate 6.10) and sometimes reddish, yellow or chorotic on a green background (Plates 6.11 and 6.12). Symptoms can darken through red to brown and even black (Plate 6.13). Chlorotic streaks and spindle shapes running parallel to the veins are occasionally seen on leaves (Plate 6.14). 'Nendran' is particularly severely affected by the disease, with leaf sheaves separating from the unusually red-coloured pseudostems of young plants. Leaves also become arranged fan-like on one plane, rather like the leaves of the traveller's palm (*Ravenala madagascariensis*) (Plate 6.15). Suckering is also suppressed and suckers that do emerge are distorted and deeply pigmented (Plate 6.16) (Anon., 1995).

In the Philippines, chlorotic streaks may be present on peduncles (L.V. Magnaye, Davao, 1998, personal observation) and a high disease incidence is associated with increased levels of malformed fruit in commercial plantations (A. Pedrosa, Davao, 1993, personal communication). In India, petioles and peduncles of 'Nendran' become brittle and fruit is only rarely carried to maturity. If fruit does mature, it is undersized. Mosaics can be seen on the fruit of other cultivars (Plate 6.17).

Initial symptoms in aphid-inoculated plants include broad, chlorotic patches along the major leaf veins, surrounded by a rusty red border and green or reddish streaks or spindle-shaped lesions on the petioles. Leaf symptoms, consisting of spindle-shaped lesions and streaks running parallel to the veins, are not always evident, but can occur on young plants that have been recently infected.

In Western Samoa, India (Tamil Nadu and Maharashtra States) and Vietnam, banana bract mosaic virus (BBrMV) has been isolated from banana plants that were showing symptoms typical of banana mosaic and lacking the characteristic symptoms on the bracts (Rodoni *et al.*, 1996, 1997). Some of these plants were shown to have a mixed infection of cucumber mosaic virus (CMV) and BBrMV.

Causal agent

Flexuous virus-like particles, each 750 nm × 11 nm, have been detected in infected banana (Muñez, 1992; Bateson and Dale, 1995; Thomas *et al.*, 1997). Purified virions (Plate 6.18) contain a major coat protein of 38–39 kDa, as estimated by sodium dodecyl sulphate-polyacrylamide gel electrophoresis (SDS-PAGE) (Bateson and Dale, 1995; Thomas *et al.*, 1997). The size of the coat protein, calculated directly from the putative coat-protein amino acid sequences of two partial clones, was 39.3 kDa (Bateson and Dale, 1995; M.L. Iskra-Caruana and J.E. Thomas, Montpellier and Brisbane, 1997, unpublished). Virions have a buoyant density in caesium chloride of 1.29–1.31 g cm^{-3} and an $A_{260/280}$ = 1.17 (Thomas *et al.*, 1997).

Nucleotide sequence analysis indicates that BBrMV is a unique, novel potyvirus.

The sequence of the 3′ end of the NIb gene, the coat protein gene and the 3′ untranslated region is available for several isolates (Bateson and Dale, 1995, 1996; M.L. Iskra-Caruana and C. Bringaud, Montpellier, 1997, personal communication; Rodoni *et al.*, 1997; Thomas *et al.*, 1997).

Isolates of BBrMV from India, Sri Lanka and the Philippines are all closely related serologically (Thomas *et al.*, 1997). Identity at the nucleotide level within the coat-protein gene was greater than 87% for isolates from the Philippines, India, Western Samoa and Vietnam (Caruana *et al.*, 1995; M.L. Iskra-Caruana, Montpellier, 1997, unpublished; Rodoni, *et al.*, 1997; Thomas *et al.*, 1997).

Weak serological relationships have been demonstrated between BBrMV and other potyviruses, including abacá mosaic (AbaMV), dasheen mosaic, maize dwarf mosaic, wheat streak mosaic, sorghum mosaic and sugar cane mosaic viruses (SCMV) (Thomas *et al.*, 1997).

Although bract mosaic-affected banana plants from India and the Philippines frequently also contain particles of banana mild mosaic virus (see Thomas *et al.*, this chapter, pp. 275–278) (M.L. Iskra-Caruana and J.E. Thomas, 1998, Montpellier and Brisbane, unpublished), BBrMV alone appears to be the causal agent of the disease. Its virions have been aphid-transmitted to healthy banana test plants in which bract mosaic disease has subsequently developed (Caruana and Galzi, 1998).

Both serological and nucleic acid-based assays are now available for BBrMV. The virus can be detected by ELISA using polyclonal (Thomas *et al.*, 1997) and/or monoclonal (J.E. Thomas, Brisbane, 1996, unpublished) antibodies. The polyclonal antiserum detects all isolates tested (Thomas *et al.*, 1997), though some individual monoclonal antibodies do not react with all isolates (J.E. Thomas and M.L. Iskra-Caruana, Brisbane and Montpellier, 1996, unpublished).

The virus can also be detected by PCR in total nucleic acid extracts from infected plants, using either virus-specific or degenerate potyvirus group primers (Bateson and Dale, 1995; Thomas *et al.*, 1997). Virion concentration in infected plants is relatively low, and the virions are usually not readily detected by direct electron microscopy of sap.

Disease cycle and epidemiology

BBrMV is transmitted by at least three species of aphids: *Aphis gossypii, Rhopalosiphum maidis* (Magnaye and Espino, 1990) and *Pentalonia nigronervosa* (Muñez, 1992; Diekmann and Putter, 1996). *Pentalonia nigronervosa* transmitted BBrMV after an acquisition access period of 1 min, indicating that transmission is of the non-persistent type (Muñez, 1992). Efficiency of transmission with the latter species was less than 10% (Caruana and Galzi, 1998).

Attempts to transmit BBrMV by sap inoculation to herbaceous indicator plants have so far been unsuccessful (Magnaye and Espino, 1990; Muñez, 1992; S. Cohen, Montpellier, 1996, personal communication; Diekmann and Putter, 1996). However, occasional sap transmission from banana to banana has been achieved (L.V. Magnaye and L. Herradura, Davao, 1998, unpublished). The virus can be transmitted through vegetative planting material, including suckers, bits and corms, and via micropropagated plantlets.

Host reaction

The natural and experimental host range of BBrMV appears to be restricted to *Musa*. The virus has been detected in a wide range of naturally infected banana cultivars and genotypes, and no resistance to the virus has been noted. Abacá is also a host (L.V. Magnaye and L. Herradura, Davao, 1997, unpublished; Thomas *et al.*, 1997).

Control

The use of indexed, virus-free planting material is the best means of control. Roguing/sanitation programmes have been introduced into commercial production areas in the Philippines (Magnaye, 1994).

Plate 6.9. Bract mosaic symptoms in a bract from the male bud of 'Seenikehel' (ABB) growing in a back garden at Gannoruwa, Sri Lanka (photo: D.R. Jones, INIBAP).

Plate 6.10. Chlorotic, spindle-shaped streak symptoms of bract mosiac disease in the abnormally red-pigmented pseudostem of 'Kippu Kadali' (AB) growing in the germplasm collection of the Tamil Nadu Agricultural University, Coimbatore, India (photo: D.R. Jones, INIBAP).

Plate 6.11. Yellow, spindle-shaped streak symptoms of bract mosaic disease on the deep green pseudostem of 'Lambi' (ABB, Bluggoe subgroup) growing in the germplasm collection at the Tamil Nadu Agricultural University, Coimbatore, India (photo: D.R. Jones, INIBAP).

Banana Mosaic

B.E.L. Lockhart and D.R. Jones

Introduction

The disease was first described in NSW, Australia, in 1930 (Magee, 1930, 1940a). It has been given a variety of names including infectious chlorosis, heart rot, virus sheath rot, cucumber mosaic and banana mosaic (Magee, 1930; Stover, 1972; Wardlaw, 1972). Mosaic is widespread in

Plate 6.12. Red spindle-shaped streak symptoms of bract mosaic disease extending up the petiole bases of 'Suwandel' (AAB) near Kandy, Sri Lanka (photo: D.R. Jones, INIBAP).

Plate 6.13. Dark brown stripe symptoms of bract mosaic on petiole leaf bases of 'Nendran' (AAB, Plantain subgroup) near Trichur, India (photo: D.R. Jones, INIBAP).

Plate 6.14. Chlorotic spindle-shaped streak symptoms of bract mosaic disease on a leaf of 'Nendran' (AAB, Plantain subgroup) near Trichur, India (photo: D.R. Jones, INIBAP).

distribution and it is assumed that it is found in most areas where banana is grown. Reports are still appearing in the scientific literature giving details of records in new locations (e.g. Pietersen *et al.*, 1998; Trindade *et al.*, 1998; Vuylsteke *et al.*, 1998).

Banana mosaic is usually only a nuisance to growers establishing new plantings using corms or suckers. However, the disease is becoming more common in some localities in plantations that are established using tissue culture-derived planting material (Tsai *et al.*, 1986; Niblett *et al.*, 1994).

Different strains of the virus pathogen are known to occur. Common strains have not been reported to cause economically important damage to banana. Mosaic symptoms may be absent or occur on only a few leaves. In contrast, severe or heart-rot strains cause significant losses, due to necrosis of the pseudostem, which results in plant death (Niblett *et al.*, 1994). The heart-rot strain is particularly destructive in banana grown under plastic in Morocco (Bouhida and Lockhart, 1990).

Symptoms

Symptoms depend on the strain of the virus pathogen and the temperature. Common or mild strains induce a diffuse mosaic (Plate 6.19) or line patterns and ring spots of the leaf lamina (Plate 6.20) (Yot-Dauthy and Bové, 1966; Lockhart, 1986; Niblett *et al.*, 1994). Occasionally, leaves can be deformed and curl (Plate 6.21). These symptoms appear sporadically and the majority of leaves may be symptomless. Mosaics sometimes appear on fruit (Plate 6.22). Symptoms are generally more severe when temperatures fall below 24°C, which occurs in the winter in the subtropics and at altitude in the tropics (Niblett *et al.*, 1994).

Severe strains of the banana mosaic virus produce more pronounced symptoms, which can include necrosis of emerging cigar leaves (Plate 6.23), leading to varying degrees of necrosis in the unfurled leaf lamina (Plate 6.24). Internal tissues of the pseudostem can also become necrotic. Leaf distortion is also more severe (see Plate 6.23). Plants with severe strains of the virus may die, especially if infected soon after planting.

Causal agent

Cucumber mosaic virus (CMV) causes banana mosaic disease (Yot-Dauthy and Bové, 1966). This virus is a member of the cucumovirus group and has spherical particles 28–30 nm in diameter (Plate 6.25), containing an ssRNA (Francki *et al.*, 1979). Most isolates of CMV have three genomic and one subgenomic RNA species (Francki *et al.*, 1979). A fifth RNA species (RNA 5) (Kaper and Waterworth, 1977) occurs in some virus isolates and has been linked to modulation of symptom expression in some plants, including banana (Gafny *et al.*, 1996). Isolates of CMV fall into two subgroups based on serological relationships. These two subgroups, designated DTL and T_0RS (Desvignes and Cardin, 1973; Piazzolla *et al.*, 1979), correspond to two classification categories based on nucleic acid hybridization. DTL serotypes are designated WT (Piazzolla *et al.*, 1979) or subgroup I (Owen *et al.*, 1990) and T_0RS serotypes are designated S (Piazzolla *et al.*, 1979) or subgroup II (Owen *et al.*, 1990). Isolates of CMV from banana have been identified as belonging to subgroup I (Hu *et al.*, 1995; Singh *et al.*, 1995; Gafny *et al.*, 1996), which is the subgroup that includes most CMV isolates from the tropics (Niblett *et al.*, 1994).

Disease cycle and epidemiology

Disease incidence in banana is determined primarily by the number of infected alternative hosts in the vicinity of the crop and the population dynamics of aphid vectors. CMV is capable of infecting over 800 plant species and is transmitted in a non-persistent manner by over 60 species of aphid, including *Aphis gossypii*, *Rhopalosiphum maidis*, *Rhopalosiphum prunifoliae* and *Myzus persicae*. As CMV is also seed-transmitted in many of its hosts, there is often a reservoir of inoculum present in the rural environment. Common and severe strains of CMV do not differ in their epidemiology.

Aphid vectors usually acquire CMV from weeds (e.g. *Commelina* (Plate 6.26), *Stellaria*, *Bryonia* and *Solanum* spp.) or crop hosts (tomato, melon (Plate 6.27), cucumber, sweet pepper) growing near or in banana fields. Viruliferous, winged aphids (alatae) migrate from these plants to infect banana. Aphid species, such as *A. gossypii* and *M. persicae*, do not normally colonize banana, but can transmit CMV during exploratory visits. Although some of these transient aphid species may be relatively inefficient vectors of CMV, their large numbers may ensure the effective transmission of the virus from alternative hosts to banana.

Movement to banana is most probably by chance, following the disturbance or destruction of preferred hosts. Disease incidence is higher in new plantings, because land is usually cleared of alternative aphid hosts, and weeds carrying CMV may later grow in great profusion in exposed soil (Niblett *et al.*, 1994). Tissue-cultured banana plants, which are low-lying and succulent, are believed to be particularly vulnerable to infection.

The means of spread of CMV contrasts with that of BBTV, which is transmitted in a persistent manner by *Pentalonia nigronervosa*, the only aphid species to colonize banana (Blackman and Eastop, 1984). Although this aphid has been reported to be unimportant in the spread of CMV (Stover, 1972; Wardlaw, 1972), it could be important in new plantations established from tissue culture-derived plants. On juvenile plants, large populations of *P. nigronervosa* develop on leaves and petioles (Plate 6.28) where aerial dissemination is more likely than from the pseudostem beneath basal leaf sheaths, as on more mature plants. This hypothesis provides an alternative explanation for the higher incidence of banana mosaic in plantations established from tissue-cultured plants. A second reason for not discounting the possible role for *P. nigronervosa* in the epidemiology of CMV is a challenge to the view that this aphid species colonizes only *Musa* (Stover, 1972; Wardlaw, 1972). In West Africa, where the incidence of CMV is high (Osei, 1995), large populations of *P. nigronervosa* have been observed colonizing aerial shoots of *Commelina diffusa* (B.E.L. Lockhart, Minnesota, 1998, personal observation), a major alternative host of CMV. These observations suggest that, in some situations, *P. nigronervosa* may be an important vector of CMV from alternative host to banana, as well as from banana to banana.

A better understanding of the epidemiology of CMV in banana requires a better understanding of the biology and ecology of the aphid vector species that colonize and/or visit banana and many other alternative hosts of the virus.

In theory, a banana plant with banana mosaic should be systemically infected with CMV. Indeed, the leaves of many suckers arising from diseased plants have mosaic symptoms. However, there are numerous instances when the leaves of suckers are symptomless. When propagated, farmers report that these suckers develop into plants that exhibit no mosaic symptoms and produce a normal bunch. Such plants need to be tested for the presence of CMV to determine if they have indeed escaped infection.

Host reaction

Many different cultivars have been reported with mosaic symptoms, but in some cases the disease may have been confused with other virus diseases (see 'Symptoms'). The cucumber mosaic diseases of plantain, recorded by Stover (1972) in Honduras, and of 'Poovan' (AAB, syn. 'Mysore'), described by Mohan and Lakshmanan (1988) in Tamil Nadu, India, were both most probably banana streak disease. This problem of diagnosis based on symptoms makes it difficult to compile lists of known host cultivars and their reaction to CMV.

It is not known if resistance to CMV occurs within *Musa*. Stover (1972) indicated that *M. balbisiana* was the only

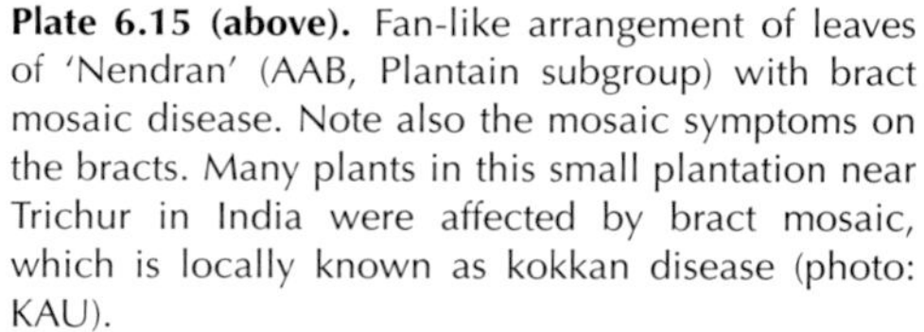

Plate 6.15 (above). Fan-like arrangement of leaves of 'Nendran' (AAB, Plantain subgroup) with bract mosaic disease. Note also the mosaic symptoms on the bracts. Many plants in this small plantation near Trichur in India were affected by bract mosaic, which is locally known as kokkan disease (photo: KAU).

Plate 6.16 (above right). Distorted and deeply red-pigmented peeper of 'Nendran' (AAB, Plantain subgroup) affected by bract mosaic disease at Kannara near Trichur, India (photo: D.R. Jones, INIBAP).

Plate 6.17 (right). Fruit of 'Karpuravalli' (ABB, syn. 'Pisang Awak') growing near Coimbatore, India, with mosaic symptoms of bract mosaic disease (photo: D.R. Jones, INIBAP).

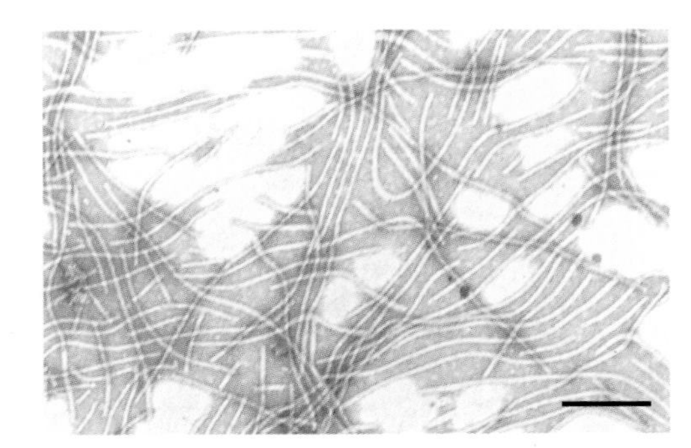

Plate 6.18. Virions of banana bract mosaic virus negatively stained with ammonium molybdate. Bar represents 200 nm (photo: J.E. Thomas, QDPI).

Plate 6.19. Chlorotic mosaic symptom of banana mosaic disease on the leaf of 'Grand Nain' (AAA, Cavendish subgroup) in St Vincent, Windward Islands (photo: D.R. Jones, SVBGA).

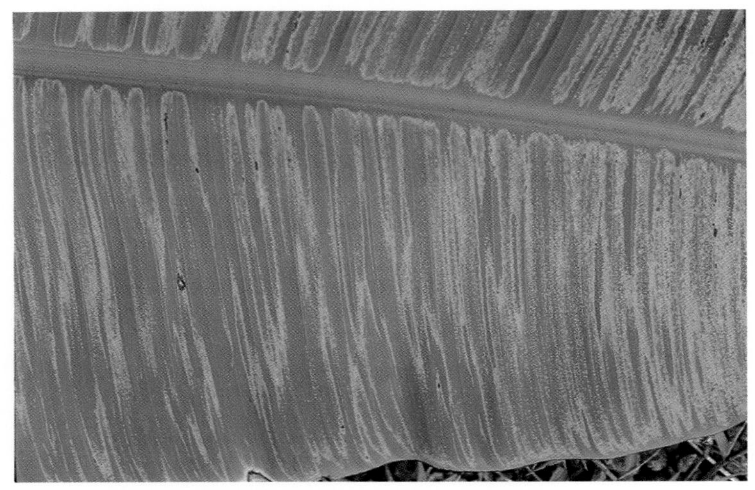

Plate 6.20. Line pattern and ring-spot symptoms of banana mosaic disease on a leaf of 'Pisang Mas' (AA, syn. 'Sucrier') growing in Melaka State, West Malaysia (photo: D.R. Jones, QDPI).

species of *Musa* that appeared free of virus symptoms in the field, although seedlings had been observed with a whitish chlorotic mottle, which disappeared as the plants matured. Incidence of mosaic has been reported to be greater in cultivars in the Cavendish subgroup (AAA) than in 'Gros Michel' (AAA) in Central America. Symptoms on Cavendish cultivars have been recorded as being more severe than on diploids (Stover, 1972).

Control

It is important to ensure that corms and suckers used for planting are healthy and not derived from plants with mosaic symptoms. Tissue cultures used for the *in vitro* multiplication of propagating material should be initiated from shoot tips excised from plants that have been tested for the presence of CMV. Although CMV has been eliminated from diseased material by heat treatment of corms followed by apical meristem culture (Berg and Bustamante, 1974), it is recommended that this technique be attempted only for exceptionally valuable germplasm, such as an important breeding line or hybrid. Otherwise, it would be prudent to discard the material and use an uninfected source of the same clone for propagation. There is some evidence to indicate that CMV may be seed-transmitted in banana (Gold, 1972), so seedlings need to be tested if they are from a plant whose virus disease status is unknown.

A variety of methods are available for diagnosing CMV infection in banana germplasm (Diekmann and Putter, 1996). These include symptomatology, indicator plant tests, serology, nucleic acid hybridization and PCR amplification. Diagnosis of CMV infection by symptomatology is the least reliable method of disease identification because of the periodicity of symptom appearance, the effect of temperature on symptom expression and the similarity of symptoms induced by other viruses.

Symptoms of banana streak (see Lockhart and Jones, this chapter, pp. 263–274) and bract mosaic (see Thomas *et al.*, this chapter. pp. 253–255) are similar to those of banana mosaic (Lockhart, 1986; Magnaye and Espino, 1990). Mosaic symptoms in the bracts are usually diagnostic for bract mosaic. Also, few, if any, leaf symptoms are associated with bract mosaic disease. Banana mosaic and banana streak are harder to differentiate because both can cause chlorotic and necrotic leaf symptoms, leaf distortion and internal necrosis of the pseudostem. However, a trained eye can usually distinguish between leaf symptoms of these two diseases, as banana streak symptoms tend to be more linear and conspicuous (Lockhart, 1986). Symptoms of banana mosaic can also be misidentified as zinc defiency (see Lahav and Israeli, Chapter 9, pp. 348–349) and the genetic abnormality known as 'mosaic' (see Israeli and Lahav, Chapter 12, p. 400).

Disease diagnosis based on symptomatology can be supported by biological assays using herbaceous indicator plants, such as tobacco, squash and cowpea (Francki *et al.*, 1979). Although indicator plant assays have been largely supplanted by other diagnostic techniques, they may still be useful in situations where other methods are unavailable. One value of this method is that it is able to distinguish the presence of CMV, which is transmissible by mechanical inoculation, from BBTV and banana streak virus (BSV), which are not transmitted by mechanical inoculation and do not infect herbaceous indicator plants.

Highly reliable methods have been developed for detection of CMV by enzyme immunoassays. Polyclonal antibodies, capable of detecting a wide range of virus strains (Diekmann and Putter, 1996), and monoclonal antibodies, which can be used to differentiate between virus subgroups (Hasse *et al.*, 1989), are available. Genome (nucleic acid)-based methods of virus identification are highly sensitive and are being used increasingly for diagnosis of CMV infection in *Musa*. Methods based on nucleic acid hybridization (Gonda and Symons, 1978) have been increasingly replaced by PCR amplification (Hu *et al.*, 1995; Singh *et al.*, 1995), which is a less cumbersome technique.

Eliminating weed hosts of CMV from plantations and surrounding areas and ensuring that susceptible crop hosts are not present within or near banana fields is another important element in the control of banana mosaic. Weeds in and alongside plantings should be controlled efficiently, especially when the crop is young and more vulnerable to infection. Banana should not be intercropped with crop hosts of CMV. In the earliest studies on the disease in New South Wales in Australia, it was reported that outbreaks occurred in plantations near to where cucurbits, tomatoes or other vegetables were growing (Magee, 1940a). In Taiwan, incidence of mosaic was much less when banana was grown next to rice than when grown next to vegetables. The disease was highest in a field intercropped with cucumber (Tsai *et al.*, 1986). Banana plantlets being acclimatized in nurseries after removal from tissue culture should also be protected from sources of infection.

Banana plants with mosaic should be removed from commercial plantations and spaces replanted with healthy material. This is because diseased plants often have a poor yield and fruit can have virus symptoms. Their removal also eliminates the chance of disease spread from banana to banana. On occasions when disease incidence is exceptionally high, fields may have to be completely replanted.

Insecticides applied to banana plants are unlikely to be effective in controlling mosaic because the major aphid vectors recognized do not colonize banana, and virus transmission occurs after only a brief probe. Nevertheless, insecticides have been used in commercial plantations in the past and, together with the removal of diseased and surrounding plants, very good control has been achieved (Adam, 1962). However, the cost-effectiveness of these operations is questionable (Jeger *et al.*, 1995) and the roguing of diseased plants alone has resulted in adequate control (Stover, 1972).

The possibility of obtaining mosaic-resistant banana clones by incorporating the gene for the coat protein of CMV into the banana genome has been suggested by Fauquet and Beachy (1993) and cross-protection using mild strains by Wu *et al.* (1997).

Because it occurs ubiquitously, CMV is not regarded as a quarantinable pathogen. This view does not take into account the fact that severe or heart-rot isolates of CMV, which are far more damaging than common isolates of the virus, do not occur in all banana-producing areas. It is therefore important to avoid the introduction of severe CMV isolates into new areas where they can cause significant damage (Bouhida and Lockhart, 1990).

Banana Streak

B.E.L. Lockhart and D.R. Jones

Introduction

Banana streak as 'la mosaïque à tirets' was first observed in the Nieky Valley in Côte d'Ivoire in 1958. Preliminary research into the problem was initiated in 1964 after a serious outbreak occurred in the region. Two years later, illustrations of symptoms of the disease, which at the time was thought to be a type of banana mosaic (see Lockhart and Jones, this chapter, pp. 256–263), were published by Yot-Dauthy and Bové (1966). In 1972, in an attempt to control another outbreak, affected plants were destroyed and work on the effect of the disease on production was begun. 'La mosaïque à tirets' was described on 'Poyo' (AAA, Cavendish subgroup) by Lassoudière (1974), who reported that symptoms could affect one leaf, but not another on the same plant, and affected plants could lose and regain symptoms between cycles. Yield losses were found to increase as the intensity of symptoms increased. Plants with severe symptoms produced few, if any, exportable bunches.

The disease was next reported from southern Morocco (Lockhart, 1986), where it was seen in almost every established planting of field-grown 'Dwarf Cavendish' (AAA, Cavendish subgroup). In some cases, incidence of banana streak exceeded 50% of plants. It was noted that the disease

Plate 6.21. Distortion of the leaf lamina of 'Grand Nain' (AAA, Cavendish subgroup) caused by cucumber mosaic virus in St Vincent, Windward Islands (photo: D.R. Jones, SVBGA).

was rarely seen in the field or under plastic where the planting material was established from imported germplasm. It was in Morocco that the causal virus was identified for the first time and its role in the aetiology of the disease confirmed (Lockhart, 1986).

Banana streak has subsequently been found to be present in different cultivars in many countries around the world (Table 6.2). Disease incidence varies greatly. In some localities in certain countries the disease appears to be serious and in other places it is confined to a few plants of certain cultivars and is not regarded as important. In southern Cameroon and southern Nigeria, plantains with banana streak have been found in just over half the villages surveyed with incidence ranging from 0.5 to 17.0% (Gauhl *et al.*, 1999a, b). Incidence is also common and widespread in plantain around in the Sula Valley in Honduras and the Los Rios province in Ecuador (D.R. Jones, Latin America, 1999, personal observation). 'Mysore' (AAB), a widely distributed dessert clone from India, is now known to be almost totally affected by banana streak (Jones, 1994; Lockhart, 1994a). The leaf-streak symptoms so com-

Plate 6.22. Yellow mosaic symptoms of banana mosaic disease on fruit of 'Robusta' (AAA, Cavendish subgroup) in St Vincent, Windward Islands (photo: D.R. Jones, SVBGA).

Plate 6.23. Necrosis of the cigar leaf and the pronounced deformation of other leaves of 'Grand Nain' (AAA, Cavendish subgroup) due to a severe strain of cucumber mosaic virus in St Vincent, Windward Islands (photo: D.R. Jones, SVBGA).

mon on 'Mysore' were until relatively recently thought to be genetic in origin (Wardlaw, 1972). In Tamil Nadu State in India, 'Mysore' is intensely cultivated and almost every plant exhibits symptoms of banana streak. 'Mysore' can only be grown for three cycles before replanting, because of the rapid decline in productivity of plantations.

The economic importance of banana streak stems from two factors: the direct effect of the disease on crop yield and the restrictions imposed on the deployment of improved hybrids due to the occurrence of the causal virus in many of these genotypes.

An analysis of the direct effect of the disease on crop yield is complex, because it is determined by a combination of variables, the most important of which are disease incidence, climatic factors and level of management. Incidence is highest when diseased source plants are used for vegetative propagation. This situation is most likely to occur in areas where growers do not recognize the disease. Climatic factors play an important role, but the components of this effect have not been clearly delineated. Seasonal fluctuations are correlated with foliar symptom expression. When

Plate 6.24. Necrosis of a leaf of 'Grand Nain' (AAA, Cavendish subgroup) caused by cucumber mosaic virus in St Vincent, Windward Islands. This symptom of banana mosaic disease can be mistaken for banana streak disease (photo: D.R. Jones, SVBGA).

Table 6.2. Distribution of banana streak disease.

Region	Country/location	Reference
Europe	Spain (Canary Islands)*	Diekmann and Putter, 1996
	Portugal (Madeira)*	Jones and Lockhart, 1993
Africa	Benin*	Diekmann and Putter, 1996
	Cameroon*	Gauhl and Pasberg-Gauhl, 1997
	Cape Verde*	Diekmann and Putter, 1996
	Côte d'Ivoire*	Yot-Dauthy and Bové, 1966
	Ghana*	Diekmann and Putter, 1996
	Guinea*	Diekmann and Putter, 1996
	Kenya†	Diekmann and Putter, 1996
	Madagascar*	Jones and Lockhart, 1993
	Malawi*	Vuylsteke *et al.*, 1996
	Mauritius*	Jones and Lockhart, 1993
	Morocco*	Lockhart, 1986
	Nigeria*	Jones and Lockhart, 1993
	Rwanda*	Sebasigari and Stover, 1988
	South Africa*	Jones and Lockhart, 1993
	Tanzania* (incl. Zanzibar)	Sebasigari and Stover, 1988
	Uganda*	Dabek and Waller, 1990
Latin America–Caribbean	Brazil†	Jones and Lockhart, 1993
	Colombia*	Reichal *et al.*, 1997
	Costa Rica*	Diekmann and Putter, 1996
	Cuba*	Jones and Lockhart, 1993
	Ecuador*	Jones and Lockhart, 1993
	Grenada†	Jones and Lockhart, 1993
	Guadeloupe*	Jones and Lockhart, 1993
	Haiti*	B.E.L. Lockhart, Minnesota, 1999, personal observation
	Honduras*	Jones and Lockhart, 1993
	Jamaica†	Jones and Lockhart, 1993
	Nicaragua*	B.E.L. Lockhart, Minnesota, 1999, personal observation
	Puerto Rico*	B.E.L. Lockhart, Minnesota, 1999, personal observation
	St Lucia*	B.E.L. Lockhart, Minnesota, 1999, personal observation
	Trinidad*	Jones and Lockhart, 1993
	USA (Florida, Virgin Islands)*	Diekmann and Putter, 1996
	Venezuela†	Jones, 1995
Asia-Pacific	Australia*	Thomas *et al.*, 1994b
	China†	Jones and Lockhart, 1993
	India*	Diekmann and Putter, 1996
	Indonesia†	Diekmann and Putter, 1996
	Malaysia†	Diekmann and Putter, 1996
	New Caledonia*	Diekmann and Putter, 1996
	Papua New Guinea†	Diekmann and Putter, 1996
	Philippines*	Diekmann and Putter, 1996
	Sri Lanka*	Diekmann and Putter, 1996
	Taiwan*	Su *et al.* (1997)
	Thailand†	Diekmann and Putter, 1996
	Tonga*	Thomas *et al.*, 1994b
	Vietnam†	Diekmann and Putter, 1996
	Western Samoa*	Thomas *et al.*, 1994b

* Confirmed by electron microscopy or serology.
† Diagnosis on symptoms only.

bunch initiation coincides with this temperature-dependent onset of increased virus replication and symptom expression, the resulting bunches are much reduced in size, do not emerge normally and may have distorted fingers. In contrast, plants that initiate flowering at a period when temperature-dependent symptom expression is repressed produce bunches of normal size and appearance, even though some leaves may show pronounced disease symptoms. Bunch size and fruit quality tend to be significantly less affected by banana streak in intensively managed plantings than under low-input management regimes, where plants also suffer from nutrient or moisture stress. In field trials in Côte d'Ivoire, banana streak has been reported to reduce stem girth of 'Poyo' by 6–53% (Lassoudière, 1974) and overall yields by 6–7% (Lassoudière, 1979). In Australia, banana streak delayed harvesting in 'Williams' (AAA, Cavendish subgroup) by 3 weeks (Daniells *et al.*, 1998).

The realization that banana streak was widespread and that the causal agent could be carried in banana without showing symptoms has caused problems for quarantine authorities and organizations involved in the international exchange of banana germplasm. Banana streak is the virus disease most commonly detected at the Virus Indexing Centres of the International Network for the Improvement of Banana and Plantain (INIBAP) in both naturally occurring landraces and hybrids from breeding programmes. Valuable germplasm resistant to major diseases of banana cannot be disseminated because of infection.

Banana streak has been mistaken for banana mosaic in the past. Some illustrations of banana mosaic in previous texts (Yot-Dauthy and Bové, 1966; Stover, 1972; Wardlaw, 1972) are now clearly recognizable as those of banana streak. Cucumber mosaic of plantains, described by Stover (1972), and cucumber mosaic of 'Poovan' (AAB, syn. 'Mysore'), described by Mohan and Lakshmanan (1988), are also most probably banana streak.

Symptoms

Symptom expression varies, depending on the isolate of the pathogen, the host cultivar and the environment (Lockhart, 1994; Gauhl and Pasberg-Gauhl, 1995; Dahal *et al.,* 1998b), and can vary from an inconspicuous chlorotic flecking to lethal necrosis. However, the most common symptoms are narrow, discontinuous and sometimes continuous, chlorotic or yellow streaks, which run from the leaf midrib to the margin (Plate 6.29). In some cases, spindle or eye-shaped patterns or blotches are present. Yellow blotches have also been associated with banana streak (Plate 6.30). Symptoms can be sparse or concentrated in their distribution. Sometimes the lamina can be distorted. Streaks later darken to orange and often become brown or black (Plate 6.31). Necrosis has also been seen on the leaf midrib and petiole (Plate 6.32). Necrosis occurs more under low temperature, short-day conditions.

Foliar symptoms appear sporadically over the course of a year. Leaves showing symptoms may be succeeded by leaves expressing few or no symptoms. It is rare to see symptoms on all leaves of the same plant. When the leaves with symptoms are shed, the plant may appear normal and it may be many months before symptoms reappear on new leaves. As mentioned in the 'Introduction', this periodicity of symptoms and their severity is associated with environmental changes and is closely correlated with temperature (Lockhart, 1995; Dahal *et al.*, 1998b). Foliar symptom expression is more regular and severe in areas where seasonal temperature fluctuations are more pronounced. Therefore, temperature fluctuation, rather than absolute temperature, may be a critical factor.

Other symptoms associated with banana streak include leaf bases falling away from the pseudostem, streaks in the pseudostem (Plate 6.32), narrow, thicker leaves and constriction of the bunch on emergence ('choking') (Lassoudière, 1979). Stunting due to a shortening of the internodes, cigar-leaf necrosis (Plate 6.33), internal necrosis

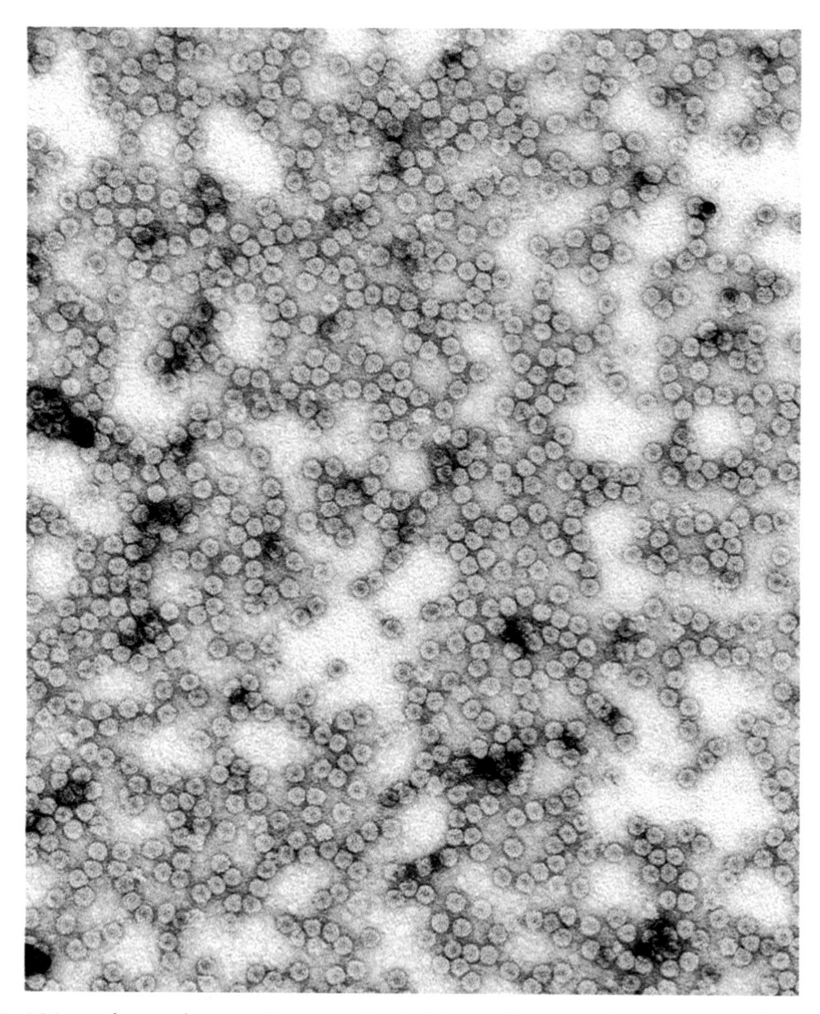

Plate 6.25. Virions of cucumber mosaic virus stained with uranyl acetate (photo: B.E.L. Lockhart, UM).

of the pseudostem (Plate 6.34), aberrant bunch emergence (Plate 6.35), reduced bunch size and distortion of fingers can also occur (Gauhl and Pasberg-Gauhl, 1994). Peel splitting and necrotic streak and spot symptoms have been observed on fingers on 'Grand Nain' (AAA, Cavendish subgroup) in Costa Rica (Plate 6.36) and Ecuador. In Australia, broad yellow lines in the leaf lamina parallel to the midrib, a purple margin to the leaf lamina, leaf twisting, grooves in the base of the pseudostem and an abnormal arrangement of the leaves similar to the traveller's palm (*Ravenala madagascariensis*) have also been associated with banana streak in 'Williams' (AAA, Cavendish subgroup) (Daniells *et al.*, 1998). Three phases of symptom expression have been recognized in commercial Cavendish plantations in Ecuador. The first is the appearance of chlorotic streaks in leaves, the second is the appearance of dark blotches on the pseudostem and midrib and the third is the splitting of the outer leaf sheaths of the pseudostem, sometimes as far up the plant as the petiole

Plate 6.26. Symptoms caused by cucumber mosaic virus on *Commelina* growing under banana in St Vincent, Windward Islands (photo: D.R. Jones, SVBGA).

Plate 6.27. Banana intercropped with melon growing under plastic in Morocco. The aphid *Aphis gossypii* is known to be a vector for severe strains of cucumber mosaic virus from the melons to the banana plants in this situation (photo: B.E.L. Lockhart, UM).

Plate 6.28. Colony of black banana aphids (*Pentalonia nigronervosa*) on a young sucker of 'Dominico-Hartón' (AAB, Plantain subgroup) in Colombia (photo: B.E.L. Lockhart, UM).

(Plate 6.37), which allows the entry of bacteria that cause a soft rot of the base of the pseudostem.

Causal agent

Banana streak virus (BSV), a member of the badnavirus group, is the cause of banana streak disease. The virions of BSV are non-enveloped, bacilliform, average 130–150 nm × 30 nm in size and contain a circular, dsDNA genome of 7.4 kb (Plate 6.38) (Lockhart and Olszewski, 1996). Together with other plant and animal viruses that encapsulate a dsDNA and replicate by reverse transcription, badnaviruses belong to the Pararetoviridae.

Another closely related virus, sugar cane bacilliform virus (SCBV) (Lockhart and Autrey, 1988), which occurs widely in sugar cane (Comstock and Lockhart, 1990), infects banana and several other monocotyledonous hosts (Bouhida *et al.*, 1993). While SCBV can be transmitted to banana, in which banana streak-like symptoms develop, BSV does not infect sugar cane (Lockhart, 1995).

A high degree of heterogeneity exists among isolates of BSV. Isolates differ serologically, genomically and biologically in that they induce different symptoms (Lockhart and Olszewski, 1996). This phenomenon has created practical problems for reliable virus detection and diagnosis. Five serological and genomically distinct isolates of BSV from Morocco, Rwanda, Trinidad, Honduras and the Philippines have so far been identified (Lockhart and Olszewski, 1993; B.E.L. Lockhart, Minnesota, 1999, unpublished).

It has recently been shown that BSV genomic sequences are integrated into genomic DNA of *Musa* and *Ensete* (LaFleur *et al.*, 1996). The presence of BSV-related sequences has been detected by PCR amplification (LaFleur *et al.*, 1996) in more than 400 *Musa* genotypes tested. While all *Musa* genotypes appear to contain integrated viral sequences, the nature of these sequences is variable. Of two such integrated sequences that have been characterized, one (LaFleur *et al.*, 1996) is incapable of giving rise to episomal BSV infection. There is good evidence, however, that a second integrated sequence is the source of *de novo* episomal BSV infection in a wide range of *Musa* genotypes, including a significant number of tetraploid hybrids which have been bred for improved yield and disease resistance (Ndowora *et al.*, 1999). The development of episomal BSV infection from integrated viral sequences is linked to *in vitro* propagation and possibly other stress factors (Ndowora *et al.*, 1999). This phenomenon has prevented the deployment of tissue cultures of improved banana and plantain hybrids, which have been found to carry episomal BSV, and has promoted research into methods of identifying the presence of expressible BSV integrants in parental *Musa* genotypes (Frison and Sharrock, 1998).

Attempts to transmit BSV by mechanical inoculation using abrasives have been unsuccessful, even when young banana plantlets are inoculated. It is not known whether this is due to some intrinsic property of the virus or whether compounds present in banana tissue inhibit initiation of infection.

Badnavirus–host interactions result in the formation of proteinaceous filamentous chains (tubules), which are visible under the electron microscope. Such chains, which are 16 nm in diameter and variable in length, are also found in association with BSV. Usually, the higher the concentration of these chains, the lower the concentration of BSV particles.

Disease cycle and epidemiology

Badnaviruses are transmitted by mealybugs and experimental transmission of BSV has been demonstrated using *Planococcus citri* (the citrus mealybug) (Plate 6.39) and *Pseudococcus* sp., which colonize the underside of banana leaves, bunches and flowers. The mode of transfer is semipersistent, as has been reported for Commelina yellow mottle virus (CoYMV), the type member of the badnavirus group. *Dysmicoccus* spp., which are found in West Africa (Matile-Ferrero and Williams, 1995) and in South America, colonize the pseudostem of banana beneath the outer leaf sheath or the rhizome just below soil level and may also be vectors. In Nigeria, *P. citri* has not been reported, but *Planococcus musae*, a new species of mealybug, has been identified in fields where incidence of banana streak is high (Dahal *et al.*, 1998a). In Ecuador, *Pseudococcus comstocki* is the most common mealybug found on banana, but transmission of BSV by this species has not yet been demonstrated. All clones of noble sugar cane have been shown to carry SCBV (Lockhart, 1994a) and the pink sugar-cane mealybug (*Saccharicoccus sacchari*) has also been shown experimentally to transmit SCBV from sugar cane to banana, producing characteristic BSV symptoms (Lockhart and Autry, 1991). In many tropical countries, banana is often found growing in close proximity to sugar cane and it is possible that sugar cane may constitute a source of inoculum for banana (Jones, 1994; Lockhart, 1994a). Detailed studies need to be undertaken to determine which mealybugs transmit banana streak and whether spread by this means is important. Field observations in many countries suggest that natural dissemination of BSV by mealybug vectors is limited in occurrence and does not play a major role in disease epidemiology (Lockhart, 1995). However, mealybugs on banana in Ecuador have been seen to be managed by ants, which may move them from plant to plant, thus spreading the disease.

Attempts to transmit BSV by mechanical inoculation using abrasives have been unsuccessful, even when young banana plantlets are inoculated. It is not known whether this is due to some intrinsic property of the virus or whether compounds present in banana tissue inhibit initiation of infection. As BSV is not transmitted mechanically, it is unlikely to be transmitted on cutting tools or during cultural operations. It is also not soil-borne. Spread of BSV is most probably by the vegetative propagation of infected source plants. All tissue cultures derived from meristems excised from diseased plants have been found to carry BSV.

Transmission through seed seems highly probable. Before banana streak was recognized in 'Mysore', the 'leaf striping' so characteristic of this cultivar was seen in Trinidad in the progeny of crosses between 'Mysore' and the tetraploid hybrid 'I.C.1' (AAAA) (Wardlaw, 1972). More recently, Daniells *et al.* (1995) have crossed diseased 'Mysore', as female parent, with 'SH-3362', an élite diploid from the Fundación Hondureña de Investigación Agricola (FHIA) breeding programme, and have shown that hybrids carry BSV. Bacilliform virus paticles have also been seen in an accession of *M. balbisiana* originally initiated from seed (J.E. Thomas, Brisbane, 1999, personal communication). The likely seed-borne nature of BSV and the fact that the virus can be present in germplasm without expressing symptoms have caused serious problems for some conventional breeding programmes.

Host reaction

Like most badnaviruses, BSV has a very restricted natural and experimental host range. BSV occurs naturally in cultivated landraces of banana and possibly enset (see Tessera and Quimio, this chapter, p. 283). No research has been undertaken to determine if BSV occurs naturally in wild *Musa* and *Ensete* species. BSV has been experimentally inoculated into *M. balbisiana* (Lockhart, 1994), but not to *M. textilis* (abacá) (B.E.L. Lockhart, Minnesota, 1999, unpublished).

There is some information on cultivars and subgroups that appear susceptible. The first reports of banana streak were from cultivars in the Cavendish subgroup (AAA) in Côte d'Ivoire (Lassoudière, 1974) and Morocco (Lockhart, 1986). Sebasigari and Stover (1988) found banana streak disease in the Kibungo Prefecture in Rwanda in 'Ney Poovan' (AB), 'Pisang Awak' (ABB) and the East African highland beer-making cultivars of the Lujugira–Mutika subgroup (AAA). In the Bukoba district of Tanzania, the disease was seen on 'Gros Michel' (AAA) and cooking cultivars in the AAA Lujugira–Mutika sub-

Plate 6.29. Chlorotic stripe, spindle and eye-shaped symptoms of banana streak disesae on a leaf of 'Mysore' (AAB) in Honduras (photo: D.R. Jones, INIBAP).

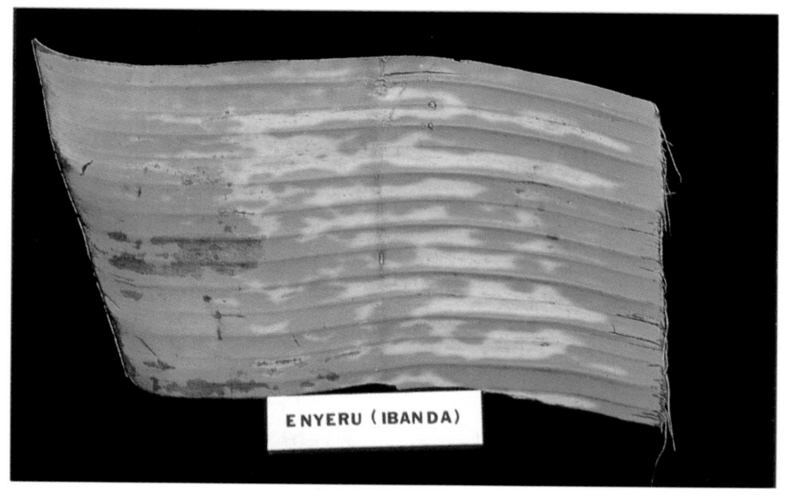

Plate 6.30. Yellow blotchy symptoms of banana streak disease on a leaf of 'Enyeru' (AAA, Lujugira–Mutika subgroup) from Uganda (photo: B.E.L. Lockhart, UM).

group. They also reported banana streak as present in Zanzibar on 'Giant Cavendish' (AAA, Cavendish subgroup) and 'Pukusa' (AAB, syn. 'Silk'). Later, Dabek and Waller (1990) recognized banana-streak symptoms in 'Mjenga' (AA, syn. 'Sucrier'), 'Kipaka' (AA, syn. 'Paka'), 'Bukoba' (AAA, syn. 'Gros Michel'), 'Paji' (AAA), 'Mzunga Mwekundu' (AAA, syn. 'Red'), Mzunga Mweupe (AAA, syn 'Green Red'), 'Mkono Wa Tembo' (AAB, 'Horn' plantain type), 'Kikonde Kenya' (AAB, syn. 'Mysore') and 'Kijakazi' (AAB, Pome subgroup) in a germplasm collection at Kizimbani in Zanzibar. All accessions of 'Kikonde Kenya' were affected and most accessions of 'Kipaka'. In Indonesia and Papua New Guinea, 'Pisang Raja' (AAB) has typical symptoms of banana streak (D.R. Jones, St Vincent, 1998, personal observation).

Plate 6.31. Black stripe symptoms of banana streak disease on a leaf of 'Mysore' (AAB) on Badu Island in the Torres Strait region of Australia (photo: D.R. Jones, QDPI).

Plate 6.32. Black stripe symptoms of banana streak disease on the pseudostem of 'Mysore' (AAB) growing in Honduras (photo: D.R. Jones, INIBAP).

The most recent reports come from the results of tests undertaken at INIBAP Virus Indexing Centres, tests undertaken by BSV research workers and personal observations of plant pathologists. These confirm the susceptibility of the Cavendish, Plantain and Lujugira–Mutika subgroups and the widespread distribution of diseased 'Mysore'. Additional clones found with BSV particles include 'Kisubi' (AB, 'Ney Poovan') from Burundi, 'Pisang Berangan' (AAA, syn. 'Lakatan') in Malaysia, 'Pisang Raja' in India and Malaysia, cultivars in the AAB Maia Maoli–Popoulu subgroup from Colombia and Australia and 'Monthan' (ABB, Bluggoe subgroup) from Côte d'Ivoire. The virus has also been found in some AAAB hybrids from breeding programmes in Guadeloupe, Honduras, Nigeria and Brazil.

Control

Since BSV appears to spread primarily by vegetative propagation, the most effective disease control strategy is to ensure that source plants used for propagation are virus-free. However, the visual diagnosis of BSV infection in banana is often difficult, as symptoms of banana streak appear sporadically during the course of the year.

Visual inspection of plants for evidence of BSV infection should therefore include an examination of all leaves present on a plant, and should be done at several intervals during the year. In Ecuador, where banana streak is a serious problem in some commercial Cavendish plantations, plants with symptoms are quickly destroyed after spraying with insecticide to kill mealybugs. Surrounding plants are also treated with insecticide in an effort to contain the outbreak. If 10 plants with symptoms are seen in a 50 m^2 area, then all plants in that area are destroyed. However, these practices have failed to stop spread and more drastic action is being contemplated.

Symptoms of banana streak can be readily mistaken for those of banana mosaic (see Lockhart and Jones, this chapter, p. 258), as both include foliar mosaic, cigar-leaf necrosis and internal pseudostem necrosis (heart rot). Although the pattern of foliar mosaic symptoms induced by BSV can, with familiarity and practice, be distinguished from that caused by CMV, this is difficult for the untrained eye.

The highly heterogeneous nature of BSV isolates (Lockhart and Olszewski, 1993) poses the most serious obstacle to the reliable detection of BSV. Serological heterogeneity has made it difficult to develop routine virus indexing protocols capable of detecting the complete range of virus isolates (Lockhart and Olszewski, 1993). A significant improvement in BSV detection by ELISA was achieved by developing assay protocols using polyclonal antibodies produced in two different animal species (Ndowora, 1998). In spite of marked improvements in the reliability of BSV detection by ELISA, there remain virus isolates that are detected weakly or not at all by this method. However, these and all other BSV isolates can be detected by immunosorbent electron microscopy (ISEM), using partially purified leaf-tissue extracts (Bouhida *et al.*, 1993). Unfortunately, this method is expensive and laborious and requires specialized equipment and skills.

Genome-based methods of BSV detection (e.g. nucleic acid hybridization, PCR amplification), which are potentially highly sensitive, are seriously compromised by both genomic heterogeneity among BSV isolates (Lockhart and Olszewski, 1993) and the occurrence of integrated BSV genomic sequences in the *Musa* genome (LaFleur *et al.*, 1996; Ndowora *et al.*, 1999). However, the use of IC-PCR (Thottappilly *et al.*, 1997) avoids false positives due to integrated viral sequences and is a potentially useful tool in diagnosing BSV infection.

In view of the problems of reliable detection of BSV discussed above, the visual inspection of all leaves over 1 year, which includes a prolonged period of cool temperatures, is advised for those without access to ISEM technology. This should be combined with the routine indexation of plants by the most reliable and sensitive method available. For those that have ISEM technology, plants established from tissue culture should be indexed after 3 and 6 months as described by Diekmann and Putter (1996).

There are no means available at present to eliminate BSV from infected material so that it can be used as a source of shoot tips for tissue-culture propagation. As the concentration of BSV decreases in leaf tissue when infected plants are grown at 28–35°C (Dahal *et al.*, 1998b), it may be possible to develop a protocol for obtaining BSV-free material through a combination of thermotherapy and other methods (Dahal *et al.*, 1998a). Inconclusive preliminary work on the effect of cryopreservation on BSV-infected meristems and cell suspensions has been reported (Muylle, 1998).

As mentioned earlier, there are no data suggesting that there is significant field spread of BSV by mealybug vectors other than in some commercial Cavendish plantations in Ecuador. However, as banana streak may significantly reduce yields, diseased plants should be eliminated and replanted with healthy material. This practice would also avoid any risk of spread, especially in fields infested by mealybugs. Mealybug populations should be reduced if incidence is high.

Banana Mild Mosaic

J.E. Thomas, B.E.L. Lockhart and M.L. Iskra-Caruana

Introduction

Uncharacterized, filamentous, virus-like particles have been noted frequently in *Musa* germplasm accessions and field specimens in recent years (Rivera *et al.*, 1992; Anon., 1993; Lockhart, 1994b; M.L. Iskra-Caruana and J.E. Thomas, Montpellier and Brisbane, 1995, personal observation; Belacázar *et al.*, 1998; Caruana and Galzi, 1998). Recent research suggests that in most cases the same single virus is present.

The economic impact of this virus is not known. Observations in Latin America indicate that yield may be affected in both banana and plantain under some conditions (B.E.L. Lockhart, Minnesota, 1998, personal observation).

The virus appears to occur widely, wherever bananas are grown, and extensively in *in vitro* germplasm collections. Field records exist from Australia, Africa, South and South-East Asia, Central and South American and the Caribbean, in a wide variety of *Musa* cultivars and genotypes (M.L. Iskra-Caruana, B.E.L. Lockhart and J.E. Thomas, Montpellier, Minnesota and Brisbane, 1998, unpublished).

Symptoms

The symptomatology of the virus, which has been named banana mild mosaic virus (BanMMV), is uncertain. In many cases, both in glasshouse-grown plants and in the field, symptomless infection occurs. 'Ducasse' (ABB, syn. 'Pisang Awak') in the field has been observed with chlorotic mosaic and streak symptoms (Plate 6.40). These symptoms, which are also often seen on young 'Ducasse' plants derived from infected tissue cultures, may disappear as the plant matures. Transitory chlorotic leaf streaks were noted on some leaves of 'Daluyao' (AAB, Plantain subgroup) infected with BanMMV alone and growing under glasshouse conditions (J.E. Thomas, Brisbane, 1998, personal observation). 'Pisang Seribu' (AAB), infected with BanMMV alone, displays silvery streaks on the leaf lamina, especially on the distal part, which are best viewed from the underside of the leaf (Plate 6.41). Plants of 'Gros Michel' (AAA) from Latin America have displayed large chlorotic blotches on the leaves, stunting and delayed bunching (B.E.L. Lockhart, Minnesota, 1998, personal observation).

Mixed infections of BanMMV and BSV are common and, in such cases, symptoms are reminiscent of those caused by BSV alone (M.L. Iskra-Caruana, B.E.L. Lockhart and J.E. Thomas, 1998, Montpellier, Minnesota, Brisbane, personal observations). In mixed infections with CMV, the presence of BanMMV seems to correlate with an additional leaf necrosis symptom (Caruana and Galzi, 1998).

Causal agent

Virions are flexuous filaments about 580 nm long and 14 nm wide (Plate 6.42) (B.E.L. Lockhart and J.E. Thomas, Minnesota and Brisbane, 1999, personal observation). The coat protein has an apparent M_r of 30 kDa in SDS-PAGE and, translated from the coat-protein gene sequence, is estimated to be 27 kDa. The concentration of virions in infected plants is variable, though generally low, with yields of purified virions in the order of 1 mg kg^{-1} tissue (C.F. Gambley and J.E. Thomas, Brisbane, 1998, unpublished).

The genome is ssRNA of size *c.* 7.4 kb, contains five ORFs (from the 5′ end: ORF1/polymerase, probably 205 kDa, triple gene block, 26, 12 and 8 kDa and ORF 5/coat protein 27 kDa), 3′ and 5′ untranslated regions and a poly-A tail. A dsRNA of *c.* 7 kbp has been detected in plants of 'Poingo' (AAB, Maia Maoli–Popoulu subgroup) and 'Kelong Mekintu' (AAB, Plantain subgroup) containing BanMMV particles (C.F. Gambley and J.E. Thomas, Brisbane, 1999, unpublished results).

BanMMV shares some, but not all, properties with the potexvirus group. It is serologically related to members of that group (B.E.L. Lockhart, Minnesota, 1999, unpublished).

Plate 6.33 (left). Cigar-leaf necrosis symptom of banana streak disease on IITA plantain hybrid 'TMPx 548–9' in Nigeria. Brown streak symptoms can also be seen on the petiole to the left of the dead cigar leaf (photo: C. Pasberg-Gauhl and F. Gauhl, IITA).

It also has a genome organization (C.F Gambley and J.E. Thomas, Brisbane, 1998, unpublished) and particle length (B.E.L. Lockhart, Minnesota, 1998, unpublished) consistent with that group. There is no information on possible strain variation for BanMMV at this stage.

Initial detection of BanMMV was based on electron microscopy of sap preparations, and later of partially purified minipreps prepared by the method of Ahlawat and Lockhart (1996). Polyclonal antisera have now been prepared to isolates from 'M'bouroukou' (AAB, Plantain subgroup) (B.E.L. Lockhart, St Paul, 1998, unpublished) and 'Ducasse' (C.F. Gambley and J.E. Thomas, Brisbane, 1998, unpublished).

(a)

(b)

Plate 6.34. Internal necrosis of the pseudostem, a symptom of banana streak disease on IITA plantain hybrid 'TMPx 597–4' in Nigeria (photo: C. Pasberg-Gauhl and F. Gauhl, IITA).

Plate 6.35 (left). Emergence of the fruit bunch through the side of the pseudostem of a banana streak disease-affected 'Grand Nain' (AAA, Cavendish subgroup) in Costa Rica (photo: S. Daillot, CIRAD).

Plate 6.36 (a and b below). Peel splitting (a) and necrotic spot symptoms (b) of banana streak disease on fingers of 'Grand Nain' in Costa Rica (photo: C. Pasberg-Gauhl and F. Gauhl, CB).

(a)

(b)

Plate 6.37 (below). Split in the outer leaf sheath of the pseudostem of a Cavendish cultivar in Ecuador infected with banana streak virus (photo: D.R. Jones, INIBAP).

Both work well in ISEM, but only the latter is suitable, albeit with limited sensitivity, for ELISA. All isolates tested with polyclonal antibodies have been shown to be serologically related. Monoclonal antibodies (MAbs) have been prepared to an isolate from 'Cardaba' (ABB) and these have been used in tissue blot and western blot assays (Caruana *et al.*, 1995). A sensitive triple antibody sandwich ELISA has been developed which uses polyclonal coating antibodies (from B.E.L. Lockhart) and these MAbs as detecting antibodies (M.L. Iskra-Caruana, Montpellier, 1999, unpublished).

PCR primers have been developed which allow detection of BanMMV in both total nucleic acid extracts from infected plants and in immunocapture reverse transcriptase (RT)-PCR, using crude sap extracts. Some variation in PCR product size has been noted and this seems to be due to variations in the length of the 3′ untranslated region (M. Sharman and J.E. Thomas, Brisbane, 1999, unpublished).

Disease cycle and epidemiology

Vegetative propagation, through both conventional planting material and tissue culture, is the only known means of transmission of BanMMV. Indeed, this virus is one of the most common contaminants of international germplasm collections, probably due to both its wide distribution and the frequently symptomless nature of infection (M.L. Iskra-Caruana and J.E. Thomas, Montpellier and Brisbane, 1999, personal observation).

BanMMV has not been transmitted mechanically with sap extracts or partially purified preparations to banana or herbaceous indicators. The banana aphid, *Pentalonia nigronervosa*, transmitted BBrMV, but not BanMMV, from banana infected with both viruses (Caruana *et al*, 1995). Under natural conditions in Guadeloupe, plants established from virus-free tissue cultures of cultivars in the Cavendish subgroup (AAA) subsequently became field-infected with CMV and BanMMV, frequently as mixed infections, and the incidence of infection increased during the cropping cycle. This implies that natural transmission of BanMMV was occurring, generally simultaneously with the aphid-transmitted CMV (Caruana and Galzi, 1998).

Host reaction

All *Musa* genotypes appear to be susceptible to BanMMV, though this observation is based only on natural infections. All attempts to transmit the virus to banana or to herbaceous indicators have been unsuccessful (Caruana and Galzi, 1998; B.E.L. Lockhart and J.E. Thomas, Minnesota and Brisbane, 1999, unpublished).

Control

The use of indexed, virus-free vegetative planting material is recommended, though the economic impact of this virus is unknown, either alone or in mixed infections with other viruses.

Banana Die-back

J.d'A. Hughes and J.E. Thomas

Symptoms characteristic of this virus disease were first seen on 'Valery' (AAA, Cavendish subgroup) in a hydromorphic plot in Ibadan, Nigeria, in 1996. Banana die-back has only been reported from a limited number of locations in Nigeria, though plants with similar symptoms have been noted in Ghana and Cameroon (Hughes *et al.*, 1998; S.K. Offei, Accra, 1998, personal communication).

Symptoms include leaf chlorosis, wrinkling, marginal necrosis and dieback of the cigar leaf (Plate 6.43). Subsequent suckers from the same mat become progressively more stunted and eventually the mother plant dies.

Isometric virions *c.* 28 nm in diameter have been purified from both infected banana plants and *Nicotiana occidentalis*. Banana dieback virus (BDBV) has been shown to be serologically related to certain nepoviruses (Hughes *et al.*, 1998).

Although limited field spread appears to occur (J.d'A. Hughes, Ibadan, 1998, personal observation), the natural vector of BDBV is unknown.

BDBV can be mechanically transmitted to a limited range of herbaceous indicator plants, including *Nicotiana clevelandii*, *N. occidentalis* and *Vigna unguiculata* cv. TVu-76.

Abacá Mosaic

J.E. Thomas and L.V. Magnaye

Introduction

The first record of abacá mosaic was on the island of Mindanao in the Philippines in 1925 (Eloja and Tinsley, 1963). Calinisan (1934) first described the disease near Davao and a few years later it was identified in Cotabato province (Celino, 1940; Kent, 1954). Abacá mosaic was initially, though incorrectly (Eloja and Tinsley, 1963), thought to be caused by cucumber mosaic virus (CMV), due to some similarities to infectious chlorosis disease of banana described from Australia by Magee (1940a).

By the late 1930s, abacá mosaic was the major factor limiting abacá production near Davao, being responsible for heavy losses of fibre and much capital investment. This caused the relocation of the industry from the Mount Apo area to more remote regions (Stover, 1972). By 1955, abacá mosaic affected 47% of the 184,000 ha under abacá cultivation. Losses of 25–50% in new plantings were common. Abacá mosaic was later reported from Luzon and certain areas in the Visayas. Today, it is found in most areas where abacá is grown in the Philippines.

Early infection renders plants unproductive and worthless. However, some fibre production is possible when older plants (1–2 years old) become infected and the irregular distribution of the virus in individual abacá clumps can allow some stalks to be temporarily disease-free (Magee, 1957; Stover, 1972).

There has been a resurgence in demand for abacá fibre in recent years and the Philippines supply over 80% of total world production. Abacá mosaic and bunchy top (see Thomas and Iskra-Caruana, this chapter, pp. 241–253) are serious constraints to production (Raymundo, 1998).

Symptoms

On inoculated abacá, symptoms become apparent after 7–21 days (Celino and Ocfemia, 1941; Eloja *et al.*, 1962; Magnaye and Eloja, 1968). Small whitish dots, which later elongate to become spindle-shaped chlorotic streaks, appear first parallel to the minor leaf veins. As these chlorotic areas expand, they may develop rusty brown borders and can extend from the midrib to the leaf margin. Subsequent leaves can develop extensive broad yellow or pale green stripes across the width of the lamina (Plate 6.44). With older infections, chlorotic, orange or dark green streaks may be present on the midribs and petioles, even when lamina symptoms are absent (Ocfemia and Celino, 1938; Eloja *et al.*, 1962; Stover, 1972). Leaves can also become distorted (Plate 6.45).

In some earlier reports, it is uncertain whether symptom descriptions pertain to abacá mosaic, as some illustrations are reminiscent of CMV infection (Calinisan, 1939). Diagnosis of the disease based on symptoms alone must be viewed with caution, as abacá in the Philippines can also be affected by BBrMV (Thomas *et al.*, 1997; this chapter, pp. 253–255). Symptoms produced by BBrMV are very similar to those of abaca mosaic.

Causal agent

Flexuous filamentous virions about 680 nm long have been detected in abacá mosaic-affected plants (Plate 6.46) (Eloja and Tinsley, 1963) and are considered to be those of the causal agent. However, the presence of this type of virus particle in abacá with mosaic symptoms does not necessarily mean that the causal agent is abacá mosaic virus (AbaMV) as BBrMV is also a potyvirus (Thomas *et al.*, 1997; Thomas *et al.*, this chapter, pp. 253–255).

On the basis of vector transmission characteristics and host range, Magee (1957) suggested that AbaMV was related to sugar cane mosaic virus (SCMV). AbaMV was subsequently shown by microprecipitin tests to be serologically related to sugar cane mosaic potyvirus (Eloja and

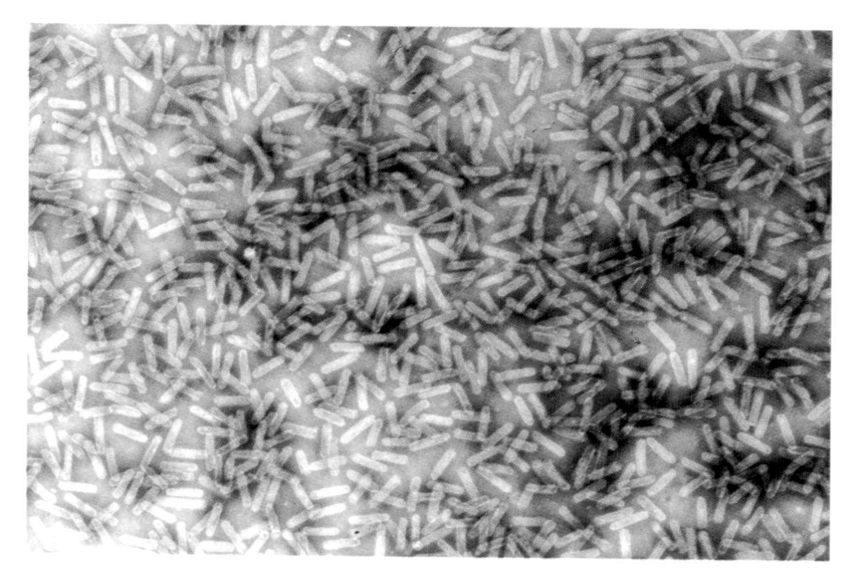

Plate 6.38. Bacilliform particles of banana streak virus stained with sodium phosphotungstate (PTA), pH 7.0 (photo: B.E.L. Lockhart, UM).

Plate 6.39. A colony of *Planococcus citri*, a mealybug vector of banana streak virus, on a banana peduncle (photo: B.E.L. Lockhart, UM).

Plate 6.40. Mosaic and chlorotic streak symptoms of banana mild mosaic disease on a leaf of 'Ducasse' (ABB, syn. 'Pisang Awak') in Queensland, Australia (photo: J.E. Thomas, QDPI).

Plate 6.41. Silver streak symptoms of banana mild mosaic disease on a leaf of 'Pisang Seribu' (AAB) in Florida (photo: B.E.L. Lockhart, UM).

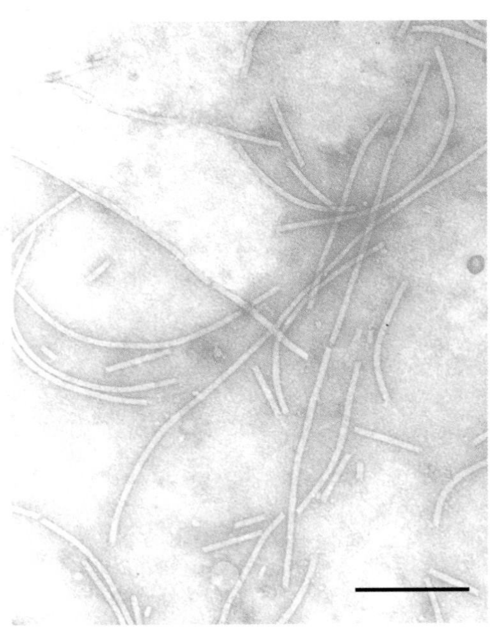

Plate 6.42. Virions of banana mild mosaic virus from 'Ducasse' (ABB, syn. 'Pisang Awak') in Australia. Bar represents 200 nm (photo: J.E. Thomas, QDPI).

Plate 6.43. Symptoms of banana dieback disease on IITA hybrid 'TMPx 7152–2' in Nigeria (photo: J.d'A. Hughes).

Tinsley, 1963). By ELISA, relationships were demonstrated with members of the sugar cane mosaic subgroup of potyviruses (including SCMV, maize dwarf mosaic, sorghum mosaic and Johnson grass mosaic viruses) and also a weak relationship with BBrMV (Gambley *et al.*, 1997; C.F. Gambley and J.E. Thomas, Brisbane, 1998, unpublished).

Sequence analysis of the coat-protein gene and 3′ untranslated region has confirmed the close relationship of AbaMV to members of the SCMV subgroup of potyviruses. AbaMV is most closely related to maize dwarf mosaic virus-B (MDMV-B, considered to be a strain of SCMV; C.F. Gambley and J.E. Thomas, Brisbane, 1998, unpublished), but these two isolates seem somewhat distantly related to other isolates in this virus cluster.

Serological and nucleic-acid-based assays are available for AbaMV. The virus can be detected by plate-trapped antigen ELISA, using antisera to SCMV or MDMV (C.F. Gambley and J.E. Thomas, Brisbane, 1998, unpublished). AbaMV can also be detected by PCR, using degenerate potyvirus group primers (Thomas *et al.*, 1997).

Disease cycle and epidemiology

AbaMV is transmitted in the non-persistent manner by a number of aphid species, including *Aphis gossypii*, *Rhopalosiphum nymphaeae* (Celino, 1940), *Rhopalosiphum maidis* (Celino and Ocfemia, 1941), *Schizaphis cyperi*, *Schizaphis graminum* (Gavarra and Eloja, 1969) and *Toxoptera citricidus* (Eloja *et al.*, 1977). *Aphis gossypii* and *R. maidis* are the principal natural vectors. The ability to transmit is rapidly lost during a series of transfers of the aphids to new hosts, consistent with a non-persistent mode of transmission (Celino, 1940). It appears that the important natural vectors of AbaMV do not colonize abacá and, indeed, the banana aphid *Pentalonia nigronervosa* appears not to be a vector of AbaMV (Celino, 1940; Ocfemia *et al.*, 1947).

There is no transmission through true seed of abacá (Calinisan, 1939), but the virus is transmitted through vegetative planting material and tissue culture.

Mechanical transmission of AbaMV is possible, though infected abacá is a poor source of inoculum (Eloja *et al.*, 1962). However, once the virus is transmitted to alternative hosts in the *Gramineae*, by either mechanical inoculation or aphid transmission, it is readily sap-transmissible, even back to abacá (Eloja *et al.*, 1962; Benigno and Del Rosario, 1965).

Host reaction

Hosts are restricted to monocotyledonous plants. Natural hosts include *Musa textilis* (abacá), *Canna indica* (*Cannaceae*) and *Maranta arundinacea* (*Marantaceae*), while the following members of the *Gramineae* are experimental hosts: *Digitaria sanguinalis*, *Echinochloa colonum*, *Imperata cylindrica*, *Rottboellia exaltata*, *Setaria palmifolia* and *Zea mays* (Velasco-Magnaye and Eloja, 1966).

The members of the *Gramineae* listed above and also *Andropogon halepensis*, *Coix lachryma-jobi*, *Panicum distachyum* and *Paspalum conjugatum* have been observed in the field to have mosaic symptoms and contain viruses transmissible to abacá with *A. gossypii* or *R. maidis*. However, the symptoms produced in abacá consist of only a few spindle-shaped chlorotic lesions on the lamina and are much less severe than those produced by typical isolates from abacá (Celino and Martinez, 1956; Velasco-Magnaye and Eloja, 1966). It is not known whether these grasses contain mild strains of AbaMV or other distinct viruses. No cross-protection was evident when abacá plants infected with the viruses from *R. exaltata* or *P. distachyum* were superinfected with typical AbaMV (Celino and Martinez, 1956).

In the *Musaceae*, *Ensete glaucum*, all commercial varieties of abacá (Bernardo and Umali, 1956) and a number of banana cultivars, which include 'Lakatan' (AAA), 'Umalag' (AAA, Cavendish subgroup), 'Grand Nain' (AAA, Cavendish subgroup) and 'Latundan' (AAB, syn. 'Silk') (L.V. Magnaye, Davao, 1998, unpublished), are susceptible to AbaMV. Resistance has been noted in *Musa balbisiana* accession 'Pacol', the *M. balbisiana* × *M. textilis* hybrid

'Canton' and *Musa ornata* (Bernardo and Umali, 1956).

Control

The aphid vectors of AbaMV do not colonize abacá. It has been recommended that maize, a host favoured by the aphid vector *R. maidis*, and other grasses should be excluded from plantations (Stover, 1972). However, studies by Celebrar *et al.* (1970) showed that total weed removal resulted in twice the incidence of AbaMV as compared with cover cropping or removal of weeds only from around the bases of the abacá plants. Insecticide spraying did not result in significant reduction in disease incidence.

A major source of infection is infected planting material. When infected corms are planted, symptoms usually start to appear within 2 months, but can take 12–18 months (Eloja *et al.*, 1962; Stover, 1972). The virus appears to be unevenly distributed in the corm, and it is possible to obtain virus-free suckers from an infected corm if the plants are severed 3–5 weeks after emergence (Pacumbaba, 1967). Roguing of infected plants and the use of virus-free planting material is considered essential to contain the spread of abacá mosaic (Magee 1957, 1960; Eloja *et al.*, 1962).

Enset Streak

M. Tessera and A.J. Quimio

Enset streak was first noticed at Hagereselem in the Sidamo zone and at the Areka Experimental Station in the North Omo zone of Ethiopia in 1991 (Tessera and Quimio, 1994). The disease was later recorded from the major growing areas (Tessera *et al.*, 1996).

Chlorotic and yellow mosaics, streaks and stripes are characteristic leaf symptoms of the disease (Plates 6.47 and 6.48). Severely affected plants also have narrow, distorted leaves and become stunted (Plate 6.49). Chlorotic areas of leaves later turn necrotic (Plate 6.49). Early infection results in a significant reduction in yield. Symptoms are more pronounced in the winter than in the summer (Tessera *et al.*, 1998).

Enset streak is believed to be caused by a badnavirus (Tessera *et al.*, 1996). Virus particles are bacilliform in shape and dimensions are given as 118–125 nm × 29.5–30.0 nm (Tessera *et al.*, 1998). The relationship of this virus with the BSV (see Lockhart and Jones, this chapter, pp. 263–274) has not been determined.

The major means of dissemination of the disease is through infected corms or suckers arising from an infected corm. A vector has not yet been identified and there is no information on clonal suscepibility, although various cultivars have been seen affected. The use of healthy corms for propagation is advocated for the control of the problem.

REFERENCES

Adam, A.V. (1962) An effective program for the control of banana mosaic. *Plant Disease Reporter* 46, 366–370.

Allen, R.N. (1978a) Epidemiological factors influencing the success of roguing for the control of bunchy top disease of bananas in New South Wales. *Australian Journal of Agricultural Research* 29, 535–544.

Allen, R.N. (1978b) Spread of bunchy top disease in established banana plantations. *Australian Journal of Agricultural Research* 29, 1223–1233.

Allen, R.N. (1987) Further studies on epidemiological factors influencing control of banana bunchy top disease, and evaluation of control measures by computer simulation. *Australian Journal of Agricultural Research* 38, 373–382.

Allen, R.N. and Barnier, N.C. (1977) The spread of bunchy top disease between banana plantations in the Tweed River District during 1975–76. In: *NSW Department of Agriculture, Biology Branch Plant Disease Survey (1975–76)*, pp. 27–28.

Plate 6.44. Chlorotic discoloration symptoms of abacá mosaic disease on an abacá leaf in the Philippines (photo: L.V. Magnaye, BPI).

Plate 6.45. Distortion symptoms of abacá mosaic disease on an abacá leaf in the Philippines (photo: L.V. Magnaye, BPI).

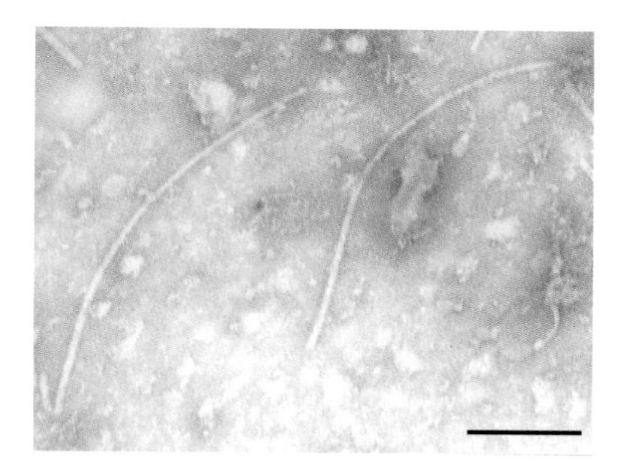

Plate 6.46. Virions of abacá mosaic virus. Bar represents 200 nm (photo: J.E. Thomas, QDPI).

Plate 6.47. Yellow leaf streak symptom of enset streak disease in Ethiopia (photo: M. Tessera, IAR).

Plate 6.49. Narrow, distorted leaves of a young enset plant affected by enset streak disease in Ethiopia. Areas of leaf tissue with chlorotic mosaics and streaks are turning necrotic (photo: A.J. Quimio, IAR).

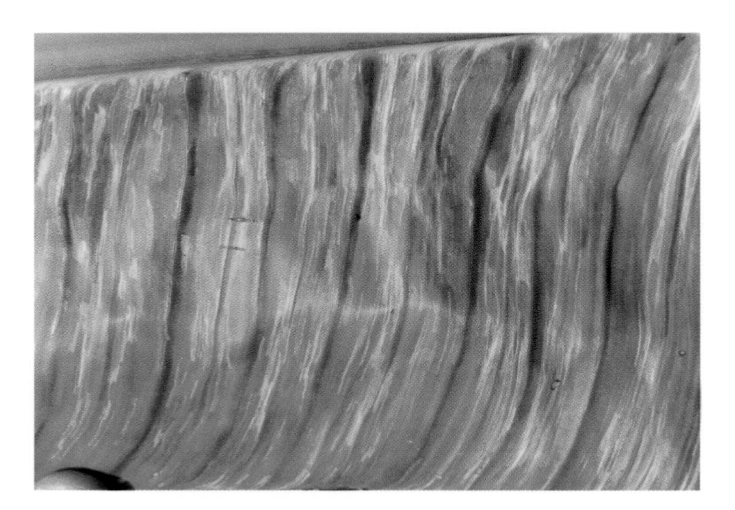

Plate 6.48. Chlorotic mosaic and leaf surface distortion symptoms of enset streak disease in Ethiopia (photo: A.J. Quimio, IAR).

Anceau, C. (1996) Identification and characterisation of viruses associated to the syndrome of banana bunchy top disease (BBTD). In: *Final Scientific Report, March 1993 to February 1996.* Faculté Universitaire des Sciences Agronomiques de Gembloux, Gembloux, Belgium.

Anon. (1992) *Bananas, Plantains and INIBAP, Annual Report 1992.* INIBAP, Montpellier, France, 31 pp.

Anon. (1993) Risks involved in the transfer of banana and plantain germplasm. In: *Bananas, Plantains and INIBAP, Annual Report 1993.* INIBAP, Montpellier, France, pp. 39–47.

Anon. (1995) *Musa*News. *Infomusa* 4(2), 26–30.

Anunciado, I.S., Balmes, L.O., Bawagan, P.Y., Benigno, D.A., Bondad, N.D., Cruz, O.J., Franco, P.T., Gavarra, M.R., Opeña, M.T. and Tabora, P.C. (1977) *The Philippines Recommends for Abaca, 1977.* Philippine Council for Agriculture and Resources Research, Los Baños, Philippines, 71 pp.

Bateson, M.F. and Dale, J.L. (1995) Banana bract mosaic virus: characterisation using potyvirus specific degenerate PCR primes. *Archives of Virology* 140, 515–527.

Beaver, R.G. (1982) Use of picloram for eradication of banana diseased with bunchy top. *Plant Disease* 66, 906–907.

Beetham, P.R., Hafner, G.J., Harding, R.M. and Dale, J.L. (1997) Two mRNAs are transcribed from banana bunchy top virus DNA-1. *Journal of General Virology* 78, 229–236.

Belalcázar, S., Reichel, H., Pérez, R., Gutierréz, T., Múnera, G. and Arévalo, E. (1998) Banana streak badnavirus infection in *Musa* plantations in Colombia. In: Frison, E. and Sharrock, S.L. (eds) (1998) *Banana Streak Virus: A Unique Virus–Musa Interaction? Proceedings of a Workshop of the PROMUSA Virology Working Group held in Montpellier, France, 19–21 January 1998.* INIBAP, Montpellier, France, pp. 55–57.

Benigno, D.B. and Del Rosario, M.S.E. (1965) Mechanical transmission of the abaca mosaic virus and some of its physical properties. *The Philippine Agriculturist* 49, 197–210.

Berg, L.A. and Bustamante, M. (1974) Heat treatment and meristem culture for the production of virus-free bananas. *Phytopathology* 64, 320–322.

Bernardo, F.S. and Umali, D.L. (1956) Possible sources of resistance to abaca mosaic and bunchy-top. *The Philippine Agriculturist* 40, 277–284.

Blackman, R.L. and Eastop, V.F. (1984) *Aphids on the World's Crops. An Identification Guide.* John Wiley and Sons, Chichester, UK, 466 pp.

Bouhida, M. and Lockhart, B.E. (1990) Increase in importance of cucumber mosaic virus infection in greenhouse-grown bananas in Morocco. *Phytopathology* 80, 981.

Bouhida, M., Lockhart, B.E.L. and Olszewski, N.E. (1993) An analysis of the complete, nucleotide sequence of a sugarcane bacilliform virus genome infections of banana and rice. *Journal of General Virology* 74, 15–22.

Bryce, G. (1921) *The Bunchy Top Plantain Disease.* Leaflet 18, Department of Agriculture, Peradeniya, Ceylon, 2 pp.

Burns, T.M., Harding, R.M. and Dale, J.L. (1995) The genome organization of banana bunchy top virus: analysis of six ssDNA components. *Journal of General Virology* 76, 1471–1482.

Calinisan, M.R. (1934) Notes on a suspected 'mosaic' of abaca in the Philippines. *Philippine Journal of Agriculture* 5, 225–258.

Calinisan, M.R. (1939) A comprehensive study on symptoms of abaca mosaic. *Philippine Journal of Agriculture* 10, 121–130.

Campbell, J.G. (1926) *Annual Report of the Fiji Department of Agriculture, 1925.* Fiji Department of Agriculture, Suva.

Caruana, M.L. and Galzi, S. (1998) Identification of uncharacterised filamentous viral particles on banana plants. *Acta Horticulturae* 490, 323–335.

Caruana, M.L., Galzi, S., Séchet, H., Bousalem, M. and Bringaud, C. (1995) Etiologie de la maladie de la mosaïque des bractées des bananiers pour un diagnostic cible. In: *Rapport d'Activités 1994–1995,* Laboratoire de Phytovirologie des Régions Chaudes, CIRAD, ORSTOM, Montpellier, pp. 6–9.

Castillo, B.S. and Martinez, A.L. (1961) Occurrence of bunchy top disease of banana in the Philippines. *FAO Plant Protection Bulletin* 9, 74–75.

Celebrar, P.R., Gonzales, C.S., Eloja, A.L. and Gavarra, M.R. (1970) Further studies on three methods of weed control and insecticide spraying as control measures against the abaca mosaic diseases. *Philippine Journal of Plant Industry* 35, 79–86.

Celino, M.S. (1940) Experimental transmission of the mosaic of abaca, or manila hemp plant (*Musa textilis* Née). *The Philippine Agriculturist* 29, 379–414.

Celino, M.S. and Martinez, A.L. (1956) Transmission of viruses from different plants to abacá (*Musa textilis* Née). *The Philippine Agriculturist* 40, 285–302.

Celino, M.S. and Ocfemia, G.O. (1941) Two additional insect vectors of mosaic of abacá, or manila hemp plant, and transmission of its virus to corn. *The Philippine Agriculturist* 30, 70–78.

Comstock, J.C. and Lockhart, B.E. (1990) Widespread occurrence of sugarcane bacilliform virus in U.S. sugarcane germplasm collections. *Plant Disease* 74, 530.

Dabek, A.J. and Waller, J.M. (1990) Black leaf streak and viral leaf streak: new banana diseases in East Africa. *Tropical Pest Management* 36, 157–158.

Dahal, G., Hughes, J.d'A. and Lockhart, B.E.L. (1998a) Status of banana streak disease in Africa: problems and future research needs. *Integrated Pest Management Reviews* 3, 85–97.

Dahal, G., Hughes, J.d'A., Thottappilly, G. and Lockhart, B.E.L. (1998b) Effect of temperature on symptom expression and reliability of banana streak badnavirus detection in naturally infected plantain and banana (*Musa* spp.). *Plant Disease* 82, 16–21.

Dale, J.L. (1987) Banana bunchy top: an economically important tropical plant virus disease. *Advances in Virus Research* 33, 301–325.

Dale, J.L. (1996) Structure and function of the banana bunchy top virus genome. In: *Abstracts of the Xth International Congress of Virology, Jerusalem, Israel, 11–16 August 1996*, Abstract W52–4, p. 76.

Dale, J.L., Phillips, D.A. and Parry, J.N. (1986) Double-stranded RNA in banana plants with bunchy top disease. *Journal of General Virology* 67, 371–375.

Daniells, J., Thomas, J.E. and Smith, B.J. (1995) Seed transmission of banana streak confirmed. *Infomusa* 4(1), 7.

Daniells, J., Geering, A. and Thomas, J.E. (1998) Banana streak virus investigations in Australia. *Infomusa* 7(2), 20–21.

Darnell-Smith, G.P. (1924) 'Bunchy top' disease in banana. *Queensland Agricultural Journal* 21, 169–179.

Darnell-Smith, G.P. and Tryon, H. (1923) Banana bunchy top disease. *Queensland Agricultural Journal* 19, 32–33.

Desvignes, J.C. and Cardin, L. (1973) Contribution à l'étude du virus de la mosaïque du concombre (CMV). IV. Essai de classification de plusieurs isolats sur la base de leur structure antigénique. *Annual Review of Phytopathology* 5, 409–430.

Diekmann, D.R. and Putter, C.A.J. (1996) Musa, 2nd edn. FAO/IPGRI Technical Guidelines for the Safe Movement of Germplasm, No. 15, Food and Agriculture Organization of the United Nations, Rome/International Plant Genetic Resources Institute, Rome, 28 pp.

Dietzgen, R.G. and Thomas, J.E. (1991) Properties of virus-like particles associated with banana bunchy top disease in Hawaii, Indonesia and Tonga. *Australasian Plant Pathology* 20, 161–165.

Drew, R.A., Moisander, J.A. and Smith, M.K. (1989) The transmission of banana bunchy-top virus in micropropagated bananas. *Plant Cell, Tissue and Organ Culture* 16, 187–193.

Eloja, A.L. and Tinsley, T.W. (1963) Abaca mosaic virus and its relationship to sugarcane mosaic. *Annals of Applied Biology* 51, 253–258.

Eloja, A.L., Velasco, L.G. and Agati, J.A. (1962) Studies on the abaca mosaic disease. 1. Sap transmission of the abaca mosaic virus. *Philippine Journal of Agriculture* 27, 75–84.

Eloja, A.L., Gavarra, M.R. and Opeña, M.T. (1977) The seasonal occurrence of aphids and its relation to abaca mosaic disease incidence in the field. *Philippine Journal of Plant Industry* 31, 56–60.

Espino, R.C., Magnaye, L.V., Johns, A.P. and Juanillo, C. (1993) Evaluation of Philippine banana cultivars for resistance to bunchy-top and fusarium wilt. In: Valmayor, R.V., Hwang, S.C., Ploetz, R., Lee, S.C. and Roa, N.V. (eds) *Proceedings: International Symposium on Recent Developments in Banana Cultivation Technology, Taiwan Banana Research Institute, Chiuju, Pingtung, Taiwan, 14–18 December 1992.* INIBAP/ASPNET, Los Baños, Philippines, pp. 89–102.

Fauquet, C.M. and Beachy, R.N. (1993) Status of coat protein-mediated resistance and its potential application for banana viruses. In: *Proceedings of the Workshop on Biotechnology Applications for Banana and Plantain Improvement held in San José, Costa Rica, 27–31 January 1992.* INIBAP, Montpellier, France, pp. 69–84.

Ferreira, S.A., Trujillo, E.E. and Ogata, D.Y. (1989) *Bunchy Top Disease of Bananas.* Commodity Fact Sheet BAN-4(A), FRUIT, information leaflet prepared by the Hawaii Cooperative Extension Service, Hawaii Institute of Tropical Agriculture and Human Resources, University of Hawaii at Manoa.

Francki, R.I.B., Mossop, D.W. and Hatta, T. (1979) *Cucumber Mosaic Virus.* Descriptions of Plant Viruses No. 213, Commonwealth Mycological Institute and Association of Applied Biologists, Kew, Surrey, England, 6 pp.

Frison, E.A. and Putter, C.A.J. (1989) *FAO/IBPGR Technical Guidelines for the Safe Movement of* Musa *Germplasm.* FAO/IBPGR, Rome, 23pp.

Frison, E. and Sharrock, S.L. (eds) (1998) *Banana Streak Virus: A Unique Virus–Musa Interaction? Proceedings of a Workshop of the PROMUSA Virology Working Group held in Montpellier, France, 19–21 January 1998.* INIBAP, Montpellier, France, 70 pp.

Gadd, C.H. (1926) Bunchy top disease of plantains (a review). *Tropical Agriculturist* 66, 3–20.

Gafny, R., Wexler, A., Mawassi, M., Israeli, Y. and Bar-Joseph, M. (1996) Natural infection of banana by a satellite-containing strain of cucumber mosaic virus: nucleotide sequence of the coat protein gene and the satellite RNA. *Phytoparasitica* 24, 49–56.

Gambley, C.F., Thomas, J.E., Geering, A.D.W. and Magnaye, L.V. (1997) Differentiation of two potyviruses infecting *Musa,* abaca mosaic virus and banana bract mosaic virus. In: *Abstracts of the 11th Biennial APPS Conference, Australasian Plant Pathology Society, Perth, 29 September–2 October 1997*, p. 108.

Gauhl, F. and Pasberg-Gauhl, C. (1997) First report of banana streak badnavirus in plantain landraces in southern Cameroon, Central Africa. *Plant Disease* 81, 1335.

Gauhl, F. and Pasberg-Gauhl, C. (1994) *Symptoms Associated with Banana Streak Virus (BSV).* Plant Health Management Division, International Institute of Tropical Agriculture, Ibadan, Nigeria, 20 pp.

Gauhl, F., Pasberg-Gauhl, C., Bopda-Waffo, A., Hughes, J.d'A. and Chen, J.S. (1999a) Occurrence of banana streak badnavirus on plantain and banana in 45 villages in southern Cameroon, Central Africa. *Zeitschrift für Pflanzenkrankheiten und Pflanzenschutz* 106, 174–180.

Gauhl, F., Pasberg-Gauhl, C., Lockhart, B.E.L., Hughes, J.d'A. and Dahal, G. (1999b) Incidence and distribution of banana streak badnavirus in the plantain production region of southern Nigeria. *International Journal of Pest Management* 44, 167–171.

Gavarra, M.R. and Eloja, A.L. (1969) Further studies on the insect vectors of the abaca mosaic virus: II. Experimental transmission of the abaca mosaic virus by *Schizaphis cyperi* (van der Groot) and *S. graminum* Rondani. *Philippine Journal of Plant Industry* 34, 89–96.

Geering, A.D.W. and Thomas, J.E. (1996) A comparison of four serological tests for the detection of banana bunchy top virus in banana. *Australian Journal of Agricultural Research* 47, 403–412.

Geering, A.D.W. and Thomas, J.E. (1997) Search for alternative hosts of banana bunchy top virus in Australia. *Australasian Plant Pathology* 26, 250–254.

Gold, A.H. (1972) Seed transmission of banana viruses. *Phytopathology* 62, 760 (abstract).

Gonda, T.J. and Symons, R.H. (1978) The use of hybridization analysis with complementary DNA to determine the RNA sequence homology between strains of plant viruses: its application to several strains of cucumoviruses. *Virology* 88, 361–370.

Hafner, G.J., Harding, R.M. and Dale, J.L. (1995) Movement and transmission of banana bunchy top virus DNA component one in bananas. *Journal of General Virology* 76, 2279–2285.

Hafner, G.J., Harding, R.M. and Dale, J.L. (1997a) A DNA primer associated with banana bunchy top virus. *Journal of General Virology* 78, 479–486.

Hafner, G.J., Stafford, M.R., Wolter, L.C., Harding, R.M. and Dale, J.L. (1997b) Nicking and joining activity of banana bunchy top virus replication protein *in vitro. Journal of General Virology* 78, 1795–1799.

Harding, R.M., Burns, T.M. and Dale, J.L. (1991) Virus-like particles associated with banana bunchy top disease contain small single-stranded DNA. *Journal of General Virology* 72, 225–230.

Harding, R.M., Burns, T.M., Hafner, G.J., Dietzgen, R.G. and Dale, J.L. (1993) Nucleotide sequence of one component of the banana bunchy top virus genome contains the putative replicase gene. *Journal of General Virology* 74, 323–328.

Hasse, A., Richter, J. and Rabenstein, F. (1989) Monoclonal antibodies for detection and serotyping of cucumber mosaic virus. *Journal of Phytopathology* 127, 129–136.

Hu, J.S., Li, H.P., Barry, K. and Wang, M. (1995) Comparison of dot blot, ELISA and RT-PCR assays for detection of two cucumber mosaic virus isolates infecting banana in Hawaii. *Plant Disease* 79, 902–906.

Hu, J.S., Wang, M., Sether, D., Xie, W. and Leonhardt, K.W. (1996) Use of polymerase chain reaction (PCR) to study transmission of banana bunchy top virus by the banana aphid (*Pentalonia nigronervosa*). *Annals of Applied Biology* 128, 55–64.

Hughes, J.d'A., Speijer, P.R. and Olatunde, O. (1998) Banana die-back virus – a new virus infecting banana in Nigeria. *Plant Disease* 82, 129.

Jeger, M.J., Eden-Green, S., Thresh, J.M., Johanson, A., Waller, J.M. and Brown, A.E. (1995) Banana Diseases. In: Gowen, S. (ed.) *Bananas and Plantains.* Chapman and Hall, London, UK, pp. 317–381.

Jones, D.R. (1994) Risks involved in the transfer of banana and plantain germplasm. In: Jones, D.R. (ed.) *The Improvement and Testing of* Musa*: a Global Partnership, Proceedings of the First Global Conference of the International Musa Testing Program held at FHIA, Honduras, 27–30 April 1994.* INIBAP, Montpellier, France, pp. 85–98.

Jones, D.R. (ed.) (1995) Black Sigatoka disease in Venezuela. *Infomusa* 4, 13–14.

Jones, D.R. and Lockhart, B.E.L. (1993) *Banana Streak Disease. Musa* Disease Fact Sheet No. 1, INIBAP, Montpellier, France.

Jose, P.C. (1981) Reaction of different varieties of banana against bunchy top disease. *Agricultural Research Journal of Kerala* 19, 108–110.

Kahlid, S. and Soomro, M.H. (1993) Banana bunchy top disease in Pakistan. *Plant Pathology* 42, 923–926.

Kahlid, S., Soomro, M.H. and Stover, R.H. (1993) First report of banana bunchy top virus in Pakistan. *Plant Disease* 77, 101.

Kaper, J.M. and Waterworth, H.E. (1977) Cucumber mosaic virus-associated RNA 5: causal agent for tomato necrosis. *Science* 196, 429–431.

Karan, M., Harding, R.M. and Dale, J.L. (1994) Evidence for two groups of banana bunchy top virus isolates. *Journal of General Virology* 75, 3541–3546.

Karan, M., Harding, R.M. and Dale, J.L. (1997) Association of banana bunchy top virus DNA components 2 to 6 with bunchy top disease. *Molecular Plant Pathology on-Line* http://www.bspp.org.uk/mppol/1997/0624karan.

Kawano, S. and Su, H.-J. (1993) Occurrence of banana bunchy top virus in Okinawa. *Annals of the Phytopathological Society of Japan* 59, 53.

Kent, G.C. (1954) Abaca mosaic. *Philippine Agriculturist* 17, 555–577.

Kenyon, L., Magnaye, L., Warburton, H., Herradura, L., Chancellor, T., Escobido, E. and Foot, C. (1996) *Epidemiology and Control of Banana Virus Diseases in the Philippines.* NRI – Department for International Development Crop Protection Programme Project A0217/X0258 Final Technical Report, Natural Resources Institute, Chatham Maritime, UK.

Kenyon, L., Brown, M. and Khonje, P. (1997) First report of banana bunchy top virus in Malawi. *Plant Disease* 81, 1096.

LaFleur, D.A., Lockhart, B.E.L. and Olszewski, N.E. (1996) Portions of the banana streak badnavirus genome are integrated in the genome of its host *Musa* spp. *Phytopathology* 86(11), S100.

Lassoudière, A. (1974) La mosaïque dite 'à tirets' du bananier 'Poyo' en Côte d'Ivoire. *Fruits* 29, 349–357.

Lassoudière, A. (1979) Mise en évidence des répercussions économiques de la mosaïque en tirets du bananier en Côte d'Ivoire. Possibilités de lutte par éradication. *Fruits* 34, 3–34.

Lockhart, B.E.L. (1986) Purification and serology of a bacilliform virus associated with a streak disease of banana. *Phytopathology* 76, 995–999.

Lockhart, B.E.L. (1994a) Banana streak virus. In: Ploetz, R.C., Gentmeyer, G.A., Nishijima, W.T., Rohrbach, K.G. and Ohr, H.D. (eds) *Compendium of Tropical Fruit Diseases.* American Phytopathological Society Press, St Paul, Minnesota, USA, 19–20.

Lockhart, B.E.L. (1994b) Development of detection methods for banana streak virus. In: *The Global Banana and Plantain Network, INIBAP Annual Report 1994.* INIBAP, Montpellier, France, pp. 20–21.

Lockhart, B.E.L. (1995) *Banana Streak Badnavirus Infection in* Musa*: Epidemiology, Diagnosis and Control.* Food and Fertilizer Technology Center Technical Bulletin 143, Food and Fertilizer Technology Center, Taipei, Taiwan, 11 pp.

Lockhart, B.E.L. and Autrey, L.J.C. (1988) Occurrence in sugarcane of a bacilliform virus related serologically to banana streak virus. *Plant Disease* 72, 230–233.

Lockhart, B.E.L. and Autrey, L.J.C. (1991) Mealybug transmission of sugarcane bacilliform and sugarcane clostero-like viruses. In: *International Society of Sugar Cane Technologists, 3rd Sugar Cane Pathology Workshop, Mauritius, 22–26 July 1991, Abstracts*, p. 17.

Lockhart, B.E.L. and Olszewski, N.E. (1993) Serological and genomic heterogeneity of banana streak badnavirus: implications for virus detection in *Musa* germplasm. In: Ganry, J. (ed.) *Breeding Banana and Plantain for Resistance to Diseases and Pests, Proceedings of the International Symposium on Genetic Improvement of Bananas for Resistance to Diseases and Pests organized by CIRAD–FLHOR, Montpellier, France, 7–9 September 1992.* CIRAD, Montpellier, France, pp. 105–113.

Magee, C.J.P. (1927) *Investigation on the Bunchy Top Disease of the Banana.* Council for Scientific and Industrial Research, Melbourne, 86 pp.

Magee, C.J.P. (1930) A new virus disease of bananas. *Agricultural Gazette of New South Wales* 41, 929.

Magee, C.J.P. (1936) Bunchy top of bananas – rehabilitation of the banana industry of N.S.W. *Journal of the Australian Institute of Agricultural Science* 2, 13–16.

Magee, C.J.P. (1939) Pathological changes in the phloem and neighbouring tissues of the banana (*Musa cavendishii* Lamb.) caused by the bunchy-top virus. *Science Bulletin, Department of Agriculture, New South Wales, Sydney*, 67, 4–32.

Magee, C.J.P. (1940a) Transmission of infectious chlorosis or heart rot of banana and its relationship to cucumber mosaic. *Journal of the Australian Institute of Agriculture* 6, 44–47.

Magee, C.J.P. (1940b) Transmission studies on the banana bunchy-top virus. *Journal of the Australian Institute of Agricultural Science* 6, 109–110.

Magee, C.J.P. (1948) Transmission of banana bunchy top to banana varieties. *Journal of the Australian Institute of Agricultural Science* 14, 18–24.

Magee, C.J.P. (1953) Some aspects of the bunchy top disease of banana and other *Musa* spp. *Journal and Proceedings of the Royal Society of New South Wales* 87, 3–18.

Magee, C.J.P. (1957) Report to the Government of the Philippines on Control of Abaca Mosaic. Report 666, FAO EPTA, 30 pp.

Magee, C.J.P. (1960) The latent infection problem in the control of abaca (Manila hemp) mosaic. In: *Report to the Sixth Commonwealth Mycological Conference*, pp. 117–119.

Magee, C.J.P. (1967) *The Control of Banana Bunchy Top.* South Pacific Commission, Noumea, New Caledonia, 13 pp.

Magnaye, L.V. (1994) Virus diseases of banana and current studies to eliminate the virus by tissue culture. In: Tangonan, N.G. (ed.) *Towards Making Pest and Disease Management Relevant to Big and Small Banana Growers. Proceedings of the 1st PPS–SMD National Symposium on Pests and Diseases in the Philippines, 23–24 April 1993, Davao City, Philippines.* Phytopathological Society, Southern Mindanao Division, Davao, Philippines, pp. 38–43.

Magnaye, L.V. and Eloja, A.L. (1968) Banana mosaic in the Philippines. I. Transmission and initial host range studies. Paper presented at the 5th Annual Conference of the Philippine Phytopathological Society, 10–12 May, 1968, Davao City.

Magnaye, L.V. and Espino, R.R.C. (1990) Note: Banana bract mosaic, a new disease of banana. I. Symptomatology. *The Philippine Agriculturist* 73, 55–59.

Manser, P.D. (1982) Bunchy top disease of plantain. *FAO Plant Protection Bulletin* 30, 78–79.

Matile-Ferrero, D. and Williams, D.J. (1995) Recent outbreaks of mealybugs on plantain (*Musa* spp.) in Nigeria including a new record for Africa and a description of a new species of *Planococcus* Ferris (Homoptera, Pseudococcidae). *Bulletin de la Société Entomologique de France* 100, 445–449.

Mohan, S. and Lakshmanan, P. (1988) Outbreak of cucumber mosaic virus on *Musa* sp. in Tamil Nadu, India. *Phytoparasitica* 16, 281–282.

Muharam, A. (1984) Test for resistance of some banana cultivars to banana bunchy-top disease. *Bulletin of Penel. Hort.* 11, 16–19.

Muñez, A.R. (1992) Symptomatology, transmission and purification of banana bract mosaic virus (BBMV) in 'Giant Cavendish' banana. MSc thesis, University of the Philippines, Los Baños, Philippines.

Muylle, H. (1998) Development of techniques for the elimination of virus diseases from *Musa*: progress report. In: Frison, E. and Sharrock, S.L. (eds) *Banana Streak Virus: A Unique Virus–Musa Interaction? Proceedings of a Workshop of the PROMUSA Virology Working Group held in Montpellier, France, 19–21 January 1998.* INIBAP, Montpellier, France, pp. 34–41.

Ndowora, T.C.R. (1998) Banana streak virus: development of an immunoenzymatic assay for detection and characterization of sequences that are integrated in the genome of the host. PhD thesis, University of Minnesota, 90 pp.

Ndowora, T., Dahal, G., LaFleur, D., Harper, G., Hull, R., Olszewski, N.E. and Lockhart, B. (1999) Evidence that badnavirus infection in *Musa* can originate from integrated pararetroviral sequences. *Virology* 255, 214–220.

Niblett, C.L., Pappu, S.S., Bird, J. and Lastra, R. (1994) Infectious chlorosis, mosaic and heart rot. In: Ploetz, R.C., Zentmyer, G.A., Nishijima, W.T., Rohrbach, K.G. and Ohr, H.D. (eds) *Compendium of Tropical Fruit Diseases.* APS Press, St Paul, Minnesota, pp. 18–19.

Ocfemia, G.O. (1926) Progress report on bunchy-top of abaca or Manila hemp. *Phytopathology* 16, 894.

Ocfemia, G.O. and Buhay, G.G. (1934) Bunchy-top of abaca, or Manila hemp: II. Further studies on the transmission of the disease and a trial planting of abaca seedlings in a bunchy-top devastated field. *Philippine Agriculturist* 22, 267–280.

Ocfemia, G.O. and Celino, M.S. (1938) Transmission of abacá mosaic. *Philippine Agriculturist* 27, 593–598.

Ocfemia, G.O., Celino, M.S. and Garcia, F.J. (1947) Further studies on transmission of bunchy-top and mosaic of abacá (manila hemp plant), separation of the two diseases, and mechanics of inoculation by *Pentalonia nigronervosa* Coquerel. *The Philippine Agriculturist* 31, 87–97.

Opina, O.S. and Milloren, H.J.L. (1996) Dynamics of banana bunchy top within commercial populations of banana (*Musa* sp.) cv Lakatan. *Philippine Phytopathology* 32, 75–82.

Osei, J.K. (1995) Cucumber mosaic virus (CMV) infects plantain in Ghana. *MusAfrica* 6, 4–5.

Owen, J., Shintaku, Aeschleman, P., Ben Tahar, S. and Palukitis, P. (1990) Nucleotide sequence and evolutionary relationships of cucumber mosaic virus (CMV) strains. CMV RNA 3. *Journal of General Virology* 71, 2243–2249.

Pacumbaba, R.P. (1967) Propagation of virus-free suckers of mosaic-infected abaca corms. *Plant Disease Reporter,* 51, 405–408.

Piazzolla, P., Diaz-Ruiz, J.R. and Kaper, J.M. (1979) Nucleic acid homologies of eighteen cucumber mosaic virus isolates determined by competition hybridization. *Journal of General Virology* 45, 361–369.

Pietersen, G., Staples, M., Kasdorf, G.C.F. and Thomas, J.E. (1998) First report of cucumber mosaic cucmovirus subgroup 1 in South Africa from banana with infectious chlorosis. *Plant Disease* 82, 1171.

Ram, R.D. and Summanwar, A.S. (1984) *Colocasia esculenta* (L) Schott. A reservoir of bunchy top disease of banana. *Current Science* 53, 145–146.

Ramos, C.S. and Zamora, A.B. (1990) Elimination of banana bunchy top infection from banana (*Musa* sp. cv. Lakatan) by heat pretreatment and meristem culture. *Philippine Journal of Crop Science* 15, 119–123.

Raymundo, A. (1998) Epidemiology and integrated management of abaca bunchy top in the Philippines. In: *Proceedings of the Regional Workshop on Disease Management of Banana and Citrus: The Use and Management of Disease-free Planting Material, 14–16 October, 1998.* International Network for the Improvement of Banana and Plantain/Food and Fertilizer Technology Center, Davao, Philippines.

Reichal, H., Belalcázar, S., Múnera, G., Pérez, R. and Avévalo, E. (1997) The presence of banana streak virus infecting sugarcane and has been confirmed in plantain (*Musa* AAB Simmonds), sugar cane (*Saccharum officinarum*) and edible canna (*Canna edulis*) in Columbia. *Infomusa* 6(1), 9–12.

Rivera, C., Ramírez, P. and Pereira, R. (1992) Preliminary characterization of viruses infecting banana in Costa Rica. In: *Biotechnology Applications for Banana and Plantain Improvement.* INIBAP, Montpellier, France, pp. 63–68.

Rodoni, B.C., Harding, R.M., Bateson, M.F. and Dale, J.L. (1996) Banana bract mosaic potyvirus (BBMV): a new and widespread virus of bananas. In: *Abstracts of the Xth International Congress of Virology, Jerusalem, Israel, 11–16 August, 1996.* Virology Division, International Union of Microbiological Societies, p. 156.

Rodoni, B.C., Ahlawat, Y.S., Varma, A., Dale, J.L. and Harding, R.M. (1997) Identification and characterization of banana bract mosaic virus in India. *Plant Disease* 81, 669–672.

Sebasigari, K. and Stover, R.H. (1988) *Banana Diseases and Pests in East Africa. Report of a Survey in November 1987.* INIBAP, Montpellier, France, 15 pp.

Shanmuganathan, N. (1980) Virus and virus-like diseases of plants in the Gilbert Islands. *FAO Plant Protection Bulletin* 28, 29–38.

Shanmugavelu, K.G., Aravindakshan, S. and Sathiamoorthy, S. (1992) *Banana Taxonomy, Breeding and Production Technology.* Metropolitan Book Co., New Delhi, India, 456 pp.

Simmonds, N.W. (1933) *Report on Visit to Samoa.* Department of Agriculture, Suva, Fiji.

Simmonds, N.W. (1962) *The Evolution of Bananas.* Longman, Green, London, UK, 170 pp.

Singh, Z., Jones, R.H.C. and Jones, M.G.K. (1995) Identification of cucumber mosaic subgroup I isolates from banana plants affected by infectious chlorosis disease using RT-PCR. *Plant Disease* 79, 713–716.

Smith, M.C., Holt, J., Kenyon, L. and Foot, C. (1998) Quantitative epidemiology of banana bunchy top virus disease and its control. *Plant Pathology* 47, 177–187.

Stover, R.H. (1972) *Banana, Plantain and Abaca Diseases.* Commonwealth Mycological Institute, Kew, Surrey, England, 316 pp.

Su, H.J., Wu, R.Y. and Tsao, L.Y. (1993) Ecology of banana bunchy-top virus disease. In: Valmayor, R.V., Hwang, S.C., Ploetz, R., Lee, S.C. and Roa, N.V. (eds) *Proceedings: International Symposium on Recent Developments in Banana Cultivation Technology, Taiwan Banana Research Institute, Chiuju, Pingtung, Taiwan, 14–18 December 1992.* INIBAP/ASPNET, Los Baños, Philippines, pp. 308–312.

Su, H.J., Hung, T.H. and Wu, M.L. (1997) First report of banana streak virus infecting banana cultivars (*Musa* spp.) in Taiwan. *Plant Disease* 81, 550.

Sulyo, Y., Duriat, A.S. and Said, A. (1978) *Peninjauan pendahuluan penyakit virus dan vektornya pada tanaman pisang.* Laporan Kegiatan Penelitian LPH, Jakarta, pp. 210–211.

Sun, S.K. (1961) Studies on the bunchy top diseases of bananas. *Special Publication of College of Agriculture, National Taiwan University* 10, 82–109.

Tessera, M. and Quimio, A.J. (1994) Research on ensat pathology. In: Herath, E. and Desalegn, L. (eds) *Proceedings of the 2nd National Horticultural Workshop of Ethiopia.* Institute of Agricultural Research, Addis Ababa, pp. 217–225.

Tessera, M., Lohuis, D. and Peters, D. (1996) A badnavirus in *Ensete.* In: Bekele, E., Abdulah, A. and Yemane, A. (eds) *Proceedings of the 3rd Annual Conference of the Crop Protection Society of Ethiopia.* Crop Protection Society of Ethiopia, Addis Ababa, pp. 143–148.

Tessera, M., Goldbach, R.W. and Peters, D. (1998) Partial purification of the virus associated with enset chlorotic leaf disease. *Pest Management Journal of Ethiopia* 2, 106–109.

Thomas, J.E. (1991) Virus indexing procedures for banana in Australia. In: Valmayor, V.V., Umali, B.E. and Besjosano, C.P. (eds) *Banana Diseases in Asia and the Pacific: Proceedings of a Technical Meeting on Diseases affecting Banana and Plantain in Asia and the Pacific, Brisbane, Australia, 15–18 April 1991.* INIBAP, Montpellier, France, pp.144–157.

Thomas, J.E. and Dietzgen, R.G. (1991) Purification, characterization and serological detection of virus-like particles associated with banana bunchy top disease in Australia. *Journal of General Virology* 72, 217–224.

Thomas, J.E. and Magnaye, L.V. (1996) *Banana Bract Mosaic Disease. Musa* Disease Fact Sheet No. 7. International Network for the Improvement of Banana and Plantain, Montpellier, France.

Thomas, J.E., Iskra-Carana, M.L. and Jones, D.R. (1994a) *Banana Bunchy Top Disease. Musa* Disease Fact Sheet No. 4, INIBAP, Montpellier, France, 2 pp.

Thomas, J.E., McMichel, L.A., Dietzgen, R.G., Searle, C., Matalevea, S. and Osasa, A. (1994b) Banana streak virus in Australia, Western Samoa and Tonga. In: *International Society of Sugar Cane Technologists, 4th Sugar Cane Pathology Workshop, Brisbane, Australia, Abstracts,* p. 40.

Thomas, J.E., Smith, M.K., Kessling, A.F. and Hamill, S.D. (1995) Inconsistent transmission of banana bunchy top virus in micropropagated bananas and its implication for germplasm screening. *Australian Journal of Agricultural Research* 46, 663–671.

Thomas, J.E., Geering, A.D.W., Gambley, C.F., Kessling, A.F. and White, M. (1997) Purification, properties and diagnosis of banana bract mosaic potyvirus and its distinction from abaca mosaic potyvirus. *Phytopathology* 87, 698–705.

Thomson, D. and Dietzgen, R.G. (1995) Detection of DNA and RNA plant viruses by PCR and RT-PCR using a rapid virus release protocol without tissue homogenisation. *Journal of Virological Methods* 54, 85–95.

Thottappilly, G., Dahal, G., Harper, G., Hull, R. and Lockhart, B.E.L. (1997) Banana streak badnavirus: development of diagnostics by ELISA and PCR. *Phytopathology* 87, S97.

Trindade, D.R., Potronier, L.S., Albuquerque, F.C., de Benchimol, R.L. and Amorim, A.M. (1998) Occurrence of cucumber mosaic virus on banana in the State of Para, Brazil. *Fitopatologia Brasileira* 23, 185.

Tsai, Y.P., Hwang, M.T., Chen, S.P. and Liu, S.S. (1986) An ecological study of banana mosaic. *Plant Protection Bulletin (Taiwan ROC)* 28, 383–387.

Vakili, N.G. (1969) Bunchy top disease of bananas in the Central Highlands of South Vietnam. *Plant Disease Reporter* 53, 634–638.

Velasco-Magnaye, L. and Eloja, A.L. (1966) Some physical properties and suscept range of the abaca mosaic virus. *Philippine Phytopathology* 2, 22–30.

Vuylsteke, D.R., Chizala, C.T. and Lockhart, B.E.L. (1996) First report of banana streak virus disease in Malawi. *Plant Disease* 80, 224.

Vuylsteke, D.R., Hughes, J.d'A. and Rajab, K. (1998) Banana streak badnavirus and cucumber mosaic virus in farmers' fields in Zanzibar. *Plant Disease* 82, 1403.

Wanitchakorn, R., Harding, R. and Dale, J.L. (1997) Banana bunchy top virus DNA-3 encodes the viral coat protein. *Archives of Virology* 142, 1673–1680.

Wardlaw, C.W. (1961) The virus diseases: bunchy top. *Banana Diseases, Including Plantains and Abaca.* Longman, Green, London, UK, pp. 68–115.

Wardlaw, C.W. (1972) *Banana Diseases including Plantains and Abaca.* Longman, Green and Co Ltd., London, UK, 648 pp.

Wu, M.L., Hung, T.H. and Su, H.J. (1997) Strain differentiation of cucumber mosaic virus associated with banana mosaic disease in Taiwan. *Annals of the Phytopathological Society of Japan* 63(3), 176–178.

Wu, R.Y. and Su, H.J. (1990a) Transmission of banana bunchy top virus by aphids to banana plantlets from tissue culture. *Botanical Bulletin of Academia Sinica* 31, 7–10.

Wu, R.Y. and Su, H.J. (1990b) Purification and characterization of banana bunchy top virus. *Journal of Phytopathology* 128, 153–160.

Wu, R.Y. and Su, H.J. (1990c) Production of monoclonal antibodies against banana bunchy top virus and their use in enzyme-linked immunosorbent assay. *Journal of Phytopathology* 128, 203–208.

Wu, R.Y. and Su, H.J. (1991) Regeneration of healthy banana plantlets from banana bunchy top virus-infected tissues cultured at high temperature. *Plant Pathology* 40, 4–7.

Wu, R.Y. and Su, H.J. (1992) Detection of banana bunchy top virus in diseased and symptomless banana plants with monoclonal antibody. *Tropical Agriculture (Trinidad)* 69, 397–399.

Wu, R.Y., You, L.R. and Soong, T.S. (1994) Nucleotide sequences of two circular single-stranded DNAs associated with banana bunchy top virus. *Phytopathology* 84, 952–958.

Xie, W.S. and Hu, J.S. (1995) Molecular cloning, sequence analysis, and detection of banana bunchy top virus in Hawaii. *Phytopathology* 85, 339–347.

Yeh, H.H., Su, H.J. and Chao, Y.C. (1994) Genome characterization and identification of viral-associated dsDNA component of banana bunchy top virus. *Virology* 198, 645–652.

Yot-Dauthy, D. and Bové, J.M. (1966) Mosaïque du bananier. Identification et purification de diverses souches du virus. *Fruits* 21, 449–466.

7
Nematode Pathogens

Burrowing Nematode

J.L. Sarah

Introduction

The burrowing nematode, which was first described by Cobb (1893) from Fiji, is one of the most important root parasites on banana in tropical areas. It is considered to be the main nematode problem where banana, especially cultivars in the Cavendish subgroup (AAA), is grown in commercial plantations (Sarah, 1989; Gowen and Quénéhervé, 1990; Stanton, 1994). The burrowing nematode is also common on plantain and cooking-banana types cultivated in the lowlands of central and eastern Africa and in Puerto Rico in the Caribbean. However, it is generally absent in plantain and cooking-banana roots in western and southern Africa and Central America (Pinochet, 1977, 1988a; Sarah, 1985, 1989; Adiko, 1988; Bridge 1988, 1993). The burrowing nematode is not found in the highest-altitude zones where banana is cultivated, such as the highlands of central and eastern Africa. It is also usually absent from the highest latitude zones where cultivation occurs, such as the Mediterranean area, Canary Islands, Madeira, the Cape Province of South Africa and Taiwan (Stover, 1972; Bridge, 1988; Sarah, 1989; Gowen and Quénéhervé, 1990), though it may be present under greenhouse cultivation. In the Philippines, the burrowing nematode causes black head rot or tip-over of abacá (Anderson and Alaban, 1968; Davide, 1972; Castillo *et al.*, 1974). The present geographical distribution of the burrowing nematode is a reflection of historical movements of infested planting material, especially corms of Cavendish cultivars, and of the temperature preference of the pathogen.

The burrowing nematode destroys root (Plate 7.1) and corm tissue, which reduces water and mineral uptake. This results in a reduction of plant growth and development and may lead to severe reduction of bunch weight and increase significantly the time period between two successive harvests (Gowen, 1975; Stanton, 1994). Furthermore, this destruction also results in a tendency for plants to uproot or topple (toppling disease), particularly during windstorms and heavy-rain periods.

Crop losses depend on several factors, including the pathogenicity of local burrowing nematode populations, associated pathogens (including other nematode species), banana cultivar, climatic conditions and soil factors, especially fertility. In commercial plantations of Cavendish cultivars in areas of Côte d'Ivoire where soils are poor and eroded, losses of over 75% have been reported, due to bunch weight reduction and uprooting (Sarah, 1989). In such instances, if there is no nematode control, banana plants become virtually unproductive after the first harvest. In the more fertile

Plate 7.1. Lesions caused by the burrowing nematode on banana roots (photo: J.L. Sarah, CIRAD).

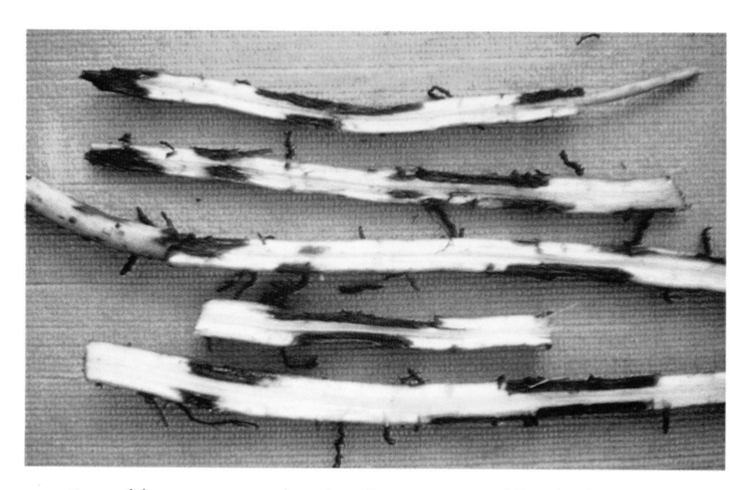

Plate 7.2. Cross-section of banana roots showing damage caused by the burrowing nematode. Lesions extend from the exterior of the root to the central cylinder (photo: J.L. Sarah, CIRAD).

peat soils of Côte d'Ivoire and in the volcanic soils of Cameroon, cumulative crop losses are generally below 30% (Sarah, 1989). In South Africa, losses may reach 75% (Jones and Milne, 1982). In Central America (Costa Rica and Panama) and South America (Colombia), crop losses estimated by counting uprooted plants fluctuate between 12 and 18%. Losses tend to be around 5% in the Sula Valley in Honduras (Pinochet, 1986).

The actual economic impact of the burrowing nematode on smallholder cultivation is difficult to estimate. In Cameroon, losses appear to be below 20% (Melin *et al.*, 1976). However, severe symptoms have been observed in plantains in Côte d'Ivoire growing near Cavendish plantations (Sarah, 1985) and damage has reached 50% in experimental plots (Sarah, 1989). In Honduras, Stover (1972) reported that there was considerably more uprooting, which resulted in complete loss of yield, in burrowing nematode-infested plots of 'Horn' plantain than in control plots.

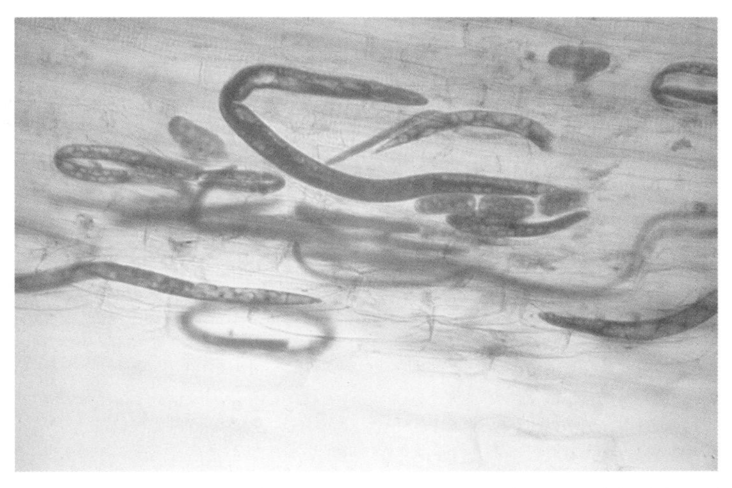

Plate 7.3. Infestation of *Radopholus similis* in banana root tissue. All stages of the life cycle of the nematode from egg to adult are present (photo: M. Boisseau, CIRAD).

Symptoms

The burrowing nematode causes a necrosis of corm and root tissues of banana. On corms, this necrosis appears as a reddish-brown discoloration, which can easily be seen after the surface has been washed free of dirt and lightly peeled. The necrosis usually begins where roots leave the corm. The size of lesions can vary from small spots to large areas of necrotic tissue. Damage caused by the banana weevil borer (*Cosmopolites sordidus*) is superficially similar to larger lesions, but extends much further into the corm as tunnels.

The symptoms of the burrowing nematode can easily be identified in roots. If a root is cleared of soil and then cut in half longitudinally, reddish-brown necrotic patches can be seen extending from the surface towards the centre (Plate 7.2). Symptoms of burrowing nematode can be distinguished from those of Fusarium wilt (see Ploetz and Pegg, Chapter 3, pp. 148–149) because symptoms of the latter are confined to vascular tissue and do not extend to the root surface.

The main impact of the burrowing nematode is to weaken the root systems of banana, so that plants are easily blown over by strong winds. Severe nematode damage can usually be seen in the corm and root tissue of toppled plants. Other symptoms include the unthrifty appearance of plants (as is evidenced by thin pseudostems and small bunches) that are otherwise growing under environmental conditions where development should be normal.

Causal agent

The Latin name for the burrowing nematode is *Radopholus similis* (Cobb) (Plate 7.3). It normally feeds in advance of necrotic areas and can be isolated from the reddish tissue at the edges of lesions. Large numbers of the nematode can be obtained by teasing affected root-cortex tissues in a dish of water. More sophisticated methods, such as centrifugal flotation and mist extraction, which allow a quantitative evaluation, are detailed by Hooper (1986).

In 1956, Ducharme and Birchfield demonstrated the existence of two 'races' of *R. similis* – one attacking banana, but not citrus, and another attacking banana and citrus. Huettel *et al.* (1984) gave specific status to the citrus parasite, which they called *Radopholus citrophilus*, based on behaviour, biochemical and, especially, karyotype differences. *Radopholus citrophilus* was described as having five chromosome pairs and *R. similis* as having four. However, the elevation of *R. citrophilus* to a species is not accepted by all, and Siddiqui (1986) has proposed a subspecific level, *R. similis* ssp. *citrophilus*, for the citrus parasite. Evidence that supports the view that *R. citrophilus* should not have separate species status has been obtained recently, following further morphological (Valette *et al.*, 1998) and genomic (Kaplan *et al.*, 1998) studies.

Using different criteria, such as chromosome number, pathogenicity, reproduction rate and host preference, three pathotypes of *R. similis* have been distinguished in populations from Central America and the Caribbean (Edwards and Wehunt, 1971; Pinochet, 1979, 1988a; Tarté *et al.*, 1981; Rivas and Roman, 1985). Recently, pathogenic diversity was reported to be worldwide and to be clearly linked to reproductive fitness in plant tissues. Isolates from Uganda, Côte d'Ivoire, Costa Rica and Guinea have been shown to have higher multiplication rates than those from Martinique, Guadeloupe, Sri Lanka and Queensland in Australia (Sarah *et al.*, 1993; Fallas and Sarah, 1995a; Fallas *et al.*, 1995; Hahn *et al.*, 1996). Enzymatic phosphoglucose isomerase, (PGI) and randomly amplified polymorphic DNA (RAPD) analyses have revealed two genomic groups of burrowing nematode that are not related to pathogenicity (Fallas *et al.*, 1996). The distribution of these genomic groups appears to be linked to historical movements of planting material.

Disease cycle and epidemiology

Penetration occurs preferentially at the root apex, but *R. similis* is able to invade any portion of the root length. The nematode migrates in and between cells in the root cortex, where it feeds on the cell cytoplasm. This results in collapsed cell walls, cavities and tunnels (Blake, 1961, 1966; Valette *et al.*, 1997). On corms, lesions begin to develop where infested roots are attached and then spread outwards. Necroses can extend to the whole cortex of corms (black-head disease) and roots, but the root stele is usually not damaged, except occasionally when very young (Mateille, 1994b; Valette *et al.*, 1997).

Radopholus similis is a migratory endoparasitic nematode, which completes its life cycle in 20–25 days in root and corm tissues under optimal conditions. Embryonic development takes 4–10 days and the four juvenile stages are completed in 10–15 days, depending on temperature (Van Weerdt, 1960; Loos, 1962). This species has a pronounced sexual dimorphism, in which males present an atrophied stylet and are considered to be non-parasitic, though they are present in plant tissue. Juveniles and adult females are actively mobile. They may migrate into the soil under adverse conditions and move towards new roots. The temperature range for *R. similis* development lies between 24 and 32°C, with optimum reproduction occurring at around 30°C (Loos, 1962; Fallas and Sarah, 1995a). It does not reproduce below 16–17°C or above 33°C (Fallas and Sarah, 1995a, b; Pinochet *et al.*, 1995).

Necrosis of root and corm tissues is accelerated if other organisms, such as fungi and bacteria, are present. Fungi commonly associated with burrowing-nematode lesions are *Cylindrocarpon musae*, *Acremonium stromaticum*, *Fusarium* spp. and *Rhizoctonia solani* (Laville, 1964; Pinochet and Stover, 1980). Fungi of the genus *Cylindrocladium* have been found to be pathogenic on banana in the French Antilles and Cameroon, causing lesions similar to those of *R. similis* (see Jones, Chapter 3, pp. 159–160). In association with the nematode, they can cause severe damage (Loridat, 1989).

Environmental factors and stages of plant development influence nematode populations. In Côte d'Ivoire, *R. similis* is the major banana nematode, except in some areas with high organic-matter content. However, as a rule, *R. similis* is less influenced by soil conditions than other species, and this may be due to its strictly endoparasitic habit (Quénéhervé, 1988). Rainfall appears to be the main factor that modulates nematode populations (Melin and Vilardebo, 1973; Jaramillo and Figueroa, 1976; Vilardebo, 1976; Jones and Milne, 1982; Hugon *et al.*, 1984; Sarah *et al.*, 1988; Quénéhervé, 1989a, b); too little and too much water suppressing nematode densities in the roots. *Radopholus similis* is generally absent in cooler banana-growing areas and this is believed to be because temperature is a limiting factor. During the crop cycle, the *R. similis* population increases gradually until after the emergence of the flower bud (Melin and Vilardebo, 1973; Vilardebo, 1976; Sarah, 1986). The increase is faster in the roots of suckers (Sarah, 1986), especially those which are not pruned (Mateille *et al.*, 1984).

Host reaction

Radopholus similis is able to attack almost all banana cultivars, as well as abacá and other seeded *Musa* species (Gowen and Quénéhervé, 1990). It has approximately 250 host plants, among which there are many weeds and some crops of economic importance, such as black pepper, coconut, tea, tuber crops, fruit trees and ornamentals (Milne and Keetch, 1976; O'Bannon, 1977; Bridge, 1987).

Parameters used to measure the reaction of *Musa* to *R. similis* have been the number of nematodes on each plant, the number of nematodes in known weights of root, the percentage of infested roots and assessments of lesion damage on roots and corms (Wehunt *et al.*, 1978; Pinochet, 1988b; Sarah *et al.*, 1992; Fallas *et al.*, 1995; Fogain, 1996; Speijer and Gold, 1996; Price and McClaren, 1996). In pot experiments using plants derived from *in vitro* propagation, root damage measured 12 weeks after inoculation correlated well with nematode infestation levels measured at 6–8 weeks (Fallas *et al.*, 1995). Trial designs, sampling strategies and methods of statistical analysis have been reviewed by Price and McClaren (1996).

An evaluation of wild *Musa* species in pots has shown that *Musa acuminata* ssp. *banksii* is quite susceptible (Wehunt *et al.*, 1978). In contrast, most accessions of *Musa balbisiana* tested have been very resistant (Fogain, 1996). *Musa acuminata* sspp. *malaccensis*, *microcarpa*, *zebrina* and *burmanniciodes* ('Calcutta 4') have been found generally to have moderate to good resistance (Wehunt *et al.*, 1978; Fogain, 1996). However, one accession of *M. acuminata* ssp. *microcarpa* has been rated as moderately susceptible (Wehunt *et al.*, 1978). *Musa textilis* is moderately resistant (Price and McClaren, 1996). Consequently, *R. similis* is not reported as a major problem in abacá crops (Anunciado *et al.*, 1977).

Diploid cultivars vary in their reaction to *R. similis*. 'Pisang Mas' (AA, syn. 'Sucrier') and 'Pisang Lidi' (AA, syn. 'Pisang Lilin') have moderate resistance (Wehunt *et al.*, 1978; Fogain, 1996). 'Pisang Batuau' (AA), 'Pisang Oli' (AA) and many accessions of 'Pisang Jari Buaya' (AA) are highly resistant (Wehunt *et al.*, 1978; Fogain, 1996). In comparison, 'Guyod' (AA) and 'Tuugia' (AA) are very susceptible (Wehunt *et al.*, 1978; Fogain, 1996). 'Safet Velchi' (AB, syn. 'Ney Poovan') appears to be very resistant (Price and McClaren, 1996).

Many cultivars in the Cavendish subgroup (AAA) have been estimated to be moderately susceptible to *R. similis*. 'Gros Michel' and its dwarf mutant 'Cocos' are less susceptible (Wehunt *et al.*, 1978; Price and McClaren, 1996). 'Yangambi Km 5' (AAA, syn. 'Ibota Bota') has very strong resistance to *R. similis* (Sarah *et al.*, 1992; Price, 1994b; Fogain, 1996; Price and McLaren, 1996; Fogain and Gowen, 1998).

Cultivars in the Plantain subgroup (AAB) are, on the whole, very susceptible to *R. similis* (Price, 1994b; Fogain, 1996; Price and McLaren, 1996). 'Popoulou' (AAB, Maia Maoli–Popoulu subgroup) is also very susceptible. This susceptibility may be linked to *M. acuminata* ssp. *banksii*, the wild banana, which may have contributed both A genomes to these clones (Carreel, 1995; Fogain, 1996). 'Focanah' (AAB, Pome subgroup), 'Figue Pomme Ekonah' (AAB, syn. 'Silk'), 'Pisang Kelat' (AAB) and 'Pisang Ceylan' (AAB, syn. 'Mysore') all have good resistance (Fogain, 1996; Price and McClaren, 1996).

Not many ABB cultivars have been screened against *R. similis*. Of those that have, 'Bluggoe' accessions would appear to be moderately susceptible and 'Pelipita' moderately resistant (Price and McLaren, 1996).

Control

Control methods, apart from the occasional paring of dead corm tissue, are not usually implemented in smallholdings. In such cropping systems, chemical control is not possible, because of the high cost and the toxicity of nematocides. As a consequence, most control methods discussed here are those which are undertaken in commercial plantations.

Preplanting measures

Reducing nematode populations in the soil before planting and the use of cleansed or nematode-free plant material are of primary importance in the control of *R. similis*. Eradication of *R. similis* from the soil is virtually impossible. After the first detection of *R. similis* in the northern Transvaal in South Africa, regulatory measures were introduced in an attempt to eradicate the nematode and prevent dissemination. All plants at infested sites were rogued and burned and the land was fumigated with methyl bromide and left as fallow for 9 months. Despite these drastic measures, *R. similis* was not eradicated, although populations were reduced almost to extinction at five out of the eight sites (Jones and Milne, 1982). In Panama, an 18-month clean fallow period did not eradicate *R. similis* (Loos, 1961).

Populations of *R. similis* may be reduced to an undetectable level by allowing non-host plants to grow on infested land for 1 year. *Panicum maximum* (*Poaceae*) has been used successfully in Queensland, Australia (Colbran, 1964) and *Chromolaena odorata* (*Asteraceae*) in West Africa (Sarah, 1989). Milne and Keetch (1976) have published a list of 44 non-host plants of *R. similis* in South Africa, among which are some cash crops, such as passion fruit (*Passiflora edulis*) and pineapple (*Ananas comosus*). This latter crop is currently used in rotation by some banana farmers in Côte d'Ivoire (Sarah, 1989). Sugar cane (*Saccharum officinarum*) is mentioned as a poor host by Milne and Keetch (1976) and has met with some success as a rotation crop in Central America (Loos, 1961).

An alternative method to fallow is soil cleansing. Loos (1961) reported that *R. similis* was eliminated after land in Honduras and Panama was flooded for 5–6 months. Flooding has also been used in Surinam (Maas, 1969). In the Côte d'Ivoire, 6–7 weeks of complete flooding was as effective as 10–12 months of fallow in reducing nematode populations (Sarah *et al.*, 1983; Mateille *et al.*, 1988). However, this method is often not practicable, as land has to be level and have a permanent water-supply.

Fumigation is quite efficient for soil cleansing. However, fumigants, such as dichloropropene and methyl bromide, are general biocides, with a detrimental effect on soil biology, and their use today is either discouraged or banned.

Nematodes may be introduced into clean soil through infested corms and suckers. Therefore, all planting material should be produced in nematode-free soils. Since this is not often the case, lightly infested corms or suckers must be treated to remove nematodes. The simplest method consists of paring the corms superficially to remove necrotic tissue. However, nematodes located deep within the cortex in tissues that appear healthy may escape removal. Storage of pared material for 2 weeks may further reduce the nematode population (Quénéhervé and Cadet, 1985b), but such techniques cannot be applied to small suckers, which are quite fragile and need to be replanted rapidly.

Paring followed by hot-water treatments (52–55°C for 15–20 min) has been a common and effective practice in Latin America and Australia (Blake, 1961; Stover, 1972; Pinochet, 1986). However, hot-water treatments are labour-intensive and require careful monitoring, since temperature and immersion times are critical. Planting material can also be disinfested using chemicals. Dipping plant material in a nematocide (phenamiphos) solution at a concentration of 2500 ppm for 30 min has been shown to be effective (Jones and Milne, 1982). Corms can also be dipped in a nematocide–mud mixture, which adheres to the surface, forming a nematocidal coat. The advantage of this method is that treated material can be removed immediately from the dip and there is much less splashing of dangerous chemicals. The dip mixture is made with 400–500 g of active nematocide ingredient either with 15 kg of bentonite in 100 l of water or with natural clay. If clay is used, the proportion that has to be mixed with water must be decided by experimentation. The technique is known as 'pralinage' (Vilardebo and Robin, 1969).

The best means of avoiding contamination is to use nematode-free plants propagated through *in vitro* techniques. Micropropagated plants are now one of the most common sources of planting material in many producing regions and should be the only type of planting material allowed when banana is grown in virgin soil.

Postplanting measures

In most cases where contaminated planting material has been used to initiate new plantations or where clean planting material has been planted in infested soil, populations of burrowing nematode will build up quite rapidly. Yield losses may be reduced through propping or guying of pseudostems to avoid toppling. Improved drainage is also an important factor in reducing nematode damage in high-rainfall regions, such as parts of Central America (Pinochet, 1986). In the same way, any measures that improve fertility and root development may increase plant tolerance to nematodes. Such measures include ploughing before planting, incorporation of organic matter in the soil, fertilization and irrigation.

Chemical control is currently the most common and efficient way for nematode control. Nematocides are generally non-volatile organophosphates or carbamates, which are applied as granules on the soil surface around the mat in a 30–40 cm radius. Emulsifiable compounds are applied as liquid sprays or through irrigation systems, such as in the Canary Islands, Martinique and Colombia (Gowen and Quénéhervé, 1990). The optimum application time, dose and frequency of applications are determined by nematocide efficiency, environmental conditions, population dynamics and the pathogenicity of local strains. In most production areas, nematocide applications vary from 2 to 3 g of active ingredient per mat and two to three applications per year (Gowen, 1979; Sarah, 1989; Gowen and Quénéhervé, 1990). To avoid problems of enhanced biodegradation induced by repetitive use of the same nematocide, alternation with different compounds is recommended (Anderson, 1988; Sarah, 1989). Nematocides currently in use for the control of *R. similis* are presented in Table 7.1.

In some banana-growing countries, nematocides are applied on a regular basis and no attempt is made to determine if the treatments are necessary or not. As well as being costly, nematocides are also very toxic and can endanger the environment if used indiscriminately. Land previously used to cultivate banana should be treated to reduce nematode populations before planting. Nematode levels should be checked from time to time to determine if an application of nematocide is warranted. The threshold for 'triggering' nematocide application will depend on local parameters such as climatic and soil conditions, as well as aggressiveness of pathotypes. For this exercise to be worthwhile, the check has to be accurate and based on a nematode count. Nematodes must be extracted from plant material and surrounding soil, using proved protocols (Sarah, 1991). A direct field estimation of the level of infestation by checking the number of lesions can be undertaken if a more precise evaluation is not possible (Speijer and Gold, 1996). The amount of visible damage to roots caused by the burrowing nematode gives an indication of the relative health of

Table 7.1. Non-volatile nematocides used to control banana nematodes.

Nematocide group	Active ingredient	Trade name and formulations
Organophosphates	Phenamiphos	Nemacur® (G, EC)
	Isazophos	Miral® (G, ME)
	Ethoprophos	Mocap® (G)
	Ebufos	Rugby® (G)
	Terbufos	Counter® (G)
Carbamates	Aldicarb	Temik® (G)
	Oxamyl	Vydate® (EC)
	Carbofuran	Furadan® (G)

G, granules; EC, emulsifiable compound; ME, microencapsulated.

the root system, but may not correlate with the actual level of infestation. It is possible that high populations of the nematode may be present in tissues that are not yet necrotic. Alternatively, previous nematocide applications may have killed the populations of nematode responsible for the formation of lesions.

Biological control

Plant-parasitic nematodes have many natural enemies and a number have been considered as possible biological control agents. The first antagonists investigated were the nematode-trapping fungi (*Arthrobotrys*, *Dactyllela*, *Dactylaria*, etc.). However, they were difficult to mass-produce, and, furthermore, their efficiency in the soil was linked to quite precise soil conditions (pH, organic-matter content, composition of microfauna/flora) (Stirling, 1991; Cayrol *et al.*, 1992; Davide, 1994). Attempts to commercialize these agents have been unsuccessful.

An industrial formulation of a parasitic fungus, *Paecilomyces lilacinus*, has recently been developed (Davide, 1994). *Paecilomyces lilacinus* is a parasite of burrowing-nematode eggs, juveniles and adults, and gave apparently promising results in the Philippines.

Strains of the bacterium *Pseudomonas fluorescens*, a fluorescent pseudomonad found in the rhizosphere of banana, and the type strain of *Pseudomonas putida* have been demonstrated to inhibit the invasion of roots of 'Grand Nain' (AAA, Cavendish subgroup) by *R. similis* (Aalten *et al.*, 1998).

Breeding for resistance

Fogain and Gowen (1997) demonstrated in field trials that population levels of *R. similis* were higher on the root systems of nematocide-treated susceptible cultivars than on an untreated resistant cultivar. Their work shows that genetic resistance can effectively control *R. similis*.

Resistance to burrowing nematode through genetic improvement has long been hindered by difficulties associated with conventional banana breeding (Menendez and Shepherd, 1975; Pinochet, 1988a). Nevertheless, progress has been made. Several clones of 'Pisang Jari Buaya' (AA) have long been recognized as an exploitable source of resistance to burrowing nematode (Pinochet and Rowe, 1978, 1979; Wehunt *et al.*, 1978; Pinochet, 1988a). Although its heritability has not been actually established (Stanton, 1994), the resistance of 'Pisang Jari Buaya' has been incorporated into breeding lines and this has led to the production of hybrids of commercial interest (Rowe and Rosales, 1994). Recent screening studies have shown that sources of resistance that may be useful in breeding programmes are present in many genotypes (Sarah *et al.*, 1992; Price, 1994b; Price and McLaren, 1996; Fogain, 1996). It has been suggested that clones with large numbers of roots may exhibit a higher tolerance to nematode attacks and selection for this character should be a worthwhile breeding objective (Gowen, 1996).

Because of differences in pathogenicity among *R. similis* populations, as well as the other nematode species able to parasitize and damage banana roots, efforts in breeding banana for broad resistance against all these pathogens will be extremely difficult (De Waele, 1996). As a first step, potentially valuable banana cultivars are being evaluated against local populations of the burrowing nematode in each ecological zone in studies coordinated by the International Network for the Improvement of Banana and Plantain (INIBAP) (Frison *et al.*, 1997). Techniques for the early screening of germplasm in small pots have been developed (Pinochet, 1988b; Sarah *et al.*, 1992; Fogain, 1996). Such methods allow susceptible germplasm to be identified very rapidly. With inferior lines eliminated, only the most promising germplasm need be retained for final evaluation in relatively costly field trials (Price and McLaren, 1996).

Recent developments in the research of resistance mechanisms to *R. similis* have shown that phenolic compounds, especially some tannins and flavonoids, could be involved reducing the inroads nematodes make into banana tissues and their multiplication within these tissues (Mateille, 1994b; Valette *et al.*, 1996, 1997). A better knowledge of mechanisms involved in the resistance of *Musa* to nematodes, leading to the identification of the dominant genes involved, would be undoubtedly helpful for the breeding programmes. New cellular and molecular banana improvement techniques may allow the natural limitations of traditional plant breeding to be circumvented in the future (De Waele, Chapter 15, pp. 492–495).

Root-lesion Nematodes

S.R. Gowen

Introduction

Root-lesion nematodes occur widely, but not universally, on banana throughout the tropics (Bridge *et al.*, 1997). Like the burrowing nematode, their distribution is likely to have been increased through the movement of infested clonal planting material. However, they are not found so commonly in commercial plantations of cultivars in the Cavendish subgroup (AAA) supplying fruit to the international market, where *Radopholus similis* is usually the dominant or most important nematode pest. Root-lesion nematodes are often reported in association with cultivars in the Plantain subgroup (AAB) (Ogier and Merry, 1970; Pinochet and Stover, 1980; Bridge *et al.*, 1995) and there is some evidence to suggest that plantain is more susceptible to these nematodes than are other banana types (Perez *et al.*, 1986). Root-lesion nematodes have been recorded on abacá (Davide, 1972). One of the root-lesion nematodes found on banana also attacks enset in Ethiopia.

Symptoms

It is possible that root-lesion nematodes are overlooked when they occur in mixed populations with *R. similis* or are mistaken for that species. The damage in roots and corms is identical to that caused by *R. similis*. Root-lesion nematodes feed on the cytoplasmic contents of cells in the cortex and migrate between and within cells. This causes the formation of cavities within the root tissue and results in characteristic, dark purple lesions and necrotic patches (Plate 7.4). Symptoms are usually confined to the cortex and the stele tissue is generally unaffected – a useful diagnostic character when examining necrotic roots. Infested plants become stunted, bunch weight is decreased and the production cycle is increased. Damage leads to a reduction in the size of the root system, the snapping of roots and the toppling of

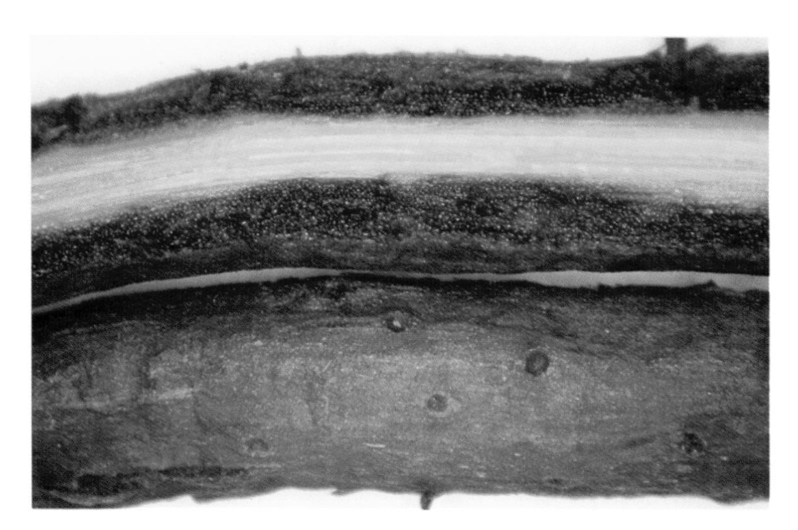

Plate 7.4. Damage to a banana root caused by *Pratylenchus goodeyi,* a root-lesion nematode (photo: B. Pembroke, UR).

plants. Plant toppling may be more prevalent in poor soils.

Reduced plant growth and toppling can increase the exposure of soils to sunlight. This results in a rise in soil temperatures and a reduction in the organic content. Nutrient leaching and erosion may also occur in soils exposed to direct rainfall (Bridge *et al.*, 1997).

Causal agent

Root-lesion nematodes are species of *Pratylenchus.* To the non-specialist, *Pratylenchus* can be confused with *R. similis* when nematodes extracted from roots or corms are viewed under a dissection microscope (× 50). However, unlike *R. similis*, *Pratylenchus* males have functional stylets. To the experienced technician, the genera can be differentiated by noting the position of the vulva, which is near to the tail in *Pratylenchus* spp. and at mid-length of the body of *R. similis* females.

Many reports in the literature do not identify *Pratylenchus* on banana to the species level. However, *Pratylenchus coffeae* (Zimmerman, 1893) Filipjev and Schuurmans Stekhaven, 1941, is the most widely reported. The only other species that is recognized as a significant pathogen of banana is *Pratylenchus goodeyi* Sher and Allen, 1953. Differences in tail morphology separate *P. coffeae* from *P. goodeyi*, but it is difficult for non-specialists to differentiate between the two nematodes.

Pratylenchus coffeae (Plate 7.5) infects a number of important crops, which include potato, yam, citrus, coffee, ginger (see Luc *et al.*, 1990), abacá and some ornamental plants. It is the most important nematode pest of banana in the Pacific and is significant in parts of South-East Asia, especially in Thailand on 'Klaui Namwa' (ABB, 'Pisang Awak'). This nematode is also reported as the most damaging on cultivars in the Cavendish subgroup in Honduras. *Pratylenchus coffeae* is widespread in South Africa and Ghana, where it may cause up to 60% losses in production of plant crops of plantain. Distribution is localized in other African countries, suggesting that it may have only recently been introduced (Bridge *et al.*, 1997).

Pratylenchus goodeyi has a more restricted distribution than *P. coffeae*, occurring in the highlands of East, Central and West Africa, the Canary Islands, Madeira, Egypt and Crete. It has also been found in subtropical areas of Australia (Anon., 1998). At present, banana and enset (O'Bannon, 1975; Peregrine and Bridge, 1992; Tessera

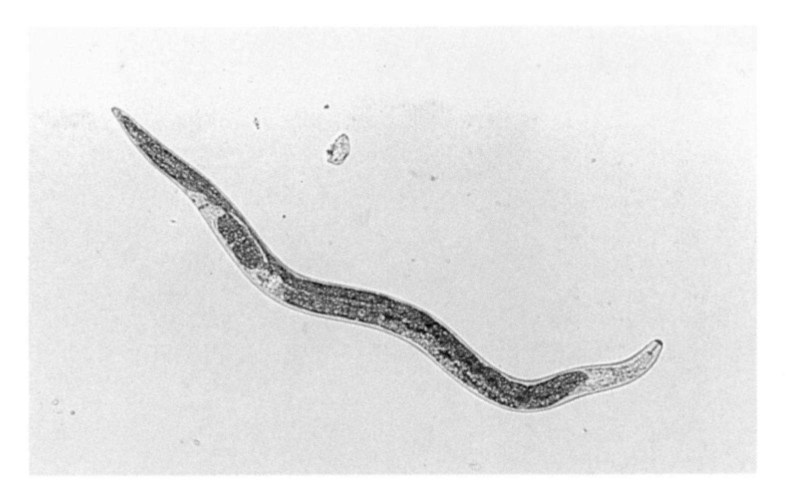

Plate 7.5. *Pratylenchus coffeae,* a root-lesion nematode that attacks banana (photo: B. Pembroke, UR).

and Quimio, 1994) are the only known hosts of economic importance. It seems to be adapted to cooler climates and is the most serious nematode pathogen at elevations above 700 m in Cameroon. In Uganda, it is often the only nematode of banana at elevations of 1500 m (Price and Bridge, 1995). The distribution of *P. goodeyi*, its general absence from commercial plantations of Cavendish cultivars in lowland areas and its presence only on smallholder banana crops suggests that it is indigenous to Africa.

Disease cycle and epidemiology

The optimum temperature for invasion and development of *P. coffeae* is 25–30°C, which is the same as for *R. similis.* With *P. goodeyi,* it is nearer 20°C. The life cycle of *P. coffeae* is completed in about 4 weeks under optimum conditions.

The question of whether there are biotypes of *Pratylenchus* spp. with different host preferences is currently under investigation. There is evidence that populations of *P. coffeae* that attack yams in Uganda and the Pacific do not infest nearby banana plants. However, this is not the case in Ghana, where both yam and banana can be infested with the same population. Diversity may also occur in *P. goodeyi.* In some areas of Uganda, very high densities of the nematode have been observed on old banana stands which still remain productive, and yet, in Tanzania, lower population levels are associated with a high incidence of plant toppling (Bridge *et al.*, 1997).

As yet there are no data that can be used to determine damage thresholds. Population densities are usually expressed on the basis of 100 g of fresh roots and results vary according to the extraction technique used. In plantations where damage is obvious, either as uprooting or on visual inspection of roots, populations greater than 10,000 nematodes for every 100 g of roots may be common.

In Cameroon, at altitudes over 900 m, the population densities of *P. goodeyi* on plantations average 15,000 with a maximum of 56,000 nematodes for every 100 g of roots (Bridge *et al.*, 1995). In Uganda, average densities on East African highland banana cultivars (AAA, Lujugira–Mutika subgroup) were 25,000 nematodes 100 g^{-1} roots at ten farms at altitudes over 1600 m, but only 680 nematodes 100 g^{-1} roots on a similar number of fields at altitudes 500 m lower (Kashaija *et al.*, 1994). However, banana plants on these lower altitude farms were suffering no less because roots were supporting densities of

32,000 *Helicotylenchus multicinctus* (spiral nematode) and 6500 *R. similis* in 100 g^{-1} roots. These two species were not present at the higher elevation. This work illustrates the complexity of determining the relative importance of nematodes in mixed populations and in different environments.

Host reaction

Pratylenchus coffeae is a significant pest on 'Pisang Awak' and cultivars in the Plantain and Cavendish subgroups. In Africa, *P. goodeyi* is an important pest on plantain and cultivars in the Lujugira–Mutika subgroup (Bridge *et al.*, 1997).

Resistance (or decreased susceptibility) to *Pratylenchus* spp. has been demonstrated in glasshouse experiments. 'Yangambi Km 5' (AAA, syn. 'Ibota Bota'), 'Paka' (AA) and 'Kunnan' (AB) have been shown to have significant resistance to *P. coffeae* (Collingborn and Gowen, 1997). 'Yangambi Km 5' has also been shown to have some resistance to *P. goodeyi* in pots (Pinochet *et al.*, 1998) and in the field in Cameroon (Fogain and Gowen, 1998).

Control

Cultural practices, such as planting nematode-free suckers or plants from tissue-culture laboratories in land free of nematodes, help control damage caused by root-lesion nematodes. Paring of suckers to remove roots and discoloured areas of the corm, followed by a hot-water treatment (20 min at 53–55°C), also helps. Host-range information is important for developing a management strategy based on clean planting material. Several crops and common weeds will support reproduction of *P. coffeae* (Gowen and Quénéhervé, 1990) and a few alternative hosts of *P. goodeyi* have been discovered in East Africa (Mbwana, 1992). In glasshouse experiments, early inoculation with arbuscular mycorrhizal fungi appeared to increase the tolerance of 'Grand Nain' (AAA, Cavendish subgroup) to infestation by *P. goodeyi* by reducing the number of lesions on roots and enhancing plant nutrition (Jaizme-Vega and Pinochet, 1997).

Nematocides that are currently used for the control of *R. similis* in commercial banana plantations (see Sarah, this chapter, pp. 300–302) are equally effective on *Pratylenchus* spp.

In the long term, conventional banana breeding, coupled with genetic transformation, should contribute towards a partial management of *Pratylenchus* spp. Sources of resistance are currently being identified by banana workers in Honduras, Cameroon, Uganda and the Canary Islands (Bridge *et al.*, 1997).

Spiral Nematode

S.R. Gowen

Introduction

There are several species known collectively as spiral nematodes, but only one is a significant pathogen on banana and occurs almost wherever the crop is grown (McSorley and Parrado, 1986). Opinions differ as to its importance on banana, but in Israel it cause serious damage (Minz *et al.*, 1960). The spiral nematode seems to be more important in countries at the edge of the climatic range for the crop.

Symptoms

Like other migratory endoparasites, the spiral nematode feeds on the cell contents in the root cortex, causing necrotic lesions. However, unlike *R. similis* and *Pratylenchus* spp., feeding is often restricted to the outer parenchymal cells of the cortex (Zuckerman and Strich-Harari, 1963; Blake, 1966; Mateille, 1994a). In roots where the spiral nematode is the only parasite, its presence may be implied by the superficial nature of the lesions (Plate 7.6). However, in severe infestations, necrosis may extend to the stele causing root death. Therefore, the uprooting of infested plants, normally associated with nematode damage, can also be caused by the spiral nematode.

Causal agent

The spiral nematode important on banana is *Helicotylenchus multicinctus* (Cobb, 1893) Golden 1956.

It occurs frequently in roots that are infested with either *R. similis* or *Pratylenchus*. In extracts from root samples, *H. multicinctus* can be readily distinguished from these other genera by comparison of the lengths of the stylet (which are longer) and by the shape of the body when killed. Dead specimens are curved in the form of a letter C, whereas other spiral nematodes die in a coiled position. *Radopholus similis* and *Pratylenchus* spp. are generally straight when at rest.

Disease cycle and epidemiology

Unlike some other spiral nematodes, *H. multicinctus* is entirely endoparasitic and, like *R. similis* and the root-lesion nematodes, it is likely to have been distributed widely on planting material. All stages (juveniles and adults) are infective and can be found in roots and in adjacent soil.

When occurring in mixed populations, numbers of *H. multicinctus* may be greater than those of *R. similis* (Kashaija *et al.*, 1994). On commercial plantations of 'Robusta' (AAA, Cavendish subgroup) in St Lucia, populations were up to 24,000 spiral nematodes 100 g^{-1} of fresh roots, which was a concentration three times greater than that of *R. similis* (Gowen, 1977b). In Venezuela, where *H. multicinctus* was found coexisting with *Meloidogyne incognita* on the roots of a cultivar in the Cavendish subgroup, populations of 35,000 spiral nematodes 100 g $roots^{-1}$ were reported (Crozzoli *et al.*, 1995).

Host reaction

There are no reports on differential susceptibility to this nematode in *Musa*. Because it is generally considered a less serious pathogen than other nematodes, little attention has been given to testing varietal responses. This may be an omission that should be corrected. Techniques for mass-culturing *H. multicinctus* are not as well established as are those for *R. similis* and *P. coffeae* and this, to a certain extent, constrains critical experimental work.

Control

Chemical treatments used to control *R. similis* (see Sarah, this chapter, pp. 300–302) will also control *H. multicinctus*.

Root-knot Nematodes

D. De Waele

Introduction

Root-knot nematodes have been found in association with banana in all producing areas. They have also been identified as infesting abacá (Ocfemia and Calinisan, 1928) and enset (O'Bannon, 1975). In general, root-knot nematodes are not considered important pathogens of banana, abacá and enset. However, this may be due to insufficient research into their occurrence and effects, especially on banana cultivars grown by smallholders. On cultivars in the Cavendish subgroup (AAA), the burrowing nematode is usually more successful and tends to dominate in situations where both types of nematode are found. Root-knot nematodes are more likely to cause problems in areas where Cavendish cultivars have not been introduced or where the climate is too cold for *Radopholus similis*.

Root-knot nematodes have been described as common and abundant on banana in countries in the Mediterranean area (e.g. Crete (Vovlas *et al.*, 1994) and Lebanon (Sikora and Schlosser, 1973)), and the subtropics (e.g. South Africa (Jones, 1979b)). In South-East Asia, they are widely distributed on local diploid and triploid dessert and cooking-banana cultivars. However, root-knot nematodes are not reported in any great numbers in countries in the Latin America–Caribbean region (e.g. Honduras (Pinochet, 1977), Colombia (Zuniga *et al.*, 1979) and the French Antilles (Kermarrec and Scotto la Massese,

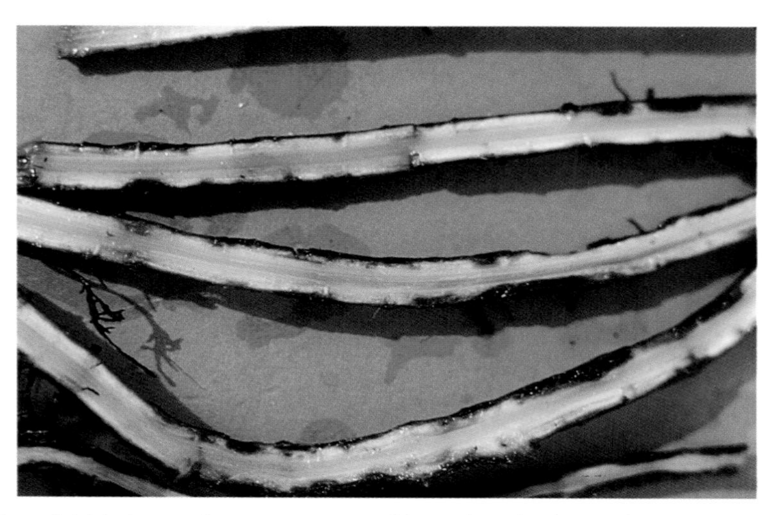

Plate 7.6. Superficial lesions on banana roots caused by *Helicotylenchus multicinctus*, a spiral nematode (photo: S.R. Gowen, UR).

1972)) or in Africa (e.g. Ghana (Afreh-Nuamah and Hemeng, 1995) and Tanzania (Nsemwa, 1991)). This may be a reflection in part of the importance of Cavendish clones in some of these countries and the dominance of the burrowing nematode. However, root-knot nematodes have been recorded as a problem in the northern Cavendish growing districts in South Africa where they are in part responsible for a condition known as false Panama disorder (Deacon *et al.*, 1985). Root-knot nematodes and *Fusarium oxysporum* (but not f. sp. *cubense*) are associated with this disorder. Treatment with nematocides can prevent the appearance of symptoms (A. Severn-Ellis, 1999, personal communication). In the Philippines, root-knot nematodes are also found on Cavendish clones. Large and widespread populations have been detected in commercial growing areas around Davao in Mindanao. The average population density was 3539 nematodes 100 g^{-1} fresh roots and 82% of all root samples examined were infected. Large root-knot nematode populations were also found on the roots of all local cultivars sampled (Davide *et al.*, 1992).

Root-knot nematodes are commonly found on local banana cultivars in Thailand (Prachasaisoradej *et al.*, 1994), Malaysia (Razak, 1994) and Indonesia (Hadisoeganda, 1994). However, they are regarded as of minor pathogenic importance in these countries. This may be because yield loss assessment data are lacking. In Malaysia, banana plants are often stunted, having thin pseudostems and small bunches, which are signs of nematode infestations (Razak, 1994). The popular cultivars 'Pisang Mas' (AA, syn. 'Sucrier'), 'Pisang Embun' (AAA, Gros Michel subgroup), 'Pisang Nangka' (AAA), 'Pisang Berangan' (AAA, syn. 'Lakatan'), 'Pisang Rastali (AAB, syn. 'Silk') and 'Pisang Tandok' (AAB, Plantain subgroup) are susceptible to root-knot. In West Malaysia, root-knot nematodes have been found to be widespread in a commercial Cavendish plantation. Plants showed extensive galls on the roots and the average number of infective, second-stage juveniles (J_2) recovered from 200 ml of soil was 2300 (Razak, 1994).

Because root-knot nematodes usually occur together with several other nematode species in banana roots, data on the damage caused by *Meloidogyne* spp. alone can only be obtained from inoculation experiments. In the Philippines, Davide and Marasigan (1985) conducted a yield loss assessment experiment in the field with a root-knot nematode. They reported that an inoculum level of 1000 J_2 $plant^{-1}$ resulted in a 26.4%

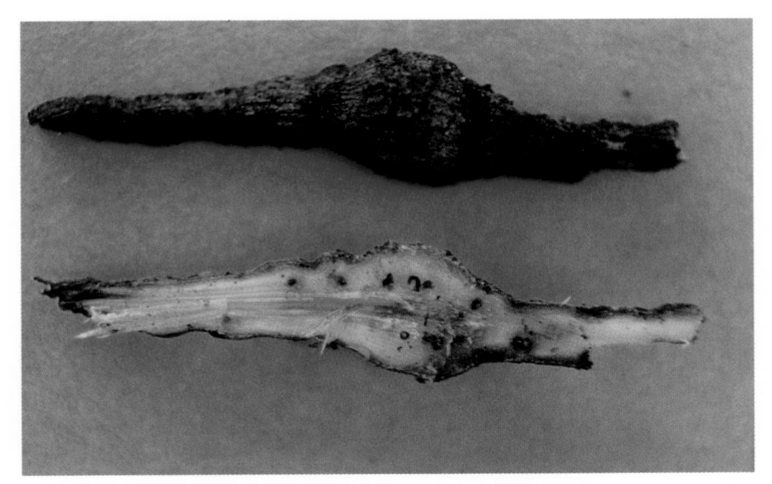

Plate 7.7. A banana root with a gall caused by root-knot nematodes. Dark-coloured, swollen females can be seen in longitudinal section (photo: K. de Jager, ITSC).

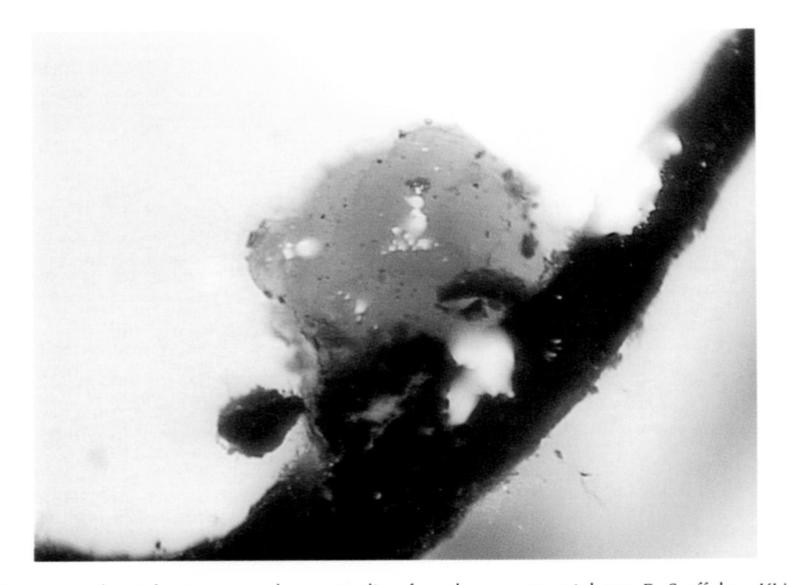

Plate 7.8. Egg mass of root-knot nematodes protruding from banana root (photo: R. Stoffelen, KUL).

yield loss in 'Giant Cavendish' (AAA, Cavendish subgroup), 10,000 J_2 in a 45.4% yield loss and 20,000 J_2 in a 57.1% yield loss, compared with the uninoculated plants.

Root-knot nematodes are very common throughout the banana-growing regions of Australia, but have never been shown to cause yield loss (Stanton, 1994).

Symptoms

On banana, galls (Plate 7.7) and swellings on primary and secondary roots are the most obvious symptoms of root-knot nematode infestation. Sometimes, the root tips are invaded and there is little or no gall formation, but root-tip growth ceases and new roots proliferate just above the infested tissues. Infested plants may have a much lower number of secondary and tertiary roots and root hairs (Claudio and Davide, 1967).

Dissection of galls reveals the typical swollen females in various stages of development (Plate 7.7). At maturity, the females are saccate. Eggs are laid singly within a gelatinous matrix to form an external egg sac or egg mass (Plate 7.8). In thick, fleshy, primary roots, the egg masses may be contained within the root. On banana roots grown under *in vitro* conditions, protruding egg masses were observed 28 days after inoculation (Coosemans *et al.*, 1994). Different species of root-knot nematode can be observed in the same gall (Pinochet, 1977). They may also colonize the outer layers of the corm up to 7 cm deep (Quénéhervé and Cadet, 1985a).

Above-ground symptoms caused on banana by root-knot nematode in Pakistan included yellowing and narrowing of leaves, stunting, reduced plant growth and less fruit production (Jabeen *et al.*, 1996). Stunted growth has also been attributed to root-knot nematodes in India (Sudha and Prabhoo, 1983) and Taiwan (Lin and Tsay, 1985).

On abacá, galls on roots have been reported to be 3–10 mm in diameter and may run together to form an irregular club-shaped body up to 5 cm long and over 1 cm in thickness. Infested roots appear normal at first, but later become brown and then change to an almost black colour. The surface of the galls crack with age and become rough to the touch. Leaves turn pale green or yellowish. The youngest leaf is generally the worst affected. Later, leaves become narrower and shorter. Plants appear stunted and leaves tend to bunch.

Galls on the primary and secondary roots of enset are associated with root-knot nematodes. Infested plants become stunted and have yellow leaves, which wilt during the dry season. Young seedlings can be seriously affected.

Causal agents

The species of root-knot nematode most commonly recorded on *Musa* are *Meloidogyne incognita* and *Meloidogyne javanica*. *Meloidogyne arenaria* and other species of *Meloidogyne* are only occasionally reported. The *Meloidogyne* species found on abacá and enset have not been identified.

Females of *Meloidogyne* spp. are sedentary. They have spherical bodies with slender necks (Plate 7.9), while the males and juveniles are vermiform. Males are rare. See Eisenback *et al.* (1981) for full descriptions of *M. incognita* and *M. javanica*.

Disease cycle and epidemiology

The life cycle of *Meloidogyne* spp. on banana is similar to its life cycle on other hosts. The endodermis is penetrated by J_2, which enter the stele and induce the vascular parenchyma or differentiating vascular cells in the central part of the stele to form multinucleate giant cells. The formation of these giant cells disturbs or blocks the surrounding xylem vessels (dos Santos and Sharma, 1978; Sudha and Prabhoo, 1983; Vovlas and Ekanayake, 1985; Jabeen *et al.*, 1996). Multiple life cycles can be completed within the same root, depending on the longevity of the root and the severity of the necrosis.

Root-knot nematodes may require much more time to become established in banana roots than root-lesion nematodes. In Cuba, *Meloidogyne* spp. needed from 24 to 30 months to establish themselves on 'Dwarf Cavendish' (AAA, Cavendish subgroup) (Fernandez and Ortega, 1982).

Root-knot nematodes are influenced by rainfall and soil conditions, such as temperature, texture and pH. After establishment, soil moisture and temperature are mainly responsible for fluctuations in populations (McSorley and Parrado, 1981; Fernandez and Ortega, 1982; Mani and Al Hinai, 1996; Youssef and Aboul-Eid, 1996). Regardless of inoculum levels, populations of *M. incognita* in the soil usually reach their highest densities during the rainy season and then decline to reach their lowest level during the dry season. The climate also affects the host. During the dry season, not enough new roots are available for nematodes to infest, resulting in low nematode populations. In Egypt, the highest population levels of *M. incognita* on banana were also positively correlated with the highest soil temperatures (26–30°C) observed at the experimental site (Youssef and Aboul-Eid, 1996). In the Philippines, Davide (1980) reported that the highest population densities of *M. incognita* were observed in sandy loam soils and at pH 5–5.6.

Meloidogyne spp. and *R. similis* can jointly infest banana. However, the root-knot nematode populations are usually reduced or completely replaced by *R. similis*, as the latter species destroys the roots, which provide the feeding sites for the root-knot nematodes (Santor and Davide, 1992).

In banana roots, *Meloidogyne* spp. often occur together with fungi capable of colonizing weakened or wounded tissue. In Yemen, Sikora (1980) observed higher levels of root rot in banana plantations where *M. incognita* and root-rot fungi (*Fusarium* and *Rhizoctonia* spp.) were present together in the soil. Synergistic effects of *M. incognita* and *Fusarium oxysporum* f. sp. *cubense*, the cause of Fusarium wilt, on roots of 'Rasthali' (AAB, syn. 'Silk') have also been reported (Jonathan and Rajendran, 1998).

Root-knot nematodes are often dispersed in run-off water and can also be spread with contaminated planting material.

Host reaction

Most of the widely grown banana cultivars are susceptible to root-knot nematode. Often, inconclusive results have been obtained when banana genotypes have been screened for resistance to *Meloidogyne* spp. on a large scale. In Brazil and India, many *Musa* genotypes were screened against *M. incognita* and *M. javanica*, but none were found to be resistant or even moderately resistant (Zem and Lordello, 1981; Patel *et al.*, 1996). In the Philippines, Davide and Marasigan (1985) screened 90 different *Musa* genotypes for reaction to *M. incognita*. They reported that the response of cultivars varied considerably, ranging from mild to severe root-gall formation. 'Viente Cohol' (AA), 'Dakdakan' (AA, syn. 'Viente Cohol'), 'Pogpogon' (AA), 'Alaswe' (AAA), 'Inambak' (AAA), 'Pastilan' (AAA), 'Sinker' (AAA), 'Maia Maole' (AAB, Maia Maoli–Popoulu subgroup) and 'Pa-a Dalaga' (ABB) showed resistance to *M. incognita*. They had a gall index of 1–2, indicating trace to slight root-gall formation, with generally few nematodes infecting the roots.

Control

Cultural practices

As noted earlier, *Meloidogyne* spp. can be disseminated with infested planting material. Risks can be minimized if the corm is peeled to remove outer tissues and then treated with hot water or a nematicide before planting (Haddad *et al.*, 1973).

Root-knot nematodes have a wide host range and associations with weeds are far more numerous than for the other nematode species associated with banana. Special attention should be given to the maintenance of weed-free fallow and the selection of cover crops in rotation systems and intercrops. In India, intercropping with *Coriandrum sativum*, *Sesamum indicum*, *Crotalaria juncea*, *Tagetes erecta* and *Acorus calamus* significantly reduced the populations of *M. incognita* on 'Robusta' (AAA, Cavendish subgroup) in field trials (Charles, 1995). The same effect on *Meloidogyne* spp. was obtained in crop rotation trials with Pangola grass, maize and sugar cane in Cuba (Stoyanov, 1971) and with *Tagetes patula* in South Africa (Milne and Keetch, 1976). Rotation with paddy rice may also drastically reduce root-knot nematode populations. This has been attributed to flooding (Sivakumar and Marimuthu, 1986). Fallowing to eradicate root-knot nematodes may, however, be ineffective, since Stoyanov (1971) reported from Cuba that *Meloidogyne* spp. persisted in soil in the absence of banana for up to 29 months.

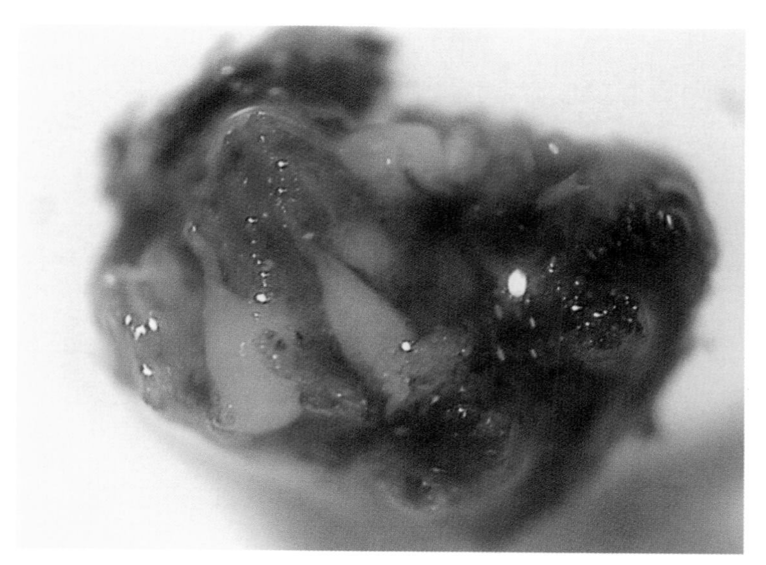

Plate 7.9. A gall on a banana root exposed to show female root-knot nematodes with swollen bodies and slender necks (photo: R. Stoffelen, KUL).

Plate 7.10. Symptoms of nematode black leaf-streak disease on the leaf of a young enset seedling (photo: M. Tessera and A.J. Quimio, IAR).

Chemical control

Numerous field experiments have shown the effectiveness of nematocides in the control of root-knot nematodes. Dipping banana corms in a solution of nematicide for 10 min before planting may protect the plants for a few months against nematode infection. Nematocides found effective include dibromochloropropane (DBCP or Nemagon®), ethoprophos and fenamiphos (organophosphates), and aldicarb and carbofuran (carbamates) (Subramanian *et al.*, 1984; Davide, 1992). Immersion of peeled corms in a 1% solution of sodium hypochlorite (NaOCl) for 5 or 10 min also controlled *Meloidogyne* spp. and is considered by Lordello *et al.* (1994) as an effective, low cost and non-toxic preplanting treatment. It has also been shown that

Plate 7.11. Leaf of an enset sucker with severe symptoms of nematode black leaf streak disease (photo: M. Tessera and A.J. Quimio, IAR).

fumigating soil with ethylene dibromide (EDB), dichloropropane-dichloropropene (D-D) or methyl bromide (Keetch *et al.*, 1976) before planting is effective. Treating soil with organophosphates (ethoprophos, cadusaphos, fenamiphos, isazophos, terbufos) and carbamates (aldicarb, carbofuran, oxamyl) several times a year has controlled root-knot nematodes in established banana plantations and improved plant growth and yield (Gupta, 1975; Valle-Lamboy and Ayala, 1976; Jones, 1979a; Caveness and Badra, 1980; Badra and Caveness, 1983; B'Chir and Horrigue-Raouani, 1991; Horrigue-Raouani and B'Chir, 1992; Crozzoli *et al.*, 1995; Araya and Cheves, 1997). By knowing the seasonal fluctuation in nematode populations, an effective nematocide application strategy can be developed. Control is most effective when treatments are timed to coincide with the build-up of nematode populations that usually occurs at the onset of the rainy season. In Puerto Rico, oxamyl applied to leaf axils of 'Giant Cavendish' four times at 30-day intervals during the growing season effectively controlled *M. incognita* (Robalino *et al.*, 1983).

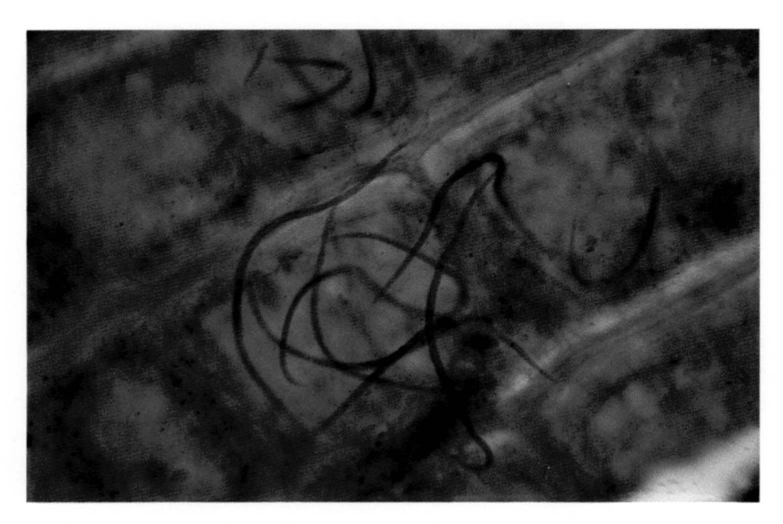

Plate 7.12. *Ektaphelenchoides* sp. in leaf tissue of enset (photo: M. Tessera and A.J. Quimio, IAR).

Hoan and Davide (1979) reported that root extracts of 11 plant species showed nematocidal effects when tested against *M. incognita* in the Philippines. Root extracts from African marigold (*T. erecta*), ipil-ipil (*Leucaena leucocephala*), Bermuda grass (*Cynodon dactylon*) and makahiya (*Mimosa pudica*) prevented eggs from hatching and controlled infestation. The performance of these root extracts was comparable to that of commercially produced chemical nematocides. Results of another study revealed that leaf extracts from Kaatoanbangkal (*Anthocephalus chinensis*) and water lily (*Eichornia crassipes*) and extracts of garlic (*Allium sativa*) and onion (*Allium cepa*) bulbs were also effective against *M. incognita* (Guzman and Davide, 1992). Characterization of the active nematocidal principle showed a phenolic aldehyde from *A. chinensis*, a carboxylic acid from *E. crassipens* and a ketone from *A. cepa*.

Biological control

Culture extracts of 17 species of microorganisms have been evaluated under laboratory and greenhouse conditions in the Philippines for nematocidal activity against *M. incognita* infesting 'Giant Cavendish' (Molina and Davide, 1986). Purified extracts of several *Penicillium* spp. (*P. oxalicum*, *P. anatolicum*) and *Aspergillus niger* showed high nematocidal activity. *Paecilomyces lilacinus* and *P. oxalicum* have been reported as being very successful in controlling *Meloidogyne* spp. and other nematodes on banana (Davide, 1994). Liquid and powder formulations containing these two fungi are marketed commercially.

Arbuscular mycorrhizal fungi are also being investigated as biological control agents. Inoculation of micropropagated 'Grand Nain' (AAA, Cavendish subgroup) with two isolates of *Glomus mosseae* suppressed gall formation and build-up of *M. incognita* in roots under greenhouse conditions. The presence of nematodes had no effect on the colonization of roots by these fungi (Jaizme-Vega *et al.*, 1997). Inoculation of the same banana cultivar with *Glomus intraradices* did not affect the build-up of *M. incognita* in the roots, but increased plant growth by enhancing plant nutrition (Pinochet *et al.*, 1997).

Fluorescent strains of *Pseudomonas* isolated from the rhizosphere have also been found to inhibit the invasion of roots of 'Grand Nain' (AAA, Cavendish subgroup) by a mixed population of *M. incognita* and *M. javanica* (Aalten *et al.*, 1998).

Acknowledgements

The information provided by Editha Lomerio on root-knot nematode disease of abacá and Mesfin Tessera and Arcadio Quimio on root-knot disease of enset is gratefully acknowledged.

Other Nematodes

D. De Waele, S.R. Gowen, M. Tessera and A.J. Quimio

In addition to the burrowing nematode, root-lesion nematodes, spiral nematodes and root-knot nematodes, many other nematode species have been found in association with banana. Often these nematodes occur in relatively low numbers in roots infested with high numbers of more pathogenic species.

Rotylenchulus reniformis Linford & Oliveira, 1940, has been found in association with banana in all producing areas throughout the world. Documented reports come from South America (Zuniga *et al.*, 1979; Crozzoli *et al.*, 1993), the Caribbean (Oramas and Roman, 1982), Africa (Adiko, 1988; Fargette and Quénéhervé, 1988), Asia (Chau *et al.*, 1997) and the Mediterranean (Aboul-Eid and Ameen, 1991). In St Lucia, populations of up to 2500 juvenile and infective immature female nematodes were found in 100 cm^3 samples of soil taken from around the fine, secondary roots in which mature adult females were permanently lodged (Gowen, 1977b). Although *R. reniformis* is sometimes mentioned as a potential threat to banana cultivation and is believed to cause damage to the root system (Edmunds, 1968), no quantitative data on the effect of this nematode species on growth and yield of banana have been reported.

Rotylenchulus reniformis penetrates the cortex of banana roots perpendicularly to the stele and establishes a permanent feeding site in the endodermis. Nematode feeding induces the fusion of endodermal, pericycle and vascular parenchymal cells to form a syncytium, with hypertrophied nuclei and prominent nucleoli (Vovlas and Ekanayake, 1985). These permanent feeding sites are generally located on the secondary roots (Ayala, 1962; Edmunds, 1968).

Rotylenchulus reniformis is usually found in association with other pathogenic nematode species. Most nematocides effective against root-knot nematodes, including oxamyl applied to leaves (Gowen, 1977a; Robalino *et al.*, 1983), were also effective against *R. reniformis*. Information on either cultural or biological control is rare. In India, intercropping with *Coriandrum sativum*, *Sesamum indicum*, *Crotalaria juncea*, *Tagetes erecta* and *Acorus calamus* significantly reduced the populations of *R. reniformis* on 'Robusta' (AAA, Cavendish subgroup) in field trials (Charles, 1995). The nematode has also been reported to infest the roots of abacá (Davide, 1972).

Hoplolaimus pararobustus (Schuurmans Stekhoven and Teunissen, 1938) Sher, 1963, which can be found in relatively high numbers on banana (1000–18,000 nematodes 100 g^{-1} roots) (Guerout, 1974; Hunt, 1977; Price, 1994b), appears to occur only in the subepidermal cortex (Mateille, 1994a). Its damage potential is thought to be minimal (Price, 1994a).

Heterodera oryzicola Rao & Jayaprakash, 1978, occurs on banana in southern India and its incidence is probably related to the cropping system, where banana is grown in rotation with paddy rice. Pathogenicity studies suggest that this nematode could cause yield loss (Charles and Venkitesan, 1993).

Nematode black leaf streak disease of enset was first recorded at Awasa Zuria in Ethiopia in 1991 (Tessera and Quimio, 1994). Later it was found in most enset-growing areas. This disease, which is caused by a species of *Ektaphelenchoides*, has the capacity to severely damage enset suckers and seedlings. The most characteristic symptom is small black streaks on leaves of suckers and seedlings (Plate 7.10). Streaks sometimes coalesce to form long necrotic stripes. Severe streaking can cause the premature death of leaves (Plate 7.11). The nematode lives and multiplies in leaf tissue (Plate 7.12) and spreads to neighbouring healthy leaves by rain splash or during watering operations. The nematode is carried to new farms on infested plants. Most enset clones seem susceptible. The early removal of infested leaves helps to control the disease and minimize the chance of spread.

REFERENCES

Aalten, P.M., Vitiur, D., Blanvillain, D., Gowen, S.R. and Sutra, L. (1998) Effect of rhizosphere fluorescent *Pseudomonas* strains on plant-parasitic nematodes *Radopholus similis* and *Meloidogyne* spp. *Letters in Applied Microbiology* 27, 357–361.

Aboul-Eid, H.Z. and Ameen, H.H. (1991) Distribution and population densities of root-knot, reniform and spiral nematodes on different banana cultivars in Egypt. *Bulletin of the Faculty of Agriculture, University of Cairo* 42, 919–928.

Adiko, A. (1988) Plant-parasitic nematodes associated with plantain, *Musa paradisiaca* (AAB), in the Ivory Coast. *Revue de Nématologie* 11, 109–113.

Afreh-Nuamah, K. and Hemeng, O.B. (1995) Nematodes associated with plantains in the eastern region of Ghana. *Musafrica* 7, 4–6.

Anderson, E.J. and Alaban, C.A. (1968) Some plant-parasitic nematodes found in the vicinities of General Santos and Davao, Mindanao. *Philippine Phytopathology* 4, 1–2 (abstract).

Anderson, J.P. (1988) Accelerated microbial degradation of nematicides and other plant protection chemicals. *Nematropica* 19, 1.

Anon. (1998) *Musa*news. *Infomusa* 7(2), 43–46.

Anunciado, I.S., Balmes, L.O., Bawagen, P.Y., Benigno, D.A., Bondad, N.D., Cruz, O.J., Franco, P.T., Gavarra, M.R., Opeña, M.T. and Tabora, P.C. (1977) *The Philippines Recommends for Abaca, 1977.* Philippines Council for Agriculture and Resources Research, Los Baños, Philippines, 71 pp.

Araya, M. and Cheves, A. (1997) Efecto de cuatro nematicidas sobre el control de nematodos en banano (*Musa* AAA). *CORBANA* 22, 35–48.

Ayala, A. (1962) Pathogenicity of the reniform nematode on various hosts. *Journal of Agriculture, University of Puerto Rico* 45, 265–299.

Badra, T. and Caveness, F.E. (1983) Effects of dosage sequence on the efficacy of nonfumigant nematicides, plantain yields, and nematode seasonal fluctuations as influenced by rainfall. *Journal of Nematology* 15, 496–502.

B'Chir, M.M. and Horrigue-Raouani, N. (1991) Particularités du problème posé par les *Meloidogyne* spp. sur bananier sous serres. *Mededelingen van de Faculteit Landbouwwetenschappen Rijksuniversiteit Gent* 56, 1271–1280.

Blake, C.D. (1961) Root rot of banana caused by *Radopholus similis* (Cobb) and its control in New South Wales. *Nematologica* 6, 295–310.

Blake, C.D. (1966) The histological changes in banana roots caused by *Radopholus similis* and *Helicotylenchus multicinctus. Nematologica* 12, 129–137.

Bridge J. (1987) Control strategies in subsistence agriculture. In: Brown, R.H. and Kerry, B.R. (eds) *Principles and Practice of Nematode Control in Crops.* Academic Press, Sydney, Australia, pp. 389–420.

Bridge, J. (1988) Plant nematode pests of banana in East Africa with particular reference to Tanzania. In: *Nematodes and the Borer Weevil in Bananas, Proceedings of an INIBAP Workshop, Bujumbura, Burundi, 1987.* INIBAP, Montpellier, France, pp. 35–39.

Bridge, J. (1993) Worldwide distribution of the major nematode parasites of bananas and plantains. In: Gold, C.S. and Gemmel, B. (eds) *Proceedings of Biological and Integrated Control of Highland Banana and Plantain Pests and Diseases in Africa, Cotonou, Bénin, 12–14 November 1991.* The Printer, Davis, California, USA, pp. 195–198.

Bridge, J., Price, N.S. and Kofi, P. (1995) Plant parasitic nematodes of plantain and other crops in Cameroon, West Africa. *Fundamental and Applied Nematology* 18, 251–260.

Bridge, J., Fogain, R. and Speijer, P. (1997) *The Root Lesion Nematodes of Banana. Musa* Pest Fact Sheet No. 2, INIBAP, Montpellier, France, 4 pp.

Carreel, F. (1995) Etude de la diversité génétique des bananiers (genre *Musa*) à l'aide des marqueurs RFLP. PhD thesis, Institut National Agronomique, Paris-Grignan, France.

Castillo, M.B., Reyes, T.T. and Davide, R.G. (1974) *Host Index of Plant Parasitic Nematodes in the Philippines.* UPLB-CA Technical Bulletin, No. 33, Los Baños, The Philippines, 32 pp.

Caveness, F.E. and Badra, T. (1980) Control of *Helicotylenchus multicinctus* and *Meloidogyne javanica* in established plantain and nematode survival as influenced by rainfall. *Nematropica* 10, 10–14.

Cayrol, J.C., Djian-Caporalino, C. and Panchaud-Mattei, E. (1992) La lutte biologique contre les nématodes phytoparasites. *Courrier de la cellule environnement de l'INRA* 17, 31–44.

Charles, J.S.K. (1995) Effect of intercropping antagonistic crops against nematodes in banana. *Annals of Plant Protection Sciences* 3, 185–187.

Charles, J.S.K. and Venkitesan, T.S. (1993) Pathogenicity of *Heterodera oryzicola* (Nemata: Tylenchina) towards banana (*Musa* AAB cv. Nendran). *Fundamental and Applied Nematology* 16, 359–365.

Chau, N.N., Thanh, N.V., De Waele, D. and Geraert, E. (1997) Plant-parasitic nematodes associated with banana in Vietnam. *International Journal of Nematology* 7, 122–126.

Claudio, M.Z. and Davide, R.G. (1967) Pathogenicity and identity of root-knot nematodes on five varieties of banana. *Philippine Agriculturist* 51, 241–251.

Cobb, N.A. (1893) Nematodes, mostly Australian and Fijian. In: *Macleay Memorial Volume.* Linnean Society of New South Wales, pp. 252–308.

Colbran, R.C. (1964) Cover crops for nematode control in old banana land. *Queensland Journal of Agricultural Science* 21, 233–236.

Collingborn, F.M.B. and Gowen, S.R. (1997) Screening of banana cultivars for resistance to *Radopholus similis* and *Pratylenchus coffeae. Infomusa* 6(2), 3.

Coosemans, J., Duchateau, K. and Swennen, R. (1994) Root-knot nematode (*Meloidogyne* spp.) infection on banana (*Musa* spp.) *in vitro. Archiv für Phytopathologie und Pflanzenschutz* 29, 165–169.

Crozzoli, P.R., Graff, R. and Rivas, G.D. (1993) Nematodos fitoparasitos asociados al cultivo del banano (*Musa acuminata* AAA) en el estado Aragua, Venezuela. *Revista de la Facultad de Agronomia, Universidad Central de Venezuela* 19, 275–287.

Crozzoli, R., Martinez, G. and Rivas, D. (1995) Manejo y fluctuaciones poblacionales de *Helicotylenchus multicinctus* y *Meloidogyne incognita* en banano en Venezuela. *Nematropica* 25, 61–66.

Davide, R.G. (1972) *Nematodes of Philippine Crops and their Control.* Philippine Phytopathology Society Bulletin No. 2, 34 pp.

Davide, R.G. (1980) Influence of cultivar, age, soil texture, and pH on *Meloidogyne incognita* and *Radopholus similis* on banana. *Plant Disease* 64, 571–573.

Davide, R.G. (1992) Evaluation of three nematicides for control of *Meloidogyne incognita* affecting banana. In: Davide R.G. (ed.) *Studies on Nematodes Affecting Bananas in the Philippines.* Philippine Agriculture and Resources Research Foundation, Los Baños, Philippines, pp. 101–110.

Davide, R.G. (1994) Biological control of banana nematodes: development of BIOCON I (BIOACT) and BIOCON II technologies. In: Valmayor, R.V., Davide, R.G., Stanton, J.M., Treverrow, N.L. and Roa, V.N. (eds) *Banana Nematodes and Weevil Borers in Asia and the Pacific: Proceedings of a Conference–Workshop on Nematodes and Weevil Borers Affecting Bananas in Asia and the Pacific, Serdang, Selangor, Malaysia, 18–22 April 1994.* INIBAP/ASPNET, Los Baños, Philippines, pp. 139–146.

Davide, R.G. and Marasigan, L.Q. (1985) Yield loss assessment and evaluation of resistance of banana cultivars to the nematodes *Radopholus similis* Thorne and *Meloidogyne incognita* Chitwood. *Philippine Agriculturist* 68, 335–349.

Davide, R.G., Gargantiel, F. and Zarate, F.A. (1992) Survey on distribution and identification of banana nematodes in the Philippines. In: Davide, R.G. (ed.) *Studies on Nematodes Affecting Bananas in the Philippines.* Philippine Agriculture and Resources Research Foundation, Los Baños, Philippines, pp. 17–36.

Deacon, J.W., Herbert, J.A. and Dames, J. (1985) Fals Panama disorder of bananas. *Subtropica* 6, 15–18.

De Waele, D. (1996) Plant resistance to nematodes in other crops: relevant research that may be applicable to *Musa.* In: Frison, E.A., Horry, J.P. and De Waele, D. (eds) *New Frontiers in Resistance Breeding for Nematode, Fusarium and Sigatoka. Proceedings of the Workshop held in Kuala Lumpur, Malaysia, 2–5 October 1995.* INIBAP, Montpellier, France, pp. 108–118.

dos Santos, A.V.P. and Sharma, R.D. (1978) Formacao das celulas gigantes em raizes de bananeira parasitadas por *Meloidogyne incognita. Ciencia e Cultura* 30, 842–848.

Ducharme, E.P. and Birchfield, W. (1956) Physiological races of the burrowing nematode. *Phytopathology* 46, 615–616.

Edmunds, J.E. (1968) Nematodes associated with bananas in the Windward Islands. *Tropical Agriculture, Trinidad* 45, 119–124.

Edwards, D.I. and Wehunt, E.J. (1971) Host range of *Radopholus similis* from banana areas of Central America with indications of additional races. *Plant Disease Reporter* 55, 415–418.

Eisenback, J.D., Hirschmann, H., Sasser, J.N. and Triantaphyllou, A.C. (1981) *A Guide to the Four Most Common Species of Root-knot Nematodes (*Meloidogyne *spp.) with a Pictorial Key.* North Carolina State University, Raleigh, USA, 48 pp.

Fallas, G. and Sarah, J.L. (1995a) Effect of temperature on the *in vitro* multiplication of seven *Radopholus similis* isolates from different banana producing zones of the world. *Fundamental and Applied Nematology* 18, 445–451.

Fallas, G. and Sarah, J.L. (1995b) Effect of storage temperature on the *in vitro* reproduction of *Radopholus similis. Nematropica* 25, 175–177.

Fallas, G., Sarah J.L. and Fargette, M. (1995) Reproductive fitness and pathogenicity of eight *Radopholus similis* isolates on banana plants (*Musa* AAA, cv. Poyo). *Nematropica* 25, 135–141.

Fallas, G., Hahn, M., Fargette, M., Burrows, P. and Sarah, J.L. (1996) Molecular and biochemical diversity among isolates of *Radopholus similis* from different areas of the world. *Journal of Nematology* 28, 422–430.

Fargette, M. and Quénéhervé, P. (1988) Populations of nematodes in soils under banana cv. Poyo, in the Ivory Coast. 1. The nematofauna occurring in the banana producing areas. *Revue de Nématologie* 11, 239–244.

Fernandez, M. and Ortega, J. (1982) Comportamiento de las poblaciones de nematodos fitoparasitos en platano Enano Cavendish. *Ciencias de la Agricultura* 13, 7–17.

Fogain, R. (1996) Screenhouse evaluation of *Musa* for suceptibility to *Radopholus similis*: evaluation of plantains AAB and diploids AA, AB and BB. In: Frison, E.A., Horry, J.P. and De Waele, D. (eds) *New Frontiers in Resistance Breeding for Nematode, Fusarium and Sigatoka. Proceedings of the Workshop held in Kuala Lumpur, Malaysia, 2–5 October 1995.* INIBAP, Montpellier, France, pp. 79–88.

Fogain, R. and Gowen, S.R. (1997) Damage to roots of *Musa* cultivars by *Radopholus similis* with and without protection of nematicides. *Nematropica* 27, 27–32.

Fogain, R. and Gowen, S.R. (1998) 'Yangambi km5' (*Musa* AAA, Ibota subgroup): a possible source of resistance to *Radopholus similis* and *Pratylenchus goodeyi. Fundamental and Applied Nematology*, 21, 75–80.

Frison, E.A., Orjeda, G. and Sharrock, S.L. (eds) (1997) *PROMUSA: A Global Programme for* Musa *Improvement. Proceedings of a Meeting held in Gosier, Guadeloupe, March 5 and 9, 1997.* INIBAP, Montpellier, France/World Bank, Washington, USA, 64 pp.

Gowen S.R. (1975) Improvement of banana yields with nematicides. In: *Proceedings of the 8th British Insecticide and Fungicide Conference, Brighton 1975.* BCPC, pp. 121–125.

Gowen, S.R. (1977a) Nematicidal effects of oxamyl applied to leaves of banana seedlings. *Journal of Nematology* 9, 158–161.

Gowen, S.R. (1977b) Studies on the control of nematodes on banana in the Windward Islands and Jamaica. PhD thesis, University of Reading, UK, 320 pp

Gowen S.R. (1979) Some considerations of problems associated with the nematode pests of bananas. *Nematropica* 9, 79–91.

Gowen S.R. (1996) The source of nematode resistance, the possible mechanisms and the potential for nematode tolerance in *Musa*. In: Frison, E.A., Horry, J.P. and De Waele, D. (eds) *New Frontiers in Resistance Breeding for Nematode, Fusarium and Sigatoka, Proceedings of the Workshop held in Kuala Lumpur, Malaysia, 2–5 October 1995.* INIBAP, Montpellier, France, pp. 108–118.

Gowen, S.R. and Quénéhervé, P. (1990) Nematode parasites of bananas, plantains and abaca. In: Luc, M., Sikora, R.A. and Bridge, J. (eds) *Plant Parasitic Nematodes in Subtropical and Tropical Agriculture.* CAB International, Wallingford, UK, pp. 431–460.

Guerout, R (1974) Quatre nématicides récents et leurs possibilités d'utilisation en bananaraie. *Fruits* 29, 339–347.

Gupta, J.C. (1975) Evaluation of various treatments and varietal resistance for the control of banana nematodes. *Haryana Journal of Horticultural Sciences* 4, 152–156.

Guzman, R.S. and Davide, R.G. (1992) Screening of various plant extracts for toxicity to *Meloidogyne incognita* and *Radopholus similis.* In: Davide, R.G. (ed.) *Studies on Nematodes Affecting Bananas in the Philippines.* Philippine Agriculture and Resources Research Foundation, Los Baños, Philippines, pp. 133–139.

Haddad, O., Meredith, J.A. and Martinez, R.G.J. (1973) Estudio preliminar sobre el control de nematodos en material de propagacion de bananos. *Nematropica* 3, 29–45.

Hadisoeganda, W.W. (1994) Status of nematode problems affecting banana in Indonesia. In: Valmayor, R.V., Davide, R.G., Stanton, J.M., Treverrow, N.L. and Roa, V.N. (eds) *Banana Nematodes and Weevil Borers in Asia and the Pacific: Proceedings of a Conference–Workshop on Nematodes and Weevil Borers affecting Bananas in Asia and the Pacific, Serdang, Selangor, Malaysia, 18–22 April 1994.* INIBAP/ASPNET, Los Baños, Philippines, pp. 63–73.

Hahn, M., Sarah, J.L, Boisseau, M., Vines, N.J., Bridge, J., Wright, D.J. and Burrows, P.R. (1996) Reproductive fitness and pathogenic potential of selected *Radopholus* populations on two banana cultivars. *Plant Pathology* 45, 223–231.

Hoan, L.T. and Davide, R.G. (1979) Nematicidal properties of root extracts of several plant species on *Meloidogyne incognita. Philippine Agriculturist* 62, 285–295.

Hooper, D.J. (1986) Extraction of nematodes from plant material. In: Southey, J.F. (ed.) *Laboratory Methods for Work with Plant and Soil Nematodes.* Ministry of Agriculture, Fisheries and Food, London, UK, pp. 51–58.

Horrigue-Raouani, N. and B'Chir, M.M. (1992) Conception de la lutte chimique contres les *Meloidogyne* associés au bananier sous serres. *Mededelingen van de Faculteit Landbouwwetenschappen Rijksuniversiteit Gent* 57, 903–912.

Huettel, R.N., Dickson, D.W. and Kaplan, D.T. (1984) *Radopholus citrophilus* sp. n. (Nematoda), a sibling species of *Radopholus similis*. *Proceedings of the Helminthological Society of Washington* 51, 32–35.

Hugon R., Ganry J. and Berthe, G. (1984) Dynamique de population du nématode *Radopholus similis* en fonction du stade de développement du bananier et du climat. *Fruits* 39, 251–253.

Hunt, D.J. (1977) Plant parasitic nematodes from the Windward Islands. *PANS* 23, 402–411.

Jabeen, S., Bilqees, F.M., Khan, A. and Khatoon, N. (1996) Pathogenicity of *Meloidogyne javanica* on banana in Pakistan. *Proceedings of Parasitology* 21, 11–76.

Jaizme-Vega, M.C. and Pinochet, J. (1997) Growth response of banana to three mycorrhizal fungi in *Pratylenchus goodeyi* infested soil. *Nematropica* 27, 69–71.

Jaizme-Vega, M.C., Tenoury, P., Pinochet, J. and Jaumot, M. (1997) Interactions between the root-knot nematode *Meloidogyne incognita* and *Glomus mosseae* in banana. *Plant and Soil* 196, 27–35.

Jaramillo, R. and Figueroa, A. (1976) Relación entre el balance hídrico de *Radopholus similis* (Cobb) Thorne en la zona bananera de Guápiles, Costa Rica. *Turrialba* 26, 187–192.

Jonathan, E.I. and Rajendran, G. (1998) Interaction of *Meloidogyne incognita* and *Fusarium oxysporum* f. sp. *cubense* on banana. *Nematologia Mediterranea* 26, 9–11.

Jones, R.K. (1979a) Control of *Helicotylenchus multicinctus* and *Meloidogyne incognita* infecting bananas by two granular systemic nematicides. *Plant Disease Reporter* 63, 744–747.

Jones, R.K. (1979b) Nematodes associated with bananas in South Africa. *Phytophylactica* 11, 79–81.

Jones, R.K. and Milne, D.L. (1982) Nematode pests of bananas. In: Keetch, D.P. and Heyns, J. (eds) *Nematology in Southern Africa*. Department of Agriculture and Fisheries, Pretoria, Republic of South Africa, pp. 30–37.

Kaplan, D.T., Vanderspool, M.C. and Opperman, C.H. (1998) Sequence tag site and host range assays demonstrate that *Radopholus similis* and *R. citrophilus* are not reproductively isolated. *Journal of Nematology* 29, 421–429.

Kashaija, I.N., Speijer, P.R., Gold, C.S. and Gowen, S.R. (1994) Occurrence, distribution and abundance of plant parasitic nematodes of bananas in Uganda. *African Crop Science Journal* 2, 99–104.

Keetch, D.P., Reynolds, R.E. and Mitchell, I.A. (1976) An evaluation of pre- and post-plant nematicides for the control of plant parasitic nematodes in bananas. *Citrus and Subtropical Fruit Journal* 506, 5–7.

Kermarrec, A. and Scotto la Massese, C. (1972) Données nouvelles sur la composition et la dynamique de la nématofaune des sols des Antilles françaises. *Annales de Zoologie Ecologie Animale* 4, 513–527.

Laville, E. (1964) Etude de la mycoflore des racines du bananier Poyo. *Fruits* 19, 435–449.

Lin, Y.Y. and Tsay, T.T. (1985) Studies on banana root-knot disease in the central area of Taiwan. *Journal of the Chinese Society for Horticultural Science* 31, 44–46.

Loos, C.A. (1961) Eradication of the burrowing nematode, *Radopholus similis*, from bananas. *Plant Disease Reporter* 29, 43–52.

Loos, C.A. (1962) Studies on the life-history and habits of the burrowing nematode *Radopholus similis* the cause of black-head disease of banana. *Proceedings of the Helminthological Society of Washington* 29, 43–52.

Lordello, R.R.A., Moreira, R.S. and Lordello, A.I.L. (1994) Hipoclorito de sodio: nova alternativa para o controle do nematoide *Radopholus similis* em mudas de bananeira. *Agronomico (Campinas)* 46, 35–40.

Loridat, P. (1989) Etude de la microflore fongique et des nématodes associés aux nécroses de l'appareil souterrain du bananier en Martinique. *Fruits* 44, 587–598.

Luc, M., Sikora, R.A. and Bridge, J. (eds) (1990) *Plant Parasitic Nematodes in Subtropical and Tropical Agriculture*. CABI, Wallingford, UK, 629 pp.

McSorley, R. and Parrado, J.L. (1981) Population fluctuations of plant-parasitic nematodes on bananas in Florida. *Proceedings of the Florida State Horticultural Society* 94, 321–323.

McSorley, R. and Parrado, J.L. (1986) *Helicotylenchus multicinctus* on bananas: an international problem. *Nematropica* 16, 73–91.

Maas, P.W.T. (1969) Two important cases of nematode infestation in Surinam. In: Peachey, J.E. (ed.) *Nematodes of Tropical Crops*. Commonwealth Agricultural Bureaux, St Albans, England, pp. 149–154.

Mani, A. and Al Hinai, M.S. (1996) Population dynamics and control of plant parasitic nematodes on banana in the Sultanate of Oman. *Nematologia Mediterranea* 24, 295–299.

Mateille, T. (1994a) Comparative host tissue reactions of *Musa acuminata* (AAA group) cvs Poyo and Gros Michel roots to three banana-parasitic nematodes. *Annals of Applied Biology* 124, 65–73.

Mateille, T. (1994b) Réactions biochimiques provoquées par trois nématodes phytoparasites dans les racines de *Musa acuminata* (groupe AAA) variétés Poyo et Gros Michel. *Fundamental and Applied Nematology* 17, 283–290.

Mateille, T., Cadet, P. and Quénéhervé, P. (1984) L'influence du recépage du bananier cv. Poyo sur le développement des populations de *Radopholus similis* et d'*Helicotylenchus multicinctus*. *Revue de Nématologie* 7, 355–361.

Mateille, T., Foncelle, B. and Ferrer, H. (1988) Lutte contre les nématodes du bananier par submersion du sol. *Revue de Nématologie* 11, 235–238.

Mbwana, A.A.S. (1992) Host range of lesion nematode, *Pratylenchus goodeyi* Sher and Allen. In: *Annual Report International Centre of Insect Physiology and Ecology (ICIPE)*, Nairobi, Kenya.

Melin, P. and Vilardebo, A. (1973) Efficacité de quelques nématicides en bananeraie dans les sols volcaniques de la région du Mungo (Cameroun). *Fruits* 28, 3–17.

Melin, P., Plaud, G. and Tézenas du Montcel, H. (1976) Influence des nématodes sur la culture du plantain. *Fruits* 31, 688–691.

Menendez, T. and Shepherd, K. (1975) Breeding new bananas. *World Crops* 27, 104–112.

Milne, D.L. and Keetch, D.P. (1976) Some observations on the host plant relationships of *Radopholus similis* in Natal. *Nematropica* 6, 13–17.

Minz, G., Ziv, D. and Strich-Harari (1960) Decline of banana plantations caused by spiral nematodes in the Jordan Valley and its control by DBCP. *Ktavim* 10, 147–157.

Molina, G.C. and Davide, R.G. (1986) Evaluation of microbial extracts for nematicidal activity against plant parasitic nematodes *Meloidogyne incognita* and *Radopholus similis*. *Philippine Agriculturist* 69, 173–186.

Nsemwa, L.T.H. (1991) Problems of banana weevil and nematodes in the Southern Highlands of Tanzania. *Fruits* 46, 541–542.

O'Bannon, J.H. (1975) *Nematode Survey*. FAO, Rome, 29pp.

O'Bannon, J.H. (1977) Worldwide dissemination of *Radopholus similis* and its importance in crop production. *Journal of Nematology* 9, 16–25.

Ocfemia, G.O. and Calinisan, M.R. (1928) The root-knot of abaca or Manila hemp. *Phytopathology* 18, 861–867.

Ogier, T.P. and Merry, C.A.A.F. (1970) Yield decline of plantains, *Musa paradisiaca*, in Trinidad associated with the nematode *Pratylenchus* sp. *Turrialba* 20, 407–417.

Oramas, D. and Roman, J. (1982) Plant parasitic nematodes associated with plantain (*Musa acuminata* × *M. balbisiana*, AAB) in Puerto Rico. *Journal of Agriculture of the University of Puerto Rico* 66, 52–59.

Patel, B.A., Vyas, R.V., Patel, D.J. and Patel, R.S. (1996) Susceptibility of banana cultivars to root-knot nematodes (*Meloidogyne* spp.). *Infomusa* 5(2), 26–27.

Peregrine, W.T.H. and Bridge, J. (1992) The lesion nematode *Pratylenchus goodeyi*, an important pest of *Ensete* in Ethiopia. *Tropical Pest Management* 38, 325–326.

Perez, J.A., Valdes, S. and Mola, Y.G. (1986) Comportamiento varietal del platano (*Musa* sp.) al ataque de los nematodos *Radopholus similis* y *Pratylenchus coffeae*. *Ciencia Tecnica Agricola, Proteccion de Plantas* 9, 13–22.

Pinochet, J. (1977) Occurence and spatial distribution of root-knot nematodes on bananas and plantains in Honduras. *Plant Disease Reporter* 61, 518–520.

Pinochet, J. (1979) Comparison of four isolates of *Radopholus similis* from Central America on Valery bananas. *Nematropica* 9, 40–43.

Pinochet, J. (1986) A note on nematode control practices on bananas in Central America. *Nematropica* 16, 197–203.

Pinochet, J. (1988a) Comments on the difficulty in breeding bananas and plantains for resistance to nematodes. *Revue de Nématologie* 11, 3–5.

Pinochet, J. (1988b) A method for screening bananas and plantains to lesion forming nematodes. In: *Nematodes and the Borer Weevil in Bananas, Proceedings of a INIBAP Workshop, Bujumbura, Burundi, 1987*. INIBAP, Montpellier, France, pp. 62–65.

Pinochet, J. and Rowe, P.R. (1978) Reaction of two banana cultivars to three different nematodes. *Plant Disease Reporter* 62, 727–729.

Pinochet, J. and Rowe, P.R. (1979) Progress in breeding for resistance to *Radopholus similis* in banana. *Nematropica* 9, 76–78.

Pinochet, J. and Stover, R.H. (1980) Fungi associated with nematode lesion on plantains in Honduras. *Nematropica* 10, 112–115.

Pinochet, J., Fernandez, C. and Sarah, J.L. (1995) Influence of temperature on *in vitro* reproduction of *Pratylenchus coffeae, P. goodeyi,* and *Radopholus similis. Fundamental and Applied Nematology* 18, 391–392.

Pinochet, J., Fernandez, C., Jaizme-Vega, M.C. and Tenoury, P. (1997) Micropropagated banana infected with *Meloidogyne javanica* responds to *Glomus intraradices* and phosphorus. *Hortscience* 32, 101–103.

Pinochet, J., Jaizme, M. del C., Fernandez, C., Jaumot, M. and De Weele, D. (1998) Screening bananas for root-knot (*Meloidogyne* spp.) and lesion nematode (*Pratylenchus goodeyi*) resistance for the Canary Islands. *Fundamental and Applied Nematology* 21, 17–23.

Prachasaisoradej, S., Chinnasri, B., Tungjitsomkid, N. and Chiemchaisri, Y. (1994) Status of nematode and weevil borer problems affecting banana in Thailand. In: Valmayor, R.V., Davide, R.G., Stanton, J.M., Treverrow, N.L. and Roa, V.N. (eds) *Banana Nematodes and Weevil Borers in Asia and the Pacific: Proceedings of a Conference–Workshop on Nematodes and Weevil Borers Affecting Bananas in Asia and the Pacific, Serdang, Selangor, Malaysia, 18–22 April 1994.* INIBAP/ASPNET, Los Baños, Philippines, pp. 115–121.

Price, N.S. (1994a) Alternate cropping in the management of *Radopholus similis* and *Cosmopolites sordidus,* two important pests of banana and plantain. *International Journal of Pest Management* 40, 237–244.

Price, N.S. (1994b) Field trial evaluation of nematode susceptibility within *Musa. Fundamental and Applied Nematology* 17, 391–396.

Price, N.S. and Bridge, J. (1995) *Pratylenchus goodeyi* (Nematoda: Pratylenchidae): a plant-parasitic nematode of the montane highlands of Africa. *Journal of African Zoology* 109, 435–442.

Price, N.S. and McLaren, C.G. (1996) Techniques for field screening of *Musa* germplasm. In: Frison, E.A., Horry, J.P. and De Waele, D. (eds) *New Frontiers in Resistance Breeding for Nematode, Fusarium and Sigatoka, Proceedings of the Workshop held in Kuala Lumpur, Malaysia, 2–5 October 1995.* INIBAP, Montpellier, France, pp. 87–107.

Quénéhervé, P. (1988) Population of nematodes in soils under banana, cv 'Poyo', in the Ivory Coast. 2. Influence of soil texture, pH and organic matter on nematode populations. *Revue de Nématologie* 11, 245–251.

Quénéhervé, P. (1989a) Population of nematodes in soils under banana, cv 'Poyo', in the Ivory Coast. 3. Seasonal dynamics of populations in mineral soil. *Revue de Nématologie* 12, 149–160.

Quénéhervé, P. (1989b) Population of nematodes in soils under banana, cv 'Poyo', in the Ivory Coast. 4. Seasonal dynamics of populations in organic soil. *Revue de Nématologie* 12, 161–171.

Quénéhervé, P. and Cadet, P. (1985a) Localisation des nématodes dans les rhizomes du bananier cv. Poyo. *Revue de Nématologie* 8, 3–8.

Quénéhervé, P. and Cadet, P. (1985b) Etude de la dynamique de l'infestation en nématodes transmis par les rhizomes du bananier cv Poyo en Côte-d'Ivoire. *Revue de Nématologie* 8, 257–263.

Razak, A.R. (1994) Plant parasitic nematodes, a potential threat to commercial cultivation of banana in Malaysia. In: Valmayor, R.V., Davide, R.G., Stanton, J.M., Treverrow, N.L. and Roa, V.N. (eds) *Banana Nematodes and Weevil Borers in Asia and the Pacific: Proceedings of a Conference–Workshop on Nematodes and Weevil Borers Affecting Bananas in Asia and the Pacific, Serdang, Selangor, Malaysia, 18–22 April 1994.* INIBAP/ASPNET, Los Baños, Philippines, pp. 34–45.

Rivas, X. and Roman, J. (1985) Oogénesis y reproducción de una población de *Radopholus similis* de Puerto Rico. *Nematropica* 15, 19–25.

Robalino, G., Roman, J. and Cordero, M. (1983) Efecto del nematicida-insecticida oxamil aplicado al suelo y a las axilas de las hojas del bananero. *Nematropica* 13, 135–143.

Rowe, P. and Rosales, F. (1994) *Musa* breeding at FHIA. In: Jones, D.R. (ed.) *The Improvement and Testing of Musa: A Global Partnership.* INIBAP, Montpellier, France, pp. 117–129.

Santor, W. and Davide, R.G. (1992) Interrelationship of *Radopholus similis* and *Meloidogyne incognita* in banana. In: Davide, R.G. (ed.) *Studies on Nematodes Affecting Bananas in the Philippines.* Philippine Agriculture and Resources Research Foundation, Los Baños, Philippines, pp. 71–78.

Sarah, J.L. (1985) Les nématodes des bananiers plantains en Côte d'Ivoire. In: *La Coopération*

Internationale pour une Recherche Efficace sur le Plantain. Proceedings of the Third Meeting of the International Association for Research on Plantain and Banana, Abidjan, Côte d'Ivoire, 27–31 Mai 1985, International Association for Research on Plantain and Banana, pp. 89–93.

Sarah, J.L. (1986) Répartition spatiale des infestations racinaires de *Radopholus similis* en relation avec la croissance et le développement du bananier 'Poyo' en Côte d'Ivoire. *Fruits* 41, 427–435.

Sarah, J.L. (1989) Banana nematodes and their control in Africa. *Nematropica* 19, 199–216.

Sarah, J.L. (1991) Estimation of nematode infestation in banana. *Fruits* 46, 643–646.

Sarah, J.L., Lassoudière, A. and Guérout, R. (1983) La jachère nue et l'inondation du sol, deux méthodes intéressantes de lutte intégrée contre *Radopholus similis* dans les sols tourbeux de Côte d'Ivoire. *Fruits* 38, 35–42.

Sarah, J.L., Kéhé, M., Beugnon, M. and Martin, P. (1988) Expérimentation avec l'aldicarbe pour lutter contre *Radopholus similis* COBB (Nematoda, Pratylenchidae) et *Cosmopolites sordidus* (GERMAR) (Coleoptera, Curculionidae) en bananeraie. 2. Expérimentation réalisée en Côte d'Ivoire. *Fruits* 43, 475–484.

Sarah, J.L., Blavignac, F., Sabatini, C. and Boisseau, M. (1992) Une méthode de laboratoire pour le criblage variétal des bananiers vis-à-vis des nématodes. *Fruits* 45, 35–42.

Sarah, J.L., Sabatini, C. and Boisseau, M. (1993) Differences in pathogenicity to banana (*Musa* sp., cv. Poyo) among isolates of *Radopholus similis* from different production areas of the world. *Nematropica* 23, 75–79.

Siddiqi, M.R. (1986) *Tylenchida Parasites of Plants and Insects.* Commonwealth Agricultural Bureaux, Farnham Royal, UK, 645 pp.

Sikora, R.A. (1980) Observations on *Meloidogyne* with emphasis on disease complexes and the effect of host plant on morphometrics. In: *Proceedings of the 2nd Research Planning Conference on Root-knot Nematodes,* Meloidogyne *spp., Region VII. Athens, Greece, 1979.* North Carolina State University, Raleigh, North Carolina, USA, pp. 93–104.

Sikora, R.A. and Schlosser, E. (1973) Nematodes and fungi associated with root systems of bananas in a state of decline in Lebanon. *Plant Disease Reporter* 57, 615–618.

Sivakumar, M. and Marimuthu, T. (1986) Population dynamics of phytonematodes associated with betelvine (*Piper betle* L.), banana and paddy rice with special reference to the crop. *Indian Journal of Nematology* 16, 277.

Speijer, P.R. and Gold, C.S. (1996) *Musa* root health assessment: a technique for the evaluation of *Musa* germplasm resistance. In: Frison, E.A., Horry, J.P. and De Waele, D. (eds) *New Frontiers in Resistance Breeding for Nematode, Fusarium and Sigatoka, Proceedings of the Workshop held in Kuala Lumpur, Malaysia, 2–5 October 1995.* INIBAP, Montpellier, France, pp. 62–78.

Stanton, J.M. (1994) Status of nematode and weevil borer problems affecting banana in Australia. In: Valmayor, R.V., Davide, R.G., Stanton, J.M., Treverrow, N.L. and Roa, V.N. (eds) *Banana Nematodes and Weevil Borers in Asia and the Pacific: Proceedings of a Conference–Workshop on Nematodes and Weevil Borers Affecting Bananas in Asia and the Pacific, Serdang, Selangor, Malaysia, 18–22 April 1994.* INIBAP/ASPNET, Los Baños, Philippines, pp. 48–56.

Stirling, G.R. (1991) *Biological Control of Plant Parasitic Nematodes.* CAB International, Wallingford, UK, 282 pp.

Stover, R.H. (1972) *Banana Plantain and Abaca Diseases.* Commonwealth Mycological Insitute, Kew, Surrey, UK, 316 pp.

Stoyanov, D. (1971) Control de los nematodos parasitos del platano por medio de rotaciones y su duracion en tierra sin hospederos. *Revista de Agricultura, Cuba* 4, 75–80.

Subramanian, D., Sathyamoorthy, P. and Velayutham, B. (1984) A note on the evaluation of phenamiphos (Nemacur 40 EC) for the control of nematodes affecting banana. *South Indian Horticulture* 32, 313–314.

Sudha, S. and Prabhoo, N.R. (1983) *Meloidogyne* (Nematoda: Meloidogynidae) induced root galls of the banana plant *Musa paradisiaca* – a study of histopathology. *Proceedings of the Indian Academy of Animal Sciences* 92, 467–473.

Tarté, R., Pinochet, J., Gabrielli, C. and Ventura, O. (1981) Differences in population increase, host preferences and frequency of morphological variants among isolates of banana race *Radopholus similis. Nematropica* 11, 43–52.

Tessera, M. and Quimio, A.J. (1994) Research on ensat pathology. In: Herath, E. and Desalegn, L. (eds) *The Proceedings of the 2nd National Horticultural Workshop of Ethiopia.* Institute of Agricultural Research, Addis Adaba, pp. 217–225.

Valette, C., Andary, C., Mondolot-Cosson, L., Boisseau, M., Geiger, J.P., Sarah, J.L. and Nicole, M. (1996) Histochemistry and cytochemistry of phenolic compounds in banana roots following infection with the nematode *Radopholus similis*. In: *Proceedings of Third International Nematology Congress. Gosier, Guadeloupe, 7–12 Juillet 1996*. Abstract.

Valette, C., Nicole, M., Sarah, J.L., Boisseau, M., Boher, B., Fargette, M. and Geiger, J.P. (1997) Ultrastructure and cytochemistry of interactions between banana and the nematode *Radopholus similis*. *Fundamental and Applied Nematology* 20, 65–77.

Valette, C., Mounport, D., Nicole, M., Sarah, J.L. and Baujard, P. (1998) Scanning electron microscope study of two African populations of *Radopholus similis* and proposal of *R. citrophilus* as a junior synonym of *R. similis*. *Fundamental and Applied Nematology* 21, 139–146.

Valle-Lamboy, S. and Ayala, A. (1976) Control of plantain nematodes with contact nematicides. *Nematropica* 6, 55–59.

Van Weerdt, L.G. (1960) Studies on the biology of *Radopholus similis* (Cobb, 1893) Thorne, 1949. III. Embriology and post-embryonic development. *Nematologica* 5, 43–51.

Vilardebo, A. (1976) Population dynamics of *Radopholus similis* in relation to climatic factors and the physiology of the plant. *Nematropica* 6, 4–5.

Vilardebo, A. and Robin, J. (1969) Nematicidal treatment of banana planting material. In: Peachey, J.E. (ed.) *Nematodes of Tropical Crops*. Commonwealth Agricultural Bureaux, St Albans, England, pp. 133–141.

Vovlas, N. and Ekanayake, H.M.R.K. (1985) Histological alterations induced by *Rotylenchus reniformis* alone or simultaneously with *Meloidogyne incognita* on banana roots. *Nematropica* 15, 9–17.

Vovlas, N., Avgelis, A., Goumas, D. and Frisullo, S. (1994) A survey of banana diseases in sucker propagated plantation in Crete. *Nematologia Mediterranea* 22, 101–107.

Wehunt, E.J., Hutchinson, D.J. and Edwards, D.I. (1978) Reaction of banana cultivars to the burrowing nematode (*Radopholus similis*). *Journal of Nematology* 10, 368–370.

Youssef, M.M.A. and Aboul-Eid, H.Z. (1996) Fluctuation of root-knot and spiral nematode populations on banana in relation to soil temperature. *Afro-Asian Journal of Nematology* 6, 67–69.

Zem, A.C. and Lordello, L.G.E. (1981) Behaviour of banana cultivars exposed to infestation by *M. incognita* and *M. javanica*. *Anais da Escola Superior de Agricultura 'Luiz de Queiroz'* 38, 875–883.

Zuckerman, B.M. and Strich-Harari, D. (1963) The life stages of *Helicotylenchus multicinctus* (Cobb) in banana roots. *Nematologica* 9, 347–353.

Zuniga, G., Ortiz, R. and de Agudelo, F.V. (1979) Nematodos asociados con el cultivo del platano (*Musa* AAB o ABB) en el Valle del Cauca. *Fitopatologia Colombiana* 8, 40–52.

8
Non-infectious Disorders of Banana

E. Lahav, Y. Israeli, D.R. Jones and R.H. Stover

INTRODUCTION

Infectious agents have not been shown to be the cause of the disorders described in this chapter. All the disorders are believed to result from the plant's response to special growing conditions, often unfavourable. They are not major problems except in localized areas and then usually for short periods. Problems affecting the plant are described first followed by disorders that specifically affect the fruit.

PLANT DISORDERS

Elephantiasis

This is a rare disorder with localized outbreaks reported from Surinam, Colombia, the Dominican Republic, Costa Rica, Panama and Honduras. It was first described from Surinam in 1911 by Essed, who attributed the cause to *Ustilaginoidella oedipigera*, which was later identified as a species of *Fusarium*. However, there was no proof that the fungus was pathogenic and the cause of elephantiasis remains unknown. Fungi are seldom present in affected tissue in Colombia and Central America, but bacteria, including *Erwinia carotovora*, are common. Inoculation with this pathogen causes some swelling and splitting at the pseudostem base, but fails to produce elephantiasis. In Honduras, the disorder was sometimes present together with yellow mat (Prescott, 1917). In 1954, about 24 ha of 'Gros Michel' (AAA) in the Dominican Republic became severely affected and were destroyed. The disorder did not reappear after ploughing and replanting.

The base of the pseudostem swells and the outer sheaths rupture just above the rhizome. Breakage continues inward until the pseudostem falls over, leaving a conical, pineapple-like stump. The breakage is often associated with a narrow band of rot, from which *E. carotovora* (see Thwaites *et al.*, Chapter 5, pp. 231–234) is obtained. There may or may not be leaf symptoms, consisting of swollen veins and midrib. Suckers may be normal or deformed and split or are easily broken off close to the ground. In severe cases, the entire mat may be killed. There are no internal symptoms in the rhizome.

Heart-leaf Unfurling Disorder of Plantain

In Central America, the heart leaves of 'French' and 'Horn' plantain (AAB,

Plantain subgroup) occasionally fail to unfurl normally. The first sign of heart-leaf abnormality is the bending over of the point of the unfurled heart leaf, which appears chlorotic. The first 30–60 cm at the tip then fails to unfurl or only one-half of the blade unfurls. The unfurled portion remains chlorotic and eventually rots. Sometimes only portions of the blades unfurl, giving the leaves a torn and tattered appearance. Affected leaves appear shorter than normal because the tip portion has rotted away. After the emergence of several abnormal heart leaves, normal leaf emergence resumes.

The cause of this disorder is unknown, but it is believed to be associated with an uneven growth rate as a result of adverse growing conditions, such as periods of drought followed by rain.

High Mat

This is a common defect in banana growth, especially in plantations several years old. The upper portion of the rhizome grows out of the soil, exposing a considerable area of root-bearing tissue. Short roots emerge above ground when the weather is moist. The base of the rhizome is usually only 1 or 2 inches below the soil surface. Such plants are not as well anchored as plants with buried rhizomes, and uprooting is more common. According to Wardlaw (1961), compact soils are a contributing factor, as well as pulling trash and soil away from the base of the plant in hoeing operations. Brouhns (1957) believed that lateral bud formation high on the rhizome was a contributing factor. Subra and Guillemot (1961) pointed out the importance of selecting axial suckers rather than lateral or terminal suckers for good anchorage. The axial sucker is the first sucker to emerge from the base of the rhizome on the side opposite to where the mother plant emerged. However, according to Charpentier (1966), all rhizomes normally tend to move upward out of the soil. Hasselo (1957) recommended earthing up by making a mound of earth around the base of the banana mat in a circle of 45 cm radius so that the root-bearing region is covered to a depth of 10–15 cm. On volcanic soils in Cameroon, this reduced uprooting losses to an insignificant level. Charpentier (1966) stated that the tendency for rhizomes to come out of the ground resulted from a natural phototropism of the vegetative parts. This occurred when a rhizome was planted at a normal depth. When rhizomes were planted excessively deep, a second corm formed above the deeply placed corm and there was some foliar deformation. By earthing up or darkening the base of the pseudostem, a second corm also formed higher up, but there was no foliar deformation. As earthing up is used to stimulate corm formation for speeding up vegetative reproduction (Barker, 1959; Charpentier, 1966), it would seem unlikely to offer a more than temporary solution to the high-mat problem.

In Central America, the high-mat condition is common in some areas and is not related to compacted soils. Also, lateral buds that emerge from well below the soil surface can develop into high mats by the time the plant has flowered. 'Gros Michel', cultivars in the Plantain subgroup and clones in the Cavendish subgroup (AAA) are affected.

Leaf-edge Chlorosis

A yellowing of the leaf margins of Cavendish cultivars was described in Central America by Stover (1972). A similar condition also occurs on the island of St Vincent in the Caribbean. Chlorosis begins at or near the edges of the leaf and spreads up the lamina towards the midrib. In severe cases, the yellowing can extend almost to the midrib (Plate 8.1). The affected tissue is sometimes invaded by weak pathogens and becomes necrotic in places. Leaves 3–6 are most commonly affected and the condition is worse in plants bearing bunches. The chlorotic areas of the leaf do not continue to expand in size as the leaf ages. The symptoms are also found in fertile plantations and their effect on photosynthesis is negligible.

Plate 8.1. Symptoms of leaf-edge chlorosis on 'Robusta' (AAA, Cavendish subgroup) in St Vincent, Windward Islands (photo: D.R. Jones, SVBGA).

The cause of this chlorosis is not known, but it may be associated with a transitional nutritional deficiency brought on by periods of stress. In St Vincent, although the symptoms resemble those caused by calcium deficiency (see Plate 9.1), tests have shown that the affected leaf tissue has sufficient calcium. However, this does not completely rule out calcium deficiency as the cause. An analysis of leaves with symptoms reveals low levels of boron, manganese, phosphorus and sulphur. Stover (1972) reported less nitrogen, manganese and potassium than normal.

Roxana Disease

This is a rare disorder, reported in 1956 from an abandoned 'Gros Michel' planting near Roxana, Limon Province, Costa Rica (Allen, 1957). About 10% of the plants were affected. Later in the year, another small outbreak occurred at Quebrada Honda, 80 km away, and also in the Sula Valley, Honduras, in 1957 (Waite, 1960). The external symptoms are very similar to those of Rayadilla disease caused by zinc deficiency in Colombia (see Lahav and Israeli, Chapter 9, pp. 348–349). In some ways, they resemble certain cases of yellow mat disease, attributed to adverse soil conditions, where chlorotic striping and narrowing of the leaves occur, along with phloem dysfunction. Leaves are erect and rosetted in appearance. The size of leaves is reduced, and laminae are irregularly scalloped and cupped upwards. Irregular, elongated, white or yellowish patches or stripes occur, principally in the interveinal areas parallel to the veins. Mottling and striping are most prominent on new leaves. Suckers may remain healthy in appearance until approaching the shooting stage, when leaf numbers are reduced from 12–16 to about six malformed, chlorotic leaves. Petioles and leaf sheaths have reddish-brown necrotic streaks in severely affected plants. Rhizomes from affected plants produce affected and stunted suckers, although severity of symptoms varies greatly. According to Allen (1957), there is phloem disorganization in leaves, with symptoms exactly as described by Magee (1927) for bunchy top. The only other disease in Central America in which the phloem is sometimes affected is yellow mat. The cause of the Roxana disease was never determined, although a virus was suspected because of its systemic nature. However, this is unlikely, as the condition never reappeared when Cavendish cultivars were planted extensively in the same province, beginning about 1963.

Yellow Mat

This disorder was first described affecting 'Gros Michel' by Prescott in 1917 and called 'Colorado disease', as it occurred commonly in the Colorado District near Tela in Honduras. Later, it was reported as most severe in the Changuinola area of Panama (Dunlap, 1923; Permar, 1925). When these areas were abandoned because of Fusarium wilt, no further records were noted until the Changuinola area was replanted with wilt-resistant cultivars in the Cavendish subgroup (AAA) in the late 1950s and early 1960s. The disorder reappeared and was called yellow mat, or 'mata amarilla' in Spanish, because of the pronounced yellowing of the foliage.

Yellow mat occurs sporadically, rarely more than 5% of the crop being affected, in localized areas along the Atlantic coast of Central America and in Surinam. Barnes (1962) described a disease with similar symptoms on the Cavendish cultivars 'Lacatan' and 'Giant Cavendish' in wet areas in Grenada. In Panama, all cultivars of dessert banana and plantain (AAA, AB, AAB) were affected. According to Dunlap (1923), yellow mat was also prevalent in abacá throughout the Changuinola area. In other banana-growing areas of Central America, yellow mat is rare or absent.

The first symptoms are a yellowing and browning of the lower leaves, usually on large plants or plants with fruit. The leaves die and the disorder progresses acropetally until only a few of the youngest leaves remain green. New leaves that emerge after symptoms appear are small and narrow. Leaves of large suckers may or may not be affected at first, depending on the severity of attack. If affected, sucker leaves may be narrow, strap-shaped and chlorotic, with chlorosis sometimes in bands. Often, leaf yellowing is the only symptom present on suckers.

Early symptoms of leaf yellowing and browning and generally poor growth are similar to those present on plants growing on heavy clay and in poorly drained areas. In yellow mat, however, suckers usually succumb. Often, young sword suckers turn black and die. The death of suckers helps to distinguish yellow-mat disease from symptoms of poor growth as a result of adverse soil texture and drainage.

Affected plants are usually stunted, the fruit is small and short-fingered and the bunch may not be vertical. Eventually, all the sword suckers attached to plants showing symptoms will die. Water suckers will continue to emerge for a few months before the entire mat is destroyed. In a minority of cases, there may be some recovery and renewal of normal growth. When symptoms first appear, an examination of the plant below soil level shows shallow rooting, extensive necrosis and a generally poor root system. Internal symptoms, when present, consist of vascular discoloration in the lower pseudostem, usually paler and less extensive than symptoms caused by Fusarium or bacterial wilts. This discoloration was shown by Reinking (quoted by Dunlap, 1923) to involve the phloem and not the xylem. On 'Gros Michel', external symptoms were often similar to those of Fusarium wilt. An examination of the lower pseudostem became a quick microscopic method of distinguishing between cases of Fusarium wilt (which affects the xylem) and yellow mat on the wilt-susceptible 'Gros Michel'. The presence of phloem discoloration was confirmed in yellow mat outbreaks in Panama in 1964. The phloem sieve tubes were plugged with a dark amorphous material. However, in some cases of yellow mat, phloem discoloration is slight or absent.

Rhizomes taken from mats with well-developed symptoms of yellow mat may or may not grow. In some cases, growth is normal for several months and then symptoms appear and the plant dies. In a minority of cases, normal growth and recovery from the disease occur. Root growth from rhizomes from plants with yellow mat is usually reduced and roots are weak in comparison with roots from normal rhizomes. The above-ground symptoms are believed to be related to a deterioration and rotting of the root system.

The precise cause of yellow mat is not known. Because of vascular discoloration, a disease caused by *Fusarium* was suspected, but Reinking (1926) showed that *Fusarium* spp. associated with the disorder were not pathogenic.

Outbreaks of yellow mat are associated with clay or light clay soils, land with drainage problems, extreme fluctuations in the water-table and rainfall well above average. There is no evidence of yellow mat spreading like a soil-borne disease. Flooding alone does not cause yellow-mat symptoms. It is believed that certain toxic soil factors are involved, which affect not only the root system but the rhizome as well, and in some way the physiology of the rhizome storage tissue is disturbed. As a result, rhizomes from affected plants usually cannot produce normal plants. The production of these toxic factors is, in some unknown way, related to soil structure, permeability and drainage. Areas where sporadic outbreaks occur persistently do not produce optimum yields of bananas. Evidence of high clay content and poor soil structure is frequently present, indicating that yellow mat is related to some soil condition that adversely affects root growth. However, the possibility of a virus being involved cannot be excluded.

Early symptoms of yellow mat resemble those of potassium deficiency. Low oxygen and high carbon dioxide levels are known to impede potassium uptake (Hammond *et al.*, 1955). It is possible that potassium uptake may also be impeded when soil conditions favourable for yellow mat are present, although there is no plant response to potassium or minor elements in yellow mat areas.

Following extensive improvements in soil drainage in Panama, yellow-mat incidence declined. Existing drains were deepened and new drains opened across contours to remove water more rapidly during periods of heavy rainfall. As a result, water-tables were lowered and root growth increased.

FRUIT DISORDERS

Alligator Skin

A rare blemish called alligator skin is occasionally found on green fruit at harvest. Portions of the peel are reddish brown and the surface becomes hard and raised and cracks into rectangular areas, somewhat resembling the skin of an alligator (Plate 8.2). The blemish usually occurs on the inner face of the fingers and usually only a few hands or fingers are affected on a stem. Sometimes a small amount is seen on the outer ridges of the fingers. The cause of this defect is unknown, but irritation of the epidermis when young is suspected. Light abrasions from leaves or bracts that rub the fruit when it is very young are probably a contributing factor. Finger-ridge scarring can be caused by abrasions from polyethylene bags when blown against the fruit in high winds.

Dark Centre of Ripe Fruit

The presence of a black centre (Plate 8.3) in otherwise normally ripened fruit is a common source of complaints from consumers. On biting into a ripe banana, the centre is found to be black and soft. There are no outward symptoms of the internal blemish. Dark centres can be induced by the impact of dropping boxes containing fruit with a colour index of 4–5. Dropping individual fingers on their flower tips six times from a height of 15 cm consistently produced dark centres without bruising. Dark centres have also been induced by applying pressures of 1 kg 14 cm^{-2} to the flower ends of fingers. As ripening advances, dark centres are less likely to develop and green fruit is not susceptible.

The condition is believed to result from rupture of latex ducts when ripening fruit

Plate 8.2. Alligator-skin blemish on a hand of green Cavendish bananas at harvest (photo: R.H. Stover).

Plate 8.3. Symptoms of dark centre in ripening fruit of a cultivar in the Cavendish subgroup (AAA) (photo: R.H. Stover).

is roughly handled. Phenolic substances are released, which, on reaction with polyphenoloxidase enzyme, turn brown or black. Latex cells are less susceptible to rupture in green fruit, as tannins, which inhibit the browning enzymes, are present.

Fruit Fall or Break Neck

In Central America, fruit of cultivars in the Cavendish subgroup sometimes falls from the plant as a result of breakage of the peduncle where it curves inward towards the pseudostem. Breakage usually occurs during or shortly after the time the fingers emerge from the flower. Sometimes the peduncle may break at some point within the pseudostem or at the point of emergence from the pseudostem. The peduncle is very brittle at this stage. There are no symptoms of disease or abnormal growth. Incidence is less than 1% and the disorder does not occur throughout the year. The reasons for the weakness of the peduncle are not known, but water stress possibly contributes. 'Lacatan' was more susceptible to break neck than other Cavendish cultivars in Honduras.

Hard Lump

Fruit of 'Rasthali' (AAB, syn. 'Silk') commonly suffers from a disorder called 'hard mass' or 'lumps' in India. These lumps, which are pinkish brown in colour and firmer than the usual soft pulp, taste like immature or unripe fruit. Hard lump is of concern because it impairs the eating qualities of affected 'Rasthali' bananas, which are highly prized in India. It has been reported that the occurrence of hard lump is seasonal and only fruit harvested from July to December have the problem. The application of 2,4-dichlorophenoxyacetic acid (2,4-D) to bunches, cut ends of peduncles or hands reduced incidence (Shanmugavelu *et al.*, 1992).

Malformed Fingers and Hands

Individual fingers and occasionally entire hands are twisted and excessively curved, destroying the symmetry of the hand. The symptoms are similar to those caused by low or high temperatures or by water stress. Such hands are discarded at the packing station because they prevent normal packing and may cause damage to other hands in the same box. A common cause of fruit malformation is the prevention of normal finger curvature and growth by the presence of leaves or bracts between the fingers or upon the young developing hands of fruit. Sometimes, fruit malformation is caused by the weight of the polyethylene bag obstructing normal curvature of the fingers. This can occur when a young bunch is covered before the fingers have curved upward. Removal of leaves and bracts that touch the young fruit and a delay in bagging of very young fruit reduce malformation. However, even in the absence of these obstructions to finger curvature, malformed fingers and hands can still be found. High density (reduced illumination) and low temperature during inflorescence development increase the rate of twisted hands in the Cavendish cultivar 'Grand Nain'. In Central America, from 1.5 to 3% of dehanded fruit at the boxing station is discarded because of malformed fingers. Malformation rate differs among cultivars of the Cavendish subgroup. Relatively more malformed fingers appear in 'Grand Nain' as compared with 'Williams' or 'Valery'.

Maturity Stain

Maturity stain, which is known as maturity bronzing in Australia, is a peel blemish that appears only on fruit with an excessive caliper grade. Longitudinal streaks and blotches appear mostly on the shoulder of outer whorl fingers of basal hands (Plate 8.4). This stain should not be confused with the reddish discoloration from red rust thrips, which is found between the fingers.

Maturity stain is stress-induced and correlated in Australia and Costa Rica with periods of heavy rains, high temperatures, high humidity and a possible potassium/magnesium imbalance. On the Pacific side of Panama and Costa Rica, the normal rejection level of fruit because of maturity stain has been reported to be 8–10%. However, this rises to 20% during periods

Plate 8.4. Maturity stain on green Cavendish fruit at harvest (photo: E. Lahav, VI).

of heavy rain (Stover and Simmonds, 1987).

A considerable amount of research on maturity stain has been undertaken in Australia. Campbell and Williams (1976) suggested that maturity stain was caused by high turgor pressure in the non-dividing cells of the peel epidermis, which resulted in the breakdown of the middle lamellae, cell rupture and oxidation of the released cell constituents. Anatomical studies on the peel of the developing fruit showed that stretching of the epidermis exceeds its elastic limit and leads to cracks and cell disruption in the peel surface and, subsequently, to formation of lesions characteristic of maturity stain (Williams *et al.*, 1990). Sealed polyethylene covers increased maturity stain, compared with open bunch covers (Daniells *et al.*, 1992). Maturity stain was found to be greatest in fruits kept in darkness and was completely prevented after continuous exposure to white light (Wade *et al.*, 1993). Maturity stain began developing on fruit of water-stressed bunches at much thinner finger diameter than in fruits of unstressed controls (Daniells *et al.*, 1987). It showed a positive association with copper in the fruit skin and calcium in leaves and a negative association with calcium and magnesium in the fruit skin (Campbell and Williams, 1978). The removal of leaves reduced the incidence of the fruit-peel disorder, while bunch trimming increased the incidence (Daniells *et al.*, 1994). However, it was concluded that the reduction in maturity stain achieved by leaf removal is of no immediate benefit to the growers. The only treatment of partial benefit is an aerial spray of calcium nitrate ($Ca(NO_3)_2$) incorporated with fungicides. Harvesting fruit at a lower grade can reduce incidence.

Mixed Ripe Fruit

Mixed ripe fruit, where some of the fruit ripens before marketing while some stays green, is a more widely spread phenomenon than premature ripening. The time that fruit ripens is related to its physiological age and not its size. Some fruit can ripen while still small if growth has been delayed due to stress conditions (Turner and Rippon, 1973; Rippon, 1975). Recording flowering time by attaching coloured ribbons to bunches on emergence and harvesting at the right physiological age reduce mixed ripening. Polyethylene sleeved fruit is known to ripen earlier than unsleeved fruit.

Marriott (1980) listed some of the factors shortening the green life of fruit and possibly inducing mixed ripening. High levels of Sigatoka disease on plants as bunches develop and mechanical damage to fruit might also stimulate earlier ripening. Generally, environmental stresses affecting fruit growth, such as water stress, drought or chilling, may stimulate premature ripening, as well as mixed ripening.

Neer Vazhai

Neer vazhai is a disorder affecting 1–2% of 'Nendran' (AAB, Plantain subgroup) grown in the Tiruchirappalli district of Tamil Nadu State in southern India. Affected plants appear normal until after flowering, when fingers on many of the hands fail to fill (Plate 8.5). Unfilled fingers tend to have persistent female flowers. When suckers from affected plants develop into mature plants, they in turn produce deformed bunches. The problem may be genetic (Shanmugavelu *et al.*, 1992).

Physiological Finger Drop

When bananas were shipped on stems (peduncles), losses were encountered as a result of fingers falling from the bunch (loose fingers). This usually occurred after ripening had begun. In the French Antilles, finger drop, or degrain, due to a physiological cause was a serious seasonal problem as stem fruit ripened (Daudin and Guyot, 1965; Guillemot and Colmet-Daage, 1965). Phillips (1970) reported a similar finger-drop problem in the Windward Islands.

Finger drop should not be confused

with finger loss resulting from mechanical injury to the necks of the finger, which may or may not be accompanied by a fungal infection, or with stem-end rot. With physiological finger drop, the fruit detaches itself from the pedicel as ripening progresses. The fibrous system of the pedicel is not involved, since the break takes place at the placental axis of the fruit, usually before the finger is fully ripe. Physiological finger drop is the premature separation of the finger from the pedicel. Fruit with finger drop tends to yellow more quickly, while the pedicel remains green. The skin is dull yellow, with a silky, soft texture.

Guillemot and Colmet-Daage (1965) found a close relationship between the pedicel ratio (pedicel length divided by width) and a tendency to finger drop. They believed that finger drop was promoted by sharp fluctuations in the amount of soil nitrogen and recommended that soil nitrogen levels should be kept high during the period following the first heavy rain. Application of nitrogen was to be avoided at the end of the dry season.

Phillips (1970) found that the pedicel ratio was adversely affected by dense planting (3000 plants ha^{-1} or more) and heavy applications of fertilizer.

Finger drop has been associated with nitrogen imbalance (Martin-Prevel and Montagut, 1966). It occurs during the hot, wet season in the tropics when there is a low potassium supply and $N\text{-}NH_4$ accumulates. The excess nitrogen delays bunch emergence and produces bunches with widely spaced hands, which are easily damaged in transport.

Short-cycle fruit (fruit reaching maturity in a short time from planting) was found to be more susceptible to finger drop than long-cycle fruit (Daudin and Guyot, 1965). However, no anatomical differences were found between short- and long-cycle fruit. It was thought that finger drop was related to seasonal factors and any other factor that reduced the maturation time of the plant crop to less than 350–365 days from planting. Vallade and Rabechault (1963) could find no differences in the anatomy of the pedicel that would explain susceptibility to finger drop. In Australia, Hicks (1934) attributed a seasonal finger dropping to ripening at high temperatures and humidity. Finger drop was reduced when ripening fruit was removed to lower humidity or green fruit was held at lower temperatures before ripening. Semple and Thompson (1988) reported that temperatures up to 30°C and ethylene increased finger drop, while relative humidity had no effect.

Some bred tetraploid cultivars have an inherent weakness in finger attachment to the pedicel at ripening. This defect has made these cultivars unsuitable for the commercial trade (New and Marriott, 1974).

Premature Field Ripening

This specific problem occurred in the Changuinola area of Panama and was common from 1955 to 1961 on 'Gros Michel' and cultivars in the Cavendish subgroup. There was an irregular development of hands, with the proximal or upper hands on the hanging bunch filling out and maturing more rapidly than the apical or lower hands. The full, proximal fingers ripened on the plant and, in advanced stages of ripeness, the fingers fell from the stem while the apical hands were still green. Outbreaks of the condition were sporadic, with up to several thousand cases $month^{-1}$ 240 ha^{-1}, followed by a decline to few or no cases. The worst outbreaks followed a drought in 1955 and a flood in 1957. Therefore, it appears that this problem is probably related to a serious check in the growth of the plant during the period when the fruit is filling. Inspections indicated that the root systems were deficient on plants with premature field ripening.

Sinkers

Fruit normally floats in water when the hands are removed from the peduncle and placed in delatexing tanks at the packing station. However, 'sinkers' appear during

Plate 8.5. Neer vazhai disorder in a bunch of 'Nendran' (AAB, Plantain subgroup) growing in India (photo: S. Sathiamoorthy, TNAU).

some periods of the year (winter in subtropical areas) and from some areas. This term is applied to fruit that sinks to the bottom of the tank, where, unless quickly removed, it remains and becomes excessively scarred due to abrasion damage.

Sinkers are most common in fruit with a slow sap flow, from areas where growing conditions have been poor, due to soil conditions, or when there is adverse weather, such as excessive rainfall. The appearance of sinkers in Honduras is seasonal and may reach 5% of the fruit (Johnson, 1979). Sinkers have relatively low levels of potassium and can be avoided by applying 400 kg of potassium ha^{-1} $year^{-1}$. Sinkers do not necessarily ripen sooner than normal fruit.

Split Peel

Split peel (Plate 8.6) results from several causes, some of which have not been defined. Split peel of green fruit in the field results from a too rapid filling out of the fingers, usually on the proximal or largest hands. The rapid expansion in pulp volume results from highly favourable growing conditions or failure to harvest fruit when the correct harvesting caliper grade is reached. Split peel may occur during transit and when ripening. Too high ripening temperatures and fruit with high caliper grade are contributing factors. Split peel was common on 'Gros Michel' ripened in November to February in Honduras. Banana growth during this period is retarded, as a result of low temperatures, and fruit maturation time is increased. Dipping the fruit in a hormone solution, such as 2,4-D, before ripening reduced the amount of split peel (Freiberg, 1955).

Sometimes split peel occurs in transit on fruit that is not overmature; the cause is unknown.

Uneven Degreening

'Uneven degreening', which becomes evident following ethylene treatment, is a serious ripening disorder of bananas harvested from some selections of cultivars in the Cavendish subgroup (AAA) in subtropical areas. In Taiwan, it affects fruit from plants of 'Pei-Chiao', which flower between December and March, the period of lowest temperatures. In mild cases, either affected fruit only partially yellows or yellowing is delayed. In severe cases, the peel remains green (Plate 8.7). The disorder mainly affects the top three hands. Herbicide contamination, virus infections and the overuse of nitrogen fertilizer have been considered as probable causes, but eliminated. Temperatures below 20°C and genetic characteristics are believed to be the major contributing factors. An integrated strategy, including the elimination of selections of 'Pei-Chiao' with high susceptibility and the use of brown paper bags instead of blue polyethylene bunch covers, is being tried to overcome the problem. Brown paper bags affect the microclimate of the bunch, as well as blocking infrared (IR) radiation and blue-green light, resulting in changes

Plate 8.6. Symptoms of split peel on green Cavendish fruit at harvest (photo: E. Lahav, VI, and Y. Israeli, JVBES).

to the configuration of epidermal cells (Chiang *et al.*, 1998; Tang, 1998).

Withered Pedicels

In boxed fruit, pedicels sometimes wither during ripening, especially following transit periods in excess of 10 days. The surface appears wrinkled and shrivelled and is dull brown or grey, in contrast to the normal greenish yellow of turgid pedicels. Clusters with shrivelled pedicels are found alongside normal, turgid pedicels. Occasionally, in some locations, up to 50% of the clusters can have severely withered pedicels. This occurs even when the fruit is under high humidity in polyethylene bags.

The cause of withered pedicels is not known. Incidence increases the longer fruit is held at ambient temperatures prior to shipboard storage at 13–15°C. Incidence is also greater in fruit from areas with poor soil or poor drainage or when growing conditions are not ideal.

Plate 8.7. Uneven degreening of bananas harvested from 'Pei-Chiao' (AAA, Cavendish subgroup). Normally degreened bananas of uniform yellow colour (right) contrast with unevenly degreened bananas (left) (photo: S.C. Chiang, TBRI).

Yellow Pulp

The colour and texture of the pulp of green fruit when harvested at the correct grade is normally dull white and firm. Sometimes, the pulp has a yellow or honey-coloured tint and is soft in the centre of the finger. Such fruit is said to have 'yellow pulp'. In some cases, a pinkish cast may be present. Colour and texture changes are most pronounced near the centre of the finger, and are usually indicative of approaching ripening or breaking of skin colour prematurely.

Any factors delaying the filling out of the fruit or increasing the time to harvest maturity can cause yellow pulp. They include loss of foliage, due to Sigatoka leaf spot or defoliation, excessive shade or drought. In Guadeloupe, yellow pulp was detected following prolonged drought by cutting open fingers at the packing station. Fruit with yellow pulp is discarded, because it would ripen in transit.

In Central America, populations of 1976–2470 robust plants per hectare on good loam soils can result in softening of fruit before harvest and yellow pulp because of excessive shading. Competition for limited light and space results in abnormally long periods for fruit to reach harvesting grade. Such fruit is 'stale' at harvest and often has a tendency to ripen in transit.

In Cameroon and Guinea, up to 70% of fruit from areas where yellow pulp was most severe has been rejected for export in the past. In these areas, high potassium/magnesium ratios, high soil-calcium levels or deficiencies in magnesium or manganese may have caused yellow pulp (Dumas and Martin-Prevel, 1958; Charpentier and Martin-Prevel, 1968). Yellow pulp can be avoided by applying high doses of sulphur to the soil which improves the balance of cations by reducing excess calcium and increasing the absorption of manganese (Marchal *et al.*, 1972).

REFERENCES

Allen, R.M. (1957) A virus-type disease of Gros Michel bananas in Costa Rica. *Turrialba* 7, 72–83.

Barker, W.G. (1959) A system of maximum multiplication of the banana plant. *Tropical Agriculture (Trinidad)* 36, 275–284.

Barnes, R.F. (1962) Grenada banana disease. In: *Report on Banana Investigations 1962.* Regional Research Centre, University of the West Indies, Trinidad, pp. 25–26.

Brouhns, G. (1957) Note sur la croissance du bananier 'Gros Michel'. *Fruits* 12, 261–268.

Campbell, S.J. and Williams, M.T. (1976) Factors associated with maturity bronzing of banana fruit. *Australian Journal for Experimental Agriculture and Animal Husbandry* 16, 428–432.

Campbell, S.J. and Williams, M.T. (1978) Mineral relationships in 'maturity bronzing' of banana fruit. *Australian Journal for Experimental Agriculture and Animal Husbandry* 18, 603–608.

Charpentier, J.M. (1966) La remontée du méristème central du bananier. *Fruits* 21, 103–119.

Charpentier, J.M. and Martin-Prevel, P. (1968) *Carences et troubles de la nutrition minéral chez le bananier. Guide de diagnostic pratique.* Institut Français de Recherches Fruitières Outre-Mer, Paris, 75 pp.

Chiang, S.C., Tang, C.Y., Chao, C.P. and Hwang, S.C. (1998) An integrated approach for the prevention of uneven degreening of bananas in Taiwan. *Acta Horticulturae* 490, 511–518.

Daniells, J.W., Watson, B.J., O'Farrel, P.J. and Mulder, J.C. (1987) Soil water stress at bunch emergence increase maturity bronzing of banana fruit. *Queensland Journal of Agriculture and Animal Sciences* 44, 97–100.

Daniells, J.W., Lisle, A.T. and O'Farrell, P.J. (1992) Effect of bunch covering methods on maturity bronzing, yield and fruit quality of bananas in North Queensland. *Australian Journal of Experimental Agriculture* 32, 121–125.

Daniells, J.W., Lisle, A.T. and Bryde, N.Y. (1994) Effect of bunch trimming and leaf removal at flowering on maturity bronzing, yield and other aspects of fruit quality of bananas in North Queensland. *Australian Journal of Experimental Agriculture* 34, 259–265.

Daudin, J. and Guyot, H. (1965) A study of finger drop affecting the Giant Cavendish banana in the French West Indies. In: *The Banana Industry and Research Development in the Caribbean.* Caribbean Organization, Puerto Rico, pp. 131–144.

Dumas, J. and Martin-Prevel, P. (1958) Contrôle de nutrition des bananeraies en Guinée (Premiers résultats). *Fruits* 13, 375–386.

Dunlap, V.C. (1923) *Weekly Reports on Banana Investigations.* United Fruit Co., Panama Division.

Essed, E. (1911) The Surinam disease: a condition of elephantiasis of the banana caused by *Ustilagioidella oedipigera. Annals of Botany* 25, 363–365.

Freiberg, S.R. (1955) Effect of growth-regulators on ripening, split peel, reducing sugars, and diastatic activity of bananas. *Botanical Gazette* 117, 113–119.

Guillemot, J. and Colmet-Daage, F. (1965) Factors affecting quality of bananas in the West Indies: finger drop – effects of variation in the nitrogen content of the soil on finger drop. In: *Banana Industry and Research Developments in the Caribbean.* Caribbean Organization, Puerto Rico, pp. 36–55.

Hammond, L.C., Allaway, W.H. and Loomis, W.E. (1955) Effects of oxygen and carbon dioxide levels on absorption of potassium by plants. *Plant Physiology* 30, 155–161.

Hasselo, H.N. (1957) Earthing up of Gros Michel bananas. *Tropical Agriculture (Trinidad)* 34, 59–64.

Hicks, E.W. (1934) Finger dropping from bunches of Australian Cavendish bananas. *Journal of the Council for Scientific and Industrial Research, Australia* 7, 3.

Johnson, T.J. (1979) Effects of potassium on buoyancy of banana fruit. *Experimental Agriculture* 15, 173–176.

Magee, C.J.P. (1927) Investigations on the bunchy-top disease of bananas. *Bulletin Australian Council for Scientific and Industrial Research* 30, 30–64.

Marchal, J., Martin-Prevel, P. and Melin, P. (1972) Le soufre et le bananier. *Fruits* 27, 167–177.

Marriott, J. (1980) Banana: physiology and biochemistry of storage and ripening for optimum quality. *CRC Critical Reviews in Food Science and Nutrition* 13, 41–88.

Martin-Prevel, P. and Montagut, G. (1966) Essais sol-plante sur bananier. 8. Dynamique de l'azote dans la croissance et le développement du végétal. *Fruits* 21, 283–294.

New, S. and Marriott, J. (1974) Post-harvest physiology of tetraploid banana fruit: response to storage and ripening. *Annals of Applied Biology* 78, 193–204.

Permar, J.H. (1925) Colorado disease in the Farm 8 Lacatan planting. In: *Annual Report United Fruit Company for Period December 27, 1924 to December 5, 1925, Panama Division.*

Phillips, G.A. (1970) Fruit quality problems of the Windward Islands' banana industry. *PANS* 16, 298–303.

Prescott, S.C. (1917) *Diseases of the Banana.* Bulletin, United Fruit Company Research Department, 35 pp.

Reinking, O.A. (1926) Fusaria inoculation experiments: relationship of various species of fusaria to wilt and Colorado disease of bananas. *Phytopathology* 16, 371–392.

Rippon, L.E. (1975) Knowing bunch age can increase growers' profits. *Banana Bulletin* 39(4), 2.

Semple, A.J. and Thompson, A.K. (1988) Influence of the ripening environment on the development of finger drop in bananas. *Journal of the Science of Food and Agriculture* 46, 139–146.

Shanmugavelu, K.G., Aravindakshan, S. and Sathiamoorthy, S. (1992) *Banana Taxonomy, Breeding and Production Technology.* Metropolitan Book Co., New Delhi, India, 456 pp.

Stover, R.H. (1972) *Banana, Plantain and Abacá Diseases.* Commonwealth Mycological Institute, Kew, Surrey, UK, 316 pp.

Stover, R.H. and Simmonds, N.W. (1987) *Bananas.* Longman Scientific and Technical, Harlow, Essex, UK, 468 pp.

Subra, P. and Guillemot, J. (1961) Contribution à l'étude du rhizome et des rejets du bananier. *Fruits* 16, 19–23.

Tang, C.Y. (1998) Response to selection of uneven degreening in the clones of a Cavendish banana cultivar (*Musa* cv. AAA) in Taiwan. *Fruits* 53, 355–363.

Turner, D.W. and Rippon, L.E. (1973) Effect of bunch covers on fruit growth and maturity in bananas. *Tropical Agriculture (Trinidad)* 50, 235–240.

Vallade, J. and Rabechault, H. (1963) Etude des caractères anatomiques des pédicelles de bananes, en corrélation avec le degrain. *Fruits* 18, 129–140.

Wade, N.L., Kavanagh, E.E. and Tan, S.C. (1993) Sunscald and ultraviolet light injury of banana fruits. *Journal of Horticultural Science* 68, 409–419.

Waite, B.H. (1960) Virus diseases of bananas in Central America. *Proceedings of the Caribbean Region, American Society of Horticultural Science* 4, 26–30.
Wardlaw, C.W. (1961) *Banana Diseases.* Longman, Green and Co. Ltd, London, pp. 40–42.
Williams, M.H., Vesk, M. and Mullins, M.G. (1990) Development of the banana fruit and occurrence of the maturity bronzing disorder. *Annals of Botany* 65, 9–19.

9
Mineral Deficiencies of Banana

E. Lahav and Y. Israeli

Introduction

For normal growth and fruit production, banana plants require high amounts of mineral nutrients, which are often only partly supplied by the soil. To establish a crop yielding 50 t ha^{-1} $year^{-1}$ of fresh fruit, about 1500 kg of potassium ha^{-1} $year^{-1}$ may be extracted from the soil. Amounts of other mineral nutrients extracted, in kg ha^{-1} $year^{-1}$, are: nitrogen – 388; phosphorus – 52; calcium – 227; magnesium – 125; sulphur – 73; chlorine – 525; sodium – 10.6; manganese – 12; iron – 6; zinc – 4.7; boron – 1.27; copper – 0.37 (Lahav, 1995). Thus large quantities of mineral nutrients have to be replaced in order to maintain soil fertility and to permit the continuous production of high yields. This is achieved by applying organic manure and/or, more efficiently, mineral fertilizers, which supply the elements in a concentrated and readily available form.

The overall requirement of mineral nutrients can be estimated from analysis of the whole plant and estimated plant growth. The grower must know the ability of the soil to meet these requirements and whether supplementary fertilizers are needed. Two approaches have been adopted to solve this problem. In one approach, field experiments can be established on a range of soil types. The results of these trials are also dependent upon local conditions of climate, the cultivar planted and the effect of pests and diseases (especially nematodes), and therefore their reliable extrapolation is limited. In order to make the results of field experiments more meaningful, plant tissues and soil samples are usually analysed to determine mineral nutrient levels, with the aim of estimating the amount of fertilizer and microelements required to optimize yields (Lacoeuilhe and Martin-Prevel, 1971; Marchal *et al.*, 1972; Warner and Fox, 1977). Symptoms are useful in diagnosing mineral nutrient imbalance and are summarized in Table 9.1.

Boron Deficiency

Boron (B) deficiency has been recorded in the field in Ecuador (Tollenaar, 1969) and in sand culture (Charpentier and Martin-Prevel, 1965; Norton, 1965; Coke and Boland, 1971). Symptoms include reduced leaf area, curling and lamina deformation. However, the most characteristic symptom is stripes perpendicular to the veins on the underside of the lamina. New leaves may have an incomplete lamina, similar to that in sulphur and calcium deficiencies. Thickening of secondary veins and inhibition of root and flower formation have also been reported. All affected roots later darken and die. In Taiwan, a heart rot of plants derived from tissue culture has been attributed to boron and calcium deficiencies.

Table 9.1. Summary of mineral deficiency symptoms.

Age of leaf	Symptoms on leaf lamina	Additional symptoms	Element
All ages	Uniform light green or yellowing coloration	Pink petioles; stunted growth	Nitrogen
		Midribs curving resulting in weeping, drooping leaves	Copper
Young leaves	Yellow to almost white coloration with intervenal chlorosis		Iron
	Pale green to yellow coloration including veins	Thickening of secondary veins; leaves deformed	Sulphur
	Streaks across veins	Leaf lamina incomplete	Boron
	Yellow stripes along veins	Reddish coloration on lower side of young leaves	Zinc
	Marginal chlorosis	Thickening veins; necrosis from margins inward; leaves deformed	Calcium
Old leaves	'Sawtooth' marginal chlorosis	Petiole breaking; dark green-purple colour of young leaves	Phosphorus
	Yellow discoloration in mid-blade; midrib and margins remain green	Limit of chlorotic borders not clearly defined, pseudostem disintegrating	Magnesium
	Dirty yellow-green		Manganese
	Yellow-orange and brown scorching along margins	Leaf bending, rapid leaf desiccation	Potassium

Symptoms were a yellowing and necrosis of leaves, degeneration of roots and the development of a blackish cavity at the apical meristem of the corm before and after planting out in the field. Plants failed to develop symptoms when sprayed with either sodium borate ($NaBO_3$) or calcium sulphate ($CaSO_4$) (Ko *et al.*, 1997). Charpentier and Martin-Prevel (1968) point out that there is no clear distinction between sulphur and boron deficiency symptoms.

Soil concentration of boron should be in the range of 0.1–1.0 p.p.m. (Walmsley and Twyford, 1976). About 12 kg ha^{-1} of borax has been suggested as an amount that should overcome boron deficiencies (Twyford and Walmsley, 1968). In India, 2 ppm of boron sprayed as boric acid (H_3BO_3) gave the best results (Srivastava, 1964).

Calcium Deficiency

Symptoms of calcium (Ca) deficiency first appear on the youngest leaves as thickened lateral veins, being especially noticeable near the midrib. Later, the tissue between veins at the leaf margin becomes chlorotic (Plate 9.1). This symptom is generally more common near the tip of the leaf. The affected tissue expands towards the midrib, turns necrotic and gradually gives the leaf a 'sawtooth' appearance (Martin-Prevel and Charpentier, 1963). Calcium deficiency symptoms also include 'spike leaves'. Here, the lamina is deformed (Plate 9.2) or almost absent, due to a temporary shortage of calcium caused by a flush of rapid growth (Lahav and Turner, 1983).

Symptoms appear mainly on soils poor in calcium, with a low pH, or as a result of large amounts of potassium (antagonist to calcium) applied with inadequate rainfall. The plants recover when roots grow after rains or when potassium is leached below root volume. 'Spike leaves' can appear in soils rich in calcium as a result of low root activity in the early spring.

Calcium is supplied to banana plantations in the carbonate form as lime or dolomite and also as a component of superphosphate (21% calcium). Carbonates are usually applied to adjust soil pH rather than to increase the supply of calcium as a plant nutrient, though the two go hand in hand. The amount of lime applied depends upon the change in soil pH required and the buffering capacity of the soil. A common rate for banana is 1–2 t ha^{-1} annually or 3–6 t ha^{-1} every 3–5 years.

Copper Deficiency

The copper (Cu) requirements of banana are small, total copper uptake being about 1% of manganese uptake (Walmsley and Twyford, 1976). Copper is actively absorbed and readily translocated within the plant (van der Vorm and van Diest, 1982). Deficiencies have been described in pot culture (Charpentier and Martin-Prevel, 1965) and in the field in Côte d'Ivoire (Moity, 1961). Deficiency symptoms appear on all leaves. They are similar to those of nitrogen deficiency in that a general uniform paleness of the laminae occurs. Petioles are not pink, but the midrib bends, giving the plant an umbrella-like appearance. Yield is greatly reduced. Deficient plants have been reported as being sensitive to fungal and virus attacks (Moity, 1961). In acute cases, the rigidity of the funnel leaf is lost and it bends over the upper leaves during unfurling.

It is preferable to correct copper deficiency with a foliar spray, and an application of up to 0.5% solution of neutralized copper sulphate ($CuSO_4$) has been advocated (Lahav, 1995). However, soil applications can also be used. Twyford and Walmsley (1968) suggested 1 kg ha^{-1} of copper sulphate as an annual rate for banana. Srivastava (1964) increased the growth of suckers by applying a nutrient solution containing 4 ppm copper as copper sulphate to the soil. Copper oxide (CuO) and copper chelates can also be applied.

Iron Deficiency

Iron (Fe) deficiency symptoms, associated with calcareous soils, have been reported in the field in Hawaii (Cooil and Shoji, 1953), Jamaica (Charpentier and Martin-Prevel, 1968) and Israel (Ziv, 1954). The most common symptom is interveinal chlorosis. In Israel, this symptom is expressed more intensively in the spring, since the soil pH is higher at that time, the roots are less active and the leaves are starting to develop. Later, the entire lamina becomes chlorotic (Plate 9.3). Leaves may become yellow-white, with an iron level as low as 3.4 ppm (Cooil and Shoji, 1953).

In Hawaii, chlorotic plants flowered when still small, producing small bunches or no bunches at all. Four applications of iron sulphate ($FeSO_4$) spray at weekly intervals controlled the chlorosis for up to 8 weeks. Application of iron salts to the soil doubled the amount of growth compared with non-treated plants.

Iron deficiency can be overcome with foliar sprays of 0.5% iron sulphate (with surfactant) or iron chelate (Fe-ethylenediaminetetra-acetic acid (EDTA)). However, lime-induced chlorosis is better corrected by a soil application of iron chelate, which can be applied directly to the soil or through irrigation water.

Magnesium Deficiency

Magnesium (Mg) deficiency has been reported from many areas where banana is grown. It usually occurs where banana has been grown for 10–20 years without magnesium fertilizer (Chalker and Turner, 1969) or where high amounts of potassium have been given for a number of years (Messing, 1974).

A large range of leaf symptoms has been

Plate 9.1. Symptoms of leaf-edge chlorosis due to calcium deficiency in banana in Costa Rica (photo: E. Lahav, VI).

Plate 9.2. Deformed lamina as a result of calcium deficiency in banana in New South Wales, Australia (photo: E. Lahav, VI).

attributed to magnesium deficiency, including marginal yellowing, extending to near the midrib, changes in phyllotaxis, purple mottling of the petioles, marginal necrosis and separation of leaf sheaths from the pseudostem. The most common symptom in the field is that the leaf margins of older leaves usually remain green, while the area between the margin and the midrib becomes chlorotic (Murray, 1959; Martin-Prevel and Charpentier, 1963; Chalker and Turner, 1969; Turner and Bull, 1970).

Imbalances in the magnesium/potassium ratio can also cause problems. The magnesium deficiency disorder called 'blue', because of the bluish and brown marbling on the petioles (Plate 9.4), may be accentuated by potassium excess. Charpentier and Martin-Prevel (1968) indicate that a 'clearing blue' disorder may be due to a potassium deficiency as a result of excess magnesium. However, the magnesium-deficient 'blue' disorder is more common, because application of potassium fertilizer is more frequent than that of magnesium. In this disorder, the fruit maturation period is increased and there is a tendency to yellow pulp. Analysis reveals that soils are only slightly saturated in bases and calcium addition could improve the potassium–magnesium imbalance. Charpentier and Martin-Prevel conclude that potassium, calcium and magnesium deficiencies involve an imbalance of at least two of the three cations. Soil and tissue analyses are therefore essential to correct the problem.

Although the expression of magnesium deficiency symptoms is not necessarily associated with reduced yields, magnesium is usually applied when symptoms occur. However, the simple addition of magnesium-containing compounds to the soil does not always give the expected rapid response (Turner and Bull, 1970). It may be

Plate 9.3. Severe chlorosis in an iron-deficient banana growing on calcareous soil in Western Galilee, Israel (photo: E. Lahav, VI).

Plate 9.4. Banana in Côte d'Ivoire with symptoms of 'blue', the magnesium deficiency disorder which is caused by a potassium/magnesium imbalance (photo: R.H. Stover).

necessary to replant and incorporate magnesium into nitrogen, phosphorus and potassium mixtures so that small, regular additions of magnesium are given to the banana plants (Messing, 1974). Soil-applied soluble magnesium fertilizers, such as magnesium sulphate ($MgSO_4$), give a more rapid response than the less soluble compounds, but they are also more expensive. Since magnesium deficiency is unlikely to become a sudden problem in the field, the regular application of small amounts of magnesium seems advisable.

Manganese Deficiency

Jordine (1962) reported manganese (Mn) deficiency in banana in Jamaica. Later, Charpentier and Martin-Prevel (1965) investigated the effects of a deficiency and excess of manganese in sand culture in Côte d'Ivoire. Though found naturally in many African peat soils, manganese deficiency can be artificially induced by adding excessive amounts of lime.

The characteristic feature of manganese deficiency is a 'tooth-comb' chlorosis in leaves. The fungus *Deightoniella torulosa* (see Jones *et al.*, Chapter 2, pp. 102–104) is usually found in these chlorotic areas. The chlorosis starts marginally on the second or third youngest leaf. Sometimes, a narrow green edge is left at the leaf margins. The chlorosis then spreads along the main veins towards the midrib, interveinal areas remaining green giving a 'tooth-comb' appearance. While normal-sized bunches are produced by deficient plants (at least in the plant crop), the fruit is covered with black spots. Poor fruit development is partly associated with the premature desiccation of the leaves.

Manganese deficiency can be corrected

by foliar sprays of a 2% solution or by the ground application of manganese sulphate ($MnSO_4$) (Jordine, 1962). A tentative annual rate of 7–11 kg ha^{-1} of manganese has been suggested (Twyford and Walmsley, 1968; Marchal and Martin-Prevel, 1971).

Nitrogen Deficiency

Growth of the banana plant is more sensitive to a lack of nitrogen (N) than of any other element. Under sand-culture conditions, nitrogen deficiency more than halves the rate of leaf production, while deficiencies of other elements have only a slight effect (Murray, 1960). For the banana, the relationship between total dry-matter production and total nitrogen uptake is close (Lahav and Turner, 1983). Therefore, the first indication of nitrogen deficiency is a reduced rate of growth, resulting in a reduction in yield. Upon application of nitrogen, there is a prompt increase in growth and fruit weight.

Nitrogen is redistributed rapidly from old banana leaves to young ones (van der Vorm and van Diest, 1982). Hence, deficiency symptoms appear quickly and soon all leaves are affected. The leaves are pale green in colour, with the midribs, petioles and leaf sheaths becoming reddish pink (Plate 9.5) (Murray, 1959). The distance between successive leaves is reduced, giving the plant a 'rosette' appearance. Nitrogen deficiency symptoms are often observed under conditions of poor rooting and when there is weed competition. In general, field symptoms of nitrogen deficiency may result from any factor that reduces growth, such as drought or poor drainage.

The banana plant cannot store nitrogen (Martin-Prevel and Montagut, 1966) and it is easily leached in the soil. Therefore, nitrogen is almost universally in short supply, even for plants grown on the very fertile soils of Central America (Butler, 1960). Nitrogen deficiency is easily corrected by a variety of nitrogen fertilizers. The most commonly used are urea ($CO(NH_2)_2$), ammonium nitrate (NH_4NO_3), ammonium sulphate ($(NH_4)_2SO_4$) and potassium nitrate (KNO_3). The form used will depend on soil pH, the presence of irrigation and the type of irrigation system (Lahav, 1995). Nitrogen is also a component of compound fertilizers, which also include phosphorus and potassium.

Phosphorus Deficiency

Deficiency symptoms are rarely seen in the field. This is because banana plants accumulate the phosphorus (P) they require over an extended period of time and a relatively small quantity of the element leaves the plant with the fruit. Phosphorus is also easily redistributed from old to young leaves (van der Vorm and van Diest, 1982), from leaves to the bunch (Lahav, 1974) and from mother plant to followers (Walmsley and Twyford, 1968). Hence, phosphorus deficiency, if observed at all, will be seen early in the development of the plant crop.

Deficiency symptoms have been recorded in the field in Dominica (Stover and Simmonds, 1987) and Guadeloupe (Lacoeuilhe and Godefroy, 1971), and in sand culture (Charpentier and Martin-Prevel, 1965). A low supply of phosphorus results in stunted plants and poor root development. Older leaves initially exhibit marginal chlorosis, in which purplish-brown flecks develop. Eventually, a 'saw-tooth' necrosis appears. The affected leaves curl, the petioles break and the younger leaves have a deep bluish-green colour. Sucker and plant growth is reduced, but bunch weight is affected only when the deficiency is severe and prolonged.

It is presumed that the banana plant obtains adequate phosphate from soils through its mycorrhizal association (Lin and Fox, 1987; Fox, 1989). Vesicular–arbuscular mycorrhizal fungi can invade banana roots and improve plant growth in soils that differ considerably in phosphorus content.

The most commonly used phosphorus fertilizers are superphosphate and rock phosphate. These fertilizers can be applied at any time of the year, but, for more soluble phosphorus fertilizers, such as phosphoric acid (H_3PO_4), application should be confined to the growing season. Phosphoric acid, because of its strong acidity, is not recommended for soils with a low pH.

Potassium Deficiency

Potassium (K) is a key element in banana nutrition. When deficient, leaves become smaller in size and midribs curve so that the tip of the leaf points towards the base of the plant (Plate 9.6.) (Murray, 1959; Martin-Prevel and Charpentier, 1963; Lahav, 1972). However, the predominant effect of potassium is on the longevity of the leaf (Murray, 1960; Lahav, 1972) and the most common symptom of potassium deficiency is the appearance of an orange-yellow colour in the oldest leaves, followed by their rapid desiccation. Other effects of potassium deficiency are choking, delay in flower initiation and reduced fruit number per hand and hand number per bunch. Fruit size is also seriously affected by a shortage of potassium (Lahav, 1995), fruit growth being restricted because of a reduction in total dry-matter production, carbohydrate translocation and photosynthesis, due to a smaller leaf area and delayed stomatal activity. Protein synthesis is also impaired and the conversion of sugars to starch restricted.

Potassium uptake is highest during the first half of the vegetative phase of the banana plant and is much reduced after bunch emergence, even when the potassium supply is abundant. As the bunch attracts potassium from other parts of the plant, it is important that the plant has an adequate supply of the element early in its development (Lopez and Espinosa, 1995). It is totally different in plantain, where more than half of the potassium is taken up after flowering (Lahav, 1995).

Potassium deficiency is common in the less fertile banana soils and there are numerous reports of positive responses of banana plants to the application of potash (Bhangoo *et al.*, 1962; Stover and Simmonds, 1987). On volcanic soils in Cameroon, Hasselo (1961) described a premature yellowing of 'Lacatan' (AAA, Cavendish subgroup), which was corrected by potash application. Similarly, in Jamaica, Tai (1959) stated that potash fertilizer reduced a premature yellowing of banana. In Honduras, banana plants on certain leached, acid soils responded to potash if nitrogen was also applied (Bhangoo *et al.*, 1962), but no response to potash was obtained in fertile alluvial soils (Butler, 1960).

Potassium chloride (KCl) is the form of potassium usually applied to banana plantations. However, if soil salinity is a problem, fertilizers containing chlorides should be avoided and the high potassium requirement of the plant satisfied by using potassium sulphate (K_2SO_4), especially if banana is grown together with avocado, which is sensitive to chlorides. Potassium nitrate (KNO_3) can also be used as a potassium source for banana.

Sulphur Deficiency

Deficiencies of sulphur (S) have been reported in the field, as well as in sand-culture experiments (Charpentier and Martin-Prevel, 1965; Messing, 1971; Marchal *et al.*, 1972). Sulphur is actively redistributed from old to young leaves (van der Vorm and van Diest, 1982). When there is a shortage, young leaves first turn yellowish white (Plate 9.7.). As the deficiency progresses, necrotic patches appear on leaf margins and a slight thickening of the veins occurs, similarly to calcium and boron deficiencies. Sometimes the morphology of the leaf is changed and a bladeless leaf appears, again similarly to calcium and boron deficiencies. Growth is also stunted and the bunch is very small or choked. Temporary sulphur deficiency is very typical in tissue-culture plants during the first 2 months after planting. The most rapid uptake of sulphur occurs between the sucker and shooting stages. After shooting, the uptake rate is reduced and sulphur

Plate 9.5. Symptoms of nitrogen deficiency in banana in St Lucia, Windward Islands. Note the pink colour of the petioles and the 'rosette' effect due to the short internodes between petioles (photo: E. Lahav, VI).

Plate 9.6. Symptoms of potassium deficiency in banana grown in sand culture in Israel. Note the orange-yellow colour of the leaf before necrosis and the downward bend of the midrib (photo: E. Lahav, VI).

needed for fruit growth comes from the leaves and pseudostem (Walmsley and Twyford, 1976).

Most sulphur is supplied to bananas as ammonium sulphate ($(NH_4)_2SO_4$), potassium sulphate (K_2SO_4) or superphosphate ($Ca(H_2PO_4) + CaSO_4$). In the Windward Islands, 127 kg of sulphur ha^{-1} is needed to establish the plant crop (Walmsley and Twyford, 1976). Subsequent losses in fruit removal amount to 23 kg ha^{-1} $year^{-1}$, which is about three times the losses in plantain (Lahav, 1995). Nitrogen–phosphorus– potassium (NPK) fertilizer, containing 4% sulphur, is used to avoid yield losses caused by sulphur deficiency (Messing,

Plate 9.7. Yellowish young leaves due to sulphur deficiency in banana in Taiwan (Photo: E. Lahav, VI).

Plate 9.9. Symptoms of zinc deficiency in mature banana plants. Note the chlorotic stripes along the secondary veins. (Photo: E. Lahav, VI.)

Plate 9.8. Anthocyanin pigmentation on the underside of young banana leaf as a result of zinc deficiency (photo: E. Lahav, VI).

1971). Sulphur-coated urea is also considered a good source (Jaramillo, 1976). Regular applications of sulphur, amounting to 50 kg ha^{-1} $year^{-1}$, are recommended to avoid deficiencies (Marchal *et al.*, 1972). If sulphur-containing fertilizers, such as ammonium sulphate, potassium sulphate and superphosphate, are not used, sulphur should be applied separately.

Zinc Deficiency

The most widely reported minor element deficiency of banana is that of zinc (Zn) (Moity, 1954; Cardenosa-Barriga, 1962; Jordine, 1962). It has often been confused with virus infection (Freiberg, 1966). Zinc deficiency is more common on naturally high-pH soils or on highly limed soils, because zinc ions in the chelate complex can be replaced by calcium ions. Zinc availability may also be restricted in organic and peat soils. In Carnarvon, Western Australia, yields declined from more than 60 t ha^{-1} at pH 4.5 to about 30 t ha^{-1} at pH 8.7. This was thought to be associated with the supply of zinc to the plants, and leaf analysis data supported this interpretation (Turner *et al.*, 1989).

Zinc is moderately redistributed from old to young leaves (van der Vorm and van Diest, 1982). The characteristic deficiency symptoms appear in young leaves, which become significantly smaller in size and more lanceolate in shape (Jordine, 1962). The emerging leaf also has a high amount of anthocyanin pigmentation on its underside (Plate 9.8), which often disappears as the leaf unfurls. The unfurled leaf has alternating chlorotic and green bands. The fruit is sometimes twisted, short and thin, with a light green colour. As with sulphur deficiency, young tissue-culture plants also show zinc deficiency symptoms immediately after planting. Zinc deficiency symptoms may appear without any apparent reduction in growth or yield, but, if the deficiency persists, plants of the next cycle are stunted (Charpentier and Martin-Prevel, 1965).

'Rayadilla' disease (Cardenosa-Barriga, 1962), now attributed to zinc deficiency, was first reported in 1944 in Colombia on plantains in the Cauca Valley and 'Gros Michel' (AAA) in the Sevilla area. In the Sevilla area, the condition is also called 'lengua de vaca', because of the narrow strap-shaped leaves. 'Rayadilla' is seasonal in appearance, being most common during dry seasons and tending to disappear during rainy months. It is worst west of the Orihueca and Rio Frio districts, where high pH and salinity problems may occur during the dry season, when plants are irrigated (Colmet-Daage and Gautheyrou, 1968). Plants have a stunted appearance, with erect, narrow leaves and small bunches of fruit. The bunch may be horizontal or partially parallel to the ground, rather than hanging perpendicularly. There is a general, uneven chlorosis, with stripes of yellow to white tissue alternating with green in young leaves (Plate 9.9). This tissue may become necrotic as the leaves mature. The narrow strap-shaped leaves may have a thickening over the veins, resulting in upward curling. These leaves become fragile and brittle. Plants approaching maturity are most affected and suckers may appear normal until they are 150 or 180 cm high, when chlorotic mottling and rosetting symptoms appear. Normal leaves may follow chlorotic, stunted leaves, or the disease symptoms may intensify with each emerging leaf. Because of the symptoms, a virus disease was originally suspected, but attempts at transmission with six species of aphids were unsuccessful (Cardenosa-Barriga, 1962). Application of zinc to chlorotic foliage resulted in the resumption of normal growth and pigmentation. Symptoms of zinc deficiency in Jamaica were very similar to those of 'Rayadilla' in Colombia (Jordine, 1962). It also seems highly likely that 'alkali chlorosis', reported in Haiti, is also a zinc deficiency disorder (Wardlaw, 1938).

The zinc requirement of banana is relatively low. On the basis of whole-plant analysis, Twyford and Walmsley (1968) suggested a rate of about 1.0 kg of zinc ha^{-1} $year^{-1}$. Field experiments in Australia on acid soils showed that double and triple this amount was required to correct the deficiency, while, in Central America and the Philippines, 5–10 kg ha^{-1} is applied. Since zinc is leached under acid conditions and fixed under alkaline conditions, it is easy to correct its deficiency with a spray of 0.5% zinc sulphate ($ZnSO_4$) (Jordine, 1962). Work in Western Australia has shown that foliar applications need to be applied every 3 months to maintain zinc concentration in the leaves above a critical level (Turner *et al.*, 1989). The long-term solution seems to be the regular addition of zinc to the soil. Zinc chelates are also sometimes applied.

REFERENCES

Bhangoo, M.S., Altman, F.G. and Karon, M.L. (1962) Investigations on the Giant Cavendish banana. I. Effect of nitrogen, phosphorus and potassium on fruit yield in relation to nutrient content of soil and leaf tissue in Honduras. *Tropical Agriculture (Trinidad)* 39, 189–201.

Butler, A.F. (1960) Fertilizer experiments with the Gros Michel banana. *Tropical Agriculture (Trinidad)* 37, 31–50.

Cardenosa-Barriga, R. (1962) La 'Rayadilla' del platano en Colombia. *Turrialba* 12, 118–127.

Chalker, F.C. and Turner, D.W. (1969) Magnesium deficiency in bananas. *Agricultural Gazette of New South Wales* 80, 474–476.

Charpentier, J.M. and Martin-Prevel, P. (1965) Culture sur milieu artificiel. Carences atténuées ou temporaires en éléments majeurs, carence en oligo-éléments chez le bananier. *Fruits* 20, 521–557.

Charpentier, J.M. and Martin-Prevel, P. (1968) *Carences et troubles de la nutrition minéral chez le bananier. Guide de diagnostic pratique.* Institut Français de Recherches Fruitières Outre-Mer, Paris, 75 pp.

Coke, L. and Boland, D.E. (1971) Boron nutrition of banana suckers. In: *Proceedings of the 2nd ACORBAT Conference, Jamaica*, pp. 59–66.

Colmet-Daage, F. and Gautheyrou, J.M. (1968) Etude préliminaire des sols de la région bananière de Santa Marta (Colombie). *Fruits* 23, 21–30.

Cooil, B.J. and Shoji, K. (1953) Studies reduce banana chlorosis. *Hawaii Farming and Science* 1, 1–8.

Fox, R.L. (1989) Detecting mineral deficiencies in tropical and temperate crops. In: Plucknett, D.L. and Sprague, H.B. (eds) *Westview Tropical Series 7.* Westview Press, Boulder, Colorado, USA, pp. 337–353.

Freiberg, S.R. (1966) Banana nutrition. In: Childers, N.F. (ed.) *Fruit Nutrition.* Horticultural Publications, New Brunswick, New Jersey, USA, pp. 77–100.

Hasselo, H.N. (1961) Premature yellowing of Lacatan bananas. *Tropical Agriculture (Trinidad)* 38, 29–34.

Jaramillo, R. (1976) Efecto de urea y de urea-azufre en la produccion de banano 'Giant Cavendish' en Guapiles, Costa Rica. *Turrialba* 26, 90–95.

Jordine, C.G. (1962) Metal deficiencies in bananas. *Nature* 194, 1160–1163.

Ko, W.H., Chen, S.P., Chao, C.P. and Hwang, S.C. (1997) Etiology and control of heart rot of banana tissue culture plantlets. *Plant Pathology Bulletin (Taiwan)* 6, 31–36.

Lacoeuilhe, J.J. and Godefroy, J. (1971) Un cas de carence en phosphore en bananeraie. *Fruits* 26, 659–662.

Lacoeuilhe, J.J. and Martin-Prevel, P. (1971) Culture sur milieu artificiel. 1. Carences en N, P, S chez le bananier, analyse foliaire. 2. Carence en K, Ca, Mg chez le bananier, analyse foliaire. *Fruits* 26, 161–167, 243–253.

Lahav, E. (1972) Effect of different amounts of potassium on the growth of the banana. *Tropical Agriculture (Trinidad)* 49, 321–335.

Lahav, E. (1974) The influence of potassium on the content of macro-elements in the banana sucker. *Agrochimica* 18, 194–204.

Lahav, E. (1995) Banana nutrition. In: Gowen, S. (ed.) *Bananas and Plantains.* Chapman and Hall, London, UK, pp. 258–315.

Lahav, E. and Turner, D.W. (1983) *Banana Nutrition.* Bulletin 7, International Potash Institute, Worblaufen-Bern, Switzerland, 62 pp.

Lin, M.L. and Fox, R.L. (1987) External and internal P requirements of mycorrhizal and non-mycorrhizal banana plants. *Journal Plant Nutrition* 10, 1341–1348.

Lopez, A.M. and Espinosa, J.M. (1995) *Manual de Nutricion y Fertilizacion del Banano*. Instituto de la Potasa y el Fosforo, Quito, Ecuador, 82 pp.
Marchal, J. and Martin-Prevel, P. (1971) Les oligo-elements Cu, Fe, Mn, Zn dans le bananier – niveaux foliaires et bilans. *Fruits* 26, 483–500.
Marchal, J., Martin-Prevel, P. and Melin, P. (1972) Le soufre et le bananier. *Fruits* 27, 167–177.
Martin-Prevel, P. and Charpentier, J.M. (1963) Culture sur milieu artificiel. Symptomes de carences en six éléments minéraux chez le bananier. *Fruits* 18, 221–247.
Martin-Prevel, P. and Montagut, G. (1966) Essais sol-plante sur bananiers. 8. Dynamique de l'azote dans la croissance et le développement du végétal. 9. Fonctions de divers organes dans l'assimilation de P, K, Ca, Mg. *Fruits* 21, 283–294, 395–416.
Messing, J.H.L. (1971) Response to sulphur in Windward Island soils. In: *Proceedings of the 2nd ACORBAT Conference, Jamaica*. Jamaican Banana Board, Kingston, pp. 51–58.
Messing, J.H.L. (1974) Long term changes in potassium, magnesium and calcium content of banana plants and soils in the Windward Islands. *Tropical Agriculture (Trinidad)* 51, 154–160.
Moity, M. (1954) La carence en zinc sur le bananier. *Fruits* 9, 354.
Moity, M. (1961) La carence en cuivre des 'tourbières du Nieky' (Côte d'Ivoire). *Fruits* 16, 399–401.
Murray, D.B. (1959) Deficiency symptoms of the major elements in the banana. *Tropical Agriculture (Trinidad)* 36, 100–107.
Murray, D.B. (1960) The effect of deficiencies of the major nutrients on growth and leaf analysis of the banana. *Tropical Agriculture (Trinidad)* 37, 97–106.
Norton, K.R. (1965) Boron deficiency in bananas. *Tropical Agriculture (Trinidad)* 42, 361–365.
Srivastava, R.P. (1964) Effect of microelements Cu, Zn, Mo, B and Mn on the growth characteristics of banana. *Science and Culture* 30, 352–355.
Stover, R.H. and Simmonds, N.W. (1987) *Bananas*. Longman Scientific and Technical, Harlow, Essex, UK, 468 pp.
Tai, E.A. (1959) *Annual Report*, 1956–57, 1957–58. Banana Board Research and Development, Kingston, Jamaica.
Tollenaar, D. (1969) Boron deficiency in sugarcane, oil palm and other monocotyledons on volcanic soils of Ecuador. *Netherlands Journal of Agricultural Science* 17, 81–91.
Turner, D.W. and Bull, J.H. (1970) Some fertilizer problems with bananas. *Agricultural Gazette of New South Wales* 81, 365–367.
Turner, D.W., Korawis, C. and Robson, A.D. (1989) Soil analysis and its relationship with leaf analysis and banana yield with special reference to a study of Carnarvon, Western Australia. *Fruits* 44, 193–203.
Twyford, I.T. and Walmsley, D. (1968) The status of some micronutrients in healthy Robusta banana plants. *Tropical Agriculture (Trinidad)* 45, 307–315.
van der Vorm, P.D.J. and van Diest, A. (1982) Redistribution of nutritive elements in a 'Gros Michel' banana plant. *Netherland Journal of Agricultural Science* 30, 286–296.
Walmsley, D. and Twyford, I.T. (1968) The uptake of 32P by 'Robusta' banana. *Tropical Agriculture (Trinidad)* 45, 223–228.
Walmsley, D. and Twyford, I.T. (1976) The mineral composition of the banana plant. 5, Sulphur, iron, manganese, boron, zinc, copper, sodium and aluminum. *Plant and Soil* 45, 595–611.
Wardlaw, C.W. (1938) Banana diseases. 12. Diseases of the banana in Haiti, with special reference to a condition described as 'plant failure'. *Tropical Agriculture (Trinidad)* 15, 276–282.
Warner, R.M. and Fox, R.L. (1977) Nitrogen and potassium nutrition of the Giant Cavendish banana in Hawaii. *Journal of the American Society for Horticultural Science* 102, 739–743.
Ziv, D. (1954) Chlorosis of bananas and other plants in the Jordan Valley due to iron deficiency. *Hassadeh* 35, 190–193 (in Hebrew).

10

Injuries to Banana Caused by Adverse Climate and Weather

Y. Israeli and E. Lahav

Introduction

Banana is a tropical crop that grows and yields best where temperatures, humidity and rainfall are high. Because of nearness to a market or for economic assistance to local agriculture, banana is sometimes grown commercially in areas where the climate is more subtropical than tropical and where temperatures are below optimum for several months of the year. This gives rise to growth and quality disorders associated with temperatures that are too low or too high for normal growth. In addition, abnormal weather such as wind, hail and excesses or deficiencies of water and sunlight can cause injury and yield reductions.

Seasonal Wind and Windstorm Damage

Banana has leaves that are extremely sensitive to tearing by winds because of their wide and thin lamina and parallel secondary veins. Skutch (1927) stated that leaf tearing is a normal phenomenon in banana. Leaf tearing might have some advantage in reducing boundary layer resistance, increasing transpiration rate and maintaining the energy balance of the leaf (Taylor and Sexton, 1972). Wind velocity of 15 km h^{-1} is assumed to have a positive effect on the energy balance in banana (Turner, 1994); however, it was proved that leaf tearing reduces rate of growth and decreases yield. Ziv (1962) found a 17% reduction in bunch weight after tearing leaves from the edge of the lamina to the petiole every 30–50 mm and Peretz (1992) measured a 10% yield reduction after tearing leaves into strips 50 mm wide. When leaves were torn to narrower strips – 12.5 mm wide – bunch weight decreased by 17.9% (Eckstein *et al.*, 1996). The negative effect of leaf tearing is explained by the decrease in photosynthesis resulting from the decrease in interception of photosynthetic active radiation (Eckstein *et al.*, 1996). In addition to leaf tearing, wind increases water loss, which results in water stress and stomata closure (Shmueli, 1953). Seasonal winds in Israel can be strong enough to destroy the total lamina (Plate 10.1) or break leaves, especially in the upper part of the pseudostem. Wind can also cause leaves, bunch covers (Plate 10.2), dust or salt particles to collide with fruit, causing bruises.

In many banana-growing areas, banana is artificially protected from the wind. In the Canary Islands, banana is commonly grown behind wind-breaks made of bricks or sheltered in greenhouses (Galan-Sauco, 1992). Artificial wind-breaks (Plate 10.3) are sometimes used in Israel (L orch, 1959; Ziv, 1962) and in New South Wales, Australia (Freeman, 1973). Closely planted trees are often employed to block strong winds. *Casuarina* is used in Egypt, *Cupressus* in

Israel (Plate 10.4), *Erythrina* in Martinique and Guadeloupe and pines in Queensland, Australia (Daniells, 1984).

A significant increase in the rate of growth and an increase of bunch weight of up to 25% have been measured in plants protected from the wind (Ziv, 1962; Peretz, 1992; Eckstein, 1994). However, wind-breaks may also shade nearby plants and thus reduce yield (Eckstein, 1994), compete with the banana plants for water and nutrients (Ziv, 1962) or host various pests and diseases (Krambias *et al.*, 1973). The cost of wind-breaks might also be high. Usually banana is naturally well protected when plant density is sufficiently high (Lorch, 1958).

Tropical storms above 50 km h^{-1}, such as hurricanes in South America and the Caribbean and cyclones in South-East Asia and Australia or tornadoes in West Africa, can cause total devastation by breaking or uprooting banana plants over large areas (Plate 10.5). Heavy rains and floods can also accompany strong winds. Toppling of banana plants occurs more often in wet soil, especially when roots and corms are damaged by nematodes or weevil borer (Gowen, 1993). Damage recovery after uprooting is slow, because of a delay in the emergence of follow-suckers (Robinson, 1996).

Storm incidence can affect the development of banana industries. Thus, commercial banana plantations in the Philippines have been sited on the island of Mindanao, which is cyclone-free, unlike Luzon island, where 19 cyclones are registered annually (Valmayor, 1990). A high-density, single-cycle method of planting was developed in Taiwan to avoid the typhoon season (Tang and Liu, 1993). Planting is timed so that only young suckers are present in the field during the typhoon season.

Rate of toppling differs among cultivars (Ahmad and Quasem, 1975; Stover and Simmonds, 1987; Daniells and Bryde, 1993, 1995; Tang and Chu, 1993) as was demonstrated by Monnet (1961) (Table 10.1). The main factor involved is stature, but pseudostem circumference is also important. Leaf petioles in tetraploid cultivars are not as strong as in triploids and the leaf breaks more easily (Stover and Simmonds, 1987). Sarma *et al.* (1995) reported that 'Bhimkal' (ABB), although a tall cultivar, has a fibrous pseudostem, which makes it resistant to toppling. The relative tolerance of 'Grand Nain' and 'Poyo' (AAA, Cavendish subgroup) was demonstrated in Côte d'Ivoire when the two cultivars grew side by side. 'Grand Nain' is shorter than 'Poyo' by 50–70 cm and suffered significantly less toppling. In St Lucia, the taller Cavendish cultivar 'Robusta' suffered 2.5 times more wind damage than the shorter 'Williams' (Holder and Gumbs, 1983a). 'FHIA-01®' (AAAB, syn. 'Goldfinger') has recently been introduced in Australia and, although taller than 'Grand Nain' or 'Williams', has a larger and stronger pseudostem. Consequently, it suffers less from strong winds (Daniells *et al.*, 1995). The effect of strong wind on three cultivars of different heights is illustrated in Plate 10.6. It can be concluded that cultivar selection and timing of planting to avoid adverse seasonal weather (Obiefuna, 1986) contribute to decreased wind damage.

Low-temperature Effects

The response of the banana to temperature has been thoroughly reviewed (Stover and Simmonds, 1987; Turner, 1994, 1995; Robinson, 1996). Therefore, this review will focus on the effect of low winter temperatures, chilling and frost on the banana plant.

Minimum temperature for meristem activity and new leaf formation is 9–11°C (Green and Kuhne, 1969; Ganry, 1973), but almost all new leaves emerge only above 14–16°C (Robinson, 1996). Leaf emergence rates are optimal between 27 and 32°C (Ganry, 1973; Turner and Lahav, 1983; Allen *et al.*, 1988). The banana grows best between 22 and 31°C.

Optimum temperature for CO_2 assimilation in the banana leaf has been found to be 28–36°C. At radiation levels above 800 μmol m^{-2} s^{-1}, the optimal temperature is 31–32°C. Assimilation of CO_2 was

Table 10.1. The effect of mature plant height on blow-down losses and yield in six different Cavendish cultivars (AAA) and 'Gros Michel' (AAA) during four crop cycles in Guinea (after Monnet, 1961).

Cultivar	Average height (cm)	Blow-down losses (%)	Yield (t ha^{-1})
'Dwarf Cavendish'	180	3.3	82.0
'Grand Nain'	210	2.4	89.1
'Seredou'	260	20.1	60.9
'Poyo'	280	35.3	23.1
'Maneah'	290	57.5	40.0
'Lacatan'	320	60.2	19.6
'Gros Michel'	360	71.6	17.4

recorded by Eckstein and Robinson (1995b) to be at a maximum at a leaf temperature of 32.5°C.

Minimum temperature for root growth is about 12°C (Robinson and Alberts, 1989). The optimum calculated by Turner (1995) is 24°C, somewhat lower than the optimum of 26.5°C found by Israeli *et al.* (1979) under controlled conditions. The lowest critical temperature for fruit growth is between 11°C (Green and Kuhne, 1975) and 14.5° (Ganry and Meyer, 1975). Turner and Barkus (1982) showed that relative fruit growth rate increases linearly between 13.3 and 21.6°C. In Israel, fruit growth continues during winter, when daily average temperatures are as low as 13–14°C and leaf emergence has almost totally stopped (Israeli and Lahav, 1986). It seems that, under such conditions, the fruit grows during the daytime, when temperatures are higher. Leaves do not emerge because the daily temperature fluctuations in the meristem zone are small. The optimum temperature for fruit growth is 30°C (Hord and Spell, 1962; Ganry and Meyer, 1975). The relationship between temperature and rate of fruit growth enables harvest dates to be predicted, but the interaction of many additional factors (water availability, nutrient supply, amount of light, strength of winds, cultivar factors, pest and disease incidence) makes the use of such models very difficult (Turner, 1995).

The relationship between temperature and other climatic factors and the banana rate of growth enables the prediction of expected flowering time (Cottin *et al.*, 1987; Israeli *et al.*, 1988). Flowering to harvest time is affected significantly by temperature. Under tropical conditions, annual variation of 30 days can be expected (Stover and Simmonds, 1987). Fruit-filling periods increased from 90 to 120 days when average temperatures decreased from 29 to 23°C (Hord and Spell, 1962). Temperatures below 20°C are quite rare in tropical lowlands, though, every 5–8 years, a cold spell may reduce temperatures to 11°C and cause chilling damage in the growing areas of Mexico, Honduras or Guatemala (Stover and Simmonds, 1987). Long fruit-filling periods resulting from low temperatures are also related to increased elevation (Daudin, 1955; Sanchez-Nieva *et al.*, 1970; Galan-Sauco *et al.*, 1984). Low winter temperatures prevailing in subtropical growing regions (north or south of latitude 20°) increase the period from flowering to harvest much more significantly. In Israel, bunches emerging in the spring are harvested after 80 days, while those emerging in the autumn may need 240 days (Oppenheimer, 1960; Smirin, 1960; Israeli and Lahav, 1986). In South Africa, this variation is between 118 and 213 days for 'Williams' (Robinson, 1981).

The first symptom of low-temperature damage, observed at 17/10°C day/night temperatures, is a shortening of the distance between petioles where they leave the pseudostem, which gives a symptom referred to as 'choking'. The leaves also point upright and have narrow laminae (Porteres, 1950; Prest and Smith, 1960;

Plate 10.1. Wind damage to banana in Israel. The leaves on some plants have been shredded until only the midribs remain (photo: Y. Israeli, JVBES).

Plate 10.2. Abrasion damage to fruit of 'Robusta' (AAA, Cavendish subgroup) in St Vincent, Windward Islands. The line scars are caused by the polyethylene bunch cover rubbing on bananas, usually the finger ridges, in the wind (photo: D.R. Jones, SVBGA).

Cann, 1964; Turner and Lahav, 1983). Fruit suffers chilling injury after several hours' exposure to temperatures below 11–14°C.

One of the most significant symptoms of low winter temperatures is 'choke throat', a condition that is characterized by the partial emergence of the infloresence from the pseudostem. The lower hands trapped in the trunk do not open normally and the whole bunch is positioned almost horizontally. This condition results from the crowding of petioles at the top of the pseudostem and the loss of elasticity of leaf sheaths during winter when growth ceases. In extreme choking, the bunch is trapped inside the pseudostem (Plate 10.7) or emerges through the side of the upper pseudostem (Plate 10.8). Such bunches are differentiated in autumn and 'choke' during the end of the winter or early spring

Plate 10.3. Artificial wind-break for banana in Israel (photo: E. Lahav, VI).

Plate 10.4. *Cupressus* trees used as a wind-break for banana in Israel (photo: Y. Israeli, JVBES).

(Oppenheimer, 1960). Choking has been described in many subtropical growing regions (Stoler, 1960; Ziv, 1962; Kuhne *et al.*, 1973; Robinson, 1993) and 'Dwarf Cavendish' (AAA, Cavendish subgroup) is especially affected. This has led to a decline in the importance of 'Dwarf Cavendish' in subtropical areas and its replacement by less susceptible Cavendish cultivars, such as 'Williams' (Robinson, 1981).

Within the Cavendish subgroup, the sensitivity to low temperatures generally decreases as plant height increases or as the length of bunch stalk increases (Kuhne, 1980a; Hill *et al.*, 1992). Although the tendency to choking decreases with height, experience gained in Israel shows that low temperatures can induce choking in the relatively tall 'Williams' and 'Grand Nain'. 'Williams' is also affected in South Africa (Robinson, 1996). In Taiwan, 'Grand Nain' suffers from 'choke throat' while the taller local cultivar 'Pei-Chiao' suffers less (Tang and Chu, 1993).

'May flowering' in the northern hemisphere (or 'November dump', as it is called in the southern hemisphere) is another condition associated with low temperatures (Summerville, 1944; Oppenheimer, 1960; Fahn *et al.*, 1961; Ziv, 1962; Cann, 1964; Blake and Peacock, 1966; Kuhne *et al.*, 1973; Israeli and Blumenfeld, 1985; Robinson, 1993, 1996; Turner, 1995). The main characteristics of this disorder are as follows.

- Fewer than five anthers in a flower.
- Fewer than three loculi in an ovary. Sometimes the loculi are totally missing. Flowers with reduced number of loculi develop small fruit scattered irregularly in the hand.
- Enlargement of the style base due to unification of several filaments. Sometimes the loculi penetrate into the style base. Such thickening forms a nipple at the tip of the finger.
- Nectary longer than normal.
- Fingers short and thick with a round cross-section and straight growth. The distal part of the finger is often larger than the proximal part.
- Variability in finger size within a single hand.
- A reduction in the number of hands to up to half the usual number.

In Israel, abnormal flowers appear during a period of 3–4 weeks on bunches emerging in June. If one assumes that there are 11 leaves hidden in the pseudostem during floral initiation (Summerville, 1944) and at the end of initiation there are still nine leaves left to emerge (Ziv, 1962), then these abnormal flowers would appear to be formed in bunches differentiated in winter or late autumn when temperatures are low. Further evidence for low temperatures being the cause of the condition comes from studying the location of the abnormal flowers on the bunch over the 'May flowering' period. At the beginning of the period, abnormal flowers are located mainly on the distal hands of emerging bunches while the basal hands are more affected on bunches emerging at the end of the period. As the basal hands complete their differentiation earlier than the distal hands, it is believed that they are not affected at the beginning of the period, because they are formed before temperatures decrease, whereas the distal hands are formed when conditions are colder. Conversely, at the end of the period, it is the basal hands that are differentiated during the colder weather, whereas the distal hands are formed later, when spring temperatures are increasing (Fahn *et al.*, 1961). These disorders may appear one at a time or simultaneously and their severity depends on the temperature. They are more frequent in 'Dwarf Cavendish', but also appear in 'Williams' and 'Grand Nain'.

The relationship between temperature at the apical meristem and quality of bunches was demonstrated in the Jordan Valley, Israel, by Stoler (1962). Inflorescences differentiated in the spring, when meristem temperatures were 20 to 24°C, flowered in July–August and produced excellent bunches, with many hands and normal fingers. Earlier-differentiated bunches, when temperatures were less than 18°C, contained few hands and these had abnormal flowers. High temperatures, ranging from 26 to 30°C, at the zone of the apical meristem resulted in bunches with few hands, but normal fingers. The average daily air temperature during bunch differentiation is related to the number of hands formed on the bunch (Israeli, 1976). Numbers of hands increase when the bunch is differentiated between 18°C and 27°C, but decrease quickly when differentiation occurs at temperatures above 27°C. A special desuckering system, aimed at avoiding bunch differentiation during spring or autumn, was developed in Israel (Oppenheimer, 1960; Ziv, 1962; Aubert, 1971; Ticho, 1971). By using this method, the actual rate of abnormal bunches is significantly reduced.

Chilling and Frost Damage

Low temperatures in tropical regions, caused by cold fronts, results mainly in cessation of growth and delay in fruit filling. Chilling damage is rare (Stover and Simmonds, 1987). However, in subtropical regions temperatures may drop much lower and chilling or frost damage may occur.

Banana plants are sensitive to low temperatures because of their anatomical, morphological and physiological characteristics (Shmueli, 1960). They have big leaves, containing large cells surrounded by large intercellular cavities, and they are highly hydrated, being very sensitive to any change in hydration level. The osmotic pressure of leaf sap is relatively low, while its freezing-point is high. The leaf is the coldest part of the plant during a bright winter night, since it is exposed to the sky. Frost damage occurs in the banana when leaf temperature drops below 0°C. Leaf temperatures during a cold night may be 2–2.5°C lower than temperatures recorded in a conventional meteorological station (Gilead and Rosenan, 1957) and are very similar to temperatures measured near the soil surface. Internal plant parts have higher temperatures, since they are more protected. On one cold night, Shmueli (1960) recorded temperatures of −1.0°C near the soil surface, −1.4°C on the surface of an upper leaf exposed to the sky, 0.4°C in the air above that leaf, 1.3°C on the surface of a lower, protected leaf and 5.6°C inside the pseudostem at a position 60 cm above the ground. At the same time, the bunch temperature under the polyethylene sleeve was 3.0°C. Frost damage occurs when sap freezes. Leaves blacken during the days following frost and this leads to complete leaf destruction (Plate 10.9).

Chilling injury takes place when temperatures drop to 5–8°C (Oppenheimer, 1960; Ticho, 1971; Kuhne and Green, 1980; Robinson, 1993; Turner, 1994). Initially, there is a gradual yellowing of the leaf (Plate 10.10), caused by the destruction of chlorophyll. Damage is more severe if there is strong radiation the following morning. This induces photoinhibition and the reduction in photosynthesis is directly related to the decrease in temperature the previous night (Eckstein and Robinson, 1995b) and to the light intensity the following morning (Smillie *et al.*, 1988). Repeated exposure to low temperatures night after night and strong illumination day after day causes bleaching. Continuous exposure to chilling temperatures also induces the destruction of chlorophyll, with leaf colour changing to pale-green, yellow, orange and finally brown. Leaf necrosis will take place after 3–4 weeks. Chilling injury is a product of time and temperature and can occur after long exposure to temperatures of 2–3°C (Robinson, 1993). Slight chilling injury might not cause irreversible damage. A young leaf affected initially by chilling and showing chlorophyll destruction may return within few weeks to normal activity if not exposed to further chilling (Shmueli, 1960). It has been demonstrated that high proportions of desaturated phosphatidylglycerol within lipids of the chloroplast membrane appear to signify chilling sensitivity of a few plant species, including banana (Roughan, 1985; Kenrick and Bishop, 1986).

Banana plantations affected by frost resulting in the loss of canopy can still produce marketable fruit if the bunches are protected and have reached the half-full filling stage. Such fruit absorbs mostly water during its last stages of development; therefore, green leaves at this stage are not critical for bunch filling and maturation (Israeli and Lahav, 1986). Bunches at a less advanced stage will not reach commercial marketing (Smith, 1973; Robinson, 1993). Frost damage on the bunch appears as water-soaked areas on the fingers (Plate 10.11). In a frost-damaged plantation, the bunch is no longer protected from the sun by leaves and symptoms of sunburn may appear on the peduncle and upper hands. Often, the peduncle breaks or snaps. Plants that have not differentiated bunches before frost have been thought to need four to six newly emerged leaves after frost to support normal bunch development (Kuhne and Green, 1980; Robinson, 1993), though ten

Plate 10.5. Storm damage to 'Grand Nain' (AAA, Cavendish subgroup) in Israel (photo: E. Lahav, VI).

Plate 10.6. Effect of plant height on losses from wind. The devastated plot in the centre contains the tall Cavendish cultivar 'Lacatan' (360 cm in height). Much less damage has been caused to the smaller Cavendish cultivar 'Valery' (290 cm in height) in the plot to the right and to 'Cocos', a dwarf mutant of 'Gros Michel', in the plot to the left (photo: R.H. Stover).

leaves may be more realistic (Smith, 1973). Bunch emergence soon after frost will probably be followed by choking and by abnormal flowers. Such bunches have no commercial value and should be eliminated. Frost damage may be related to temperatures prevailing before the frost. When frost follows a period of high temperatures and intensive growth, more severe damage can be expected. On the other hand, field observations suggest that a gradual decrease in temperatures over several weeks before a frost may reduce the amount of damage significantly.

Sensitivity to chilling and frost varies among cultivars and clones. Cultivars con-

Plate 10.7. Bunch trapped inside the pseudostem as a result of early flowering and low temperatures (photo: Y. Israeli, JVBES).

Plate 10.8. A bunch that has emerged through the side of the pseudostem (photo: Y. Israeli, JVBES).

Plate 10.9. Severe frost damage to banana in Western Galilee, Israel (photo: E. Lahav, VI).

taining the B genome inherited from *Musa balbisiana* are more resistant than those based solely on the A genome inherited from *Musa acuminata*. Liang *et al.* (1994) tested the effect of chilling temperatures on potted AAA and ABB banana cultivars. The AAA banana clones were significantly injured, while no damage was observed on the ABB clones. Peroxidase activity was initially much higher in leaves of ABB plants than in the AAA plants. It remained high in the ABB banana after exposure to chilling, but decreased in the AAA plants. Paclobutrazol treatment reduced the damage caused by the low temperature. The new 'FHIA 01®' hybrid is relatively more frost-resistant than 'Grand Nain' (Rowe and Rosales, 1993; Z.C. de Beer, South Africa, 1995, personal communication). Within the Cavendish subgroup, the clones that are low in stature suffer more than the tall ones. However, their rate of growth and development is faster and they recover quickly. This height-related difference in frost sensitivity is well demonstrated in Israel, where short cultivars, such as 'Dwarf Cavendish', and taller ones, such as 'Williams', often grow side by side.

Frost damage can be reduced by several methods, the most important one being the selection of a frost-free area for planting. Before planting in a region where there are risks, a topographical/climatic survey should be undertaken (MacHattie and Schnelle, 1974; Gat, 1984) for the selection of optimal sites (Lomas *et al.*, 1989). In frost-hazardous zones, heavy clay soils with no air drainage are generally not recommended for planting banana.

Methods for predicting frost and protection against frost have been reviewed for many crops (Bagdonas *et al.*, 1978; Barak and Israeli, 1989; Kalma *et al.*, 1992), including banana (Shmueli *et al.*, 1959; Oppenheimer, 1960). Conditions favourable for the development of a radiational frost include initial low air temperatures, clear skies, low air humidity and no wind. During the night, the soil loses heat through long-wave radiation. This radiation heats the plants above, which slows the cooling process. Therefore, keeping the soil surface exposed without weeds or trash will help heat convection (Shmueli *et al.*, 1959) and reduce frost damage. A high-density plantation will be less damaged than a low-density plantation, since relatively less of its canopy is exposed to the sky. Banana plants in plantations should not be cut down immediately after harvest if there is a risk of frost. The lower parts of vegetative wind-breaks should be pruned so that air can move freely over the soil surface. Frost in subtropical regions can affect the following cycle as well as the current crop, due to loss of leaves, destruction of some followers and delayed flowering.

Frost can be prevented by using heaters or solid-fuel fires, overhead sprinklers to provide artificial rain, misters to provide artificial fog or wind machines or helicopters that mix the lower cool air layers with higher hotter layers. All these methods have been tested in banana plantations (Shmueli *et al.*, 1959; Oppenheimer, 1960; Barak and Israeli, 1989), but none are used commercially.

The most efficient means for protection against environmental stresses, including chilling and frost, is to grow banana under polyethylene (Plate 10.12). This technique is widely used in Morocco (Janick and Ait-Oubahau, 1989; Choukr-Allah, 1990), Canary Islands (Galan-Sauco *et al.*, 1992) and southern Turkey (Cevik *et al.*, 1984) and to a smaller extent in Sardinia, Sicily, Crete, South Korea and Israel (Ahn *et al.*, 1990; Pala and Ovadia, 1990; Colombo *et al.*, 1993). The quality of fruit is much improved and bananas produced by these methods can compete successfully in the market with those from the tropics.

Hail Damage

Hail damage is a rare event in tropical lowlands (Stover and Simmonds, 1987), but is typical in many subtropical growing regions, especially South Africa (Kuhne and Green, 1980; Robinson, 1993; Eckstein and Fraser, 1996). Hail can tear banana leaves to narrow shreds. These later turn brown and desiccate (Plate 10.13). Leaf shredding by hail is probably more damaging than leaf tearing by wind. Additional damage is caused when hail accumulates in the funnel leaf and the upper petioles. Plant tissue that is in direct contact with ice freezes and desiccates within a few days. Hail also damages fruit, the points of impact being characterized by sunken scars (Plate 10.14). Unprotected fruit is totally damaged and, in extreme cases, fingers can break (Kuhne and Green, 1980). Polyethylene sleeves (especially when more than 0.1 mm in thickness) give fruit partial protection (Kuhne and Green, 1980; Eckstein and Fraser, 1996).

After hailstorms, each plant should be evaluated for damage. Plants with no leaf area and severely damaged bunches should be cut down. Good-quality, mature bunches from severely affected plants should be harvested immediately. Undamaged bunches close to harvest may complete their development, even when most of the leaves on the plant are severely damaged. For plants that have not bunched, the damaged leaves should be cut and removed from plants in order to accelerate new growth and development from reserves stored in the pseudostem. If four or more new leaves emerge after the hailstorm, a commercial bunch may be obtained. If flowering is expected before a sufficient number of new leaves emerge, the plant should be sacrificed for the sake of the follower (Kuhne and Green, 1980; Robinson *et al.*, 1992; Robinson, 1993, 1996).

Heat Stress

Banana can stand high temperatures and grow in hot and semi-arid areas, such as Egypt, Israel and Western Australia, given an adequate water-supply. Banana leaf growth stops at an ambient temperature of 38–40°C (Stover and Simmonds, 1987; Robinson, 1996). Photosynthesis decreases at temperatures above 35°C (Eckstein and Robinson, 1995b) and leaf burning occurs at 37°C (Turner and Lahav, 1983).

A study of the energy balance of the banana in tropical lowlands showed that the temperature of a leaf in a horizontal position exposed to direct sunlight may reach 12°C above ambient temperature. Under conditions of reduced transpiration, as a consequence of limited water-supply or after artificial stomata closure using petroleum jelly, temperatures rise about 20°C above air temperature (Stoutjesdijk, 1970). Taylor and Sexton (1972), working with *Heliconia latispatha*, recorded a temperature of 46°C in a leaf exposed to direct sun during the dry season, but a temperature of only 40°C in a leaf with a torn lamina. They demonstrated in *Strelizia nicolai*, another relative of the banana, under controlled conditions, that, when the leaf is exposed during a short time to 47.5°C, it will turn necrotic a few days later. This is, therefore, probably the lethal temperature for banana leaves. They concluded that, in the *Musaceae*, tearing of the leaf lamina along the secondary veins is a natural process, reducing the boundary-layer resistance, increasing the transpiration rate and helping to remove excess heat.

The last unfurled leaf is the most sensitive to overheating (Plate 10.15). In this leaf, the stomata are still not fully functioning and therefore water conductivity and transpiration are limited (Eckstein and Robinson, 1995a). The last expanded leaf shows burns at an ambient temperature of 42–43°C, even when the water-supply is plentiful (Robinson, 1993). The cigar or funnel leaf is also very sensitive. Under high temperatures, it fails to unfurl normally and necrotic spots caused by sunburn are observed on the side facing the sun. The cigar leaf is most affected by heat during periods of fast growth or when there is insufficient hardening. Under these conditions, it is often distorted and sometimes bends horizontally. Plants close to bunch emergence are especially vulnerable to heat stress, which results in thin and degenerated fingers (Plate 10.16). When banana corms were heated for 6 weeks to 32–37°C, chlorotic leaves with a deformed lamina or with no lamina at all emerged (Anon., 1976). Similar symptoms have been observed in the field after a heat wave (Stover and Simmonds, 1987).

Plate 10.10. Advanced stage of leaf chilling resulting in a loss of chlorophyll. Note necrosis of the leaf tissue next to the petiole because of direct exposure to the sky (photo: Y. Israeli, JVBES).

Plate 10.11. Fingers with water-soaked areas after frost damage to cell membranes (photo: Y. Israeli, JVBES).

Heat may also damage the underground parts of the plant. Israeli *et al.* (1979) grew banana plants in nutrient solutions at mean temperatures of 16.2, 25.0, 26.2 and 33.6°C. At the highest temperature treatment, dry matter and potassium content declined in the corm and roots and the root/shoot dry-matter ratio decreased. Ramcharan *et al.* (1995) grew banana plants in pots for 6–10 weeks under controlled conditions, with root zone temperatures of 28, 33, 38 and 43°C (6 h daily). As temperatures increased, the dry weight of the roots decreased. In this experiment, leaf width decreased at 38 and 43°C. At 43°C, roots were brown and suberized and lacked root tips. Lack of root tips may affect the hormone balance of the plant and thus the distribution of photosynthates.

Ingram and Ramcharan (1988) investi-

Plate 10.12. Polyethylene shelter used to protect banana from frost in Tenerife, Canary Islands (photo: Y. Israeli, JVBES).

Plate 10.13. Leaf shredding as a result of hail (photo: Y. Israeli, JVBES).

gated the physiological response of banana roots exposed to a range of temperatures between 30 and 60°C. Damage to cell membranes was measured by electrolyte leakage and was first detected at 45°C. At 48°C, 225 ± 36 min exposure was needed to cause 50% of the maximal electrolyte leakage, while, at 57°C, 7 ± 4 min were needed. Heat injury, therefore, is an interaction between time and temperature.

Lahav *et al.* (1993) measured the effect of polyethylene soil covers on soil temperatures in a banana plantation in the Jordan Valley, Israel. During hot afternoons, the covers raised the soil temperature from 30.6 to 37.9°C. This rise caused partial stomata closure, increased leaf temperatures from 35.7 to 37°C and decreased photosynthesis. The number of hands per bunch was also reduced and bunch weight decreased by 28%.

Above-optimum temperatures at the apical meristem during the time of bunch differentiation result in bunches with few hands (Stoler, 1962). The number of hands per bunch declines rapidly, from 11.7 hands at 26.5°C to 9.4 hands at 30°C. Stapleton and Garza-Lopez (1988) covered soil around banana in Mexico with transparent and black polyethylene and recorded soil temperatures. At a depth of 20 cm, the maximum temperature was 44°C under transparent polyethylene, 40°C under black polyethylene and 38°C in uncovered soil. Banana plants in the covered rows were stunted and chlorotic and survived for only a few weeks. Soil covers should, therefore, be used with care in banana plantations.

Another aspect to be considered is the effect of heat stress on fruit. Bananas maturing under high temperatures do not ripen normally and they soften quickly, the pulp having less aroma and taste (Tang and Liu, 1993). High temperatures in ripening rooms delay the normal decrease in chlorophyll concentration or yellowing (Seymour *et al.*, 1987) and result in oversoft, green fruit, called 'cooked fruit' (Thompson *et al.*, 1972; Rippon and Trochoulias, 1976). Increasing temperatures in ripening rooms from 20°C to 35°C accelerates the softening processes, ethylene production and the accumulation of sugars, but above 30°C the degreening process is inhibited. At 40°C, a total cessation of the ripening process was recorded (Yoshioka *et al.*, 1978). This is related to decreased protein synthesis, decreased acid phosphatase activity and suppression of isoamyl acetate synthesis (Yoshioka *et al.*, 1980a, b, 1982).

High temperatures in the field may interfere with fruit development. Stover and Simmonds (1987) described a heat wave in Panama where daily maximum temperatures of 33–35°C, high radiation and a limited amount of rain caused changes in latex vessels in the peel. Latex globules exploded and cell walls and cells turned reddish brown. This resulted in symptoms of under-peel discoloration, similar to those caused by chilling. Under-peel discoloration caused by heat stress is quite common in Western Australia (Robinson, 1993). The internal breakdown of fruit after ripening has been linked to temperatures of 45°C at bunch emergence (Robinson, 1993).

Radiation Damage

The natural habitat of the wild banana is in open spaces in the tropical forest. Under shade, the banana declines rapidly (Simmonds, 1962). However, in many banana-growing areas in tropical lowlands, sunlight hours are limited to 4–5 h day^{-1}, because of cloudiness. Nevertheless, very good crops can be obtained under such conditions. Stover (1984) showed that 14–18% of incident radiation normally passes through the canopy to reach the ground in plantations in Central America. When plant density was increased so that only 10% of the light reached the ground, plants affected by reduced radiation could be found. Israeli *et al.* (1995) showed that shading slows the rate of growth of plants and time of appearance of the first ratoon. The visible symptoms of heavy shade are dark green leaves (which is the result of an increase in the concentration of chlorophyll), reduced growth, reduced height, 'rosetting' of leaves, a reduction in the size and number of suckers, a reduction in bunch size and small, defective fingers.

Excessive radiation can also bring problems. Exposed parts of the peduncle and fingers, especially those on the upper hand of the bunch, can yellow and blacken due to sunburn (Plate 10.17). The damage is more severe in bunches covered with transparent polyethylene. Bunch covers made from non-transparent, white polyethylene significantly reduce the effects of strong sunlight, as does paper placed over the upper hands (Rippon and Turner, 1970; Kuhne, 1980b; Ke *et al.*, 1981). Sometimes, sunburn can be prevented by folding a leaf over the bunch. In plantations, most bunches are protected by the leaf canopy, but those on plants next to roadways are vulnerable. As it is believed that banana plants gradually become acclimatized to high radiation, severe sunburn may result from the sudden exposure of a protected bunch to direct sunlight.

Wade *et al.* (1993) found that ultraviolet (UV-C) radiation (but not UV-A) induced bronzing spots on the fruit. When white light was applied immediately afterwards, its effect was much reduced.

Another type of radiation damage can occur in subtropical areas during the winter months, when strong sunlight during the day is often combined with low night temperatures, low humidity, closed leaf stomata and an only partially active plant root system. Under these conditions, the temperature of a leaf exposed to direct sunlight increases significantly, because transpiration is minimal. This results in the leaf bleaching due to chlorophyll destruction. Symptoms are commonly found on the half of the leaf lamina facing the afternoon sun (Robinson, 1996).

Drought Stress

Milburn (1993), Turner (1994, 1995) and Robinson (1996) have reviewed the physiological and agronomical changes that take place in banana as a result of water deficit. Banana has a high water consumption (Ghavami, 1972; Sarraf and Bovee, 1973; Meyer and Schoch, 1976; Israeli and Nameri, 1987; Turner, 1987; Robinson and Alberts, 1989) and high sensitivity to water stress (Shmueli, 1953; Aubert, 1968; Doorenbos and Kassam, 1979; Eckstein and Robinson, 1996). It also grows much better under irrigation (Arscott *et al.*, 1965a, b; Trochoulias, 1973; Krishnan and Shanmugavelu, 1980; Meyer, 1980; Holder and Gumbs, 1983b; Robinson and Alberts, 1986). Banana needs an ample supply of water, because it has a large vegetative mass (a significant proportion of which is leaf tissue), a high hydration level, broad, thin leaves, which absorb much energy and transpire water easily, and a sparse and shallow root system (Shmueli, 1953; Olsson *et al.*, 1984; Robinson, 1985). Recently, Eckstein (1994) showed that the first symptoms of water stress (partial closure of stomata and reduction in carbon fixation) in the field are expressed 4 days after cessation of irrigation. Twelve days without water resulted in a 79% reduction in CO_2 assimilation, caused mainly by stomata closure and partially by reduced photosynthesis. The reduced assimilation results in reduced growth and production. The first symptom in the crop, even when no other symptoms are observed, can be a reduced finger length. If the fruit is affected during its initial stages of development, several weeks or even months may pass between the water stress and the emergence of the affected bunch.

The effect of a lack of water is especially noticeable during periods of high temperatures and otherwise optimal growing conditions (Arscott *et al.*, 1965a; Robinson and Green, 1981). Water stress slows down the rate of fruit growth and at the same time stimulates maturation (Arscott *et al.*, 1965a; Srikul and Turner, 1995). The fruit will be affected even when a limited period of water stress occurs between bunch emergence and harvest (Hegde and Srinivas, 1989).

The first symptoms of water stress are usually expressed in leaves as a change to a pale green colour and the continuous folding of the margins when temperatures are high. Later, leaves become yellowish green, followed by necrosis at the margins (Plate 10.18). The tissue on both sides of the midrib, where the lamina is more exposed to the sun, also becomes necrotic. These symptoms are much more pronounced in leaves sprayed with oil to control leaf-spot disease. During continuous water stress, the petioles are crowded at the top of the pseudostem and the bunch is choked on emergence, similar to the 'choke throat' caused by low temperatures (Kuhne and Green, 1980; Holder and Gumbs, 1983a). Water stress also affects bunch shape, such as non-uniform distances between hands, large differences in finger length between the upper and the lower hands and extreme curving of fingers towards the bunch axis or in the opposite direction to form an open hand (Holder and Gumbs,

Plate 10.14. Hail damage to uncovered fruit (photo: Y. Israeli, JVBES).

Plate 10.15. Heat stress damage to a young banana leaf in the Jordan Valley, Israel (photo: Y. Israeli, JVBES).

Plate 10.16. Thin fingers and premature ripening resulting from high temperatures during flower differentiation. Only the first hands, usually the largest, were affected, since they were formed in unfavourable conditions (photo: Y. Israeli, JVBES).

Plate 10.17. Symptoms of sunburn injury on fruit of 'Grand Nain' (AAA, Cavendish subgroup) in St Vincent, Windward Islands (photo: D.R. Jones, SVBGA).

Plate 10.18. Leaf burn resulting from water stress at Santa Marta, Colombia (photo: Y. Israeli, JVBES).

1983a). Leaves can also age earlier, followed by leaf folding and desiccation (Wardlaw, 1961a; Kallarackal *et al.*, 1990). Later, plants become stunted and small, with a reduced leaf area. Finally, growth ceases and plants have small, malformed bunches (Wardlaw, 1961a; Stover and Simmonds, 1987; Robinson, 1996). Hill *et al.* (1992) reported up to 35% unmarketable bunches due to pseudostem breakage and snapping of the peduncle in Carnarvon, Western Australia, from October to March, a period of strong winds, high temperature and low humidity, which cause extreme moisture stress. Bunches are also frequently lost in the Windward Islands during the dry season, due to the breakage of drought-weakened pseudostems (Plate 10.19).

The B genome of *M. balbisiana* increases relative resistance to drought. Banana clones with the AAB genome are more resistant than those with the AAA genome. Cultivars in the ABB group are the most drought-resistant (Stover and Simmonds, 1987; Valmayor, 1990). Ekanayake *et al.* (1995) showed that differences occur between clones in the same group according to their ability to close their stomata under high vapour-pressure deficit during the afternoon.

Within the AAA group, the Cavendish subgroup seems to be more sensitive than 'Gros Michel' (Stover and Simmonds, 1987). Variations also exist within the Cavendish subgroup. 'Grand Nain' seems to be sensitive to water stress, expressed as choking and fruit deformation, in climatic conditions where 'Williams' is symptomless. 'Williams' on the other hand is more sensitive than 'Robusta', showing a 14% reduction in the number of hands under water stress, as compared with 7% in 'Robusta' (Holder and Gumbs, 1983a). Ploidy level may also influence reaction to water stress. Triploids were found to be more susceptible than diploids and tetraploids (Ramadass and Sheriff, 1993).

In banana-growing areas throughout the world, drought is a major factor limiting production (Shou, 1990; Valmayor, 1990; Akomeah *et al.*, 1995). Most new commercial plantations have irrigation, even in areas where annual precipitation is high. In North Queensland, Australia, for example, annual rainfall is 2000–4000 mm, with some months of only 100 mm. Irrigation during the dry months increases production by 15–20% and improves fruit quality (Daniells, 1984; Behncken, 1990).

Excess-water Stress

Too much water is as big a problem as too little water in many banana plantations in tropical lowlands. Drainage canals and ditches are prepared before planting in order to avoid waterlogging. Ideally, such drainage systems should allow excess water to be channelled away as early as possible (2 h after rain) and lower the water-table to below 1 m about 24 h later (Stover and Simmonds, 1987). In some areas, drainage pumps are required (Lassoudier and Martin, 1974). Drainage problems are more severe in clay soils with low hydraulic conductivity. If planting in such soils is unavoidable, drainage ditches should be quite close to one another.

Growth ceases in plants flooded for 48 h. First, yellowing symptoms become apparent in leaves at this time. Longer periods of flood (72–96 h) result in irreversible damage (Plate 10.20). Flood damage is more severe when water is standing than when water is flowing. This may be related to the lower oxygen content of standing water.

Irizarry *et al.* (1980) showed that lowering the water-table from 12 to 36 cm increased production in the plantain cultivar 'Maricongo' from 5.6 to 37.8 t ha^{-1} and that this was correlated with an increase in root production. Ghavami (1976), working with the Cavendish cultivar 'Valery', found that damage first became evident when the water-table rose above 60 cm and was more significant at the second and third cycles, probably because excess water not only caused root necrosis but also prevented root formation.

Both Ghavami (1976) and Holder and Gumbs (1983c) related significant yield decrease to rises in the water-table. Floods associated with tropical storms may cause more damage than stormy winds (Stover and Simmonds, 1987; Tang and Liu, 1993).

In addition to the direct damage caused by waterlogging, flood waters also weaken the resistance of plants to diseases such as Fusarium wilt (Stover and Malo, 1972; Valmayor, 1990; Shivas *et al.*, 1995).

The construction of ridges is an efficient method to reduce the damage to banana plants in areas of high water-table (Avilan *et al.*, 1982; Daniells, 1984). In Thailand, banana is grown successfully on raised embankments in redundant paddy rice fields (Plate 10.21).

Lightning Injury

This is a rare disorder, but it has been reported from Jamaica, Surinam (Maas, 1967) and Honduras. Symptoms include the yellowing, browning and dieback of the leaf tip. The midrib also collapses about 20–30 cm from the leaf tip, causing it to hang down. These collapsed yellow leaf tips or yellow 'flags' attract attention to otherwise normal green plants in plantations. A purple-brown or reddish discoloration is present in the midrib from the point of leaf collapse to the pseudostem. The entire leaf eventually turns brown and collapses. Discoloration advances from the collapsed leaves along the midrib and petiole into the pseudostem and leaf sheaths. The inner leaf sheaths show brown, elongated lesions up to 2.5 cm in length on the outer epidermis. The vascular tissue may turn brown. The pseudostem begins to rot from the top downwards and collapses as a soft rotten mass, but the rhizome is not affected. The condition appears in localized areas of 0.5–2 ha and affects up to 50% of the plants. Symptoms appear rapidly over a period of about a week and then gradually cease after 10–14 days. Badly affected plants should be cut down and replanted.

Under-peel Discoloration

Under-peel discoloration consists of a reddish-brown streaking in the vascular tissue just below the epidermis of the fruit (Plate 10.22). Symptoms are visible when the epidermis is peeled back with a knife. Where under-peel discoloration is present, latex flow is reduced or stopped and the peel tends to adhere strongly to subepidermal layers.

Under-peel discoloration usually indicates that the fruit was subjected to chilling temperatures. Less frequently, it can also be a symptom of fruit desiccation, ageing at low humidity, with or without high temperatures, and bruising. Fruit with under-peel discoloration will ripen more slowly than normal fruit and will have a dull yellow colour and, if the discoloration is severe, fruit may not yellow, but may become dull grey. Also, the placenta may be hardened. Eating quality is affected only when chilling injury is severe.

Under-peel discoloration from chilling may occur in the field when temperatures drop below 12.8°C for several hours as a result of cold fronts. More commonly, it occurs in transit, due to improper refrigeration or ventilation in ships and vans. Fruit exposed to winter temperatures in

Table 10.2. Time–temperature relationship to chilling injury of 'Valery' (AAA, Cavendish subgroup) and 'Cocos' (AAA, a dwarf 'Gros Michel') fruit.

		Number of hours of exposure to temperature to induce		
Cultivar	Temperature (°C)	Light injury*	Moderate injury†	Severe injury‡
'Valery'	13.3	100	350–370	650
	12.8	10–20	70	70
	12.2	5–10	24–26	56–59
	11.7	1	3–5	25–27
	11.1	1	1	18–20
'Cocos'	11.1	4–6	20–22	46–48
	10.6	1	1	33–35
	10.0	1	1	1

*0–10 lesions in each cm^2.
†10–25 lesions in each cm^2.
‡25 lesions or over in each cm^2.

Plate 10.19. Breakage of a drought-weakened pseudostem of 'Robusta' (AAA, Cavendish subgroup) in St Vincent, Windward Islands. Propping can help prevent such losses (photo: D.R. Jones, SVBGA).

Plate 10.20. Yellowing of leaves of 'Kluai Namwa' (ABB, syn. 'Pisang Awak') due to floods along the Mekong River in Thailand (photo: D.R. Jones, INIBAP).

non-tropical areas while in transit will also be chilled.

Chilling in transit is avoided by maintaining strict control over the temperature of inflowing air. Air temperatures should not be below 13.2°C for fruit of cultivars in the Cavendish subgroup (AAA) and 12.2°C for fruit of 'Gros Michel' (AAA). Chilling is most likely to occur on fruit stored near air delivery ducts. Periods of storage beyond 14 days can also lead to under-peel discoloration, even at 13.3°C (Table 10.2). Under-

Plate 10.21. Irrigation of banana growing on embankments in old paddy rice fields near Bangkok, Thailand (photo: D.R. Jones, INIBAP).

Plate 10.22. Under-peel discoloration caused by chilling temperatures. The latex ducts are discoloured and such fruit ripens to a dull greyish-yellow colour (photo: R.H. Stover).

peel discoloration associated with long storage or transit periods is probably related to desiccation, as well as low temperature.

Von Loesecke (1950), Wardlaw (1961b), Palmer (1971), Mattei (1978) and Marriott (1980) have reviewed the physiology of postharvest chilling of banana fruit. Recently, the use of a colorimeter has enabled changes in peel colour resulting from chilling, either before or after ripening, to be evaluated objectively and quantitatively (Hewage *et al.*, 1996). Thus chilling damage can now be identified even before external symptoms are visible and after very moderate chilling temperatures (11°C), resulting in very light chilling symptoms. Another parameter used to assess damage is electrical conductivity of the peel (Deullin, 1980).

Chilling injury is related to temperature and exposure time. Genotypes also vary in their response to different temperatures (Marriott and New, 1975; Broughton and Wu, 1979; Mohammed and Campbell, 1993). Fruit of Cavendish cultivars is more sensitive than fruit of 'Gros Michel', its shorter sport 'Cocos' (Table 10.2) and cultivars in the AAB and ABB genomic groups (Satyan *et al.*, 1992).

Chilling injury is less if the chilling process is gradual, when oxygen is reduced and relative humidity increases in the ambient atmosphere (Mattei, 1978). Rate of under-peel discoloration may decrease when the fruit is protected by vegetable oil, mineral oil or dimethylpolysiloxane solution. Treating fruit with these chemicals enabled fruit to be stored at 9°C for 24 h with no significant damage (Jones *et al.*, 1978).

Chilling injury is related to a decrease in activity of peroxidases (Toraskar and Modi, 1984), oxidation of phenols, especially dopamine (Abd El-Wahab and Nawwar, 1977), and an increase in the level of α-farnesene, which precedes the external chilling symptoms (Wills *et al.*, 1975). A possible effect of chilling on phase transition in membrane structural lipids (Lyons, 1973) was demonstrated by the identification of saturated lipid acids, typical to sensitive plants, in banana (Roughan, 1985; Kernick and Bishop, 1986). Also, chilling reduces unsaturated lipid acids, especially linolenic acid (Wang and Gemma, 1994), and changes membrane permeability (Gemma *et al.*, 1994).

REFERENCES

Abd El-Wahab, F.K. and Nawwar, M.A.M. (1977) Physiological and biochemical studies on chilling-injury of banana. *Scientia Horticulturae* 7, 373–376.

Ahmad, K.U. and Quasem, A. (1975) Effect of cyclone on some banana varieties. *Bangladesh Horticulture* 3, 48–49.

Ahn, S.B., Cho, W.S., Lim, S.E., Kim, T.J., Lee, M.H. and Choi, K.M. (1990) Banana insect pest species and their damage in a vinyl house on Cheju Island. *Korean Journal of Applied Entomology* 29, 6–13.

Akomeah, F., Ohemeng-Appiah, L. and Adomako, D. (1995) Country reports: Ghana. *MusAfrica* 7, 11–13.

Allen, R.N., Belinda Dettmann, E., Johns, G.G. and Turner, D.W. (1988) Estimation of leaf emergence rates of bananas. *Australian Journal of Agricultural Research* 39, 53–62.

Anon. (1976) Etude des effets de la température sur la croissance. Station de Neufchateau (Guadeloupe). *Fruits* 31, 270–271.

Arscott, T.G., Bhangoo, M.S. and Karon, M.L. (1965a) Irrigation investigations of the Giant Cavendish banana. iii. Banana production under different water regimes and cultivation practices. *Tropical Agriculture (Trinidad)* 42, 210–216.

Arscott, T.G., Bhangoo, M.S. and Karon, M.L. (1965b) Irrigation investigations of the Giant Cavendish banana – a note on the relationship between bunch weight and quantity of supplemented irrigation. *Tropical Agriculture (Trinidad)* 42, 367–368.

Aubert, B. (1968) Etude préliminaire des phénomènes de transpiration chez le bananier. *Fruits* 23, 357–381.

Aubert, B. (1971) Action du climat sur le comportement du bananier en zones tropicale et subtropicale. *Fruits* 26, 175–187.

Avilan, R.L., Meneses, R.L. and Sucre, R.E. (1982) Banana root distribution under different soil management systems. *Fruits* 37, 103–110.

Bagdonas, A., Georg, J.C. and Gerber, J.F. (1978) *Techniques of Frost Prediction and Methods of Frost and Cold Protection.* World Meteorological Organization, Technical Note No.157, Geneva, 160 pp.

Barak, A. and Israeli, A. (1989) *Means for Frost Protection.* Special Publication of the Meteorological Service and Agricultural Extension Service, Ministry of Agriculture, Tel Aviv, Israel, 30 pp. (in Hebrew).

Behncken, G.M. (1990) The banana industry in Australia. In: Valmayor, R.V. (ed.) *Banana and Plantain R & D in Asia and the Pacific: Proceedings of a Regional Consultation on Banana and Plantain R & D Networking, Manila and Davao, 20–24 November 1989.* INIBAP, Montpellier, France, pp. 22–40.

Blake, J.R. and Peacock, B.C. (1966) Some aspects of the abnormal fruit of November-flowering banana. *Queensland Journal of Agricultural and Animal Sciences* 23, 449–452.

Broughton, W.J. and Wu, K.F. (1979) Storage conditions and ripening of two cultivars of banana. *Scientia Horticulturae* 10, 83–93.

Cann, H.J. (1964) How cold weather affects banana growing in N.S.W. *Agricultural Gazette of New South Wales* 75, 1012–1019.

Cevik, B., Kaska, N., Tekinel, O. and Dinc, U. (1984) The effects of drip and basin irrigation on growth, yield and quality of bananas grown in greenhouse conditions with various mulching materials. *Doga Bilim Dergisi, D2 Tarim ve Ormancilik* 8, 265–275 (in Turkish).

Choukr-Allah, R. (1990) Problems of greenhouse winter production in Morocco. *Acta Horticulturae* 263, 39–46.

Colombo, A., Campo, G. and Calabro, M. (1993) Control of *Tetranychus urticae* on banana in the greenhouse using predators. *Informatore Fitopatologico* 43(12), 56–61.

Cottin, R., Melin, P. and Ganry, J. (1987) Modelling of banana production: the effect of some parameters in Martinique. *Fruits* 42, 691–701, 763.

Daniells, J.W. (1984) The banana industry in north Queensland. *Queensland Agricultural Journal,* September–October, 282–290.

Daniells, J. and Bryde, N. (1993) Yield and plant characteristics of seven banana hybrids from Jamaica and Honduras in north Queensland. *Infomusa* 2(1), 18–20.

Daniells, J. and Bryde, N. (1995) Semi-dwarf mutant of Yangambi Km5. *Infomusa* 4(2), 16–17.

Daniells, J., Pegg, K., Searle, C., Smith, M., Whiley, T., Langdon, P., Bryde, N. and O'Hare, T. (1995) Goldfinger in Australia: a banana variety with potential. *Infomusa* 4(1), 5–6.

Daudin, J. (1955) Conseils pratiques à un planteur de bananes Martiniquais. *Bulletin de l'Institut Français Argumes Coloniaux* 13, 55.

Deullin, R. (1980) Electrical conductivity of the skin of the banana, a physical characteristic which can be used for improved assessment of fruit development. *Fruits* 35, 273–281.

Doorenbos, J. and Kassam, A.H. (1979) Yield response to water: banana. In: *FAO Irrigation and Drainage Paper No. 33.* FAO, Rome, pp. 73–76.

Eckstein, K. (1994) Physiological responses of banana (*Musa* AAA; Cavendish subgroup) in the subtropics. PhD thesis, Institut für Obstbau und Gemusebau, Universitat Bonn, Germany, 203 pp.

Eckstein, K. and Fraser, C. (1996) Effect of hail on banana fruit grown with or without bunch covers. *Citrus and Subtropical Fruit Research Institute Information Bulletin* 284, 1–2.

Eckstein, K. and Robinson, J.C. (1995a) Physiological responses of banana (*Musa* AAA; Cavendish subgroup) in the subtropics. 1. Influence of internal plant factors on gas exchange of banana leaves. *Journal of Horticultural Science* 70, 147–156.

Eckstein, K. and Robinson, J.C. (1995b) Physiological responses of banana (*Musa* AAA; Cavendish subgroup) in the subtropics. 2. Influence of climatic conditions on seasonal and diurnal variations in gas exchange of banana leaves. *Journal of Horticultural Science* 70, 157–167.

Eckstein, K. and Robinson, J.C. (1996) Physiological responses of banana (*Musa* AAA; Cavendish subgroup) in the subtropics. 6. Seasonal responses of leaf gas exchange to short-term water stress. *Journal of Horticultural Science* 71, 679–692.

Eckstein, K., Robinson, J.C. and Fraser, C. (1996) Physiological responses of banana (*Musa* AAA; Cavendish sub-group) in the subtropics. 5. Influence of leaf tearing on assimilation potential and yield. *Journal of Horticultural Science* 71, 503–514.

Ekanayake, I.J., Ortiz, R. and Vuylsteke, D.R. (1995) Physiological factors in drought tolerance of various *Musa* genotypes. *IITA Research* 11, 7–10.

Fahn, A., Klarman Kislev, N. and Ziv, D. (1961) The abnormal flower and fruit of May-flowering Dwarf Cavendish bananas. *Botanical Gazette* 123, 116–125.

Freeman, B. (1973) Artificial windbreaks. *Agricultural Gazette of New South Wales* 84, 176–180.

Galan-Sauco, V. (1992) *Tropical Fruit Crops in the Subtropics. II. Banana.* Ediciones Mundi-Prensa, Madrid, 173 pp.

Galan-Sauco, V., Garcia-Samarin, J. and Carbonell, E. (1984) Study of the practice of disbudding and of the phenology of banana (*Musa acuminata* Colla (AAA), cv. Pequena enana) on the island of Tenerife. III. Phenology of banana in Tenerife. *Fruits* 39, 595–605.

Galan-Sauco, V., Cabrera-Cabrera, J. and Hernandez-Delgado, P.M. (1992) Phenological and production differences between greenhouse and open-air bananas (*Musa acuminata* Colla AAA cv. Dwarf Cavendish) in Canary Islands. *Acta Horticulturae* 296, 97–111.

Ganry, J. (1973) Etude du développement du systeme foliaire du bananier en fonction de la température. *Fruits* 28, 499–516.

Ganry, J. and Meyer, J.P. (1975) Recherche d'une loi d'action la température sur la croissance des fruits du bananier. *Fruits* 30, 375–392.

Gat, Z. (1984) Topoclimatology in Israel: methods and applications. *Alon Hanotea* 38, 421–427 (in Hebrew).

Gemma, H., Matsuyama, Y. and Wang, H.G. (1994) Ripening characteristics and chilling injury of banana fruit. I. Effect of storage temperature on respiration, ethylene production and membrane permeability of peel and pulp tissues. *Japanese Journal of Tropical Agriculture (Trinidad)* 38, 216–220.

Ghavami, M. (1972) Determining the water needs of the banana plant. *Transactions of the ASAE* 16, 598–600.

Ghavami, M. (1976) Banana plant response to water table levels. *Transactions of the ASAE* 19, 675–677.

Gilead, M. and Rosenan, N. (1957) Micro-climatological observations in a banana plantation. *Indian Journal of Meteorology and Geophysics* 8, 1–7.

Gowen, S.R. (1993) Yield losses caused by nematodes on different banana varieties and some management techniques appropriate for farmers in Africa. In: *Biological and Integrated Control of Highland Banana and Plantain Pests and Diseases: Proceedings of a Research Coordination Meeting, Cotonou, Benin, 12–14 November 1991.* IITA, Ibadan, Nigeria. pp. 199–208.

Green, G.C. and Kuhne, F.A. (1969) Growth of the banana plant in relation to winter air temperature fluctuations. *Agroplantae* 1, 157–162.

Green, G.C. and Kuhne, F.A. (1975) Estimating the state of maturation of a banana bunch from meteorological and supporting data. *Agrochemophysica* 7, 27–32.

Hegde, D.M. and Srinivas, K. (1989) Effect of soil moisture stress on fruit growth and nutrient accumulation in banana cultivar 'Robusta'. *Fruits* 44, 135–138.

Hewage, K.S., Wainwright, H. and Wijeratnam, R.S.W. (1996) Quantitative assessment of chilling injury in bananas using a colorimeter. *Journal of Horticultural Science* 71, 135–139.

Hill, T.R., Bissell, A.R.J. and Burt, J.R. (1992) Yield, plant characteristics, and relative tolerance to bunch loss of four banana varieties (*Musa* AAA Group, Cavendish subgroup) in the semi-arid subtropics of Western Australia. *Australian Journal of Experimental Agriculture* 32, 237–240.

Holder, G.D. and Gumbs, F.A. (1983a) Agronomic assessment of the relative suitability of the banana cultivars 'Robusta' and 'Giant Cavendish' (Williams hybrid) to irrigation. *Tropical Agriculture (Trinidad)* 60, 17–24.

Holder, G.D. and Gumbs, F.A. (1983b) Effects of irrigation on the growth and yield of banana. *Tropical Agriculture (Trinidad)* 60, 25–30.

Holder, G.D. and Gumbs, F.A. (1983c) Effects of waterlogging on the growth and yield of banana. *Tropical Agriculture (Trinidad)* 60, 111–116.

Hord, H.H.V. and Spell, D.P. (1962) Temperature as a basis for forecasting banana production. *Tropical Agriculture (Trinidad)* 29, 219–223.

Ingram, D.L. and Ramcharan, C. (1988) 'Grande Naine' banana and *Dracaena marginata* 'Tricolor' root cell membrane heat tolerance. *Fruits* 43, 29–33.

Irizarry, H., Silva, S. and Vicente-Chandler, J. (1980) Effect of water table level on yield and root system of plantains. *Journal of Agriculture of the University of Puerto Rico* 64, 33–36.

Israeli, Y. (1976) The effect of air temperature at the time of floral initiation on the number of female hands in banana in the Jordan Valley. *Alon Hanotea* 31, 682–685 (in Hebrew).

Israeli, Y. and Blumenfeld, A. (1985) *Musa*. In: Halevy, A. (ed.) *Handbook of Flowering*, Vol. 3. CRC Press, Boca Raton, Florida, pp. 390–409.

Israeli, Y. and Lahav, E. (1986) Banana. In: Monselise, S.P. (ed.) *Handbook of Fruit Set and Development*. CRC Press, Boca Raton, Florida, pp. 45–73.

Israeli, Y. and Nameri, N. (1987) Seasonal changes in banana plant water use. *Water and Irrigation Review (Israel)*, January, 10–14.

Israeli, Y., Ziv, D. and Kafkafi, U. (1979) The rate of leaf growth, water uptake and mineral content of banana plants grown in nutrient solution at various temperatures. *Alon Hanotea* 34, 39–45 (in Hebrew).

Israeli, Y., Nameri, N., Gat, Z. and Burd, P. (1988) Relating banana flowering distribution in the Jordan Valley (Israel) to climate and phenology. *Agricultural and Forest Meteorology* 43, 109–119.

Israeli, Y., Plaut, Z. and Schwartz, A. (1995) Effect of shade on banana morphology, growth and production. *Scientia Horticulturae* 62, 45–56.

Janick, J. and Ait-Oubahau, A. (1989) Greenhouse production of banana in Morocco. *Hortscience* 24, 22–27.

Jones, R.L., Freebairn, H.T. and McDonnell, J.F. (1978) The prevention of chilling injury, weight loss reduction, and ripening retardation in banana. *Journal of the American Society for Horticultural Science* 103, 219–221.

Kallarackal, J., Milburn, J.A. and Baker, D.A. (1990) Water relations of the banana. III. Effects of controlled water stress on water potential, transpiration, photosynthesis and leaf growth. *Australian Journal of Plant Physiology* 17, 79–90.

Kalma, J.D., Laughlin, G.P., Caprio, J.M. and Hamer, P.J. (1992) *The Bioclimatology of Frost, Its Occurrence, Impact and Protection*. In: Advances in Bioclimatology 2, Springer-Verlag, New York, 144 pp.

Ke, L.S., Weng, M.H., Wangin, S.L. and Ke, D.F. (1981) Effect of PE bunch covers with different blue colour gradings on sunburning, appearance and ripening of banana fruit. *Journal of the Chinese Society for Horticultural Science* 27, 177–186.

Kernick, J.R. and Bishop, D.G. (1986) Phosphatidylglycerol and sulphoquinovosyldiacylglycerol in leaves and fruits of chilling-sensitive plants. *Phytochemistry* 25, 1293–1295.

Krambias, A., Zyngas, J.P. and Shiakides, T. (1973) Lepidopterous pest on banana. *FAO Plant Protection Bulletin* 21, 64–66.

Krishnan, B.M. and Shanmugavelu, K.G. (1980) Studies on water requirements of banana cv. Robusta: total consumptive use and water use efficiency. *Mysore Journal of Agricultural Sciences* 14, 27–31.

Kuhne, F.A. (1980a) Plant selection in bananas. *Farming in South Africa. Bananas* C4, 1–3.

Kuhne, F.A. (1980b) The use of bunch covers in banana cultivation. *Farming in South Africa. Bananas* G5, 1.

Kuhne, F.A. and Green, G.C. (1980) Cultivation of bananas under unfavourable weather conditions. *Farming in South Africa. Bananas* B3, 1–3.

Kuhne, F.A., Kruger, J.J. and Green, G.C. (1973) Phenological studies of the banana plant. *Citrus and Subtropical Fruit Journal* 472, 12–16.

Lahav, E., Israeli, Y., Weizman, Z., Zalach, Y. and Meirav, N. (1993) The effect of soil polyethylene covers on bananas. *Alon Hanotea* 47, 494–502 (in Hebrew).

Lassoudiere, A. and Martin, P. (1974) Drainage problems in the organic soils of banana plantations in Agneby, Ivory Coast. *Fruits* 29, 255–266.

Liang, L.F., Wang, Z.H., Zhou, B.Y. and Huang, H.B. (1994) Effects of low temperature and paclobutrazol on the activities and isozymes of peroxidase in banana leaves. *Journal of South China Agricultural University* 15, 65–70.

Lomas, J., Gat, Z., Borsuk, Z. and Raz, Z. (1989) *Frost Atlas of Israel (10 Regional Maps)*. Survey of Israel, Israel Meteorological Service, Bet-Dagan, Israel.

Lorch, J. (1958) Analysis of windbreak effects. *Bulletin of the Research Council of Israel* 6, 211–219.

Lorch, J. (1959) Windbreaks and windbreak effects in the banana plantations of the Jordan Valley, Israel. *Israel Meteorotogical Service Meteorological Notes* 17, 1–25.

Lyons, M.J. (1973) Chilling injury in plants. *Annual Review of Plant Physiology* 24, 445–466.
Maas, P.W.T. (1967) Lightning damage to Congo bananas in Surinam. *Surinaamse Landbouw* 15, 44–46.
MacHattie, L.B. and Schnelle, F. (1974) *An Introduction to Agrotopoclimatology.* World Meteorological Organization Technical Note No.133, Geneva, 131 pp.
Marriott, J. (1980) Banana: physiology and biochemistry of storage and ripening for optimum quality. *CRC Critical Reviews in Food Science and Nutrition* 13, 41–88.
Marriott, J. and New, S. (1975) Storage physiology of bananas from new tetraploid clones. *Tropical Science* 17, 155–163.
Mattei, A. (1978) La frisure (chilling) de la banane. *Fruits* 33, 51–56.
Meyer, J.P. (1980) Besoins en eau et irrigation des bananiers aux Antilles Françaises. *Fruits* 35, 95–99.
Meyer, J.P. and Schoch, P.G. (1976) Besoin en eau du bananier aux Antilles. Mesure de l'évapotranspiration maximale. *Fruits* 31, 3–9.
Milburn, J.A. (1993) Advances in studying water relations in the banana and other plants. In: Valmayor, R.V., Hwang, S.C., Ploetz, R., Lee, S.W. and Roa, N.V. (eds) *Proceedings of the International Symposium on Recent Developments in Banana Cultivation Technology, Chiuju, Pingtung, Taiwan, 14–18 December 1992.* INIBAP/ASPNET, Los Baños, Laguna, Philippines, pp. 114–132.
Mohammed, M. and Campbell, R.J. (1993) Quality changes in 'Lacatan' and 'Gros Michel' bananas stored in sealed polyethylene bags with an ethylene absorbent. XXXIX Annual Meeting of the Interamerican Society for Tropical Horticulture, Santo Domingo, Dominican Republic, 22–27 August, 1993. *Proceedings of the Interamerican Society for Tropical Horticulture* 37, 67–72.
Monnet, J. (1961) Comparaison de quelques varietes de bananiers. *Fruits* 16, 74–76.
Obiefuna, J.C. (1986) The effect of monthly planting on yield, yield patterns and yield decline of plantains (*Musa* AAB). *Scientia Horticulturae* 29, 47–54.
Olsson, K.A., Cary, P.R. and Turner, D.W. (1984) Fruit crops – banana. In: Pearson, C.J. (ed.) *Control of Crop Productivity.* Academic Press, Sydney, pp. 230–237.
Oppenheimer, C. (1960) *The Influence of Climatic Factors on Banana Growing in Israel.* Publication of the National and University Institute of Agriculture, Series No. 350-B, Rehovot, 8 pp.
Pala, M. and Ovadia, R. (1990) Initial evaluation of greenhouse banana cultivation. *Colture Protette* 19(12), ci–civ.
Palmer, J.K. (1971) The banana. In: Hulme, A.C. (ed.) *The Biochemistry of Fruits and their Products.* Academic Press, London, pp. 65–105.
Peretz, E.C. (1992) The effect of banana leaf tearing on vegetative growth and production. Graduate student thesis, Beit Yerach Regional High School, 36 pp. (in Hebrew).
Porteres, R. (1950) La maladie physiologique de l'éventail foliaire et de l'engorgement du stipe des bananiers dans l'Ouest Africain. *Fruits* 5, 208–213.
Prest, R.L. and Smith, A.E. (1960) Spike leaf in banana. *Queensland Agricultural Journal* 86, 283–285.
Ramadass, R. and Sheriff, M.M. (1993) Effect of moisture stress on the yield and yield components in banana as influenced by their genome and ploidy. *Madras Agricultural Journal* 80, 130–133.
Ramcharan, C., Ingram, D.L., Nell, T.A. and Barrett, J.E. (1995) Interactive effects of root-zone temperature and irrigation volume on banana vegetative growth in two environments. *Fruits* 50, 225–232.
Rippon, L.E. and Trochoulias, T. (1976) Ripening responses of bananas to temperature. *Australian Journal of Experimental Agriculture and Animal Husbandry* 16, 140–144.
Rippon, L.E. and Turner, D.W. (1970) Summer bunch covers for bananas. *Agricultural Gazette of New South Wales* 81, 344.
Robinson, J.C. (1981) Studies on the phenology and production potential of Williams banana in a subtropical climate. *Subtropica* 2, 12–16.
Robinson, J.C. (1985) Root depth in bananas. *Citrus and Subtropical Fruit Research Institute Information Bulletin* 155, 6–8.
Robinson, J.C. (1993) Climate-induced problems of growing bananas in the subtropics. *Citrus and Subtropical Fruit Institute Information Bulletin* 245, 16–22.
Robinson, J.C. (1996) *Bananas and Plantains.* Crop Production Science in Horticulture 5, CAB International, Wallingford, UK, 238 pp.
Robinson, J.C. and Alberts, A.J. (1986) Growth and yield responses of banana (cultivar 'Williams') to

drip irrigation under drought and normal rainfall conditions in the subtropics. *Scientia Horticulturae* 30, 187–202.

Robinson, J.C. and Alberts, A.J. (1989) Seasonal variations in the crop water-use coefficient of banana (cultivar 'Williams') in the subtropics. *Scientia Horticulturae* 40, 215–225.

Robinson, J.C. and Green, G.C. (1981) Plant bananas where the temperature is right. *Farming in South Africa. Bananas* B2, 1–4.

Robinson, J.C., Anderson, T. and Eckstein, K. (1992) The influence of functional leaf removal at flower emergence on components of yield and photosynthetic compensation in banana. *Journal of Horticultural Science* 67, 403–410.

Roughan, P.G. (1985) Phosphatidylglycerol and chilling sensitivity in plants. *Plant Physiology* 77, 740–746.

Rowe, P. and Rosales, F. (1993) Diploid breeding at FHIA and the development of Goldfinger (FHIA-01). *Infomusa* 2(2), 9–11.

Sanchez-Nieva, F., Colom-Covas, G., Hernandez, I., Bueso de Vinas, C., Guadalupe, R. and Torres, A. (1970) Effect of zone and climate on yields, quality and ripening characteristics of Montecristo bananas grown in Puerto Rico. *Journal of Agriculture, University of Puerto Rico* 54, 195–210.

Sarma, R., Prasad, S. and Mohan, N.K. (1995) Bhimkal, description and uses of a seeded edible banana of northeastern India. *Infomusa* 4(1), 8.

Sarraf, S. and Bovee, A.C.J. (1973) *Evapotranspiration of the Banana.* Magon, Serie Technique, Institut de Recherches Agronomiques, Tel-Amara, Liban, 18 pp.

Satyan, S.H., Scott, K.J. and Best, D.J. (1992) Effects of storage temperature and modified atmospheres on cooking bananas grown in New South Wales. *Tropical Agriculture (Trinidad)* 69, 263–267.

Seymour, G.B., Thompson, A.K. and John, P. (1987) Inhibition of degreening in the peel of bananas ripened at tropical temperatures. 1. Effect of high temperature on changes in the pulp and peel during ripening. *Annals of Applied Biology* 110, 145–151.

Shivas, R.G., Wood, P.M., Darcey, M.W. and Pegg, K.G. (1995) First record of *Fusarium oxysporum* f.sp. *cubense* on Cavendish bananas in Western Australia. *Australasian Plant Pathology* 24, 38–43.

Shmueli, E. (1953) Irrigation studies in the Jordan Valley. I. Physiological activity of the banana in relation to soil moisture. *Bulletin of the Research Council of Israel* 3, 228–247.

Shmueli, E. (1960) Chilling and frost damage in banana leaves. *Bulletin of the Research Council of Israel* 8D, 225–238.

Shmueli, E., Manes, E., Ravitz, E. and Mills, D. (1959) Prevention of frost damage in banana groves. In: *Special Bulletin No. 23, Project 18/22.* Division of Publications, Beit-Dagan, Israel, pp. 1–17 (in Hebrew).

Shou, B.H. (1990) Agroclimatology of the major fruit production in China: a review of current practice. *Agricultural and Forest Meteorology* 53, 125–142.

Simmonds, N.W. (1962) *The Evolution of the Bananas.* Longman, Green, London, 170 pp.

Skutch, A.F. (1927) Anatomy of leaf of banana *M. sapientum* L var hort Gros Michel. *Botanical Gazette* 84, 337–391.

Smillie, R.M., Hetherington, S.E., He, J. and Nott, R. (1988) Photoinhibition at chilling temperatures. *Australian Journal of Plant Physiology* 15, 207–222.

Smirin, S. (1960) Banana growing in Israel. *Tropical Agriculture (Trinidad)* 37, 87–95.

Smith, I.E. (1973) Recovery of frost damaged bananas. *Hortus* 19, 23–25.

Srikul, S. and Turner, D.W. (1995) High N supply and soil water deficits change the rate of fruit growth of bananas (cv 'Williams') and promote tendency to ripen. *Scientia Horticulturae* 62, 165–174.

Stapleton, J.J. and Garza-Lopez, J.G. (1988) Mulching of soils with transparent (solarization) and black polyethylene films to increase growth of annual and perennial crops in southwestern Mexico. *Tropical Agriculture (Trinidad)* 65, 29–33.

Stoler, S. (1960) The banana. In: *Notes and Studies.* Sifriat Hassadeh, Tel-Aviv, pp. 75–165 (in Hebrew).

Stoler, S. (1962) Banana pseudostem temperatures. In: Ziv, D. (ed.) *Investigations on the Banana and the Rotation of Irrigated Crops.* Sifriat Hassadeh, Tel-Aviv, pp. 177–186 (in Hebrew).

Stoutjesdijk, P. (1970) Some measurements of leaf temperatures of tropical and temperate plants and their interpretation. *Acta Botanica Neerlandica* 19, 373–384.

Stover, R.H. (1984) Canopy management in Valery and Grand Nain using leaf area index and photosynthetically active radiation measurements. *Fruits* 39, 89–93.

Stover, R.H. and Malo, S.E. (1972) The occurrence of fusarial wilt in normally resistant 'Dwarf Cavendish' banana. *Plant Disease Reporter* 56, 1000–1003.

Stover, R.H. and Simmonds, N.W. (1987) *Bananas*, 3rd edn. Longman Scientific and Technical, Harlow, Essex, UK, 468 pp.

Summerville, W.A.T. (1944) Studies on nutrition as qualified by development in *Musa cavendishii* L. *Queensland Journal of Agricultural Science* 1, 1–127.

Tang, C.Y. and Chu, C.K. (1993) Performance of semi-dwarf banana cultivars in Taiwan. In: Valmayor, R.V., Hwang, S.C., Ploetz, R., Lee, S.W. and Roa, N.V. (eds) *Proceedings: International Symposium on Recent Developments in Banana Cultivation Technology, Chiuju, Pingtung, Taiwan, 14–18 December 1992.* INIBAP/ASPNET, Los Baños, Laguna, Philippines, pp. 43–52.

Tang, C.Y. and Liu, C.C. (1993) Banana-based farming system in Taiwan. In: Valmayor, R.V., Hwang, S.C., Ploetz, R., Lee, S.W. and Roa, N.V. (eds) *Proceedings: International Symposium on Recent Developments in Banana Cultivation Technology, Chiuju, Pingtung, Taiwan, 14–18 December 1992.* INIBAP/ASPNET, Los Baños, Laguna, Philippines, pp. 14–30.

Taylor, S.E. and Sexton, O.J. (1972) Some implications of leaf tearing in Musaceae. *Ecology* 53, 141–149.

Thompson, A.K., Been, B.O. and Perkins, C. (1972) Handling, storage and marketing of plantains. *Proceedings of the Tropical Region, American Society for Horticultural Science* 16, 205–212.

Ticho, R.J. (1971) The banana industry in Israel. *Tropical Science* 13, 289–301.

Toraskar, M.V. and Modi, V.V. (1984) Peroxidase and chilling injury in banana fruit. *Journal of Agricultural and Food Chemistry* 32, 1352–1354.

Trochoulias, T. (1973) The yield response of bananas to supplementary watering. *Australian Journal of Experimental Agriculture and Animal Husbandry* 13, 470–472.

Turner, D.W. (1987) Nutrient supply and water use of bananas in a subtropical environment. *Fruits* 42, 89–93.

Turner, D.W. (1994) Bananas and plantains. In: Schaffer, B. and Andersen, P.C. (eds) *Handbook of Environmental Physiology of Fruit Crops.* Vol. II. *Sub-tropical and Tropical Crops.* CRC Press, Boca Raton, Florida, USA, pp. 37–64.

Turner, D.W. (1995) The response of the plant to the environment. In: Gowen, S. (ed.) *Bananas and Plantains.* Chapman and Hall, London, pp. 206–229.

Turner, D.W. and Barkus, B. (1982) Yield, chemical composition, growth and maturity of 'Williams' banana fruit in relation to supply of potassium, magnesium and manganese. *Scientia Horticulturae* 16, 239–252.

Turner, D.W. and Lahav, E. (1983) The growth of banana plants in relation to temperature. *Australian Journal of Plant Physiology* 10, 43–53.

Valmayor, R.V. (1990) Bananas and plantains in the Philippines. In: Valmayor, R.V. (ed.) *Banana and Plantain R & D in Asia and the Pacific: Proceedings of a Regional Consultation on Banana and Plantain R & D Networking, Manila and Davao, 20–24 November 1989.* INIBAP, Montpellier, France, pp. 87–120.

Von Loesecke, H.W. (1950) *Bananas*, 2nd edn. Interscience, New York, 189 pp.

Wade, N.L., Kavanagh, E.E. and Tan, S.C. (1993) Sunscald and ultraviolet light injury of banana fruits. *Journal of Horticultural Science* 68, 409–419.

Wang, H.G. and Gemma, H. (1994) Ripening characteristics and chilling injury of banana fruit. II. Changes in degree of unsaturation of fatty acids during exposure to chilling. *Japanese Journal of Tropical Agriculture* 38, 246–250.

Wardlaw, C.W. (ed.) (1961a) Non-infectious diseases. In: *Banana Diseases Including Plantains and Abaca.* Longman, Green, London, pp. 32–67.

Wardlaw, C.W. (ed.) (1961b) Causes and extent of wastage in harvested fruit. In: *Banana Diseases Including Plantains and Abaca.* Longman, Green, London, pp. 466–487.

Wills, R.B.H., Bailey, W.M. and Scott, K.J. (1975) Possible involvement of alpha-farnesene in the development of chilling injury in bananas. *Plant Physiology* 56, 550–551.

Yoshioka, H., Ueda, Y. and Ogata, K. (1978) Effect of elevated temperatures on ripening of banana fruit. *Journal of Japanese Society of Food Science and Technology* 25, 607–611.

Yoshioka, H., Ueda, Y. and Chachin, K. (1980a) Effect of high temperature (40 deg C) on changes in acid-phosphatase activity during ripening in banana. *Journal of Japanese Society of Food Science and Technology* 27, 511–516.

Yoshioka, H., Ueda, Y. and Chachin, K. (1980b) Inhibition of banana fruit ripening and decrease of protein synthesis at high temperature (40 deg C): physiological studies of fruit ripening in relation to heat injury. Part III. *Journal of Japanese Society of Food Science and Technology* 27, 610–615.

Yoshioka, H., Ueda, Y. and Iwata, T. (1982) Development of the isoamyl acetate biosynthetic pathway in banana fruits during ripening and suppression of its development at high temperature. *Journal of Japanese Society of Food Science and Technology* 29, 333–339.

Ziv, D. (1962) *Investigations on the Banana and the Rotation of Irrigated Crops.* Sifriat Hassadeh, Tel-Aviv, 330 pp. (in Hebrew).

11
Chemical Injury to Banana

E. Lahav and Y. Israeli

INTRODUCTION

Cultural and biological control practices reduce the need to use agricultural chemicals to sustain plant growth and protect crops against insects and diseases. However, commercial banana production requires the input of large quantities of chemicals to maintain optimum soil fertility, control weeds and combat diseases and pests. Occasionally these materials are either applied in excess or improperly and injury to the banana plant or fruit can occur. Banana plants can also be damaged if grown in soil containing minerals in toxic concentrations or in environments where the air is polluted. Damage associated with chemical injury to banana is described in this chapter.

INJURY CAUSED BY AGRICULTURAL CHEMICALS

Injury Caused by Fertilizers

Excessive fertilizer application may result in a marginal chlorosis of the lower leaves, followed by necrosis. The first symptom, which appears on leaves 2–5, is a whitish discoloration or brown band, similar in appearance to sunburn. Brown spots may appear parallel to the edge of the leaf, but not at distal or proximal positions. These spots become necrotic and the tissue disintegrates. The leaf becomes thin and greyish white and tends to puncture and fragment (Charpentier and Martin Prevel, 1968). These symptoms usually disappear when the fertilizer is leached from the soil by rain or irrigation. If the symptoms persist, this is most probably the result of a lack of drainage or the presence of excessive salts associated with saline soils.

Injury Caused by Pest Control Chemicals

Oil, used against banana leaf-spot diseases since the late 1950s, initially caused significant injury. External symptoms of oil toxicity are leaf burns, usually around the central vein, and black necrotic spots. This damage was much reduced with the change to oils of a narrow distillation range (346–354°C), containing less than 12% aromatic compounds and above 90% unsulphonated residues (Calpouzos *et al.*, 1961; Calpouzos, 1968). Sprayed plants, which have no external injury symptoms, can carry bunches 8% lighter than untreated control plants (Israeli *et al.*, 1993).

Tridemorph, a fungicide used to control Sigatoka leaf spot and black leaf streak diseases, may sometimes cause leaf yellowing. The yellowing is stronger in mature leaves

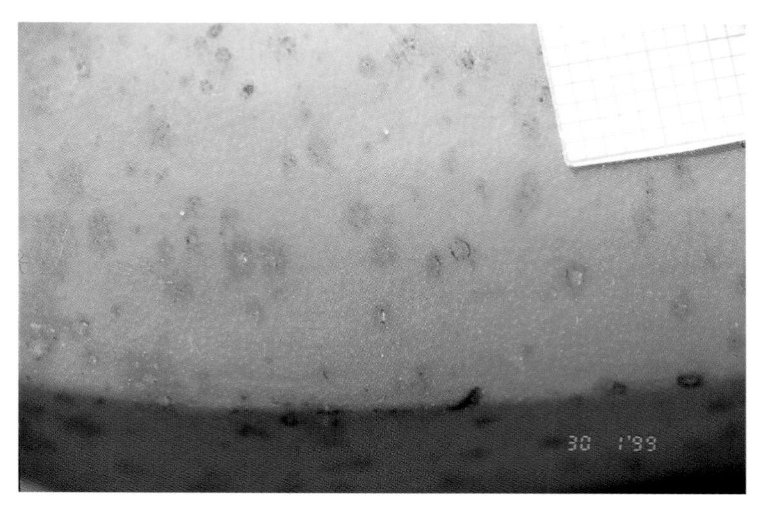

Plate 11.1. Water-soaked speckling caused by a combination of fungicides, oil and water on fruit of 'Grand Nain' (AAA, Cavendish subgroup) in Costa Rica (photo: C. Pasberg-Gauhl and F. Gauhl, CB).

Plate 11.2. Necrotic speckling caused by a combination of fungicides, oil and water on fruit of 'Grand Nain' (AAA, Cavendish subgroup) in Costa Rica (photo: C. Pasberg-Gauhl and F. Gauhl, CB).

exposed to the sun (Cronshaw, 1987), in water-stressed plants, when radiation is high and after the use of some herbicides. Tridemorph is always applied with oil and the phytotoxic reaction is a result of this combination. Damage to fruit of 'Grand Nain' (AAA, Cavendish subgroup) as a result of sprays containing different fungicides, oil and water is illustrated in Plates 11.1 and 11.2.

Insecticides approved for use on banana (like carbaryl, chlorpyrifos, diazinon) are rarely phytotoxic. Often damage is due to the carrier or poor application techniques, especially when these result in large droplets.

Quaternary ammonium compounds used for disinfecting tools to prevent the spread of Moko disease are phytotoxic if applied to fruit.

Plate 11.3. Paraquat herbicide-induced, necrotic spots on the leaf of a sucker of 'Robusta' (AAA, Cavendish subgroup) in St Vincent, Windward Islands (photo: D.R. Jones, SVBGA).

Exceptionally high levels of chlorine in water used for removing latex from dehanded fruit can cause a reddish flecking if fruit is left in contact with the chlorinated water for more than 30 min.

Injury Caused by Herbicides

The use of herbicides in dessert banana and plantain plantations was reviewed by Lassoudiere and Pinon (1971), Feakin (1972), Lassoudiere (1972), Irizarry (1987), Stover and Simmonds (1987), Tezenas du Montcel (1987), Galan-Sauco (1992), Shanmugavelu *et al.* (1992) and Robinson (1996). Earlier works dealing specifically with phytotoxicity were reviewed by Lassoudiere (1972) and by Soto (1985). More recent publications on this subject are by Guillemot (1975), Velez-Ramos and Vega-Lopez (1977), Ramadass *et al.* (1980), Liu *et al.* (1985), Cronshaw (1987), Liu and Rodriguez-Garcia (1988), Liu and Singmaster (1990) and Gonzalez and Piedrahita (1994). Symptoms associated with herbicide damage are summarized in Table 11.1.

There is much evidence that weed competition affects the rate of growth and yield of banana and that mechanical or chemical weed control is necessary for maximal production (Mishra and Das, 1984). Mechanical weed control might give better results than chemical control (Venero, 1980), but in most cases this method is more expensive (Irizarry, 1987). Therefore, herbicides are commonly used. The most popular contact herbicides are: paraquat, dalapon, glyphosate and glufosinate (Stover and Simmonds, 1987; Shanmugavelu *et al.*, 1992; Achard, 1993; Dave, 1993). Simazine, diuron, oxyfluorfen and ametryne are contact herbicides with residual effect (Liu *et al.*, 1985; Irizarry, 1987; Liu and Santiago-Cordova, 1991; Dave, 1993). Many other herbicides are in limited use worldwide.

All herbicides, if not properly used, might delay growth and induce symptoms of damage in banana leaves. However, the degree of risk differs greatly according to the compound in use and is affected by the stage of plant development, cultivar, climate, soil type, method of application, agrotechniques and presence of other chemicals. Phytotoxicity is only one factor

Table 11.1. Characteristics of herbicides in use and their phytotoxic symptoms on banana.

Name of herbicide (a.i.)	Commercial product and content of a.i.	Pre- or post-emergence	Rate (g a.i. ha^{-1})	Mode of action	Phytotoxicity hazard	Phytotoxicity symptoms	References
Paraquat	Gramoxone® 200 g l^{-1}	Post	200–1000	Contact	Low	Local chlorosis and necrosis on direct contact with leaves or live leaf sheaths	Lassoudiere and Pinon, 1971; Liu, 1990; Achard, 1993
Dalapon	Dowpon® 85%	Post	3000–12,000	Systemic, residual	Low	Not reported	Irizarry, 1987; Shanmugavelu *et al.*, 1992
Glyphosate	Round Up® 480 g l^{-1}	Post	500–1500	Systemic	High	Reduced growth rate, typical chlorosis and lamina deformations on young leaves	Guillemot, 1975; Achard, 1993
Glufosinate	Basta® 200 or 600 g l^{-1}	Post	300–600	Mainly contact	Low, moderate on young, *in vitro*-propagated plants	Local yellowing and browning after direct contact, spreading towards the cigar leaf of young, *in vitro*-propagated plants	Perkins, 1990; Gonzalez and Piedrahita, 1994; Achard, 1993; Y. Israeli, unpublished
Ametryne	Gesapax® 500 g l^{-1}	Pre and post	800–5000	Contact, systemic, residual	Low	Temporary chlorosis on young leaves; reduced fruit size	Guillemot, 1975; Soto, 1985; Cronshaw, 1987
Oxyfluorfen	Goal® 240 g l^{-1}	Pre and post	500–1200	Contact, systemic, residual	Moderate	Leaf necrosis after direct contact; marginal leaf necrosis when affected by gaseous form	Escorriola *et al.*, 1979; Liu and Singmaster, 1990; Y. Israeli, unpublished
Diuron	Karmex® 80%	Pre	1000–4800	Residual	High	Yellowing along midrib of older leaves, spreading towards the margins; reduced growth rate	Guillemot, 1975; Velez-Ramos and Vega-Lopez, 1977; Ramadass *et al.*, 1980; Cronshaw, 1987; Shanmugavelu *et al.*, 1992
Simazine	Gesatop® 500 g l^{-1}	Pre	500–3000	Residual	Moderate	Marginal leaf necrosis	Soto, 1985; Y. Israeli, unpublished

a.i., active ingredient.

to be taken into account for herbicide selection. Efficiency and cost are other major factors to be considered.

Paraquat is not translocated systemically. Yellowing, chlorosis and necrosis only appear where the herbicide has been in contact with the leaf tissue (Lassoudiere, 1972; Liu and Singmaster, 1990) (Plate 11.3). Therefore, it is recommended, as with other contact herbicides, that it be used only when banana suckers are larger than 1 m (Feakin, 1972; Stover and Simmonds, 1987; Galan-Sauco, 1992; Robinson, 1996). Manual weed control is recommended when suckers are smaller. Planting material derived from tissue culture is extremely sensitive to paraquat, as well as other herbicides, as it is initially low-lying with dense foliage (Kwa and Ganry, 1990; Marie *et al.*, 1993). Drift is the main risk in using paraquat. The risk of drift might be reduced if special equipment, such as a low-pressure spray, special spray nozzle and mechanical shield, is used. Keeping the spray nozzle close to the ground and the use of specific adjuvants may also help. Under no circumstances should herbicides be sprayed when conditions are windy (Feakin, 1972; Liu and Singmaster, 1990).

Glufosinate is a partially systemic (Perkins, 1990), relatively new, contact herbicide for banana (Gonzalez and Piedrahita, 1994). Chlorotic and necrotic spots are induced when glufosinate is in direct contact with the lamina. Young plants derived from tissue culture are especially sensitive and contact with exposed leaves results in chlorosis and necrosis of the funnel leaf several days later (Plate 11.4).

Oxyfluorfen is both a contact and a systemic herbicide, with a residual effect in the soil. Because of its high volatility, it may also act in gaseous form. Brown necrotic spots are formed on leaves following direct contact with oxyfluorfen and marginal necrosis is a result of systemic effects (Escorriola *et al.*, 1979). Usually the damage is limited and temporary and can be controlled by reducing drift and volatility.

Glyphosate is a contact herbicide with long-range, systemic activity. After contact with leaves or pseudostem, it is absorbed and translocated to the meristems, resulting in interveinal chlorosis and deformations to young leaves (Plate 11.5). Sometimes the lamina is totally absent (Guillemot, 1975; Achard, 1993). The major danger from glyphosate is its delayed action and the cessation of growth even when no external symptoms are observed. Frequent applications are recommended to reduce thick stands of weeds and spraying should be as close to the ground as possible (Liu and Rodriguez-Garcia, 1988). Banana suckers are highly sensitive to glyphosate and it has been used in high concentration to destroy diseased and abandoned plantations (Anon., 1983).

Ametryne is one of the oldest herbicides in use, as both a contact and a pre-emergence herbicide (Liu *et al.*, 1985; Stover and Simmonds, 1987; Dave, 1993; Robinson, 1996). Ametryne is usually not phytotoxic, but chlorosis might be induced when high concentrations are absorbed by banana roots (Lassoudiere, 1972). The range of efficiency of ametryne is limited and it is usually mixed with other herbicides, such as paraquat, simazine or other triazynes (Soto, 1985). Such combinations might induce phytotoxic symptoms in banana when applied at double-strength concentrations (Guillemot, 1975). Because of its persistence in soil, ametryne, as with other residual herbicides, may be displaced downhill during heavy rains. Displacement depends on the slope and rain intensity (Liu *et al.*, 1985).

Simazine is a well-known residual herbicide, which can be used alone as a pre-emergent (Lassoudiere and Pinon, 1971; Tosh *et al.*, 1982; Dave, 1993; Robinson, 1996) or in combination with other herbicides, such as ametryne or paraquat. When used at recommended concentrations, no phytotoxic symptoms are observed, but, when high concentrations are used, browning and necrosis appear at leaf margins (Lassoudiere, 1972; Guillemot, 1975; Soto, 1985). Such symptoms were seen in Israel after 2000 g active ingredient (a.i.) ha^{-1} of simazine were applied. Phytotoxicity

Plate 11.4. Chlorosis and necrosis of young leaves of tissue culture-derived plants due to glufosinate herbicide in Martinique, French Antilles (photo: Y. Israeli, JVBES).

Plate 11.5. Interveinal chlorosis and lamina deformation caused by glyphosate herbicide in Martinique, French Antilles (photo: Y. Israeli, JVBES).

symptoms are followed by a significant retardation of growth. Stronger phytotoxic symptoms occur in the presence of amitrol.

Diuron was one of the most popular residual herbicides of the past (Feakin, 1972; Lassoudiere, 1972; Tosh *et al.*, 1982; Soto, 1985; Irizarry, 1987; Stover and Simmonds, 1987) and is still popular today (Soto, 1985; Shanmugavelu *et al.*, 1992; Dave, 1993; Robinson, 1996). It is widely used alone or with contact herbicides. Phytotoxicity symptoms have been reported by Lassoudiere (1972), Guillemot (1975), Velez-Ramos and Vega-Lopez (1977), Ramadass *et al.* (1980), Tosh *et al.* (1982) and Shanmugavelu *et al.* (1992). Initial symptoms are yellowing along the central midrib and later towards the margins of mature leaves (Guillemot, 1975; Ramadass *et al.*, 1980). The bunch is also greatly damaged (Plate 11.6). Phytotoxic effects occur more frequently in plants growing during the dry season in light soils, where the organic matter is low (Lassoudiere, 1972). Lower concentrations of diuron should be used under these conditions. Phytotoxicity is more pronounced when the systemic fungicide tridemorph is used to control Sigatoka leaf spot or black

Plate 11.6. Bunch showing symptoms of excessive diuron herbicide application in Western Galilee, Israel (photo: E. Lahav, VI).

leaf streak diseases (Cronshaw, 1987). Plantain is more sensitive to diuron than other banana types (Tezenas du Montcel, 1987).

Bromacil and other related compounds are residual herbicides. When bromacil is applied to banana roots, leaves quickly yellow and become necrotic. Suckers can be killed (Lassoudiere and Pinon, 1971; Guillemont, 1975; Tosh *et al.*, 1982). Therefore, because of the high sensitivity of banana suckers to bromacil, it is recommended not to use it in banana plantations.

The hormone herbicide 2,4-dichlorophenozyacetic acid (2,4-D) and its derivatives were used in the past as weed killers in banana plantations (Feakin, 1972; Ramadass *et al.*, 1980; Stover and Simmonds, 1987; Shanmugavelu *et al.*, 1992). Not only banana leaves, but also the pseudostem and follow suckers are known to be extremely sensitive to 2,4-D. Even 2,4-D in gaseous form will damage banana suckers (Lassoudiere, 1972). The herbicide is a systemic auxin that affects the apical meristem, resulting in deformation of leaves and total death of the whole plant. When applied to banana plants after bunch emergence, 2,4-D interferes with the normal growth of the fruit and may cause finger malformations. It has been replaced in banana plantations by paraquat, glyphosate

Plate 11.7. Twisting, breaking and splitting of the leaf sheaths of a sucker caused by 2,4-D herbicide used for desuckering (photo: E. Lahav, VI).

and other contact herbicides. However, because of the high sensitivity of banana to 2,4-D, it is sometimes used for pruning unwanted suckers (Plate 11.7) (Chundawat and Patel, 1992).

Injury Caused by Desuckering Agents

Chemicals used for thinning suckers, such as 2,4-D and kerosene, might cause damage when applied at too high a concentration (Stover and Simmonds, 1987; Chundawat and Patel, 1992; Robinson, 1996). In young plantations established from plants derived from tissue culture, a relatively large number of suckers are produced, which need to be destroyed. If paraffin is used, the corm of the mother plant might be injured and the incidence of *Erwinia carotovora* frequently increases (see Thwaites *et al.*, Chapter 5, pp. 231–234). The dose of paraffin should not exceed 2.5 ml $plant^{-1}$.

Containers used for herbicides should not be used for the storage of desuckering chemicals. Even small amounts of paraquat or glyphosate injected into suckers might significantly affect the mother plant and result in loss of the bunch.

INJURY CAUSED BY ELEMENTS IN TOXIC CONCENTRATIONS

Symptoms caused by an excess of various elements are summarized in Table 11.2.

Aluminium Toxicity

Aluminium (Al) is not an essential element for growth, but can cause toxicity problems, especially when the soil pH is below 4.0 (Godefroy *et al.*, 1978; Lopez and Espinosa, 1995). The only direct symptom of aluminium toxicity described in banana is deformed roots (Lopez and Espinosa, 1995). Recently, the *in vitro* propagation method was used for selection of aluminium-tolerant variants of 'Nanicão' (AAA, Cavendish subgroup) (Matsumoto and Yamaguchi, 1990). Protocorm-like bodies were irradiated by gamma rays and cultured on a high-aluminium liquid medium at low pH to simulate natural conditions where aluminium toxicity is expected. One variant showed higher stress tolerance and was transferred to the field for evaluation.

Arsenic Toxicity

Arsenic (As) toxicity has been reported by Fergus (1955) in Australia. Leaves of all ages had chlorotic stripes and bunch development was poor. An analysis of affected plants showed 0.25–2.00 ppm of arsenic as dry leaf matter, while healthy plants had 0–0.50 ppm. Herbicides were thought to be the source of the excess arsenic. The condition can be overcome by spreading 50–100 kg of clean soil, which adsorbs arsenic and renders it unavailable to banana, around each plant.

Boron Toxicity

Marginal paling and necrosis of leaves have been reported as toxicity symptoms in plants with more than 850 ppm of boron (B) in the leaf lamina (Lahav and Turner, 1983).

Calcium Toxicity

Marginal leaf necrosis (Plate 11.8) resulting from calcium (Ca) in excess of 4% dry weight was reported from the Jordan Valley, Israel (Israeli, 1981).

Copper Toxicity

The accumulation of copper (Cu) in some acid soils of Central America as a result of

Table 11.2. Summary of symptoms on banana caused by excess minerals.

Symptoms on	Description of symptoms	Element
Petioles	Blue coloration	Magnesium
Leaf	Marginal browning followed by necrosis	Salinity (sodium chloride)
	Irregular chlorosis followed by necrosis	Magnesium
	Marginal chlorosis followed by necrosis	Sodium, boron, calcium
	Marginal blackening followed by necrosis	Iron, manganese
	Chlorotic stripes	Arsenic
	Water-soaked lesions followed by necrosis	Fluorine
Fruit	Not filled	Arsenic
	Weak bunch, widely spaced hands, long peduncles	Nitrogen
Roots	Growth inhibited	Copper
	Club-like root tips	Aluminium

applications of Bordeaux spray for Sigatoka control over a period of 15–20 years has affected root growth of cultivars in the Cavendish subgroup (AAA). Copper injury is not evident when the crop is first planted, but appears after the second or third cycle and becomes worse as the plantation ages. This toxicity is believed to result from: (i) the uptake of copper by the roots and its release when the roots die, resulting in an accumulation of the element in the root zone close to the rhizome; (ii) a gradually increasing soil acidity in the surface soil; and (iii) a 'high mat' condition (the base of the rhizome is only 5–10 cm below the soil surface 3–4 years after planting). Copper levels in the top 5–10 cm of the root zone of affected plants varied from 600 to 2000 ppm and pH varied from 5 to 6. Root injury was reduced after deep ploughing to dilute copper concentrations in the upper soil layer and liming to raise the pH.

Fluorine Toxicity

Symptoms of toxicity resulting from excess airborne fluorine (F) have been reported in Taiwan (Su *et al.*, 1978). Symptoms were described as typical marginal scorching, characterized by water-soaked lesions, developing into necrosis at the leaf margin (Plate 11.9). The content of fluorine in leaves with symptoms was 117–2119 ppm, as compared with 36–82 ppm in leaves from plants without symptoms. The corresponding values in fruit from affected and unaffected plants were 450 and 105 ppm, respectively.

Iron Toxicity

A black, necrotic marginal scorch on older leaves has been associated with iron (Fe) concentrations of up to 800 ppm in the Canary Islands (Lahav and Turner, 1983). Marginal scorching can sometimes be found in fertile plantations, but it is not necessarily an indication of a problem, unless associated with severe necrosis.

Manganese Toxicity

Worldwide, manganese (Mn) excess is considered to be a greater problem than manganese deficiency. Butler (1960) reports experiments in Jamaica where 20 kg of manganese ha^{-1} $year^{-1}$ reduced yields by 5%. However, in sand-culture experiments, high levels of manganese have not reduced yield (Charpentier and Martin Prevel, 1965; Turner and Barkus, 1983). Black spots and marginal necrosis on leaves have been associated with high concentrations of manganese in Australia (Lahav and Turner,

Plate 11.8. Marginal necrosis resulting from calcium excess in the Jordan Valley, Israel (photo: Y. Israeli, JVBES).

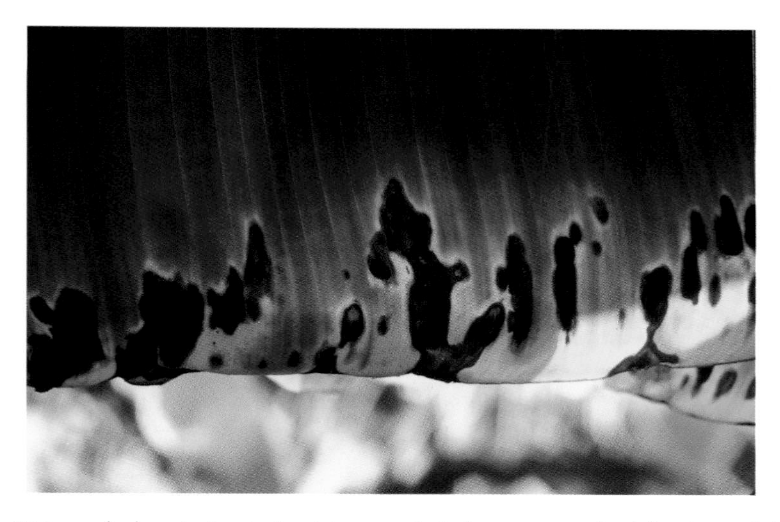

Plate 11.9. Water-soaked spots developing into necrotic lesions as a result of fluorine toxicity in Taiwan (photo: E. Lahav, VI).

1983). Toxic concentrations may be the result of natural high concentrations of available manganese in the soil or from a build-up of the element applied in fungicides, such as mancozeb. However, the tolerance of banana to high concentrations of manganese in the soil solution is high.

Manganese toxicity observed in the field may be attributed to indirect effects, such as low calcium, magnesium and zinc availability in acid soils, rather than to a high concentration of manganese. High manganese levels in the soil are corrected by liming.

Plate 11.10. Marginal necrosis due to salt toxicity in Santa Marta, Colombia (photo: E. Lahav, VI).

Sodium and Chlorine Toxicity

The sensitivity of banana to salinity or sodium chloride (NaCl) toxicity depends on environmental conditions, especially soil characteristics and evaporative demand. Problems do occur in India, the Philippines, Haiti, the Dominican Republic, Ecuador, Colombia, Jamaica, Israel, Jordan, Egypt and the Canary Islands. Dunlap and McGregor (1932) found that plants grew satisfactorily at 100–500 ppm total soluble salts, whereas plants and fruit were visibly affected from 500 to 1000 ppm. Above 1000 ppm, plants were stunted or dead. Saline soils produce marginal chlorosis, leading to necrosis (Plate 11.10), stunted growth and thin, deformed fruit. The adverse effect of salinity was demonstrated in some parts of Israel where banana plants were irrigated with water containing 500–600 ppm of chlorides. When the salty water was replaced by fresh water, a significant improvement of growth and production was recorded. Under saline conditions, sucker growth is reduced and fruits do not fill. Plants growing very near to the ocean in areas affected by sea spray also show signs of salt toxicity.

The effect of salinity on banana is assumed to be due to an increase in osmotic pressure of the soil solution, which results in difficulties in water uptake. The banana plant may be able to stand a higher than normal salinity level when evaporative demand is not too high, water-supply is adequate and drainage is functioning well. However, the plant may be adversely affected when evaporative demand increases or water-supply decreases. Leaching the salts with fresh water can also reduce salinity problems.

In most cases, it is difficult to distinguish between the effects of salinity, excess sodium (Na) and excess chlorine (Cl). However, symptoms of sodium toxicity have been said to start as an initial marginal chlorosis, which becomes marginal necrosis in older leaves. Eventually about one-third of the leaf is affected. Excess sodium adversely affects growth and yield and causes flowering to be delayed for 35 days, a 31% drop in bunch weight and a 23% decrease in finger weight (Israeli *et al.*, 1986).

Under saline conditions, sodium and chlorine levels in the lamina and petiole are only slightly affected, while sodium levels in the roots greatly increase. It has been suggested that sodium is toxic in the lamina at a concentration of 0.5% (Adinarayana *et al.*, 1986), but the roots seem to be a better indicator for sodium and chlorine levels. Tentative toxicity levels in the roots are 1.0% for sodium and 3.3% for chlorine (Israeli *et al.*, 1986).

When the exchangeable sodium percentage is high, water movement is restricted. This condition might be corrected by gypsum ($CaSO_4.2H_2O$) or sulphur treatments. Soils in Santa Marta, Colombia, were treated this way and sodium toxicity damage was remarkably reduced.

REFERENCES

Achard, R. (1993) Weed control in banana plantations in Cameroon. *Fruits* 48, 101–105.

Adinarayana, V.D., Balaguravaiah, Y., Narasimha, Rao, P. and Subba Rao, I.V. (1986) Potassium and sodium disorders in banana. *Journal of Potassium Research* 2, 102–107.

Anon. (1983) Using glyphosate to kill bananas in Panama wilt-affected plantations. *Citrus and Subtropical Fruit Research Institute Information Bulletin* No.131, 9.

Butler, A.F. (1960) Fertilizer experiments with the Gros Michel banana. *Tropical Agriculture (Trinidad)* 37, 31–50.

Calpouzos, L. (1968) Oils. In: Torgason, D.C. (ed.) *Fungicides: An Advanced Treatise,* Vol. 2. Academic Press, New York, pp. 367–393.

Calpouzos, L., Delfel, N.E., Colberg, C. and Theis, T. (1961) Relation of petroleum oil composition to phytotoxicity and sigatoka disease control on banana leaves. *Phytopathology* 51, 317–321.

Charpentier, J.M. and Martin Prevel, P. (1965) Culture sur milieu artificiel. Carences atténuées ou temporaires en éléments majeurs, carence en oligo-elements chez le bananier. *Fruits* 20, 521–557.

Charpentier, J.M. and Martin Prevel, P. (1968) *Carences et troubles de la nutrition mineral chez le bananier. Guide de diagnostic pratique.* Institut Français de Recherches Fruitières Outre Mer, Paris, 75 pp.

Chundawat, B.S. and Patel, N.L. (1992) Studies on chemical desuckering in banana. *Indian Journal of Horticulture* 49, 218–221.

Cronshaw, D.K. (1987) The role of calixin in phytotoxic and stress-induced symptoms on banana leaves. In: *Proceedings of the 7th ACORBAT Conference, 1985, San José, Costa Rica.* Centro Agronomico Tropical de Investigación y Enseñanza (CATIE), Turrialba, Costa Rica, pp. 171–173.

Dave, B. (1993) Weed control in banana plantations of Martinique. *Fruits* 48, 95–99.

Dunlap, V.C. and McGregor, J.D. (1932) *The Relationship Between Soil Alkalinity and Banana Production in St. Catherine District, Jamaica.* United Fruit Company Bulletin 45.

Escorriola, J.M., Arce, J. and Beltran, A. (1979) Oxyfluorfen: nueva herramienta con gran potencial herbicida para la industria bananera. In: *Proceedings of the 4th ACORBAT Conference, 1979, Panama.* Union de Paices Exportadores de Banano (UPEB), Panama, pp. 267–268.

Feakin, S.D. (1972) Weeds. In: Feakin, S.D. (ed.) *Pest Control in Bananas.* PANS Manual No. 1., Centre for Overseas Pest Research, Foreign and Commonwealth Office, Overseas Development Administration, London, pp. 5–14.

Fergus, I.F. (1955) A note on arsenic toxicity in some Queensland soils. *Queensland Journal of Agricultural Science* 12, 95–100.

Galan-Sauco, V. (1992) Control de malas hierbas. In: *Los Frutales Tropicales en los Subtropicos II: Platano (Banano).* Ediciones Mundi-Prensa, Madrid, pp. 87–88.

Godefroy, J., Lassoudiere, A., Loissois, P. and Penel, J.P. (1978) Action du chaulage sur les caractéristiques physico-chimiques, et la productivité d'un sol tourbeux en culture bananière. *Fruits* 33, 77–90.

Gonzalez, S. and Piedrahita, W. (1994) Glufosinate: new molecule for the weed control on banana. In: *Proceedings of the 10th Conference ACORBAT, 1991, Villahermosa, Mexico.* Corporación Bananera Nacional (CORBANA), San José, Costa Rica, pp. 117–122.

Guillemot, J. (1975) Tests sur l'efficacité et la phytotoxicité de quelques herbicides en bananeraie. *Fruits* 30, 75–81.

Irizarry, H. (1987) Intensive plantain production in the humid mountains of Puerto Rico. In: *Proceedings 3rd Meeting of the International Association for Research on Plantain and Bananas, Abidjan, Ivory Coast, 1985*, pp. 55–59.

Israeli, Y. (1981) A case of calcium excess in Israeli bananas. *International Banana Nutrition Newsletter* 3, 11–12.

Israeli, Y., Lahav, E. and Nameri, N. (1986) The effect of salinity and sodium absorption ratio in the irrigation water, on growth and productivity of banana under drip irrigation conditions. *Fruits* 41, 297–301.

Israeli, Y., Shabi, E. and Slabaugh, W.R. (1993) Effect of banana spray oil on banana yield in the absence of Sigatoka (*Mycosphaerella* sp.). *Scientia Horticulturae* 56, 107–117.

Kwa, M. and Ganry, J. (1990) Utilisation agronomique des vitroplants de bananier. *Fruits*, Numéro special, 107–111.

Lahav, E. and Turner, D.W. (1983) *Banana Nutrition.* International Potash Institute Bulletin 7, Worblaufen-Bern, Switzerland.

Lassoudiere, A. (1972) Utilisation de herbicides en culture bananiere. Etude bibliographique. *Fruits* 27, 87–105.

Lassoudiere, A. and Pinon, A. (1971) Indications préliminaires sur des essais de desherbage chimique en bananeraie. *Fruits* 26, 333–348.

Liu, L.C. and Rodriguez-Garcia, J. (1988) Optimum time interval and frequency of glyphosate application for weed control in plantain (*Musa* sp.). *Journal of Agriculture of the University of Puerto Rico* 72, 297–300.

Liu, L.C. and Santiago-Cordova, M. (1991) Persistence of three herbicides in a drip-irrigated banana field. *Journal of Agriculture of the University of Puerto Rico* 75, 19–23.

Liu, L.C. and Singmaster, J.A. (1990) Herbicide drift control in plantains and taniers. *Journal of Agriculture of the University of Puerto Rico* 74, 471–475.

Liu, L.C., Fernandez-Horta, D. and Santiago-Cordova, M. (1985) Diuron and ametryn runoff from a plantain field. *Journal of Agriculture of the University of Puerto Rico* 69, 177–183.

Lopez, A.M. and Espinosa, J.M. (1995) *Manual de Nutricion y Fertilizacion del Banano.* Instituto de la Potasa y el Fosforo, Quito, Ecuador, 82 pp.

Marie, P., Dave, B. and Cote, F. (1993) Use of banana vitroplants in the West Indies: assets and constraints. *Fruits* 48, 89–94.

Matsumoto, K. and Yamaguchi, H. (1990) Selection of aluminium-tolerant variants from irradiated protocorm-like bodies in banana. *Tropical Agriculture (Trinidad)* 67, 229–232.

Mishra, A.K. and Das, P.C. (1984) Weed control in banana. *South Indian Horticulture* 32, 61–64.

Perkins, G.R. (1990) Basta – a new herbicide for horticulture. In: *Proceedings of the 9th Australian Weeds Conference*, pp. 544–547.

Ramadass, R., Vaithialingam, R., Bhakthavathsalu, C.M. and Veerannan, L. (1980) Control of weeds in banana with the aid of herbicides. In: *Proceedings of the National Seminar on Banana Production and Technology, Tamil Nadu, India.* Tamil Nadu Agricultural University, Coimbatore, India, pp. 93–97.

Robinson, J.C. (1996) Weed control. In: *Bananas and Plantains.* CAB International, Wallingford, UK, pp. 155–156.

Shanmugavelu, K.G., Aravindakshan, K. and Sathiamoorthy, S. (1992) Weeding. In: *Banana Taxonomy, Breeding and Production Technology.* Metropolitan Book Co., New Delhi, India, pp. 97–100.

Soto, M.B. (1985) Manejo y control de las malas hierbas. In: *Cultivo y Comercializacion de Banano.* Litografia e Impresa LIL, San Jose, Costa Rica, pp. 242–265.

Stover, R.H. and Simmonds, N.W. (1987) *Bananas*, 3rd edn. Longman, Scientific and Technical, Harlow, Essex, UK, 468 pp.

Su, H.J., Ko, W.H., Chuang, S.Y., Huang, M.T. and Hwang, S.C. (1978) *Etiological Studies on Marginal Scorch of Banana, with Special Reference to Air-polluted Fluoride Associated with the Disease.* Banana Research Institute, Special Issue No. 21, Taiwan Banana Research Institute, Pingtung, Taiwan, 20 pp.

Tezenas du Montcel, H. (1987) Weeding. In: *Plantain Bananas.* Macmillan Publishers, London, pp. 51–53.

Tosh, G.C., Mohanty, D.C. and Nanda, K.C. (1982) Herbicides for the control of weeds in interplanted pineapples and bananas in India. *Tropical Pest Management* 28, 431–432.

Turner, D.W. and Barkus, B. (1983) The uptake and distribution of mineral nutrients in the banana in response to supply of K, Mg and Mn. *Fertilizer Research* 4, 89–99.

Velez-Ramos, A. and Vega-Lopez, J.A. (1977) Chemical weed control in plantains (*Musa acuminata* × *Musa balbisiana*, AAB). *Journal of Agriculture of the University of Puerto Rico* 18, 411–416.

Venero, R. (1980) Chemical weed control in banana and its influence on yields. *Cultivos Tropicales* 2, 127–137.

12

Genetic Abnormalities of Banana*

Y. Israeli and E. Lahav

Introduction

Vegetative mutations have contributed greatly to diversity in banana (Champion, 1967; Stover and Simmonds, 1987). Mutations affecting plant stature, foliage and fruit pigmentation, waxiness, hairiness, bract persistence, finger length and shape, and absence of female flowers in banana or male part of the inflorescence in plantain appear to be harmless and have survived, sometimes giving rise to new clones. Many mutations do not survive in commercial plantations because of human selection. However, mutations may survive if they appear in botanical collections or on smallholders' farms and if they are not deleterious. Some mutations are characterized by morphological or physiological disadvantages, such as extreme dwarfism, abnormal leaf emission, extreme mosaic or a mutant with difficulty in iron uptake, and will not survive without intentional care.

Since the introduction of *in vitro* propagation techniques for banana, the occurrence of off-type plants has greatly increased, as has been reported from many different places in the world (Israeli *et al.*, 1995). The frequency of off-types in banana vary from an extremely high 91% (Daniells and Smith, 1993) and 25% (Stover, 1987) to a low 3% (Hwang and Ko, 1987) and 1% (Arias and Valverde, 1987). In plantain, variable rates of off-types have also been reported (Sandoval *et al.*, 1991; Vuylsteke *et al.*, 1991; Krikorian *et al.*, 1993). The appearance of these plants is, therefore, a common phenomenon.

Off-type plants might differ permanently or temporarily from the source plants. The latter, the result of an epigenetic effect, is characterized by a non-heritable change and is reversible. The permanent off-type, referred to as a somaclonal variant, is heritable and is an expression of pre-existing variation in the source plant or is due to the creation of *de novo* variation via undetermined genetic mechanisms (Larkin and Scowcroft, 1981). More general information about somaclonal variation, its occurrence and causes may be found in reviews by Scowcroft (1984), Lee and Phillips (1988), Bajaj (1990), Swartz (1991) and Côte *et al.* (1993). Very little is known about the causes of somaclonal variation in banana (Reuveni and Israeli, 1990; Reuveni *et al.*, 1996) and the identification of an off-type as a somaclonal variant requires observations over several ratoon cycles (Scowcroft, 1984). The type and rate of variation is specific to the genotype (Stover, 1987; Smith, 1988; Robinson and Nel, 1989; Israeli *et al.*, 1991; Vuylsteke *et al.*, 1991) and this is true not only for different cultivars, but also for different selections of the same

*This chapter is dedicated to the memory of Dr Oded Reuveni.

cultivar (Israeli *et al.*, 1996). The micropropagation procedure may also affect the rate of variation (Drew and Smith, 1990; Israeli *et al.*, 1995).

Variations in Plant Stature

The mutation of genes controlling the height of banana is one of the more common mutations and was described by Cheesman in 1933 and Champion in 1952. In the Cavendish subgroup (AAA), Cheesman described tall mutants of 'Dwarf Cavendish' and suggested that 'Grand Nain' was probably derived from a mutant of 'Dwarf Cavendish'. Stover and Simmonds (1987) listed nine cultivars with dwarf counterparts and ten dwarfs have been reported at the subgroup level (J. Daniells, South Johnstone, 1997, personal communication). According to Richardson (1961), 'Gros Michel' (AAA) has mutated to the 'dwarf variant' form in Jamaica ('Highgate'), in Panama ('Cocos') and in Honduras ('Lowgate'). Richardson also reported on the reverse mutation of dwarf 'Cocos' to the tall 'Gros Michel'. He noted that changes from short to tall plants are more frequent than the reverse. Over a 3-year period, 50–70 tall reversions appeared among 28,000 'Cocos' mats. Richardson (1961) believes that this change was controlled by the somatic loss of the dominant dwarf allele, rather than by the reverse gene mutation. On the other hand, with micropropagated Cavendish cultivars 'Williams' and 'Grand Nain', tall variants are rare and many dwarfs (similar, but not necessarily identical to 'Dwarf Cavendish') are observed (Israeli *et al.*, 1991). It seems that the genetic factor determining height in the Cavendish subgroup is unstable, in both conventional and in micropropagated plants, but the pattern of change differs *in vitro* and *in situ*. Height variation has also been recorded in the ABB group (Ventura *et al.*, 1988).

The fact that the *in vitro* condition enhances dwarfism means that a plant produced from a culture that has been subcultured many times could be a dwarf or an 'extra dwarf', the latter originating from the former. The dwarf variant is the most frequent in the Cavendish subgroup, but does not appear in the micropropagated 'Red' (AAA). However, the mosaic variant is found in micropropagated 'Red', which indicates that different genomes and cultivars differ in the kind and rate of variants generated in culture (Israeli *et al.*, 1995). A similar conclusion was reached by Vuylsteke *et al.* (1988).

In the Cavendish subgroup, dwarfs are very similar to 'Dwarf Cavendish' and vary significantly from the original cultivar. Their height, leaf size, distance between petioles (Plate 12.1), bunch length, bunch mass and finger dimensions are smaller than those of the original cultivar. The rate of growth of dwarf plants is fast and they flower earlier than the original cultivars from which they are derived (Israeli and Nameri, 1985; Smith and Drew, 1990). There is no difference in number of hands or in pseudostem and bunch stalk circumference. Some of the flowers have a persistent perianth and the male flowers and bracts are persistent on the male axis (Plate 12.2). Their fruits differ in size and shape from those of the original cultivars. All these are typical characteristics of 'Dwarf Cavendish'. It was recently demonstrated that dwarf off-types show improved tolerance to low-temperature-induced photoinhibition, as compared with the normal 'Williams' (Damasco *et al.*, 1997). Most dwarfs produced by Cavendish clones resemble each other in their basic characteristics, whether or not they originated from different cultivars, propagation batches or growth plots.

Tissue culture-derived dwarfs can be detected during their growth period in the nursery, before transplanting to the field. Differences in height, distance between petioles and leaf size can be distinguished if all plants are grown under uniform conditions. When the height of normal plants from the ground to the base of the youngest petiole is 35–40 cm, dwarfs of 'Grand Nain' are about 5 cm shorter and dwarfs of 'Williams' about 10 cm shorter. Because of the greater height difference, the 'Williams' dwarfs can be screened out more efficiently at this stage (Israeli *et al.*, 1991).

Efficient detection of dwarfs in the field, based on the above parameters, is achieved during periods of uniform growth. Differences in height, leaf index (length–width ratio) and distance between petioles is pronounced. The leaf index is less than 2.0 in more than 90% of dwarf 'Williams' plants and 2.2 or greater among normal 'Williams' plants (Israeli *et al.*, 1991). Inflorescence parameters, such as shape and size of fingers and persistence of bracts and male flowers on the axis, are the most reliable means of characterizing dwarfs.

Smith and Hamil (1993) showed that, in the Cavendish subgroup, dwarfs can be recognized and removed when plantlets are 7 weeks old and 20 cm high. The best morphological characteristics for screening at this stage are length of lamina and petiole and the ratio between them. Reuveni (1990) demonstrated that dwarfs may be identified at the tissue-culture propagation stage by adding gibberellic acid (GA_3) to the growth medium. Normal plants respond with intensive growth, while the reaction of dwarfs is very slow. The same response is achieved as a result of spraying GA_3 after deflasking (Damasco *et al.*, 1996a). Significant differences are found in endogenous GA_3 levels between dwarf, normal and giant 'Grand Nain' (Sandoval *et al.*, 1995). The level of GA_3 was 3.6 times higher in the normal 'Grand Nain' than in the dwarf off-type and 4.6-fold in giant 'Grand Nain' as compared with the dwarf. The differences in GA_{20} were 4.6 and 7.3, respectively.

Ortiz and Vuylsteke (1995) investigated the inheritance of dwarfism in plantain and concluded that a single recessive gene controls the trait. Recently, banana cultivars and clones have been identified by modern methods of molecular analysis (Howell *et al.*, 1994; Bhat and Jarret, 1995). These methods have also been used for identification of dwarf off-types. Damasco *et al.* (1996b) identified a random amplified polymorphic DNA (RAPD) marker specific to the dwarf off-type produced during *in vitro* propagation of 'New Guinea Cavendish' and 'Williams' cultivars. The marker provided a reliable means with which to detect dwarf variants in tissue culture. The detection of dwarf off-types in plantain using the RAPD technique is also reported (Ford-Lloyd *et al.*, 1993).

An extra-dwarf variant is an extremely low-stature plant, about 1 m tall. The leaves are greatly condensed, forming a 'rosette', and are very short and broad. The bunch stalk is very short and the bunch becomes choked in the pseudostem on emergence. As a result, the bunch tends to lie horizontally and is sometimes even orientated vertically. The fingers are very short and the fruit has no commercial value. These characteristics are typical of the 'Extra-Dwarf Cavendish' cultivar described by Stover and Simmonds (1987). The incidence of the extra-dwarf variant is very high in 'Nathan' (Israeli selection of 'Dwarf Cavendish'). Very few extra-dwarfs arise naturally in Cavendish cultivars 'Williams' and 'Grand Nain'. 'Dwarf Parfitt' (Plate 12.3), an extra-dwarf Cavendish variant found in Australia, has been shown to be resistant to Fusarium wilt race 4. Compared with 'Williams', it also has a higher rate of photosynthesis and a better tolerance to low temperatures (Moore *et al.*, 1993).

The extra-dwarf variants, contrary to other stature variants, can be detected in tissue culture by their compact appearance and the formation of rosette-like leaves. They can be recognized and removed easily at later hardening and nursery stages (Israeli *et al.*, 1991).

Giant variants are rare. The plants are similar to the tall cultivars of the Cavendish subgroup, such as 'Valery' and 'Lacatan'. Giants can be detected in the nursery, as they are approximately 30% taller than normal plants. These variants are also observed in conventionally propagated banana.

Abnormal Foliage

A 'drooping-leaf variant' was observed by Smith (1988) and Vuylsteke *et al.* (1988) in plantain (AAB) and by Israeli *et al.* (1995)

Plate 12.1. A dwarf variant (right) that has arisen in tissue culture compared with a normal plant (left). Note the short internodes, the short petioles and the different leaf length–width ratio on the dwarf variant (photo: Y. Israeli, JVBES, and E. Lahav, VI).

Plate 12.2. Fruiting dwarf variant of 'Grand Nain' (AAA, Cavendish subgroup) (right) compared with normal 'Grand Nain' (left). Note the persistent perianth parts and bracts in the dwarf mutant (photo: Y. Israeli, JVBES, and E. Lahav, VI).

Plate 12.3. Extra-dwarf Cavendish variant 'Dwarf Parfitt' in Queensland, Australia. Note that the bunch has 'choked' on emergence (photo: QDPI).

Plate 12.4 (above left). Drooping-leaf variant of 'Grand Nain' (AAA, Cavendish subgroup) (photo: Y. Israeli, JVBES, and E. Lahav, VI).

Plate 12.5 (above). Pronounced variegation in the leaves of 'Pelipita' (ABB) growing in Nigeria. Note that the lead follower is not variegated, but the smaller sucker appears affected (photo: D.R. Jones, INIBAP).

Plate 12.6 (left). 'Variegated-leaf' variant with variegated leaf and fruit. The white tissue lacks chlorophyll (photo: Y. Israeli, JVBES, and E. Lahav, VI).

in the Cavendish cultivar 'Grand Nain' (Plate 12.4). This variation is associated with slow growth, delayed flowering and agronomically inferior bunches. The bunches of the plantain variant were of the 'False Horn' type, but extremely small. The origin of this variation is suggested to be due to a high ploidy level.

Random pale patches without chlorophyll in the leaf lamina characterizes the 'variegated-leaf' variant. To the untrained eye, it can be mistaken for symptoms of cucumber mosaic virus. Variegation is irregular and differs from plant to plant. Sometimes it appears as a red colour on a pale background. Variegation is very likely a sectorial chimera. A very pronounced variegation is illustrated in Plate 12.5.

In some plants, variegation disappears after several months and the newly emerged leaves are symptomless and the followers normal. However, other variegated plants retained the symptoms and can produce variegated fruit (Plate 12.6). In many cases, both normal and variegated suckers arise from variegated mother plants. This type of variant occurs in both tissue cultured and conventionally propagated plants of 'Williams' and 'Grand Nain' in Israel (Israeli *et al.*, 1991).

The first detection of variegated leaves depends on the intensity of symptoms. Extremely variegated plants can be detected in tissue-cultured plants at the test-tube stage or at the very beginning of the hardening stage. Less intense symptoms can be detected at the nursery stage before transplanting to the field.

The leaf blade of the 'deformed-lamina variant' is deformed or lobed (Plate 12.7) and in some cases the affected part has a bright mosaic. The deformed portion of the leaf is usually close to the petiole. This variant was found in *in vitro* plants of banana (Israeli *et al.*, 1991) and plantain (Vuylsteke *et al.*, 1996). It was only once recorded in conventionally propagated banana in Israel (Israeli *et al.*, 1991).

Affected tissue-cultured plants can be detected easily in the nursery before planting in the field. In most cases, the deformation disappears gradually in the field, but in a few cases it is inherited for many cycles.

The most common genetic abnormality in foliage is the 'mosaic' variant, which is also called 'Masada' in Central America. The variably thick, rubbery, narrow leaves, with different degrees of pale green mottling, may resemble banana mosaic disease (see Lockhart and Jones, Chapter 6, pp. 256–263) to the inexperienced eye. The mottling, which is irregular and not restricted by secondary veins, can be seen more clearly when the lamina is lit from the back. The upper leaf surface shows a pattern of bright spots and is covered with depressions and protuberances (Plate 12.8). In addition, very narrow and irregular lamina, wavy margins and longitudinal bright stripes along the midrib and petiole down to the leaf sheaths are sometimes observed. The plants grow relatively slowly and inflorescence emergence is delayed. The number of hands is normal, but the peduncle is thin and the fingers are very small, with no commercial value. Reuveni (1990) recorded an exceptionally low number of stomata on a mosaic leaf.

Mosaic variants have appeared so far in all clones of the Cavendish subgroup and also in 'Red', but usually in small numbers. They can be detected at the nursery stage, when tissue-cultured plants reach a height of 25–30 cm, with a leaf length exceeding 20 cm. The symptoms observed on the eighth to tenth leaves are more easily distinguished than those on the older leaves. One case only of a mosaic sucker originating from a normal mother plant was recorded in 'Williams' in conventionally propagated plants (Israeli *et al.*, 1991). It is, therefore, very typical of *in vitro*-propagated plants. An extreme mosaic, with more pronounced characteristics, is also known (Plate 12.9). It is very droopy in appearance and has very narrow laminae (only a few centimetres wide) and very intense mottling. The extreme mosaic can be detected at the hardening stage or immediately after potting.

Mosaic plants have been found to be aneuploids (Reuveni *et al.*, 1986; Sandoval *et al.*, 1996). It is clear that the *in vitro* procedure promotes their development. A change in chromosome number has also been reported for somaclonal variants of other species (Lee and Phillips, 1988). The mosaic-variant plants are similar, but not identical, to each other, which points to a difference in the level of aneuploidy, as confirmed by Sandoval *et al.* (1996). The extreme mosaic might have a larger number of extra chromosomes than the common mosaic variant.

The connection between various off-types and chromosome number in meristematic cells of root tips of 'Grand Nain' was investigated by Sandoval *et al.* (1996). About 5% of cells in plants propagated conventionally had abnormal chromosome numbers, while 14% of cells were affected in normal *in vitro*-propagated plants. Dwarf variants had 22% abnormal cells and mosaic variants 35%.

An 'iron-deficient' variant was found in a 'Dwarf Cavendish' plantation in the Jordan Valley, Israel. This mutant had reduced ability for iron uptake. It had almost totally yellow leaves unless iron-chelate was applied. The physiology of this abnormal variation is unknown.

Variations in Pseudostem Pigmentation

The 'reddish pseudostem' variant is characterized by a decrease in black pigmentation and the appearance of a reddish colour on the leaf sheaths and petioles. The red colour is pronounced, especially in young plants, either in the nursery or in the field. These plants also have bright green leaf sheaths, petioles, midribs and fruit peel. Another typical characteristic of the reddish variant is an elongated funnel leaf, which does not open normally during the hot midsummer months.

Reddish pseudostem variants are quite rare in both conventional and *in vitro* plants. They are easily distinguished when they exceed a height of 30 cm in the nursery. Detection is more reliable under conditions of high light intensity, as opposed to partial shade, where normal pseudostems are also pale and diagnosis is more difficult. As a consequence, detection under field conditions is simple. Variants with a reddish pseudostem produce satisfactory commercial fruit.

The pseudostem of the 'black pseudostem' variant is almost completely black. Coloration is also found in the petioles and central veins. The characteristic persists during many cycles (Israeli *et al.*, 1991). The black pseudostem variant occurs in conventionally propagated plants, as well as in tissue-cultured plants. 'Pisang Klutuk Wulung', the black stemmed *Musa balbisiana* from Indonesia, is illustrated in Plate 14.2.

Other pigmentation variants are greenish (hardly any anthocyanin in the lamina and leaf sheaths), various degrees of striping, black spotting (where the fruit is also spotted) and red-green variation, which is common in 'Red' (Israeli *et al.*, 1991). The 'Red' to 'Green Red' phenomenon can be explained by the periclinical chimera structure of the mother plants (Stover and Simmonds, 1987). Waxiness is another pseudostem variation. Its inheritance was investigated by Ortiz *et al.* (1995), who found that waxiness is controlled by a recessive gene, with a significant dosage effect.

Chimeras

In fruit, chimeras are strips of abnormally pigmented tissue, which stand out against the normal green background. They often consist of reddish or brown, well-defined bands on the peel. Occasionally, part of an entire bunch is affected. Chimeras are not common and the fruit is rejected at the packing station.

Several variations of *in vitro*-propagated plants, such as variegated leaves and deformed laminae, seem to have the characteristics of chimeras. In many cases, after one to four subsequent cycles in the field, normal followers emerge alongside the symptom-carrying suckers. In the Plantain subgroup (AAB), the high percentage of

Plate 12.7. Leaves of a 'deformed-lamina' variant (photo: Y. Israeli, JVBES, and E. Lahav, VI).

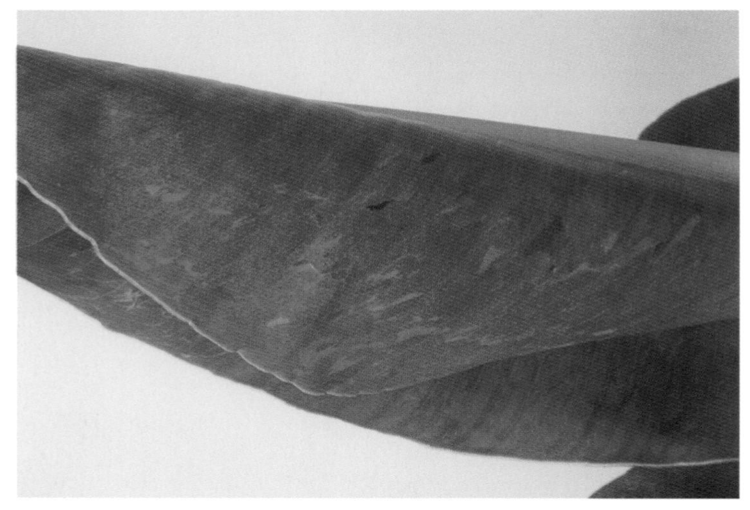

Plate 12.8. Leaf of 'mosaic' variant covered with irregular bright spots. The leaf surface is also covered with depressions and protuberances and the margins are wavy (photo: Y. Israeli, JVBES, and E. Lahav, VI).

'French' plantain-type inflorescence reversion found in micropropagated 'False Horn' plantain was also related to a chimeric constitution (Krikorian *et al.*, 1993). According to Stover and Simmonds (1987), the red peel of the fruit of 'Red' is a chimera, consisting of a 'red skin' on a 'green core', and frequent sectoring and variegation result from accidents in development, rather than mutation.

Inflorescence and Fruit Variations

Variations have been documented in flower, fruit and bunch characteristics (Stover and Simmonds, 1987; Côte *et al.*, 1993).

With a 'persistent flower' variant, perianths of female flowers do not abscise after flowering, but remain attached to the ovaries. There is no nectar secretion and bracts remain attached to the cushion and

Plate 12.9. 'Extreme mosaic' variant (photo: Y. Israeli, JVBES, and E. Lahav, VI).

Plate 12.10. 'Split-finger' variant (photo: Y. Israeli, JVBES).

Plate 12.11. The fusion of fingers in the proximal hand is one of the most common genetic defects (photo: Y. Israeli, JVBES).

are delayed in their withering. The fruits are less curved and take an 'open-hand' form. A large part of the male axis is covered with residual male flowers. This mutant is known in both tissue-cultured and conventionally propagated plants (Israeli, 1977).

The 'inflorescence with no female flowers' variant is caused by interference to the parthenocarpic development of the ovary. It is quite a common off-type in conventional plantations, where it occurs in various degrees of severity, but almost none at all are found in micropropagated plants. Such inflorescences have been found in 'Dwarf Cavendish', 'Williams' and 'Grand Nain'. The suckers of such plants grow rapidly, as a result of lack of competition for nutrients from developing fruit.

The 'French reversion' variant in plantain is characterized by the appearance of a typical 'French' inflorescence in 'False Horn' plantain. The plant produces relatively small fruit and persistent intermediate flowers and a male bud is present at maturity. All these traits are totally different from the original plant. *In vitro*-propagated plantains produce a high rate of this variation. Krikorian *et al.* (1993) suggested a chimera breakdown mechanism as an explanation for this high variation rate. The 'Monganga'-type is another variation in plantain, producing a degenerated 'False Horn' inflorescence (Vuylsteke *et al.*, 1996).

In a plant that is a 'split-finger' variant fingers split longitudinally on their external side (Plate 12.10). Similar fruit has been found in conventionally propagated plants. Fingers in some ratoon bunches can be normal.

The fusion of fingers in the proximal hand is one of the most common genetic defects (Plate 12.11). In a 'fused finger' variant, up to four fingers may be fused together. About 5% of bunches may have fused fingers and up to 0.5% of the weight of banana clusters can be rejected at the packing station because of this defect. The fused finger condition is common in 'Grand Nain', but much less frequent in the tall or dwarf Cavendish cultivars. Therefore, it has a genetic basis.

A plant that is a 'multi-bunching' variant has multiple pseudostems, peduncles and male flower buds. This condition has been described by Cheesman (1933), Champion (1952, 1967), Nayar *et al.* (1958), Gill (1968) and Stover and Simmonds (1987). Davis (1984) observed about 35 different cases of abnormal bunching. In 1991, a variant of 'Giant Cavendish' with three bunches was found in a farmer's field in southern Taiwan and propagated *in vitro* (Tang, 1995). When plants arising from this tissue culture were planted in the field, 21.1% were observed to be multi-bunching variants. These variants were described as a single bunch with multiple male buds (Plate 12.12), multiple bunches on a single peduncle and multiple bunches on independent peduncles. In the ratoon crop, the number of multi-bunching variants rose to 61.1% and a new type with multiple pseudostems appeared. This variation suggests that the abnormal division of the meristematic tissue occurred at various stages of inflorescence development.

A bunch that bends horizontally and has widely spaced hands has been observed on a tissue culture variant of 'Williams' (AAA, Cavendish subgroup) (Plate 12.13). Other bunch and fruit variants are a long tapering bunch with no male bud (Plate 12.14), waxy fingers and blunter fruit tips. These variations appear mostly in cultivars in the Plantain subgroup and ABB group (Stover and Simmonds, 1987). Daniells and Smith (1993) have noted variants with pointed-tip fingers (susceptible to mokillo disease), severe fruit scarring, long peduncles, longer fruit, short fingers, incomplete fruit that does not fill, different-coloured male buds and bunches in which the two top hands fail to fill.

The genetic abnormalities described in this chapter have been primarily off-types (natural and *in vitro*-induced) in the Cavendish subgroup. This reflects the documented information and the major economic importance of Cavendish cultivars. However, genetic abnormalities exist in other clones and subgroups, such as plantain, which are important food sources. Many cultivars have not been multiplied *in vitro* to any great extent and it is possible that, when they are, new variants will arise. The variations listed above are mainly deleterious. However, variants waiting to be discovered may have resistance to pests and diseases and improved horticultural performance and could be of great economic importance.

Acknowledgement

J.W. Daniells is thanked for his important contribution to this chapter.

REFERENCES

Arias, O. and Valverde, M. (1987) Production y variacion somaclonal de plantas de banano variedad Grande Naine producida por cultivo de tejidos. *Revista de la Asociacion Bananera Nacional (ASBANA) Costa Rica* 28, 6–11.

Bajaj, Y.P.S. (1990) Somaclonal variation-origin, induction, cryopreservation and implications in plant breeding. In: Bajaj, Y.P.S. (ed.) *Somaclonal Variation in Crop Improvement. I. Biotechnology in Agriculture and Forestry.* Springer Verlag, Berlin, pp. 3–48.

Bhat, K.V. and Jarret, R.L. (1995) Random amplified polymorphic DNA and genetic diversity in Indian *Musa* germplasm. *Genetic Resources and Crop Evolution* 42, 107–118.

Champion, J. (1952) Note sur quelques anomalies et mutations du bananier Nain (*Musa sinensis*). *Fruits* 7, 478–480.

Champion, J. (1967) *Les Bananiers et leur culture.* Tome I. *Botanique et genetique.* Institut Français de Recherches Fruitières Outre-Mer, Paris, pp. 192–194.

Cheesman, E.E. (1933) Mutant types of the dwarf banana. *Tropical Agriculture (Trinidad)* 10, 4–5.

Côte, F.X., Sandoval, J.A., Marie, P. and Fluboiron, E. (1993) Variations in micropropagated bananas and plantains: literature survey. *Fruits* 48, 11–18.

Damasco, O.P., Godwin, I.D., Smith, M.K. and Adkins, S.W. (1996a) Gibberellic acid detection of dwarf off-types in micropropagated Cavendish bananas. *Australian Journal of Experimental Agriculture* 36, 237–241.

Damasco, O.P., Graham, G.C., Henry, R.J., Adkins, S.W., Smith, M.K. and Godwin, I.D. (1996b) Random amplified polymorphic DNA (RAPD) detection of dwarf off-types in micropropagated Cavendish (*Musa* spp. AAA) bananas. *Plant Cell Reports* 16, 118–123.

Damasco, O.P., Smith, M.K., Godwin, I.D., Adkins, S.W., Smillie, R.M. and Hetherington, S.E. (1997) Micropropagated dwarf off-type Cavendish bananas (*Musa* spp. AAA) show improved tolerance to suboptimal temperatures. *Australian Journal of Agricultural Research* 48, 377–384.

Daniells, J.W. and Smith, M.K. (1993) Somatic mutations of bananas – their stability and potential. In: Valmayor, R.V., Hwang, S.C., Ploetz, R., Lee, S.W. and Roa, V.N. (eds) *Proceedings of the International Symposium on Recent Developments in Banana Cultivation Technology, Chiuju, Pingtung, Taiwan, 14–18 December 1992.* INIBAP/ASPNET, Los Baños, Laguna, Philippines, pp. 162–171.

Davis, T.A. (1984) A multi-bunching banana sport. *Newsletter from IBPGR Regional Committee for Southeast Asia* 8(4), 21–23.

Drew, R.A. and Smith, M.K. (1990) Field evaluation of tissue-cultured bananas in south-eastern Queensland. *Australian Journal of Experimental Agriculture* 30(4), 569–574.

Ford-Lloyd, B.V., Howell, E. and Newbury, H.J. (1993) An evaluation of random amplified polymorphic DNA (RAPD) as a tool for detecting genetic instability in *Musa* germplasm stored *in vitro.* In: Ganry, J. (ed.) *Breeding Banana and Plantain for Resistance to Diseases and Pests. Proceedings of the International Symposium on Genetic Improvement of Bananas for Resistance to Diseases and Pests, organized by CIRAD-FLHOR, Montpellier, France, 7–9 September 1992.* CIRAD, Montpellier, France, p. 375.

Gill, M.M. (1968) A note on dichotomy of the inflorescence in the plantain (*Musa paradisiaca* Linn.). *Tropical Agriculture (Trinidad)* 45, 337–341.

Plate 12.12. 'Multi-bunching' variant with multiple male buds arising from a single bunch (photo: E. Lahav, VI).

Plate 12.13. Tissue-culture variant of 'Williams' (AAA, Cavendish subgroup) with widely spaced hands and a bunch tending to the horizontal (photo: D.R. Jones, QDPI).

Plate 12.14. 'Tapering bunch' tissue-culture variant of 'Chinese Cavendish' (AAA, Cavendish subgroup) (photo: D.R. Jones, INIBAP).

Howell, E.C., Newbury, H.J., Swennen, R.L., Withers, L.A. and Ford-Lloyd, B.V. (1994) The use of RAPD for identifying and classifying *Musa* germplasm. *Genome* 37, 328–332.

Hwang, S.C. and Ko, W.H. (1987) Somaclonal variation of bananas and screening for resistance to *Fusarium* wilt. In: Persley, G.J. and De Langhe, E.A. (eds) *Banana and Plantain Breeding Strategies, Proceedings of an International Workshop, Cairns, Australia.* ACIAR Proceedings No. 21, Australian Centre for International Agricultural Research, Canberra, Australia, pp. 151–156.

Israeli, Y. (1977) Deciduous and persistent flowers in the Cavendish banana. MSc thesis, Hebrew University of Jerusalem, Israel, 60 pp. (Hebrew, with English summary).

Israeli, Y. and Nameri, N. (1985) Off-types of banana plants multiplied *in vitro.* In: Report on Observations and Experiments in Bananas in the Jordan Valley in the Years 1978–84. *Report, Banana Experimental Station, Jordan Valley, Israel* 24, 50–59 (in Hebrew).

Israeli, Y., Reuveni, O. and Lahav, E. (1991) Qualitative aspects of somaclonal variations in banana propagated by *in vitro* techniques. *Scientia Horticulturae* 48, 71–88.

Israeli, Y., Lahav, E. and Reuveni, O. (1995) *In vitro* culture of bananas. In: Gowen S. (ed.) *Bananas and Plantains.* Chapman and Hall, London, pp. 147–178.

Israeli, Y., Ben Bassat, D. and Reuveni, O. (1996) Selection of stable banana clones which do not produce dwarf somaclonal variants during in vitro culture. *Scientia Horticulturae* 67(3–4), 197–205.

Krikorian, A.D., Irizarry, H., Cronauer Mitra, S.S. and Rivera, E. (1993) Clonal fidelity and variation in plantain (*Musa* AAB) regenerated from vegetative stem and floral axis tips *in vitro. Annals of Botany* 71, 519–535.

Larkin, P.J. and Scowcroft, W.R. (1981) Somaclonal variation: a novel source of genetic variability from cell cultures for improvement. *Theoretical Applied Genetics* 60, 197–214.

Lee, M. and Phillips, R.L. (1988) The chromosomal basis of somaclonal variation. *Annual Review Plant Physiology* 39, 413–437.

Moore, N.Y., Pegg, K.G., Langdon, P.W. and Whiley, A.W. (1993) Current research on Fusarium wilt of banana in Australia. In: Ploetz, R.C., Valmayor, R.V., Hwang, S.C., Ploetz, R., Lee, S.W. and Roa, N.V. (eds) *Proceedings, International Symposium on Recent Developments in Banana Cultivation Technology, Chiuju, Pingtung, Taiwan, 14–18 December 1992.* INIBAP/ASPNET, Los Baños, Laguna, Philippines, pp. 270–284.

Nayar, T.G., Seshadri, V.S. and Bakthavathasalu, C.M. (1958) Abnormalities in bananas – III. *Indian Journal Agricultural Science* 28, 401–402.

Ortiz, R. and Vuylsteke, D. (1995) Inheritance of dwarfism in plantain (*Musa* spp. AAB group). *Plant Breeding* 114, 466–468.

Ortiz, R., Vuylsteke, D. and Ogburia, N.M. (1995) Inheritance of pseudostem waxiness in banana and plantain (*Musa* spp.). *Journal of Heredity* 86, 297–299.

Reuveni, O. (1990) Methods for detecting somaclonal variants in 'Williams' bananas. In: Jarret, R.L. (ed.) *Identification of Genetic Diversity in the Genus* Musa. *Proceedings of an International Workshop held at Los Baños, Philippines, 5–10 September 1988.* INIBAP, Montpellier, France, pp. 108–113.

Reuveni, O. and Israeli, Y. (1990) Measures to reduce somaclonal variation in *in vitro* propagated bananas. *Acta Horticulturae* 275, 307–313.

Reuveni, O., Israeli, Y., Eshdat, Y. and Degani, H. (1986) *Genetic Variability of Banana Plants Multiplied via* in vitro *Techniques.* Final report submitted to IBPGR (No. PR 3/11), Agricultural Research Organization, The Volcani Center, Bet-Dagan, Israel.

Reuveni, O., Israeli, Y. and Lahav, E. (1996) Somaclonal variation in banana and plantain (*Musa* species). In: Bajaj, Y.P.S. (ed.) *Somaclonal Variation in Crop Improvement II.* Springer, New York, pp. 174–196.

Richardson, D.L. (1961) Note on the reversion of the dwarf banana 'Cocos' to 'Gros Michel'. *Tropical Agriculture (Trinidad)* 38, 35–37.

Robinson, J.C. and Nel, D.J. (1989) Mutations in tissue-culture field plantings of banana. *Citrus and Subtropical Fruit Research Institute Information Bulletin* 200, 7.

Sandoval, J.A., Tapia, F.A.C., Miller, L. and Villa Lobos, A.B. (1991) Observations about the variability encountered in micropropagated plants of *Musa* cv. 'False Horn' AAB. *Fruits* 46, 533–539.

Sandoval, J., Kerbellec, F., Cote, F. and Doumas, P. (1995) Distribution of endogenous gibberellins in dwarf and giant off-types banana (*Musa* AAA, cv. 'Grand Nain') plants from *in vitro* propagation. *Plant Growth Regulation* 17, 219–224.

Sandoval, J.A., Côte, F.X. and Escoute, J. (1996) Chromosome number variations in micropropagated

true-to-type and off-type banana plants (*Musa* AAA 'Grande Naine' cv.). In Vitro *Cellular and Development Biology Plant* 32, 14–17.
Scowcroft, W.R. (1984) Genetic Variability in Tissue Culture: Impact on Germplasm Conservation and Utilization. Technical Report to the International Board for Plant Genetic Resources (IBPGR), Rome (84/152), 41 pp.
Smith, M.K. (1988) A review of factors influencing the genetic stability of micropropagated bananas. *Fruits* 43, 219–223.
Smith, M.K. and Drew, R.A. (1990) Growth and yield characteristics of dwarf off-types recovered from tissue-cultured bananas. *Australian Journal of Experimental Agriculture* 30, 575–578.
Smith, M.K. and Hamill, S.D. (1993) Early detection of dwarf off-types from micropropagated Cavendish bananas. *Australian Journal of Experimental Agriculture* 33, 639–644.
Stover, R.H. (1987) Somaclonal variation in 'Grand Nain' and 'Saba' bananas in the nursery and field. In: Persley, G.J. and De Langhe, E.A. (eds) *Bananas and Plantain Breeding Strategies. Proceedings of an International Workshop, Cairns, Australia.* ACIAR Proceedings No. 21, Australian Centre for International Agricultural Research, Canberra, Australia, pp. 136–139.
Stover, R.H. and Simmonds, N.W. (1987) *Bananas.* Longman Scientific and Technical, Harlow, Essex, UK, 468 pp.
Swartz, H.J. (1991) Post culture behaviour: genetic and epigenetic effects and related problems. In: Debergh, G.J. and Zimmerman, R.H. (eds) *Micropropagation: Technology and Application.* Kluwer Academic Publishers, Dordrecht, pp. 95–112.
Tang, C.Y. (1995) Variation of bunch types in mericlones of a multi-bunching banana. *Infomusa* 4(2), 17–18.
Ventura, J.de-la-C, Rojas, M.E., Yera, E.C., Lopez, J. and Rodriguez-Nodals, A.A. (1988) Somaclonal variation in micropropagated bananas (*Musa* spp.). *Ciencia y Tecnica en la Agricultura, Viandas Tropicales* 11(1), 7–16.
Vuylsteke, D., Swennen, R.L., Wilson, G.F. and De Langhe, E.A. (1988) Phenotypic variation among *in vitro* propagated plantain (*Musa* sp. cultivar 'AAB'). *Scientia Horticulturae* 36, 79–88.
Vuylsteke, D.R., Swennen, R.L. and De Langhe, E.A. (1991) Somaclonal variation in plantains (*Musa* spp., AAB group) derived from shoot-tip culture. *Fruits* 46, 429–439.
Vuylsteke, D.R., Swennen, R.L. and De Langhe, E.A. (1996) Field performance of somaclonal variants of plantain (*Musa* spp., AAB group). *Journal, American Society Horticultural Science* 121, 42–46.

13

Quarantine and the Safe Movement of *Musa* Germplasm

D.R. Jones and M. Diekmann

Introduction

The sheaths of wild banana must have been useful to early humans as a source of fibre and the leaves as wrapping materials and temporary thatch. The male buds, male flowers and immature fruit may also have been utilized as vegetables, as they still are in some areas of South-East Asia today. Early cultivators would have eagerly transplanted suckers of banana with fruit showing the first signs of parthenocarpy and sterility, as this feature would have added to the attractiveness of an already versatile plant. Superior edible types would have been propagated and distributed widely among and between ethnic groups. In this way, banana cultivars would have spread throughout tropical Asia. Early migrants and traders are thought to be responsible for the later transfer of planting material from Asia to Africa. Relatively recently in historic terms, Polynesians distributed banana throughout the Pacific and Europeans took the plant across the Atlantic to the New World (Simmonds, 1962).

Although *Musa* is affected by a number of viruses, bacteria, fungi and nematodes, only a few of the disease problems known today seem to have been disseminated during the first movements of banana out of South-East Asia. The distribution of germplasm during the 20th century seems to be responsible for the long-distance spread of most important banana pathogens from their centres of origin. *Fusarium oxysporum* f. sp. *cubense*, the cause of Fusarium wilt (see Ploetz and Pegg, Chapter 3, pp. 143–159) is a notable exception to this rule. Intercontinental spread of the pathogen is considered to have occurred long before the disease was described for the first time in 1876 (Stover, 1962). It is fortuitous that populations of the Fusarium wilt fungus capable of attacking Cavendish clones in the tropics did not also spread outside Asia at this time.

The present distribution of serious diseases of banana is outlined in Table 13.1. One of these diseases – blood disease (see Thwaites *et al.*, Chapter 5, pp. 221–223) – is confined to certain areas of Indonesia, while others, such as bract mosaic (see Thomas *et al.*, Chapter 6, pp. 253–255) have only been found in a few countries. Some have a regional or near-global distribution. Factors controlling the rate of spread are numerous. They include the life cycle of the pathogen, the effectiveness of quarantine measures, the presence or absence of insect vectors, climate and geographical features, such as oceans, deserts, forests and high mountain ranges, which may act as natural barriers.

Mycosphaerella fijiensis is an example of a pathogen with a high capability for spread. Although it may have been present in certain localities in the Pacific long before its recognition in Fiji in 1963, its

Table 13.1. Known distribution of some important pathogens of *Musa*.

Pathogen	Disease	Mainland Asia	Pacific*	Latin America	Caribbean	Africa
Fungi						
Mycosphaerella fijiensis	Black leaf streak	L	W	W	L	W
Mycosphaerella musicola	Sigatoka	W	W	W	W	W
Mycosphaerella sp.	Septoria leaf spot	W	–	–	–	L
Guignardia musae	Freckle	W	W	–	?	?
Fusarium oxysporum f. sp. *cubense* (race 1)	Fusarium wilt	W	L	W	W	W
Fusarium oxysporum f. sp. *cubense* (tropical race 4)	Fusarium wilt	L	L	–	–	–
Bacteria						
Ralstonia solanacearum	Moko	–	L	W	L	–
Ralstonia solanacearum	Bugtok	–	L	–	–	–
Ralstonia sp.	Blood disease	–	L	–	–	–
Viruses						
Banana bunchy top virus	Bunchy top	W	W	–	–	L
Banana bract mosaic virus	Bract mosaic	L	L	–	–	–
Banana streak virus	Banana streak	W	W	W	W	W
Cucumber mosaic virus	Banana mosaic	W	W	W	W	W
Abacá mosaic virus	Abacá mosaic	–	L	–	–	–
Nematodes						
Rhadopholus similis	Burrowing nematode	W	W	W	W	W
Pratylenchus coffeae	Root lesion nematode	W	W	W	W	W
Pratylenchus goodeyi	Root lesion nematode	–	L[†]	–	–	L

* Includes Australia, East Malaysia, Indonesia, Papua New Guinea, the Philippines and Taiwan lying off the Asian mainland.
[†] Subtropical Australia.
L, distribution limited; W, distribution widespread; –, absent; ?, records questionable.

Plate 13.1. Quarantine sign at airport on Horn Island in the Torres Strait area of Australia. In order to prevent the spread of black leaf streak disease, air passengers are warned that it is prohibited under quarantine legislation to carry banana material (photo: D.R. Jones, QDPI).

subsequent rapid movement to new areas has been dramatic (see Pasberg-Gauhl *et al.*, Chapter 2, pp. 37–44). As a fungal disease of banana foliage, black leaf streak can be carried long distances with leaf trash, fresh leaves used to wrap produce and leaves attached to suckers. Airborne spores are also thought to carry the pathogen considerable distances and to ensure that, once established in a new region, a steady and inexorable spread to all surrounding banana-growing districts and countries is inevitable. Ascospores carried by winds from Cuba are the most likely explanation for the appearance of the disease in Jamaica, approximately 140 km away.

When black leaf streak was first found on islands in the Torres Strait and on the tip of Cape York Peninsula in Australia (Jones and Alcorn, 1982), attempts were made to eradicate the disease and prevent the spread of the fungal pathogen to other parts of Australia (Plate 13.1). When eradication failed (Jones, 1984), most probably as a result of inoculum carried from Papua New Guinea, susceptible cultivars growing in the area where the disease was found were replaced with resistant ones to reduce inoculum levels (Jones, 1990). Since then, black leaf streak has been discovered only occasionally on susceptible banana plants growing in isolated communities on Cape York Peninsula, to the south of the affected zone. These outbreaks have been eradicated as they occur. Because of the vigilance of the quarantine authorities and the natural barrier to spread afforded by the sparsely populated Cape York Peninsula, the commercial banana-growing areas of Australia are still free of black leaf streak. This is the one success story in the otherwise relentless spread of the disease to all areas where banana is cultivated.

The spread of other diseases has been less dramatic than that of black leaf streak. Bunchy top is increasing its area of distribution at a much slower pace. The long-distance dissemination of the banana bunchy top virus (BBTV) relies primarily on the movement of infected planting material from one country to another. The unrestricted movement of banana suckers, corms and tissue cultures across international borders is against the quarantine regulations of many banana-growing countries, but undoubtedly occurs from time to time as legitimate channels of entry are circumvented. Although the risk of spread is not very high, because propagating material, including meristems used to initiate *in vitro* cultures, would not normally be taken from plants with obvious bunchy-top symptoms, there is a possibility that germplasm could be selected for propagation after infection but before symptom expression. This may be how long-distance spread has occurred in the past. In countries where bunchy top has become established, total eradication has proved impossible so it is important to prevent introduction.

Almost all selections of 'Mysore' (AAB), a cultivar believed to have originated in India, where it is known as 'Poovan', are infected with banana streak virus (BSV) and the virus most probably spread with this clone when planting material of the clone was introduced into other countries. Symptoms of BSV in 'Mysore' are periodic, and effects can be much more subtle than those of BBTV. As a result, the virus has been widely disseminated. The 'peculiarities' of 'Mysore' were only relatively recently suspected of being a disease and were not associated with BSV infection until 1993 (Anon., 1993). Further spread of BSV from 'Mysore' to other clones via mealybug vectors may be an uncommon event in many countries, as 'Mysore' is often the only clone with BSV symptoms in long-established field germplasm collections.

It is important that care is taken to slow or prevent the further dissemination of pathogens of *Musa*. The greatest risk to commercial production in the Latin American–Caribbean region would be the introduction of populations of *F. oxysporum* f. sp. *cubense* that are capable of attacking Cavendish clones in the tropics. If this occurred and a wilt-resistant replacement for Cavendish were not found, then the damage already caused in West Malaysia (Ong, 1996) and elsewhere would be repeated on a much larger scale. Major threats to smallholders and subsistence farmers come from the continued spread of bunchy top (see Thomas and Iskra-Caruana, Chapter 6, pp. 241–253), bract mosaic (see Thomas *et al.*, Chapter 6, pp. 253–255), Moko and blood diseases (see Thwaites *et al.*, Chapter 5, pp. 213–223). However, it is not only the most serious diseases that may pose problems. A disease that may cause minor damage in one environment may cause significant damage in another. It is prudent, therefore, to try and exclude all exotic pathogens from gaining entry to countries where they do not occur. Even if a disease is present in a country, it is good policy not to be indifferent to its reintroduction, because of possible strain or race differences that could increase problems.

The basis of good quarantine policy relies on a knowledge of the diseases present in a country and of exotic diseases that threaten crops in that country. An understanding of the epidemiology of exotic diseases and how they may gain entry is also important. Without this information, it is very difficult to develop regulations and guidelines for plant importations. For this reason, it is crucial that countries have the capability of diagnosing disease problems. Basic taxonomic training of plant pathologists should be encouraged and supported, especially in developing countries.

Movement of Pathogens

Most banana diseases can be spread from one location to another with planting material. Until fairly recently, banana planting material consisted of rhizomes, with or without attached leaf and pseudostem tissue (Plate 13.2). The potential for moving pathogens with rhizomes is high, as adhering soil can harbour root-attacking organisms, such as nematodes, fungi and bacteria. In addition, the rhizome tissue itself could be infected with all three and also viruses. Any attached leaf tissue and adhering leaf trash could also carry foliar pathogens. Risks can be significantly reduced if the rhizome is washed to remove soil and pared to trim off roots, the outermost tissue layers, discoloured areas and any leaf material. However, risks are not eliminated and there is still the possibility that some pathogens, which may not show obvious symptoms, could be present.

It is known that *F. oxysporum* f. sp. *cubense* is disseminated with rhizomes and infested soil. Bancroft (1876) first recognized the importance of planting uninfected rhizomes to control the disease. However, as corms can carry the disease without showing obvious symptoms, his recommendations were easier said than done. Nematodes have also been introduced to uninfested land with planting material and accompanying soil. The appearance of the burrowing nematode in some new areas has been associated with shipments of rhizomes of highly susceptible Cavendish cultivars. It has been suggested that the bacterium that causes Moko disease was introduced to Mindanao in the Philippines on Cavendish rhizomes imported from Central America (Buddenhagen, 1986). BBTV was almost certainly introduced into Australia with planting material from Fiji (Dale, 1987).

Some pathogens may move with true seed. Seed is produced in fruit of wild banana species (Plate 13.3), sometimes in fruit of edible cultivars, such as 'Pisang Awak' (ABB), which retains a high degree of fertility, and in fruit of plants following artificial pollination in breeding programmes. There is strong evidence that BSV is seed-borne (Daniells *et al.*, 1995) and it is possible that cucumber mosaic virus may also be seed-borne (Gold, 1972; Stover, 1972). Although not absolutely proved, it is likely that the bacterial agents of Moko, bugtok and blood diseases, which attack fruit, may also move with seed as surface contaminants.

In many parts of the banana-growing world, banana leaves are used to wrap food and pack produce, which may be carried or taken long distances by travellers and traders. Although it is unlikely that heavily diseased leaves would be deliberately utilized for this purpose, it is possible that an occasional leaf with an odd lesion may be selected. Banana leaves are not allowed entry into Australia, because of the risk of introducing foliar pathogens (Jones, 1991). Dried leaf tissue could also harbour foliar and fruit disease pathogens (Stover, 1977).

Banana fruit can carry crown-rot fungi and other pathogens. Fungi causing crown rot are found everywhere, and quarantine measures to exclude them are not warranted. However, bananas can also be affected by serious bacterial diseases and not show obvious external symptoms. Often, the presence of these diseases is only discovered when fingers are peeled or cut to expose the pulp. Diseased fruit could be moved from one location to another and then discarded in the garden when found to be rotten. This may allow the bacterial pathogen to spread to nearby banana plants. A quarantine regulation restricting the movement of bananas in Indonesia, as well as vegetative parts of banana, may well have delayed the spread of blood disease from Sulawesi to West Java until 1987 (Eden-Green, 1994). There is also a risk that leaf trash with viable fruiting bodies of fungal pathogens may also accompany bananas packed in cartons. For these reasons, the risks associated with the movement of banana fruit from one banana-growing country to another need to be carefully evaluated.

Some banana pathogens can affect alternative hosts, which then become sources of infection. Not much is known about the host range of many banana pathogens so risks have not been totally defined. However, some threats have been identified. *Heliconia* is known to pose a significant risk, as it has long been recognized as a host of the bacterium that causes Moko disease of banana. In 1989/90, a survey of 20 commercial Heliconia farms in Hawaii found five with plants affected by Heliconia wilt. Of 17 isolates of *Ralstonia solanacearum* obtained from diseased Heliconia, five were pathogenic on banana. Fortunately for the banana industry in Hawaii, these banana-attacking isolates had not spread from *Heliconia* to banana. The results of the survey suggested that quarantine regulations governing the import of *Heliconia* into Hawaii and intrastate movement of vegetative cuttings or rhizomes were needed (Ferreira *et al.*, 1991).

Plate 13.2. Suckers of 'Klaui Namwa' (ABB, syn. 'Pisang Awak') on sale in northern Thailand. This planting material comes from an area near the Mekong River, where Fusarium wilt is common, and some suckers would undoubtedly carry the disease (photo: D.R. Jones, INIBAP).

Plate 13.3. Fruit of *Musa acuminata* ssp. *banksii* showing seed (photo: D.R. Jones, QDPI).

At the time of the Hawaiian survey, the threat that ornamental *Heliconia* posed to the banana industry in Australia had long been recognized and quarantine regulations were in place that required imported *Heliconia* to be held in post-entry quarantine before release. The quarantine period was 3 months in a private post-entry quarantine house for plants from areas free of Moko and 9 months in a government post-entry quarantine facility for those from an area where the disease was known to occur. All needed to be inspected regularly by a government plant pathologist before release. However, after learning of the detection of Heliconia wilt in Hawaii, the Australian quarantine authorities immediately organized a survey of *Heliconia* from

Plate 13.4. Banana plantlets in tissue culture (photo: QDPI).

Hawaii recently imported into Australia. A number of plants were found to be wilting in a private post-entry quarantine house in a banana-growing area. Some were affected with strains of *R. solanacearum* that were not pathogenic on banana, but two plants were shown to carry *R. solanacearum* race 2, strain SFR. Fifty-five plants already released from quarantine because of an absence of obvious wilt symptoms during the 3-month quarantine period had been planted outside in the importer's nursery. Two were found to be diseased and removed. The nursery was quarantined, unaffected *Heliconia* returned to quarantine, banana plants on the property destroyed and soil that might have been contaminated removed and treated with formalin (Jones, 1991; Akiew and Hyde, 1993). Luckily, the pathogen was contained and eradicated, due to prompt quarantine action. This incident highlights the need for extreme caution if *Heliconia* is allowed entry to banana-growing countries.

Sugar cane bacilliform virus (SCBV) is closely related serologically to BSV (see

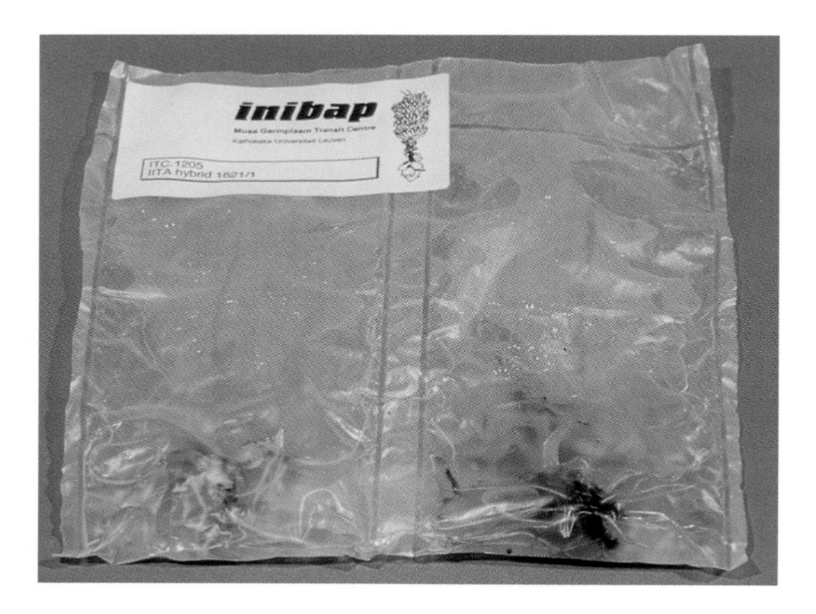

Plate 13.5. Polyethylene bags are now preferred by INIBAP to distribute rooted plantlets (photo: I. Van den Houwe, KUL).

Lockhart and Jones, Chapter 6, pp. 268–270) and they may be the same virus (Lockhart and Autrey, 1991). SCBV has been transmitted experimentally from *Saccharum officinarum* to banana by the pink sugar-cane mealybug (*Saccharicoccus sacchari*) to produce typical streak-disease symptoms (Lockhart and Autrey, 1991). Almost all noble canes and many interspecific hybrids tested in Mauritius carry SCBV (Autry *et al.*, 1991) and sugar cane must be looked on as a potential source of inoculum to infect banana. This movement of SCBV from sugar cane to banana may have occurred many times in tropical countries at locations where the two crops were grown next to one another. These events offer an alternative explanation to the widespread distribution of BSV. The revelation that BSV sequences are integrated into the genome of banana (LaFleur *et al.*, 1996) leads to speculation that the virus and *Musa* may have had a long association.

A number of alternative hosts outside the *Musaceae* have been suggested for BBTV (see Thomas and Iskra-Caruana, Chapter 6, pp. 250–251), but the reports need confirmation.

Principles of Safe Movement of *Musa* Germplasm

The safe movement of *Musa* germplasm was reviewed by Stover (1977), who outlined the danger of disease and pest transfer on traditional planting material and emphasized the advantages of using tissue-cultured plantlets for the first time.

The movement of conventional propagating material of *Musa*, such as rhizomes and suckers, from one banana-growing country to another should no longer take place. Even when accompanied by certification that the material has been inspected and found free of disease symptoms, there is still a risk, because certain pathogens may be carried without exhibiting symptoms. Only if the importing country has the capability of immediately initiating shoot-tip cultures from the imported material and testing plants derived from these cultures for virus pathogens, which may be carried as latent infections, should this be contemplated.

Seed of *Musa* and *Heliconia* imported into Australia is submerged in a solution of sodium hypochlorite (1% available chlorine) for 10 min as a precaution against the introduction of the Moko disease pathogen (Jones, 1991). If seeds of species in these genera need to be imported, they should also be sown in post-entry quarantine and plants observed regularly for disease symptoms.

The safest way of introducing new germplasm is as tissue culture (Plate 13.4). This eliminates the risks posed by fungi, bacteria and nematode pathogens, as these would almost certainly contaminate the culture medium if present. Artificial media used to culture germplasm introductions should not contain antibiotics or colouring, which might inhibit or mask contaminants.

The germplasm exchange system currently provided and administered by the International Network for the Improvement of Banana and Plantain (INIBAP) utilizes *in vitro* cultures (Van den Houwe and Jones, 1994). Proliferating tissue cultures are dispatched in tubes for those who need to further multiply the germplasm *in vitro*. Plantlets with roots that are ready for immediate planting are distributed in polyethylene bags (Plate 13.5).

Although the risks are significantly reduced using tissue cultures, they are not eliminated. The possibility of moving virus pathogens with tissue cultures of *Musa* still remains. In recent years, efforts have been made to develop and improve tests to detect *Musa* viruses in germplasm held in collections and in quarantine.

Development of a System for the Safe Movement of *Musa* Germplasm

Until relatively recently, the detection of banana viruses relied almost entirely on the inspection of plants for symptoms. Before the mid-1980s, the only threats recognized were from BBTV and CMV.

Introduced banana was usually grown for 9–12 months in quarantine and examined at regular intervals for disease symptoms. As it had been reported that some strains of CMV could be carried as latent infections (Stover, 1972), the routine inoculation of cotyledons of indicator plants, such as cowpea (local lesion host) and cucumber (systemic host), with sap from young banana leaves was advocated.

Soon after its creation in November 1984, INIBAP established an *in vitro* banana germplasm collection at the Laboratory of Tropical Crop Husbandry of the Katholieke Universiteit Leuven (KUL), Belgium. The function of this collection was not only to conserve valuable germplasm, but to actively distribute superior clones to departments of agriculture, regional centres, international institutes and universities undertaking banana improvement programmes. To reflect this aspect of its work, the facility was named the INIBAP Transit Centre (ITC). The storage, multiplication and transfer of tissue-cultured material began in 1985 and it quickly became obvious that germplasm health was an important issue that needed to be addressed. One of the major concerns was to ensure that BBTV was not introduced to the western hemisphere through germplasm exchange. Another problem emerging worldwide with the development of tissue-culture techniques was how to ensure that material selected as a source of shoot tips for culture initiation and subsequent mass propagation was free of virus disease.

It was now even more important than before to develop quick, reliable tests that would detect banana viruses (Dale, 1988). Virology laboratories in Australia, France, the Philippines and Taiwan became involved in projects to characterize BBTV and to develop rapid indexing tests that would allow for its early detection in germplasm in quarantine and material selected for *in vitro* propagation.

In 1988, the Food and Agriculture Organization of the United Nations (FAO), the International Bureau of Plant Genetic Resources (IBPGR) and INIBAP organized a meeting of banana virologists and quarantine experts at the University of the Philippines, Los Baños (UPLB), to define protocols for the international transfer of *Musa* germplasm. The meeting reaffirmed that all *Musa* movement should be by tissue culture and that the only quarantine risk from *in vitro* material was the possible spread of virus diseases. The virus diseases identified were bunchy top, banana mosaic, banana streak and bract mosaic, which had recently been recognized in the Philippines as something different from mosaic. The detection of banana virus diseases by visual observation was seen as inadequate, following reports proving that BBTV could be carried in banana tissue cultures without expressing obvious symptoms (Drew *et al.*, 1989). Unfortunately, although herbaceous indicator and serological tests were available to detect CMV, reliable diagnostic tests for other banana viruses were only just being developed. The spread of BBTV was perceived to be the biggest threat associated with the movement of banana in tissue culture, and protocols advocated at this time for international transfer and quarantine reflected this concern. It was thought important that the quarantine environment should be optimum for the expression of BBTV symptoms and that, as a consequence, plants in quarantine should be grown at 32°C under conditions of high natural light, if necessary supplemented with artificial light. The recommendations of the meeting were published as the *FAO/IBPGR Technical Guidelines for the Safe Movement of* Musa *Germplasm* (Frison and Putter, 1989).

As a result of the UPLB meeting, an international system of safe germplasm transfer was initiated under the auspices of INIBAP. INIBAP established two virus indexing centres, utilizing the banana virology expertise available at the Queensland Department of Primary Industries (QDPI) in Brisbane, Australia, and the Centre de coopération internationale en recherche agronomique pour le développement (CIRAD) in Montpellier, France. The task of these centres was to quarantine and test accessions held by INIBAP at the ITC so

Plate 13.6. Young banana plants established from introduced tissue cultures in the banana quarantine facility in Brisbane, Australia (photo: D.R. Jones, QDPI).

that they could be cleared for international movement. Approximately 20 tubes of proliferating tissue cultures of each *Musa* germplasm accession are held at the ITC. All this material is derived from a single shoot tip. Following recommendations made in the *FAO/IBPGR Technical Guidelines for the Safe Movement of* Musa *Germplasm*, a representative sample of five plantlets per accession was dispatched to an INIBAP virus indexing centre. Four were grown in quarantine for 1 year (or longer if the eighth-leaf growth stage had not been reached after 1 year), inspected weekly for virus symptoms and indexed for CMV. The fifth plant was held in reserve in case of loss of a test plant. As an extra precaution against the spread of BBTV, accessions originating from continents where bunchy top was present were sent for testing at both of INIBAP's virus indexing centres. When an accession had been cleared, INIBAP could distribute tissue cultures derived from stocks remaining at the ITC.

The next step forward came in 1990, when INIBAP organized a meeting of bunchy-top specialists at the laboratories of CIRAD's Institut de recherche sur les fruits et agrumes (IRFA) in Montpellier, France. The objective of this meeting was to compare the various diagnostic tests for BBTV that had been developed and make recommendations on indexing requirements. An addendum to the *FAO/IBPGR Technical Guidelines on the Safe Movement of* Musa *Germplasm* was published as a result of the meeting. This addendum detailed agreed changes to the BBTV indexing protocol, which now included the use of BBTV-specific monoclonal antibodies (for use in enzyme-linked immunosorbent assay (ELISA) tests) and complementary DNA (cDNA) probes. Petiole sap was recommended to test for BBTV, as virus concentration was high in phloem tissue, and it was thought that tests should be conducted twice during the quarantine period. It was deemed important that ten new leaves should be produced by each plant in quarantine, the youngest being at least 1 m in length, to allow for the expression of bunchy top symptoms (Iskra-Caruana, 1994). The 'state-of-the-art' quarantine protocols being used in Australia at this time to prevent the entry of banana pathogens reflect the leading role that virologists and quarantine specialists in this country were playing in determining international standards (Plate 13.6) (Jones, 1991; Thomas, 1991).

Plate 13.7. *FAO/IPGRI Technical Guidelines for the Safe Movement of Germplasm No. 15* Musa *spp.*, 2nd edn, edited by Diekmann and Putter (1996) (photo: D.R. Jones, INIBAP).

Over the next 5 years, it became evident that BBTV, although still important, was not the main cause for concern. Not once was BBTV found in ITC germplasm indexed at INIBAP virus indexing centres. In fact, accessions that were originally collected because they were infected with BBTV also indexed negative, indicating a possible loss of the virus after years of storage and subculturing. Work in Australia had also suggested that BBTV is not always passed on to all plantlets derived from infected tissue cultures (Drew *et al.*, 1992).

BSV was the pathogen being detected more and more frequently at INIBAP virus indexing centres in germplasm from many countries and also from banana-breeding programmes. Previously, BSV had only been found in a few African countries, but it now appeared to have a worldwide distribution. Streak symptoms were often slow to develop in plants in quarantine and sometimes they failed to appear during the quarantine period. The most reliable technique for detecting BSV at the time was the examination of leaf sap under the electron microscope for bacilliform particles, and this became mandatory at INIBAP virus indexing centres. Antisera also became available as a result of research into SCBV and BSV that was being undertaken at the University of Minnesota by B.E.L. Lockhart. However, there were serious doubts as to whether they could detect all strains of the virus. At this time, as part of its International *Musa* Testing Programme sponsored by the United Nations Development Programme (UNDP), INIBAP commissioned Dr Lockhart to develop a reliable test that would detect all strains of BSV.

Increasingly, through the routine use of the electron microscope to screen ITC and other germplasm, isometric and filamentous particles that could not be identified were being found in banana sap (Jones, 1994) (see Thomas *et al.*, Chapter 6, pp. 247–278). Sometimes symptoms were associated with the presence of these viruses, but often no obvious symptoms could be seen. The significance of the viruses was unknown and work was needed to clarify the situation. Up until 1995, the Philippines was the only known location of bract mosaic disease. In 1995, it was confirmed that banana bract mosaic virus (BBrMV) was the probable cause of kokkan disease of 'Nendran' (AAB, Plantain subgroup) in India. BBrMV was also detected in banana specimens from Sri Lanka (Thomas *et al.*, 1997). It became apparent that the distribution of this serious pathogen was greater than had been previously thought and that a reliable diagnostic test for BBrMV was urgently needed. INIBAP proposed and later funded research to characterize, investigate and develop diagnostic tests for BBrMV and the unknown viruses seen under the electron microscope.

Because INIBAP had been adjusting its indexing protocols and requirements in line with new knowledge on the detection and distribution of *Musa* viruses, the procedures being advocated by INIBAP in 1994 to screen banana germplasm for virus diseases differed significantly from those outlined in the *FAO/IBPGR Technical Guidelines on the Safe Movement of* Musa *Germplasm* (Jones and Iskra-Caruana, 1994). It became obvious that another meeting of banana virus experts was necessary to update the *FAO/IBPGR Technical Guidelines* of 1989 and its addendum of 1990.

Current Recommendations for the Safe Movement of *Musa* Germplasm

The recommendations of the meeting of banana virologists that took place under the auspices of FAO and the International Plant Genetic Resources Institute (IPGRI) in Rome in June 1995 closely followed the protocols for virus detection followed by INIBAP (Jones, 1996). However, modifications advocated included the testing of a larger proportion of a tissue-cultured accession held at the ITC, to reduce the chances of the sample being unrepresentative of the health status of the 'mother stock', and a shortening of the quarantine period, because of increased confidence in indexing tests.

The recommendations, as the new *FAO/IPGRI Technical Guidelines for the Safe Movement of Germplasm* Musa *spp.* (Diekmann and Putter, 1996) (Plate 13.7), are available on request from IPGRI. The guidelines offer the best possible advice at the time of publication for institutions and commercial companies involved in the international transfer of *Musa*. They also include descriptions of abacá mosaic, banana mosaic, bract mosaic, bunchy top and banana streak diseases and the protocols of tests to detect the virus pathogens.

Since the 1995 Rome meeting, it has been recognized that the high rate of BSV infection in bred tetraploids may be related to the *de novo* synthesis of the virus from BSV genomic sequences integrated into the DNA of *Musa*. This synthesis seemed to be associated with *in vitro* culture and possibly other stress factors related to embryo rescue (see Lockhart and Jones, Chapter 6, pp. 263–274). In addition, banana die-back, a new virus disease problem, has been reported from West Africa (see Hughes and Thomas, Chapter 6, p. 278). Further advances in our knowledge of banana virus problems will undoubtedly continue to be made that will eventually require further revision to the *Technical Guidelines* of 1996.

General recommendations of the 1996 Technical Guidelines

Germplasm should be obtained from the safest source possible, such as the ITC (Laboratory of Tropical Crop Improvement, Katholieke Universiteit Leuven, Kardinal Mercierlaan 92, 3001 Heverlee, Belgium) and the indexed collection of the International Institute of Tropical Agriculture (IITA) (Plantain and Banana Improvement Programme, IITA, c/o L.W. Lambourn & Co., Carolyn House, 26 Dingwall Road, Croydon CR9 3EE, UK).

All germplasm should be moved in the form of tissue culture. If this is not possible, full quarantine measures must be taken until the vegetative material or seed is cultured *in vitro*.

Germplasm should be tested for all viruses known to affect *Musa*, according to the protocols specified in the *Technical Guidelines*. However, in some instances, tests may be omitted if there is strong, reliable evidence that particular viruses are not present in the country of origin of the germplasm.

Indexing procedures and results should be documented, e.g. in a germplasm health statement.

Technical protocols – vegetative material

1. Select a sucker from a plant without symptoms of systemic infection.
2. Soil, roots and any other extraneous material should be removed from the sucker. The corm should then be trimmed down to a small block of tissue containing the meristem. About 10 cm of tissue should lie above the meristem. The dimensions of the tissue block should be about 20 cm high and 10–15 cm in diameter. The block should be air-dried for 2–3 days and

wrapped in newspaper. The material should be labelled and dispatched in a cardboard box. No plastic should be used for wrapping.

3. The material should be sent for tissue-culturing in the country of origin. However, if this is not possible, the material should be sent to an appropriate tissue-culture laboratory in another country, but preferably one not in a banana-growing area.

4. Meristems should be excised, surface-disinfected and cultured.

5. The meristem culture should be cloned to seven plantlets, of which five should be sent to an indexing facility and two should remain in culture for future multiplication.

6. At the indexing facility, four plants should be established in a vector-free, insect-proof greenhouse under conditions conducive to vigorous plant growth. The fifth plant serves as a backup in case of the death of a test plant.

7. After 3 months of growth, tissue samples should be taken from the three youngest expanded leaves and indexed for viruses.

8. Three months later, tissue samples should again be taken from the three youngest expanded leaves and indexed for virus. In addition, electron-microscopic observations should be undertaken to look for the presence of other viruses.

9. If all the tests are negative, the four indexed plants may be released and the cultures derived from the two remaining subclones may be further propagated and distributed *in vitro*. For the movement of *in vitro* material, neither charcoal, fungicides nor antibiotics should be added to the medium. *In vitro* cultures should be shipped in transparent tubes and visually inspected for bacteria, fungi and arthropods. Contaminated material should be destroyed.

Technical protocols – seed

Seed should be free of pulp, air-dried, inspected for the absence of insect pests and fumigated, if necessary. The seed should be sent for tissue-culturing in the country of origin. However, if this is not possible, the seed should be sent to an appropriate tissue-culture facility in another country, but preferably one in a non-banana-growing area.

1. Seed should be surface-disinfected with 0.5% sodium hypochlorite for 10 min at room temperature to eliminate pathogens carried on the seed surface.

2. The seed-coat should be removed before culturing *in vitro*.

3. Seedlings should be indexed in the same way as material derived from meristem culture.

REFERENCES

Akiew, E. and Hyde, K.D. (1993) First detection of *Pseudomonas solanacearum* race 2, strain SFR, on *Heliconia* in Australia. *Plant Disease* 77, 319.

Anon. (1993) *Musa*News. *Infomusa* 2(2), 21–23.

Autry, L.J.C., Boolell, S. and Lockhart, B.E. (1991) Sugar cane bacilliform virus: symptomology, diagnosis and distribution in cane germplasm in Mauritius. In: *International Society of Sugar Cane Technologists, 3rd Sugar Cane Pathology Workshop, Mauritius, 22–26 July 1991, Abstracts*, p. 15.

Bancroft, J. (1876) Report of the board appointed to enquire into the causes of disease affecting livestock and plants, Queensland, 1876. *Votes and Proceedings* 1877(3), 1011–1038.

Buddenhagen, I.W. (1986) Bacterial wilt revisited. In: Persley, G.J. (ed.) *Bacterial Wilt Disease in Asia and the South Pacific.* ACIAR Proceedings No. 13, ACIAR, Canberra, Australia, pp. 126–143.

Dale, J.L. (1987) Banana bunchy top, an economically important tropical plant virus disease. *Advances in Virus Research* 33, 301–325.

Dale, J.L. (1988) The status of disease indexing and the international distribution of banana germplasm. In: *Conservation and Movement of Vegetatively Propagated Germplasm:* In Vitro *Culture and Disease Aspects.* International Board for Plant Genetic Resources, Rome, Italy, pp. 43–46.

Daniells, J., Thomas, J.E. and Smith, M. (1995) Seed transmission of banana streak virus confirmed. *Infomusa* 4(1), 7.

Diekmann, M. and Putter, C.A.J. (1996) *FAO/IPGRI Technical Guidelines for the Safe Movement of Germplasm, No. 15,* Musa *spp.*, 2nd edn. Food and Agriculture Organization of the United Nations, Rome/International Plant Genetic Resources Institute, Rome, 26 pp.

Drew, R.A., Moisander, J.A. and Smith, M.K. (1989) The transmission of banana bunchy top virus in micropropagated bananas. *Plant Cell, Tissue and Organ Culture* 16, 187–193.

Drew, R.A., Smith, M.K. and Anderson, D.W. (1992) Field evaluation of micropropagated bananas derived from plants containing banana bunchy top virus. *Plant Cell, Tissue and Organ Culture* 28, 203–205.

Eden-Green, S.J. (1994) *Banana Blood Disease. Musa* Disease Fact Sheet No. 3, INIBAP, Montpellier, France.

Ferreira, S., Pitz, K., Alvarez, A. and Isherwood, M. (1991) Heliconia wilt in Hawaii. *Phytopathology* 81, 1159.

Frison, E.A. and Putter, C.A.J. (eds) (1989) *FAO/IBPGR Technical Guidelines for the Safe Movement of* Musa *Germplasm.* FAO/IBPGR, Rome, Italy, 23 pp.

Gold, A.H. (1972) Seed transmission of banana viruses. *Phytopathology* 62, 760.

Iskra-Caruana, M.L. (ed.) (1994) *Second 'Hands On' Workshop on Banana Bunchy Top Disease, Report, Montpellier, France, 30 July–10 August 1990.* INIBAP, Montpellier, France, 38 pp.

Jones, D.R. (1984) Failure of the black Sigatoka eradication campaign in the Torres Strait region. *Australasian Plant Pathology* 13, 57–58.

Jones, D.R. (1990) Black Sigatoka – a threat to Australia. In: Fullerton, R.A. and Stover, R.H. (eds) *Sigatoka Leaf Spot Diseases of Bananas, Proceedings of an International Workshop held at San José, Costa Rica, 28 March–1 April 1989.* INIBAP, Montpellier, France, pp. 38–46.

Jones, D.R. (1991) Banana quarantine in Australia. In: Valmayor, V.V., Umali, B.E. and Besjosano, C.P. (eds) *Banana Diseases in Asia and the Pacific: Proceedings of a Technical Meeting on Diseases Affecting Banana and Plantain in Asia and the Pacific, Brisbane, Australia, 15–18 April 1991.* INIBAP, Montpellier, France, pp. 133–143.

Jones, D.R. (1994) Risks involved in the transfer of banana and plantain germplasm. In: Jones, D.R. (ed.) *The Improvement and Testing of* Musa*: A Global Partnership, Proceedings of the First Global Conference of the International* Musa *Testing Program held at FHIA, Honduras, 27–30 April 1994.* INIBAP, Montpellier, France, pp. 85–98.

Jones, D.R. (1996) The safe movement of *Musa* germplasm. In: Croft, B.J., Piggin, C.M., Wallis, E.S. and Hogarth, D.M. (eds) *Sugarcane Germplasm Conservation and Exchange, Report of an International Workshop held in Brisbane, Queensland, Australia, 28–30 June 1995.* ACIAR Proceedings No. 67, Australian Centre for International Agricultural Research, Canberra, Australia, pp. 85–89.

Jones, D.R. and Alcorn, J.L. (1982) Freckle and black Sigatoka diseases of banana in far north Queensland. *Australasian Plant Pathology* 11, 7–9.

Jones, D.R. and Iskra-Caruana, M.-L. (1994) Screening banana and plantain at INIBAP's virus indexing centres. In: Jones, D.R. (ed.) *The Improvement and Testing of Musa: A Global Partnership, Proceedings of the First Global Conference of the International Musa Testing Program held at FHIA, Honduras, 27–30 April 1994.* INIBAP, Montpellier, France, pp. 99–105.

LaFleur, D.A., Lockhart, B.E.L. and Olszewski, N.E. (1996) Portions of the banana streak badnavirus genome are integrated in the genome of its host *Musa* spp. *Phytopathology* 86(11), S100.

Lockhart, B.E.L. and Autrey, L.J.C. (1991) Mealybug transmission of sugarcane bacilliform and sugarcane clostero-like viruses. In: *International Society of Sugar Cane Technologists, 3rd Sugar Cane Pathology Workshop, Mauritius, 22–26 July 1991, Abstracts*, p. 17.

Ong, K.P. (1996) Fusarium wilt of Cavendish banana in a commercial farm in Malaysia. In: Frison, E.A., Horry, J.P. and De Waele, D. (eds) *New Frontiers in Resistance Breeding for Nematode, Fusarium and Sigatoka: Proceedings of an International Banana Breeding Workshop held in Serdang, Selangor, Malaysia, 2–5 October, 1995.* INIBAP, Montpellier, France, pp. 211–217.

Simmonds, N.W. (1962) *The Evolution of the Bananas.* Longmans, Green, London, 170 pp.

Stover, R.H. (1962) *Fusarial Wilt (Panama Disease) of Bananas and other* Musa *Species.* Commonwealth Mycological Institute, Kew, Surrey, UK, 117 pp.

Stover, R.H. (1972) *Banana, Plantain and Abaca Diseases.* Commonwealth Mycological Institute, Kew, Surrey, UK, 319 pp.

Stover, R.H. (1977) Banana (*Musa* spp.). In: Hewitt, W.B. and Chiarappa, L. (eds) *Plant Health and Quarantine in International Transfer of Genetic Resources.* CRC Press, Boca Raton, Florida, USA, pp. 71–79.

Thomas, J.E. (1991) Virus indexing procedures for banana on Australia. In: Valmayor, V.V., Umali, B.E. and Besjosano, C.P. (eds) *Banana Diseases in Asia and the Pacific: Proceedings of a Technical Meeting on Diseases Affecting Banana and Plantain in Asia and the Pacific, Brisbane, Australia, 15–18 April 1991.* INIBAP, Montpellier, France, pp. 144–157.

Thomas, J.E., Geering, A.D.W., Gambley, C.F., Kessling, A.F. and White, M. (1997) Purification, properties and diagnosis of banana bract mosaic potyvirus and its distinction from abaca mosaic potyvirus. *Phytopathology* 87, 698–705.

Van den Houwe, I. and Jones, D.R. (1994) *Musa* germplasm distribution from the INIBAP Transit Center. In: *The Improvement and Testing of* Musa*: a Global Partnership, Proceedings of the First Global Conference of the International Musa Testing Program held at FHIA, Honduras, 27–30 April 1995.* INIBAP, Montpellier, France, pp. 99–105.

14

Banana Breeding for Disease Resistance

HISTORY OF BANANA BREEDING

D.R. Jones

Introduction

Commercial banana growing for export has always been a monoculture based firstly on 'Gros Michel' (AAA) and, since the 1960s, on the almost genetically identical members of the Cavendish subgroup (AAA). The dangers of monoculture are well known. Disease epidemics can arise that result in the simultaneous decimation of a crop in entire regions.

Fusarium wilt and its catastrophic inroads into the 'Gros Michel' plantations in the Latin American–Caribbean region was the stimulus that prompted the development of the first banana-breeding programmes. Planting material used to establish new farms invariably introduced the disease and it wasn't long before levels of Fusarium wilt increased to a point where production became uneconomic. As the pathogen could survive for long periods in infested soil, there was no recourse but to move to new land. By the 1920s, export banana production had become shifting cultivation. A solution was needed to overcome the problem.

Beginnings in the West Indies

Banana breeding began at the Imperial College of Tropical Agriculture (ICTA) in Trinidad in 1922 (Cheesman, 1931) and later in Jamaica in 1924 (Larter, 1947). The goal of these two collaborating programmes was to produce a 'Gros Michel' resistant to Fusarium wilt. Both centres formed germplasm collections and shared interesting accessions. Much of the early work on taxonomy, cytology and cytogenetics of *Musa* spp. and cultivars was undertaken at ICTA (Shepherd, 1968). An important finding was that the basic haploid chromosome number of banana in the *Eumusa* series is 11. 'Gros Michel' and many other cultivars were shown to be triploids, with 33 chromosomes, and others, including wild species in the *Eumusa* section, were revealed to be diploids, with 22 chromosomes. It was also discovered that, when pollen from a diploid was applied to female flowers of 'Gros Michel', a few seeds were produced that yielded tetraploid hybrids, with 44 chromosomes. This led to a pathway being opened for breeding, as it was realized that it might be possible for these hybrids, with their pronounced 'Gros Michel' features, to inherit useful characteristics, such as disease resistance, from the diploid parent. In about 1928, 'I.C.2', the first AAAA hybrid banana released by ICTA, was obtained from 'Gros Michel' crossed with a wild *Musa acuminata*. It was distributed widely, but succumbed to Fusarium wilt in Honduras, where it was known as 'Golden Beauty', and was not grown commercially.

Soon after the arrival of Sigatoka in Trinidad in 1933, accessions in the banana collection and ICTA tetraploids were evaluated for their reaction to the disease. A high degree of resistance was found among wild species, including *M. acuminata* subspecies used as pollen sources in the breeding programme. Many hybrids were also rated as resistant to Sigatoka, unlike 'Gros Michel', their female parent. It was concluded that the best hope of obtaining an export banana with resistance to Sigatoka lay in the same breeding strategy as had been developed to obtain resistance to Fusarium wilt (Cheesman and Wardlaw, 1937).

Few tetraploid hybrids of 'Gros Michel' were produced at ICTA, this function being transferred to Bodles in Jamaica. After the Second World War, ICTA concentrated on the genetic improvement of diploids by crossing and selection, as it was believed that, with a genetically fixed female contribution, the success of hybrids must depend totally on the positive genetic qualities transmitted by the pollen (Dodds, 1943). Collecting expeditions were made to East Africa in 1948 and the Pacific and South-East Asia in 1954/55 to augment the germplasm available for diploid breeding.

In the 1950s, AAAA tetraploids were generated in Jamaica by crossing 'Gros Michel', and later its dwarf variant 'Highgate', with unimproved diploids. 'Bodles Altafort' or '1847', which was a 'Gros Michel' × 'Pisang Lilin' cross, and '2390–2', a cross between 'Highgate' and 'Pisang Lilin', were released for distribution as Fusarium wilt-resistant 'Gros Michel' types during this decade. '2390–2' was, and still is, grown by a few banana farmers in Australia, but was too short-fruited to be considered seriously for export from the Latin American–Caribbean region.

The West Indies programme, with K. Shepherd as leading breeder, became centred at Bodles and managed by the Banana Board of Jamaica in 1960. Male parents were initially selected from crosses between *M. acuminata* ssp. *banksii* (collected in Western Samoa and named 'Samoa') and 'Paka', an AA diploid collected in Zanzibar. AAAA tetraploids bred when these pollen donors were crossed with 'Highgate' included 'T6' ('61–86') and 'T8' ('61–882–1'). Unfortunately, resistance to Sigatoka and Fusarium wilt was not high and fruit size differed significantly between early and late hands. Because of these defects, they were not adopted commercially. However, tests in the Cook Islands in the early 1980s showed that both 'T6' and 'T8' had very high resistance to black leaf streak (Gonsalves, 1987). 'T8' has been supplied to small-scale farmers in the Torres Strait–Cape York region of Australia to act as a buffer against the further spread of black leaf streak in Australia (Jones, 1990) and was introduced to countries in the South Pacific as part of a crop improvement programme (Jones, 1993).

In the early 1960s, the problem of Fusarium wilt became less important, because the Fusarium wilt-resistant cultivars in the Cavendish subgroup were replacing 'Gros Michel' as the banana of the export trade. However, Cavendish cultivars, such as 'Gros Michel', were susceptible to Sigatoka and resistance to this serious disease was seen as being of increasing importance. Unfortunately, Cavendish cultivars are sterile and conventional breeding methods could not be used to improve this subgroup. Therefore, the West Indies programme still looked to improve 'Gros Michel' as a means of breeding a better dessert banana for the export market.

A wild-type, but parthenocarpic, male parent derived from a cross between *M. acuminata* ssp. *malaccensis* (collected in Kedah State in West Malayasia) and *M. acuminata* ssp. *banksii* ('Samoa') was used for breeding in the mid-1960s. AAAA tetraploids, which were obtained in crosses with 'Highgate', had quite high resistance to Sigatoka and Fusarium wilt and fruit with an acceptable taste. Unfortunately, simulated shipment trials revealed some undesirable horticultural features. Fruit had a short green life and ripened rapidly and fingers had weak pedicels, which caused them to drop from crowns when ripe. The hybrids were never as commercially viable as cultivars in the Cavendish subgroup, which became a standard by which new hybrids for the export trade were measured. Nevertheless, 'Calypso' ('65–3405–1') was tested in the Philippines as a possible export cultivar, because of its reported moderate resistance to black leaf streak in the Cook Islands (Gonsalves, 1987). In addition, 'Buccaneer' ('65–168–12'), which was found to be highly resistant to black leaf streak in the Cook Islands (Gonsalves, 1987), was acquired by Brazil and Australia for possible use as a buffer cultivar to control the spread of black leaf streak.

Most Jamaican tetraploids in this second wave produced bananas which had excellent cooking qualities when green and made tasty snacks, similar in texture and flavour to potato, when cut thin and fried in hot fat to make chips. 'Ambrosia,' ('64–2596') rated very highly in taste tests when ripe, but was susceptible to black leaf streak, a trait which has limited its potential for local production (K. Shepherd, Portugal, 1997, personal communication).

The Banana Breeding Scheme of Jamaica continued until financing problems in the 1970s forced work to slow down and eventually almost cease by 1980. In the later stages of the programme, diploids were further improved by crossing the *M. acuminata* ssp. *malaccensis* × *banksii* ('Samoa') diploids with various others, including *M. acuminata* ssp. *banksii* ('Samoa') × 'Paka'. The breeding diploid 'M53', which is now being utilized in plantain improvement in Cameroon, was one successful product. Superior diploids were crossed with 'Highgate' and the last batch of tetraploids evaluated included '72–1242', which had resistance to Sigatoka and Fusarium wilt and uniform hands and fingers, with only a slight tendency to finger drop. This was later released to Jamaican growers for local consumption.

The Jamaican programme continued after 1980 under the direction of R.A. Gonsalves, Director of the Banana Board, although output was greatly reduced because of limited funds. Tetraploids were crossed with improved diploids to obtain secondary triploids (see 'Triploid Breeding' below). One AAA triploid, designated 'RG1', is reputed to have a satisfactory green life, tolerance to nematodes and resistance to Sigatoka leaf spots and Fusarium wilt. It can also be cooked green. However, one drawback of 'RG1'is its long fingers, which make the fruit unsuitable for packing in regular cartons. 'RG1' is being multiplied in tissue culture for release to Jamaican growers and for further trials (R.A. Gonsalves, Jamaica, 1998, personal communication).

Simmonds (1966), Shepherd (1968, 1974), Menendez and Shepherd (1975) and Gonsalves and Dixon (1990) have recorded the accomplishments of the pioneer banana-breeding programmes in the Caribbean.

Honduras

The United Fruit Company had a brief breeding programme in Panama in the 1920s. Germplasm was initially obtained from Cuba, Costa Rica and Panama, but this was augmented by 134 accessions of wild species and cultivars collected in South-East Asia, India, Australia and New Guinea between 1921 and 1927 by O.A. Reinking, a German expert on *Fusarium*. With the onset of the depression, the programme was cancelled and, in 1930, the germplasm was transferred to Tela in Honduras (Rowe and Richardson, 1975).

In the mid-1950s, the United Fruit Company tested Fusarium wilt-resistant clones with commercial potential from Reinking's collection as possible replacements for 'Gros Michel'. 'Valery' (AAA, Cavendish subgroup), which had been collected in Vietnam in 1925, was selected for commercial production.

The breeding programme of United Fruit was resurrected in the late 1950s and based at an experimental farm at La Lima, Honduras. The long-term objective of the programme was to combine all possible desirable qualities, such as disease and insect resistance, improved growth habit, better fruit quality and other favourable characteristics, in one commercial banana. To provide a broad genetic base for this venture, 72 of the original clones in the Tela collection were moved to La Lima in 1958 and more material added following expeditions undertaken by P.H. Allen in the Western Pacific and South-East Asia in 1959–1961. Rhizomes collected overseas were at first quarantined at the US Department of Agriculture (USDA) facilities in Maryland before being sent to Utila Island off the coast of Honduras for a further quarantine period. Later, accessions were sent directly to Utila. By 1964, 850 accessions had been assembled at La Lima, but a number were obvious duplicates. Losses occurred due to Fusarium wilt and Moko disease and, after a revision, the collection had been reduced to 574 accessions by 1972 (Rowe and Richardson, 1975).

Accessions were evaluated for their reaction to the tomato strain of *Ralstonia solanacearum* (Buddenhagen, 1962), burrowing nematode (Wehunt *et al.*, 1965, 1978), Sigatoka (Vakili, 1968) and Moko (Stover, 1972). Studies on the inheritance of resistance to disease, such as Fusarium wilt (Vakili, 1965) and Sigatoka (Vakili, 1968), were also conducted.

Prior to the1960s, the goal of the breeding programme was to produce a Fusarium wilt-resistant 'Gros Michel' type of banana. As in Jamaica, disease-resistant AA diploids, such as 'Pisang Lilin', were crossed with 'Gros Michel' to produce AAAA tetraploids. However, during the early 1960s, Cavendish cultivars were rapidly becoming the industry standard. Much higher yields per hectare could be achieved with Cavendish cultivars than with 'Gros Michel'. This was believed to be because Cavendish cultivars could be planted more densely and were faster suckering. A dwarf habit was also becoming an important requirement to lessen losses due to wind damage, and tetraploids derived from 'Gros Michel' were tall. Less tall variants of 'Gros Michel', such as the semi-dwarf 'Cocos', the shorter 'Highgate' from Jamaica and the even more dwarf 'Lowgate' found in a plantation in Honduras, were used as the female parent in an endeavour to produce hybrids that were nearer the new ideal height. It was now obvious that, for a hybrid derived from a 'Gros Michel'-type background to be commercially successful, it must combine all the desirable features of Cavendish cultivars with resistance to important diseases. This was not going to be an easy task. The best yield of a Fusarium wilt-resistant 'Cocos' × 'Pisang Lilin' cross was only 85% of 'Valery' (Rowe and Richardson, 1975). It became obvious that the production of a commercially viable hybrid could only be attempted if the pollen donor was not only resistant to disease, but also agronomically superior to those diploids currently available for breeding.

The synthesis of agronomically improved diploids with resistance to Fusarium wilt, burrowing nematode and black leaf streak, after its arrival in Central America in 1972, became the prime concern of the United Fruit breeding programme. The time from pollination until a bred diploid could be evaluated and utilized took 3 years. This time-consuming exercise was begun in the early 1960s by D.L. Richardson and was later continued by P.R. Rowe.

In 1984, a discouraged United Brands Company (formally United Fruit) donated their breeding programme to the Fundación Hondureña de Investigación Agrícola (FHIA). With funding from international donor agencies, the breeding emphasis shifted from the so far unsuccessful undertaking of trying to develop an export banana superior to Cavendish cultivars to producing disease-resistant hybrids that could be used by subsistence farmers in developing countries. In this endeavour, the improved diploids, which had been under development since the 1960s, were to play a pivotal role. Promising new tetraploid hybrids have emerged in recent times (Rowe and Rosales, 1994; see Rowe and Rosales, this chapter, pp. 440–448).

In 1991, FHIA became the first breeding programme to test promising hybrids for reaction to black leaf streak disease at various locations around the world under the first phase of an International *Musa* Testing

Programme (IMTP) coordinated by the International Network for the Improvement of Banana and Plantain (INIBAP) and funded by the United Nations Development Programme (UNDP). Three tetraploids, designated 'FHIA-01'® (AAAB), 'FHIA-02' (AAAA) and 'FHIA-03' (AABB), were shown to be highly resistant (Jones, 1994b) and distributed widely for further testing and evaluation (Jones, 1994c). 'FHIA-01'®, which is also resistant to *Fusarium oxysporum* f. sp *cubense* race 4 in Australia, has fruit with a sweet-acid taste, similar to 'Pome' (AAB). 'FHIA-03', a robust cooking banana with drought tolerance, has fruit that can also be eaten fresh when ripe and may have an important role to play in Africa (Plate 14.1). The long-term diploid breeding programme was beginning to bear fruit.

Since IMTP phase 1, other FHIA black leaf streak-resistant hybrids have gained favour. These include 'FHIA-21'® (AAAB), a plantain-type banana with a large bunch, which has found an important niche in several Central American countries, and 'FHIA-18' (AAAB), a sibling of 'FHIA-01'®, with fruit of a more agreeable flavour, which is very popular in Brazil.

More information on the achievements of the FHIA breeding programme is presented later (see Rowe and Rosales, this chapter, pp. 435–449).

Brazil

Apart from a small-scale endeavour that began in India in 1949 (Sathiamoorthy, 1994), the only major breeding programmes up until 1980 were those centred in Jamaica and Honduras. However in 1982, a programme was launched in Brazil under the auspices of Empresa Brasiliera de Pesquisa Agropecuaria–Centro Nacional de Pesquisa de Mandioca e Fruticultura (EMBRAPA–CNPMF) and based at Cruz das Almas in Bahia State. The aim of the programme was to improve the locally popular 'Prata' (AAB, Pome subgroup) and 'Maçã' (AAB, syn. 'Silk') cultivars, particularly in respect to resistance to Fusarium wilt, Sigatoka and also black leaf streak, which, it was realized, would one day spread to Brazil. K. Shepherd, formerly of the Jamaican programme, was recruited as consultant and he began by establishing the genetic base necessary for conventional breeding work. Germplasm was obtained on a collecting trip to India, the Philippines, Papua New Guinea and Hawaii in 1982 and was supplemented by other accessions from Ecuador in 1983 and Guadeloupe, Thailand and Indonesia in 1985.

The primary task was to produce disease-resistant, tetraploid hybrids with characteristics of 'Prata', this being the most common type of banana in Brazil. 'Prata' has poor agronomic traits and increases in yield were also expected. As in the West Indies and Honduras, improvement of the diploid pollen parent was seen as the key to success.

Testing diploid cultivars, synthetic diploids and hybrids for resistance to Fusarium wilt and Sigatoka has been undertaken at Cruz das Almas. Tetraploids have been tested against black leaf streak in Costa Rica and are now being evaluated elsewhere under INIBAP's IMTP. One tetraploid clone, designated 'PA 12–03' (AAAB), was derived from a 'Prata Anã' (a dwarf 'Prata') × 'Pisang Lilin' cross. It was found to have useful resistance to Sigatoka and a higher yield potential than other 'Prata' types in Brazil. 'PA 12–03' has been released for distribution to farmers in Brazil under the name 'Pioniera'.

Progress in the breeding programme has been summarized by Shepherd *et al.* (1994) and de Oliveira e Silva *et al.* (1997a, b).

Africa

The 1980s saw a revival in interest in banana research, especially banana breeding, mainly as a result of the spread of black leaf streak disease to West Africa, which threatened the livelihood of subsistence plantain growers. The growing susceptibility of Cavendish cultivars to Fusarium wilt in subtropical areas, including South Africa, was a further stimulus.

Plate 14.1. FHIA Breeder P.R. Rowe beside a bunch of 'FHIA-03' at La Lima, Honduras (photo: D.R. Jones, INIBAP).

Plate 14.2. CIRAD Breeder C. Jenny beside 'Pisang Klutuk Wulung', a black-stemmed variant of *Musa balbisiana*, in the CRBP germplasm collection in Nyombe, Cameroon (photo: D.R. Jones, INIBAP).

Plate 14.3. IITA Breeders R. Ortiz (left) and D. Vuylsteke (right) beside a display at Onne in Nigeria, which shows bunches of 'Obino l'Ewai' (AAB, Plantain subgroup) (left) and *Musa acuminata* ssp. *burmannicoides* accession 'Calcutta 4' (right), the parents of the black leaf streak-resistant, plantain-like tetraploid 'TMPx 6930–1' (centre) (photo: D.R. Jones, INIBAP).

The situation in developing countries in Africa was perceived as serious and resulted in the formation in 1984 of INIBAP by a group of international donors. INIBAP was to coordinate and stimulate international banana research for the benefit of the small-scale banana grower. In 1986, INIBAP, the Australian Centre for International Agricultural Research (ACIAR) and the Queensland Department of Primary Industries (QDPI) organized a workshop in Australia, which brought together banana breeders and banana experts from around the world. The result of the meeting was the identification of global and regional research and breeding needs (Persley and De Langhe, 1987). The improvement of cultivars in the Plantain subgroup (AAB), locally important AAB dessert banana types, such as 'Silk' and 'Pome', and export Cavendish were high on the agenda. Since then, important breeders' meetings have been held in France in 1992 (Ganry, 1993) and Honduras in 1994 (Anon., 1994; Jones 1994a) and breeding goals redefined (Jones, 1994d).

Collections of cultivars in the Plantain subgroup were assembled and characterized by the Centre régional bananiers et plantains (CRBP) at Nyombe in Cameroon and by the International Institute of Tropical Agriculture (IITA) at Onne in Nigeria. Conventional breeding programmes, aimed at improving the resistance of plantain to black leaf streak, were later established at these institutes. The initial strategy of both programmes was to produce disease-resistant, plantain-like tetraploids pollinating with fertile plantain cultivars with pollen from diploids. At CRBP, improved Jamaican diploids, obtained from EMBRAPA–CNPMF, are being repeatedly selfed to increase homozygosity and then used as the male parent in crosses with plantain cultivars (Jenny *et al.*, 1994) (Plate 14.2). At IITA, black leaf streak-resistant tetraploids have been produced using 'Obino l'Ewai' from Nigeria and 'Bobby Tannap' from Cameroon as the triploid female parents and *M. acuminata* ssp. *burmannicoides* accession 'Calcutta 4' as the diploid male parent (Plate 14.3) (Ortiz and Vuylsteke, 1994). As well as true plantain cultivars, the plantain-like 'Laknau' (AAB) has also been utilized as a female parent (Ortiz and Vuylsteke, 1998). In addition to tetraploids, black leaf streak-resistant triploids have been developed from crosses between bred tetraploids and bred diploids (Ortiz *et al.*, 1998) (see 'Triploid Breeding' below). Breeders at Onne are also interested in increasing the vigour of plantain by improving the root system. A new IITA breeding project, aimed at improving ABB cooking-banana types and East African highland banana cultivars (AAA, Lujugira–Mutika subgroup) has been initiated at Namulonge in Uganda.

Triploid Breeding

The wisdom of breeding tetraploids as commercial cultivars has been questioned (Stover and Buddenhagen, 1986). Generally, tetraploids have droopier leaves with weaker petioles than triploids, which make them more liable to break in high wind. They are also much more prone to seed set if pollinated, an undesirable characteristic in edible banana. It has been argued that these problems are not serious enough to exclude tetraploids from consideration as a final breeding product (Shepherd, 1994). Nevertheless, few tetraploids are found in nature and triploids are by far the most common group of edible banana. It was suggested that, as triploidy has evolved in *Musa* as the most productive ploidy level, breeding programmes should aim to produce triploid hybrids.

Triploid hybrid production had been attempted in Honduras by crossing bred tetraploids with improved diploids to produce progeny that approached commercial acceptance (Rowe and Richardson, 1975). An advantage of this scheme was that bred diploids, carrying perhaps different resistance genes, could be utilized twice: once to cross a triploid to produce a tetraploid (3x × 2x = 4x) and secondly to cross a tetraploid to produce a secondary triploid (4x × 2x = 3x). A disadvantage was that, unlike tetraploid breeding, which resulted in progeny that retained many of the desirable characteristics of the triploid female parent, secondary triploid breeding resulted in progeny that had totally new combinations of genes and could lose desirable characteristics.

A new triploid breeding scheme, however, envisages using colchicine to double the chromosomes of desirable synthetic diploids in order to produce autotetraploids, which could then be crossed with other improved diploids to form triploids (Vakili, 1967; Stover and Buddenhagen, 1986). In theory the use of the autotetraploid leads to the retention of the genetic make-up of the diploid in the triploid, thus conserving desirable traits. Another perceived advantage is that the synthesis of triploids is flexible, due to the high level of diversity available in diploid germplasm. Genetic variability of the triploid in the new scheme comes from three diploid parents and, at any time, new criteria for selection to respond to new objectives, such as resistance to a different strain of pathogen, can be quickly incorporated into progeny.

Guadeloupe

The triploid breeding scheme outlined above forms the basis of a breeding programme that the Institut de recherche sur les fruits et agrumes (IRFA), now the Département des productions fruitières et horticoles (FLHOR) of the Centre de coopération internationale en recherché agronomique pour le développement (CIRAD), established at Neufchâteau in Guadeloupe in the early 1980s. The long-term strategy of this programme is to improve diploid populations (Tezenas du Montcel *et al.*, 1996). Pure lines are being produced by repeated autofertilization, so that desired characteristics will be in a homozygous form. Innovative research, made possible by new tissue-culture techniques, has included the initiation of haploid cultures from immature pollen, so that homozygous diploids can be produced after chromosome doubling. Different pure lines are now being crossed to form élite diploids, which can be used to produce autotetraploids. These, in turn, will be used in crosses with other pure lines to obtain the final triploid. It is envisaged that this method will give more control over the genetic make-up of the final triploid product. CIRAD–FLHOR intends to improve locally consumed cooking- and dessert banana types by these methods (Bakry and Horry, 1994; Tezenas du Montcel *et al.*, 1994).

Coupled with this work, research is being undertaken by CIRAD to identify the wild progenitors of subgroups and cultivars. With this knowledge, breeding disease-resistant, high-yielding replacements will be on a more scientific basis. For example, it is possible that new disease-resistant triploids to replace cultivars in the disease-susceptible Pome subgroup could be assembled from diploids containing the genetic elements needed to form the desired agronomic characters that distinguish members of this subgroup from other cultivars. Integrated with the breeding programme are other important CIRAD projects, such as mapping the *Musa* genome to locate resistance genes, investigating the genetic diversity of *Mycosphaerella fijiensis*, identifying pathogenic variability in *Radopholus similis* and learning more of the virus problems that are constraints on the safe movement of *Musa* germplasm.

The CIRAD and CRBP breeding programmes and associated projects are closely linked under a francophone umbrella and are a significant force in the global banana improvement effort. Much is expected in the future from this comprehensive approach to banana improvement.

Biotechnological Innovations

The first biotechnological innovations to have an impact on banana occurred after the advent of tissue-culture propagation methods. Aseptic shoot-tip culture became a means of speeding up the multiplication of banana clones and was quickly adopted by breeding programmes. Associated with this development was the routine use of aseptic techniques to rescue and culture embryos from seed that would not normally germinate. More recently, the goal has been to develop protocols for the production of plants from pollen (see 'Guadeloupe' above), cell suspensions, single cells and protoplasts, as a prelude to more technologically advanced breeding techniques.

As mentioned previously, cultivars in the important Cavendish subgroup are not amenable to improvement through conventional breeding, because of their sterility. As a consequence, many of the newer biotechnological methods have been directed towards improving Cavendish. The somaclonal variation of banana in tissue culture has been exploited at the Taiwan Banana Research Institute (TBRI) near Pingtung to produce clones of Cavendish that are more resistant to *Fusarium oxysporum* f. sp. *cubense* race 4 (Tang and Hwang, 1994; see Hwang and Tang, this chapter, pp. 449–458). Banana in Taiwan is usually grown as an annual. Planting at the same time every year ensures that the crop avoids serious damage during the typhoon season and that fruit is harvested when best financial returns can be expected. The replanting of banana every year using tissue-cultured material also helps reduce disease incidence. The use of the new resistant clones has ensured that losses caused by Fusarium wilt are kept to a minimum.

Breeding by inducing beneficial mutations in banana in tissue culture by using the chemical mutagen ethyl methanesulphonate and gamma-irradiation was undertaken in the late 1970s by the Del Monte Corporation in the Philippines in an attempt to produce a Cavendish clone that was resistance to Fusarium wilt (Epp, 1987). At this time, Fusarium wilt was killing the occasional plant in Cavendish plantations and was perceived to be a potential threat to Del Monte's export industry in Mindanao. Four clones of 'Umalag' (AAA, Cavendish subgroup) were identified that, although not immune, had a higher level of resistance than control 'Umalag' plants. Unfortunately, work was terminated in the early 1980s, when the problem was re-evaluated and considered not to warrant continued research.

Mutation breeding by gamma-irradiation has been an important part of the programme of the International Atomic Energy Agency (IAEA) at their laboratories at Seibersdorf in Austria (Novak, 1991; Afza *et al.*, 1994; Jamaluddin, 1994; Roux *et al.*, 1994) and has been attempted elsewhere (de Beer and Visser, 1994; Smith *et al.*, 1994; Bhagwat and Duncan, 1998). Although no superior, disease-resistant banana clones derived from mutation breeding techniques have been released to the industry to date, this method may have potential for banana improvement and work continues.

A number of workers have attempted to produce disease-resistant clones of banana by treating tissue cultures with culture filtrates of fungal pathogens containing phytotoxins. This is a relatively simple and therefore attractive method of trying to improve banana and has been advocated as a technique for applying selection pressure to callus (Escalant, 1990) and for screening material for resistance following mutagenic treatment (Novak *et al.*, 1993). Fusaric acid, which may be produced by *F. oxysporum* f. sp. *cubense* (Matsumoto *et al.*, 1995), and phytotoxic exudates from *M. fijiensis* (Okole and Schulz, 1993, 1997; Garcia *et al.*, 1997) have been utilized in this work. Surviving banana cultures that can tolerate high concentrations of phytotoxin have been used to produce plants. However, despite claims that clones selected for tolerance to phytotoxins have been shown to be more resistant to disease, no banana selected by these methods has yet been released to the industry.

Daub (1986) reviewed the problems associated with the use of tissue-culture techniques for the selection of resistance to pathogens and concluded that more needed to be known about the role that fungal metabolites play in host–pathogen interactions before the technology could be utilized to its fullest extent. Toxins produced in culture by *M. fijiensis* have been used to evaluate banana for reaction to black leaf streak (Molina and Krausz,

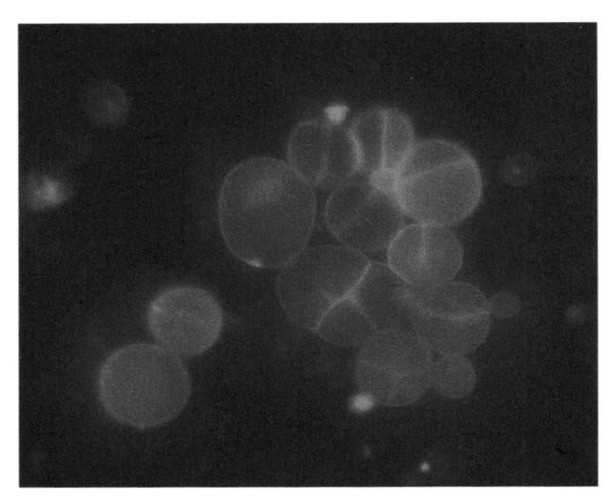

Plate 14.4. Cell-wall regeneration and first division of protoplasts of 'Bluggoe' (ABB) 4 days after inoculation on to a culture medium. Stained with Calcofluor white and observed under ultraviolet illumination (photo: B. Panis, KUL).

1989), but this is not proof of their involvement in pathogenicity (Lepoivre *et al.*, 1993). Harelimana *et al.* (1997) report that toxins of *M. fijiensis* are probably not involved in infection initiation or in the hypersensitive reaction of highly resistant cultivars. The effects of toxins on chlorophyll fluorescence indicated to these authors that their site of action is the chloroplast and therefore *in vitro* cultures of banana may not be suitable for screening for the effects of toxins. It was considered that, at most, toxins may act as secondary determinants of pathogenicity, contributing to lesion expansion in cultivars exhibiting partial resistance.

Genetic engineering for improving banana offers hope for the future. Protocols have been developed that enable banana plants to be regenerated from protoplasts (Plate 14.4) (Megia *et al.*, 1993), cell suspensions (Dhed'a *et al.*, 1991) and somatic embryos (Escalant *et al.*, 1994b) that makes transformation of the whole plant and not just parts of plants possible. Particle bombardment (biolistics) of embryogenic cell suspensions or somatic embryos is the method being developed at the Katholieke Universiteit Leuven (KUL). Banana plants carrying reporter genes or genes encoding antimicrobial proteins in all cells have been regenerated from transformed tissue. Work is now under way to incorporate genes into the banana genome that code for other proteins with antifungal activity or may prevent the development of banana bract mosaic virus (see Sági, Chapter 15, pp. 471–472, 492). Strategies of a francophone group engaged in banana improvement through biotechnological methods have been reviewed by Escalant *et al.* (1994a).

Although genetic engineering offers opportunities to improve Cavendish clones, which cannot be bred conventionally, it can also be seen as a tool to complement conventional breeding methods. Useful genes for disease resistance could be engineered into breeding diploids, which would then be used to produce disease-resistant triploids and tetraploids by more traditional techniques.

The impact that biotechnology is having on banana improvement and possibilities for the future have been reviewed by Dale (1990), Novak (1992), Anon. (1993), Sági *et al.* (1995) and Sági *et al.* (1998). However, the recent emphasis on biotechnological approaches to crop improvement has been criticized by some banana workers who argue that scarce financial resources would be better utilized supporting conventional breeding programmes that are now showing results (Simmonds, 1997; Rowe, 1998).

CONVENTIONAL BANANA BREEDING IN HONDURAS

P.R. Rowe and F.E. Rosales

Introduction

Conventional breeding for disease resistance began in the Caribbean in 1922 at a time when race 1 of *Fusarium oxysporum* f. sp. *cubense* was widespread on the export cultivar 'Gros Michel' (AAA) (Shepherd, 1968). The gradual, almost universal, destruction of 'Gros Michel' by Fusarium wilt (Stover, 1962) is a classic example of the dangers inherent in a monoculture. Switching to *F. oxysporum* f. sp. *cubense* race 1-resistant cultivars of the Cavendish subgroup (AAA) in the 1950s solved the problem. However, since all the Cavendish clones are very closely related genetically, plantations producing bananas for export are once again vulnerable. No other naturally occurring clones with the required disease resistance and agronomic characters for growing in large commercial plantations are known.

In anticipation of the eventual appearance of new diseases that could destroy Cavendish, the United Fruit Company began an extensive breeding programme in Honduras in 1959. This programme was donated to the new FHIA in 1984. With financing from international donor agencies and the governments of Honduras and Ecuador, the breeding objectives in the FHIA programme were expanded to include development of disease-resistant plantain and other cooking-banana hybrids. Rowe and Richardson (1975), Rowe (1984) and Rowe and Rosales (1996) have reported the developments in this programme.

Three comprehensive accounts of banana breeding have recently been published (Ortiz, 1994; Ortiz *et al.*, 1994; Rowe and Rosales, 1996) they include details of germplasm collections and techniques, such as seed extraction and embryo culture, employed by modern conventional breeders. It is not intended to repeat this information. The emphasis in this report is that conventional breeding has proved to be successful for *Musa* improvement – just as the pioneers in the early programmes prophesied. Now, more than 70 years after banana breeding was begun, the first bred disease-resistant hybrids have been planted for commercial cultivation. The development of these first commercial hybrids and the expectations from current and subsequent genetic improvement activities are outlined in the following discussions.

Basis of Early Banana Breeding

A genetic abnormality makes conventional banana breeding possible. The triploid 'Gros Michel' produces an average of about two seeds per bunch when pollinated by diploids (Shepherd, 1954). It was found that 'Gros Michel' does not undergo normal meiosis during sexual reproduction, but instead produces unreduced triploid gametes. Diploid pollen parents undergo normal meiosis, and most progenies from crosses with 'Gros Michel' are tetraploids (Cheesman, 1932; Cheesman and Larter, 1935; Larter, 1935). Some useless heptaploid progenies are also produced, due to double restitution during the formation of the 'Gros Michel' gametes. These heptaploids are readily identified by their thick leaves and slow growth, and are discarded as seedlings.

A few of the early tetraploids derived from 'Gros Michel' had bunch features that approached commercial acceptability. Subsequent use of these primary tetraploids in 4x × 4x and 2x × 4x crosses resulted in inferior tetraploid and triploid progenies. This lack of desirable traits in secondary polyploids has been explained by Dodds (1943). The primary tetraploids contained all three 'Gros Michel' genomes intact, whereas the secondary tetraploids and triploids were the products of meioses in which the 'Gros Michel' chromosome sets had participated. The diploids that were available for these cross-pollinations had inadequate bunch qualities and these genetic deficiencies were readily expressed with the breakdown of the 'Gros Michel' genomic complement.

Dodds concluded that the future of banana breeding depended upon first developing improved diploid male parents and then seeking commercial hybrids among the primary tetraploids produced from crossing with 'Gros Michel'. The ideal pollen parents should have combinations of disease resistance, be parthenocarpic and seed-sterile, have good agronomic features and produce viable pollen.

The Challenges

The first formidable obstacle that faced breeders was to develop diploids with combinations of disease resistance and desirable agronomic qualities. Even though an abundance of genetically diverse natural clones were available for diploid improvement schemes, all had inferior bunch features or were practically sterile when used in cross-pollinations (Rowe and Rosales, 1996).

A second challenge was to identify additional seed-fertile triploids for use in breeding other types of banana. This need for the development of disease-resistant plantain and other cooking-banana types became critical after the black leaf streak disease reached the major producing areas and began causing yield reductions of up to 50% (see Pasberg-Gauhl *et al.*, Chapter 2, pp. 37–44).

Both of these initial problems have been sufficiently solved to permit the development of the first commercial-type hybrids, which are now being planted to replace natural cultivars in certain areas. Since the genetic improvements of these hybrids were dependent upon the breakthroughs in diploid breeding, the development of these diploid parental lines for the subsequent crosses with seed-fertile triploids will be discussed first.

Breeding Diploid Parental Lines

A bred diploid resistant to black leaf streak

Many wild subspecies of *Musa acuminata* Colla are resistant to black leaf streak disease and some have been used as parental lines in attempts to incorporate this resistance into bred diploids with improved bunch features (Rowe and Rosales, 1996). The accession that has given the best results from these cross-pollinations is the *M. acuminata* ssp. *burmannica* accession 'Calcutta 4'. This clone was collected in Burma and has been described by Simmonds (1956). The Honduran programme obtained 'Calcutta 4' from the ICTA in Trinidad in 1962 and it was given the code 'IV-9' in the germplasm collection.

The resistance of 'IV-9' to black leaf streak is its main feature of interest, since its non-parthenocarpic bunch, weighing less than 2.0 kg, is among the poorest of all the natural, seeded diploids. In efforts to produce resistant hybrids with improved bunch traits, many different bred diploids with advanced agronomic qualities were crossed with 'IV-9' and about 2500 hybrids were planted in the field for evaluation. Only one hybrid, 'SH-2989' (SH = selected hybrid), which is resistant to black leaf streak and has a much better bunch than 'IV-9', merited selection from this huge population.

By the time 'SH-2989' was selected, several diploids with exceptional bunch sizes had been developed in the Honduran programme for use in breeding for combinations of agronomic excellence and disease resistance in subsequent diploid improvement efforts. From crosses of these agronomically superior, bred diploids with 'SH-2989', about 500 hybrids were planted in the field for evaluation. One hybrid, 'SH-3437', which has a high level of resistance to black leaf streak and bunch weights of up to 35.0 kg, was selected from these segregating populations. The bunch features of 'IV-9', 'SH-2989' and 'SH-3437' in this progressive agronomic improvement of diploids resistant to black leaf streak are shown in Plate 14.5. The other parental line of 'SH-3437' is 'SH-3217', which remains one of the major accomplishments in breeding for diploids with large bunches in the Honduran programme. Rowe and Rosales (1996) describe the extensive cross-pollinations and selections that were involved in the development of 'SH-3217'. The special significance of 'SH-3437' is that it is a disease-resistant diploid that has resulted in a commercial-type hybrid when crossed with seed-fertile triploids. An exceptional tetraploid plantain hybrid, which was derived from these 3x × 2x cross-pollinations, will be discussed under the section on breeding plantain.

A bred diploid resistant to burrowing nematode

Root and corm rot, caused by the invasion of nematodes, followed by certain fungi and bacteria, became a serious disease with the switch from 'Gros Michel' to the Cavendish cultivars. The burrowing nematode (*Radopholus similis*) is the most widespread of the nematodes that attack banana and the application of nematocide to soil is required to control this pest in most producing countries (Stover, 1972; see Sarah, Chapter 7, pp. 295–303).

Accessions of the AA cultivar 'Pisang Jari Buaya' (13 apparently identical clones collected from different places) were shown to be resistant to burrowing nematode shortly after the initiation of the Honduran programme (Wehunt *et al.*, 1978). However, these clones are male-sterile and most pollinated bunches remain seedless. Efforts to produce sufficient numbers of hybrids were further complicated by poor seed germination. Only about 2.0% of the seeds germinated, even when cultured as embryos.

To compensate for these limiting factors, some 10,000 bunches of 'Pisang Jari Buaya' were pollinated to secure an adequate seed supply. From the 35 resultant segregating hybrids, only one had bunch qualities that merited its selection for further evaluation. This hybrid, 'SH-3142', was also found to be highly resistant to the burrowing nematode (Pinochet and Rowe, 1979). Bunch features of the 'Pisang Jari Buaya' parental line and the 'SH-3142' hybrid are shown in Plate 14.6.

Unlike 'Pisang Jari Buaya', 'SH-3142' is readily fertile as a seed and pollen parent and it has been used to pollinate seed-fertile triploids. Tetraploid hybrids of a banana and a plantain with 'SH-3142' parentage, both of which are now being planted commercially, are discussed under the sections on breeding dessert banana and breeding plantain, respectively.

A bred diploid resistant to F. oxysporum *f. sp.* cubense *race 4*

The designation of race 4 is given to populations of *F. oxysporum* f. sp. *cubense* capable of overcoming Cavendish resistance (see Ploetz and Pegg, Chapter 3, pp. 149–150). Observations by Ploetz and Correll (1988) that race 4 isolates from different places belong to at least three different vegetative compatibility groups indicate genetic diversity among strains of this race. For several years after the first report of race 4 in Taiwan in 1967 (Su *et al.*, 1977), it was thought to be a disease problem only in subtropical areas. Now that there are confirmations of Cavendish being attacked in tropical regions, the possibility exists that this race of Fusarium wilt could eventually eliminate Cavendish from commercial cultivation, just as race 1 eliminated 'Gros Michel'. Hwang and Tang (1996) have described somaclonal variants of the 'Pei-Chiao' (Cavendish subgroup) that remain resistant to race 4 in Taiwan, where banana is cultivated as an annual crop (see also Hwang and Tang, this chapter, pp. 449–458). It is not yet known how this resistance would hold up in perennial plantings. While there are still control measures for other banana diseases, the only control for Fusarium wilt is genetic resistance.

Most of the diploid accessions used in the FHIA programme have demonstrated field resistance to race 1 of the Fusarium wilt pathogen. The widely used *M. acuminata* ssp. *malaccensis* and *M. acuminata* ssp. *burmannica* have shown

Plate 14.5. Progress in breeding for better bunch qualities in diploids resistant to black leaf streak: 'IV-9', the *M. acuminata* ssp. *burmannica* ('Calcutta 4') source of resistance (left), 'SH-2989' progeny of 'IV-9' (centre) and 'SH-3437' progeny of 'SH-2989' (right) (photo: FHIA).

Plate 14.6. Bunch features of the burrowing nematode-resistant 'Pisang Jari Buaya' (AA) accession (left) and the 'SH-3142' hybrid, which was derived from crosses with this accession (right). 'SH-3142' is also resistant to the burrowing nematode and, unlike the almost sterile 'Pisang Jari Buaya', is readily fertile both as a male and as a female parental line in cross-pollinations (photo: FHIA).

resistance to both races 1 and 2 (Vakili, 1965). Evaluations for resistance to races 1 and 2 (race 2 attacks the ABB 'Bluggoe' cooking banana) can be undertaken in Honduras, but testing for resistance to race 4 must be conducted outside banana-growing areas or in countries where this race already exists.

In 1980, three advanced diploids ('SH-2095', 'SH-2989' and 'SH-3142'), along with cultivars 'Valery' and 'Grand Nain' as Cavendish controls, were evaluated for reaction to *Fusarium oxysporum* f. sp. *cubense* race 4 from Taiwan in containers in South Florida (Stover and Buddenhagen, 1986). These three diploids were chosen because of their diverse pedigrees and extensive use as parental lines in the diploid breeding scheme. The ratings from this test were: 'SH-2095', highly susceptible; 'Valery' and 'Grand Nain', susceptible; 'SH-2989', moderately resistant; 'SH-3142', highly resistant.

With the finding that 'SH-3142' was resistant to *Fusarium oxysporum* f. sp. *cubense* race 4 from Taiwan in this preliminary screening, this diploid was sent to cooperating scientists in Australia for evaluation of reaction to the local race 4. From subsequent crosses of 'SH-3142' × 'SH-3217', the 'SH-3362' diploid was selected and also sent for testing. Langdon and Pegg (1988) reported that several plants of 'SH-3142' succumbed in the field, but 'SH-3362' proved to be highly resistant.

These findings suggested that: (i) *Fusarium oxysporum* f. sp. *cubense* race 4 in Taiwan differs from the race 4 found in Australia; and (ii) the genetics of resistance may be complex. The 'SH-3217' parent of

Plate 14.7. Bunch characteristics of the 'SH-3362' diploid (right), which is resistant to *Fusarium oxysporum* f. sp. *cubense* race 4 in Australia. 'Pisang Jari Buaya' (left) and 'SH-3142' (middle) diploids are in the pedigree of 'SH-3362' (photo: FHIA).

Plate 14.8. Plant and bunch features of the 'SH-3649' tetraploid, which was derived from 'Highgate' × 'SH-3362'. 'SH-3649' has a larger bunch than its 'Highgate' parent and is resistant to *Fusarium oxysporum* f. sp. *cubense* race 1. The first-crop plant shown has an ideal short plant height, but ratoon crop plants are about 1.0 m taller (photo: FHIA).

'SH-3362' is a progeny of the highly susceptible 'SH-2095' × 'SH-2766'. The pedigrees of 'SH-2095' and 'SH-2766' differ only in that 'Sinwobogi', an AA cultivar from Papua New Guinea, is not in the lineage of 'SH-2766' (Rowe and Rosales, 1996), so it is unlikely that 'SH-3217' contributed to the resistance of 'SH-3362'.

'SH-3362' has an excellent bunch size (Plate 14.7) and, like the 'SH-3437' and 'SH-3142' diploids, it has been crossed with seed-fertile triploids. The reaction of tetraploids with 'SH-3362' parentage to *Fusarium oxysporum* f. sp. *cubense* race 4 is not yet known. However, two selected tetraploid hybrids from 'Highgate' × 'SH-3362' have proved to be resistant to race 1. These two tetraploids, one of which is already being grown commercially, will be discussed further in the next section on breeding dessert banana substitutes.

Breeding Tetraploids

Breeding disease-resistant AAA dessert banana substitutes

For many years, breeding for improved export dessert banana types was the only objective of the two oldest programmes in Jamaica and Honduras. While 'Gros Michel' was used in the early 3x × 2x crosses in Jamaica, the 'Highgate' spontaneous dwarf mutant of 'Gros Michel', which was discovered in Jamaica (Larter,

1934), became the fixed female parent for crossing with bred diploids in both programmes.

The early expectations were that eventually the bred tetraploids would have the disease resistance of the improved diploid parental lines and have bunch sizes equal to those of 'Highgate'. As anticipated, with progress in developing diploids with better agronomic qualities, the bunch sizes of the tetraploids derived from them were correspondingly better. Now, tetraploids derived from crossing the recently developed superior diploids with 'Highgate' often have larger bunches than those of 'Highgate'.

A breakthrough in the development of bred diploids, which results in tetraploids with both disease resistance and agronomic qualities better than those of 'Highgate', is illustrated by the 'SH-3649' tetraploid, which was derived from 'Highgate' × 'SH-3362'. A plant of 'SH-3649' with a first-crop 52.0 kg bunch is shown in Plate 14.8. It is already known that this tetraploid is resistant to *Fusarium oxysporum* f. sp. *cubense* race 1 in Honduras (M. Rivera, 1995, unpublished results) and trials are under way to evaluate its reaction to race 4 in Australia (K.G. Pegg, Brisbane, 1996, personal communication).

Unfortunately, tetraploids derived from 'Highgate' have plant heights of up to 3.7 m in the second and subsequent fruit production cycles. This height was acceptable in export banana types when 'Valery' and 'Giant Cavendish', which have similar plant heights, were the main varieties cultivated. However, with the switch from these two semi-dwarf Cavendish clones to the shorter (2.7 m) and thus more wind-resistant 'Grand Nain' as the favoured export cultivar in the late 1970s, this shorter plant height became the desired standard.

'Highgate' continues to be used in 3x × 2x crosses in the Honduran programme, because plant height sometimes becomes a secondary consideration if the tetraploid hybrids are otherwise attractive alternatives to the standard Cavendish cultivars. For example, the 'FHIA-23' tetraploid, which was derived from the same cross and is very similar to the 'SH-3649' hybrid just described, is now being cultivated on about 2000 ha in Cuba for domestic markets. In addition to being resistant to race 1 of *Fusarium oxysporum* f. sp. *cubense*, 'FHIA-23' is tolerant to black leaf streak and has proved to be twice as productive as the Cavendish cultivars when no fungicides are applied to control this disease (J.M. Alvarez, Cuba, 1998, personal communication).

In endeavours to develop new tetraploids with a more desirable, shorter plant height, another triploid is now being evaluated as the female parental line in breeding for a new export banana. This triploid is the 'Lowgate' dwarf mutant of 'Gros Michel', which was discovered in Honduras (Rowe and Richardson, 1975). Plant and bunch features of this mutant, which has a plant height about 1.0 m shorter than that of 'Highgate', are shown in Plate 14.9. 'Lowgate' has not previously been widely used, because of its inferior bunch size and shorter finger length when compared with 'Highgate'. Instead, this clone has been held in reserve, awaiting the development of superior diploids that could compensate for its agronomic deficiencies. Now that it has been demonstrated that the current diploids used in crosses with 'Highgate' frequently result in tetraploids with bunches larger than those of this triploid parental clone, it is expected that crossing these diploids with 'Lowgate' will provide tetraploid hybrids with the desired bunch sizes. Extensive pollinations of this shortest dwarf mutant of 'Gros Michel' have recently been initiated in the FHIA programme.

An obstacle in these crosses with 'Lowgate' is that this clone produces very few seeds (about one seed per 50 pollinated bunches). While this very low seed fertility is a handicap for producing large numbers of hybrids for evaluation and selection, this characteristic should be valuable in breeding commercial tetraploids. With the female sterility genes of 'Lowgate' combined with those of a parthenocarpic seedless, or almost seedless, diploid pollen parent, it is expected that tetraploids can be bred that will remain seedless if planted on a large scale. Pollen from tetraploids is much less efficient than that of diploids when used in pollinations for seed production and this inefficiency in fertilization would also contribute to the desired seedlessness.

Eating qualities of tetraploids with 'Lowgate' parentage are anticipated to be equal to those of the Cavendish clones. In a taste panel test of 24 tetraploids derived from 'Highgate' in Jamaica, some were similar in acceptability to the 'Valery' Cavendish cultivar (Baldry *et al.*, 1981). One undesirable trait that has often been observed in tetraploids is 'weak neck', the premature detachment of fruit from the cluster crown when fully ripe (New and Marriott, 1974). However, this is not an inborn defect of the ploidy level. Tetraploids with 'strong neck' are readily produced by selection of diploid pollen parents that have strong pedicels.

Breeding disease-resistant AAB dessert banana substitutes

While the historical emphasis in banana breeding has been on the development of a disease-resistant dessert cultivar for the production of export bananas, the advanced diploids developed in these endeavours have also now proved to be useful in breeding for disease-resistant hybrids of other banana types. Although Cavendish cultivars are grown as a source of both dessert and green cooking bananas for local markets in many countries, a large number of consumers prefer dessert bananas with the sweet-acid flavour characteristic of the 'Pome', 'Silk' and 'Mysore' clones of the AAB genomic group. For example, Brazil produces more than 5,000,000 tonnes of bananas annually for internal distribution. About two-thirds of this production is with the 'Prata' (Pome subgroup) and 'Maçã' (syn. 'Silk') cultivars, even though these two clones are susceptible to Fusarium wilt and less productive than Cavendish (Alves *et al.*, 1987). 'Poovan' (syn. 'Mysore') is the most important banana cultivar in India, while 'Pisang Rastali' (syn. 'Silk') is highly prized in Malaysia and Indonesia (Simmonds, 1966).

Attempts to improve these three tall AAB dessert banana cultivars by crossing them with bred diploids have not been successful in the FHIA programme. They are seed-fertile, but the hybrids produced have not had the desired plant and bunch features. However, crosses with 'Prata Anã', a dwarf 'Pome' from Brazil, have resulted in a tetraploid hybrid, 'SH-3481', which has exceptional agronomic and disease-resistance characteristics. A 41.0 kg bunch of 'SH-3481' and representative bunches of its 'Prata Anã' triploid and 'SH-3142' diploid parents are shown in Plate 14.10. 'SH-3481' was given the code name of 'FHIA-01®' for inclusion in international trials to determine its reaction to black leaf streak disease (see Jones, this chapter, pp. 428–429). It was shown to have a very high level of resistance in Colombia, Honduras, Costa Rica, Nigeria and Burundi (Jones, 1994b).

In addition to this resistance to black leaf streak, it has been found that 'FHIA-01®' is also resistant to *F. oxysporum* f. sp. *cubense* race 4 in Australia (K.G. Pegg, Queensland, 1994, personal communication). After conducting marketing trials in 1995 to determine consumer acceptance of fruit and receiving a favourable reaction, the Australian Banana Growers' Council recommended 'FHIA-01®' for commercial production in the subtropical growing regions of southern Queensland and New South Wales under the name 'Goldfinger'.

For unknown reasons, 'Goldfinger' performs better in the subtropics than in the tropics, especially in regard to ripe fruit quality. The pulp of tropically grown fruit is too soft when ripened by the standard techniques with ethylene. This softness of pulp is not as pronounced when the fruit is allowed to ripen naturally, so it is thought that this new hybrid would still be a suitable shelf-ripened banana in several tropical countries where disease pressures are limiting the availability of fruit from the traditional banana cultivars.

At present, 'Goldfinger' is the only bred

Plate 14.9. Plant and bunch features of 'Lowgate', the shortest of the dwarf mutants of 'Gros Michel'. 'Lowgate' is currently being used as a parental line in the FHIA programme to breed tetraploid hybrids with shorter plant heights than those obtainable by using the taller 'Highgate' (photo: FHIA.)

banana that is known to be resistant to both *F. oxysporum* f. sp. *cubense* race 4 and black leaf streak diseases. Bunches of 'Goldfinger' are about twice as large as those of 'Prata Anã' and provide optimistic expectations that forthcoming tetraploids from further crosses with this AAB clone will have equally improved bunch sizes.

Breeding disease-resistant AAB plantain substitutes

Plantain breeding has only recently received attention, and thus no background approaches to the development of new hybrids of this crop were initially available to serve as guidelines. The key finding was that the triploid 'French' plantain has the same genetic abnormality as 'Gros Michel', in that it produces seeds when pollinated with diploids (De Langhe, 1969; Rowe, 1987). As with the progenies of 'Gros Michel', the progenies from 'French' plantain × diploid crosses consist mostly of normal tetraploids and useless heptaploids. Vuylsteke and Ortiz (1995) found that diploid hybrids are also sometimes produced from these 3x × 2x crosses.

Plate 14.10. Bunch features of the 'FHIA-01®' ('SH-3481') tetraploid hybrid (right) as compared with those of its 'Prata Anã' triploid (left) and 'SH-3142' diploid (middle) parental lines. 'FHIA-01®' has been named 'Goldfinger' and in 1995 was recommended as a new disease-resistant commercial variety for the subtropical banana-growing regions of Australia (photo: FHIA).

Plate 14.11. Bunch features of the 'FHIA-21®' tetraploid hybrid (right) as compared with those of the 'AVP-67' ('French' plantain-type) (left) and 'SH-3142' diploid (middle) parental lines. This 37.0 kg bunch of 'FHIA-21®' was dehanded shortly after emergence, to leave only the first five hands for subsequent development (photo: FHIA).

The widely grown cultivars of the 'False Horn' plantain type have a male bud that degenerates very shortly after the bunch is formed. In contrast, the commercially less popular 'French' plantain cultivars have a normal male bud, which continues to grow for about 3 months while the fruits are maturing. 'French' plantain variants, which commonly appear as low-frequency, spontaneous, stable mutations in plantings of 'False Horn', also differ from 'False Horn' by having more slender fingers, but the eating qualities of the two are very similar. It is not understood why 'French' plantain is seed-fertile while 'False Horn' plantain is practically sterile (12 seeds and no useful hybrid plants were obtained from pollinating about 3000 bunches of the latter in the FHIA programme).

With the ready availability of advanced bred diploids, rapid progress has been made in the genetic improvement of plantain. The code for the 'French' plantain parental line that has been used extensively in cross-pollinations in the FHIA programme is 'AVP-67'. There are differences among the various 'French' plantain cultivars in regard to bunch features and female fertility, and 'AVP-67' is considered the best of the several clones of this type of plantain in the Honduran germplasm collection.

Both the 'SH-3142' and 'SH-3437' diploids have been used for breeding tetraploid plantain-like hybrids with resistance to black leaf streak. One of these tetraploids, 'FHIA-21®', which was derived from 'AVP-67' × 'SH-3142', is now being grown for commercial production in Honduras. In experimental trials with no chemical control of black leaf streak, the number of functional leaves at harvest is four to five for this plantain hybrid, while it is none to one for the 'False Horn' standard clone. In all trials, whether under marginal or excellent conditions, the yield of 'FHIA-21®' has been at least double that of 'False Horn'. Consumer taste panel tests with both green and ripe fruit have also shown a preference for this new hybrid over the traditional plantain.

The other agronomic characteristics of 'FHIA-21®', such as plant height, time from planting to harvest and speed of subsequent fruiting cycles, are very similar to those of 'False Horn'. However, bunches of this hybrid are like those of the 'French' plantain cultivars, and have up to ten hands with the typical slender fingers of this type of plantain when left to develop naturally. To promote production of the thicker and longer fingers which are preferred by consumers in both domestic and export markets, it is recommended that bunches of 'FHIA-21®' be dehanded soon after emergence to leave only the first five hands for development. It has been found that the weights of these five-hand bunches are only slightly less than those of unpruned bunches of this hybrid, but that the fingers on these five hands are similar in thickness and length to those of 'False Horn' at maturity. Even with this recommended dehanding, the number of fingers on these five-hand bunches of 'FHIA-21®' has been more than twice the number on entire bunches of 'False Horn' in the same trials.

This dehanding also serves another useful purpose, in that the male bud is removed at the same time, before the staminate flowers are exposed. 'FHIA-21®', like most other tetraploid plantain hybrids, produces abundant pollen, and leaving the male buds to develop naturally would result in an occasional fruit with seeds.

Bunch features of the 'AVP-67' and 'SH-3142' parental lines as compared with a 37.0 kg bunch of the 'FHIA-21®' hybrid, which was dehanded to five hands shortly after emergence, are shown in Plate 14.11. Bunches of this size for 'FHIA-21®' are common under good growing conditions, even with no chemical control of black leaf streak.

One precaution that must be taken for exported fruit of 'FHIA-21®' is to harvest bunches by age control. Fruit at full maturity (100–105 days after the first visible tip of the emerging bunch appears) ripens rather quickly after harvest, but slightly younger fruit (90–95 days) has the necessary green life for shipping without problems with premature ripening. This 90–95-day fruit has the required fruit size, and it is now being exported from Honduras to the USA.

While plantains are an important food in Latin America and the Caribbean, more than half the world's plantains are produced and consumed in West and Central Africa. In view of its excellent eating qualities and productivity, it is expected that 'FHIA-21®' could also become a popular replacement of 'False Horn' in family food plots in these regions of Africa. However, for the domestic open markets common in all countries where plantain is cultivated, 'FHIA-21®' fruit would have a shorter shelf-life (both green and ripe) than fruit of 'False Horn'.

Another plantain hybrid, 'FHIA-20', which was derived from the 'AVP-67' × 'SH-3437' cross, now appears to have the fruit qualities that would permit it to be handled in essentially the same way as 'False Horn' is currently treated. 'FHIA-20' has a fruit shape that is not as typically plantain-like (the flower ends of the fingers are not as pointed as are those of 'FHIA-21®'), but it has a large bunch size and good eating qualities, like 'FHIA-21®'. The advantages of 'FHIA-20' are that it has a long green life, similar to that of 'False Horn', both before and after harvest, and its ripe fruit does not become soft as rapidly as 'FHIA-21®'. 'FHIA-20' is still being evaluated, but an additional positive preliminary observation is that the black leaf-streak resistance of this hybrid appears to be equal to that of 'FHIA-21®'. Bunch characteristics of both 'FHIA-20' and 'False Horn' are compared with those 'FHIA-21®' in Plate 14.12.

These discussions of activities and results in plantain breeding have centred on the experiences in the FHIA programme. Vuylsteke *et al.* (1993a, b, 1995) and Ortiz and Vuylsteke (1994) describe the development and performance of black leaf streak-resistant plantain hybrids in the IITA's programme in Nigeria. Progress in the Centre de recherches régionales sur bananiers et plantains (CRBP) programme in Cameroon is outlined by Jenny *et al.* (1994).

Breeding disease-resistant AAA and ABB cooking-banana substitutes

While seed-fertile triploid natural clones have been identified for use in breeding tetraploid dessert banana and plantain hybrids, triploid parental lines for breeding improved East African highland cooking- and beer-making banana clones (AAA, Lujugira–Mutika subgroup) have not been forthcoming from among natural cultivars. Several of these cooking-banana clones have been found to be seed-fertile, but the hybrids produced from them have not had desirable plant and bunch features. However, a secondary triploid hybrid, which originated from initial crosses with an ABB cooking-banana clone, now appears to be an exceptional parental line in breeding disease-resistant hybrids for East Africa. This secondary triploid was developed by pursuing a crossing scheme that previously had only been a theoretical approach to genetic improvement.

Theoretically, it is possible to breed genetically diverse, disease-resistant tetraploid and triploid *Musa* hybrids in abundance. The process begins by crossing diploids with seed-fertile triploid natural clones, and this leads to selection of primary tetraploids from the resultant segregating hybrids. Then, by crossing diploids with these selected primary tetraploids, an array of secondary triploids is produced. Selected seed-fertile secondary triploids would serve for succeeding 3x × 2x cross-pollinations and a new generation of tetraploids, which, in turn, would provide the parental lines for subsequent 4x × 2x crosses and the production of tertiary triploids. The attractiveness of this continuous sequence of cross-pollinations and selections is that four different disease-resistant bred diploids could be introduced into the pedigrees of these tertiary triploids. Schematically, this series of crosses is as follows:

(primary) 3x × 2x(a) = (primary)
4x × 2x(b) = (secondary) 3x × 2x(c) =
(secondary) 4x × 2x(d) = (tertiary) 3x

This theoretical approach to the development of genetically diverse hybrids remains a theory in breeding for new dessert banana and plantain types. The secondary triploid dessert banana hybrids, with both 'Highgate' and 'Prata Anã' in their pedigrees, produced so far in the FHIA programme have had inferior bunch features, which eliminated them from consideration as parental lines for further cross-pollinations. Secondary triploids resulting from 4x × 2x crosses in the plantain breeding scheme have lacked the plantain cooking qualities and, likewise, have been deemed to be useless as subsequent parental lines. However, in breeding for new ABB cooking-banana types, this theory had already proved to be an excellent practical approach for producing genetically improved hybrids. The key breeding achievement in this sequence was the development of the dwarf 'SH-3386' secondary triploid (Plate 14.13). 'SH-3386' was selected from among the segregating hybrids derived from crossing a bred dwarf diploid (AA) with a selected primary tetraploid (ABBB). The pedigree of this primary tetraploid parental line of 'SH-3386' is 'Gaddatu' (ABB, syn.'Gubao') × *Musa balbisiana* (BB).

It is noteworthy that 'SH-3386' was the only dwarf triploid that merited selection from extensive earlier efforts to incorporate the gene for dwarfness into several ABB clones. Now, the exceptional 'FHIA-03' tetraploid, which was derived from crossing a black leaf streak-resistant bred diploid with 'SH-3386', is the first cooking-banana hybrid to be grown commercially.

This 'FHIA-03' secondary tetraploid hybrid has received ready acceptance by consumers in Grenada and Cuba, the two countries where it was tested first outside Honduras. In Grenada, 'Bluggoe' (ABB) is a staple food crop, but this cultivar is extremely susceptible to Moko disease (*Ralstonia solanacearum*) (see Thwaites *et al.*, Chapter 5, pp. 219–220), which limits production. 'FHIA-03' was planted in Moko-infested localities in Grenada for evaluating its reaction under intense natural disease pressure. These original plantings of this hybrid have remained free of Moko infection through five cycles of fruit production to date (H. Fagan, St Lucia, 1995, personal communication). In Cuba, more than 3900 ha of 'FHIA-03' have now been planted and bunches with weights of

Plate 14.12. Bunch features of the 'FHIA-20' (left) and 'FHIA-21®' (right) plantain hybrids as compared with those of the standard 'False Horn' plantain cultivar (middle). This 'False Horn' bunch weighed 16.5 kg. The 'FHIA-20' and 'FHIA-21®' bunches, which had been dehanded to five hands when first formed, to promote the development of longer and thicker fingers, weighed 38.5 and 37.0 kg, respectively (photo: FHIA).

Plate 14.13. Plant and bunch features of the dwarf 'SH-3386' secondary triploid, which has 'Gaddatu' (ABB), *Musa balbisiana* (BBw), and a bred dwarf diploid (AA) in its pedigree. 'SH-3386' has been the key seed-fertile triploid parental line in crosses with diploids for breeding new cooking-banana hybrids (photo: FHIA).

up to 70.0 kg have been harvested (J.M. Alvarez, Cuba, 1998, personal communication). Bunch features of 'Bluggoe', as compared with those of 'FHIA-03', are shown in Plate 14.14.

Boiled green fruit of 'FHIA-03' has a very good flavour and texture, and the time required for cooking is about half that required for cooking 'Bluggoe'. However, a weakness of this hybrid is that the fruit has a relatively short green life after the bunch is harvested. For this reason, it is recommended primarily for home gardens, where the fruit can be harvested a hand at a time as needed from bunches which remain on the plants. Since the plants are strong and short, this progressive partial dehanding of hanging bunches is easily accomplished with a ladder and a sharp knife. This technique preserves the green life of the remaining hands for up to a month after the first hand is harvested, and the fruits of these remnant hands continue to enlarge while awaiting removal.

While 'FHIA-03' has shown excellent potential as a new robust, dwarf, black leaf streak-resistant, drought-resistant and productive cooking banana for domestic consumption in countries where boiled green bananas or plantains are popular dishes, its

Plate 14.14. The bunch sizes of the 'FHIA-03' tetraploid cooking-banana hybrid (right) have been twice as large as those of the 'Bluggoe' (ABB) natural cooking-banana clone (left) in replicated trials in Honduras. This new hybrid is being grown in Grenada and Cuba (photo: FHIA).

Plate 14.15. Bunches of 'Nyamwihogora' (AAA, Lujugira–Mutika subgroup) (left) and the black leaf streak-resistant 'SH-3748' tertiary triploid hybrid (right), which illustrate that new productive cooking-banana hybrids can be bred with bunch configurations and finger shapes similar to those of the East African highland cultivars. This comparison also shows the small bunch sizes of the East African cultivars in Honduras, as a result of severe defoliation by black leaf streak (photo: FHIA).

high levels of male and female sterility have prevented it from being readily usable in further cross-pollinations. However, another dwarf tetraploid, 'SH-3648', which was also derived from crossing a diploid (in this case, *F. oxysporum* f. sp. *cubense* race 4-resistant 'SH-3362') with 'SH-3386', is fertile as both a male and female parental line.

The availability of 'SH-3648' as a seed-fertile secondary tetraploid has permitted making the 4x × 2x crosses for producing tertiary triploid hybrids (as described in the theoretical sequence of crosses presented earlier). While both 'FHIA-03' and 'SH-3648' have the angular finger shapes of the ABB cooking bananas, some of the tertiary triploid hybrids produced by crossing with 'SH-3648' have the plump, rounded fruit shapes typical of the East African highland cooking-banana cultivars. Bunch features of the 'SH-3748' tertiary triploid (which was derived from crossing the black leaf streak-resistant 'SH-3437' diploid with 'SH-3648'), as compared with those of the 'Nyamwihogora' cultivar (Lujugira–Mutika subgroup), are shown in Plate 14.15. This comparison also shows the small bunch sizes of the East African highland cultivars in Honduras, as a result of severe defoliation by black leaf streak.

'SH-3748' is a dwarf, vigorous plant and has a high level of resistance to black leaf streak. This hybrid produces bunches which weigh up to 45.0 kg, and its boiled green fruit has a flavour and texture similar to those of the East African highland cooking bananas. 'SH-3748' and other similar selected tertiary triploids are being sent to the INIBAP Transit Centre at the Katholieke Universiteit Leuven for virus indexing. Once found to be virus-free, they will be sent to East African countries for further evaluation of their potential as new cooking bananas for this region of Africa.

The prospects for continued breeding of new disease-resistant, productive hybrids that could substitute for the East African highland cooking-banana cultivars are very promising. By crossing several different disease-resistant diploids with 'SH-3386' to produce additional seed-fertile secondary tetraploids, and then crossing the various improved diploids with these new tetraploids, large numbers of genetically diverse, tertiary triploids could be produced for evaluation and selection.

Conclusions

The diploid, triploid and tetraploid selected hybrids described above represent the major achievements in the FHIA programme. Other endeavours, including attempts to improve the bunch qualities of several alternative diploid sources of resistance to black leaf streak and to develop dwarf plantain hybrids by using dwarf 'French' plantain clones in the plantain breeding scheme, although not covered here, are important ongoing activities of the FHIA breeding programme.

In some instances, such as was the case with 'Pisang Jari Buaya' in breeding for nematode resistance, the absence of pollen has limited the usefulness of certain diploids. This is the case now with the black leaf streak-resistant 'SH-3723' diploid, which has exceptional bunch features (Plate 14.16), but which, because of lack of pollen, cannot be used in the 3x × 2x crossing schemes. 'SH-3723' does produce a few seeds when pollinated, so it is now being used as a female parent in crosses with the 'SH-3142', 'SH-3362' and 'SH-3437' diploids. Just as the 'SH-3142' progeny of 'Pisang Jari Buaya' has abundant pollen, it is expected that the diploid progenies from these crosses with 'SH-3723' will also produce pollen. 'SH-3723' already has the black leaf streak-resistant 'SH-2989', the burrowing nematode-resistant 'SH-3142' and the *F. oxysporum* f. sp. *cubense* race 4-resistant 'SH-3362' in its pedigree, so some of the progenies from the crosses currently being made with 'SH-3723' should have multiple disease resistance. Such diploids with 'SH-3723' parentage are anticipated to be exceptional pollen parents for subsequently crossing with the various seed-fertile triploids.

In retrospect, the destruction of the 'Gros Michel' export banana by *F. oxysporum* f. sp. *cubense* race 1 had a positive aspect. As a direct result of this devastation of 'probably the most important single fruit variety in the world in terms of volume produced' (Popenoe, 1941), banana breeding programmes were started. Without the actual and pending accomplishments in these programmes, there would be little hope of practical solutions for the current disease problems affecting the world's various types of banana. Now, as has been discussed, four bred, disease-resistant hybrids

Plate 14.16. Bunch features of the 'SH-3723' diploid, which was derived from a series of crosses among diploids individually resistant to black leaf streak, the burrowing nematode and *F. oxysporum* f. sp. *cubense* race 4. Lack of pollen limits this hybrid for use as a female parent only, but it is anticipated that progenies from crossing the black leaf streak-resistant 'SH-3437', the burrowing nematode-resistant 'SH-3142' and the *F. oxysporum* f. sp. *cubense* race 4-resistant 'SH-3362' with 'SH-3723' will have multiple disease resistance and produce pollen. Such progenies would be exceptional parental lines for subsequently crossing with the various seed-fertile triploids (photo: FHIA).

are already being planted commercially in certain areas.

Conventional breeding has been the way by which almost all cultivated plants have historically been improved. Banana was recalcitrant in this regard, but it has now joined the long list of crops that have been improved by cross-pollinations and selections. Indeed, with the parental lines now available for subsequent breeding endeavours, the possibilities are almost limitless, just as Simmonds (1962) predicted.

UNCONVENTIONAL BANANA BREEDING IN TAIWAN

S.C. Hwang and C.Y. Tang

Introduction

The cultivar 'Pei-Chiao' (AAA, Cavendish subgroup) was introduced into Taiwan more than 240 years ago and is now the most important export banana cultivar in the country. In 1967, a total of 26 million boxes of bananas (16 kg of fruit box^{-1}) were exported to Japan. However, the banana industry has declined dramatically since 1970, due to the competition from the Philippines and the occurrence of Fusarium wilt caused by *F. oxysporum* f. sp. *cubense* race 4 (Su *et al.*, 1986) (see Ploetz and Pegg, Chapter 3, pp. 143–159). Recently, it was estimated that about 1500 ha out of 5000 ha of land used for the production of banana in Taiwan were infested with the disease.

Measures such as field sanitation, soil

fumigation and liming were not successful in controlling Fusarium wilt in Taiwan (Hwang, 1985). The rotation of banana with paddy rice was shown to be effective, but for only 1–2 years in the same field. Thus, it was believed that the identification or development of resistant cultivars and their adoption by industry would be the most effective means of overcoming the problem.

Several approaches were considered. The first was to introduce new cultivars from other countries in the hope that a suitable resistant cultivar would be found. Exotic germplasm was imported, but, of 22 Cavendish and related cultivars collected from various parts of the world, none showed resistance to *F. oxysporum* f. sp. *cubense* race 4 (Hwang *et al.*, 1984a). The second approach contemplated was conventional breeding. However, Cavendish cultivars are highly sterile and, consequently, are very difficult to improve by hybridization. As the chances of obtaining a resistant cultivar with commercial potential by using this approach were thought to be remote, an innovative breeding scheme was recommended for their improvement.

In 1985, a programme was initiated at the TBRI to screen somaclonal variants of 'Pei-Chiao' for resistance to *F. oxysporum* f. sp. *cubense* race 4 (Hwang and Ko, 1987). More than ten resistant clones were subsequently identified (Hwang and Ko, 1988). Further improvement of these resistant clones through somaclonal variation techniques has proved to be effective in selecting Cavendish cultivars with commercial potential (Hwang and Ko, 1989). The first resistant cultivar, 'Tai-Chiao No. 1' was released for commercial production in 1992 (Hwang *et al.*, 1992, 1993).

Somaclonal Variation in *Musa*

Natural somatic mutations have been described in various banana cultivars at different locations around the world (see Israeli and Lahav, Chapter 12, pp. 395–408). Over 50 of these, resulting in changes in plant stature and habit, colour and bunch and fruit characters, were identified by Stover and Simmonds (1987).

After the development of micropropagation techniques for banana using meristem culture (Ma and Shii, 1972; Hwang *et al.*, 1984b), somaclonal variation was commonly found in tissue-cultured plantlets (Hwang, 1986; Reuveni and Israeli, 1990; Israeli *et al.*, 1991; Vuylsteke *et al.*, 1991; Daniells and Smith, 1993). In a comprehensive review (Israeli *et al.*, 1995), 29 cases of somaclonal variation were reported in various types of banana and plantain, with an incidence ranging from 0 to 69.1%.

Hwang (1986) reported that the percentage of off-types detectable among 30,000 young tissue-cultured plantlets derived from 'Pei-Chiao' was 0.37%. Variants identified were those with variegated leaves (0.25%) and those with narrow, pointed leaves (0.12%). When 46,260 mature 'Pei-Chiao' plants derived from tissue culture were inspected, 2.43% were found to be off-types. Variations consisted of changes in plant stature (1.44%), leaf shape (0.46%), pseudostem colour (0.11%) and bunch characters (0.42%).

Daniells and Smith (1993) reported 28 distinct off-types among several banana cultivars in Australia. As found in Taiwan, these off-types could be classified into variants in plant stature, leaf morphology, pseudostem colour and bunch characteristics. Between 1989 and 1992, 6–38% of off-types were detected among tissue cultured-plants of Cavendish cultivars.

Most somaclonal variants of 'Pei-Chiao' are genetically stable (Hwang, 1986). Some of the variants and their stability in the subsequent generations are listed in Table 14.1. With the exception of the mutant with variegated leaves, it was found that about 95% of the asexual progenies of most variants retained the selected character. In the mutant with variegated leaves, 63–69% reversion was observed. Similar results were reported in a study of somatic mutation in 'Giant Cavendish' in Australia (Daniells and Smith, 1993). Genetic stability of somaclonal variants is important in the improvement of banana cultivars,

Table 14.1. Genetic stability of somaclonal variants of 'Pei-Chiao'.

Clone	Selected trait	Genetic stability (%)*		
		1st gen.	2nd gen.	3rd gen.
	Stature			
GCTCV-215	Taller stature	98.6	100	100
TC1–229[†]	Shorter stature	94.3	100	100
	Foliage			
GCTCV-23	Upright leaves	97.3	98.6	100
GCTCV-40	Drooping leaves	100	100	100
	Plant colour			
GCTCV-43	Darker pseudostem	100	100	100
GCTCV-15	Variegated leaves	37.5	31.7	35.8
	Fruit characteristics			
GCTCV-105	Compact bunch with short fingers	99.6	100	100
GCTCV-119	Starchy pulp	100	100	100

* Genetic stability as measured by the percentage of plants in the subsequent asexual generations retaining the selected feature. Data were taken from about 200 plants per variant per generation. 1st generation was established from tissue-cultured plantlets, while 2nd and 3rd generations were established from suckers.
[†] A tertiary variant derived from GCTCV-215–1.

because it ensures the usefulness of the selected traits.

Most reported somaclonal variants are not advantageous and, consequently, not useful for commercial production. However, some beneficial off-types, such as longer finger length and a sturdier pseudostem with erect leaves, have been observed (Daniells and Smith, 1993) and deemed worthy of further evaluation. With a suitable screening technique, selection for disease resistance induced through somaclonal variation becomes an effective approach for the improvement of banana cultivars.

Methods for Screening Somaclonal Variants for Resistance to Fusarium Wilt

Both suckers and plants derived from tissue culture can be used as planting material for screening for resistance to Fusarium wilt. However, tissue-cultured plants are preferred because of their higher susceptibility level. It was shown that disease incidence in tissue-cultured plants was 88% in heavily infested fields, while that of suckers was 51% (Anon., 1984). Also, the time for appearance of disease symptoms was almost 2–3 months earlier in tissue-cultured plants than in plants grown from suckers. The following methods of screening for resistance were used in different programmes.

Field evaluation using suckers

A prerequisite of this method is a field with a uniform and heavy infestation of the Fusarium wilt pathogen (500–2000 propagules g^{-1} of soil). This concentration can be reached by augmenting natural levels of the pathogen by ploughing large amounts of diseased plant tissue into the soil. Suckers are planted in the infested field for observation for several months until the appearance of disease symptoms. Diseased plants are eliminated and suckers from surviving plants used for further evaluation. Decisions on further selection should be based on the overall performance of a clone rather than on that of individual plants.

This method is applicable if planting material cannot be produced by tissue culture. However, the chance of disease escape is high. Su *et al.* (1986) reported that eight clones selected for resistance by this method showed disease symptoms only in the sixth generation. Thus, results from this screening technique should be considered as tentative. Confirmation of resistance would require the planting of a large population of tissue-cultured plants.

Field evaluation using plants derived from tissue culture

Plants derived from shoot-tip culture or other tissue-culture techniques are preferred for screening for resistance to Fusarium wilt in the field. Two-month-old plants are planted at high density in a heavily infested field. After 3–4 months, depending on seasonal temperatures, the rhizomes of surviving plants are unearthed, cut and examined for vascular discoloration. Any plants without disease symptoms are further evaluated after multiplication.

Pot evaluation using inoculum in sand–maize meal

De Beer and Visser (1994) in South Africa developed this method. A sterile, sand–maize-meal mixture is inoculated with the Fusarium wilt fungus. After 2 weeks' incubation, the mixture with fungal inoculum is placed at the bottom of planting bags. Three-month-old plants derived from tissue culture are then replanted on top of the sand–maize-meal mixture. After 4 weeks, the plants are removed and examined for internal symptoms of infection. This screening technique is especially effective in the cool season.

Pot evaluation using fungal spore suspensions and young, tissue-cultured plantlets

This screening technique is a modification of a rapid method for determining pathogenicity of *F. oxysporum* f. sp. *cubense* developed by Sun and Su (1984). A crude suspension of conidial spores is obtained by flooding fungal cultures after 7 days' growth on potato dextrose agar under light. The inoculum is prepared by filtering the extract through two layers of cheesecloth and adjusting the suspension to 3×10 spores ml^{-1}. Roots of young tissue-cultured plantlets, which have been hardened following removal from the culture flask, are washed and dipped into the spore suspension for 1 min. After inoculation, the roots are dried by evaporation under shade and the plantlets replanted in potting mix in pots. About 6–8 weeks later, surviving plants are removed and inspected by cutting the rhizome area and looking for discoloration of the vascular tissue. Vakili (1965) used a similar method to screen seedlings of *Musa* species for reaction to Fusarium wilt.

Fusarium wilt of banana is a soil-borne disease. The fungus gains entry into the xylem elements of adventitious roots and then spreads into the rhizome stele and pseudostem. The successful invasion of the plant by the pathogen depends on many factors. In screening for resistance, misleading results may be obtained, due to disease escape. In order to ensure that the selected clones have true resistance to the disease, an exhaustive evaluation procedure should be followed. Initially, single plants are identified as being potentially resistant. In the next stage of evaluation, resistance should be confirmed using clonal derivatives, such as suckers and/or tissue-cultured plants. At this stage, plants of one or two known susceptible cultivars should also be planted for reference. The final evaluation should be undertaken in several diseased fields, with many hundred tissue-cultured plants of each selected clone in each field.

Table 14.2. Horticultural characteristics of *F. oxysporum* f. sp. *cubense* race 4-resistant clones derived from 'Pei-Chiao' by somaclonal variation.

Clone	Resistance	Horticultural characteristics
GCTCV-40	High	Tall and slender pseudostem; weak petioles with narrow, drooping leaves
GCTCV-44	High	Short and slender pseudostem; weak petiole and drooping leaves; bunch normal, but weak pedicel
GCTCV-46	Moderate	Black spots on pseudostem and leaf sheath; upright leaves; small bunch with short fingers
GCTCV-53	Moderate	Dark green pseudostem; drooping leaves; elongate male bud; small bunch with short fingers
GCTCV-62	Moderate	Pale green pseudostem; slow growth of suckers; small bunch and fingers
GCTCV-104	High	Pale green pseudostem; fewer fingers; long growing cycle
GCTCV-105	High	Upright leaves; compact bunch and shorter fingers
GCTCV-119	High	Very tall and slender pseudostem; wavy leaves; long growing cycle
GCTCV-201	Moderate	Robust pseudostem; short fruit stalk
GCTCV-215	Moderate	Tall and slender pseudostem; leaf-tip curl and splitting; normal bunch
GCTCV-216	Moderate	Very tall and strong pseudostem; very large and heavy bunch
GCTCV-217	High	Erect leaves; compact, but heavy bunch

GCTCV, Giant Cavendish tissue-culture variant; High, disease incidence of less than 10%; Moderate, disease incidence of 10–30%.

Identification of Resistant Clones

Field screening by growing tissue-cultured plants of susceptible cultivars at high density in a disease nursery was the major method used to search for variants resistant to *F. oxysporum* f. sp. *cubense* race 4 at TBRI.

From 1984 to 1986, about 30,000 tissue-cultured plants of 'Pei-Chiao' were screened and ten potentially useful clones selected, following final evaluation (Hwang and Ko, 1988, 1989, 1992). From 1992 to 1995, two resistant clones were found among 11,180 tissue-cultured plants of 'Tai-Chiao No. 2', a semi-dwarf Cavendish cultivar. The efficiency of selection was about two to three resistant clones for every 10,000 plants screened.

Among the ten selected clones derived from 'Pei-Chiao', five were highly resistant, with a disease incidence of less than 10%, while the other five were moderately resistant, with an infection rate of 10–30%. In addition to the normal screening procedure, putative resistant plants derived from 'Pei-Chiao' were occasionally found in farmer's fields or in the disease nursery at TBRI. Clones 'GCTCV-216' and 'GCTCV-217' were selected in this way. The resistance levels and characteristics of the 12 'Pei Chiao'-derived clones are presented in Table 14.2. 'GCTCV-217' has only recently been selected and shows promise for commercial production.

'GCTCV-217'

'GCTCV-217' is a highly resistant clone with a distinct erect leaf type (Plate 14.17). Plant height is about 270 cm and its growing cycle is 12 months. The bunch is slightly more compact than the bunch of 'Pei-Chiao', but with normal finger length. The bunch weight is about 30 kg, which is 3 kg more than 'Pei-Chiao'. Because of its erect leaf habit, the leaves shade a circular area of land, with a diameter of about 347 cm, which is approximately 50 cm less than that of 'Pei-Chiao' (see Table 14.3). This gives 'GCTCV-217' the advantage of tolerating higher planting densities.

Plate 14.17. 'GCTCV-217' (left) has high resistance to *F. oxysporum* f. sp. *cubense* race 4 compared to 'Pei-Chiao' (right), which is very susceptible (photo: S.C. Hwang, TBRI).

Plate 14.18. 'GCTCV-215–1', which was selected as a secondary variant of 'GCTCV-215' in 1988. The clone, as 'Tai-Chiao No. 1', was officially released for commercial production in Taiwan in 1992. One of the characteristics of 'GCTCV-215–1' is its susceptibility to marginal scorch, which appears as necrosis along leaf margins (photo: S.C. Hwang, TBRI).

Evaluation of fruit taste and marketability is under way.

Changes in Fruit Quality

Fruit quality is the key to the success of a new commercial cultivar. When 'Pei-Chiao', the traditional standard for flavour and other fruit characteristics, was compared with its *F. oxysporum* f. sp. *cubense* race 4-resistant variants, differences in fruit quality became apparent. Bananas from some, such as 'GCTCV-46' and 'GCTCV-215', were found to be similar to 'Pei-Chiao' in terms of taste and texture, but bananas from others, such as 'GCTCV-40' and 'GCTCV-216', were inferior (see Table 14.4). An added bonus for 'GCTCV-215' was that fruit degreened uniformly after ripening was induced, compared with fruit of 'Pei-Chiao', which is moderately affected by a ripening disorder (see Lahav *et al.*, Chapter 8, pp. 334–335). Bananas from 'GCTCV-119' and 'GCTCV-105' were much sweeter and the pulp more granular than the fruit of 'Pei-Chiao'. 'GCTCV-53' was the worst variant in terms of fruit quality. It was lower in sugar content, less aromatic and had a higher frequency of uneven ripening. Thus possible changes in fruit quality should be taken into consideration during the selection process.

Table 14.3. Resistance levels to *F. oxysporum* f. sp. *cubense* race 4 and horticultural traits of promising clones compared to 'Pei-Chiao'.

Trait	GCTCV-217	GCTCV-215–1	GCTCV-105–1	TC1–229	'Pei Chiao'
Resistance level	High	Moderate	High	Moderate	Susceptible
Plant height (cm)	270	290	260	240	270
Girth (cm)	70	70	70	68	75
Crown coverage diameter (cm)	347	445	365	342	395
Growing cycle (months)	11–12	12–13	11–12	12–13	11–12
Bunch weight (kg)	30.1	24.1	24.0	22.4	27.4
Bunch shape	Compact	Loose	Compact	Loose	Loose
Finger length	Medium	Long	Short	Medium	Long

Table 14.4. Fruit quality of variants of 'Pei-Chiao' selected because of their resistance to *F. oxysporum* f. sp. *cubense* race 4.

Clone	Resistance level	Taste	Texture	Ripening disorder*	Finger characteristics
GCTCV-40	High	Fair	Soft	–	Short
GCTCV-44	High	Fair	Soft	–	Long and slender
GCTCV-46	Moderate	Good	Normal	++	Short
GCTCV-53	Moderate	Fair	Normal	+++	Short
GCTCV-62	Moderate	Fair	Normal	–	Fewer fingers
GCTCV-104	High	Good	Normal	–	Fewer fingers
GCTCV-105	High	Very good	Granular	++	Short
GCTCV-119	High	Very good	Granular	+++	Normal
GCTCV-201	Moderate	Good	Normal	++	Normal
GCTCV-215	Moderate	Good	Normal	–	Slender
GCTCV-216	Moderate	Fair	Soft	++	Long
GCTCV-217	High	Good	Normal	++	Normal, blunt tip
Pei-Chiao	Susceptible	Good	Normal	++	Normal

* Uneven ripening process during April and May.
–, no disorder; ++, moderate level of disorder; +++, serious level of disorder.

Improvement of Resistant Clones

All resistant clones selected so far have been morphologically different from the original cultivars, 'Pei-Chiao' and 'Tai-Chiao No. 2'. Most of them carry inferior horticultural traits (Hwang and Ko, 1988), such as reduced bunch weight, increased height, long growing cycle or poor fruit quality (Tables 14.3 and 14.4). However, from tissue-cultured plants of these clones, secondary somaclonal variants of various forms have been identified (Hwang and Ko, 1989). The frequency of secondary variation has been found to range from 0.2 to 10.1%, depending on the clone.

Some secondary variants have shown improved horticultural traits, such as a larger pseudostem, shorter plant height, shorter growing cycle or bigger bunch and have retained their resistance to *F. oxysporum* f. sp. *cubense* race 4. 'GCTCV-215–1', 'GCTCV-105–1' and 'TC1–229', which is a tertiary variant, were selected because of their potential for commercial production.

'GCTCV-215–1'

A moderately resistant clone, 'GCTCV-215–1' (Plate 14.18), was selected in 1988 and is a secondary variant of 'GCTCV-215' (Hwang, 1991; Hwang *et al.*, 1992, 1993). Extensive evaluation of this clone was carried out in farmers' diseased fields in 1990–1992. Its horticultural traits, productivity and fruit quality are comparable to those of 'Pei-Chiao', its parental cultivar.

'GCTCV-215–1' is moderately resistant to Fusarium wilt, having a susceptibility rate of 17.6% when planted as tissue-cultured plants in very seriously diseased plots, as compared with 74.6% with 'Pei-Chiao'. When planted as suckers, the infection rate was reduced to 5.2%, as compared with 77.8% with 'Pei-Chiao'.

'GCTCV-215–1' is slightly taller than 'Pei-Chiao' and has a more slender pseudostem and longer growing cycle (see Table 14.3). Although the bunch weight of 'GCTCV-215–1' is about 10% lighter than that of 'Pei-Chiao', the bunch shape is more uniform and ripening is more even, giving better-quality fruit.

'GCTCV-215–1' was officially released for commercial production as 'Tai-Chiao No.1' in 1992 (Hwang, 1991; Hwang *et al.*, 1992, 1993). It has been estimated that 'Tai-Chiao No.1' is now planted on about 1500 ha in southern Taiwan every year. Surveys in infested fields planted with 'Tai-Chiao No. 1' showed that the percentage of wilt incidence averaged 6.5% in 1994 and 5.1% in 1995. The incidences for 'Pei-Chiao' were 69.0% and 42.6%. The release of 'Tai-Chiao No.1' has boosted banana production remarkably in recent years, thus saving the Taiwan banana industry from devastation by Fusarium wilt.

'GCTCV-105–1'

The clone 'GCTCV-105–1' was selected in 1991 from the clone 'GCTCV-105', which was highly resistant, but low-yielding. Its plant height and growing cycle are similar to that of 'Pei-Chiao', its parental cultivar, but it is highly resistant to *F. oxysporum* f. sp. *cubense* race 4, with a disease incidence of less than 5%. The major characteristic of this clone is its short finger length. Fingers are about 19–20 cm long, which is approximately 3 cm shorter than in 'Pei-Chiao'. The bunch of 'GCTCV-105–1' is compact and the number of fingers per hand is higher than in 'Pei-Chiao'. However, due to shorter fingers, the bunch weight of 'GCTCV-105–1' is slightly less than that of its parental cultivar (see Table 14.3). In field trials, disease incidence averaged 1.6% in 1994 and 1.9% in 1995.

'TC1–229'

Wind damage is one of the major problems facing the Taiwanese banana industry and plants of smaller stature would help reduce yield losses, as well as being easier to manage. In 1992, a semi-dwarf mutant of 'GCTCV-215–1', designated 'TC1–229', was found in a farmer's orchard (Tang and Hwang, 1994, 1998). After 2 years' evaluation, the semi-dwarf trait was shown to be stable, with a reduction in plant height of 50–70 cm compared with 'GCTCV-215–1' (see Table 14.3). Other traits, such as growth cycle, bunch weight and level of resistance to *F. oxysporum* f. sp. *cubense* race 4, were not significantly different from those of 'GCTCV-215–1'.

Discussion

Somaclonal variation has been reported in many crops, such as maize, rice and tomato (Larkin and Scowcroft, 1981). As in banana, the potential of somaclonal variation for crop improvement has been thoroughly explored in potato and sugar cane, which are also propagated vegetatively (Liu and Chen, 1978; Shephard *et al.*, 1980; Liu *et al.*, 1983). As well as an improvement in agronomic traits, changes in disease resistance have been reported. For example, increased resistance to early blight and late blight has been found in potato and to Fiji disease, downy mildew and eyespot in sugar cane (Bright *et al.*, 1986).

In the work undertaken at TBRI on the improvement of Cavendish banana for resistance to *F. oxysporum* f. sp. *cubense* race 4, different levels of host–pathogen interactions, ranging from moderate resistance to high resistance, were detected in somaclonal variants. Variants that have improved agronomic traits, such as shorter stature, precocious flowering and higher bunch weight have also been found (Tang and Hwang, 1994). Since vegetative propagation is the major means of reproduction in edible banana, any beneficial and stable genetic changes occurring in somaclonal variation can be transmitted to the subsequent generations. Work at TBRI has shown that the improved resistance of selected clones can be maintained in the field for more than 10 years. Since its selection in 1988 and commercial registration in 1992, 4.6 million tissue-cultured plants of 'GCTCV-215–1' have been propagated and distributed to farmers throughout the southern part of the country. The horticultural traits and resistance level of 'GCTCV-215–1' have been shown to be consistent over the years.

The major drawback of using somaclonal variants to improve resistance to *F. oxysporum* f. sp. *cubense* race 4 is the pleiotropic effect of the resistance trait. All resistant clones isolated so far are morphologically different from the parental cultivars. Very often, they are inferior to the parent in terms of the agronomic characters necessary for commercial acceptance. Two approaches can be used to remedy such a drawback. One is to increase the sample size used for selection. A routine screening procedure could be devised to search for resistant clones with good horticultural characteristics at the same time as screening for disease resistance. However, an efficient tissue-culture propagation system and the availability of large areas of infested land for planting or the development of biotechnological screening procedures are prerequisites for this approach. Another approach is to search for improved variants among tissue-culture progenies of agronomically inferior resistant clones. This method is often still not attractive to many research workers, because large numbers of plants derived from resistant clones with inferior traits have to be grown in the field in order to search for improved variants. However, this improvement procedure has proved useful at TBRI and has led to the selection of 'GCTCV-215–1', 'GCTCV-105–1' and 'TC1–229'.

It is important that improved clones are held under conditions that minimize the chances of further somaclonal variation. If precautions are not taken, the selected clone may change unnoticed to another genotype and could lose a desirable agronomic trait or resistance to disease. At least 20 plants of each improved cultivar should be maintained in soil, preferably in a protected environment, such as an insect-proof net house, to reduce the chances of transmission of banana virus diseases. The soil should be infested with the Fusarium wilt pathogen if it is a resistant cultivar. It is important that this *mother stock* is not maintained in tissue culture. Suckers from mother stock should be used to establish a nursery for the maintenance of *multiplication stock*. The size of the nursery may vary from 200 to 500 plants, depending on commercial demand for the cultivar. Again, for a resistant cultivar, the nursery should be in infested soil. Any diseased plants and/or abnormal plants different from the original type should be rogued out from the nursery. Suckers from multiplication stock should be used as a source of shoot tips for mass tissue-culture propagation.

Conclusion

Exploiting somaclonal variation is a simple, but effective approach to crop improvement. Though the procedure suffers from unpredictability and requires directional screening and systematic searching, it has proved useful in Taiwan in improving the resistance of Cavendish banana to *F. oxysporum* f. sp. *cubense* race 4. The somaclonal variation approach to improving the resistance of Cavendish to *F. oxysporum* f. sp. *cubense* race 4 is also being tried in South Africa (de Beer and Visser, 1994). In addition to Cavendish, it is believed that this method of improvement would be applicable to other types of banana and for inducing resistance to other types of disease.

REFERENCES

Afza, R., Roux, N., Brunner, H., van Duren, M. and Morpurgo, R. (1994) In-vitro mutation techniques for *Musa*. In: Jones, D.R. (ed.) *The Improvement and Testing of* Musa*: a Global Partnership, Proceedings of the First Global Conference of the International Testing Program held at FHIA, Honduras, 27–30 April 1994.* INIBAP, Montpellier, France, pp. 207–212.

Alves, E.J., Shepherd, K. and Dantas, J.L.L. (1987) Cultivation of bananas and plantains in Brazil and needs for improvement. In: Persley, G.J. and De Langhe, E.A. (eds) *Banana and Plantain Breeding Strategies: Proceedings of an International Workshop held in Cairns, Australia, 13–17 October, 1986.* ACIAR Proceedings No. 21, Australian Centre for International Agricultural Research, Canberra, Australia, pp. 44–49.

Anon. (1984) *Taiwan Banana Research Institute Research Report 1983.* TBRI, Pingtung, Taiwan, 60 pp.

Anon. (1993) *Proceedings of a Workshop on Biotechnology Applications for Banana and Plantain Improvement held in San José, Costa Rica, 27–31 January 1992.* INIBAP, Montpellier, France, 251 pp.

Anon. (1994) *Banana and Plantain Breeding: Priorities and Strategies, Proceedings of the First Meeting of the* Musa *Breeders' Network held in La Lima, Honduras, 2–3 May 1994.* INIBAP, Montpellier, France, 55 pp.

Bakry, F. and Horry, J.-P. (1994) *Musa* breeding at CIRAD–FLHOR. In: Jones, D.R. (ed.) *The Improvement and Testing of* Musa*: a Global Partnership, Proceedings of the First Global Conference of the International Testing Program held at FHIA, Honduras, 27–30 April 1994.* INIBAP, Montpellier, France, pp. 168–175.

Baldry, J., Coursey, D.G. and Howard, G.E. (1981) The comparative consumer acceptability of triploid and tetraploid banana fruit. *Tropical Science* 23, 33–66.

Bhagwat, B. and Duncan, E.J. (1998) Mutation breeding of banana cv. Highgate (*Musa* spp., AAA group) for tolerance to *Fusarium oxysporum* f. sp. *cubense* using chemical mutagens. *Scientia Horticulturae* 73, 11–22.

Bright, S.W.J., Ooms, G., Foulger, D., Karp, A. and Evans, N. (1986) Mutation and tissue culture. In: Whiters, L.A. and Alderson, P.G. (eds) *Plant Tissue Culture and its Agricultural Applications.* Butterworths, London, pp. 431–449.

Buddenhagen, I.W. (1962) Bacterial wilt of certain seed-bearing *Musa* spp. caused by the tomato strain of *Pseudomonas solanacearum. Phytopathology* 52, 286.

Cheesman, E.E. (1931) *Banana Breeding at the Imperial College of Tropical Agriculture.* Empire Marketing Board Report 47, 35 pp.

Cheesman, E.E. (1932) Genetic and cytological studies of *Musa*. I. Certain hybrids of the 'Gros Michel' banana. *Journal of Genetics* 26, 291–316.

Cheesman, E.E. and Larter, L.N.H. (1935) Genetical and cytological studies of *Musa*. III. Chromosome numbers in the Musaceae. *Journal of Genetics* 30, 31–52.

Cheesman, E.E. and Wardlaw, C.W. (1937) Specific and varietal susceptibility of bananas to Cercospora leaf spot. *Tropical Agriculture (Trinidad)* 14, 335–336.

Dale, J.L. (1990) Banana and plantain. In: Persley, G.J. (ed.) *Agricultural Biotechnology, Opportunities for International Development.* CAB International, Wallingford, UK, pp. 225–240.

Daniells, J.W. and Smith, M.K. (1993) Somatic mutation of bananas–their stability and potential. In: Valmayor, R.V., Hwang, S.C., Ploetz, R., Lee, S.W. and Rao, N.V. (eds) *Proceedings: International Symposium on Recent Developments in Banana Cultivation Technology, Taiwan Banana Research Institute, Chiuju, Pingtung, Taiwan, 14–18 December 1992.* INIBAP/ASPNET, Los Baños, Philippines, pp. 162–171.

Daub, M.E. (1986) Tissue culture and the selection of resistance to pathogens. *Annual Review of Phytopathology* 24, 159–186.

de Beer, Z.C. and Visser, A.A. (1994) Mutation breeding of banana in South Africa. In: Jones, D.R. (ed.) *The Improvement and Testing of* Musa*: a Global Partnership, Proceedings of the First Global Conference of the International Testing Program held at FHIA, Honduras, 27–30 April 1994.* INIBAP, Montpellier, France, pp. 243–247.

De Langhe, E. (1969) Bananas. In: Ferwerda, F.P. and Wit, F. (eds) *Outlines of Perennial Crop Breeding in the Tropics.* H. Veenman and Zonen N.V., Wageningen, the Netherlands, pp. 53–78.

de Oliveira e Silva, S., de Matos, A.P., Alves, E.J. and Shepherd, K. (1997a) Breeding diploid banana (AA) at EMBRAPA/CNPMF. *Infomusa* 6(2), 4–6.

de Oliveira e Silva, S., de Matos, A.P., Alves, E.J. and Shepherd, K. (1997b) Breeding 'Prata' ('Pome') and 'Maçã' ('Silk') banana types: current achievements and opportunities. *Infomusa* 6(2), 7–10.

Dhed'a, D., Dumortier, F., Panis, B., Vuylsteke, D. and De Langhe, E. (1991) Plant regeneration in cell suspension cultures of the cooking banana cv. Bluggoe (*Musa* spp., ABB group). *Fruits* 46, 125–135.

Dodds, K.S. (1943) The genetic system of banana in relation to banana breeding. *Empire Journal of Experimental Agriculture* 11, 89–98.

Epp, M.D. (1987) Somaclonal variation in bananas: a case study with Fusarium wilt. In: Persley, G.J. and De Langhe, E.A. (eds) *Banana and Plantain Breeding Strategies, Proceedings of an International Workshop held in Cairns, Australia, 13–17 October 1986.* ACIAR Proceedings No. 21, Australian Centre for International Agricultural Research, Canberra, pp.140–150.

Escalant, J.V. (1990) Using tissue culture technques for production of new varieties of plantain and early screening against black Sigatoka. In: Fullerton, R.A. and Stover, R.H. (eds) *Sigatoka Leaf Spot Diseases of Banana, Proceedings of an International Workshop held at San José, Costa Rica, March 28-April 1, 1989.* INIBAP, Montpellier, France, pp. 338–348.

Escalant, J.V., Côte, F., Grappin, A., Aubiron, E., Legavre, T., Lagoda, P., Carreel, F., Bakry, F., Horry, J.-P., Kerbelec, F., Teisson, C. and Tezenas du Montcel, H. (1994a) Biotechnologies for *Musa* improvement – strategies and results. In: INIBAP (ed.) *Banana and Plantain Breeding: Priorities and Strategies, Proceedings of the First Meeting of the Musa Breeders' Network held in La Lima, Honduras, 2–3 May 1994.* INIBAP, Montpellier, France, pp. 22–26.

Escalant, J.V., Teisson, C. and Côte, F. (1994b) Amplified somatic embryogenisis from male flowers of triploid banana and plantain cultivars (*Musa* sp.). In Vitro *Cell Development Biology* 30P, 181–186.

Ganry, J. (ed.) (1993) *Breeding Banana and Plantain for Resistance to Diseases and Pests, Proceedings of the International Symposium on Genetic Improvement of Bananas for Resistance to Diseases and Pests, Organized by CIRAD–FLHOR, Montpellier, France, 7–9 September 1992.* CIRAD, Montpellier, France, 393 pp.

Garcia, L., Herrera, L., Bermudez, I., Veitia, N., Clavero, J., Acosta Mayra, C. and Romero, C. (1997) Method for the *in vitro* selection of *Mycosphaerella fijiensis* Morelet in the banana. *Infomusa* 6(1), 14–15.

Gonsalves, R.A. (1987) Reactions of breeding lines of banana from Jamaica to black leaf streak. In: Jaramillo, R. and Mateo, N. (eds) *Memoria de la Reunion Regional de INIBAP para America Latina y el Caribe, San José, Costa Rica.* INIBAP, Montpellier, France, pp. 224–228.

Gonsalves, R.A. and Dixon, J.A. (1990) Advances of the Jamaica Banana Board breeding scheme. In: Fullerton, R.A. and Stover, R.H. (eds) *Sigatoka Leaf Spot Diseases of Bananas: Proceedings of an International Workshop held at San José, Costa Rica, March 28–April 1, 1989.* INIBAP, Montpellier, France, pp. 267–269.

Harelimana, G., Lepoivre, P., Jijakli, H. and Mourichon, X. (1997) Use of *Mycosphaerella fijiensis* toxins for the selection of banana cultivars resistant to black leaf streak. *Euphytica* 96, 125–128.

Hwang, S.C. (1985) Ecology and control of fusarial wilt of banana. *Plant Protection Bulletin (Taiwan)* 27, 233–245.

Hwang, S.C. (1986) Variation in banana plants propagated through tissue culture. *Journal of the Chinese Society of Horticultural Science* 32, 117–125.

Hwang, S.C. (1991) Somaclonal resistance in Cavendish banana to Fusarium wilt. *Plant Protection Bulletin (Taiwan)* 33, 124–132.

Hwang, S.C. and Ko, W.H. (1987) Somaclonal variation of bananas and screening for resistance of Fusarium wilt. In: Persley, G.J. and De Langhe, E.A. (eds) *Banana and Plantain Breeding*

Strategies, Proceedings of an International Workshop held at Cairns, Australia, 13–17 October 1986. ACIAR Procceedings No. 21, Australian Centre for International Agricultural Research, Canberra, Australia, pp. 157–160.

Hwang, S.C. and Ko, W.H. (1988) Mutants of Cavendish banana resistant to race 4 of *Fusarium oxysporum* f. sp. *cubense*. *Plant Protection Bulletin (Taiwan)* 30, 386–392.

Hwang, S.C. and Ko, W.H. (1989) Improvement of fruit quality of Cavendish banana mutants resistant to race 4 of *Fusarium oxysporum* f. sp. *cubense*. *Plant Protection Bulletin (Taiwan)* 31, 131–138.

Hwang, S.C. and Ko, W.H. (1992) Breeding for resistance to fusarial wilt of Cavendish banana by using tissue culture method. In: *Proceedings of the SABRAO International Symposium on the Impact of Biological Research on Agricultural Productivity held at Taichung, Taiwan,* 10–13 March 1992. TDAIS/SABRAO, Taichung, Taiwan, pp. 229–237.

Hwang, S.C. and Tang, C.Y. (1996) Somaclonal variation and its use for improving Cavendish (AAA dessert) bananas in Taiwan. In: Frison, E.A., Horry, J.P. and De Waele, D. (eds) *New Frontiers in Resistance Breeding for Nematode, Fusarium and Sigatoka: Proceedings of an International Banana Breeding Workshop held in Serdang, Selangor, Malaysia, 2–5 October, 1995.* INIBAP, Montpellier, France, pp. 173–181.

Hwang, S.C., Chen, C.C. and Wu, F.L. (1984a) An investigation on susceptibility of banana clones to fusarial wilt, freckle and marginal scorch disease in Taiwan. *Plant Protection Bulletin (Taiwan)* 26, 155–156.

Hwang, S.C., Chen, S.L., Lin J.C. and Lin, H.L. (1984b) Cultivation of banana using plantlets from meristem culture. *Horticultural Science* 19, 231–233.

Hwang, S.C., Ko, W.H. and Chao, C.P. (1992) Control of fusarial wilt of Cavendish banana by planting a resistant clone derived from breeding. In: Proceedings of the Symposium on the Non-chemical Control of Crop Diseases and Pests. *Plant Protection Bulletin (Taiwan)*, Special Publication, 259–280.

Hwang, S.C., Ko, W.H. and Chao, C.P. (1993) GCTCV-215–1: a promising Cavendish clone resistant to race 4 of *Fusarium oxysporum* f. sp. *cubense*. In: Valmayor, R.V., Hwang, S.C., Ploetz, R., Lee, S.W. and Rao, N.V. (ed) *Proceedings: International Symposium on Recent Developments in Banana Cultivation Technology, Taiwan Banana Research Institute, Chiuju, Pingtung, Taiwan, 14–18 December 1992.* INIBAP/ASPNET, Los Baños, Philippines, pp. 43–57.

Israeli, Y., Reuveni, O. and Lahav, E. (1991) Qualitative aspects of somaclonal variations in banana propagated by *in vitro* techniques. *Scientia Horticulturae*, 48, 71–88.

Israeli, Y., Lahav, E. and Reuveni, O. (1995) *In vitro* culture of bananas. In: Gowen, S. (ed.) *Bananas and Plantains.* Chapman and Hall, London, pp. 147–178.

Jamaluddin, S.H. (1994) Mutation breeding of banana in Malaysia. In: Jones, D.R. (ed.) *The Improvement and Testing of* Musa*: a Global Partnership, Proceedings of the First Global Conference of the International Testing Program held at FHIA, Honduras, 27–30 April 1994.* INIBAP, Montpellier, France, pp. 228–232.

Jenny, C., Auboiron, A. and Beveraggi, A. (1994) Breeding plantain-type hybrids at CRBP. In: Jones, D.R. (ed.) *The Improvement and Testing of* Musa*: a Global Partnership, Proceedings of the First Global Conference of the International Testing Program held at FHIA, Honduras, 27–30 April 1994.* INIBAP, Montpellier, France, pp. 176–187.

Jones, D.R. (1990) Black Sigatoka – a threat to Australia. In: Fullerton, R.A. and Stover, R.H. (eds) *Sigatoka Leaf Spot Diseases of Bananas, Proceedings of an International Workshop held at San José, Costa Rica, March 28-April 1, 1989.* INIBAP, Montferrier sur Lez, France, pp. 38–46.

Jones, D.R. (1993) Evaluating banana and plantain for reaction to black leaf streak in the South Pacific. *Tropical Agriculture (Trinidad)* 70, 39–44.

Jones, D.R. (ed.) (1994a) *The Improvement and Testing of* Musa*: a Global Partnership, Proceedings of the First Global Conference of the International Testing Program held at FHIA, Honduras, 27–30 April 1994.* INIBAP, Montpellier, France, 303 pp.

Jones, D.R. (1994b) International *Musa* Testing Program Phase 1. In: Jones, D.R. (ed.) *The Improvement and Testing of* Musa*: a Global Partnership, Proceedings of the First Global Conference of the International Testing Program held at FHIA, Honduras, 27–30 April 1994.* INIBAP, Montpellier, France, pp. 12–20.

Jones, D.R. (1994c) Distribution of recommended FHIA hybrids. In: Jones, D.R. (ed.) *The Improvement and Testing of* Musa*: a Global Partnership, Proceedings of the First Global*

Conference of the International Testing Program held at FHIA, Honduras, 27–30 April 1994. INIBAP, Montpellier, France, pp. 21–22.

Jones, D.R. (1994d) Report on the meeting. In: INIBAP (ed.) *Banana and Plantain Breeding: Priorities and Strategies, Proceedings of the First Meeting of the* Musa *Breeders' Network held in La Lima, Honduras, 2–3 May 1994.* INIBAP, Montpellier, France, pp. 7–15.

Krikorian, A.D. and Scott, M.E. (1995) Somatic embryogenesis in bananas and plantains (*Musa* clones and species). In: Bajaj, Y.P.S. (ed.) *Biotechnology in Agriculture and Forestry, Vol. 31. Somatic Embryogenesis and Synthetic Seed 11.* Springer Verlag, Berlin, pp. 183–195.

Langdon, P.W. and Pegg, K.G. (1988) *Field Screening for Resistance to* Fusarium *wilt.* Research Report, ACIAR Banana Improvement Project for the Pacific and Asia – Fusarium Wilt (Panama Disease) Studies, QDPI, Brisbane, 12 pp.

Larkin, P.J. and Scowcroft, W.R. (1981) Somaclonal variation – a novel source of variability from cell cultures for plant improvement. *Theoretical Applied Genetics* 60, 197–214.

Larter, L.N.H. (1934) Sports of the 'Gros Michel'. *Journal of the Jamaica Agricultural Society* 38, 461–465.

Larter, L.N.H. (1935) Hybridism in *Musa.* I. Somatic cytology of certain Jamaican seedlings. *Journal of Genetics* 31, 297–316.

Larter, L.N.H. (1947) *Report on Banana Breeding.* Bulletin 34, Department of Agriculture, Kingston, Jamaica, 24pp.

Lepoivre, P., Acuna, C.P. and Riveros, A.S, (1993) Screening procedures for improving resistance to banana black leaf streak disease. In: Ganry, J. (ed.) (1993) *Breeding Banana and Plantain for Resistance to Diseases and Pests, Proceedings of the International Symposium on Genetic Improvement of Bananas for Resistance to Diseases and Pests, organized by CIRAD–FLHOR, Montpellier, France, 7–9 September 1992.* CIRAD, Montpellier, France, pp. 213–220.

Liu, M.C. and Chen, W.H. (1978) Tissue and cell culture as aids to sugarcane breeding. II. Performance and yield potential of sugarcane callus derived lines. *Euphytica* 27, 273–282.

Liu, M.C., Yeh, H.S. amd Chen, W.H. (1983) Tissue and cell culture as aids to sugarcane breeding. IV. A high-sucrose and vigorously growing calliclone 71–4829. *Report of the Taiwan Sugarcane Research Institute* 102, 1–12.

Ma, S.S. and Shii, C.T. (1972) *In vitro* formation of adventitous buds in banana shoot apex following decapitation. *Journal of the Chinese Society of Horticultural Science* 18, 135–142.

Matsumoto, K., Luz Barbosa, M., Copati Souza, L.A. and Batista Teixeira, J. (1995) Race 1 Fusarium wilt tolerance on banana plants selected by fusaric acid. *Euphytica* 84, 67–71.

Megia R., Haicour, R., Tizroutine, S., Bui Trang, V., Rossignol, L., Sihachakr, D. and Schwendiman, J. (1993) Plant regeneration from cultured protoplasts of the cooking banana cv. Bluggoe (*Musa* spp., ABB group). *Plant Cell Reports* 13, 41–44.

Menendez, T. and Shepherd, K. (1975) Breeding new bananas. *World Crops*, May–June, 104–112.

Molina, G.C. and Krausz (1989) A phytotoxic activity in extracts of broth cultures of *Mycosphaerella fijiensis* var. *difformis* and its use to evaluate host resistance to black Sigatoka. *Plant Disease* 73, 142–144.

New, S. and Marriott, J. (1974) Post-harvest physiology of tetraploid banana fruit: response to storage and ripening. *Annals of Applied Biology* 78, 193–204.

Novak, F.J. (1991) *In vitro* mutation system for crop improvement. In: *Proceedings of an International Symposium on the Contribution of Plant Mutation Breeding to Crop Improvement jointly organized by the International Atomic Energy Agency and the Food and Agriculture Organization of the United Nations and held in Vienna, 18–22 June 1990*, Vol. 2. IAEA, Vienna, pp. 327–342.

Novak, F.J. (1992) *Musa* (bananas and plantains). In: Hammerschlag, F.A. and Litz, R.E. (eds) *Biotechnology of Perennial Fruit Crops.* CABI, Wallingford, UK, pp. 449–488.

Novak, F.J., Brunner, H., Afza, R., Morpungo, R.K., Van Duren, M., Sacchi, M., Sitti Hawa, J., Khatri, A., Kahl, G., Kaemmer, D., Ramser, J. and Weising, K. (1993) Improvement of *Musa* through biotechnology and mutation breeding. In: INIBAP (ed.) *Biotechnology Applications for Banana and Plantain Improvement.* INIBAP, Montpellier, France, pp. 143–158.

Okole, B.N. and Schulz, F.A. (1993) Selection of banana and plantain (*Musa* spp.) resistant to toxins produced by *Mycosphaerella* species using in-vitro culture techniques. In: Ganry, J. (ed.) *Breeding Banana and Plantain for Resistance to Diseases and Pests, Proceedings of the International Symposium on Genetic Improvement of Bananas for Resistance to Diseases and Pests, Organized by CIRAD–FLHOR, Montpellier, France, 7–9 September 1992.* CIRAD, Montpellier, France, p. 378.

Okole, B.N. and Schulz, F.A. (1997) Selection of *Mycosphaerella fijiensis*-resistant cell lines from micro-cross sections of banana and plantain. *Plant Cell Reports* 16, 339–343.

Ortiz, R. (1994) *Musa* genetics. In: Gowen, S. (ed.) *Bananas and Plantains.* Chapman and Hall, London, pp. 84–109.

Ortiz, R. and Vuylsteke, D. (1994) Plantain breeding at IITA. In: Jones, D.R. (ed.) *The Improvement and Testing of* Musa*: a Global Partnership, Proceedings of the First Global Conference of the International Testing Program held at FHIA, Honduras, 27–30 April 1994.* INIBAP, Montpellier, France, pp. 130–156.

Ortiz, R. and Vuylsteke, D. (1998) 'BITA-3': a starchy banana with partial resistance to black sigatoka and tolerance to streak virus. *HortScience* 33, 358–359.

Ortiz, R., Ferris, R.S.B. and Vuylsteke, D. (1994) Banana and plantain breeding. In: Gowen, S. (ed.) *Bananas and Plantains.* Chapman and Hall, London, pp. 110–146.

Ortiz, R., Vuylsteke, D., Crouch, H. and Crouch, J. (1998) TMSx: triploid black-sigatoka-resistant *Musa* hybrid germplasm. *HortScience* 33, 362–365.

Persley, G.J. and De Langhe, E.A. (eds) (1987) *Banana and Plantain Breeding Strategies, Proceedings of an International Workshop held in Cairns, Australia, 13–17 October 1986.* ACIAR Proceedings No. 21, Australian Centre for International Agricultural Research, Canberra, 187 pp.

Pinochet, J. and Rowe, P.R. (1979) Progress in breeding for resistance to *Radopholus similis* on bananas. *Nematropica* 9, 76–78.

Ploetz, R.C. and Correll, J.C. (1988) Vegetative compatibility among races of *Fusarium oxysporum* f. sp. *cubense. Plant Disease* 72, 325–328.

Popenoe, W. (1941) Banana culture around the Caribbean. *Tropical Agriculture (Trinidad)* 18, 8–12.

Reuveni, O. and Israeli, Y. (1990) Measures to reduce somaclonal variation in vitro propagated bananas. *Acta Horticulturae* 275, 307–313.

Roux, N., Afza, R., Brunner, H., Morpungo, R. and van Duren, M. (1994) Complementary approaches to cross-breeding and mutation breeding for *Musa* improvement. In: Jones, D.R. (ed.) *The Improvement and Testing of* Musa*: a Global Partnership, Proceedings of the First Global Conference of the International Testing Program held at FHIA, Honduras, 27–30 April 1994.* INIBAP, Montpellier, France, pp. 213–218.

Rowe, P.R. (1984) Breeding bananas and plantains. In: Janick, J. (ed.) *Plant Breeding Reviews,* Vol. 2. AVI Publishing Company, Westport, Connecticut, pp. 135–155.

Rowe, P.R. (1987) Breeding plantains and cooking bananas. In: *International Cooperation for Effective Plantain and Banana Research, Proceedings of the Third Meeting of IARPB held in Abidjan, Côte d'Ivoire, 27–31 May, 1985.* INIBAP, Montpellier, France, pp. 21–23.

Rowe, P.R. (1998) A banana breeder's response to 'The Global Programme for *Musa* Improvement'. *Infomusa* 7(1), I–IV.

Rowe, P.R. and Richardson, D.L. (1975) *Breeding Bananas for Disease Resistance, Fruit Quality and Yield.* Bulletin 2, Tropical Agriculture Research Services (SIATSA), La Lima, Honduras, 41 pp.

Rowe, P. and Rosales, F. (1994) *Musa* breeding at FHIA. In: Jones, D.R. (ed.) *The Improvement and Testing of* Musa*: a Global Partnership, Proceedings of the First Global Conference of the International Testing Program held at FHIA, Honduras, 27–30 April 1994.* INIBAP, Montpellier, France, pp. 117–129.

Rowe, P.R. and Rosales, F.E. (1996) Bananas and plantains. In: Janick, J. and Moore, J.N. (eds) *Fruit Breeding,* Vol. I: *Tree and Tropical Fruits.* John Wiley and Sons, New York, pp. 167–211.

Ságí, L., Remy, S., Verelst, B., Swennen, R. and Panis, B. (1995) Genetic transformation in *Musa* species (banana). In: Bajaj, Y.P.S. (ed.) *Biotechnology in Agriculture and Forestry, Vol. 34, Plant Protoplasts and Genetic Engineering VI.* Springer, Berlin, pp. 214–227.

Ságí, L. Remy, S. and Swennen, R. (1998) Genetic transformation for the improvement of bananas – a critical assessment. In: *Networking Banana and Plantain, INIBAP Annual Report 1997.* INIBAP, Montpellier, France, pp. 33–36.

Sathiamoorthy, S. (1994) *Musa* improvement in India. In: Jones, D.R. (ed.) *The Improvement and Testing of* Musa*: a Global Partnership, Proceedings of the First Global Conference of the International Testing Program held at FHIA, Honduras, 27–30 April 1994.* INIBAP, Montpellier, France, pp. 188–200.

Shephard, J.F., Bidney, D. and Shahin, E. (1980) Potato protoplast in crop improvement. *Science* 208(4439), 17–24.

Shepherd, K. (1954) Seed fertility of the 'Gros Michel' banana in Jamaica. *Journal of Horticultural Science* 29, 1–11.

Shepherd, K. (1968) Banana breeding in the West Indies. *Pest Articles and News Summary, Section B* 14, 370–379.

Shepherd, K. (1974) Banana research at ICTA. *Tropical Agriculture (Trinidad)* 51, 482–490.

Shepherd, K. (1994) History and methods of banana breeding. *Infomusa* 3 (1), 10–11.

Shepherd, K., Dantas, J.L.L. and de Oliveira e Silva, S. (1994) Breeding Prata and Maça cultivars in Brazil. In: Jones, D.R. (ed.) *The Improvement and Testing of* Musa*: a Global Partnership, Proceedings of the First Global Conference of the International Testing Program held at FHIA, Honduras, 27–30 April 1994.* INIBAP, Montpellier, France, pp. 157–168.

Simmonds, N.W. (1956) Botanical results of the banana collecting expedition, 1954–1955. *Kew Bulletin* 3, 463–489.

Simmonds, N.W. (1962) *The Evolution of the Bananas.* Longman, Green, London, 170 pp.

Simmonds, N.W. (1966) *Bananas,* 2nd edn. Longman, Green, London, 512 pp.

Simmonds, N.W. (1997) Pie in the sky. *The Planter* 73, 615–623.

Smith, M.K., Hamill, S.D., Langdon, P.W. and Pegg, K.G. (1994) Mutation breeding for banana improvement in Australia. In: Jones, D.R. (ed.) *The Improvement and Testing of* Musa*: a Global Partnership, Proceedings of the First Global Conference of the International Testing Program held at FHIA, Honduras, 27–30 April 1994.* INIBAP, Montpellier, France, pp. 233–242.

Stover, R.H. (1962) *Fusarial Wilt (Panama Disease) of Banana and Other* Musa *Species.* CMI Phytopathological Paper No. 4, Commonwealth Mycological Institute, Kew, Surrey, UK, 117 pp.

Stover, R.H. (1972) *Banana, Plantain and Abaca Diseases.* Commonwealth Mycological Institute, Kew, Surrey, UK, 316 pp.

Stover, R.H. and Buddenhagen, I.W. (1986) Banana breeding: polyploidy, disease resistance and productivity. *Fruits* 41, 175–191.

Stover, R.H. and Simmonds, N.W. (1987) *Bananas,* 3rd edn. Longman Scientific and Technical, Harlow, Essex, 468 pp.

Su, H.J., Chuang, T.Y. and Kong, W.S. (1977) *Physiological Race of Fusarial Wilt Fungus Attacking Cavendish Banana of Taiwan.* Special Publication No. 2, Taiwan Banana Research Institute, Pingtung, Republic of China, 21 pp.

Su, H.J., Hwang, S.C. and Ko, W.H. (1986) Fusarial wilt of Cavendish banana in Taiwan. *Plant Disease* 70, 814–818.

Sun, E.J. and Su, H.J. (1984) Rapid method for determining differential pathogenicity of *Fusarium oxysporum* f. sp. *cubense* using banana plantlets. *Tropical Agriculture (Trinidad)* 61, 7–8.

Tang, C.Y. and Hwang, S.C. (1994) *Musa* mutation breeding in Taiwan. In: Jones, D.R. (ed.) *The Improvement and Testing of* Musa*: a Global Partnership, Proceedings of the First Global Conference of the International Testing Program held at FHIA, Honduras, 27–30 April 1994.* INIBAP, Montpellier, France, pp. 219–227.

Tang, C.Y. and Hwang, S.C. (1998) Selection and asexual inheritance of a dwarf variant of Cavendish banana resistant to race 4 of *Fusarium oxysporum* f. sp. *cubense. Australian Journal of Experimental Agriculture* 38, 189–194.

Tezenas du Montcel, H., Bakry, F. and Horry, J.-P. (1994) Breeding for the improvement of banana and plantain. In: INIBAP (ed.) *Banana and Plantain Breeding: Priorities and Strategies, Proceedings of the First Meeting of the Musa Breeders' Network held in La Lima, Honduras, 2–3 May 1994.* INIBAP, Montpellier, France, pp. 27–28.

Tezenas du Montcel, H., Carreel, F. and Bakry, F. (1996) Improve the diploids; the key for banana breeding. In: Frison, E.A., Horry, J.-P. and De Waele, D. (eds) *New Frontiers in Resistance Breeding for Nematode, Fusarium and Sigatoka. Proceedings of the Workshop held in Kuala Lumpur, Malaysia, 2–5 October 1995.* INIBAP, Montpellier, France, pp. 119–128.

Vakili, N.G. (1965) *Fusarium* wilt resistance in seedlings and mature plants of *Musa* species. *Phytopathology* 55, 135–140.

Vakili, N.G. (1967) The experimental formation of polyploidy and its effect in the genus *Musa. American Journal of Botany* 54, 24–36.

Vakili, N.G. (1968) Responses of *Musa acuminata* species and edible cultivars to infection by *Mycosphaerella musicola. Tropical Agriculture (Trinidad)* 45, 13–22.

Vuylsteke, D. and Ortiz, R. (1995) Plantain-derived diploid hybrids (TMP2x) with black Sigatoka resistance. *HortScience* 30, 147–149.

Vuylsteke, D., Swennen, R. and De Langhe, F. (1991) Somaclonal variation in plantains (*Musa* spp, AAB group) derived from shoot-tip culture. *Fruits* 46, 429–439.

Vuylsteke, D.R., Swennen, R.L. and Ortiz, R. (1993a) Development and performance of black Sigatoka-resistant tetraploid hybrids of plantains (*Musa* spp., AAB group). *Euphytica* 65, 33–42.

Vuylsteke, D.R., Swennen, R.L. and Ortiz, R. (1993b) Registration of 14 improved tropical *Musa* plantain hybrids with black Sigatoka resistance. *HortScience* 28, 957–959.

Vuylsteke, D., Ortiz, R., Ferris, S. and Swennen, R. (1995) 'PITA-9': A black Sigatoka-resistant hybrid from the 'False Horn' plantain gene pool. *HortScience* 30, 395–397.

Wehunt, E.J., Edwards, D.I. and Hutchison, D.J. (1965) Reaction of *Musa acuminata* to *Radopholus similis*. *Phytopathology* 55, 1082.

Wehunt, E.J., Hutchison, D.J. and Edwards, D.I. (1978) Reaction of banana cultivars to the burrowing nematode (*Radopholus similis*). *Journal of Nematology* 10, 368–370.

15

Genetic Engineering of Banana for Disease Resistance – Future Possibilities

Introduction

L. Sági

Molecular plant improvement is the term now widely used for the unconventional breeding of crop plants using a range of molecular biological techniques. These techniques include the identification, isolation and characterization of novel genes encoding agronomically important traits and their introduction into cultivars via a combination of genetic transformation and *in vitro* regeneration. The transfer of novel genes into the plant genome may result in increased abiotic stress tolerance or protection against pathogens and pests or may improve nutritional or physiological quality or storage properties. Molecular improvement should not be regarded as an alternative to classical breeding. It is more of a complementary tool. Molecular biologists and traditional breeders need to work closely together to identify genes of interest and to field-test transgenic cultivars (Michelmore, 1995).

The first priority in banana improvement is the generation of cultivars with resistance to diseases and pests. However, classical breeding, which is based on sexual hybridization, is seriously hampered by a high degree of male and female sterility in the most important edible cultivars. Conventional breeding is possible in most instances, but many hundreds of pollinations are necessary to obtain very few seeds. Cultivars in the commercially important Cavendish subgroup (AAA) are not amenable to conventional breeding because of their almost total female sterility. In addition, the male inflorescence of many cultivars in the Plantain subgroup (AAB) is either completely absent or gradually degenerates. Therefore, the introduction of resistance traits by genetic transformation has the potential of being a valuable banana improvement technique.

Genetic transformation can be defined as the introduction and stable integration of genes into the nuclear genome and their expression in a transgenic plant. During the last decade, a wide range of transformation techniques have been tested and it is now evident that cells capable of regenerating plants can be transformed by one of just two methods: either by using particle bombardment or through cocultivation with *Agrobacterium tumefaciens*. Particle bombardment (or biolistic transformation) uses accelerated heavy-metal microparticles coated with DNA to penetrate and deliver foreign genes into plant cells, which are then selected and regenerated into plants. *Agrobacterium tumefaciens*, a soil bacterium, transforms its plant hosts by integrating a segment (T-DNA) of its tumour-inducing plasmid into the plant nuclear genome. The transfer of this T-DNA is regulated by a complex process, involving numerous bacterial genes (the majority called virulence genes) that are located outside the T-DNA. This mechanism allows T-DNA to be used as a vehicle to

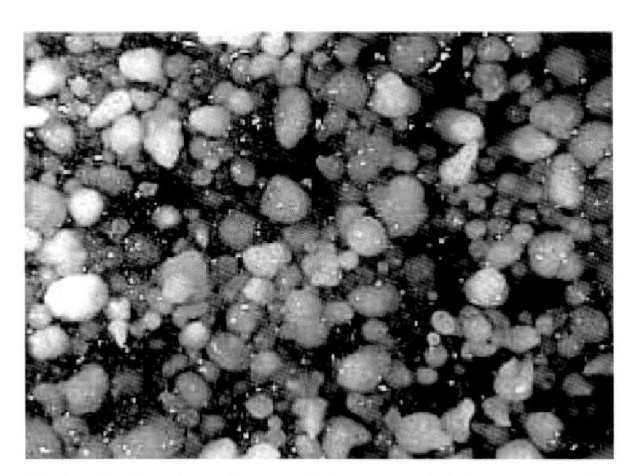

Plate 15.1. A culture of 'transgenic' banana cells ('Three Hand Planty' (AAB, Plantain subgroup)) expressing the *gusA* reporter gene (blue coloration after treatment with X-Gluc) in the Laboratory of Tropical Crop Improvement, Katholieke Universiteit Leuven, Belgium (photo: L. Sági, KUL).

introduce virtually any novel gene into a plant cell.

Both of these systems have been used successfully on banana. Transgenic plants have been produced by means of particle bombardment, using a particle-gun device, optimized for transformation of banana embryogenic cell-suspension cultures (Plate 15.1) (Sági *et al.*, 1995b, c; Remy *et al.*, 1998) and by *Agrobacterium*-mediated transformation of *in vitro* meristems (May *et al.*, 1995) or cell-suspension cultures (Pérez Hernández *et al.*, 1998, 1999). In addition, DNA has been introduced into regenerable, banana cell suspension-derived protoplasts by electroporation (Sági *et al.*, 1994). The status of research on the genetic transformation of banana has been reviewed by Sági *et al.* (1995a, 1998, 1999).

Foreign genes introduced by genetic transformation may not always confer a high level of resistance to the transgenic plant, as is the ideal in traditional plant breeding, but may protect a plant sufficiently to enable the plant to tolerate a given pathogen. Strategies for genetic engineering for resistance to fungal, bacterial, viral and nematode pathogens of plants, together with advances to date, are reviewed below. Relevant applications to banana are also discussed in detail.

Engineering Resistance to Diseases Caused by Viruses

L. Sági

Introduction

Banana is afflicted by a number of serious virus diseases (see Chapter 6, pp. 241–293). Attempts can be made to protect banana against these viruses by engineering a variety of genes into the banana genome. Genes of non-viral origin, some of which have the potential for protecting a plant against a broad range of viruses, can be used, but there is a possibility that plant morphology may be affected. Strategies using genes or sequences of the virus pathogen (called pathogen-derived resistance (PDR)) or a related virus can also be utilized. These genes can encode structural components, such as the coat/capsid protein, or non-structural components involved in virus replication or virus movement. In addition, these sequences can be engineered to be translatable, untranslatable or expressed in sense or antisense orientation, which can assist in understanding the mechanism of resistance in question. Although the spectrum of protection may be limited to only one virus in this PDR strategy, a higher degree of resistance could be expected.

Genes of non-viral origin

Virus disease resistance genes

Among the several ways to create virus resistance in plants by molecular improvement, the most obvious is the isolation of plant genes imparting resistance to virus diseases and their transfer into susceptible cultivars. Initial work in this area with the resistance gene *N* of tobacco, which induces a hypersensitive response to a number of tobamoviruses, including tobacco mosaic virus (TMV) and tomato mosaic virus (ToMV), and which was isolated by transposon tagging (Whitham *et al.*, 1994), has been encouraging. Sequence analysis of the gene has shown that it encodes a 131 kDa protein with domains, such as the leucine-rich repeat and the nucleotide-binding site, homologous to the cytoplasmic region of animal receptors and to other recently isolated resistance genes that are also thought to act as receptors (see Sági, this chapter, p. 481). The presence of these domains indicates that the *N* resistance gene is involved in a signal transduction pathway. More recently, Whitham *et al.* (1996) have transferred the *N* resistance gene to tomato and found that it functions in a similar way to that in tobacco, rendering transgenic tomato plants resistant to TMV and ToMV. These results not only suggest that the presumed signal transduction pathway is conserved (at least in the *Solanaceae*), but also demonstrate the potential that this resistance gene has for expression in other plant species.

Not much is known about the elicitors that are able to trigger the action of this resistance gene. Several viral products have been reported to be involved in the induction of the *N*-mediated hypersensitive response: the coat protein (Saito *et al.*, 1989; Culver and Dawson, 1991; Pfitzner and Pfitzner, 1992), a replicase-associated protein (Padgett and Beachy, 1993) and the cell-to-cell movement protein (Calder and Palukaitis, 1992). More recently, Taraporewala and Culver (1996) have identified a region in the three-dimensional structure of the TMV coat protein that is responsible for the elicitor activity. The coat protein has also been implicated as a hypersensitivity elicitor in other plant–virus interactions (Bendahmane *et al.*, 1995; de la Cruz *et al.*, 1997). In cowpea, the hypersensitive response-mediated resistance to cucumber mosaic virus (CMV) is elicited by the *2a* gene, which encodes the viral polymerase (Kim and Palukaitis, 1997).

The hypersensitive defence mechanism mediated by virus disease resistance genes remained unclear until the recent work of Zhang and Klessig (1998). They have found that (post)translational activation of a wound-induced protein kinase, as well as transcriptional activation of the corresponding gene, after local and systemic infection by TMV, depends on the presence of the *N* gene. This observation confirms that virus disease resistance genes probably act like bacterial disease-resistance genes in a signal transduction pathway and induce the expression of a series of genes that eventually leads to a hypersensitive response.

A large number of genes conferring resistance or tolerance to a broad range of plant-pathogenic viruses have been identified in various crop plants (Fraser, 1986, 1990). The isolation of these genes for genetic transformation may be cumbersome, due to a lack of detailed physical maps and tools for map-based cloning in most of these crops. An alternative approach can be the use of *Arabidopsis* as a model plant for heterologous viruses, because identified loci can be more easily mapped and isolated from the *Arabidopsis* genome than from the original host. Using this approach, Lee *et al.* (1996) identified and mapped a single incompletely dominant locus in *Arabidopsis* that conferred tolerance to tobacco ringspot nepovirus. It should be noted, however, that resistance to plant viruses may not always be under simple genetic control, i.e. involving only a single gene and with completely dominant or recessive character.

Plant genes encoding antiviral proteins

RIBOSOME-INACTIVATING PROTEINS Plants have long been known to produce proteins with antiviral activity. The best-characterized group of these proteins belongs to type I ribosome-inactivating proteins (RIPs), which are single-chain, basic proteins with a mass of about 30 kDa. RIPs can enzymatically depurinate a specific adenine residue in a highly conserved region of the large subunit of both prokaryotic and eukaryotic ribosomal RNA (Lord *et al.*, 1991; Stirpe *et al.*, 1992; Barbieri *et al.*, 1993; Hartley *et al.*, 1996). Their antiviral activity appears to be non-specific (Chen *et al.*, 1991) and is believed to prevent early events during replication of both DNA and RNA viruses (Chen *et al.*, 1993). One group of type I RIPs has been shown to be active on conspecific ribosomes (Bonness *et al.*, 1994), but, because RIPs are localized in the cell-wall matrix, antiviral activity is blocked by spatial separation (Ready *et al.*, 1986). Upon mechanical infection or aphid penetration, these RIPs may enter the cytosol by selective endocytosis and inactivate the host ribosomes. RIPs make attractive candidates for a potentially universal control of plant virus diseases.

The best-known examples of plant antiviral RIPs are four pokeweed (*Phytolacca americana*) antiviral proteins (PAPs) (Irvin *et al.*, 1980; Barbieri *et al.*, 1982). Complementary DNA (cDNA) and genomic DNA (gDNA) clones were identified and sequenced for three PAPs, which form part of a multi-gene family (Lin *et al.*, 1991) and are expressed either in the leaves, seeds or cell cultures. When a PAP cDNA clone was introduced into tobacco and potato and expressed in the intercellular fluid, Lodge *et al.* (1993) observed resistance in the transgenic plants to infection by three unrelated viruses (including CMV) that are transmitted either mechanically or by aphids. Expression of RIPs in transgenic plants frequently results in aberrant phenotypes and disturbed organ development (Lodge *et. al.*, 1993; Görschen *et al.*, 1997). However, Wang, P. *et al.* (1998) have recently found that transformation of the cDNA corresponding to the leaf-specific PAP II into tobacco results in normal phenotype, while the transgenic plants still confer resistance to TMV and potato virus X (PVX). Tumer *et al.* (1997) have also shown that enzymatic *N*-glycosidase activity is required for the antiviral effect of PAP, as expression of an active-site mutant PAP did not provide local resistance to viral infection in transgenic tobacco. Using the same mutant PAP gene in grafting experiments with transgenic tobacco rootstocks and wild-type scions, Smirnov *et al.* (1997) found that the enzymatic activity of PAP was also responsible for systemic resistance to infections by TMV.

Another highly active 28 kDa RIP was isolated from the four o'clock plant (*Mirabilis jalapa*) with activities similar to PAPs. This activity was significantly increased by genetically engineering the protein structure (Habuka *et al.*, 1991). Yet another RIP, dianthin from *Dianthus caryophyllus*, has recently been expressed in transgenic *Nicotiana benthamiana* plants (Hong *et al.*, 1996), which proved to be resistant to infection with the African cassava mosaic geminivirus.

An important condition for agricultural applications of these antiviral proteins is that the expression of RIPs should be precisely timed and tightly controlled. This can be achieved either by compartmentally regulated or by virus-inducible expression of the antiviral gene(s) (Lodge *et al.*, 1993; Hong *et al.*, 1996).

INDUCERS OF SYSTEMIC RESISTANCE OR HYPERSENSITIVE RESPONSE Another group of plant antiviral proteins, called systemic resistance inducers (SRIs), appear to induce resistance to a wide range of mechanically and vector-transmitted RNA viruses (Verma *et al.*, 1998). The SRIs, which were first isolated in the mid-1970s (mainly from *Bougainvillea spectabilis*, *Boerhaavia diffusa* and *Clerodendrum aculeatum*), are highly basic glycoproteins, with an apparent molecular weight of 30–34 kDa. It appears that SRIs induce the fast synthesis of a systemic host protein, rather than acting on the virus directly (Verma *et al.*, 1996). However, the precise nature of the SRIs remained unknown until several research groups demonstrated that SRIs isolated from the above species possess *N*-glycosidase activity that is characteristic of RIPs (Olivieri *et al.*, 1996; Bolognesi *et al.*, 1997; Balasaraswathi *et al.*, 1998). Independently, Kumar *et al.* (1997) also found that the cDNA corresponding to the 34 kDa SRI of *C. aculeatum* was homologous to a large number of known RIPs and that both the natural SRIs and a recombinant SRI were indeed RIPs. Whether SRIs and RIPs can be unified to a common group of antiviral proteins in the future will depend on the similarity of their mode of action, which remains to be clarified. Nevertheless, an important feature of the SRIs is that they are not toxic and they appear to induce host defence mechanisms. Therefore, it may be worthwhile expressing such SRIs in transgenic plants for testing their potential to achieve resistance to a wide range of viruses.

An unexpectedly discovered connection between two unrelated pathways may also have significant applications in obtaining broad-range virus disease resistance. Bachmair *et al.* (1990) overexpressed a mutant ubiquitin, which inhibits polyubiquitin-dependent proteolytic degradation, in transgenic tobacco and observed the development of spontaneous necrotic lesions under mild stress conditions. The similarity of these symptoms to hypersensitivity indicates that the hypersensitive response in plants is suppressed by the protein degradation programme, and disruption of this programme allows its manifestation. Becker *et al.* (1993) found that the transgenic tobacco plants showed partial resistance to TMV infection and a decrease in virus replication. In addition, TMV infection also induced the expression of the gene encoding the pathogenesis-related protein PR-1. The importance of the role that ubiquitin can play in the hypersensitive response has been confirmed by Karrer *et al.* (1998), who isolated tobacco genes whose products elicit the hypersensitive response after a challenge with TMV. They found that one of the cDNAs that activated a pathogenesis-related protein gene also encoded ubiquitin. The ubiquitin produced appeared to disrupt protein degradation by cosuppression and a consequence of this was the activation of a hypersensitive response. The connection of these two important pathways may be further exploited for generating disease resistance in transgenic plants.

A connection between two independent signal transduction pathways was also observed by Sano *et al.* (1994) during their work on ectopic expression in tobacco of a rice gene encoding a small GTP-binding protein. The abnormal phenotype of the transgenic plants was associated with an elevated level of endogenous cytokinins, as well as with activation of pathogenesis-related genes after wounding. More importantly, these changes in the transgenic plants resulted in a hypersensitive-like response and increased resistance to TMV infection. Since wound- and pathogen-induced pathways are normally considered to be distinct, this apparent cross-signalling could have been caused by altered cytokinin biosynthesis or by the action of the overexpressed small GTP-binding protein as a molecular switch. Though it is not known at this stage, it is possible that GTP-binding proteins mediate a signal that is involved in disease resistance.

Yang and Klessig (1996) have isolated and characterized a TMV-induced tobacco gene with a significant homology to the human oncogene *myb*. It appears that the new gene product is able to bind to the promoter of pathogenesis-related protein PR-1 gene as well as to induce systemic resistance. When the gene has been overexpressed in transgenic tobacco, many plants have become highly resistant to TMV infection, indicating that this *myb* homologue is somehow involved in disease resistance (Yang and Klessig, 1998).

ANTIBODIES An alternative strategy for utilizing proteins with antiviral effect is the expression of antibodies in transgenic plants (also called plantibodies) that bind specifically to particular virus components. The basis for this approach is the establishment of various successful methods to produce functional full-size or single-chain antibodies in plants (Hiatt *et al.*, 1989; van Engelen *et al.*, 1994; see Sági, this chapter, p. 477). When a single-chain antibody specific to a part of the coat protein of a tombusvirus was expressed in tobacco, it resulted in delayed symptoms and a reduced virus titre in systemically infected leaves (Tavladoraki *et al.*, 1993). A full-size antibody specific to epitopes on the surface of TMV virions can also be effective in reducing necrotic lesions after infection, as demonstrated by Voss *et al.* (1995). That virus-specific plantibodies may have a potential in distinct virus groups was shown by Fecker *et al.* (1997), who observed that a single-chain antibody specific to the coat protein of beet necrotic yellow-vein virus, a furovirus, conferred protection from early and late stages of infection in transgenic tobacco. Such antibodies could be used additionally against other essential components, such as viral replicases, proteases (or their cleavage site) and movement proteins. The specificity of these antibodies can be tailored so that they recognize conserved parts of the target viral component. It is expected that antibodies would be more effective during the early stages of infection, when there would be few viruses. The simultaneous expression of antibodies that recognize different components of the same virus may result in more durable and a higher degree of resistance than the use of single antibodies.

Non-plant genes for virus inactivation

A membrane-interacting, 26-residue peptide from bee venom, melittin (Dempsey, 1990), which has been reported to have activity against mammalian viruses, including human immunodeficiency virus (HIV) (Wachinger *et al.*, 1998), is structurally similar to a functionally important region of the TMV coat protein (Marcos *et al.*, 1995). Various melittin analogues were found to exhibit high antiviral activity, which was correlated with the degree of similarity between the coat protein and the melittin analogues. This result indicates the potential use of membrane-interactive peptides against viruses, as well as bacteria and fungi (see Sági, this chapter, pp. 474–477).

Most plant-pathogenic viruses are single-stranded RNA (ssRNA) viruses and, although they follow widely different replication strategies, all these viruses should have a double-stranded RNA (dsRNA) form at some point in their replication cycle. Therefore, it is conceivable that a double-stranded RNA-dependent ribonuclease (dsRNase) would be able to interfere with these replication intermediates and might confer broad-spectrum resistance. Several genes encoding dsRNA-binding proteins have been isolated from various organisms. Of these, the *rnc* gene for a dsRNA-dependent endoRNase III from *Escherichia coli* has been expressed in tobacco by Langenberg *et al.* (1997) and more recently by Zhang, L. *et al.* (1998). The transgenic tobacco plants were resistant to infections by three multipartite viruses, but not against two other viruses that contained only one component. Another gene encoding dsRNase from yeast, *pac1*, has been more extensively used (Köhn *et al.*, 1995; Watanabe *et al.*, 1995; Sano *et al.*, 1997) and found to exert activity in tobacco and potato against several viruses (including CMV and TMV) and potato spindle tuber viroid.

The potential also exists for various components of the mammalian antiviral defence system to be utilized in plants. In higher vertebrates, a family of cytokine proteins, called interferons, which are responsible for an antiviral state in the cell after virus infection or in the presence of the replication intermediate dsRNA molecules, are known to stimulate the expression of various proteins. One of these proteins is the 2′,5′-oligoadenylate (2–5A) synthetase enzyme, which adenylates the ends of dsRNA molecules present in the cell. The oligoadenylated dsRNA activates a ribonuclease L, which quickly degrades both viral and host cellular RNA, causing cell death and thus virus control (Nielsen *et al.*, 1982). Kulaeva *et al.* (1992) demonstrated that human interferon or 2–5A applied to tobacco or wheat plants increased their cytokinin content and induced the production of pathogenesis-related proteins. More recently, potato and tobacco have been transformed with the cDNA of 2–5A synthetase alone (Truve *et al.*, 1993) or in combination with that of RNase L (Mitra *et al.*, 1996; Ogawa *et al.*, 1996) and enzymatic activities in the transgenic plants have been confirmed. In laboratory and field tests, the transgenic plants have been protected against diseases caused by several viruses. In one case, virus infection induced rapid cell death (Ogawa *et al.*, 1996).

A related interferon-induced mechanism for virus control inhibits protein translation in mammalian cells. Besides the 2–5A system, production of a dsRNA-dependent protein kinase (PKR) enzyme is also stimulated in infected cells. In the presence of abundant dsRNA and upon binding to it, this PKR becomes activated by autophosphorylation and then inactivates a eukaryotic translation initiation factor, also by phosphorylation. This reaction inhibits polypeptide chain initiation to ensure that, if any undegraded RNA is left after the action of the 2–5A system, it will not be translated (Wu and Kaufman, 1996). A plant PKR homologue has also been identified and shown to be similar to the mammalian one (Langland *et al.*, 1995). So far, neither of these enzymes has been expressed in plants. It is probable that they would only have a major effect on plant virus control in combination with the 2–5A system.

Genes of viral origin

Pathogen-derived resistance

It has long been known that tobacco plants infected with a mild strain of tobacco mosaic virus became resistant to subsequent infection by a more severe strain of the same virus (McKinney, 1929). This naturally occurring cross-protection, though not fully understood, has been successfully applied in the past to control plant virus diseases (Ponz and Bruening, 1986; Gonsalves and Garnsey, 1989). However, classical cross-protection has limitations or even risks, for a number of reasons (Fulton, 1986). Firstly, protecting mild strains are not available for most agriculturally important viruses and, even where available, they may provide only a limited and often unpredictable level of protection. Secondly, protecting mild strains of a virus may produce disease symptoms that can result in a significant yield loss and also spread to other hosts with more severe consequences. Thirdly, the protecting virus strain may recombine or act synergistically with the subsequent challenging strain or with other viruses to produce a more severe disease than normal. The protecting strain may also mutate into a more severe form, causing a destructive disease.

There has been a long-standing discussion on the mechanism of cross-protection and on the viral products, such as the coat protein, that are involved (de Zoeten and Fulton, 1975; Zaitlin, 1976). In 1980, it was postulated that, if interference of viral gene products was responsible for cross-protection, then expression of these viral genes in transgenic plants might also result in protection against plant viruses (Hamilton, 1980). Based on the results of Sherwood and Fulton (1982), Sequiera (1984) proposed that the coat-protein gene could possibly be used for this purpose. The general concept of pathogen-derived resistance (PDR) was later proposed by Sanford and Johnston (1985). The idea was that the expression of key functions of a pathogen (in this case a virus) in the host plant in a dysfunctional form, either in excess or at the wrong stage during the infection cycle, could interfere with replication and/or movement of the virus and disrupt or modulate infection. Since then, PDR has been reviewed by a number of authors (Gadani *et al.*, 1990; Hull and Davies, 1992; Register and Nelson, 1992; Fitchen and Beachy, 1993; Scholthof *et al.*, 1993; Wilson, 1993; Baulcombe, 1994; Beachy, 1997). Though little is known about the precise mechanism of either cross-protection (Wilson and Watkins, 1985) or PDR, it is likely that any structural or non-structural part of a plant viral genome may give rise to PDR (Lomonossoff, 1995). Several viral components have since proved to be capable of conferring resistance and are discussed below.

Coat protein-mediated resistance

The coat protein (CP) was the first viral component to demonstrate the validity of the PDR concept (Powell Abel *et al.*, 1986) and viral CP genes have been extensively utilized to produce transgenic plants with CP-mediated resistance (CPMR). During the last 12 years, CPMR has been reported in plants transformed with CP genes from most RNA virus groups, including tobamoviruses, cucumoviruses, potyviruses, potexviruses, carlaviruses, furoviruses, ilarviruses, luteoviruses, nepoviruses, tenuiviruses, tobraviruses, tombusviruses and tospoviruses. However, only limited success has been achieved with DNA viruses. Though efficient transformation methods for the most important monocotyledons have become available in the past 5 years, many reported transformation experiments have been undertaken with tobacco, potato and tomato. However, the first transgenic squash cultivar resistant to CMV, watermelon mosaic potyvirus and zucchini yellow mosaic potyvirus (Tricoli *et al.*, 1995) has been released to industry. In general, CPMR has been shown to be effective against strains and isolates of the same virus and closely related viruses. It appears that the resistance spectrum depends on the degree of CP homology among the different members of a virus group (Nejidat and Beachy, 1990). The level of CPMR may be quite variable and the frequency of highly resistant plants appears to depend on a number of factors. One such factor is the plant species (and even the cultivar) and another is the specific construct used for the transformation, including the type of the promoter and other signals that regulate gene expression. The variation in CPMR levels may be caused by different mechanisms of action against viruses.

In general, the use of antisense CP genes (sequences coding for a transcript that is complementary to the original CP RNA) has not proved to be effective in obtaining resistance. This can be due to the low concentration of antisense RNA molecules in the cytoplasm and to poor accessibility of the uncoated viral RNA to the antisense RNA. Therefore, antisense CPMR is expected to work with viruses that occur at a low titre and are confined to specific plant tissues (Kawchuk *et al.*, 1991), which is the case for the DNA viruses infecting banana. It may be more profitable to incorporate antisense RNA into a ribozyme (see p. 474) and thus target the cleavage activity of the ribozyme (Baulcombe, 1989).

The use of CPMR in banana is hampered because the most important viruses are DNA viruses, where success with CPMR is not well documented. A first attempt to develop a banana with resistance to a virus, therefore, is being attempted using the CP gene of banana bract mosaic virus (Bateson and Dale, 1995), which is an RNA virus belonging to the potyvirus group, where CPMR has been successfully applied.

Interference with virus replication

REPLICASE-MEDIATED RESISTANCE Replicase-mediated resistance (RMR), which was discovered by Golemboski *et al.* (1990), is associated with the expression of various mutated, truncated or full-length genes that encode the viral replicase. RMR has been successful in providing protection against viruses in the same groups as are controlled by CPMR and also in a few DNA viruses (Palukaitis and Zaitlin, 1997). Plants with RMR are usually extremely resistant to the target virus isolate. However, RMR is highly strain-specific and transgenic plants may be susceptible to other isolates of the same virus. Full-length replicase genes may provide a broader RMR than the modified replicase genes. The mechanism of RMR is unclear. In certain cases replicase protein expression is necessary to obtain resistance, while in other cases RMR is mediated by RNA.

RMR has been shown to be effective against various CMV isolates, as well as against potyviruses (Audy *et al.*, 1994), and may have application in banana. Singh and Sansavini (1998) have reported that expression of a defective replicase gene from CMV in tobacco conferred high resistance to various Australian isolates of CMV from banana.

DEFECTIVE INTERFERING VIRUS AND VIRAL SATELLITE-MEDIATED RESISTANCE Defective interfering (DI) viruses and viral satellites are 'parasites' of viruses. DI viruses have significant homology with the viral genome, but viral satellites show little sequence similarity. DI viruses and viral satellites depend on and interfere with the replication of their host or helper virus. This frequently results in attenuated disease symptoms (Roux *et al.*, 1991). It is this feature that makes them interesting candidates for virus control.

Since DI viruses have not been identified in virus groups related to banana viruses, they will not be discussed further. Satellite RNAs, on the other hand, have been described in CMV and used for biocontrol in the field (Tien and Wu, 1991). Transgenically expressed satellite RNA has also conferred resistance to CMV in tobacco (Harrison *et al.*, 1987), tomato and sweet pepper (Kim *et al.*, 1997). However, there are risks associated with the extensive use of this technology. Palukaitis and Roossinck (1996) have found that satellite RNAs that normally attenuate CMV isolates may become plant pathogens by a single point mutation. Though this and similar observations (Sleat and Palukaitis, 1992) do not necessarily mean that this approach to developing virus-resistant transgenic plants has no future, they certainly highlight the potential risks.

RESISTANCE TO POTYVIRUSES BY PROTEASES The genome of potyviruses, a virus group that includes banana bract mosaic virus, encodes a polyprotein that is (post)translationally processed to at least seven different gene products. This process is catalysed by the coordinated action of three proteases (Riechmann *et al.*, 1992). Two independent reports (Maiti *et al.*, 1993; Vardi *et al.*, 1993) have shown that expression of potyvirus protease genes confers virus resistance. However, it has not yet been demonstrated whether the observed resistance was based on interference with the processing of polyprotein. At present, this approach remains a future possibility.

Interference with virus movement

An important condition for full manifestation of viral diseases in plants is systemic infection of the host. Consequently, blocking virus movement would be an attractive disease-resistance strategy because the virus would be contained in the initially infected cells. Virus spread within plants takes place in two steps and depends on two different mechanisms: cell-to-cell movement through plasmodesmata and long-distance movement in the vascular system. In order to be able to move to an adjacent cell, viruses produce a specific movement protein, which interacts with a protein complex within plasmodesmata to increase their pore size locally (Wolf *et al.*, 1989) and/or modifies the virus structure to enhance transport (Deom *et al.*, 1992). Long-distance movement involves the CP (Saito *et al.*, 1990), possibly the replicase and perhaps other viral proteins required for an interaction with endogenous proteins in plant sieve elements (Gilbertson and Lucas, 1996).

The significant role that these proteins play in viral movement has been confirmed by their expression in transgenic tobacco plants. Here, transgenic movement proteins are localized in the plasmodesmata. However, their dysfunctional mutants did not accumulate in the plasmodesmata and they inhibited cell-to-cell as well as long-distance movement of several tobamoviruses (Lapidot *et al.*, 1993). Unlike plants transformed with the functional protein, plants with dysfunctional mutants also inhibited the systemic spread of several other viruses, including a caulimovirus and a potyvirus (Cooper *et al.*, 1995). Tacke *et al.* (1996) have found that transgenic potato plants expressing mutant movement proteins of potato leafroll luteovirus showed resistance against this virus and also to infection by two other unrelated potato viruses. These results indicate that movement protein-mediated resistance may be effective against a broad range of viruses, compared with other strategies using genes of viral origin.

Viral proteins are also actively involved in the transmission of viruses from plant to plant by mechanical means and by fungal, insect or nematode vectors. The role of CP and a helper component protein in aphid transmission is especially well established in potyviruses (Flasinski and Cassidy, 1998; Wang, R.Y. *et al.*, 1998) and can be used to decrease the rate of virus spread in plant populations. Strategies using transgenic plants expressing pesticidal crystal toxins (Schnepf *et al.*, 1998) or insect protease inhibitors (Duan *et al.*, 1996) and lectins (Rao *et al.*, 1998; Zhou *et al.*, 1998) may also be adapted as an indirect measure for this purpose.

Ribozymes

Ribozymes or RNA enzymes catalyse the specific intra- or intermolecular cleavage of RNA substrates in simple organisms. Ribozymes with heterologous cleavage activity can be engineered as specific virus-cutting tools if they can be targeted. This is achieved by combining the enzymatic activity of ribozymes with antisense-specific recognition. Ribozymes are, therefore, an improvement in antisense technology. There are two major types of ribozymes: the hammerhead class (Porta and Lizardi, 1995) and the hairpin structures (Hampel *et al.*, 1993). Both types can be engineered for targeting and have shown *in vitro* cleavage activity. However, their *in vivo* activity has been significantly lower. The hammerhead class is used much more frequently in plant research because of its versatility (Haseloff and Gerlach, 1988). Ribozyme genes against plant viruses have been introduced into plants by Lamb and Hay (1990) and later by de Feyter *et al.* (1996), but they were not active *in planta*. However, ribozyme expression in transgenic potato was effective in inducing strong resistance to potato spindle tuber viroid (Yang *et al.*, 1997). If high *in vivo* activity can be achieved, application of ribozymes could be a very attractive technology, because multiple sites in the viral genome can be targeted simultaneously.

Engineering Resistance to Diseases Caused by Bacteria

L. Sági

Introduction

The development of crops resistant to bacterial pathogens through genetic engineering has not been progressing as rapidly as in other areas, which is surprising given the information that is available on interactions between plants and bacteria. The first avirulence genes were cloned from bacteria (Staskawicz *et al.*, 1984) and the very first plant resistance genes defined and cloned in tomato (Martin *et al.*, 1993; Salmeron *et al.*, 1996) and *Arabidopsis* (Bent *et al.*, 1994; Mindrinos *et al.*, 1994; Grant *et al.*, 1995) conferred resistance specifically to strains of *Pseudomonas syringae*. Plant–bacterial interactions have thus provided ideal systems for studying plant–pathogen relationships and the molecular basis of plant disease resistance.

Banana is affected by several serious bacterial diseases (see Thwaites *et al.*, Chapter 5, pp. 213–239). In principle, many of the engineering strategies outlined below may be applied to control these diseases. However, before transformation experiments can proceed, detailed biochemical and genetic studies of well-characterized strains of virulent bacterial pathogens will be necessary. Initial investigations should concentrate on the mode of action and structure of extracellular polysaccharides and toxins.

Resistance through the expression of antibacterial peptides or proteins

Most antibacterial peptides interact with the bacterial membrane by forming transient ion channels and pores, by blocking the membrane's own ion channels or by inhibiting the synthesis of membrane proteins. Some of these peptides are active against a wide range of organisms, which include viruses and various eukaryotes, while others have specific activity against certain groups of bacteria, e.g. Gram-negative bacteria.

Antibacterial agents from insects

Probably the best-known antibacterial peptides of insect origin are cecropins, which accumulate in the haemolymph of the giant silk moth (*Hyalophora cecropia*), the silkworm (*Bombyx mori*) and *Drosophila* as a response to infection. These short, linear peptides (31–39 amino acids) interact with the outer phospholipid membranes of both

Gram-negative and Gram-positive bacteria and modify them by forming a large number of transient ion channels (Durell *et al.*, 1992). Native and synthetic cecropins are active *in vitro* against a wide range of plant-pathogenic bacteria (Nordeen *et al.*, 1992; Mills and Hammerschlag, 1993; Kaduno-Okuda *et al.*, 1995) and, therefore, have been considered as potential candidates for protecting plants against bacterial pathogens. Unfortunately, apart from an early positive report (Jaynes *et al.*, 1993), accumulating evidence has shown that the expression of cecropins does not result in resistance to bacterial pathogens. This is due to low levels of the peptide being produced *in situ* (Hightower *et al.*, 1994) and to early proteolysis by endogenous proteases in transgenic plants (Mills *et al.*, 1994; Florack *et al.*, 1995). However, Owens and Heutte (1997) found that a synthetic mutant form of cecropin B with one amino acid substitution became, on average, three times more resistant to degradation by various endogenous peptidases and proteases from the intercellular fluid of plant leaves. Low expression in plants of this and other foreign proteins from lower organisms may also be caused by a difference in codon usage (Perlak *et al.*, 1991) and in potential regulatory sequences that determine intracellular localization or intercellular secretion of the foreign protein. When the same mutant cecropin gene was fused to an α-amylase secretion signal and expressed under the control of a proteinase inhibitor gene promoter in transgenic tobacco plants (Huang *et al.*, 1997), there was either no symptom development or a delay in symptom expression after *P. syringae* pv. *tabaci* was infiltrated into leaves. Huang *et al.* (1996) also introduced a different synthetic cecropin B gene into two japonica rice varieties and found it to be transcriptionally active in the transgenic plants, with some lines showing improved resistance to bacterial blight and streak diseases.

A group of similar insect defensins has been purified from flesh fly (*Sarcophaga peregrina*), including three families of sarcotoxins and the family of sapecins (Natori, 1994). Sarcotoxins are active against a wide range of Gram-negative bacteria and the gene coding for sarcotoxin IA has been introduced into potato by *Agrobacterium*-mediated transformation (Galun *et al.*, 1996). Okamoto *et al.* (1998) have found that the low expression of sarcotoxin IA, perhaps due to the instability of this short (39-residue) peptide in transgenic tobacco, can be significantly increased if it is expressed as a fusion protein or under the control of a strong recombinant promoter, which results in enhanced resistance to *P. syringae* pv. *tabaci* and *Erwinia carotovora* ssp. *carotovora* (Ohshima *et al.*, 1999).

Sapecins are 40-residue peptides with six half-cystine residues, which are essential for the antibacterial activity against mainly Gram-positive bacteria. They show a significant structural and functional similarity to charybdotoxin from scorpion venom in that they are able to bind to calcium-activated potassium ion channels (Yamada and Natori, 1991). New toxins with homology to the sapecin family, such as royalisin from the royal jelly of the honey-bee (Fujiwara *et al.*, 1990) or phormicin from *Phormia terranovae*, have also been isolated. However, these antimicrobial peptides have not yet been tested against plant-pathogenic bacteria.

The same insects also synthesize attacins, which belong to another family of six 20 kDa antibacterial proteins, in response to bacterial infection (Hultmark *et al.*, 1983). Attacins alter the structure and permeability of prokaryotic membranes by binding to lipopolysaccharide in the bacterial envelope (Carlsson *et al.*, 1998) and inhibiting the synthesis of the outer-membrane proteins (Carlsson *et al.*, 1991). Increased *in vitro* and greenhouse resistance to fire blight caused by *Erwinia amylovora* by the expression of attacin E in transgenic apple plants was reported by Norelli *et al.* (1994). This work is continuing (Borejsza-Wysocka *et al.*, 1997), but the gene encoding attacin is now being used in combination with the T4 lysozyme gene from bacteriophage. Chen and Kuehnle (1996) have also demonstrated the expression of attacin in calluses induced by transformed *Anthurium*. A number of distinct, antibacterial peptides or proteins have been described in other insects, including apidaecins from honey-bees (Casteels *et al.*, 1989) and moricin from silkworm (Hara and Yamakawa, 1995), but have not yet been used in plants.

Antibacterial agents from invertebrates

Tachyplesins are a family of antimicrobial peptides first isolated from acid extracts of haemocytes of the Japanese horseshoe crab (*Tachypleus tridentatus*). These strongly basic 2.3 kDa peptides (17–18 residues), with two disulphide bridges, primarily inhibit the growth of both Gram-negative and Gram-positive bacteria by forming a complex with bacterial lipopolysaccharides (Nakamura *et al.*, 1988) or with phospholipid membranes (Oishi *et al.*, 1997). Allefs *et al.* (1996) fused the sequence encoding a synthetic tachyplesin I gene with that of the barley hordothionin signal peptide. A low expression of this chimeric gene in three potato cultivars revealed slight inhibitory effects on *E. carotovora* ssp. *atroseptica*. Tachyplesin was also found to be very effective in controlling the growth of bacteria that are typically found in vase water, such as *Bacillus*, *Enterobacter* and *Pseudomonas* spp. (Florack *et al.*, 1996).

Antibacterial agents from higher animals

The expression in plants of antibacterial proteins from higher animals has scarcely been investigated. The best-known toxins, such as the pore-forming magainins (Bevins and Zasloff, 1990), bombinins (Simmaco *et al.*, 1991), brevinins and esculentin (Simmaco *et al.*, 1993, 1994), rugosins (Suzuki *et al.*, 1995) and temporins (Simmaco *et al.*, 1996) isolated from frog skin (Barra and Simmaco, 1995), have not yet been expressed successfully in plants for use against bacteria.

Lactoferrin, a member of a family of iron-binding glycoproteins found in human milk, has been reported to have antibacterial properties in transgenic plants. Mitra and Zhang (1994) observed that the introduction of a human lactoferrin cDNA into tobacco resulted in the expression of a truncated lactoferrin protein, which exhibited enhanced activity against several pathovars of *P. syringae*, *Xanthomonas campestris* pv. *phaseolii* and *Clavibacter flaccumfaciens* pv. *flaccumfaciens*, and more recently to *Ralstonia solanacearum* (Zhang, Z. *et al.*, 1998). These observations may not be surprising since Bellamy *et al.* (1992) found that lactoferricin, an acid-pepsin cleavage product of lactoferrin, had activity against a broad range of bacteria.

Lysozyme

Another source of antibacterial proteins has been lysozyme, either from bacteriophage or from hen eggs. These enzymes attack the murein layer of bacterial peptidoglycan resulting in cell-wall weakening and eventually leading to lysis of both Gram-negative and Gram-positive bacteria. Hippe *et al.* (1989) reported that the expression of a bacteriophage T4 lysozyme with a plant signal peptide in transgenic tobacco plants was localized in the intercellular space. The application of this strategy in potato led to increased resistance to infection by *E. carotovora* ssp. *atroseptica* in the greenhouse (Düring *et al.*, 1993). A lysozyme gene from bacteriophage T7 has also been used to construct expression vectors for plant transformation (Huang *et al.*, 1994). Initial experiments with hen-egg lysozyme for bacterial resistance met with little success (Trudel *et al.*, 1992), due to a low level secretion of the lysozyme with its own signal peptide to the intercellular space. However, high extracellular secretion of hen-egg lysozyme in transgenic tobacco resulted in growth inhibition of *Clavibacter michiganense* and *Micrococcus luteus* (Trudel *et al.*, 1995) in laboratory assays. More recently, a synthetic human lysozyme gene has also been expressed in transgenic tobacco plants and found to inhibit the growth of *P. syringae* pv. *tabaci* (Nakajima *et al.*, 1997).

Antibacterial agents from plants

Higher plants are also a rich source of various proteins with lysozyme activity, such as a chitinase from cucumber (Métraux *et al.*, 1989) or hevamine from *Hevea brasiliensis* (Jekel *et al.*, 1991). However, these proteins have not been expressed so far in transgenic plants in order to control bacterial infections. The best-characterized plant antimicrobial proteins are thionins (Florack and Stiekema, 1994), which are able to inhibit a broad range of pathogenic bacteria *in vitro* (Molina *et al.*, 1993a). Of these, the expression of a barley α-thionin gene in transgenic tobacco enhanced resistance to two pathovars of *P. syringae* in laboratory assays (Carmona *et al.*, 1993), while synthetic hordothionin genes were not secreted into the intercellular space in transgenic tobacco plants (Florack *et al.*, 1994). Unfortunately, most thionins can be toxic to animal (Carrasco *et al.*, 1981) and plant cells (Reimann-Philipp *et al.*, 1989) and thus may not be ideal for expression in transgenic plants. The recently discovered antimicrobial peptides called fabatins, which are active against both Gram-negative and Gram-positive bacteria (Zhang and Lewis, 1997), a defensin-type pseudothionin from potato (Moreno *et al.*, 1994) or plant defensins from various seed plants (Broekaert *et al.*, 1995; see Sági, this chapter, p. 484) may be more suitable for this purpose. Based on the first positive results obtained in transgenic tobacco against *P. syringae* pv. *tabaci* (Molina and García-Olmedo, 1997), lipid transfer proteins (see Sági, this chapter, p. 484) may be good candidates for use against some plant-pathogenic bacteria.

Many of the above toxic peptides may be useful for the control of bacterial pathogens in plants and they should be screened for activity in laboratory assays to determine if they have potential for use in transgenic plants. In addition, more efficient synthetic compounds, designed by combining different protein domains responsible for toxicity to bacteria (Powell *et al.*, 1995; Desnottes, 1996), could also be tested.

Antibodies

A promising strategy for control of bacteria may lie in the expression in transgenic plants of genes coding for monoclonal antibodies that target specific bacterial pathogenicity factors (Düring, 1996). Specific binding of antibodies to one or more bacterial factors, such as secreted lytic enzymes or extracellular proteins, is expected to compromise bacterial pathogenicity. The feasibility of this concept has already been demonstrated for antiviral strategies (see Sági, this chapter, pp. 469–470). The hybridoma technology for production of antibodies is laborious and involves the maintenance and care of immunized animals. However, the recently developed phage-display technique (Scott and Smith, 1990; Clackson *et al.*, 1991) offers a more convenient and faster means of producing single-chain variable-fragment (scFv) antibodies. This new method is based on bacterial recombination and gene expression techniques and it also allows the ready isolation of the gene encoding the desired antibody. For the antibody strategy to be successful, antibodies must be correctly expressed and secreted in plants. Although full-size as well as scFv antibodies have been successfully expressed in transgenic plants for various applications (Conrad and Fiedler, 1994; Whitelam and Cockburn, 1996), the obvious advantage of recombinant scFv antibodies over full-size antibodies is that, with the former, only correct folding is required and not the assembly of the light and heavy chains, as with the latter. It should be stressed, however, that transgenes encoding antibodies, just like many other foreign genes, can be inactivated by various mechanisms in the plant, such as gene silencing (De Neve *et al.*, 1999).

Resistance due to the inactivation of bacterial toxins

Pathogenic organisms produce various toxins, which can be classified as either host-specific toxins or non-host-specific toxins. The former account for the specificity of the host–pathogen interaction at the molecular level, while the latter are able to affect a wider range of host and non-host organisms (Mitchell, 1984; Gross,

1991). Only non-host-specific toxins can be considered as targets for the control of plant-pathogenic bacteria, because host-selective toxins have so far been identified only in fungal pathogens.

Bacterial toxins target certain enzymes that are present in both the host plant and the bacterial pathogen producing the toxin. Two major strategies have been adopted by bacteria to protect themselves from their own toxin. One is resistance based on the production of target enzymes that are insensitive to the toxin (Durbin and Langston-Unkefer, 1988) and the other is resistance by the production of enzymes that detoxify the toxin (De la Fuente-Martínez and Herrera-Estrella, 1993). The best-characterized bacterial toxins originate from various pathovars of *P. syringae* (Table 15.1). Consequently, transgenic research has so far only been undertaken with this family of toxins.

Ornithine carbamoyltransferase (OCTase) is a key enzyme in the arginine and polyamine biosynthetic pathway in plastids. It is inhibited by phaseolotoxin, a tripeptide toxin produced by *P. syringae* pv. *phaseolicola*. Phaseolotoxin-producing bacterial strains, however, synthesize a different OCTase, which is resistant to the toxin and its derivatives (Mosqueda *et al.*, 1990). When this toxin-insensitive OCTase was expressed in chloroplasts in transgenic tobacco plants, *in vitro* and biological assays demonstrated its effect in complementing the action of toxin-sensitive OCTase, which led to the hypersensitive defence reaction of cells to *P. syringae* pv. *phaseolicola* (De la Fuente-Martínez *et al.*, 1992; Hatziloukas and Panopoulos, 1992).

Tabtoxin, another phytotoxic dipeptide from *P. syringae* pv. *tabaci*, inhibits the target enzyme glutamine synthetase and causes chlorotic symptoms (wildfire disease) in tobacco. The toxin-producing strains are insensitive to the toxin, due to inactivation by an acetyltransferase enzyme encoded by the *ttr* gene. Transgenic tobacco plants expressing this tabtoxin acetylase showed high levels of resistance to the purified toxin and to infection by *P. syringae* pv. *tabaci* (Anzai *et al.*, 1989).

Albicidins are a family of phytotoxins and antibiotics, produced by the xylem-invading bacterium *Xanthomonas albilineans*, which specifically block prokaryotic DNA replication (Birch *et al.*, 1990). They are involved in leaf-scald disease of sugar cane by causing chlorosis in invaded host plants. Several genes that confer resistance to albicidins have been cloned from heterologous bacteria, such as *Klebsiella oxytoca* (Walker *et al.*, 1988), *Alcaligenes denitrificans* (Basnayake and Birch, 1995) and *Pantoea dispersa* (syn. *Erwinia herbicola*) (Zhang and Birch, 1997) and from *X. albilineans* (Wall and Birch, 1997). A certain level of homology between some of these genes at the protein level suggests a common functional domain. Therefore, these genes may be useful candidates for transfer into the sugar-cane genome.

Resistance through a reduction in the effects of bacterial polysaccharides

Many plant-pathogenic or symbiotic bacteria, including several species of *Clavibacter*, *Erwinia*, *Pseudomonas*, *Rhizobium* and *Xanthomonas*, produce large amounts of extracellular polysaccharide (EPS) during growth and during pathogenesis (Denny, 1995). EPSs have multiple functions and

Table 15.1. List of characterized bacterial toxins from pathovars of *Pseudomonas syringae*.

Name	Pathovar	Biochemical target
Coronatine	*glycinea, tomato*	Ethylene biosynthesis?
Phaseolotoxin	*phaseolicola*	Ornithine carbamoyltransferase (OCTase)
Syringomycin	*syringae*	Plasma-membrane H^+-ATPase
Tabtoxin	*tabaci, coronafaciens*	Glutamine synthetase
Tagetitoxin	*tagetis*	Eukaryotic (plastid) RNA polymerase III

appear to provide a selective advantage for bacteria during their epiphytic or saprophytic existence, in that they protect bacteria from desiccation, concentrate nutrients and enhance attachment to surfaces. During pathogenesis, EPSs regulate and minimize interaction with plant cells, thereby reducing the effect of host defence responses (Király *et al.*, 1997) and contact with toxic substances, while promoting colonization. In addition, EPSs may play a primary role in the development of disease symptoms, e.g. wilting caused by plugging of xylem vessels.

At least three gene clusters, which are organized more or less similarly in various phytopathogenic bacteria, are important for EPS biosynthesis. In *Ralstonia solanacearum*, an 18 kb operon (*eps*), with at least nine genes, is responsible for the acidic component of EPSs (Huang and Schell, 1995). Another gene cluster (*ops*), which contains at least seven structural genes, seems to be necessary for nucleotide sugar components of both EPSs and lipopolysaccharides (Kao and Sequeira, 1991). More recently, Kao *et al.* (1994) have identified a regulator gene, whose overexpression resulted in decreased EPS production and reduced virulence of *R. solanacearum*. Though expression of the *eps* operon in *R. solanacearum* appears to be controlled by a complex regulatory network (Huang *et al.*, 1995), which is environmentally responsive, it may be interesting to express either the regulator gene or another mutated regulatory system in the apoplast of transgenic plants to see whether it affects the EPS production and virulence of *R. solanacearum*.

An attractive way to understand the role of EPS in more depth and eventually increase disease resistance by reduced EPS production in plant-pathogenic bacteria could be through the use of polysaccharide depolymerase enzymes. Hartung *et al.* (1986) described the isolation of a polysaccharide depolymerase gene from a bacteriophage of *E. amylovora*. The corresponding enzyme when expressed in *Escherichia coli* lysed the EPS of *E. amylovora*, indicating that correct expression of the gene in plants may be interesting for testing this approach to the control of bacterial diseases.

It should be mentioned that numerous bacteriophages of bacterial plant pathogens have been described in the past (Okabe and Goto, 1963), including phages of *R. solanacearum* (Hayward, 1964) and its banana-attacking strain (Buddenhagen, 1960).

Resistance through the use of bacterial avirulence genes and plant disease resistance genes

Gene-for-gene interactions

The gene-for-gene hypothesis (Flor, 1971; Keen, 1990) explains host–pathogen recognition events on the assumption that single genes at particular loci in the plant and single genes at particular loci in the pathogen together determine whether the plant–pathogen interaction will be compatible (resulting in infection and disease development) or incompatible (resulting in the initiation of a hypersensitive defence response and disease resistance). Biological races of pathogens that contain a set of genes with alleles for virulence and avirulence will therefore induce a specific pattern of compatible and incompatible reactions when inoculated into a collection of host-plant cultivars that contain a set of genes with alleles for resistance and susceptibility. Incompatibility results when the resistance allele at a particular gene locus in the plant complements the avirulence allele at the corresponding gene locus in the pathogen. All other combinations result in compatibility. This concept presupposes that plants with a particular resistance gene recognize a pathogen-produced elicitor, a direct or indirect product of the corresponding avirulence gene (Gabriel and Rolfe, 1990). It is possible that, when the interaction is compatible, the direct or indirect product of many avirulence genes may contribute to the pathogen's virulence (Dangl, 1994).

It has been found that a gene cluster called *hrp* (hypersensitive response and pathogenicity) determines the expression of avirulence genes (Huynh *et al.*, 1989) in a wide range of plant-pathogenic bacteria. The gene cluster probably controls secretion and transport of products of avirulence genes outside bacterial cells, since at least eight of the known 20 Hrp genes show remarkable homology to proteins involved either in protein secretion pathways (Van Gijsegem *et al.*, 1993; Bogdanove *et al.*, 1996) or in flagellum biogenesis (Rosqvist *et al.*, 1995) of animal-pathogenic bacteria. In addition, Pirhonen *et al.* (1996) demonstrated that non-pathogenic *E. coli* cells transformed with a functional *hrp* gene cluster were able to trigger a genotype-specific hypersensitive response. The *hrp* gene cluster itself encodes and determines the secretion of a protein called harpin, which is also able to elicit the hypersensitive response when infiltrated into plant leaves (He *et al.*, 1993).

Bacterial avirulence genes

During the last decade, over 30 bacterial avirulence genes, mainly from strains of *P. syringae* and *X. campestris,* have been cloned. More recently, several plant disease-resistance genes corresponding to avirulence genes have also been cloned and characterized at a molecular level. However, little is known about the gene products, their function and the mechanism by which they act during incompatible plant–bacterial interactions. So far only an avirulence gene product of *P. syringae* has been shown to function as an enzyme that is involved in the synthesis of syringolides, which are able to elicit the hypersensitive response (Keen *et al.*, 1990). More recent results indicate that an avirulence gene product of *X. campestris* may also be an enzyme (Swords *et al.*, 1996) and act within plant cells (Yang and Gabriel, 1995; Gopalan *et al.*, 1996). The predicted sequence of the protein produced by this avirulence gene shows similarity to enzymes involved in the synthesis or hydrolysis of phosphodiester linkages, such as agrocinopine synthase from *Agrobacterium tumefaciens*. The indirect evidence that avirulence gene proteins may be delivered into plant cells via the *hrp* system would indicate that they could be involved in bacterial pathogenesis, rather than merely controlling genotype-specific elicitation of the hypersensitive response (Collmer, 1996).

Although it would be tempting to express an avirulence (or a harpin) gene in transgenic plants in order to create bacterial disease resistance by the elicitation of the hypersensitive response, there are a number of reasons why this should not be considered for immediate practical use. Firstly, as seen before, avirulence genes determine race-specific plant-recognition events; thus a whole set of genes would be necessary to provide protection against a wide range of pathogenic races. Secondly, as has already been reported for a fungal avirulence gene (Joosten *et al.*, 1994), the introduced gene may mutate, which would prevent the induction of the hypersensitive response. Thirdly, since the function of most avirulence genes is not clear yet, expression of an avirulence transgene alone might not induce the hypersensitive response, as has already been observed when bacteria harbouring various avirulence genes were infiltrated into plant leaves (Dangl, 1994; Gopalan *et al.*, 1996). Finally, expression of avirulence genes in plants would require very tightly regulated (preferably pathogen-inducible) promoters; otherwise, background or leaky expression of the avirulence transgene may lead to an uncontrolled hypersensitive response (Michelmore, 1995). Nevertheless, the feasibility of this approach has been demonstrated by McNellis *et al.* (1998), who expressed an avirulence gene of *P. syringae* under the control of a chemical-inducible promoter in transgenic *Arabidopsis* plants that contained the complementary resistance gene. After chemical induction of the transgenic plants, transcription of the introduced avirulence gene was detected in 30 min, its protein product was detected within 2 h and a hypersensitive response was observed in 6 h.

Disease-resistance genes from plants

Six plant resistance genes conferring resistance to bacterial diseases have been cloned to date (Martin *et al.*, 1993; Bent *et al.*, 1994; Mindrinos *et al.*, 1994; Grant *et al.*, 1995; Song *et al.*, 1995; Salmeron *et al.*, 1996; Yoshimura *et al.*, 1998). All have been isolated by map-based cloning, which includes genetic localization, marker saturation in the region localized, isolation of large genomic (bacterial artificial chromosome (BAC) or yeast artificial chromosome (YAC)) clones, identification of cDNAs and complementation. Though different classes of resistance genes may exist (Michelmore, 1995), the genes cloned have structures related not only to each other, but also to other plant resistance genes (Dangl, 1995). A variable number of leucine-rich repeats, which indicate a function for protein–protein interactions (Kobe and Deisenhofer, 1994), is one common feature. Another is a P-loop motif, with a common nucleotide-binding site, in diverse proteins with ATP/GTP-binding activity (Saraste *et al.*, 1990), which suggests that binding of these nucleotide triphosphates to the proteins is essential for their functioning (Traut, 1994). Basically, all these genes seem to encode components of receptor systems and probably form part of a signal transduction pathway, which triggers some general defence reactions, such as reinforcement of the cell wall, synthesis of phytoalexins, oxidation of phenolic compounds, activation of defence-related genes and the hypersensitive response. That the majority of the plant disease-resistance genes isolated so far belong to the receptor-related class may suggest that many of the classical resistance genes will fall into this category (Michelmore, 1995). Furthermore, while significant sequence similarities were found among the different disease-resistance genes, this has not been the case for avirulence genes. These observations indicate that pathogen recognition and response mechanisms in plants may be similar, but are activated by a wide range of specific elicitors (Dangl, 1994). This hypothesis was given credence when the first cloned disease-resistance gene *Pto* (a tomato resistance gene against *P. syringae* pv. *tomato*) was found to function in both *Nicotiana tabacum* (Thilmony *et al.*, 1995) and *Nicotiana benthamiana* (Rommens *et al.*, 1995), suggesting that disease-resistance functions are conserved in a wide range of plant species. A gene has recently been identified in *Arabidopsis* at a recessive locus for resistance to *R. solanacearum* (Deslandes *et al.*, 1998), a pathogen of banana (see Thwaites *et al.*, Chapter 5, pp. 213–221). If this gene can be isolated and transferred to banana, then it is possible that the transgenic plants may be resistant to Moko disease.

Nothing was known about how plant disease-resistance genes trigger defence reactions until Zhou *et al.* (1997) identified several classes of cDNAs encoding proteins that physically and specifically interact with the Pto protein. The sequence of these proteins shared significant homology with a wide range of transcription factors, some of them with the ability to bind to a core sequence (pathogenesis related (PR) box) present in the promoters of a large number of plant defence-related genes. Zhou *et al.* (1997) also showed that the interaction between the Pto protein and the corresponding avirulence gene product correlated with the early induction and increased expression of defence-related genes that contained the PR box. This provided indirect evidence for a direct connection between a disease-resistance gene and the specific activation of plant defence-related genes. Zhou *et al.* (1995) showed previously that another Pto-interacting protein is also involved in the induction of the hypersensitive response. Thus, at least two pathways of plant defence reactions can now be directly linked to a particular plant disease-resistance gene.

Another important event associated with plant defence mechanisms is a rapid and transient release of different active oxygen species (AOS), such as the superoxide anion radical (O_2^-), hydroxyl radical (OH^-) and hydrogen peroxide (H_2O_2). Of these AOS, production of hydrogen peroxide appears to occur early and is involved in a direct oxidative reaction with the pathogen, in biosynthesis of phytoalexins and in activating defence-related genes, as well as in the induction of acquired disease resistance. Evidence for a direct role of hydrogen peroxide in plant defence was provided by Wu *et al.* (1995), when they expressed the *Aspergillus niger* gene for glucose oxidase, which catalyses the oxidation of glucose to gluconic acid and hydrogen peroxide, in transgenic potato plants. An increased level of hydrogen peroxide was observed, which resulted in a simultaneously increased resistance to *E. carotovora* ssp. *carotovora*.

Engineering Resistance to Diseases Caused by Fungi

L. Sági

Introduction

Fungi are far more complex organisms than viruses or bacteria and have developed numerous strategies to survive in nature, which include saprotrophy, necrotrophy, hemibiotrophy and biotrophy. Interactions between plant-pathogenic fungi and their hosts are particularly complex and involve different mechanisms, such as the production of fungal toxins and the synthesis of enzymes that degrade the plant cell wall and the formation of occlusions in vascular tissue (Rodriguez and Redman, 1997). It is impossible, therefore, to design a simple defence strategy, such as the use of a single gene or a few genes from the genome of the pathogen, as in the case of viruses (see Sági, this chapter, pp. 471–474). The development of efficient antifungal strategies requires a detailed genetic, cytological and biochemical characterization of the particular host–fungal pathogen interaction. Higher plants and other organisms, including fungi, apply a wide range of defence reactions to protect themselves against fungal invasion and a variety of mechanisms might be utilized in gene transfer technologies.

Host-derived resistance strategies

Inhibition of Fungal Penetration

One of the first barriers plant pathogens, in particular necrotrophs, encounter during penetration and subsequent colonization is the plant cell wall. Fungal pathogens secrete a number of enzymes to degrade the major plant cell-wall polymers. The major enzymes used for this purpose are cutinases, endopolygalacturonases and pectatelyases. In response, higher plants apply a number of strategies to inhibit penetration of the pathogen through the cell wall.

One strategy is cell-wall reinforcement. This is a complex process, involving the rapid synthesis of phenolic compounds, which leads to the accumulation of lignins in the cell wall, and the synthesis of hydroxyproline-rich and other glycoproteins to strengthen the extracellular matrix. Another protective measure is initiated by the degradation of cell walls. Cell-wall components, such as sugar oligomers, which are released by fungal pectinolytic enzymes, can serve as elicitors to activate plant defence reactions (see below). It is possible, therefore, that the engineered expression of a pectinolytic enzyme may result in the transgenic plant having an activated defence status. Wegener *et al.* (1996) found that, when the pectate-lyase gene from the bacterium *Erwinia carotovora* was expressed in potato tubers, the transgenic plants were resistant to infection by this bacterium. However, this strategy has not yet been tested against necrotrophic fungal pathogens.

Plants also synthesize proteins that inhibit fungal enzymes that degrade plant cell-wall homogalacturonans to small monomeric uronides. One such inhibitor is the polygalacturonase-inhibiting protein (PGIP), which is specific to fungal endopolygalacturonase and is secreted into the extracellular matrix. It is believed that PGIPs only slighty inhibit fungal polygalacturonases and that this leads to the production of longer, oligomeric degradation products, which are large enough to act as elicitors of plant defence responses

(Cervone *et al.*, 1989). A gene encoding PGIP has been isolated from bean (Toubart *et al.*, 1992) and it is now possible to test if PGIP expression in transgenic plants does confer resistance. However, the transgenic tomato expressing the bean PGIP was not resistant to *Fusarium oxysporum*, *Botrytis cinerea* or *Alternaria solani* and the purified transgenic PGIP did not inhibit fungal polygalacturonases (Desiderio *et al.*, 1997). The authors concluded that the expression of more than one PGIP may be required to confer resistance to fungi.

Expression of phytoalexins

Phytoalexins are low-molecular-weight, antimicrobial (primarily antifungal), plant products that are synthesized or accumulated as a response to infection or stress related to infection. More than 350 phytoalexins have been reported to be present in vegetative and generative parts of higher plants in about 30 botanical families (Kuć, 1995). While a third of them have been isolated from leguminous plants, banana also appears to contain at least 25 different phytoalexins, including resveratrol (Holscher and Schneider, 1996), musanolones (Luis *et al.*, 1996) and phenalenone-type phytoalexins (Hirai *et al.*, 1994; Binks *et al.*, 1997; Kamo *et al.*, 1998).

Although the production of phytoalexins is induced by infection, their involvement in disease resistance is by no means certain. However, two lines of evidence suggest that some phytoalexins may play a role in resistance in certain plant–fungus interactions. Firstly, numerous fungal pathogens have active detoxifying systems against phytoalexins (VanEtten *et al.*, 1995). Pisatin, a phytoalexin from pea, is enzymatically inactivated by a pisatin demethylase from *Nectria haematococca*. When the corresponding fungal gene was transferred to the maize pathogen *Cochliobolus heterostrophus*, a non-pathogen of pea, the fungus showed a limited virulence on pea (Schäfer *et al.*, 1989). This indicates that pisatin might function as an antifungal agent in pea. Secondly, transgenic expression of a phytoalexin has been found to confer resistance to pathogens, and suppression of expression resulted in susceptibility (Maher *et al.*, 1994). Stilbene synthase catalyses the conversion of *para*-coumarate to the phytoalexin resveratrol. The expression of stilbene synthase in rice (Stark-Lorenzen *et al.*, 1997), tomato (Thomzik *et al.*, 1997) and tobacco (Hain *et al.*, 1993) and barley (Leckband and Lörz, 1998) results in increased resistance to *Pyricularia oryzae*, *Phytophthora infestans* and *B. cinerea*, respectively.

Though resveratrol has been found in banana rhizomes (Holscher and Schneider, 1996), it may be absent in leaves and fruit. High and constitutive expression of stilbene synthase in banana fruit may confer resistance to preharvest and postharvest diseases.

Expression of antimicrobial peptides

Several reviews have been published recently on the characterization and application of various classes of plant proteins with distinct antimicrobial activities (Broekaert *et al.*, 1997; Shewry and Lucas, 1997; Yun *et al.*, 1997). Antimicrobial peptides (AMPs) have a broad-spectrum antimicrobial activity against fungi, as well as bacteria, and most are non-toxic to plant and mammalian cells. Only the transgenic expression of plant AMPs, which have an upper size limit of 10 kDa, will be discussed here (for other antimicrobial proteins, see pp. 484–487).

THIONINS Besides their antibacterial activity (see Sági, this chapter, p. 477), thionins inhibit the growth *in vitro* of about 20 different fungal plant pathogens, including *Fusarium* spp., *B. cinerea*, *P. infestans* and *Rhizoctonia solani* (Cammue *et al.*, 1992; Molina *et al.*, 1993a). Thionins, which have a mass of 5 kDa and are six or eight cysteine-containing basic peptides, are divided into five classes. Type III thionins are viscotoxins from the plant semiparasite *Viscum album* L. Holtorf *et al.* (1998) have expressed viscotoxin A3 in *Arabidopsis thaliana* and transgenic plants showed increased resistance to infections of the clubroot pathogen *Plasmodiophora brassicae*. Epple *et al.* (1997) also observed that constitutive overexpression of an endogenous thionin in transgenic *Arabidopsis* resulted in enhanced resistance against *F. oxysporum* f. sp. *matthiolae*. These results indicate that thionins are defence proteins.

PLANT DEFENSINS The number of known plant defensins, which are structurally related to insect defensins (Broekaert *et al.*, 1995) and contain eight disulphide-linked cysteines, is steadily increasing. At least 20 plant species are now known to contain defensins, including those recently discovered in bell pepper (Meyer *et al.*, 1996), broad bean (Zhang and Lewis, 1997), spinach (Segura *et al.*, 1998) and maize (Kushmerick *et al.*, 1998). In laboratory assays, several defensins have been found to be toxic to *Mycosphaerella fijiensis* and *F. oxysporum* f. sp. *cubense* (Cammue *et al.*, 1993). So far, only one defensin has been constitutively expressed in transgenic plants. This conferred partial resistance to the tobacco pathogen *Alternaria longipes* (Terras *et al.*, 1995). The cDNAs encoding three different defensins have recently been transferred to banana (Remy *et al.*, 1998).

NON-SPECIFIC, LIPID-TRANSFER PROTEINS Non-specific lipid-transfer proteins (nsLTPs) in plants are highly basic, 9–10 kDa peptides containing eight disulphide-linked cysteines (Kader, 1996). A few nsLTPs with moderate or high antifungal activity to a broad range of fungi, including *Fusarium* spp., *B. cinerea* and *P. oryzae*, have been isolated from radish (Terras *et al.*, 1992), onion (Cammue *et al.*, 1995) and cereals (Molina *et al.*, 1993b; Segura *et al.*, 1993; García-Olmedo *et al.*, 1995). One of these nsLTPs has recently been expressed in transgenic banana (Plate 15.2) (S. Remy and L. Sági, 1998, unpublished results).

NON-ENZYMATIC, CHITIN-BINDING PROTEINS In addition to class I endochitinases, which contain a chitin-binding domain, non-enzymatic, chitin-binding proteins, such as hevein from the rubber tree, a lectin from stinging-nettle (Raikhel *et al.*, 1993) and AMPs from seeds of amaranth (De Bolle *et al.*, 1993) and *Pharbitis nil* L. (Koo *et al.*, 1998) are also inhibitory *in vitro* to a diverse range of pathogenic fungi. The gene encoding the amaranth AMP has been introduced into tobacco, but no increased resistance to *A. longipes* was observed. This may have been because of the sensitivity of this AMP to the presence of antagonistic cations (De Bolle *et al.*, 1996). Lectins may be better suited for the control of insects (Peumans and Van Damme, 1995).

OTHER PLANT AMPS Distinct AMPs containing four disulphide-linked cysteine residues have been isolated from seeds of maize (Duvick *et al.*, 1992) and *Impatiens balsamina* (Tailor *et al.*, 1997). Another distinct, highly basic AMP, with a mass of 8.1 kDa and six cysteine residues, has been purified from nut kernels of *Macadamia integrifolia* (Marcus *et al.*, 1997) and expressed in *Escherichia coli* (Harrison *et al.*, 1999). Snakin, which contains 12 cysteines, is a new type of AMP from potato tubers (Segura *et al.*, 1999). These AMPs have been inhibitory to a wide range of pathogenic fungi and to Gram-positive bacteria, but have not been expressed yet in transgenic plants.

The potential of antimicrobial plant proteins

PATHOGENESIS-RELATED PROTEINS A large number of diverse plant proteins are synthesized *de novo* after infection with viruses or bacterial and fungal pathogens or after treatment with biotic or chemical elicitors. These pathogenesis-related proteins (PRPs) differ in their structure, expression, spectrum of antimicrobial activity and modes of action. Five major types of PRPs can be distinguished (Shewry and Lucas, 1997; Yun *et al.*, 1997; Kitajima and Sato, 1999).

PR-1 proteins are the most abundant PRPs and have a mass of 14–16 kDa. They have been isolated from both monocots (barley and maize) and dicots (tobacco and *Arabidopsis*), but their precise mode of action is still unknown.

Plate 15.2. Transgenic banana plant ('Williams' (AAA, Cavendish subgroup)), which has an integrated and expressed non-specific lipid-transfer protein gene, in the glasshouse at the Laboratory of Tropical Crop Improvement, Katholieke Universiteit Leuven, Belgium (photo: W. Dillemans, KUL).

PR-2 and PR-3 proteins have endohydrolytic enzyme activity and are present in a wide range of plant species. PR-2 proteins are β-1,3-endoglucanases and may induce defence reaction by releasing elicitors from the fungal cell wall. PR-3 proteins are endochitinases with a similar function, but catalysing the hydrolysis of the other major component in the fungal cell wall. A combination of PR-2 and PR-3 proteins has a synergistic inhibitory effect on pathogenic fungi, both *in vitro* and in transgenic plants (Table 15.2). PR-2 proteins alone can also be effective against pathogens that belong to the *Oomycetes*, in which the cell wall does not contain chitin.

PR-4 proteins are a heterogeneous group of PRPs, many of which have chitin-binding activity. Some of these proteins show homology to products of wound-inducible (*win*) genes, which have been isolated from potato, barley and wheat. PR-4 proteins have not yet been expressed in transgenic plants and even their *in vitro* antifungal activity needs to be confirmed.

PR-5 proteins comprise a group of proteins, such as permatin zeamatin and linusitin, that are homologous to salt-induced osmotins and the super-sweet protein thaumatin. Nevertheless, they have a clear antifungal activity against a wide range of fungi *in vitro*, as well as in transgenic plants. This activity is believed to be caused by permeabilization of the fungal plasma membrane.

Within each of the above PRP groups, except for PR-4 proteins, several classes exist, according to structure and expression in the cell. The two major classes, class I and class II, contain basic and acidic proteins, respectively. Acidic PRPs are secreted in the extracellular space, while the much more potent basic homologues are targeted to the vacuole. The class I (vacuolar) PR-3 proteins have an N-terminal, cysteine-rich, chitin-binding domain, which is homologous to the chitin-binding region of lectins (see above). Expression of acidic PRP genes is mediated by salicylic acid and active oxygen species (see below), whereas ethylene and methyl jasmonate are known to mediate the expression of basic PRP genes. Surprisingly, PR-1-like and PR-5-like proteins or genes have also been identified in the animal kingdom, though their functions are not yet known.

Recently, six new groups of PRPs have been classified: proteinase inhibitors (PR-6), proteinases (PR-7), chitinases with lysozyme activity (PR-8), extracellular peroxidases (PR-9) and other PRPs (PR-10 and PR-11). The expression of a peroxidase from the tropical legume *Stylosanthes humilis* in transgenic tobacco has resulted in reduced symptom development after infection with *Phytophthora parasitica* var. *nicotianae*. In canola, the same peroxidase reduced symptoms caused by *Leptosphaeria maculans* (Kazan *et al.*, 1998a).

Table 15.2. Examples of transgenic resistance to fungal diseases conferred by genes encoding pathogenesis-related proteins.

PRP gene (origin)	Fungal pathogen	Transgenic plant species	Observed effect	Reference
Class II PR-1a (tobacco)	*Peronospora tabacina* *Phytophthora parasitica*	Tobacco	Disease symptoms reduced	Alexander *et al.* (1993)
Class I PR-2 (soybean)	*Phytophthora parasitica* *Alternaria longipes*	Tobacco	Disease symptoms reduced*	Yoshikawa *et al.* (1993)
Class II PR-2 (alfalfa)	*Phytophthora megasperma*	Alfalfa	Disease symptoms reduced	Masoud *et al.* (1996)
	Cercospora nicotianae	Tobacco	Disease symptoms reduced	Zhu *et al.* (1994)
Class I PR-3 (bean)	*Rhizoctonia solani*	Tobacco, oilseed rape	Disease symptoms reduced	Broglie *et al.* (1991)
Class I PR-3 (tobacco)	*Rhizoctonia solani*	*Nicotiana sylvestris*	Disease symptoms reduced	Vierheilig *et al.* (1993)
	Botrytis cinerea, R. solani *Sclerotinium rolfsii*	Carrot	Disease symptoms reduced	Punja and Raharjo (1996)
Chimeric class I PR-3 (tomato, tobacco)	*Sclerotinia sclerotiorum* *Phoma lingam*	Oilseed rape	Disease symptoms reduced	Grison *et al.* (1996)
Class I PR-3 (rice)	*Cercospora nicotianae*	Tobacco	Disease symptoms reduced	Zhu *et al.* (1994)
	Rhizoctonia solani	Rice (indica)	Disease symptoms reduced	Lin *et al.* (1995)
	Botrytis cinerea	Cucumber	Disease symptoms reduced*	Tabei *et al.* (1998)
	Diplocarpon rosae	Rose	Disease symptoms reduced	Marchant *et al.* (1998)
Class II PR-2 (alfalfa)/ class I PR-3 (rice)	*Cercospora nicotianae*	Tobacco	Disease symptoms reduced*	Zhu *et al.* (1994)
Class II PR-2/PR-3 (barley)	*Rhizoctonia solani*	Tobacco	Disease symptoms reduced*	Jach *et al.* (1995)
Basic PR-5 (tobacco)	*Phytophthora infestans*	Potato	Disease symptoms reduced	Liu *et al.* (1994)

* Disease symptoms reduced significantly.

Table 15.2 lists several PRPs that have been expressed in transgenic plants and conferred increased resistance to various plant-pathogenic fungi. A possible deleterious side-effect of PRPs in transgenic plants may be their action against mycorrhizal fungi, which are important for stimulating the growth of many plants, including banana (Declerck *et al.*, 1995; Jaizme-Vega and Azcón, 1995). When class I (vacuolar) PR-2 or PR-3 proteins, the most commonly used PRPs for transformation, have been expressed in transgenic plants, no harmful effect has been observed on the symbiotic bacterium *Rhizobium meliloti* (Masoud *et al.*, 1996) or on the vesicular–arbuscular mycorrhizal fungus *Glomus mosseae* (Vierheilig *et al.*, 1995). However, a delay of colonization by *G. mosseae* was observed in tobacco plants expressing a class II (extracellular) PR-2 protein (Vierheilig *et al.*, 1995), indicating that transgenic expression of antifungal proteins in the apoplast, which is preferred for efficient protection, may have adverse effects on endophytic symbionts.

RIBOSOME-INACTIVATING PROTEINS The RNA *N*-glycosidase activity of RIPs has been successfully applied to generate virus-resistant crops (see Sági, this chapter, pp. 467–468), but RIPs isolated from cereals are also active on fungal ribosomes (Stirpe and Hughes, 1989). The inhibitory effect of RIPs on the growth of fungi *in vitro* has been demonstrated (Leah *et al.*, 1991), in combination with chitinase or glucanase. A barley RIP has been expressed in transgenic tobacco either alone under the control of a wound-inducible promoter from a potato gene (Logemann *et al.*, 1992) or in combination with a barley class II chitinase (Jach *et al.*, 1995). In both cases, increased resistance to *R. solani* was observed, with enhanced protection when the combined expression strategy was used. Since then, a maize RIP (Maddaloni *et al.*, 1997) and a mutant of the pokeweed RIP, which is non-toxic to plants, have been used in transgenic tobacco to confer resistance to *R. solani* (Zoubenko *et al.*, 1997).

OTHER PLANT ANTIMICROBIAL PROTEINS Though 2S albumins, which contain two subunits of about 9 kDa and 4 kDa, function primarily as seed storage proteins, those from seeds of plants in the *Brassica* genus are able to inhibit fungal growth. Although antagonized by inorganic cations, these 2S albumins could be enhanced by a synergistic action with thionins (Terras *et al.*, 1993). 2S albumin proteins have not yet been evaluated for disease resistance in transgenic plants.

A new type of antifungal protein has been isolated from pearl millet seeds (Joshi *et al.*, 1998). This basic, 24 kDa protein was a cysteine protease inhibitor and showed *in vitro* antifungal activity against important plant-pathogenic fungi, such as *Alternaria* spp., *Fusarium* spp. and *Helminthosporium* spp. In general, protease inhibitors are not strictly antimicrobial in character and are more effective against insect or nematode pests (see De Waele, this chapter, p. 493). However, Lorito *et al.* (1994) have also found that the application of a mixture of trypsin and chymotrypsin inhibitors from tobacco inhibited the spore germination of *B. cinerea* and *Fusarium solani*. The significance of this finding has yet to be determined.

Induction of an oxidative burst and hypersensitive response

As described for defence mechanisms activated by bacterial infections (see Sági, this chapter, pp. 481–482), fungal attack also results in an oxidative burst in plants, i.e. the release of active oxygen species (Lamb and Dixon, 1997). Generation of hydrogen peroxide by the expression of a fungal glucose oxidase in potato has also resulted in enhanced resistance to late blight disease caused by *P. infestans* (Wu *et al.*, 1995). This resistance has been linked to an increase in lignin content and the accumulation of peroxidases plus a class II PR-3 protein (Wu *et al.*, 1997). Increased resistance to *Cercospora nicotianae* has also been correlated with the induction of the PR-1a gene (Kazan *et al.*, 1998b). It is possible that plant genes, such as germin, which encode an oxalate oxidase (Berna and Bernier, 1997), can also be used for this purpose.

Deák *et al.* (1999) have described the transgenic use of ferritin, an iron-binding protein, to reduce cell damage caused by a variety of environmental stress conditions. It is proposed that overexpression of ferritin in transgenic plants would give protection to oxidative stress, which is caused, among other things, by pathogen infection. Overexpression of alfalfa ferritin in transgenic tobacco has resulted in increased resistance to infections of *Alternaria alata* and *B. cinerea*.

Mittler *et al.* (1995) have introduced into tobacco a gene from *Halobacterium halobium*, which encodes a proton pump called bacterio-opsin (bO), and observed a disease lesion-mimic phenotype (a hypersensitive response in the absence of a pathogen) in various organs of the transgenic plants. These plants turned out to be resistant to infections with TMV or *Pseudomonas syringae* pv. *tabaci*, but fungal pathogens were not tested. More recently, the same bO gene was transferred to potato and the same lesion-mimic phenotype was again observed (Abad *et al.*, 1997). The transgenic plants were resistant to an isolate of *P. infestans*, but not to another isolate of the same pathogen, and were susceptible to infections with potato virus X and the bacterial pathogen *E. carotovora*. This result indicates that such induced defence reactions may be specific to certain pathogens or distinct plant–pathogen interactions.

Plant resistance genes and fungal avirulence genes

Nine fungal disease-resistance genes have been cloned so far by either transposon tagging or map-based cloning (Table 15.3). Many of these genes show structural similarity, indicating a similar function, which is related to signalling a pathogen attack. Of particular interest for banana is the *I2* gene from tomato, which confers resistance to *F. oxysporum* (Ori *et al.*, 1997). In tomato, three loci have been mapped, which confer resistance to *F. oxysporum* f. sp. *lycopersici* race 1, race 2 and races 1, 2 and 3, respectively. The *I2* locus that confers resistance to race 2 contains a cluster of at least four homologous genes. Transgenic expression of antisense genes in plants containing the *I2* locus conferred a high susceptibility to race 2, but the plants remained completely resistant to race 1. This result indicates that the *I2* genes are indeed specific to race 2 and may not be a promising candidate for transformation into banana in an attempt to confer resistance to *F. oxysporum* f. sp. *cubense*.

To date, eight fungal avirulence genes have been cloned. Five of these genes direct race-specific interactions with specific plant resistance genes, as predicted by the gene-for-gene hypothesis (Laugé and De Wit, 1998). However, the other three genes direct a broader, species-specific plant–pathogen interaction. Interestingly, the low-molecular-weight protein products of these avirulence genes appear to be elicitins, a proteinaceous class of elicitors that are able to trigger a broad-spectrum defence reaction. This indicates that at least some of the elicitins may have functions related to avirulence and perhaps to virulence. One of the most active elicitins is cryptogein, which is produced by *Phytophthora cryptogea*. Recently, Keller *et al.* (1999) have generated transgenic tobacco plants containing a fusion between a pathogen-inducible tobacco promoter and the cryptogein gene. Infection with *Phytophthora parasitica* turned on cryptogein production in the transgenic plants, which induced a localized hypersensitive response and activation of defence-related genes. As a result, these plants displayed a broad-spectrum resistance to several unrelated fungal pathogens, such as *B. cinerea* and *Erysiphe cichoracearum*.

Non-host-derived resistance strategies

Expression of non-plant antimicrobial peptides and proteins

NON-PLANT ANTIMICROBIAL PEPTIDES The antibacterial properties of cecropin are well known (see Sági, this chapter pp. 474–475),

Table 15.3. Cloned fungal disease-resistance genes.

R gene	Plant	Pathogen	*Avr* gene	Structure	Cloning method	Reference
Hm	Maize	*Cochliobolus carbonum*	None	Enzyme	Transposon tagging	Johal and Briggs, 1992
L6	Flax	*Melampsora lini*	*AL6*	NBS-LRR	Transposon tagging	Lawrence *et al.*, 1995
M	Flax	*M. lini*	*AM*	NBS-LRR	Transposon tagging	Anderson *et al.*, 1997
RPP5	*Arabidopsis*	*Peronospora parasitica*	*avrPp5*	NBS-LRR	Map-based cloning	Parker *et al.*, 1997
I2	Tomato	*Fusarium oxysporum*	?	NBS-LRR	Map-based cloning	Ori *et al.*, 1997
Cf-9	Tomato	*Cladosporium fulvum*	*Avr9*	LRR-TM	Transposon tagging	Jones *et al.*, 1994
Cf-4	Tomato	*C. fulvum*	*Avr4*	LRR-TM	Map-based cloning	Thomas *et al.*, 1997
Cf-2	Tomato	*C. fulvum*	*Avr2*	LRR-TM	Map-based cloning	Dixon *et al.*, 1996
Mlo	Barley	*Erysiphe graminis*	?	New	Map-based cloning	Büschges *et al.*, 1997

R gene, resistance gene; *Avr* gene, avirulence gene; NBS, nucleotide binding site; LRR, leucine-rich repeat; TM, transmembrane region.

but its potential for fungus control in plants has been discovered only recently. Both cecropin (DeLucca *et al.*, 1997) and its derivatives and hybrid peptides with melittin have been found to inhibit growth of several important fungal pathogens, including *B. cinerea*, *P. infestans*, *Verticillium albo-atrum* and numerous formae speciales of *F. oxysporum* (Qui *et al.*, 1995; Cavallarin *et al.*, 1998). Similarly, Kristyanne *et al.* (1997) have demonstrated that magainin, another bacterial membrane-interactive peptide, interfered with membrane integrity and inhibited the growth of the plant–pathogenic fungi *F. oxysporum*, *R. solani* and *Verticillium dahliae*. More recently, transgenic poinsettia plants expressing a magainin analogue displayed increased resistance to powdery mildew, apparently caused by reduced infection and conidial production of the pathogen (Smith *et al.*, 1998).

On the basis of their broad-spectrum activity against fungal as well as bacterial pathogens, expression of cecropin, magainin and their derivatives in banana may result in increased resistance to some pathogens. However, problems related to *in planta* expression and stability still need to be solved. Synthetic peptides from combinatorial libraries can also be effective in controlling fungal plant pathogens (*F. oxysporum* and *R. solani*) as demonstrated by Reed *et al.* (1997).

HYDROLYTIC ENZYMES OF NON-PLANT ORIGIN

Chitin is commonly found in cell walls of most fungi or in the exoskeletons of insects. These organisms need to synthesize chitinases in order to outcompete other species and to regulate their own growth and development. Other organisms, in particular bacteria or entomopathogenic nematodes, also contain chitinases in order to degrade the chitin polymer for food. Though a number of fungal chitinase genes, especially from mycoparasitic fungi (Blaiseau and Lafay, 1992; Hayes *et al.*, 1994), have been available for several years, the first gene (from *Rhizopus oligosporus*) was only recently introduced into plants (Terakawa *et al.*, 1997). The transgenic tobacco plants produced were resistant to *Sclerotinia sclerotiorum* and *B. cinerea*. The activity of chitinolytic endochitinases from the mycoparasite *Trichoderma harzianum* has been compared to that of plant chitinases. These enzymes exerted a much stronger activity against almost all pathogenic fungi tested than the plant chitinases, which showed only weak activity against a few fungal species (Lorito *et al.*, 1993). This result indicated that a single chitinase might produce a broad-spectrum resistance to plant-pathogenic fungi. This was confirmed by transgenic expression of the corresponding gene, whose *in planta* activity conferred resistance not only to the foliar pathogens *B. cinerea, Venturia inaequalis* and two *Alternaria* species, but also to the soil-borne pathogen *R. solani* (Bolar *et al.*, 1997; Lorito *et al.*, 1998). No effect was observed on the life cycle of the symbiotic mycorrhiza *Gigantus margaritus* (Lorito *et al.*, 1997a).

Chitinase genes of bacterial origin have not been used extensively. This may be because they are exochitanases and presumed to be less active against fungi (Roberts and Selitrennikoff, 1988). However, when Jones (1988) and Suslow *et al.* (1988) transferred the chitinase gene of *Serratia marcescens* into tobacco, the transgenic plants exhibited elevated chitinase activity and increased resistance to *A. longipes*. Later, Howie *et al.* (1994) found that the same gene provided tolerance to *R. solani* in the field. The protective role of bacterial chitinases was also confirmed by Toyoda *et al.* (1991), who injected chitinase from *Streptomyces griseus* into barley epidermis cells, and found that the enzyme digested the haustoria of *Erysiphe graminis*, the powdery mildew pathogen. Other sources, such as chitinases from nematode-associated bacteria (Chen *et al.*, 1996) or insects (Ding *et al.*, 1998), may also be effective against fungal infections.

Glucan is another component of the fungal cell wall. Glucanases have also been purified from numerous bacteria, as well as from fungi, including mycoparasites (de la Cruz *et al.*, 1995; Lorito *et al.*, 1997b), and may be as effective as chitinases. However, bacterial glucanase genes transferred to plants (Darbinyan *et al.*, 1996; Jensen *et al.*, 1996; Monzavi-Karbassi *et al.*, 1998) have not yet been tested for resistance to plant-pathogenic fungi.

Chitosan, which is a deacetylated derivative of chitin, is another important polymer component of the fungal cell wall, especially of species in the class *Zygomycetes*. Chitosan and its oligomers have been shown to have antifungal activity (El Ghaouth *et al.*, 1992) and to elicit defence reactions (Kendra *et al.*, 1989). Therefore, genes encoding chitosanase, a chitosan-degrading enzyme, may be useful in transgenic plants. El Quakfaoui *et al.* (1995) found that the recombinant chitosanase of *Streptomyces* sp. inhibits the growth of *Rhizopus nigricans* (a zygomycete) and also *F. oxysporum* and *V. albo-atrum*. The enzyme was also active in transgenic tobacco.

Human lysozyme, unlike some animal and bacteriophage lysozymes, is able to degrade both bacterial peptidoglycan and chitin. This raises the possibility that it may be an efficient tool for simultaneous control of both fungal and bacterial pathogens. A synthetic human lysozyme gene has been transferred to tobacco, and the transgenic plants exhibited enhanced resistance to *E. cichoracearum*, as well as the bacterium *P. syringae* (Nakajima *et al.*, 1997).

Inactivation of fungal toxins

HOST-SELECTIVE TOXINS Host-selective toxins (HSTs) are produced by approximately 20 species of pathogenic fungi and are believed to be primary determinants of host range and pathogenicity (Walton and Panaccione, 1993). HSTs are toxic only to hosts susceptible to the fungus. In theory, HST-resistant plants should be less susceptible to HST-producing pathogens. Most HSTs, such as victorin, which is produced by *Cochliobolus victoriae* and is probably the most phytotoxic and most selective compound known, are highly specific. Victorin is active against sensitive oat cultivars at 10 ng l^{-1}, while resistant oats are not affected even at 10 g l^{-1}. Due to this specificity, some HSTs may turn out to function as elicitors of plant defence reactions (Walton, 1996). Victorin has recently been found to induce a specific proteolytic cleavage of the large subunit of the ribulose-1,5-bisphosphate carboxylase/oxygenase enzyme, which is encoded in the chloroplasts (Navarre and Wolpert, 1999). In susceptible plants, victorin results in chlorophyll loss, which is also characteristic of banana leaves treated with a crude toxic extract from cultures of *M. fijiensis* (Harelimana *et al.*, 1997).

The banana pathogen *M. fijiensis* is reported to synthesize 2,4,8-trihydroxytetralone, which is believed to be an HST (Stierle *et al.*, 1991; Okole and Schulz, 1997). However, its biosynthesis, role in pathogenesis and any plant resistance mechanism to the toxin remain to be determined. As several HSTs appear to be immunogenic and anti-HST antibodies are available (Akimitsu *et al.*, 1992), it may be worthwhile to express antibodies that bind to the toxin in a transgenic banana to see if it confers resistance to the toxin and the pathogen.

NON-HOST-SPECIFIC TOXINS Non-host-specific toxins are active on both host and non-host species, but they may nevertheless have a significant role during pathogenesis in particular plant–pathogen interactions. A large number of plant-pathogenic fungi synthesize non-host-specific toxins, including *Fusarium* spp., which produce fusaric acid (5-*n*-butyl-2-pyridine-carboxylic acid) and *Cercospora* spp., which produce cercosporin. After light activation, cercosporin generates reactive oxygen, which is toxic to plants, many bacteria and fungi, but not to the fungus synthesizing the toxin. Cercosporin resistance genes have recently been cloned from several *Cercospora* species and one of them has been transformed into tobacco (Upchurch *et al.*, 1997). It is not known whether *Paracercospora fijiensis*, the imperfect stage of *M. fijiensis*, produces cercosporin or a related toxin and has resistance genes that could be exploited in a similar manner.

Like many other *Fusarium* spp., it is possible that *F. oxysporum* f. sp. *cubense*, the causative agent of Fusarium wilt of banana, also synthesizes fusaric acid or its derivatives. If it does, then the isolation of toxin-resistance genes from these fungi and their expression in banana may contribute to a better understanding of the role of the toxin in pathogenesis.

Expression of antifungal toxins

Several isolates of *Ustilago maydis*, a fungal pathogen of maize, are infected with strains of a dsRNA virus that code for antifungal killer toxins, which are polypeptides with a mass of 7–14 kDa. These fungal isolates, which are resistant to the toxin, are able to gain selective advantage by killing other susceptible isolates. At least some of these toxins appear to act by the formation of ion channels in cellular membranes, and two of the three known killer toxins have been expressed in transgenic tobacco without an adverse effect on the plants (Kinal *et al.*, 1995; Park *et al.*, 1996). If these killer toxins are found to be active against a broad range of plant-pathogenic fungi, they could be a novel means for control.

Applications in banana

Fungi are the cause of some of the most serious diseases of banana and thus are of the greatest economic concern to the commercial banana industry. Until genes conferring resistance to fungal pathogens are identified in banana (Wiame *et al.*, 1999), heterologous genes encoding proteins with antifungal activity are the primary targets for transgenic expression. Perhaps the most promising candidates are AMPs of plant origin (Broekaert *et al.*, 1997), as they display high *in vitro* activity against *M. fijiensis* and *F. oxysporum* f. sp. *cubense*, while they exert no toxicity towards human or banana cells.

Particle-bombardment technology (Sági *et al.*, 1995b) has enabled several hundred transgenic lines, which express different genes encoding defensin-type AMPs and nsLTPs (see Sági, this chapter, p. 484), alone or in combinations, to be produced from 'Three Hand Planty' (AAB, Plantain subgroup) and 'Williams' (AAA, Cavendish subgroup) at the Katholieke Universiteit Leuven. A large-scale molecular analysis by polymerase chain reaction (PCR), reverse transcription (RT)-PCR, Southern and Northern hybridization and biochemical characterization has confirmed that a vast majority of these lines contain and express the introduced genes. Approximately 100 transgenic lines have been micropropagated for testing the stability and effect of the transgenes in different field environments. In addition, about 200 individual plants have been transferred to the glasshouse and a simple and reproducible leaf-disc bioassay has been developed for the evaluation of resistance to fungal pathogens. A differential disease response has been observed among transgenic plants, which ranges from no resistance to a high degree of resistance, compared with control plants. Computer image capturing and software-based area calculation allowed for the precise measurement of the infected leaf area and for the classification of the level of resistance of each transformant (Remy *et al.*, 1999). To date, no transgenic banana has been field-tested. This is mainly due to the lack of biosafety guidelines and/or regulatory bodies in many tropical countries, where these genetically modified banana plants need to be grown in the field under natural disease pressures. It is hoped that this obstacle to progress can soon be overcome.

Engineering Resistance to Diseases Caused by Nematodes

D. De Waele

Introduction

The science of genetically engineering plants for resistance to nematodes is in its infancy. Consequently, there have been no reports to date of a transgenic banana that is either less susceptible or less sensitive to nematode pathogens. However, various options are under consideration that hold promise for the future.

In general, there are three main approaches to the molecular improvement of plants that will make them less susceptible or less sensitive to nematodes. These are the development of strategies that: (i) directly target the nematode (either inside or outside plant tissue); (ii) utilize and enhance existing plant defence mechanisms; or (iii) interfere with the nematode–plant interaction (De Waele, 1993; De Waele *et al.*, 1994). The most suitable strategy will depend upon the life cycle of the nematode, its feeding behaviour and its interaction with the host plant.

Strategies that directly affect the nematode

Strategies having a direct effect on the nematode can be divided into two groups: those that target external structures and those that target internal structures.

Strategies that target external structures have been focused on the degradation of chitin and collagen, which are important components of, respectively, the nematode eggshell (Bird, 1976) and the cuticle of juvenile and adult nematodes (Reddigari *et al.*, 1986). However, although transgenic plants expressing chitinase and (human) collagenases have been constructed (Broglie *et al.*, 1991; Havstad *et al.*, 1991), plants that resist nematodes have not been reported. As more details on the complex structure of the nematode eggshell and nematode cuticle become known, there is a growing awareness that co-expression of several lytic enzymes – not only chitinases or collagenases, but also proteases and lipases – will be necessary to lethally damage the eggshell or cuticle.

Strategies targeting internal nematode structures have concentrated on the utilization of proteins whose expression in transgenic plants results in enhanced resistance against a variety of insects. Examples of such proteins are protease inhibitors, lectins and the insecticidal crystal proteins produced by *Bacillus thuringiensis* isolates.

In transgenic potatoes, the cowpea trypsin inhibitor (CpTI), a protease inhibitor isolated from cowpea, reduced the fecundity of root-knot nematodes (*Meloidogyne* spp.) and shifted the sex ratio of *Globodera pallida*, the potato cyst nematode, to favour the production of males (Atkinson *et al.*, 1995). In hairy root cultures of tomato and in transgenic *Arabidopsis thaliana*, Urwin *et al.* (1995, 1997) showed that an engineered cystine protease inhibitor derived from rice (oryzacystatin-I) markedly affected the development of *G. pallida* and the size and fecundity of females of both *Heterodera schachtii*, the beet cyst nematode, and *Meloidogyne incognita*.

In transgenic oilseed rape, co-expression of the snowdrop lectin (GNA – *Galanthus nivalis* agglutinin) and CpTI reduced the number of *H. schachtii* females and *Pratylenchus neglectus* (all developmental stages) by approximately 25 and 75%, respectively. GNA expressed alone conferred moderate levels of resistance to *P. neglectus*, with 55% fewer nematodes being recovered relative to the controls. In transgenic potatoes, GNA reduced the female numbers of *G. pallida* by 50% (Burrows and De Waele, 1997).

Although several genes coding for crystal proteins with nematocidal activity against juveniles and adults of some bacteriophagous nematodes and *Pratylenchus scribneri* were claimed in a series of patent applications (see Edwards *et al.*, 1989; Narva *et al.*, 1991) and many transgenic plants expressing these proteins must have been constructed and tested, no nematode-resistant transgenic plants have been reported so far. Detailed observations on the nematocidal activity, mode of action and the nematode developmental stage and species specificity of several nematocidal *B. thuringiensis* compounds were recently reported (Leyns *et al.*, 1995; Borgonie *et al.*, 1996a, b). Although the source of the observed nematocidal activity was most probably a protein, there were no indications that the crystal proteins were the source of the nematocidal activity.

Strategies that utilize and enhance existing plant defence mechanisms

Strategies aimed at utilizing or enhancing existing plant defence mechanisms have focused on the cloning and transfer of natural, single, dominant, resistance genes. However, to date, only two genes ($Hs1^{pro-1}$, Mi-1), conferring resistance to *H. schachtii* and *M. incognita* respectively have been cloned (Cai *et al.*, 1997; Milligan *et al.*, 1998). Resistance to *Radopholus similis* is present in 'Pisang Jari Buaya' (AA) (Pinochet and Rowe, 1978; Wehunt *et al.*, 1978) and in 'Yangambi Km 5' (AAA, syn. 'Ibota Bota') (Sarah *et al.*, 1992; Price, 1994). This latter cultivar is also resistant to *Pratylenchus goodeyi* (Fogain and Gowen, 1998). The number and nature of the genes involved are still unknown.

Strategies that interfere with the nematode–plant interaction

Strategies aimed at interfering with the nematode–plant interaction have concentrated on the search for proteins that block nematode–plant recognition and inhibit or destroy the specialized feeding structures induced by the sedentary endoparasitic nematodes.

Strategies aimed at blocking nematode–plant recognition focus on the interference of the action of carbohydrates, which apparently play an important role in nematode sensory perception and which are associated with the head region of plant-parasitic nematodes (Zuckerman, 1983). In a compatible interaction, plant-parasitic nematodes are able to recognize and locate suitable host plants during the prepenetration phase and to enter the host tissues during the penetration phase, often at specific sites. Then, they migrate towards and recognize those plant cells that are suitable for feeding or for the induction of the specialized feeding structures. It is generally believed that the sense organs situated in the nematode head region, such as the paired amphids, are involved in this recognition process. Thus, the expression of lectins or enzymes in transgenic plants blocking the functioning of these sense organs could lead to failure of the nematodes to establish a compatible interaction.

There are two major approaches to the inhibition or destruction of the specialized feeding structures that are induced by sedentary endoparasitic nematodes and which are vital for nematode survival. The first focuses on the identification of plant genes whose expression is strongly upregulated by nematodes during the induction, formation and maintenance of the specialized feeding structures. These genes are usually not expressed in the same way in the absence of the nematodes. Several of these plant genes have already been identified (see, for example, Gurr *et al.*, 1991; Niebel *et al.*, 1993; Van der Eycken *et al.*, 1996). It is assumed that the cloning of these genes and their regulatory sequences or promoters could lead to the development of nematode-resistant transgenic plants, either through the expression of antisense RNA downregulating these genes or through the expression of cytotoxic proteins inside the specialized feeding structures (Gheysen *et al.*, 1996, 1998). Transgenic tobacco plants with a gene inserted in their genomes that is only expressed inside the giant cells upon infection by root-knot nematodes have been reported (Opperman *et al.*, 1994). Inhibition, destruction or physiological changes of the specialized feeding structures will lead to the death of the nematodes, as these structures are vital for their survival.

The second approach concentrates on the identification of proteins that are secreted by the sedentary endoparasitic nematodes and which are essential for the induction, formation or maintenance of the specialized feeding structures. It is believed that the expression in plants of antibodies directed against these proteins, the so-called plantibodies, will inhibit their biological activity (Stiekema *et al.*, 1997).

Constraints

One of the problems in developing a strategy for engineering resistance or tolerance to migratory endoparasitic nematodes, such as *R. similis* (see Sarah, Chapter 7, pp. 295–303), *Pratylenchus coffeae*, *P. goodeyi* and *Helicotylenchus multicinctus* (see Gowen, Chapter 7, pp. 303–307), which are the most important nematode pathogens of banana, is that they do not have such an intimate relationship with their hosts as do the sedentary endoparasitic nematodes. This makes it very difficult to develop strategies that distrupt the host–nematode interaction, as reported above, and limits the possibilities for control by genetic engineering. Strategies aimed at the inhibition or destruction of the specialized feeding structures that sedentary endoparasitic nematodes induce plant cells to form and which are necessary for survival have so far been the most promising for nematode control. Therefore, control of the less important, root-knot nematode pathogens of banana, such as *M. incognita* and *Meloidogyne javanica* (see De Waele, Chapter 7, pp. 307–314), has more chance of success through genetic engineering in the short term.

Conclusion

L. Sági

A number of the strategies reviewed above have the potential to confer increased resistance to many plant pathogens and pests. Some strategies may help control a number of similar types of pathogen or unrelated pathogens, such as several different bacteria and fungal species. Future research will result in elevating individual gene expression levels. Pyramiding of different genes acting in different ways on the pathogen could lead to enhanced resistance to both major and minor diseases of banana. This work will also increase our knowledge of the basic underlying mechanisms contolling banana–pathogen interactions.

Many plants transformed with different genes show resistance to plant pathogens under laboratory conditions and in the greenhouse. However, little is published on the reaction of transgenic plants in the field. There are numerous reasons why many promising genetically modified plants do not reach the stage of field trials. These include a lack of interaction between the genetic breeder, who tends to be a laboratory-based scientist, and the conventional breeder, who usually works in a farm environment, quite detached from biotechnologists. In addition, laboratory-based scientists often have little expertise in plant pathology, while field specialists may not always be aware of how to manage transgenic plants. A lack of continued financial support and other issues, such as intellectual properties and biosafety concerns, also delay or prevent the further testing of promising material.

Due to environmental factors and to changes in the genetic make-up of pathogens, it is often difficult for plant breeders to accurately determine the degree of resistance in their own lines, let alone the resistance of transgenic plants, where changes in levels of resistance may be extremely subtle. The same evaluation methods used on testing élite lines may not be applicable to transgenic plants. The need for a standardized and rational evaluation of transgenic plants in the laboratory and field is becoming more urgent with the increasing number of lines available for testing. A constant exchange of information between laboratory and field scientists is necessary if novel, integrated pest-management strategies are to be realized.

The development of transgenic banana clones with increased resistance to major pathogens is still in its infancy. However, the technology to introduce foreign genes into the banana genome has now been mastered and rapid progress is expected within the next few years. Though many strategies will undoubtedly fail, others will bring benefits in the long term.

REFERENCES

Abad, M.S., Hakimi, S.M., Kaniewski, W.K., Rommens, C.M., Shulaev, V., Lam, E. and Shah, D.M. (1997) Characterization of acquired resistance in lesion-mimic transgenic potato expressing bacterio-opsin. *Molecular Plant–Microbe Interactions* 10, 635–645.

Akimitsu, K., Hart, L.P., Walton, J.D. and Hollingsworth, R. (1992) Covalent binding sites of victorin in oat tissues detected by anti-victorin polyclonal antibodies. *Plant Physiology* 98, 121–126.

Alexander, D., Goodman, R.M., Gut-Rella, M., Glascock, C., Weymann, K., Friedrich, L., Maddox, D., Ahl-Goy, P., Luntz, Y., Ward, E. and Ryals, J. (1993) Increased tolerance to two oomycete pathogens in transgenic tobacco expressing pathogenesis-related protein 1a. *Proceedings of the National Academy of Sciences USA* 90, 7327–7331.

Allefs, S.J.H.M., De Jong, E.R., Florack, D.E.A., Hoogendoorn, C. and Stiekema, W.J. (1996) *Erwinia* soft rot resistance of potato cultivars expressing antimicrobial peptide tachyplesin I. *Molecular Breeding* 2, 97–105.

Anderson, P.A., Lawrence, G.J., Morrish, B.C., Ayliffe, M.A., Finnegan, E.J. and Ellis, J.G. (1997) Inactivation of the flax rust resistance gene *M* associated with loss of a repeated unit within the leucine-rich repeat coding region. *Plant Cell* 9, 641–651.

Anzai, H., Yoneyama, K. and Yamaguchi, I. (1989) Transgenic tobacco resistant to a bacterial disease by the detoxification of a pathogenic toxin. *Molecular and General Genetics* 219, 492–494.

Atkinson, H.J., Urwin, P.E., Hansen, E. and McPherson, M.J. (1995) Designs for engineering resistance to root parasitic nematodes. *Trends in Biotechnology* 13, 369–374.

Audy, P., Palukaitis, P., Slack, S. and Zaitlin, M. (1994) Replicase-mediated resistance to potato virus Y in transgenic tobacco plants. *Molecular Plant–Microbe Interactions* 7, 15–22.

Bachmair, A., Becker, F., Masterson, R.V. and Schell, J. (1990) Perturbation of the ubiquitin system causes leaf curling, vascular tissue alterations and necrotic lesions in a higher plant. *EMBO Journal* 9, 4543–4549.

Balasaraswathi, R., Sadasivam, S., Ward, M. and Walker, J.M. (1998) An antiviral protein from *Bougainvillea spectabilis* roots; purification and characterisation. *Phytochemistry* 47, 1561–1565.

Barbieri, L., Aron, G.M., Irvin, J.D. and Stirpe, F. (1982) Purification and partial characterization of another form of the antiviral protein from the seeds of *Phytolacca amaricana* L. (pokeweed). *Biochemical Journal* 203, 55–59.

Barbieri, L., Battelli, M.G. and Stirpe, F. (1993) Ribosome-inactivating proteins from plants. *Biochimica Biophysica Acta* 1154, 237–282.

Barra, D. and Simmaco, M. (1995) Amphibian skin: a promising resource for antimicrobial peptides. *Trends in Biotechnology* 13, 205–209.

Basnayake, W.V. and Birch, R.G. (1995) A gene from *Alcaligenes nitrificans* that confers albicidin resistance by reversible antibiotic binding. *Microbiology* 141, 551–560.

Bateson, M.F. and Dale, J.L. (1995) Banana bract mosaic virus: characterisation using potyvirus specific degenerate PCR primers. *Archives of Virology* 140, 515–527.

Baulcombe, D. (1989) Strategies for virus resistance in plants. *Trends in Genetics* 5, 56–60.

Baulcombe, D. (1994) Novel strategies for engineering virus resistance in plants. *Current Opinion in Biotechnology* 5, 117–124.

Beachy, R.N. (1997) Mechanisms and applications of pathogen-derived resistance in transgenic plants. *Current Opinion in Biotechnology* 8, 215–220.

Becker, F., Buschfeld, E., Schell, J. and Bachmair, A. (1993) Altered response to viral infection by tobacco plants perturbed in ubiquitin system. *Plant Journal* 3, 875–881.

Bellamy, W., Takase, M., Wakabayashi, H., Kawase, K. and Tomita, M. (1992) Antibacterial spectrum of lactoferricin B, a potent bactericidal peptide derived from the N-terminal region of bovine lactoferrin. *Journal of Applied Bacteriology* 73, 472–479.

Bendahmane, A., Köhm, B.A., Dedi, C. and Baulcombe, D.C. (1995) The coat protein of potato virus X is a strain-specific elicitor of *Rx1*-mediated resistance in potato. *Plant Journal* 8, 933–941.

Bent, A.F., Kunkel, B.N., Dahlbeck, D., Brown, K.L., Schmidt, R., Giraudat, J., Leung, J. and Staskawicz, B.J. (1994) *RPS2* of *Arabidopsis thaliana*: a leucine-rich repeat class of plant disease resistance genes. *Science* 265, 1856–1860.

Berna, A. and Bernier, F. (1997) Regulated expression of a wheat germin gene in tobacco: oxalate oxidase activity and apoplastic localization of the heterologous protein. *Plant Molecular Biology* 33, 417–429.

Bevins, C.L. and Zasloff, M. (1990) Peptides from frog skin. *Annual Review of Biochemistry* 59, 395–414.

Binks, R.H., Greenham, J.R., Luis, J.G. and Gowen, S.R. (1997) A phytoalexin from roots of *Musa acuminata* var. Pisang sipulu. *Phytochemistry* 45, 47–49.

Birch, R.G., Pemberton, J.M. and Basnayake, W.V. (1990) Stable albicidin resistance in *Escherichia coli* involves an altered outer-membrane nucleoside uptake system. *Journal of General Microbiology* 136, 51–58.

Bird, A.F. (1976) The development and organization of skeletal structures in nematodes. In: Croll, N.A. (ed.) *The Organization of Nematodes.* Academic Press, New York, USA, pp. 107–137.

Blaiseau, P.L. and Lafay, J.F. (1992) Primary structure of a chitinase-encoding gene (*chi1*) from the filamentous fungus *Aphanocladium album*: similarity to bacterial chitinases. *Gene* 120, 243–248.

Bogdanove, A.J., Wei, Z.M., Zhao, L. and Beer, S.V. (1996) *Erwinia amylovora* secretes harpin via a type III pathway and contains a homolog of yopN of *Yersinia* spp. *Journal of Bacteriology* 178, 1720–1730.

Bolar, J.P., Aldwinckle, H.S., Harman, G.E., Norelli, J.L. and Brown, S.K. (1997) Endochitinase-transgenic McIntosh apple lines have increased resistance to scab. *Phytopathology* 87, S10.

Bolognesi, A., Polito, L., Olivieri, F., Valbonesi, P., Barbieri, L., Battelli, M.G., Carusi, M.V., Benvenuto, E., Del Vecchio Blanco, F., Di Maro, A., Parente, A., Di Loreto, M. and Stirpe, F. (1997) New ribosome-inactivating proteins with polynucleotide:adenosine glycosidase and antiviral activities from *Basella rubra* L. and *Bougainvillea spectabilis* Willd. *Planta* 203, 422–429.

Bonness, M.S., Ready, M.P., Irvin, J.D. and Mabry, T.J. (1994) Pokeweed antiviral protein inactivates pokeweed ribosomes: implications for the antiviral mechanism. *Plant Journal* 5, 173–183.

Borejsza-Wysocka, E.E., Norelli, J.L., Ko, K. and Aldwinckle, H.S. (1997) Transformation of M26 apple rootstock with lytic proteins for enhanced fire blight resistance. *Phytopathology* 87, S10.

Borgonie, G., Claeys, M., Leyns, F., Arnaut, G., De Waele, D. and Coomans, A. (1996a) Effect of nematicidal *Bacillus thuringiensis* strains on free-living nematodes. 1. Light microscopic observations, species and biological stage specificity and identification of resistant mutants of *Caenorhabditis elegans. Fundamental and Applied Nematology* 19, 391–398.

Borgonie, G., Claeys, M., Leyns, F., Arnaut, G., De Waele, D. and Coomans, A. (1996b) Effect of nematicidal *Bacillus thuringiensis* strains on free-living nematodes. 2. Ultrastructural analysis of the intoxication process in *Caenorhabditis elegans. Fundamental and Applied Nematology* 19, 407–414.

Borkowska, M., Krzymowska, M., Talarczyk, A., Awan, M.F.M., Yakovleva, L., Kleczkowski, K. and Wielgat, B. (1998) Transgenic potato plants expressing soybean β-1,3-endoglucanase gene exhibit an increased resistance to *Phytophthora infestans. Zeitschrift für Naturforschung* 53, 1012–1016.

Broekaert, W.F., Terras, F.R.G., Cammue, B.P.A. and Osborn, R.W. (1995) Plant defensins: novel antimicrobial peptides as components of the host defense system. *Plant Physiology* 108, 1353–1358.

Broekaert, W.F., Cammue, B.P.A., De Bolle, M., Thevissen, K., De Samblanx, G. and Osborn, R.W. (1997) Antimicrobial peptides from plants. *Critical Reviews in Plant Sciences* 16, 297–323.

Broglie, K., Chet, I., Holliday, M., Cressman, R., Biddle, P., Knowlton, S., Mauvais, C.J. and Broglie, R. (1991) Transgenic plants with enhanced resistance to the fungal pathogen *Rhizoctonia solani. Science* 254, 1194–1197.

Buddenhagen, I.W. (1960) Strains of *Pseudomonas solanacearum* in indigenous hosts in banana plantations of Costa Rica, and their relationship to bacterial wilt of bananas. *Phytopathology* 50, 660–664.

Burrows, P.R. and De Waele, D. (1997) Engineering resistance against plant parasitic nematodes using anti-nematode genes. In: Fenoll, C., Grundler, F.M.W. and Ohl, S.A. (eds) *Cellular and Molecular Aspects of Plant–Nematode Interactions.* Developments in Plant Pathology, Vol. 10, Kluwer Academic Publishers, Dordrecht, pp. 217–236.

Büschges, R., Hollricher, K., Panstruga, R., Simons, G., Wolter, M., Frijters, A., van Daelen, R., van der Lee, T., Diergaarde, P., Groenendijk, J., Töpsch, S., Vos, P., Salamini, F. and Schulze-Lefert, P. (1997) The barley *Mlo* gene: a novel control element of plant pathogen resistance. *Cell* 88, 695–697.

Cai, D., Kleine, M., Kifle, S., Harloff, H.J., Sandal, N.N., Marcker, K.A., Klein-Lankhorst, R.M., Salentijn, E.M.J., Lange, W., Stiekema, W.J., Wyss, U., Grundler, F.M.W. and Jung, C. (1997) Positional cloning of a gene for nematode resistance in sugar beet. *Science* 275, 832–834.

Calder, V.L. and Palukaitis, P. (1992) Nucleotide sequence analysis of the movement genes of resistance breaking strains of tomato mosaic virus. *Journal of General Virology* 74, 1157–1162.

Cammue, B.P.A., De Bolle, M.F.C., Terras, F.R.G., Proost, P., Van Damme, J., Rees, S.B., Vanderleyden, J. and Broekaert, W.F. (1992) Isolation and characterization of a novel class of plant antimicrobial peptides from *Mirabilis jalapa* L. seeds. *Journal of Biological Chemistry* 267, 2228–2233.

Cammue, B.P.A., De Bolle, M.F.C., Terras, F.R.G. and Broekaert, W.F. (1993) Fungal disease control in *Musa*: application of new antifungal proteins. In: Ganry, J. (ed.) *Breeding Banana and Plantain for Resistance to Diseases and Pests, Proceedings of the International Symposium on Genetic Improvement of Bananas for Resistance to Diseases and Pests, organized by CIRAD–FLHOR, Montpellier, France, 7–9 September 1992*. CIRAD, Montpellier, France, pp. 221–225.

Cammue, B.P.A., Thevissen, K., Hendriks, M., Eggermont, K., Goderis, I.J., Proost, P., Van Damme, J., Osborn, R.W., Guerbette, F., Kader, J.-C. and Broekaert, W.F. (1995) A potent antimicrobial protein from onion (*Allium cepa* L.) seeds showing sequence homology to plant lipid transfer proteins. *Plant Physiology* 109, 445–455.

Carlsson, A., Engstrom, P., Palva, E.T. and Bennich, H. (1991) Attacin, an antibacterial protein from *Hyalophora cecropia*, inhibits synthesis of outer membrane proteins in *Escherichia coli* by interfering with *omp* gene transcription. *Infection and Immunity* 59, 3040–3045.

Carlsson, A., Nystrom, T., de Cock, H. and Bennich, H. (1998) Attacin – an insect immune protein – binds LPS and triggers the specific inhibition of bacterial outer-membrane protein synthesis. *Microbiology* 144, 2179–2188.

Carmona, M.J., Molina, A., Fernandez, J.A., López-Fando, J.J. and García-Olmedo, F. (1993) Expression of the alpha-thionin gene from barley in tobacco confers enhanced resistance to bacterial pathogens. *Plant Journal* 3, 457–462.

Carrasco, L., Vazquez, D., Hernandez-Lucas, C., Carbonero, P. and García-Olmedo, F. (1981) Thionins: plant peptides that modify membrane permeability in cultured mammalian cells. *European Journal of Biochemistry* 116, 185–189.

Casteels, P., Ampe, C., Jacobs, F., Vaeck, M. and Tempst, P. (1989) Apidaecins: antibacterial peptides from honeybees. *EMBO Journal* 8, 2387–2391.

Cavallarin, L., Andreu, D. and San Segundo, B. (1998) Cecropin A-derived peptides are potent inhibitors of fungal plant pathogens. *Molecular Plant–Microbe Interactions* 11, 218–227.

Cervone, F., Hahn, M.G., De Lorenzo, F., Darvill, F. and Albersheim, P. (1989) Host–pathogen interactions. XXXIII. A plant protein converts a fungal pathogenesis factor into an elicitor of plant defense responses. *Plant Physiology* 90, 542–548.

Chen, F.C. and Kuehnle, A.R. (1996) Obtaining transgenic *Anthurium* through *Agrobacterium*-mediated transformation of etiolated internodes. *Journal of the American Society for Horticultural Science* 121, 47–51.

Chen, G., Zhang, Y., Li, J., Dunphy, G.B., Punja, Z.K. and Webster, J.M. (1996) Chitinase activity of *Xenorhabdus* and *Photorhabdus* species, bacterial associates of entomopathogenic nematodes. *Journal of Invertebrate Pathology* 68, 101–108.

Chen, Z.C., White, R.F., Antoniw, J.F. and Lin, Q. (1991) Effect of pokeweed antiviral protein (PAP) on the infection of plant viruses. *Plant Pathology* 40, 612–620.

Chen, Z.C., Antoniw, J.F. and White, R.F. (1993) A possible mechanism for the antiviral activity of pokeweed antiviral protein. *Physiological and Molecular Plant Pathology* 42, 249–258.

Clackson, T., Hoogenboom, H.R., Griffiths, A.D. and Winter, G. (1991) Making antibody fragments using phage display libraries. *Nature* 352, 624–628.

Collmer, A. (1996) Bacterial avirulence proteins: where's the action? *Trends in Plant Science* 1, 209–210.

Conrad, U. and Fiedler, U. (1994) Expression of engineered antibodies in plant cells. *Plant Molecular Biology* 26, 1023–1030.

Cooper, B., Lapidot, M., Heick, J.A., Dodds, J.A. and Beachy, R.N. (1995) A defective movement protein of TMV in transgenic plants confers resistance to multiple viruses whereas the functional analogue increases susceptibility. *Virology* 206, 307–313.

Culver, J.N. and Dawson, W.O. (1991) Tobacco mosaic virus elicitor CP genes produce a hypersensitive phenotype in transgenic *Nicotiana sylvestris* plants. *Molecular Plant–Microbe Interactions* 4, 458–463.

Dangl, J.L. (1994) The enigmatic avirulence genes of phytopathogenic bacteria. *Current Topics in Microbiology and Immunology* 192, 99–118.

Dangl, J.L. (1995) Pièce de résistance: novel classes of plant disease resistance genes. *Cell* 80, 363–366.

Darbinyan, N.S., Popov, Y.G., Mochulskii, A.V., Oming, G., Piruzyan, E.S. and Vasilevko, V.T. (1996) Construction of *Nicotiana tabacum* L. plants expressing the bacterial gene for β-1,3-glucanase. 2. Transgenic plants expressing the β-1,3-glucanase gene of *Clostridium thermocellum* represent a model for studying differential expression of stress response-related genes. *Genetika* 32, 204–210.

Deák, M., Horváth, G.V., Davletova, S., Török, K., Sass, L., Vass, I., Barna, B., Király, Z. and Dudits, D. (1999) Plants ectopically expressing the iron-binding protein, ferritin, are tolerant to oxidative damage and pathogens. *Nature Biotechnology* 17, 192–196.

De Bolle, M.F.C., David, K.M.M., Rees, S.B., Vanderleyden, J., Cammue, B.P.A. and Broekaert, W.F. (1993) Cloning and characterization of a cDNA encoding an antimicrobial chitin-binding protein from amaranth, *Amaranthus caudatus. Plant Molecular Biology* 22, 1187–1190.

De Bolle, M.F.C., Osborn, R.W., Goderis, I.J., Noe, L., Acland, D., Hart, C.A., Torrekens, S., Van Leuven, F. and Broekaert, W.F. (1996) Antimicrobial peptides from *Mirabilis jalapa* and *Amaranthus caudatus*: expression, processing, localization and biological activity in transgenic tobacco. *Plant Molecular Biology* 31, 993–1008.

Declerck, S., Plenchette, C. and Strullu, D.G. (1995) Mycorrhizal dependency of banana (*Musa acuminata*, AAA group) cultivar. *Plant and Soil* 176, 183–187.

de Feyter, R., Young, M., Schroeder, K., Dennis, E.S. and Gerlach, W. (1996) A ribozyme gene and an antisense gene are equally effective in conferring resistance to tobacco mosaic virus on transgenic tobacco. *Molecular and General Genetics* 250, 329–338.

de la Cruz, A., López, L., Tenllado, F., Díaz-Ruíz, J.R., Sanz, A.I., Vaquero, C., Serra, M.T. and García-Luque, I. (1997) The coat protein is required for the elicitation of the *Capsicum* L^2 gene-mediated resistance against the tobamoviruses. *Molecular Plant–Microbe Interactions* 10, 107–113.

de la Cruz, J., Pintor-Toro, J.A., Benitez, T., Llobell, T. and Romero, L.C. (1995) A novel endo-β-1,3-glucanase, BGN13.1, involved in the mycoparasitism of *Trichoderma harzianum. Journal of Bacteriology* 177, 6937–6945.

De la Fuente-Martínez, J.M. and Herrera-Estrella, L. (1993) Strategies to design transgenic plants resistant to toxins produced by pathogens. *AgBiotech News and Information* 5, 295N–299N.

De la Fuente-Martínez, J.M., Mosqueda-Cano, G., Alvarez-Morales, A. and Herrera-Estrella, L. (1992) Expression of a bacterial phaseolotoxin-resistant ornithyl transcarbamylase in transgenic tobacco confers resistance to *Pseudomonas syringae* pv. *phaseolicola. Bio/Technology* 10, 905–909.

DeLucca, A.J., Bland, J.M., Jacks, T.J., Grimm, C., Cleveland, T.E. and Walsh, T.J. (1997) Fungicidal activity of cecropin A. *Antimicrobial Agents in Chemotherapy* 41, 481–483.

Dempsey, C.E. (1990) The actions of melittin on membranes. *Biochimica Biophysica Acta* 1031, 143–161.

De Neve, M., De Buck, S., De Wilde, C., Van Houdt, H., Strobbe, I., Jacobs, A., Van Montagu, M. and Depicker, A. (1999) Gene silencing results in instability of antibody production in transgenic plants. *Molecular and General Genetics* 260, 582–592.

Denny, T.P. (1995) Involvement of bacterial polysaccharides in plant pathogenesis. *Annual Review of Phytopathology* 33, 173–197.

Deom, C.M., Lapidot, M. and Beachy, R.N. (1992) Plant virus movement proteins. *Cell* 69, 221–224.

Desiderio, A., Aracri, B., Leckie, F., Mattei, B., Salvi, G., Tigelaar, H., Van Roekel, J.S.C., Baulcombe, D.C., Melchers, L.S., De Lorenzo, G. and Cervone, F. (1997) Polygalacturonase-inhibiting proteins (PGIPs) with different specificities are expressed in *Phaseolus vulgaris. Molecular Plant–Microbe Interactions* 10, 852–860.

Deslandes, L., Pileur, F., Liaubet, L., Camut, S., Can, C., Williams, K., Holub, E., Beynon, J., Arlat, M. and Marco, Y. (1998) Genetic characterization of RRS1, a recessive locus in *Arabidopsis thaliana* that confers resistance to the bacterial soilborne pathogen *Ralstonia solanacearum. Molecular Plant–Microbe Interactions* 11, 659–667.

Desnottes, J.F. (1996) New targets and strategies for the development of antibacterial agents. *Trends in Biotechnology* 14, 134–140.

De Waele, D. (1993) Potential of gene transfer for engineering resistance against nematode attack. In: *Proceedings of the Workshop on Biotechnology Applications for Banana and Plantain Improvement, San José, Costa Rica, January 27–31, 1992*. INIBAP, Montpellier, France, pp. 116–124.

De Waele, D., Sági, L. and Swennen, R. (1994) Prospects to engineer nematode resistance in banana.

In: Valmayor, R.V., Davide, R.G., Stanton, J.M. and Treverrow, J.M. (eds) *Banana Nematodes and Weevil Borers in Asia and the Pacific. Proceedings of a Conference–Workshop on Nematodes and Weevil Borers Affecting Banana in Asia and the Pacific, Serdang, Selangor, Malaysia, 18–22 April 1994.* INIBAP/ASPNET, Los Baños, Laguna, Philippines, pp. 204–216.

de Zoeten, G.A. and Fulton, R.W. (1975) Understanding generates possibilities. *Phytopathology* 65, 221–222.

Ding, X., Gopalakrishnan, B., Johnson, L.B., White, F.F., Wang, X., Morgan, T.D., Kramer, K.J. and Muthukrishnan, S. (1998) Insect resistance of transgenic tobacco expressing an insect chitinase gene. *Transgenic Research* 7, 77–84.

Dixon, M.S., Jones, D.A., Keddie, J.S., Thomas, C.M., Harrison, K. and Jones, J.D.G. (1996) The tomato *Cf-2* disease resistance locus comprises two functional genes encoding leucine-rich repeat proteins. *Cell* 84, 451–459.

Duan, X., Li, X., Xue, Q., Abo-el-Saad, M., Xu, D. and Wu, R. (1996) Transgenic rice plants harboring an introduced potato proteinase inhibitor II gene are insect resistant. *Nature Biotechnology* 14, 494–498.

Durbin, R.D. and Langston-Unkefer, P.J. (1988) The mechanisms for self-protection against bacterial phytotoxins. *Annual Review of Phytopathology* 26, 313–329.

Durell, S.R., Raghunathan, G. and Guy, H.R. (1992) Modeling the ion channel structure of cecropin. *Biophysical Journal* 63, 1623–1631.

Düring, K. (1996) Genetic engineering for resistance to bacteria in transgenic plants by introduction of foreign genes. *Molecular Breeding* 2, 297–305.

Düring, K., Porsch, P., Fladung, M. and Lörz, H. (1993) Transgenic potato plants resistant to the phytopathogenic bacterium *Erwinia carotovora*. *Plant Journal* 3, 587–598.

Duvick, J.P., Rood, T., Rao, G.A. and Marshak, D.R. (1992) Purification and characterization of a novel antimicrobial peptide from maize (*Zea mays* L.) kernels. *Journal of Biological Chemistry* 267, 18814–18820.

Edwards, D.L., Payne, J. and Soares, G.G. (1989) Novel isolates of *Bacillus thuringiensis* having activity against nematodes. European Patent Application 88307309.0.

El Ghaouth, A., Arul, J., Asselin, A. and Benhamou, N. (1992) Antifungal activity of chitosan on post-harvest pathogens: induction of morphological and cytological alterations in *Rhizopus stolonifer*. *Mycological Research* 96, 769–779.

El Quakfaoui, S., Potvin, C., Brzezinski, R. and Asselin, A. (1995) A *Streptomyces* chitosanase is active in transgenic tobacco. *Plant Cell Reports* 15, 222–226.

Epple, P., Apel, K. and Bohlmann, H. (1997) Overexpression of an endogenous thionin enhances resistance of Arabidopsis against *Fusarium oxysporum*. *Plant Cell* 9, 509–520.

Fecker, L.F., Koenig, R. and Obermeier, C. (1997) *Nicotiana benthamiana* plants expressing beet necrotic yellow vein virus (BNYVV) coat protein-specific scFvs are partially protected against the establishment of the virus in the early stages of infection and its pathogenic effects in the late stages of infection. *Archives of Virology* 142, 1857–1863.

Fitchen, J.H. and Beachy, R.N. (1993) Genetically engineered protection against viruses in transgenic plants. *Annual Review of Microbiology* 47, 739–763.

Flasinski, S. and Cassidy, B.G. (1998) Potyvirus aphid transmission requires helper component and homologous coat protein for maximal efficiency. *Archives of Virology* 143, 2159–2172.

Flor, H.H. (1971) Current status of the gene-for-gene concept. *Annual Review of Phytopathology* 9, 275–296.

Florack, D.E. and Stiekema, W.J. (1994) Thionins: properties, possible biological roles and mechanisms of action. *Plant Molecular Biology* 26, 25–37.

Florack, D., Allefs, S., Bollen, R., Bosch, D., Visser, B. and Stiekema, W. (1995) Expression of giant silkmoth cecropin B genes in transgenic tobacco. *Transgenic Research* 4, 132–141.

Florack, D.E.A., Dirkse, W.G., Visser, B., Heidekamp, F. and Stiekema, W.J. (1994) Expression of biologically active hordothienins in tobacco. Effects of pre- and pro-sequences at the amino acid carboxyl termini of the hordothienin precursor on mature protein expression and sorting. *Plant Molecular Biology* 24, 83–96.

Florack, D.E.A., Stiekema, W.J. and Bosch, D. (1996) Toxicity of peptides to bacteria present in vase water of cut roses. *Postharvest Biology and Technology* 8, 285–291.

Fogain, R, and Gowen, S.R. (1998) 'Yangambi km5' (*Musa* AAA, Ibota subgroup): a possible source of resistance to *Radopholus similis* and *Pratylenchus goodeyi. Fundamental and Applied Nematology* 21, 75–80.

Fraser, R.S.S. (1986) Genes for resistance to plant viruses. *Critical Reviews in Plant Sciences* 3, 257–294.

Fraser, R.S.S. (1990) The genetics of resistance to plant viruses. *Annual Review of Phytopathology* 28, 179–200.

Fujiwara, S., Imai, J., Fujiwara, M., Yaeshima, T., Kawashima, T. and Kobayashi, K. (1990) A potent antibacterial protein in royal jelly: purification and determination of the primary structure of royalisin. *Journal of Biological Chemistry* 265, 11333–11337.

Fulton, R.W. (1986) Practices and precautions in the use of cross protection for plant virus disease control. *Annual Review of Phytopathology* 26, 67–81.

Gabriel, D.W. and Rolfe, B.G. (1990) Working models of specific recognition in plant–microbe interaction. *Annual Review of Plant Physiology* 28, 365–391.

Gadani, F., Mansky, L.M., Medici, R., Miller, W.A. and Hill, J.H. (1990) Genetic engineering of plants for virus resistance. *Archives of Virology* 115, 1–21.

Galun, E., Wolf, S. and Mailer-Slasky, Y. (1996) Defence against plant pathogenic bacteria in transgenic potato plants. In: *Abstracts of the Third International Symposium on* In Vitro *Culture and Horticultural Breeding, June 16–21 1996, Jerusalem*, p. 10.

García-Olmedo, F., Molina, A., Segura, A. and Moreno, M. (1995) The defensive role of nonspecific lipid-transfer proteins in plants. *Trends in Microbiology* 3, 72–74.

Gheysen, G., Van der Eycken, W., Barthels, N., Karimi, M. and Van Montagu, M. (1996) The exploitation of nematode-responsive plant genes in novel nematode control methods. *Pesticide Science* 47, 95–101.

Gheysen, G., Karimi, M., De Meutter, J., Barthels, N. and Van Montagu, M. (1998) The exploitation of nematode-responsive plant genes to engineer nematode resistance. *Mededelingen Faculteit Landbouwwetenschappen Universiteit Gent* 63, 1633–1640.

Gilbertson, R.L. and Lucas, W.J. (1996) How do viruses traffick on the 'vascular highway'? *Trends in Plant Science* 1, 260–268.

Golemboski, D.B., Lomonossoff, G.P. and Zaitlin, M. (1990) Plants transformed with a tobacco mosaic virus nonstructural gene sequence are resistant to the virus. *Proceedings of the National Academy of Sciences USA* 87, 6311–6315.

Gonsalves, D. and Garnsey, S.M. (1989) Cross-protection techniques for control of plant virus diseases in the Tropics. *Plant Disease* 73, 592–597.

Gopalan, S., Bauer, D.W., Alfano, J.R., Loniello, A.O., He, S.Y. and Collmer, A. (1996) Expression of the *Pseudomonas syringae* avirulence protein AvrB in plant cells alleviates its dependence on the hypersensitive response and pathogenicity (Hrp) secretion system in eliciting genotype-specific hypersensitive cell death. *Plant Cell* 8, 1095–1105.

Görschen, E., Dunaeva, M., Hause, B., Reeh, I., Wasternack, C. and Parthier, B. (1997) Expression of the ribosome-inactivating protein JIP60 from barley in transgenic tobacco leads to an abnormal phenotype and alterations on the level of translation. *Planta* 202, 470–478.

Grant, M.R., Godiard, L., Straube, E., Ashfield, T., Lewald, J., Sattler, A., Innes, R.W. and Dangl, J.L. (1995) Structure of the Arabidopsis *RPM1* gene enabling dual specificity of disease resistance. *Science* 269, 843–846.

Grison, R., Grezes-Besset, B., Schneider, M., Lucante, N., Olsen, L., Leguay, J.J. and Toppan, A. (1996) Field tolerance to fungal pathogens of *Brassica napus* constitutively expressing a chimeric chitinase gene. *Nature Biotechnology* 14, 643–646.

Gross, D.C. (1991) Molecular and genetic analysis of toxin production by pathovars of *Pseudomonas syringae. Annual Review of Phytopathology* 29, 247–278.

Gurr, S.J., McPherson, M.J., Scollan, C., Atkinson, H.J. and Bowles, D.J. (1991) Gene expression in nematode-infected plant roots. *Molecular and General Genetics* 226, 361–366.

Habuka, N., Miyano, M., Kataoka, J., Tsuge, H., Ago, H. and Noma, M. (1991) Substantial increase of the inhibitory activity of *Mirabilis* antiviral protein by an elimination of the disulfide bond with genetic engineering. *Journal of Biological Chemistry* 266, 23558–23560.

Hain, R., Reif, H.J., Krause, E., Langebartels, R., Kindl, H., Vornam, B., Wiese, W., Schmelzer, E., Schreier, P.H., Stöcker, R.H. and Stenzel, K. (1993) Disease resistance results from foreign phytoalexin expression in a novel plant. *Nature* 361, 153–156.

Hamilton, R.I. (1980) Defenses triggered by previous invaders: viruses. In: Horsfall, J.G. and Cowling, E.B. (eds) *Plant Disease: an Advanced Treatise*, Vol. 5. Academic Press, New York, pp. 279–303.
Hampel, A., Nesbitt, S., Tritz, R. and Altschuler, M. (1993) The hairpin ribozyme. *Methods* 5, 37–42.
Hara, S. and Yamakawa, M. (1995) Moricin, a novel type of antibacterial peptide isolated from the silkworm, *Bombyx mori*. *Journal of Biological Chemistry* 270, 29923–29927.
Harelimana, G., Lepoivre, P., Jijakli, H. and Mourichon, X. (1997) Use of *Mycosphaerella fijiensis* toxins for the selection of banana cultivars resistant to black leaf streak. *Euphytica* 96, 125–128.
Harrison, B.D., Mayo, M.A. and Baulcombe, D.C. (1987) Virus resistance in transgenic plants that express cucumber mosaic virus satellite RNA. *Nature* 328, 799–802.
Harrison, S.J., McManus, A.M., Marcus, J.P., Goulter, K.C., Green, J.L., Nielsen, K.J., Craik, D.J., Maclean, D.J. and Manners, J.M. (1999) Purification and characterization of a plant antimicrobial peptide expressed in *Escherichia coli*. *Protein Expression and Purification* 15, 171–177.
Hartley, M.R., Chaddock, J.A. and Bonness, M.S. (1996) The structure and function of ribosome-inactivating proteins. *Trends in Plant Science* 1, 254–260.
Hartung, J.S., Fulbright, D.W. and Klos, E.J. (1986) Cloning of a bacteriophage polysaccharide depolymerase gene and its expression in *Erwinia amylovora*. *Molecular Plant–Microbe Interactions* 1, 87–93.
Haseloff, J. and Gerlach, W.L. (1988) Simple RNA enzymes with new and highly specific endoribonuclease activities. *Nature* 334, 585–591.
Hatziloukas, E. and Panopoulos, N.J. (1992) Origin, structure and regulation of *argK*, encoding the phaseolotoxin-resistant ornithine carbamoyltransferase in *Pseudomonas syringae* pv. *phaseolicola*, and functional expression of *argK* in transgenic tobacco. *Journal of Bacteriology* 174, 5895–5909.
Havstad, P., Sutton, D., Thomas, S., Sengupta-Gopalan, C. and Kemp, J. (1991) Collagenase expression in transgenic plants: an alternative to nematicides. In: *Third International Congress of Plant Molecular Biology held in Tucson, Arizona, USA*, Abstract 345.
Hayes, C.K., Klemsdal, S., Lorito, M., Di Pietro, A.D., Nakas, J.P., Tronsmo, A. and Harman, G.E. (1994) Isolation and characterization of an endochitinase-encoding gene from a cDNA library of *Trichoderma harzianum*. *Gene* 138, 143–148.
Hayward, A.C. (1964) Characteristics of *Pseudomonas solanacearum*. *Journal of Applied Bacteriology* 27, 265–277.
He, S.Y., Huang, H.-C. and Collmer, A. (1993) *Pseudomonas syringae* pv. *syringae* harpin$_{Pss}$: a protein that is secreted via the Hrp pathway and elicits the hypersensitive reponse in plants. *Cell* 73, 1255–1266.
Hiatt, A., Cafferkey, R. and Bowdish, K. (1989) Production of antibodies in transgenic plants. *Nature* 342, 76–78.
Hightower, R., Baden, C., Penzes, E. and Dunsmuir, P. (1994) The expression of cecropin peptide in transgenic tobacco does not confer resistance to *Pseudomonas syringae* pv *tabaci*. *Plant Cell Reports* 13, 295–299.
Hippe, S., Düring, K. and Kreuzaler, F. (1989) *In situ* localization of a foreign protein in transgenic plants by immunoelectron microscopy following high pressure freezing, freeze substitution and low temperature embedding. *European Journal of Cell Biology* 50, 230–234.
Hirai, N., Ishida, H. and Koshimizu, K. (1994) A phenalenone-type phytoalexin from *Musa acuminata*. *Phytochemistry* 37, 383–385.
Holscher, D. and Schneider, B. (1996) A resveratrol dimer from *Anigozanthes preissii* and *Musa cavendish*. *Phytochemistry* 43, 471–473.
Holtorf, S., Ludwig-Müller, J., Apel, K. and Bohlmann, H. (1998) High-level expression of a viscotoxin in *Arabidopsis thaliana* gives enhanced resistance against *Plasmodiophora brassicae*. *Plant Molecular Biology* 36, 673–680.
Hong, Y., Saunders, K., Hartley, M.R. and Stanley, J. (1996) Resistance to geminivirus infection by virus-induced expression of dianthin in transgenic plants. *Virology* 220, 119–127.
Howie, W., Joe, L., Newbigin, E., Suslow, T. and Dunsmuir, P. (1994) Transgenic tobacco plants which express the *chiA* gene from *Serratia marcescens* have enhanced tolerance to *Rhizoctonia solani*. *Transgenic Research* 3, 90–98.
Huang, D.N., Zhu, B., Yang, W., Xue, R., Xiao, H., Tian, W.Z., Li, L.C. and Dai, S.H. (1996) Introduction of cecropin B gene into rice (*Oryza sativa* L.) by particle bombardment and analysis of transgenic plants. *Science in China* 39, 652–661.

Huang, J. and Schell, M.A. (1995) Molecular characterization of the *eps* locus of *Pseudomonas solanacearum* and its transcriptional regulation at a single promoter. *Molecular Microbiology* 16, 977–989.

Huang, J., Carney, B.F., Denny, T.P., Weissinger, A.K. and Schell, M.A. (1995) A complex network regulates expression of *eps* and other virulence genes of *Pseudomonas solanacearum*. *Journal of Bacteriology* 177, 1259–1267.

Huang, W., Cui, X., Tian, Y., Lin, M. and Peng, X. (1994) Cloning of T7 lysozyme gene and construction of the vector for transgenic plants resistant to bacterial infection. *Wei Sheng Wu Hsueh Pao* 34, 261–265.

Huang, Y., Nordeen, R.O., Di, L., Owens, L.D. and McBeath, J.H. (1997) Expression of an engineered cecropin gene cassette in transgenic tobacco plants confers disease resistance to *Pseudomonas syringae* pv. *tabaci*. *Phytopathology* 87, 494–499.

Hull, R. and Davies, J.W. (1992) Approaches to nonconventional control of plant virus diseases. *Critical Reviews in Plant Sciences* 11, 17–33.

Hultmark, D., Engstrom, A., Andersson, K., Steiner, H., Bennich, H. and Boman, H.G. (1983) Insect immunity. Attacins, a family of antibacterial proteins from *Hyalophora cecropia*. *EMBO Journal* 2, 571–576.

Huynh, T.V., Dahlbeck, D. and Staskawicz, B.J. (1989) Bacterial blight of soybean: regulation of a pathogen gene determining host cultivar specificity. *Science* 245, 1374–1377.

Irvin, J.D., Kelly, T. and Robertus, J.D. (1980) Purification and properties of a second antiviral protein from *Phytolacca americana* which inactivates eukaryotic ribosomes. *Archives of Biochemistry and Biophysics* 200, 418–425.

Jach, G., Görnhardt, B., Mundy, J., Logemann, J., Pinsdorf, E., Leah, R., Schell, J. and Maas, C. (1995) Enhanced quantitative resistance against fungal disease by combinatorial expression of different barley antifungal proteins in transgenic tobacco. *Plant Journal* 8, 97–109.

Jaizme-Vega, M.C. and Azcón, R. (1995) Responses of some tropical and subtropical cultures to endomycorrhizal fungi. *Mycorrhiza* 5, 213–217.

Jaynes, J.M., Nagpala, P., Destéfano-Beltrán, L., Huang, J.H., Kim, J.H., Denny, T. and Cetiner, S. (1993) Expression of a Cecropin B lytic peptide analog in transgenic tobacco confers enhanced resistance to bacterial wilt caused by *Pseudomonas solanacearum*. *Plant Science* 89, 43–53.

Jekel, P.A., Hartmann, B.H. and Beintema, J.J. (1991) The primary structure of hevamine, an enzyme with lysozyme/chitinase activity from *Hevea brasiliensis* latex. *European Journal of Biochemistry* 200, 123–130.

Jensen, L.G., Olsen, O., Kops, O., Wolf, N., Thomsen, K.K. and von Wettstein, D. (1996) Transgenic barley expressing a protein-engineered, thermostable (1,3–1,4)-β-glucanase during germination. *Proceedings of the National Academy of Sciences USA* 93, 3487–3491.

Johal, G.S. and Briggs, S.P. (1992) Reductase activity encoded by the *Hm1* disease resistance gene in maize. *Science* 258, 985–987.

Jones, D.A., Thomas, C.M., Hammond-Kosack, K.E., Balint-Kurti, P.J. and Jones, J.D.G. (1994) Isolation of the tomato *Cf-9* gene for resistance to *Cladosporium fulvum* by transposon tagging. *Science* 266, 789–793.

Jones, J.D.G. (1988) Expression of bacterial chitinase protein in tobacco leaves using two photosynthetic gene promoters. *Molecular and General Genetics* 212, 536–542.

Joosten, M.H., Cozijsen, T.J. and De Wit, P.J. (1994) Host resistance to a fungal tomato pathogen lost by a single base-pair change in an avirulence gene. *Nature* 367, 384–386.

Joshi, B.N., Sainani, M.N., Bastawade, K.B., Gupta, V.S. and Ranjekar, P.K. (1998) Cysteine protease inhibitor from pearl millet: a new class of antifungal protein. *Biochemical and Biophysical Research Communications* 246, 382–387.

Kader, J.C. (1996) Lipid-transfer proteins in plants. *Annual Review of Plant Physiology and Plant Molecular Biology* 47, 627–654.

Kaduno-Okuda, K., Taniai, K., Kato, Y., Kotani, E. and Yamakaula, M. (1995) Effects of synthetic *Bombyx mori* cecropin B on growth of plant pathogenic bacteria. *Journal of Invertebrate Pathology* 65, 309–319.

Kamo, T., Kato., Hirai, N., Tsuda, M., Fujioka, D. and Ohigashi, H. (1998) Phenylphenalenone-type phytoalexins from unripe Buñgulan banana fruit. *Bioscience and Biotechnological Biochemistry* 62, 95–101.

Kao, C.C. and Sequeira, L. (1991) A gene cluster required for coordinated biosynthesis of lipopolysaccharide and extracellular polysaccharide also affects virulence of *Pseudomonas solanacearum*. *Journal of Bacteriology* 173, 7841–7847.

Kao, C.C., Gosti, F., Huang, Y. and Sequeira, L. (1994) Characterization of a negative regulator of exopolysaccharide production by the plant-pathogenic bacterium *Pseudomonas solanacearum*. *Molecular Plant–Microbe Interactions* 7, 121–130.

Karrer, E.E., Beachy, R.N. and Holt, C.A. (1998) Cloning of tobacco genes that elicit the hypersensitive response. *Plant Molecular Biology* 36, 681–690.

Kawchuk, L.M., Martin, R.R. and McPherson, J. (1991) Sense and antisense RNA-mediated resistance to potato leafroll virus in Russet Burbank potato plants. *Molecular Plant–Microbe Interactions* 4, 247–253.

Kazan, K., Goulter, K.C., Way, H.M. and Manners, J.M. (1998a) Expression of a pathogenesis-related peroxidase of *Stylosanthes humilis* in transgenic tobacco and canola and its effect on disease development. *Plant Science* 136, 207–217.

Kazan, K., Murray, F.R., Goulter, K.C., Llewellyn, D.J. and Manners, J.M. (1998b) Induction of cell death in transgenic plants expressing a fungal glucose oxidase. *Molecular Plant–Microbe Interactions* 11, 555–562.

Keen, N.T. (1990) Gene-for-gene complementarity in plant–pathogen interactions. *Annual Review of Genetics* 24, 447–463.

Keen, N.T., Tamaki, S., Kobayashi, D., Gerhold, D., Stayton, M., Shen, H., Gold, S., Lorang, J., Thordal-Christensen, H., Dahlbeck, D. and Staskawicz, B. (1990) Bacteria expressing avirulence gene D produce a specific elicitor of the soybean hypersensitive reaction. *Molecular Plant–Microbe Interactions* 3, 112–121.

Keller, H., Pamboukdjian, N., Ponchet, M., Poupet, A., Delon, R., Verrier, J.L., Roby, D. and Ricci, P. (1999) Pathogen-induced elicitin production in transgenic tobacco generates a hypersensitive response and nonspecific disease resistance. *Plant Cell* 11, 223–236.

Kendra, D.F., Christian, D. and Hadwiger, L.A. (1989) Chitosan oligomers from *Fusarium solani*/pea interactions, chitinase/β-glucanase digestion of sporelings and from fungal wall chitin actively inhibit fungal growth and enhance disease resistance. *Physiological and Molecular Plant Pathology* 35, 215–230.

Kim, C.H. and Palukaitis, P. (1997) The plant defense response to cucumber mosaic virus in cowpea is elicited by the viral polymerase gene and affects virus accumulation in single cells. *EMBO Journal* 16, 4060–4068.

Kim, S.J., Lee, S.J., Kim, B.D. and Paek, K.H. (1997) Satellite-RNA-mediated resistance to cucumber mosaic virus in transgenic plants of hot pepper (*Capsicum annuum* cv. Golden Tower). *Plant Cell Reports* 16, 825–830.

Kinal, H., Park, C.M., Berry, J.O., Koltin, Y. and Bruenn, J.A. (1995) Processing and secretion of a virally encoded antifungal toxin in transgenic tobacco plants: evidence for a Kex2p pathway in plants. *Plant Cell* 7, 677–688.

Király, Z., Elzahaby, H.M. and Klement, Z. (1997) Role of extracellular polysaccharide (EPS) slime of plant pathogenic bacteria in protecting cells to reactive oxygen species. *Journal of Phytopathology* 145, 59–68.

Kitajima, S. and Sato, F. (1999) Plant pathogenesis-related proteins: molecular mechanisms of gene expression and protein function. *Journal of Biochemistry* 125, 1–8.

Kobe, B. and Deisenhofer, J. (1994) The leucine-rich repeat: a versatile binding motif. *Trends in Biochemical Sciences* 19, 415–421.

Köhn, B., Guerra, D.G., Mowry, T.M. and Berger, P.H. (1995) A yeast-derived RNase that confers resistance to multiple plant viruses. *Phytopathology* 85, 1142.

Koo, J.C., Lee, S.Y., Chun, H.J., Cheong, Y.H., Choi, J.S., Kawabata, S., Miyagi, M., Tsunasawa, S., Ha, K.S., Bae, D.W., Han, C.D., Lee, B.L. and Cho, M.J. (1998) Two hevein homologs isolated from the seed of *Pharbitis nil* L. exhibit potent antifungal activity. *Biochimica Biophysica Acta* 1382, 80–90.

Kristyanne, E.S., Kim, K.S. and Stewart, J.McD. (1997) Magainin 2 effects on the ultrastructure of five plant pathogens. *Mycologia* 89, 353–360.

Kuć, J. (1995) Phytoalexins, stress metabolism, and disease resistance in plants. *Annual Review of Phytopathology* 33, 275–297.

Kulaeva, O.N., Fedina, A.B., Burkhanova, E.A., Karavaiko, N.N., Karpeisky, M.Ya., Kaplan, I.B., Taliansky, M.E. and Atabekov, J.G. (1992) Biological activities of human interferon and 2′-5′ oligoadenylates in plants. *Plant Molecular Biology* 20, 383–393.

Kumar, D., Verma, H.N., Tuteja, N. and Tewari, K.K. (1997) Cloning and characterisation of a gene encoding an antiviral protein from *Clerodendrum aculeatum* L. *Plant Molecular Biology* 33, 745–751.

Kushmerick, C., de Souza Castro, M., Santos Cruz, J., Bloch, C. and Beirao, P.S. (1998) Functional and structural features of γ-zeathionins, a new class of sodium channel blockers. *FEBS Letters* 440, 302–306.

Lamb, C. and Dixon, R.A. (1997) The oxidative burst in plant disease resistance. *Annual Review of Plant Physiology and Plant Molecular Biology* 48, 251–275.

Lamb, J.W. and Hay, R.T. (1990) Ribozymes that cleave potato leafroll virus RNA within the coat protein and polymerase genes. *Journal of General Virology* 71, 2257–2264.

Langenberg, W.G., Zhang, L., Court, D.L., Giunchedi, L. and Mitra, A. (1997) Transgenic tobacco plants expressing the bacterial *rnc* gene resist virus infection. *Molecular Breeding* 3, 391–399.

Langland, J.O., Jin, S., Jacobs, B.L. and Roth, D.A. (1995) Identification of a plant-encoded analog of PKR, the mammalian double-stranded RNA-dependent protein kinase. *Plant Physiology* 108, 1259–1267.

Lapidot, M., Gafny, R., Ding, B., Wolf, S., Lucas, W.J. and Beachy, R.N. (1993) A dysfunctional movement protein of tobacco mosaic virus that partially modifies the plasmodesmata and limits virus spread in transgenic plants. *Plant Journal* 4, 959–970.

Laugé, R. and De Wit, P.J.G.M. (1998) Fungal avirulence genes: structure and possible functions. *Fungal Genetics and Biology* 24, 285–297.

Lawrence, G.J., Finnegan, E.J., Ayliffe, M.A. and Ellis, J.G. (1995) The *L6* gene for flax rust resistance is related to the Arabidopsis bacterial resistance gene *RPS2* and the tobacco viral resistance gene *N*. *Plant Cell* 7, 1195–1206.

Leah, R., Tommerup, H., Svendsen, I. and Mundy, J. (1991) Biochemical and molecular characterization of three anti-fungal proteins from barley seed. *Journal of Biological Chemistry* 266, 1564–1573.

Leckband, G. and Lörz, H. (1998) Transformation and expression of a stilbene synthase gene of *Vitis vinifera* L. in barley and wheat for increased fungal resistance. *Theoretical and Applied Genetics* 96, 1004–1012.

Lee, J.M., Hartman, G.L., Domier, L.L. and Bent, A.F. (1996) Identification and map location of TTR1, a single locus in *Arabidopsis thaliana* that confers tolerance to tobacco ringspot nepovirus. *Molecular Plant–Microbe Interactions* 9, 729–735.

Leyns, F., Borgonie, G., Arnaut, G. and De Waele, D. (1995) Nematicidal activity of *Bacillus thuringiensis* isolates. *Fundamental and Applied Nematology* 18, 211–218.

Lin, Q., Chen, Z.C., Antoniw, J.F. and White, R.F. (1991) Isolation and characterization of a cDNA clone encoding the anti-viral protein from *Phytolacca americana*. *Plant Molecular Biology* 17, 609–614.

Lin, W., Anuratha, C.S., Datta, K., Potrykus, I., Muthukrishnan, S. and Datta, S.K. (1995) Genetic engineering of rice for resistance to sheath blight. *Biotechnology* 13, 686–691.

Liu, D., Raghothama, K.G., Hasegawa, P. and Bressan, R.A. (1994) Osmotin overexpression in potato delays development of disease symptoms. *Proceedings of the National Academy of Sciences USA* 91, 1888–1892.

Lodge, J.K., Kaniewski, W.K. and Tumer, N.E. (1993) Broad-spectrum virus resistance in transgenic plants expressing pokeweed antiviral protein. *Proceedings of the National Academy of Sciences USA* 90, 7089–7093.

Logemann, J., Jach, G., Tommerup, H., Mundy, J. and Schell, J. (1992) Expression of a barley ribosome-inactivating protein leads to increased fungal protection in transgenic tobacco plants. *Biotechnology* 10, 305–308.

Lomonossoff, G.P. (1995) Pathogen-derived resistance to plant viruses. *Annual Review of Phytopathology* 33, 323–343.

Lord, J.M., Hartley, M.R. and Roberts, L.M. (1991) Ribosome inactivating proteins of plants. *Seminars in Cell Biology* 2, 15–22.

Lorito, M., Harman, G.E., Hayes, C.K., Broadway, R.M., Tronsmo, A., Woo, S.L. and Di Pietro, A. (1993) Chitinolytic enzymes produced by *Trichoderma harzianum*: antifungal activity of purified endochitinase and chitobiosidase. *Phytopathology* 83, 302–307.

Lorito, M., Broadway, R.M., Hayes, C.K., Woo, S.L., Noviello, C., Williams, D.L. and Harman, G.E.

(1994) Proteinase inhibitors from plants as a novel class of fungicides. *Molecular Plant–Microbe Interactions* 7, 525–527.

Lorito, M., Muccifora, S., Woo, S.L., Bonfante, P., Bianciotto, V., Gori, P., Filippone, E. and Scala, F. (1997a) Transgenic enzyme localization and mycorrhiza infection of plants expressing a *Trichoderma harzianum* endochitinase which improves plant disease resistance. *Phytopathology* 87, S59.

Lorito, M., Woo, S.L., Zoina, A. and Scala, F. (1997b) Purification, characterization and antifungal activity of cell wall degrading enzymes from a hyperproducer strain of the mycoparasite *Sepedonium chrysospermum. Phytopathology* 87, S59.

Lorito, M., Woo, S.L., Garcia Fernandez, I., Colucci, G., Harman, G.E., Pintor-Toro, J.A., Filippone, E., Muccifora, S., Lawrence, C.B., Zoina, A., Tuzun, S. and Scala, F. (1998) Genes from mycoparasitic fungi as a source for improving plant resistance to fungal pathogens. *Proceedings of the National Academy of Sciences USA* 95, 7860–7865.

Luis, J.G., Quiñones, W., Echeverri, F., Grillo, T.A., Kishi, M.P., Garcia-Garcia, F., Torres, F. and Cardona, G. (1996) Musanolones: four 9-phenylphenalenones from rhizomes of *Musa acuminata. Phytochemistry* 41, 753–757.

McKinney, H.H. (1929) Mosaic diseases in the Canary Islands, West Africa, and Gibraltar. *Journal of Agricultural Research* 39, 557–578.

McNellis, T.W., Mudgett, M.B., Li, K., Aoyama, T., Horvath, D., Chua, N.-H. and Staskawicz, B.J. (1998) Glucocorticoid-inducible expression of a bacterial avirulence gene in transgenic *Arabidopsis* induces hypersensitive cell death. *Plant Journal* 14, 247–257.

Maddaloni, M., Forlani, F., Balmas, V., Donini, G., Stasse, L., Corazza, L. and Motto, M. (1997) Tolerance to the fungal pathogen *Rhizoctonia solani* AG4 of transgenic tobacco expressing the maize ribosome-inactivating protein b-32. *Transgenic Research* 6, 393–402.

Maher, E.A., Bate, N.J., Ni, W., Elkind, Y., Dixon, R.A. and Lamb, C.J. (1994) Increased disease susceptibility of transgenic tobacco plants with suppressed levels of preformed phenylpropanoid products. *Proceedings of the National Academy of Sciences USA* 91, 7802–7806.

Maiti, I.B., Murphy, J.F., Shaw, J.G. and Hunt, A.G. (1993) Plants that express a potyvirus proteinase gene are resistant to virus infection. *Proceedings of the National Academy of Sciences USA* 90, 6110–6114.

Marchant, R., Davey, M.R., Lucas, J.A., Lamb, C.J., Dixon, R.A. and Power, J.B. (1998) Expression of a chitinase transgene in rose (*Rosa hybrida* L.) reduces development of blackspot disease (*Diplocarbon rosae* Wolf). *Molecular Breeding* 4, 187–194.

Marcos, J.F., Beachy, R.N., Houghten, R.A., Blondelle, S.E. and Pérez-Payá, E. (1995) Inhibition of a plant virus infection by analogs of melittin. *Proceedings of the National Academy of Sciences USA* 92, 12466–12469.

Marcus, J.P., Goulter, K.C., Green, J.L., Harrison, S.J. and Manners, J.M. (1997) Purification, characterisation and cDNA cloning of an antimicrobial peptide from *Macadamia integrifolia. European Journal of Biochemistry* 244, 743–749.

Martin, G.B., Brommonschenkel, S.H., Chunwongse, J., Frary, A., Ganal, M.W., Spivey, R., Wu, T., Earle, E.D. and Tanksley, S.D. (1993) Map-based cloning of a protein kinase gene conferring disease resistance in tomato. *Science* 262, 1432–1436.

Masoud, S.A., Zhu, Q., Lamb, C. and Dixon, R.A. (1996) Constitutive expression of an inducible β-1,3-glucanase in alfalfa reduces severity caused by the oomycete pathogen *Phytophthora megasperma* f. sp. *medicaginis*, but does not reduce disease severity of chitin-containing fungi. *Transgenic Research* 5, 313–323.

May, G.D., Rownak, A., Mason, H., Wiecko, A., Novak, F.J. and Arntzen, C.J. (1995) Generation of transgenic banana (*Musa acuminata*) plants via *Agrobacterium*-mediated transformation. *Bio/Technology* 13, 486–492.

Métraux, J.P., Burkhart, W., Moyer, M., Dincher, S., Middlesteadt, W., Williams, S., Payne, G., Carnes, M. and Ryals, J. (1989) Isolation of a complementary DNA encoding a chitinase with structural homology to a bifunctional lysozyme/chitinase. *Proceedings of the National Academy of Sciences USA* 86, 896–900.

Meyer, B., Houlné, G., Pozueta-Romero, J., Schantz, M.L. and Schantz, R. (1996) Fruit-specific expression of a defensin-type gene family in bell pepper: upregulation during ripening and upon wounding. *Molecular and General Genetics* 259, 504–510.

Michelmore, R. (1995) Molecular approaches to manipulation of disease resistance genes. *Annual Review of Phytopathology* 15, 393–427.

Milligan, S.B., Bodeau, J., Yaghoobi, J., Kaloshian, I., Zabel, P. and Williamson, V.M. (1998) The root knot nematode resistance gene Mi from tomato is a member of the leucine zipper, nucleotide binding, leucine-rich repeat family of plant genes. *Plant Cell* 10, 1307–1319.

Mills, D. and Hammerschlag, F.A. (1993) Effect of cecropin B on peach pathogens, protoplasts, and cells. *Plant Science* 93, 143–150.

Mills, D., Hammerschlag, F.A., Nordeen, R.O. and Owens, L.D. (1994) Evidence for the breakdown of cecropin B by proteinases in the intercellular fluid of peach leaves. *Plant Science* 104, 17–22.

Mindrinos, M., Katagiri, F., Yu, G.-L. and Ausubel, F.M. (1994) The *Arabidopsis thaliana* disease resistance gene *RPS2* encodes a protein containing a nucleotide-binding site and leucine-rich repeats. *Cell* 78, 1089–1099.

Mitchell, R.E. (1984) The relevance of non-host-specific toxins in the expression of virulence by pathogens. *Annual Review of Phytopathology* 22, 215–245.

Mitra, A. and Zhang, Z. (1994) Expression of a human lactoferrin cDNA in tobacco cells produces antibacterial protein(s). *Plant Physiology* 106, 977–981.

Mitra, A., Higgins, D., Langenberg, W.G., Nie, H., Sengupta, D.N. and Silverman, R.H. (1996) A mammalian 2–5A system functions as an antiviral pathway in transgenic plants. *Proceedings of the National Academy of Sciences USA* 93, 6780–6785.

Mittler, R., Shulaev, V. and Lam, E. (1995) Coordinated activation of programmed cell death and defense mechanism in transgenic tobacco plants expressing a bacterial proton pump. *Plant Cell* 7, 29–42.

Molina, A. and García-Olmedo, F. (1997) Enhanced tolerance to bacterial pathogens caused by the transgenic expression of barley lipid transfer protein LTP2. *Plant Journal* 12, 669–675.

Molina, A., Ahl Goy, P., Fraile, A., Sánchez-Monge, R. and García-Olmedo, F. (1993a) Inhibition of bacterial and fungal pathogens by thionins of types I and II. *Plant Science* 92, 169–177.

Molina, A., Segura, A. and García-Olmedo, F. (1993b) Lipid transfer proteins (nsLTPs) from barley and maize leaves are potent inhibitors of bacterial and fungal plant pathogens. *FEBS Letters* 316, 119–122.

Monzavi-Karbassi, B., Goldenkova, I.V., Darbinian, N.S., Kobets, N.S., Vasilevko, V.T. and Piruzian, E.S. (1998) The effective secretion of bacterial β-glucanase into the intercellular space in *Nicotiana tabacum* transgenic plants. *Genetika* 34, 475–479.

Moreno, M., Segura, A. and García-Olmedo, F. (1994) Pseudothionin-St1, a potato peptide active against potato pathogens. *European Journal of Biochemistry* 223, 135–139.

Mosqueda, G., Van den Broeck, G., Saucedo, O., Bailey, A.M., Alvarez-Morales, A. and Herrera-Estrella, L. (1990) Isolation and characterization of the gene from *Pseudomonas syringae* pv. *phaseolicola* encoding the phaseolotoxin-insensitive ornithine carbamoyltransferase. *Molecular and General Genetics* 222, 461–466.

Nakajima, H., Muranaka, T., Ishige, F., Akutsu, K. and Oeda, K. (1997) Fungal and bacterial disease resistance in transgenic plants expressing human lysozyme. *Plant Cell Reports* 16, 674–679.

Nakamura, T., Furunaka, H., Miyata, T., Tokunaga, F., Muta, T., Iwanaga, S., Niwa, M., Takao, T. and Shimonishi, Y. (1988) Tachyplesin, a class of antimicrobial peptide from the hemocytes of the horseshoe crab (*Tachypleus tridentatus*). Isolation and chemical structure. *Journal of Biological Chemistry* 263, 16709–16713.

Nakamura, Y., Sawada, H., Kobayashi, S., Nakajima, I. and Yoshikawa, M. (1999) Expression of soybean β-1,3-endoglucanase cDNA and effect on disease tolerance in kiwifruit plants. *Plant Cell Reports* 18, 527–532.

Narva, K.E., Payne, J., Schwab, G.E., Hickle, L.A., Galasan, T. and Sick, A.J. (1991) Novel *Bacillus thuringiensis* microbes active against nematodes, and genes encoding novel nematode-active toxins cloned from *Bacillus thuringiensis* isolates. European Patent Application 91305047.2.

Natori, S. (1994) Function of antimicrobial proteins in insects. *Ciba Foundation Symposia* 186, 123–132.

Navarre, D.A. and Wolpert, T.J. (1999) Victorin induction of an apoptotic/senescence-like response in oats. *Plant Cell* 11, 237–249.

Nejidat, A. and Beachy, R.N. (1990) Transgenic tobacco plants expressing a coat protein gene of tobacco mosaic virus are resistant to some other tobamoviruses. *Molecular Plant–Microbe Interactions* 3, 247–251.

Niebel, A., de Almeida Engler, J., Tiré, C., Engler, G., Van Montagu, M. and Gheysen, G. (1993)

Induction patterns of an extensin gene in tobacco upon nematode infection. *The Plant Cell* 5, 1697–1710.
Nielsen, T.W., Maroney, P.A., Robertson, H.D. and Baglioni, C. (1982) Heterogeneous nuclear RNA promotes synthesis of (2′,5′)oligo-adenylate-activated endoribonuclease. *Molecular and Cellular Biology* 2, 154–160.
Nordeen, R.O., Sinden, S.L., Jaynes, J.M. and Owens, L.D. (1992) Activity of cecropin SB37 against protoplasts from several plant species and their bacterial pathogens. *Plant Science* 82, 101–107.
Norelli, J.L., Aldwinckle, H.S., Destéfano-Beltrán, L. and Jaynes, J.M. (1994) Transgenic 'Malling 26' apple expressing the attacin E gene has increased resistance to *Erwinia amylovora. Euphytica* 77, 123–128.
Ogawa, T., Hori, T. and Ishida, I. (1996) Virus-induced cell death in plants expressing the mammalian 2′,5′ oligoadenylate system. *Nature Biotechnology* 14, 1566–1569.
Ohshima, M., Mitsuhara, I., Okamoto, M., Sawano, S., Nishiyama, K., Kaku, H., Natori, S. and Ohashi, Y. (1999) Enhanced resistance to bacterial diseases of transgenic tobacco plants overexpressing sarcotoxin IA, a bactericidal peptide of insect. *Journal of Biochemistry* 125, 431–435.
Oishi, O., Yamashita, S., Nishimoto, E., Lee, S., Sugihara, G. and Ohno, M. (1997) Conformations and orientations of aromatic amino acid residues of tachyplesin I in phospholipid membranes. *Biochemistry* 36, 4352–4359.
Okabe, N. and Goto, M. (1963) Bacteriophages of plant pathogens. *Annual Review of Phytopathology* 1, 397–418.
Okamoto, M., Mitsuhara, I., Ohshima, M., Natori, S. and Ohashi, Y. (1998) Enhanced expression of an antimicrobial peptide sarcotoxin IA by GUS fusion in transgenic tobacco plants. *Plant and Cell Physiology* 39, 57–63.
Okole, B.N. and Schulz, F.A. (1997) Selection of *Mycosphaerella fijiensis*-resistant cell lines from micro-cross sections of banana and plantain. *Plant Cell Reports* 16, 339–343.
Olivieri, F., Prasad, V., Valbonesi, P., Srivastava, S., Ghosal-Chowdhury, P., Barbieri, L., Bolognesi, A. and Stirpe, F. (1996) A systemic antiviral resistance-inducing protein isolated from *Clerodendrum inerme* Gaertn. is a polynucleotide:adenosine glycosidase (ribosome-inactivating protein). *FEBS Letters* 396, 132–134.
Opperman, C.H., Taylor, C.G. and Conkling, M.A. (1994) Root-knot nematode directed expression of a plant root-specific gene. *Science* 263, 221–223.
Ori, N., Eshed, Y., Paran, I., Presting, G., Aviv, D., Tanksley, S., Zamir, D. and Fluhr, R. (1997) The *I2C* family from the wilt disease resistance locus *I2* belongs to the nucleotide binding, leucine-rich repeat superfamily of plant resistance genes. *Plant Cell* 9, 521–532.
Owens, L.D. and Heutte, T.M. (1997) A single amino acid substitution in the antimicrobial defense protein cecropin B is associated with diminished degradation by leaf intercellular fluid. *Molecular Plant–Microbe Interactions* 10, 525–528.
Padgett, H.S. and Beachy, R.N. (1993) Analysis of a tobacco mosaic virus capable of overcoming N gene-mediated resistance. *Plant Cell* 5, 577–586.
Palukaitis, P. and Roossinck, M.J. (1996) Spontaneous change of a benign satellite RNA of cucumber mosaic virus to a pathogenic variant. *Nature Biotechnology* 14, 1264–1268.
Palukaitis, P. and Zaitlin, M. (1997) Replicase-mediated resistance to plant virus disease. *Advances in Virus Research* 48, 349–377.
Park, C.M., Berry, J.O. and Bruenn, J.A. (1996) High-level secretion of a virally encoded anti-fungal toxin in transgenic plants. *Plant Molecular Biology* 30, 359–366.
Parker, J.E., Coleman, M.J., Szabó, V., Frost, L.N., Schmidt, R., van der Biezen, E.A., Moores, T., Dean, C., Daniels, M.J. and Jones, J.D.G. (1997) The Arabidopsis downy mildew resistance gene *RPP5* shares similarity to the Toll and interleukin-1 receptors with *N* and *L6*. *Plant Cell* 9, 879–894.
Pérez Hernández, J.B., Remy, S., Galán Saúco, V., Swennen, R. and Sági, L. (1998) Chemotaxis, attachment and transgene expression in the *Agrobacterium*-mediated banana transformation system. *Mededelingen Faculteit Landbouwwetenschappen Universiteit Gent* 63(4b), 1603–1606.
Pérez Hernández, J.B., Remy, S., Galán Saúco, V., Swennen, R. and Sági, L. (1999) Chemotactic movement and attachment of *Agrobacterium tumefaciens* to banana cells and tissues. *Journal of Plant Physiology* 155, 245–250.
Perlak, F.J., Fuchs, R.L., Dean, D.A., McPherson, S.L. and Fischoff, D.A. (1991) Modification of the coding sequence enhances plant expression of insect control protein genes. *Proceedings of the National Academy of Sciences USA* 88, 3324–3328.

Peumans, J. and Van Damme, E.J.M. (1995) Lectins as plant defense proteins. *Plant Physiology* 109, 347–352.

Pfitzner, U.M. and Pfitzner, A.J. (1992) Expression of a viral avirulence gene in transgenic plants is sufficient to induce the hypersensitive defense response. *Molecular Plant–Microbe Interactions* 5, 318–321.

Pinochet, J. and Rowe, P.R. (1978) Progress in breeding for resistance to *Radopholus similis* on bananas. *Nematropica* 9, 76–78.

Pirhonen, M.U., Lidell, M.C., Rowley, D.L., Lee, S.W., Jin, S., Liang, Y., Silverstone, S., Keen, N.T. and Hutcheson, S.W. (1996) Phenotypic expression of *Pseudomonas syringae avr* genes in *E. coli* is linked to the activities of the *hrp*-encoded secretion system. *Molecular Plant–Microbe Interactions* 9, 252–260.

Ponz, F. and Bruening, G. (1986) Mechanisms of resistance to plant viruses. *Annual Review of Phytopathology* 24, 355–381.

Porta, H. and Lizardi, P.M. (1995) An allosteric hammerhead ribozyme. *Bio/Technology* 13, 161–164.

Powell, W.A., Catranis, C.M. and Maynard, C.A. (1995) Synthetic antimicrobial peptide design. *Molecular Plant–Microbe Interactions* 8, 792–794.

Powell Abel, P., Nelson, R.S., De, B., Hoffmann, N., Rogers, S.G., Fraley, R.T. and Beachy, R.N. (1986) Delay of disease development in transgenic plants that express the tobacco mosaic virus coat protein gene. *Science* 232, 738–743.

Price, N.S. (1994) Field trial evaluation of nematode susceptibility within *Musa*. *Fundamental and Applied Nematology* 17, 391–396.

Punja, Z.K. and Raharjo, S.H.T. (1996) Response of transgenic cucumber and carrot plants expressing different chitinase enzymes to inoculation with fungal pathogens. *Plant Disease* 89, 999–1005.

Qui, X., Wu, Y., Jaynes, J., Goodwin, P. and Erickson, L.R. (1995) Effect of a membrane interactive peptide on plant cells of canola (*Brassica napus*) and two fungal pathogens. *Plant Cell Reports* 15, 115–118.

Raikhel, N.W., Lee, H.I. and Broekaert, W.F. (1993) Structure and function of chitin-binding proteins. *Annual Review of Plant Physiology and Plant Molecular Biology* 44, 591–615.

Rao, K.V., Rathore, K.S., Hodges, T.K., Fu, X., Stoger, E., Sudhakar, D., Williams, S., Christou, P., Bharathi, M., Bown, D.P., Powell, K.S., Spence, J., Gatehouse, A.M. and Gatehouse, J.A. (1998) Expression of snowdrop lectin (GNA) in transgenic rice plants confers resistance to rice brown planthopper. *Plant Journal* 15, 469–477.

Ready, M.P., Brown, D.T. and Robertus, J.D. (1986) Extracellular localization of pokeweed antiviral protein. *Proceedings of the National Academy of Sciences USA* 83, 5053–5056.

Reddigari, S.R., Jansma, P.L., Premachandran, D. and Hussey, R.S. (1986) Cuticular collagenous proteins of second-stage juvenile and adult females of *Meloidogyne incognita*; isolation and partial characterization. *Journal of Nematology* 18, 294–302.

Reed, J.D., Edwards, D.L. and Gonzalez, C.F. (1997) Synthetic peptide combinatorial libraries: a method for the identification of bioactive peptides against phytopathogenic fungi. *Molecular Plant–Microbe Interactions* 10, 537–549.

Register, J.C., III and Nelson, R.S. (1992) Early events in plant virus infection: relationships with genetically engineered protection and host gene resistance. *Seminars in Virology* 3, 441–451.

Reimann-Philipp, U., Schrader, G., Martinoia, E., Barkholt, V. and Apel, K. (1989) Intracellular thionins of barley: a second group of leaf thionins closely related to but distinct from cell wall-bound thionins. *Journal of Biological Chemistry* 264, 8978–8984.

Remy, S., François, I., Cammue, B.P.A., Swennen, R. and Sági, L. (1998) Co-transformation as a potential tool to create multiple and durable resistance in banana (*Musa* spp.). *Acta Horticulturae* 461, 361–365.

Remy, S., Deconinck, I., Swennen, R. and Sági, L. (1999) Assessment of fungus tolerance in transgenic banana to a leaf disc assay. In: *Abstracts of the International Symposium on the Molecular Biology and Cellular Biology of Banana, March 22–25, 1999, Ithaca*, p. 38.

Riechmann, J.L., Laín, S. and García, J.A. (1992) Highlights and prospects of potyvirus molecular biology. *Journal of General Virology* 73, 1–16.

Roberts, W.K. and Selitrennikoff, C.P. (1988) Plant and bacterial chitinases differ in antifungal activity. *Journal of General Microbiology* 134, 169–176.

Rodriguez, R.J. and Redman, R.S. (1997) Fungal life-styles and ecosystem dynamics: biological

aspects of plant pathogens, plant endophytes and saprophytes. *Advances in Botanical Research* 24, 169–193.
Rommens, C.M.T., Salmeron, J.M., Oldroyd, G.E. and Staskawicz, B.J. (1995) Intergeneric transfer and functional expression of the tomato disease resistance gene *Pto*. *Plant Cell* 7, 1537–1544.
Rosqvist, R., Hakansson, S., Forsberg, A. and Wolf-Watz, H. (1995) Functional conservation of the secretion and translocation machinery for virulence proteins of yersiniae, salmonellae and shigellae. *EMBO Journal* 14, 4187–4195.
Roux, L., Simon, A.E. and Holland, J.J. (1991) Effects of defective interfering viruses on virus replication and pathogenesis *in vitro* and *in vivo*. *Advances in Virus Research* 40, 181–211.
Sági, L., Remy, S., Panis, B., Swennen, R. and Volckaert, G. (1994) Transient gene expression in electroporated banana (*Musa* spp., cv. 'Bluggoe', ABB group) protoplasts isolated from regenerable embryogenic cell suspensions. *Plant Cell Reports* 13, 262–266.
Sági, L., Remy, S., Verelst, B., Swennen, R. and Panis, B. (1995a) Genetic transformation in *Musa* species (banana). In: Bajaj, Y.P.S. (ed.) *Biotechnology in Agriculture and Forestry*, Vol 34. Plant Protoplasts and Genetic Engineering VI, Springer, Berlin, Heidelberg, New York, pp. 214–227.
Sági, L., Panis, B., Remy, S., Schoofs, H., De Smet, K., Remy, S., Swennen, R. and Cammue, B.P.A. (1995b) Genetic transformation of banana and plantain (*Musa* spp.) via particle bombardment. *Bio/Technology* 13, 481–485.
Sági, L., Remy, S., Verelst, B., Panis, B., Cammue, B.P.A., Volckaert, G. and Swennen, R. (1995c) Transient gene expression in transformed banana (*Musa* spp., cv. 'Bluggoe') protoplasts and embryogenic cell suspensions. *Euphytica* 85, 89–95.
Sági, L., May, G.D., Remy, S. and Swennen, R. (1998) Recent developments in biotechnological research on bananas (*Musa* spp.). *Biotechnology and Genetic Engineering Reviews* 15, 313–327.
Sági, L., Remy, S., Pérez Hernández, J.B., Cammue, B.P.A. and Swennen, R. (1999) Transgenic banana (*Musa* species). In: Bajaj, Y.P.S. (ed.) *Biotechnology in Agriculture and Forestry*, Vol. 47. Transgenic Crops II, Springer, Berlin, Heidelberg and New York (in press).
Saito, T., Yamanaka, K., Watanabe, Y., Takamatsu, N., Meshi, T. and Okada, Y. (1989) Mutational analysis of the coat protein gene of tobacco mosaic virus in relation to hypersensitive response in tobacco plants with the N′ gene. *Virology* 173, 11–20.
Saito, T., Yamanaka, K. and Okada, Y. (1990) Long distance movement and viral assembly of tobacco mosaic virus mutants. *Virology* 176, 329–336.
Salmeron, J.M., Oldroyd, G.E.D., Rommens, C.M.T., Scofield, S.R., Kim, H.S., Lavelle, D.T., Dahlbeck, D. and Staskawicz, B.J. (1996) Tomato *Prf* is a member of the leucine-rich repeat class of plant disease resistance genes and lies embedded with the *Pto* kinase gene cluster. *Cell* 86, 123–133.
Sanford, J.C. and Johnston, S.A. (1985) The concept of parasite-derived resistance – deriving resistance genes from the parasite's own genome. *Journal of Theoretical Biology* 113, 395–405.
Sano, H., Seo, S., Orudgev, E., Youssefian, S., Ishizuka, K. and Ohashi, Y. (1994) Expression of the gene for a small GTP binding protein in transgenic tobacco elevates endogenous cytokinin levels, abnormally induces salicylic acid in response to wounding, and increases resistance to tobacco mosaic virus infection. *Proceedings of the National Academy of Sciences USA* 91, 10556–10560.
Sano, T., Nagayama, A., Ogawa, T., Ishida, I. and Okada, Y. (1997) Transgenic potato expressing a double-stranded RNA-specific ribonuclease is resistant to potato spindle viroid. *Nature Biotechnology* 15, 1290–1294.
Sarah, J.L., Blavignac, F., Sabatini, C. and Boisseau, M. (1992) Une méthode de laboratoire pour le criblage variétal des bananiers vis-à-vis de la résistance aux nématodes. *Fruits* 47, 559–564.
Saraste, M., Sibbald, P.R. and Wittinghofer, A. (1990) The P-loop – a common motif in ATP- and GTP-binding proteins. *Trends in Biochemical Sciences* 15, 430–434.
Schäfer, W., Straney, D., Ciuffetti, L., VanEtten, H.D. and Yoder, O.C. (1989) One enzyme makes a fungal pathogen, but not a saprophyte, virulent on a new host plant. *Science* 246, 247–249.
Schnepf, E., Crickmore, N., Van Rie, J., Lereclus, D., Baum, J., Feitelson, J., Zeigler, D.R. and Dean, D.H. (1998) *Bacillus thuringiensis* and its pesticidal crystal proteins. *Microbiology and Molecular Biology Reviews* 62, 775–806.
Scholthof, K.B.G., Scholthof, H.B. and Jackson, A.O. (1993) Control of plant virus diseases by pathogen-derived resistance in transgenic plants. *Plant Physiology* 102, 7–12.
Scott, J.K. and Smith, G.P. (1990) Searching for peptide ligands with an epitope library. *Science* 249, 386–390.

Segura, A., Moreno, M. and García-Olmedo, F. (1993) Purification and antipathogenic activity of lipid transfer proteins (LTPs) from the leaves of *Arabidopsis* and spinach. *FEBS Letters* 332, 243–246.

Segura, A., Moreno, M., Molina, A. and García-Olmedo, F. (1998) Novel defensin subfamily from spinach (*Spinacia oleracea* L.). *FEBS Letters* 435, 159–162.

Segura, A., Moreno, M., Madueno, F., Molina, A. and García-Olmedo, F. (1999) Snakin-1, a peptide from potato that is active against plant pathogens. *Molecular Plant–Microbe Interactions* 12, 16–23.

Sequiera, L. (1984) Cross protection and induced resistance: their potential for plant disease control. *Trends in Biotechnology* 2, 25–29.

Sherwood, J.L. and Fulton, R.W. (1982) The specific involvement of coat protein in tobacco mosaic virus cross protection. *Virology* 119, 150–158.

Shewry, P.R. and Lucas, J.A. (1997) Plant proteins that confer resistance to pests and pathogens. *Advances in Botanical Research* 26, 135–192.

Simmaco, M., Barra, D., Chiarini, F., Noviello, L., Melchiorri, P., Kreil, G. and Richter, K. (1991) A family of bombinin-related peptides from the skin of *Bombina variegata*. *European Journal of Biochemistry* 199, 217–222.

Simmaco, M., Mignogna, G., Barra, D. and Bossa, F. (1993) Novel antimicrobial peptides from skin secretion of the European frog *Rana esculenta*. *FEBS Letters* 324, 159–161.

Simmaco, M., Mignogna, G., Barra, D. and Bossa, F. (1994) Antimicrobial peptides from skin secretions of *Rana esculenta*. Molecular cloning of cDNAs encoding esculentin and brevinins and isolation of new active peptides. *Journal of Biological Chemistry* 269, 11956–11961.

Simmaco, M., Mignogna, G., Canofeni, S., Miele, R., Mangoni, M.L. and Barra, D. (1996) Temporins, antimicrobial peptides from the European red frog *Rana temporaria*. *European Journal of Biochemistry* 242, 788–792.

Singh, Z. and Sansavini, S. (1998) Genetic transformation and fruit crop improvement. *Plant Breeding Reviews* 16, 87–134.

Sleat, D.E. and Palukaitis, P. (1992) A single nucleotide change within a plant virus satellite alters the host specificity of disease induction. *Plant Journal* 2, 43–49.

Smirnov, S., Shulaev, V. and Tumer, N.E. (1997) Expression of pokeweed antiviral protein in transgenic plants induces virus resistance in grafted wild-type plants independently of salicylic acid accumulation and pathogenesis-related protein synthesis. *Plant Physiology* 114, 1113–1121.

Smith, F.D., Gadoury, D.M., Van Eck, J.M., Blowers, A., Sanford, J.C., Van der Meij, J. and Eisenreich, R. (1998) Enhanced resistance to powdery mildew in transgenic poinsettia conferred by antimicrobial peptides. *Phytopathology* 88, S83.

Song, W.Y., Wang, G.L., Chen, L.L., Kim, H.S., Pi, L.Y., Holsten, T., Gardner, J., Wang, B., Zhai, W.X., Zhu, L.-H., Fauquet, C. and Ronald, P. (1995) A receptor kinase-like protein encoded by the rice resistance gene, *Xa21*. *Science* 270, 1804–1806.

Stark-Lorenzen, P., Nelke, B., Hänler, G., Mühlbach, H.P. and Thomzik, J.E. (1997) Transfer of a stilbene synthase gene to rice (*Oryza sativa* L.). *Plant Cell Reports* 16, 668–673.

Staskawicz, B.J., Dahlbeck, D. and Keen, N.T. (1984) Cloned avirulence gene of *Pseudomonas syringae* pv. *glycinea* determines race-specific incompatibility on *Glycine max* (L.) Merr. *Proceedings of the National Academy of Sciences USA* 81, 6024–6028.

Stiekema, W.J., Bosch, D., Wilmink, A., De Boer, J.M., Schouten, A., Roosien, J., Goverse, A., Smant, G., Stokkermans, J., Gommsers, F.J., Schots, A. and Bakker, J. (1997) Towards plantibody-mediated resistance against nematodes. In: Fenoll, C., Grundler, F.M.W. and Ohl, S.A. (eds) *Cellular and Molecular Aspects of Plant–Nematode Interactions*. Developments in Plant Pathology, Vol. 10, Kluwer Academic Publishers, Dordrecht, pp. 262–271.

Stierle, A.A., Upadhyay, R., Hershenhorn, J., Strobel, G.A. and Molina, G. (1991) The phytotoxins of *Mycosphaerella fijiensis*, the causative agent of black Sigatoka disease of bananas and plantains. *Experientia* 47, 853–859.

Stirpe, F. and Hughes, R.C. (1989) Specificity of ribosome-inactivating proteins with RNA *N*-glycosidase activity. *Biochemical Journal* 262, 1001–1002.

Stirpe, F., Barbieri, L., Battelli, L.G., Soria, M. and Lappi, D.A. (1992) Ribosome-inactivating proteins from plants: present status and future prospects. *Bio/Technology* 10, 405–412.

Suslow, T.V., Matsubara, D., Jones, J., Lee, R. and Dunsmuir, P. (1988) Effect of expression of bacterial chitinase on tobacco susceptibility to leaf brown spot. *Phytopathology* 78, 1556.

Suzuki, S., Ohe, Y., Okube, T., Kakegawa, T. and Tatemoto, K. (1995) Isolation and characterization of novel antimicrobial peptides, rugosins A, B and C, from the skin of the frog, *Rana rugosa*. *Biochemical and Biophysical Research Communications* 212, 249–254.

Swords, K.M., Dahlbeck, K., Kearney, B., Roy, M. and Staskawicz, B.J. (1996) Spontaneous and induced mutations in a single open reading frame alter both virulence and avirulence in *Xanthomonas campestris* pv. *vesicatoria avrBs2*. *Journal of Bacteriology* 178, 4661–4669.

Tabei, Y., Kitade, S., Nishizawa, Y., Kikuchi, N., Kayano, T., Hibi, T. and Akutsu, K. (1998) Transgenic cucumber plants harboring a rice chitinase gene exhibit enhanced resistance to gray mold (*Botrytis cinerea*). *Plant Cell Reports* 17, 159–164.

Tacke, E., Salamini, F. and Rohde, W. (1996) Genetic engineering of potato for broad-spectrum protection against virus infection. *Nature Biotechnology* 14, 1597–1601.

Tailor, R.H., Acland, D.P., Attenborough, S., Cammue, B.P.A., Evans, I.J., Osborn, R.W., Ray, J.A., Rees, S.B. and Broekaert, W.F. (1997) A novel family of small cysteine-rich antimicrobial peptides from seed of *Impatiens balsamina* is derived from a single precursor protein. *Journal of Biological Chemistry* 272, 24480–24487.

Taraporewala, Z.F. and Culver, J.N. (1996) Identification of an elicitor active site within the three-dimensional structure of the tobacco mosaic tobamovirus coat protein. *Plant Cell* 8, 169–178.

Tavladoraki, P., Benvenuto, E., Trinca, S., De Martinis, D., Cattaneo, A. and Galeffi, P. (1993) Transgenic plants expressing a functional single-chain Fv antibody are specifically protected from virus attack. *Nature* 366, 469–472.

Terakawa, T., Takaya, N., Horiuchi, H., Koike, M. and Takagi, M. (1997) A fungal chitinase gene from *Rhizopus oligosporus* confers antifungal activity to transgenic tobacco. *Plant Cell Reports* 16, 439–443.

Terras, F.R.G., Goderis, I.J., Van Leuven, F., Vanderleyden, J., Cammue, B.P.A. and Broekaert, W.F. (1992) *In vitro* antifungal activity of a radish (*Raphanus sativus* L.) seed protein homologous to nonspecific lipid transfer proteins. *Plant Physiology* 100, 1055–1058.

Terras, F.R.G., Schoofs, H.M.E., Thevissen, K., Osborn, R.W., Vanderleyden, J., Cammue, B.P.A. and Broekaert, W.F. (1993) Synergistic enhancement of the antifungal activity of wheat and barley thionins by radish and oilseed rape 2S albumins and barley trypsin inhibitors. *Plant Physiology* 103, 1311–1319.

Terras, F.R.G., Eggermont, K., Kovaleva, V., Raikhel, N., Osborn, R., Kester, A., Rees, S.B., Vanderleyden, J., Cammue, B.P.A. and Broekaert, W.F. (1995) Small cysteine-rich antifungal proteins from radish (*Raphanus sativus* L.): their role in host defense and their constitutive expression in transgenic tobacco leading to enhanced resistance to a fungal disease. *Plant Cell* 7, 573–588.

Thilmony, R.L., Chen, Z., Bressan, R.A. and Martin, G.B. (1995) Expression of the tomato *Pto* gene in tobacco enhances resistance to *Pseudomonas syringae* pv *tabaci* expressing *avrPto*. *Plant Cell* 7, 1529–1536.

Thomas, C.M., Jones, D.A., Parniske, M., Harrison, K., Balint-Kurti, P.J., Hatzixanthis, K. and Jones, J.D.G. (1997) Characterization of the tomato *Cf-4* gene for resistance to *Cladosporium fulvum* identifies sequences that determine recognitional specificity in Cf-4 and Cf-9. *Plant Cell* 9, 2209–2224.

Thomzik, J.E., Stenzel, K., Stöcker, R.H., Schreier, P.H., Hain, R. and Stahl, D.J. (1997) Synthesis of a grapevine phytoalexin in transgenic tomatoes (*Lycopersicon esculentum* Mill.) conditions resistance against *Phytophthora infestans*. *Physiological and Molecular Plant Pathology* 51, 265–278.

Tien, P. and Wu, G. (1991) Satellite RNA for the biocontrol of plant disease. *Advances in Virus Research* 39, 321–339.

Toubart, P., Desiderio, A., Salvi, G., Cervone, F., Daroda, L., De Lorenzo, G., Bergmann, C., Darvill, A.G. and Albersheim, P. (1992) Cloning and characterization of the gene encoding the endopolygalacturonase-inhibiting protein (PGIP) of *Phaseolus vulgaris* L. *Plant Physiology* 96, 390–397.

Toyoda, H., Matsuda, Y., Yamaga, T., Ikeda, S., Morita, M., Tamai, T. and Ouchi, S. (1991) Suppression of the powdery mildew pathogen by chitinase microinjected into barley coleoptile epidermal cells. *Plant Cell Reports* 10, 217–220.

Traut, T.W. (1994) The functions and consensus motifs of nine types of peptide segments that form different types of nucleotide-binding sites. *European Journal of Biochemistry* 222, 9–19.

Tricoli, D.M., Carney, K.J., Russell, P.F., McMaster, J.R., Groff, D.W., Hadden, K.C., Himmel, P.T., Hubbard, J.R., Boeshore, M.L., Reynolds, J.F. and Quemada, H.D. (1995) Field evaluation of transgenic squash containing single or multiple virus coat protein gene constructs for resistance to cucumber mosaic virus, watermelon mosic virus 2, and/or zucchini yellow mosaic virus. *Bio/Technology* 13, 1458–1465.

Trudel, J., Potvin, C. and Asselin, A. (1992) Expression of active hen egg white lysozyme in transgenic tobacco. *Plant Science* 87, 55–67.

Trudel, J., Potvin, C. and Asselin, A. (1995) Secreted hen lysozyme in transgenic tobacco: recovery of bound enzyme and *in vitro* growth inhibition of plant pathogens. *Plant Science* 106, 55–62.

Truve, E., Aaspöllu, A., Honkanen, J., Puska, R., Mehto, M., Hassi, A., Teeri, T.H., Kelve, M., Seppänen, P. and Saarma, M. (1993) Transgenic potato plants expressing mammalian 2′-5′ oligoadenylate synthetase are protected from potato virus X infection under the field conditions. *Bio/Technology* 11, 1048–1052.

Tumer, N.E., Hwang, D.J. and Bonness, M. (1997) C-terminal deletion mutant of pokeweed antiviral protein inhibits viral infection but does not depurinate host ribosomes. *Proceedings of the National Academy of Sciences USA* 94, 3866–3871.

Upchurch, R.G., Rose, M.S., Allen, G.C. and Zuo, W.-N. (1997) Expression of cercosporin resistance gene in tobacco. *Phytopathology* 87, S99.

Urwin, P.E., Atkinson, H.J., Waller, D.A. and McPherson, M.J. (1995) Engineered oryzacystatin-I expressed in transgenic hairy roots confers resistance to *Globodera pallida*. *The Plant Journal* 8, 121–131.

Urwin, P.E., Lilley, C.J., McPherson, M.J. and Atkinson, H.J. (1997) Resistance to both cyst and root-knot nematodes conferred by transgenic *Arabidopsis* expressing a modified plant cystatin. *The Plant Journal* 12, 455–461.

Van der Eycken, W., de Almeida Engler, J., Inzé, D., Van Montagu, M. and Gheysen, G. (1996) A molecular study of root-knot nematode-induced feeding sites. *The Plant Journal* 9, 45–54.

van Engelen, F.A., Schouten, A., Molthoff, J.W., Roosien, J., Salinas, J., Dirkse, W.G., Schots, A., Bakker, J., Gommers, F.J., Jongsma, M.A., Bosh, D. and Stiekema, W.J. (1994) Coordinated expression of antibody subunit genes yields high levels of functional antibodies in roots of transgenic tobacco. *Plant Molecular Biology* 26, 1701–1710.

VanEtten, H.D., Sandrock, R.W., Wasmann, C.C., Soby, S.D., McCluskey, K. and Wang, P. (1995) Detoxification of phytoanticipins and phytoalexins by phytopathogenic fungi. *Canadian Journal of Botany* 73, S518–S525.

Van Gijsegem, F., Genin, S. and Boucher, C.A. (1993) Conservation of secretion pathways for pathogenicity determinants of plant and animal bacteria. *Trends in Microbiology* 1, 175–180.

Vardi, E., Sela, I., Edelbaum, O., Livneh, O., Kuznetsova, L. and Stram, Y. (1993) Plants transformed with a cistron of a potato virus Y protease (NIa) are resistant to virus infection. *Proceedings of the National Academy of Sciences USA* 90, 7513–7517.

Verma, H.N., Srivastava, S., Varsha and Kumar, D. (1996) Induction of systemic resistance in plants against viruses by a basic protein from *Clerodendrum aculeatum* leaves. *Phytopathology* 86, 485–492.

Verma, H.N., Baranwal, V.K. and Srivastava, S. (1998) Antiviral substances of plant origin. In: Hadidi, A., Khetarpal, R.K. and Koganezawa, H. (eds) *Plant Virus Disease Control.* American Phytopathological Society Press, St Paul, Minnesota, pp. 154–162.

Vierheilig, H., Alt, M., Neuhaus, J.-M., Boller, T. and Wiemken, A. (1993) Colonization of transgenic *Nicotiana sylvestris* plants, expressing different forms of *Nicotiana tabacum* chitinase, by the root pathogen *Rhizoctonia solani* and by the mycorrhizal fungus *Glomus mosseae*. *Molecular Plant–Microbe Interactions* 6, 261–264.

Vierheilig, H., Alt, M., Lange, J., Gut-Rella, M., Wiemken, A. and Boller, T. (1995) Colonization of transgenic tobacco constitutively expressing pathogenesis-related proteins by the vesicular-arbuscular mycorrhizal fungus *Glomus mosseae*. *Applied and Environmental Microbiology* 61, 3031–3034.

Voss, A., Niersbach, M., Hain, R., Hirsch, H.J., Yu, C.L., Kreuzaler, F. and Fischer, R. (1995) Reduced virus infectivity in *N. tabacum* secreting a TMV-specific full-size antibody. *Molecular Breeding* 1, 39–50.

Wachinger, M., Kleinschmidt, A., Winder, D., von Pechmann, N., Ludvigsen, A., Neumann, M., Holle, R., Salmons, B., Erfle, V. and Brack-Werner, R. (1998) Antimicrobial peptides melittin and cecropin inhibit replication of human immunodeficiency virus 1 by suppressing viral gene expression. *Journal of General Virology* 79, 731–740.

Walker, M.J., Birch, R.G. and Pemberton, J.M. (1988) Cloning and characterization of an albicidin resistance gene from *Klebsiella oxytoca*. *Molecular Microbiology* 2, 443–454.

Wall, M.K. and Birch, R.G. (1997) Genes for albicidin biosynthesis and resistance span at least 69 kb in the genome of *Xanthomonas albilineans*. *Letters of Applied Microbiology* 24, 256–260.

Walton, J.D. (1996) Host-selective toxins: agents of compatibility. *Plant Cell* 8, 1723–1733.

Walton, J.D. and Panaccione, D.G. (1993) Host-selective toxins and disease specificity: perspectives and progress. *Annual Review of Phytopathology* 31, 275–303.

Wang, P., Zoubenko, O. and Tumer, N.E. (1998) Reduced toxicity and broad spectrum resistance to viral and fungal infection in transgenic plants expressing pokeweed antiviral protein II. *Plant Molecular Biology* 38, 957–964.

Wang, R.Y., Powell, G., Hardie, J. and Pirone, T.P. (1998) Role of the helper component in vector-specific transmission of potyviruses. *Journal of General Virology* 79, 1519–1524.

Watanabe, Y., Ogawa, T., Takahashi, H., Ishida, I., Takeuchi, Y., Yamamoto, M. and Okada, Y. (1995) Resistance against multiple plant viruses in plants mediated by double stranded RNA-specific ribonuclease. *FEBS Letters* 372, 165–168.

Wegener, C., Bartling, S., Olsen, O., Weber, J. and von Wettstein, D. (1996) Pectate lyase in transgenic potatoes confers preactivation of defence against *Erwinia carotovora*. *Physiological and Molecular Plant Pathology* 49, 359–376.

Wehunt, E.J., Hutchinson, D.J. and Edwards, D.I. (1978) Reaction of banana cultivars to the burrowing nematode *Radopholus similis*. *Journal of Nematology* 10, 368–370.

Whitelam, G.C. and Cockburn, W. (1996) Antibody expression in transgenic plants. *Trends in Plant Science* 1, 268–272.

Whitham, S., Dinesh-Kumar, S.P., Choi, D., Hehl, R., Corr, C. and Baker, B. (1994) The product of the tobacco mosaic virus resistance gene N: similarity to toll and the interleukin-1 receptor. *Cell* 78, 1101–1115.

Whitham, S., McCormick, S. and Baker, B. (1996) The *N* gene of tobacco confers resistance to tobacco mosaic virus in transgenic tomato. *Proceedings of the National Academy of Sciences USA* 93, 8776–8781.

Wiame, I., Swennen, R. and Sági, L. (1999) Towards PCR-based cloning of candidate disease resistance genes from banana (*Musa acuminata*). *Acta Horticulturae* (in press).

Wilson, T.M.A. (1993) Strategies to protect crop plants against viruses: pathogen-derived resistance blossoms. *Proceedings of the National Academy of Sciences USA* 90, 3134–3141.

Wilson, T.M.A. and Watkins, P.A.C. (1985) Influence of exogenous coat protein on the co-translational disassembly of tobacco mosaic virus (TMV) particles *in vitro*. *Virology* 149, 132–135.

Wolf, S., Deom, C.M., Beachy, R.N. and Lucas, W.J. (1989) Movement protein of tobacco mosaic virus modifies plasmodesmatal size exclusion limit. *Science* 276, 377–379.

Wu, G., Shortt, B.J., Lawrence, E.B., Levine, E.B., Fitzsimmons, K.C. and Shah, D.M. (1995) Disease resistance conferred by expression of a gene encoding H_2O_2-generating glucose oxidase in transgenic potato plants. *Plant Cell* 7, 1357–1368.

Wu, G.S., Shortt, B.J., Lawrence, E.B., Leon, J., Fitzsimmons, K.C., Levine, E.B., Raskin, I. and Shah, D.M. (1997) Activation of host defense mechanisms by elevated production of H_2O_2 in transgenic plants. *Plant Physiology* 115, 427–435.

Wu, S.Y. and Kaufman, R.J. (1996) Double stranded (ds) RNA binding and not dimerization correlates with the activation of dsRNA-dependent protein kinase (PKR). *Journal of Biological Chemistry* 271, 1756–1763.

Yamada, K. and Natori, S. (1991) Purification, sequence and antibacterial activity of two novel sapecin homologues from *Sarcophaga* embryonic cells: similarity of sapecin B to charybdotoxin. *Biochemical Journal* 291, 275–279.

Yang, X., Yie, Y., Zhu, F., Liu, Y. Kang, L., Wang, X. and Tien, P. (1997) Ribozyme-mediated high resistance against potato spindle viroid in transgenic plants. *Proceedings of the National Academy of Sciences USA* 94, 4861–4865.

Yang, Y. and Gabriel, D.W. (1995) *Xanthomonas* avirulence/pathogenicity gene family encodes functional nuclear targeting signals. *Molecular Plant–Microbe Interactions* 8, 627–631.

Yang, Y. and Klessig, D.F. (1996) Isolation and characterization of a tobacco mosaic-inducible *myb* oncogene homolog from tobacco. *Proceedings of the National Academy of Sciences USA* 93, 14972–14977.

Yang, Y. and Klessig, D.F. (1998) Alteration of *myb* gene expression in tobacco affects disease resistance and susceptibility. *Phytopathology* 88, S101.

Yoshikawa, M., Tsuda, M. and Takeuchi, Y. (1993) Resistance to fungal diseases in transgenic tobacco plants expressing the phytoalexin elicitor-releasing factor, β-1,3-endoglucanase, from soybean. *Naturwissenschaften* 80, 417–420.

Yoshimura, S., Yamanouchi, U., Katayose, Y., Toki, S., Wang, Z.-X., Kono, I., Kurata, N., Yano, M., Iwata, N. and Sasaki, T. (1998) Expression of *Xa1*, a bacterial blight-resistance gene in rice, is induced by bacterial inoculation. *Proceedings of the National Academy of Sciences USA* 95, 1663–1668.

Yun, D.J., Bressan, R.A. and Hasegawa, P.M. (1997) Plant antifungal proteins. *Plant Breeding Reviews* 14, 39–88.

Zaitlin, M. (1976) Viral cross protection: more understanding is needed. *Phytopathology* 66, 382–383.

Zhang, L. and Birch, R.G. (1997) Mechanisms of biocontrol by *Pantoea dispersa* of sugar cane leaf scald disease caused by *Xanthomonas albilineans*. *Journal of Applied Microbiology* 82, 448–454.

Zhang, L., Mitra, A. and Langenberg, W.G. (1998) Expression of a bacterial *rnc70* gene reduces accumulation of barley stripe mosaic virus in transgenic wheat. *Phytopathology* 88, S119.

Zhang, S. and Klessig, D.F. (1998) Resistance gene N-mediated *de novo* synthesis and activation of a tobacco mitogen-activated protein kinase by tobacco mosaic virus infection. *Proceedings of the National Academy of Sciences USA* 95, 7433–7438.

Zhang, Y. and Lewis, K. (1997) Fabatins: new antimicrobial plant peptides. *FEMS Microbiology Letters* 149, 59–64.

Zhang, Z., Coyne, D.P., Vidaver, A.K. and Mitra, A. (1998) Expression of human lactoferrin cDNA confers resistance to *Ralstonia solanacearum* in transgenic tobacco plants. *Phytopathology* 88, 730–734.

Zhou, J., Loh, Y.T., Bressan, R.A. and Martin, G.B. (1995) The tomato gene *Pti1* encodes a serine/threonine kinase that is phosphorylated by Pto and is involved in the hypersensitive response. *Cell* 83, 925–935.

Zhou, J., Tang, X. and Martin, G.B. (1997) The Pto kinase conferring resistance to tomato bacterial speck disease interacts with proteins that bind a *cis*-element of pathogenesis-related genes. *EMBO Journal* 16, 3207–3218.

Zhou, Y., Tian, Y., Wu, B. and Mang, K. (1998) Inhibition effect of transgenic tobacco plants expressing snowdrop lectin on the population development of *Myzus persicae*. *Chinese Journal of Biotechnology* 14, 9–16.

Zhu, Q., Maher, E., Masoud, S., Dixon, R.A. and Lamb, C.J. (1994) Enhanced protection against fungal attack by constitutive co-expression of chitinase and glucanase genes in transgenic tobacco. *Biotechnology* 12, 807–812.

Zoubenko, O., Uckun, F., Hur, Y., Chet, I. and Tumer, N. (1997) Plant resistance to fungal infection induced by nontoxic pokeweed antiviral protein mutants. *Nature Biotechnology* 15, 992–996.

Zuckerman, B.M. (1983) Hypothesis and possibilities of intervention in nematode chemoresponses. *Journal of Nematology* 15, 173–182.

Index of *Musa* and *Ensete* Species, Banana Subgroups and Clone Sets, and Banana, Abacá and Enset Cultivars

Musa and *Ensete* Species

Banana Subgroups

Banana, Abacá and Enset Cultivars Including Bred Hybrids and Wild *Musa* Accessions

*Abacá cultivars.
†Enset cultivars.
‡Illustration.

General Index